T0181279

Springer Optimization and Its Applications

Volume 131

Aims and Scope

Optimization has been expanding in all directions at an astonishing rate during the last few decades. New algorithmic and theoretical techniques have been developed, the diffusion into other disciplines has proceeded at a rapid pace, and our knowledge of all aspects of the field has grown even more profound. At the same time, one of the most striking trends in optimization is the constantly increasing emphasis on the interdisciplinary nature of the field. Optimization has been a basic tool in all areas of applied mathematics, engineering, medicine, economics and other sciences.

The series *Springer Optimization and Its Applications* publishes undergraduate and graduate textbooks, monographs and state-of-the-art expository works that focus on algorithms for solving optimization problems and also study applications involving such problems. Some of the topics covered include nonlinear optimization (convex and nonconvex), network flow problems, stochastic optimization, optimal control, discrete optimization, multi-objective programming, description of software packages, approximation techniques and heuristic approaches.

More information about this series at http://www.springer.com/series/7393

Nicholas J. Daras • Themistocles M. Rassias
Editors

Modern Discrete Mathematics and Analysis

With Applications in Cryptography, Information Systems and Modeling

Editors
Nicholas J. Daras
Department of Mathematics
Hellenic Military Academy
Vari, Greece

Themistocles M. Rassias
Department of Mathematics
National Technical University of Athens
Athens, Greece

ISSN 1931-6828 ISSN 1931-6836 (electronic)
Springer Optimization and Its Applications
ISBN 978-3-030-08964-1 ISBN 978-3-319-74325-7 (eBook)
https://doi.org/10.1007/978-3-319-74325-7

Mathematics Subject Codes (2010): 06-XX, 11T71, 11YXX, 14H52, 14G50, 34L25, 34A34, 41A05, 41-XX, 49M30, 60G25, 65C20, 68M10, 68M12, 68P25, 68P30, 68U20, 74J20, 78A46, 81U40, 81P94, 90B10, 90B20, 90C90, 94A05, 94A15, 94A60, 97R30, 97R60, 97U70

This Springer imprint is published by the registered company Springer Nature Switzerland AG
The registered company address is: Gewerbestrasse 11, 6330 Cham, Switzerland

Preface

Modern Discrete Mathematics and Analysis with Applications in Cryptography consists of contributions written by research scientists from the international mathematical community.

The chapters are devoted to significant advances in a number of theories and problems of mathematical analysis both continuous and discrete with applications.

These chapters focus on both old and modern developments of mathematics. The volume also presents a few survey papers which discuss the progress made in broader areas which could be particularly useful to graduate students and young researchers.

We would like to express our deepest thanks to the contributors of chapters as well as to the staff of Springer for their excellent collaboration in the presentation of this publication.

Vari, Greece
Athens, Greece
October 2017

Nicholas J. Daras
Themistocles M. Rassias

Preface

Contents

Contributors

Deepika Agrawal Department of Mathematics, Netaji Subhas Institute of Technology, New Delhi, India

Argyrios Alexopoulos Croisy Sur Seine, France

Muhammad Uzair Awan GC University, Faisalabad, Pakistan

Hassen Aydi Department of Mathematics, College of Education in Jubail, Imam Abdulrahman Bin Faisal University, Jubail, Saudi Arabia

Wing Sum Cheung Department of Mathematics, The University of Hong Kong, Pokfulam, Hong Kong

Stefan Czerwik Institute of Mathematics, Silesian University of Technology, Gliwice, Poland

Nicholas J. Daras Department of Mathematics, Hellenic Military Academy, Vari Attikis, Greece

E. Daravigkas Department of Informatics, Aristotle University of Thessaloniki, Thessaloniki, Greece

Konstantinos Demertzis School of Engineering, Department of Civil Engineering, Faculty of Mathematics Programming and General Courses, Democritus University of Thrace, Xanthi, Greece

Silvestru Sever Dragomir Mathematics, College of Engineering & Science, Victoria University, Melbourne City, MC, Australia

DST-NRF Centre of Excellence in the Mathematical and Statistical Sciences, School of Computer Science & Applied Mathematics, University of the Witwatersrand, Johannesburg, South Africa

K. A. Draziotis Department of Informatics, Aristotle University of Thessaloniki, Thessaloniki, Greece

Paraskevi Fika National and Kapodistrian University of Athens, Department of Mathematics, Athens, Greece

Vijay Gupta Department of Mathematics, Netaji Subhas Institute of Technology, New Delhi, India

Mehdi Hassani Department of Mathematics, University of Zanjan, Zanjan, Iran

Lazaros Iliadis School of Engineering, Department of Civil Engineering, Faculty of Mathematics Programming and General Courses, Democritus University of Thrace, Xanthi, Greece

Hassan Azadi Kenary Department of Mathematics, College of Sciences, Yasouj University, Yasouj, Iran

Christos P. Kitsos Technological Educational Institute of Athens, Athens, Greece

Nicholas Kolokotronis Department of Informatics and Telecommunications, University of Peloponnese, Tripoli, Greece

Dimitrios Kotanidis School of Pure and Applied Sciences, Open University of Cyprus, Latsia, Cyprus

Konstantina Lantitsou Democritus University of Thrace, Xanthi, Greece

Jung Rye Lee Department of Mathematics, Daejin University, Pocheon, Republic of Korea

Konstantinos Limniotis School of Pure and Applied Sciences, Open University of Cyprus, Latsia, Cyprus

Hellenic Data Protection Authority, Athens, Greece

Neha Malik Department of Mathematics, Netaji Subhas Institute of Technology, New Delhi, India

Konstantinos Mattas Democritus University of Thrace, Department of Civil Engineering, Xanthi, Greece

Vassilios C. Moussas Laboratory of Applied Informatics, Department of Civil Engineering, University of West Attica, Egaleo-Athens, Greece

Khalida Inayat Noor COMSATS Institute of Information Technology, Islamabad, Pakistan

Muhammad Aslam Noor Mathematics Department, King Saud University, Riyadh, Saudi Arabia

COMSATS Institute of Information Technology, Islamabad, Pakistan

Basil K. Papadopoulos Democritus University of Thrace, Department of Civil Engineering, Xanthi, Greece

A. Papadopoulou Department of Mathematics, Aristotle University of Thessaloniki, Thessaloniki, Greece

I. N. Parasidis Department of Electrical Engineering, TEI of Thessaly, Larissa, Greece

Choonkil Park Research Institute for Natural Sciences, Hanyang University, Seoul, Republic of Korea

E. Providas Department of Mechanical Engineering, TEI of Thessaly, Larissa, Greece

Themistocles M. Rassias Department of Mathematics, National Technical University of Athens, Athens, Greece

Teerapong Suksumran Center of Excellence in Mathematics and Applied, Department of Mathematics, Faculty of Science, Chiang Mai University, Chiang Mai, Thailand

Apostolos Syropoulos Greek Molecular Computing Group, Xanthi, Greece

Thomas L. Toulias Technological Educational Institute of Athens, Egaleo, Athens, Greece

Mihai Turinici A. Myller Mathematical Seminar, A. I. Cuza University, Iaşi, Romania

Bicheng Yang Department of Mathematics, Guangdong University of Education, Guangzhou, Guangdong, People's Republic of China

Chang-Jian Zhao Department of Mathematics, China Jiliang University, Hangzhou, People's Republic of China

Theodoros T. Zygiridis Department of Informatics and Telecommunications Engineering, University of Western Macedonia, Kozani, Greece

Fixed Point Theorems in Generalized b-Metric Spaces

Hassen Aydi and Stefan Czerwik

Abstract In the paper we present fixed point theorems in generalized b-metric spaces both for linear and nonlinear contractions, generalizing several existing results.

Keywords Fixed point · Generalized type contraction · Metric-like space

Introduction and Preliminaries

The idea of b-metric space is due is due to the second author of this paper (for details see [4] and [2]).

Now we extend this idea in the following way. Let X be a set (nonempty).

A function $d\colon X \times X \to [0, \infty]$ is said to be a generalized b-metric (briefly gbm) on X, provided that for $x, y, z \in X$ the following conditions hold true:

(a) $d(x, y) = 0$ if and only if $x = y$,
(b) $d(x, y) = d(y, x)$,
(c) $d(x, y) \leq s[d(x, z) + d(z, y)]$, where $s \geq 1$ is a fixed constant.

Then (X, d) is called a generalized b-metric space with generalized b-metric d.

As usual, by $\mathbb{N}$, $\mathbb{N}_0$, $and \mathbb{R}_+$, we denote the set of all natural numbers, the set of all nonnegative integers, or the set of all nonnegative real numbers, respectively.

H. Aydi
Department of Mathematics, College of Education in Jubail, Imam Abdulrahman Bin Faisal University, Jubail, Saudi Arabia
e-mail: hmaydi@uod.edu.sa

S. Czerwik (✉)
Institute of Mathematics, Silesian University of Technology, Gliwice, Poland

N. J. Daras, Th. M. Rassias (eds.), *Modern Discrete Mathematics and Analysis*, Springer Optimization and Its Applications 131,
https://doi.org/10.1007/978-3-319-74325-7_1

If $f: X \to X$, by f^n, we denote the n-th iterate of f:

$$f^0(x) = x, \quad x \in X; \quad f^{n+1} = f \circ f^n.$$

Here the symbol $\varphi \circ f$ denotes the function $\varphi[f(x)]$ for $x \in X$.

Linear Quasi-Contractions

We start with the following theorem

Theorem 2.1 *Let (X, d) be a complete generalized b-metric space. Assume that $T: X \to X$ is continuous and satisfies the condition*

$$d(T(x), T^2(x)) \leq \alpha d(x, T(x)), \tag{1}$$

for $x \in X$, such that $d(x, T(x)) < \infty$, and

$$\alpha s = q < 1. \tag{2}$$

Let $x \in X$ be an arbitrarily fixed. Then the following alternative holds either:

(A) *for every nonnegative integer $n \in \mathbb{N}_0$,*

$$d(T^n(x), T^{n+1}(x)) = \infty,$$

or

(B) *there exists an $k \in \mathbb{N}_0$ such that*

$$d(T^k(x), T^{k+1}(x)) < \infty.$$

In (B)

(i) *the sequence $\{T^m(x)\}$ is a Cauchy sequence in X;*
(ii) *there exists a point $u \in X$ such that*

$$\lim_{m\to\infty} d(T^m(x), u) = 0 \quad \text{and} \quad T(u) = u.$$

Proof From (1) we get (in case (B))

$$d(T^{k+1}(x), T^{k+2}(x)) \leq \alpha d(T^k(x), T^{k+1}(x)) < \infty$$

and by induction

$$d(T^{k+n}(x), T^{k+n+1}(x)) \leq \alpha^n d(T^k(x), T^{k+1}(x)), \quad n = 0, 1, 2, \dots . \tag{3}$$

Consequently, for $n, v \in \mathbb{N}_0$, by (3), we obtain

$$\begin{aligned} d(T^{k+n}(x), T^{k+n+v}(x)) &\leq sd(T^{k+n}(x), T^{k+n+1}(x)) + \ldots + s^{v-1}d(T^{k+n+v-2}(x), T^{k+n+v-1}(x)) \\ &+ s^v d(T^{k+n+v-1}(x), T^{k+n+v}(x)) \\ &\leq s\alpha^n d(T^k(x), T^{k+1}(x)) + \ldots + s^{v-1}\alpha^{n+v-2} d(T^k(x), T^{k+1}(x)) \\ &+ \alpha^{n+v-1} s^v d(T^k(x), T^{k+1}(x)) \\ &\leq s\alpha^n [1 + (s\alpha) + \ldots + (s\alpha)^{v-1}] d(T^k(x), T^{k+1}(x)) \\ &\leq s\alpha^n \sum_{m=0}^{\infty} (s\alpha)^m d(T^k(x), T^{k+1}(x)) \\ &\leq \frac{s\alpha^n}{1 - s\alpha} d(T^k(x), T^{k+1}(x)). \end{aligned}$$

Finally,

$$d(T^{k+n}(x), T^{k+n+v}(x)) \leq \frac{s\alpha^n}{1 - s\alpha} d(T^k(x), T^{k+1}(x)) \tag{4}$$

for $n, v \in \mathbb{N}_0$.
By (4) it follows that $\{T^n(x)\}$ is a Cauchy sequence of elements of X.
Since X is complete, so there exists $u \in X$ with

$$\lim_{n \to \infty} d(T^n(x), u) = 0.$$

Because T is continuous with respect to d (see the assumptions), therefore

$$u = \lim_{n \to \infty} T^{n+1}(x) = T(\lim_{n \to \infty} T^n(x)) = T(u),$$

and u is a fixed point of T, which ends the proof. □

Remark 2.1 In the space X, T may have more than one fixed point (for consider $T(x) = x$).

Remark 2.2 If additionally d is a continuous function (as a function of one variable), then we have the estimation (see (4)):

$$d(T^{k+n}(x), u) \leq \frac{s\alpha^n}{1 - s\alpha} d(T^k(x), T^{k+1}(x)). \tag{5}$$

Remark 2.3 A function d may not be continuous even as a function of one variable (e.g., see [10]).

Remark 2.4 An operator T satisfying (1) may not be continuous (see [4]).
In fact, if T satisfies

$$d(T(x), T(y)) \le \frac{\alpha}{2}[d(x, T(x)) + d(y, T(y))], \quad x, y \in X, \tag{6}$$

then T satisfies also (1).

Remark 2.5 Theorem 2.1 generalizes theorems Diaz, Margolis ([7]), Luxemburg ([6, 9]), Banach ([3]), Czerwik, Król ([1]), and others (see also [2, 8, 11–14]).

Nonlinear Contractions

In this section we present the following result.

Theorem 3.1 *Assume that (X, d) is a complete generalized b-metric space, and $T: X \to X$ satisfies the condition*

$$d(T(x), T(y)) \le \varphi[d(x, y)] \tag{7}$$

for $x, y \in X$, $d(x, y) < \infty$, where $\varphi\colon [0, \infty) \to [0, \infty)$ is nondecreasing and

$$\lim_{n\to\infty} \varphi^n(z) = 0 \quad for\ z > 0. \tag{8}$$

Let $x \in X$ be arbitrarily fixed. Then the following alternative holds either:

(C) *for every nonnegative integer $n \in \mathbb{N}_0$*

$$d(T^n(x), T^{n+1}(x)) = \infty,$$

or

(D) *there exists an $k \in \mathbb{N}_0$ such that*

$$d(T^k(x), T^{k+1}(x)) < \infty.$$

In (D)

(iii) *the sequence $\{T^m(x)\}$ is a Cauchy sequence in X;*
(iv) *there exists a point $u \in X$ such that*

$$\lim_{n\to\infty} d(T^n(x), u) = 0 \quad and \quad T(u) = u;$$

(v) *u is the unique fixed point of T in* $B := \{t \in X : d(T^k(x), t) < \infty\}$;
(vi) *for every* $t \in B$,

$$\lim_{n\to\infty} d(T^n(t), u) = 0.$$

If, moreover, d is continuous (with respect to one variable) and

$$\sum_{k=1}^{\infty} s^k \varphi^k(t) < \infty \quad for \quad t > 0,$$

then for $t \in B$

$$d(T^m(t), u) \le \sum_{k=0}^{\infty} s^{k+1} \varphi^{m+k}[d(t, T(t))], \quad m \in \mathbb{N}_0. \tag{9}$$

Proof The proof will consist with a few steps.

1^0) Let's take $x \in X$ and $\varepsilon > 0$. Take $n \in \mathbb{N}$ such that

$$\varphi^n(\varepsilon) < \frac{\varepsilon}{2s},$$

and put $F = T^n$, $\alpha = \varphi^n$ and $x_m = F^m(x)$ for $m \in \mathbb{N}$. Then for all $x, y \in X$ such that $d(x, y) < \infty$, one gets

$$d(F(x), F(y)) \le \varphi^n[d(x, y)] = \alpha[d(x, y)]. \tag{10}$$

2^0) One can prove that (B, d) is a complete b-metric space. Clearly, $T^k(x),\ T^{k+1}(x) \in B$.

3^0) Now we observe that $T : B \to B$. For if $t \in B$, i.e., $d(T^k(x), t) < \infty$, then

$$\begin{aligned} d(T^k(x), T(t)) &\le s[d(T^k(x), T^{k+1}(x)) + d(T^{k+1}(x), T(t))] \\ &\le s\left[\varepsilon_1 + \varphi[d(T^k(x), t)]\right] \\ &\le s[\varepsilon_1 + \varepsilon_2] < \infty, \end{aligned}$$

where $\varepsilon_1,\ \varepsilon_2$ some positive numbers. Also $F : B \to B$.

4^0) For $t \in B$ we have $\{F^m(t)\} \subset B$, for all $m \in \mathbb{N}_0$. We verify that $\{F^m(t)\}$ is a Cauchy sequence. In fact, putting $y_m = F^m(t)$, $m \in \mathbb{N}_0$, we get

$$d(F(t), F^2(t)) \le \alpha[d(t, F(t))]$$

and by induction

$$d(F^m(t), F^{m+1}(t)) \leq \alpha^m[d(t, F(t))]$$

i.e.,

$$d(y_m, y_{m+1}) \leq \alpha^m[d(t, F(t))],$$

whence $d(y_m, y_{m+1}) \to 0$ as $m \to \infty$. Let m be such that

$$d(y_m, y_{m+1}) < \frac{\varepsilon}{2s}.$$

Then for every $z \in K(y_m, \varepsilon) := \{y \in X : d(y_m, y) \leq \varepsilon\}$ we obtain

$$d(F(z), F(y_m)) \leq \alpha[d(z, y_m)] \leq \alpha(\varepsilon) = \varphi^n(\varepsilon) < \frac{\varepsilon}{2s},$$

$$d(F(y_m), y_m) = d(y_{m+1}, y_m) < \frac{\varepsilon}{2s},$$

whence

$$d(F(z), y_m) \leq s\left(\frac{\varepsilon}{2s} + \frac{\varepsilon}{2s}\right) = \varepsilon,$$

which means that F maps $K(y_m, \varepsilon)$ into itself. Therefore

$$d(y_r, y_l) \leq 2s\varepsilon \quad for \quad r, l \geq m,$$

so the sequence $\{y_r\} = \{F^r(t)\}$ is a Cauchy sequence.

5^0) Since B is complete, so there exists $u \in B \subset X$ such that $y_r \to u$ as $r \to \infty$. Furthermore, by the continuity of F (arising from (10))

$$F(u) = \lim_{r\to\infty} F(y_r) = \lim_{r\to\infty} y_{r+1} = u,$$

i.e., u is a fixed point of F. Taking into consideration that $\alpha(t) = \varphi^n(t) < t$ for any $t > 0$, it is obvious that F has exactly one fixed point in B.

Moreover, by (7) T is continuous on B, so we have

$$T(u) = \lim_{r\to\infty} T(F^r(t)) = \lim_{r\to\infty} F^r(T(t)) = u$$

and hence u is a fixed point of T as well. Obviously by (7) such point is only one in B. Eventually, since for every $t \in B$ and every $r = 0, 1, \ldots, n-1$,

$$T^m(t) = T^{nl+r}(t) = F^l[T^r(t)] \to u \quad as \quad l \to \infty,$$

so

$$d(T^m(t), u) \to 0 \quad as \quad m \to \infty$$

for every $t \in B$, whence we get *(vi)*.

6^0) For any $t \in B$ and $m, n \in \mathbb{N}_0$, we have

$$\begin{aligned} d(T^m(t), T^{m+n}(t)) &\leq s[d(T^m(t), T^{m+1}(t)) + d(T^{m+1}(t), T^{m+n}(t))] \\ &\leq sd(T^m(t), T^{m+1}(t)) + \ldots + s^n d(T^{m+n-1}(t), T^{m+n}(t)) \\ &\leq s\varphi^m[d(t, T(t))] + \ldots + s^n \varphi^{m+n-1}[d(t, T(t))] \\ &\leq \sum_{r=0}^{\infty} s^{r+1} \varphi^{m+r}[d(t, T(t))], \end{aligned}$$

consequently

$$d(T^m(t), T^{m+n}(t)) \leq \sum_{r=0}^{\infty} s^{r+1} \varphi^{m+r}[d(t, T(t))].$$

Therefore, if $n \to \infty$ and d is continuous (with respect to one variable), then for $t \in B, m \in \mathbb{N}_0$

$$d(T^m(t), u) \leq \sum_{r=0}^{\infty} s^{r+1} \varphi^{m+r}[d(t, T(t))]$$

which concludes the proof. □

If X is bms, then $B = X$, and we have from Theorem 3.1:

Corollary 3.1 *Let (X, d) be a complete b-metric space and $T: X \to X$ satisfy*

$$d(T(x), T(y)) \leq \varphi[d(x, y)], \quad x, y \in X,$$

where $\varphi: \mathbb{R}_+ \to \mathbb{R}_+$ is nondecreasing function such that $\lim_{n\to\infty} \varphi^n(t) = 0$ *for each $t > 0$. Then T has exactly one fixed point $u \in X$, and*

$$\lim_{n\to\infty} d(T^n(x), u) = 0$$

for each $x \in X$.

If, moreover, d is continuous (with respect to one variable) and the series of iterates

$$\sum_{k=1}^{\infty} s^k \varphi^k(t) < \infty \quad for \quad t > 0,$$

then for $z \in X$ and $m \in \mathbb{N}_0$

$$d(T^m(z), u) \leq \sum_{k=0}^{\infty} s^{k+1} \varphi^{m+k}[d(z, T(z))]. \tag{11}$$

Remark 3.1 Corollary is contained in [4] for $s = 2$ (see also [2]).

Remark 3.2 It would be interesting to consider a convergence of series of iterates:

$$\sum_{n=1}^{\infty} \varphi^n(t).$$

For example, one of the sufficient conditions is the following

$$\limsup_{n\to\infty} \frac{\varphi^{n+1}(t)}{\varphi^n(t)} = q(t) < 1, \quad t > 0. \tag{12}$$

Let's also note another one such conditions:

$$\liminf_{n\to\infty} \left[-\frac{\ln \varphi^n(t)}{\ln n} \right] = \alpha(t) > 1, \quad t > 0. \tag{13}$$

For more details, see [5].

References

1. Afshari, H., Aydi, H., Karapinar, E.: Existence of fixed points of set-valued mappings in b-metric spaces. East Asian Math. J. **32**(3), 319–332 (2016)
2. Aydi, H., Felhi, A., Sahmim, S.: On common fixed points for (α, ψ)-contractions and generalized cyclic contractions in b-metric-like spaces and consequences. J. Nonlinear Sci. Appl. **9**, 2492–2510 (2016)
3. Aydi, H., Bota, M.F., Karapinar, E., Mitrovic, S.: A fixed point theorem for set-valued quasicontractions in b-metric spaces. Fixed Point Theory Appl. **2012**, 88 (2012)
4. Aydi, H., Bota, M.F., Karapinar, E., Moradi, S.: A common fixed point for weak phicontractions on b-metric spaces. Fixed Point Theory **13**(2), 337–346 (2012)
5. Banach, S.: Sur les opérations dans les ensembles abstraits et leur application aux équations intégrales. Fund. Math. **3**, 133–181 (1922)
6. Czerwik, S.: What Cauchy has not said about series (submitted)
7. Czerwik, S.: Contraction mappings in b-metric spaces. Acta Math. et Informatica Univ. Ostraviensis **1**, 5–11 (1993)
8. Czerwik, S.: Nonlinear set-valued contraction mappings in b-metric spaces. Atti Sem. Mat. Fis. Univ. Modena. **XLVI**, 263–276 (1998)
9. Czerwik, S., Krol, K.: Fixed point theorems in generalized metric spaces. Asian-European J. Math. **10**(1), 1–9 (2017)
10. Diaz, J.B., Margolis, B.: A fixed point theorem of the alternative, for contractions on a generalized complete metric space. Bull. Am. Math. Soc. **74**, 305–309 (1968)

11. Dugundji, J., Granas, A.: Fixed Point Theory, pp. 5–209. Polish Scientific Publishers, Warszawa (1982)
12. Luxemburg, W.A.J.: On the convergence of successive approximations in the theory of ordinary differential equations. II, Koninkl, Nederl. Akademie van Wetenschappen, Amsterdam, Proc. Ser. A (5) 61; Indag. Math. (5) **20**, 540–546 (1958)
13. Luxemburg, W.A.J.: On the convergence of successive approximations in the theory of ordinary differential equations. III, Nieuw. Arch. Wisk. (3) **6**, 93–98 (1958)
14. Roshan, J.R., Shobkolaei, N., Sedghi, S., Abbas, M.: Common fixed point of four maps in b-metric spaces. Hacet. J. Math. Stat. **43**(4), 613–624 (2014)

Orlicz Dual Brunn-Minkowski Theory: Addition, Dual Quermassintegrals, and Inequalities

Chang-Jian Zhao and Wing Sum Cheung

Abstract We further consider the Orlicz dual Brunn-Minkowski theory. An Orlicz radial harmonic addition is introduced, which generalizes the L_p-radial addition and the L_p-harmonic addition to an Orlicz space, respectively. The variational formula for the dual mixed quermassintegrals with respect to the Orlicz radial harmonic addition is proved, and the new Orlicz dual quermassintegrals generalizes the L_p-dual quermassintegrals. The fundamental notions and conclusions of the dual quermassintegrals and the Minkoswki and Brunn-Minkowski inequalities for the dual quermassintegrals are extended to an Orlicz setting. The new Orlicz-Minkowski and Brunn-Minkowski inequalities in special case yield the Orlicz dual Minkowski inequality and Orlicz dual Brunn-Minkowski inequality, which also imply the L_p-dual Minkowski inequality and L_p-dual Brunn-Minkowski inequality for the dual quermassintegrals. As application, a dual log-Minkowski inequality is proved.

Introduction

One of the most important operations in geometry is vector addition. As an operation between sets K and L, defined by

$$K + L = \{x + y : x \in K, y \in L\},$$

it is usually called Minkowski addition, and combine volume plays an important role in the Brunn-Minkowski theory. During the last few decades, the theory has

C. -J. Zhao
Department of Mathematics, China Jiliang University, Hangzhou, People's Republic of China
e-mail: chjzhao@163.com; chjzhao@aliyun.com

W. S. Cheung (✉)
Department of Mathematics, The University of Hong Kong, Pokfulam, Hong Kong
e-mail: wscheung@hku.hk

N. J. Daras, Th. M. Rassias (eds.), *Modern Discrete Mathematics and Analysis*,
Springer Optimization and Its Applications 131,
https://doi.org/10.1007/978-3-319-74325-7_2

been extended to L_p-Brunn-Minkowski theory. The first, a set called as L_p addition, introduced by Firey in [6] and [7]. Denoted by $+_p$, for $1 \le p \le \infty$, defined by

$$h(K +_p L, x)^p = h(K, x)^p + h(L, x)^p, \tag{1}$$

for all $x \in \mathbb{R}^n$ and compact convex sets K and L in $\mathbb{R}^n$ containing the origin. When $p = \infty$, (1) is interpreted as $h(K +_\infty L, x) = \max\{h(K, x), h(L, x)\}$, as is customary. Here the functions are the support functions. If K is a nonempty closed (not necessarily bounded) convex set in $\mathbb{R}^n$, then

$$h(K, x) = \max\{x \cdot y : y \in K\},$$

for $x \in \mathbb{R}^n$, defined the support function $h(K, x)$ of K. A nonempty closed convex set is uniquely determined by its support function. L_p addition and inequalities are the fundamental and core content in the L_p Brunn-Minkowski theory. For recent important results and more information from this theory, we refer to [17–19, 21, 27, 32–36, 39–41, 46–49] and the references therein. In recent years, a new extension of L_p-Brunn-Minkowski theory is to Orlicz-Brunn-Minkowski theory, initiated by Lutwak et al. [37] and [38]. In these papers the notions of L_p-centroid body and L_p-projection body and the fundamental inequalities were extended to an Orlicz setting. The Orlicz centroid inequality for star bodies was introduced in [55] which is an extension from convex to star bodies. Recently, the Orlicz addition of convex bodies was introduced and extend the L_p Brunn-Minkowski inequality to the Orlicz Brunn-Minkowski inequality (see [50]). A dual Orlicz-Brunn-Minkowski theory was introduced (see [56]). The other articles advance the theory, and its dual theory can be found in literatures [20, 23, 25, 43, 53], and [54].

Very recently, Gardner et al. [12] constructed a general framework for the Orlicz-Brunn-Minkowski theory and made clear for the first time the relation to Orlicz spaces and norms. They introduced the Orlicz addition $K +_\varphi L$ of compact convex sets K and L in $\mathbb{R}^n$ containing the origin, implicitly, by

$$\varphi\left(\frac{h(K, x)}{h(K +_\varphi L, x)}, \frac{h(L, x)}{h(K +_\varphi L, x)}\right) = 1, \tag{2}$$

for $x \in \mathbb{R}^n$, if $h(K, x) + h(L, x) > 0$, and by $h(K +_\varphi L, x) = 0$, if $h(K, x) = h(L, x) = 0$. Here φ is the set of convex function $\varphi : [0, \infty)^2 \to [0, \infty)$ that are increasing in each variable and satisfy $\varphi(0, 0) = 0$ and $\varphi(1, 0) = \varphi(0, 1) = 1$. Orlicz addition reduces to L_p addition, $1 \le p < \infty$, when $\varphi(x_1, x_2) = x_1^p + x_2^p$, or L_∞ addition, when $\varphi(x_1, x_2) = \max\{x_1, x_2\}$.

The radial addition $K \widetilde{+} L$ of star sets (compact sets that is star shaped at o and contains o) K and L can be defined by

$$K \widetilde{+} L = \{x \widetilde{+} y : x \in K, y \in L\},$$

where $x\widetilde{+}y = x + y$ if x, y and o are collinear, $x\widetilde{+}y = o$, otherwise, or by

$$\rho(K\widetilde{+}L, \cdot) = \rho(K, \cdot) + \rho(L, \cdot),$$

where $\rho(K, \cdot)$ denotes the radial function of star set K. The radial function of star set K is defined by

$$\rho(K, u) = \max\{c \geq 0 : cu \in K\},$$

for $u \in S^{n-1}$. Hints as to the origins of the radial addition can be found in [9, p. 235]. If $\rho(K, \cdot)$ is positive and continuous, K will be called a star body. Let $\mathscr{S}^n$ denote the set of star bodies in $\mathbb{R}^n$. When combined with volume, radial addition gives rise to another substantial appendage to the classical theory, called the dual Brunn-Minkowski theory. Radial addition is the basis for the dual Brunn-Minkowski theory (see, e.g., [3, 11, 15, 16, 24, 26, 30], and [48] for recent important contributions). The original is originated from the Lutwak [28]. He introduced the concept of dual mixed volume laid the foundation of the dual Brunn-Minkowski theory. In particular, the dual Brunn-Minkowski theory can count among its successes the solution of the Busemann-Petty problem in [8, 11, 29, 45], and [51].

More generally, for any $p < 0$ (or $p > 0$), the p-radial addition $K\widetilde{+}_pL$ defined by (see [10])

$$\rho(K\widetilde{+}_pL, x)^p = \rho(K, x)^p + \rho(L, x)^p, \tag{3}$$

for $x \in \mathbb{R}^n$ and $K, L \in \mathscr{S}^n$. When $p = \infty$ or $-\infty$, (3) is interpreted as $\rho(K\widetilde{+}_\infty L, x) = K \cup L$ or $\rho(K\widetilde{+}_{-\infty}L, x) = K \cap L$ (see [14]). In 1996, the harmonic p-addition for star bodies was defined by Lutwak: If K, L are star bodies, for $p \geq 1$, the harmonic p-addition defined by (see [32])

$$\rho(K\widehat{+}_pL, x)^{-p} = \rho(K, x)^{-p} + \rho(L, x)^{-p}. \tag{4}$$

For convex bodies, the harmonic p-addition was first investigated by Firey [6].

The operations of the p-radial addition, the harmonic p-addition, and the L_p-dual Minkowski, Brunn-Minkwski inequalities are fundamental notions and inequalities from the L_p-dual Brunn-Minkowski theory. For recent important results and more information from this theory, we refer to [3, 5, 10, 13–16], and [26] and the references therein.

In the paper, we consider the Orlicz dual Brunn-Minkowski theory, an idea suggested by Gardner et al. [12]. The fundamental notions and conclusions of the radial addition, the p-radial addition, the harmonic addition, the harmonic L_p-addition, the harmonic Blaschke addition, the p-harmonic Blaschke addition, the dual mixed quermassintegral, the p-dual mixed quermassintegral, the L_p-dual quermassintegral, the L_p-dual mixed quermassintegral, and inequalities are extended to an Orlicz setting. It represents a generalization of the L_p-dual-Brunn-Minkowski theory, analogous to the way that Orlicz spaces generalize L_p spaces.

In section "Orlicz Radial Harmonic Addition", we introduce a notion of Orlicz radial harmonic addition $K\widehat{\widetilde{+}}_\varphi L$ of $K, L \in \mathscr{S}^n$, by

$$\varphi\left(\frac{\rho(K,x)}{\rho(K\widehat{\widetilde{+}}_\varphi L,x)}, \frac{\rho(L,x)}{\rho(K\widehat{\widetilde{+}}_\varphi L,x)}\right) = 1. \tag{5}$$

Here $\varphi \in \Phi_2$, the set of convex function $\varphi : [0,\infty)^2 \to (0,\infty)$ that are decreasing in each variable and satisfy $\varphi(0,0) = \infty$, $\varphi(0,x_2) = \varphi(x_1,0) = \infty$, and $\varphi(0,1) = \varphi(1,0) = \infty$.

When $p < 0$ and $\varphi(x_1,x_2) = x_1^p + x_2^p$, the Orlicz radial harmonic addition $\widehat{\widetilde{+}}_\varphi$ becomes the p-radial addition $\widetilde{+}_p$. When $p \geq 1$ and $\varphi(x_1,x_2) = x_1^{-p} + x_2^{-p}$, the Orlicz radial harmonic addition $\widehat{\widetilde{+}}_\varphi$ becomes the harmonic p-radial addition $\widehat{+}_p$. This shows that (5) is a generalization of (3) and (4).

In section "Preliminaries", we introduce the basic definitions and notions and shows the dual mixed p-quermassintegral and the dual mixed harmonic p-quermassintegral.

In section "Orlicz Dual Brunn-Minkowski Inequality", we establish an Orlicz dual Brunn-Minkowski inequality for the Orlicz radial harmonic addition. If $K, L \in \mathscr{S}^n$, $0 \leq i < n$ and $\varphi \in \Phi_2$, then

$$1 \geq \varphi\left(\left(\frac{\widetilde{W}_i(K)}{\widetilde{W}_i(K\widehat{\widetilde{+}}_\varphi L)}\right)^{1/(n-i)}, \left(\frac{\widetilde{W}_i(L)}{\widetilde{W}_i(K\widehat{\widetilde{+}}_\varphi L)}\right)^{1/(n-i)}\right). \tag{6}$$

If φ is strictly convex, equality holds if and only if K and L are dilates. Here, $\widetilde{W}_i(K)$ denotes the usual dual quermassintegral of star body K.

When $p < 0$ and $\varphi(x_1,x_2) = x_1^p + x_2^p$ in (6), (6) becomes the following result. If $K, L \in \mathscr{S}^n$, $0 \leq i < n$ and $p < 0$, then

$$\widetilde{W}_i(K\widetilde{+}_p L)^{p/(n-i)} \geq \widetilde{W}_i(K)^{p/(n-i)} + \widetilde{W}_i(L)^{p/(n-i)}, \tag{7}$$

with equality if and only if K and L are dilates. Taking $i = 0$ in (7), this yields Gardner's Brunn-Minkowski inequality for p-radial addition (see [10]). If $K, L \in \mathscr{S}^n$ and $p < 0$, then

$$V(K\widetilde{+}_p L)^{p/n} \geq V(K)^{p/n} + V(L)^{p/n}, \tag{8}$$

with equality if and only if K and L are dilates.

When $p \geq 1$ and $\varphi(x_1,x_2) = x_1^{-p} + x_2^{-p}$ in (6), (6) becomes the following result. If $K, L \in \mathscr{S}^n$, $0 \leq i < n$ and $p \geq 1$, then

$$\widetilde{W}_i(K\widehat{+}_p L)^{-p/(n-i)} \geq \widetilde{W}_i(K)^{-p/(n-i)} + \widetilde{W}_i(L)^{-p/(n-i)}, \tag{9}$$

with equality if and only if K and L are dilates. Taking $i = 0$ in (9), this yields Lutwak's L_p-Brunn-Minkowski inequality for harmonic p-addition (see [32]). If $K, L \in \mathscr{S}^n$ and $p \geq 1$, then

$$V(K\widehat{+}_pL)^{-p/n} \geq V(K)^{-p/n} + V(L)^{-p/n}, \tag{10}$$

with equality if and only if K and L are dilates. If $i = 0$ and only when $\widehat{\widetilde{+}}_\varphi$ is $\widehat{+}_\varphi$, (6) changes Zhu, Zhou, and Xu's result [53].

The p-dual mixed volume, $\widetilde{V}_p(K, L)$, for $p < 0$ (or $p > 0$), defined by (see [52])

$$\widetilde{V}_p(K, L) = \frac{1}{n}\int_{S^{n-1}} \rho(K, u)^{n-p}\rho(L, u)^p dS(u). \tag{11}$$

Let $K, L \in \mathscr{S}^n$, $0 \leq i < n$ and $p < 0$ (or $p > 0$), dual mixed p-quermassintegral of star bodies K and L, defined by

$$\widetilde{W}_{p,i}(K, L) = \frac{1}{n}\int_{S^{n-1}} \rho(K \cdot u)^{n-i-p}\rho(L \cdot u)^p dS(u). \tag{12}$$

Obviously, when $i = 0$, (12) reduces (11).

The L_p-dual mixed volume, for $p \geq 1$, $\widetilde{V}_{-p}(K, L)$, defined by (see [32])

$$\widetilde{V}_{-p}(K, L) = \frac{1}{n}\int_{S^{n-1}} \rho(K, u)^{n+p}\rho(L, u)^{-p} dS(u). \tag{13}$$

Let $K, L \in \mathscr{S}^n, 0 \leq i < n$ and $p \geq 1$, dual mixed harmonic p-quermassintegral of star bodies K and L, defined by

$$\widetilde{W}_{-p,i}(K, L) = \frac{1}{n}\int_{S^{n-1}} \rho(K \cdot u)^{n-i+p}\rho(L \cdot u)^{-p} dS(u). \tag{14}$$

Obviously, when $i = 0$, (14) reduces (13).

In section "Orlicz Radial Harmonic Addition" we introduce also a notion of Orlicz dual linear combination, by means of an appropriate modification of (5). Unlike the L_p dual case, an Orlicz scalar multiplication cannot generally be considered separately. The particular instance of interest corresponds to using (5) with $\varphi(x_1, x_2) = \varphi_1(x_1) + \varepsilon\varphi_2(x_2)$ for $\varepsilon > 0$ and some $\varphi_1, \varphi_2 \in \Phi$, where the sets of convex functions $\varphi_1, \varphi_2 : [0, \infty) \to (0, \infty)$ that are decreasing and satisfy $\varphi_1(0) = \varphi_2(0) = \infty$ and $\varphi_1(1) = \varphi_2(1) = 1$.

In section "Orlicz Dual Mixed Quermassintegral", a notion of Orlicz dual mixed quermassintegral is introduced. We prove that there exist a new dual mixed volume an call Orlicz dual mixed quermassintegral, obtaining the equation

$$\frac{(\varphi_1)'_l(1)}{n-i}\lim_{\varepsilon\to 0^+}\frac{\tilde{W}_i(K\widehat{\widetilde{+}}_{\varphi,\varepsilon}L) - \tilde{W}_i(K)}{\varepsilon} = \frac{1}{n}\int_{S^{n-1}} \varphi_2\left(\frac{\rho(L, u)}{\rho(K, u)}\right)\rho(K, u)^{n-i} dS(u),$$

where $0 \leq i < n$, $\varphi \in \Phi_2$ and $\varphi_1, \varphi_2 \in \Phi$. Denoting by $\widetilde{W}_{\varphi,i}(K, L)$, the integral on the right-hand side with φ_2 replaced by φ, we see that either side of the above equation is equal to $\widetilde{W}_{\varphi_2,i}(K, L)$. Hence a new Orlicz-dual mixed volume $\widetilde{W}_{\varphi,i}(K, L)$ is found.

$$\widetilde{W}_{\varphi,i}(K, L) = \frac{1}{n}\int_{S^{n-1}} \varphi\left(\frac{\rho(L, u)}{\rho(K, u)}\right) \rho(K, u)^{n-i} dS(u). \tag{15}$$

When $\varphi_1(t) = \varphi_2(t) = t^p$ and $p < 0$, (15) becomes (12). When $\varphi_1(t) = \varphi_2(t) = t^{-p}$ and $p \geq 1$, (15) becomes (14). This shows, Orlicz-dual mixed quermassintegral $\widetilde{W}_{\varphi,i}(K, L)$ changes to dual mixed p-quermassintegral $\widetilde{W}_{p,i}(K, L)$ $(p < 0)$ and dual mixed harmonic p-quermassintegral $\widetilde{W}_{-p,i}(K, L)$ $(p \geq 1)$, respectively.

In section "Orlicz Dual Minkowski Inequality", we establish an Orlicz dual-Minkowski inequality: If $K, L \in \mathscr{S}^n$, $0 \leq i < n$ and $\varphi \in \Phi$, then

$$\widetilde{W}_{\varphi,i}(K, L) \geq \widetilde{W}_i(K)\varphi\left(\left(\frac{\widetilde{W}_i(L)}{\widetilde{W}_i(K)}\right)^{1/(n-i)}\right). \tag{16}$$

If φ is strictly convex, equality holds if and only if K and L are dilates.

When $\varphi_1(t) = \varphi_2(t) = t^p$ and $p < 0$, (16) becomes the following inequality. If $K, L \in \mathscr{S}^n$, $0 \leq i < n$ and $p < 0$, then

$$\widetilde{W}_{p,i}(K, L)^{n-i} \geq \widetilde{W}_i(K)^{n-i-p}\widetilde{W}_i(L)^p, \tag{17}$$

with equality if and only if K and L are dilates. Taking $i = 0$ in (17), this yields the p-dual Minkowski inequality (see [52]): If $K, L \in \mathscr{S}^n$ and $p < 0$, then

$$\widetilde{V}_p(K, L)^n \geq V(K)^{n-p}V(L)^p, \tag{18}$$

with equality if and only if K and L are dilates.

When $\varphi_1(t) = \varphi_2(t) = t^{-p}$ and $p \geq 1$, (16) becomes the following inequality. If $K, L \in \mathscr{S}^n$, $0 \leq i < n$ and $p \geq 1$, then

$$\widetilde{W}_{-p,i}(K, L)^{n-i} \geq \widetilde{W}_i(K)^{n-i+p}\widetilde{W}_i(L)^{-p}, \tag{19}$$

with equality if and only if K and L are dilates. Taking $i = 0$ in (19), this yields Lutwak's L_p-dual Minkowski inequality is following: If $K, L \in \mathscr{S}^n$ and $p \geq 1$, then

$$\widetilde{V}_{-p}(K, L)^n \geq V(K)^{n+p}V(L)^{-p}, \tag{20}$$

with equality if and only if K and L are dilates.

In 2012, Böröczky et al. [4] proposed a conjecture: For all $n \geq 2$ and origin-symmetric convex bodies K and L, is it true that

$$\int_{S^{n-1}} \log\left(\frac{h(L,u)}{h(K,u)}\right) dV_n(K,u) \geq \frac{1}{n}\log\left(\frac{V(L)}{V(K)}\right)\ ? \tag{21}$$

They proved only when $n = 2$. The conjecture was called as log-Minkowski inequality. Here $dV_n(K,u)$ denotes a Borel measure of convex body K in S^{n-1}, called as normalized cone measure, defined by (see, e.g., [12])

$$dV_n(K,u) = \frac{h(K,u)}{nV(K)} dS(K,u),$$

where $S(K,\cdot)$ is a Boel measure of convex body K on S^{n-1}.

In section "Dual Log-Minkowski Inequality", for all $n \geq 2$, we prove a dual log-Minkowski inequality. If $K, L \in \mathscr{S}^n$ such that $L \subset \mathrm{int}K$, then

$$\int_{S^{n-1}} \log\left(\frac{\rho(L,u)}{\rho(K,u)}\right) d\tilde{V}_n(K,u) \leq \frac{1}{n}\log\left(\frac{V(L)}{V(K)}\right), \tag{22}$$

with equality if and only if K and L are dilates. Here $d\tilde{V}_n(K,u)$ denotes a Borel measure of star body K in S^{n-1}, call as dual normalized cone measure, defined by

$$d\tilde{V}_n(K,u) = \frac{\rho(K,u)^n}{nV(K)} dS(u).$$

Preliminaries

The setting for this paper is n-dimensional Euclidean space $\mathbb{R}^n$. A body in $\mathbb{R}^n$ is a compact set equal to the closure of its interior. For a compact set $K \subset \mathbb{R}^n$, we write $V(K)$ for the (n-dimensional) Lebesgue measure of K and call this the volume of K. Let $\mathscr{K}^n$ denote the class of nonempty compact convex subsets containing the origin in their interiors in $\mathbb{R}^n$. Associated with a compact subset K of $\mathbb{R}^n$, which is star shaped with respect to the origin and contains the origin, its radial function is $\rho(K,\cdot) : S^{n-1} \to [0,\infty)$, defined by

$$\rho(K,u) = \max\{\lambda \geq 0 : \lambda u \in K\}.$$

Note that the class (star sets) is closed under unions, intersection, and intersection with subspace. The radial function is homogeneous of degree -1, that is,

$$\rho(K,ru) = r^{-1}\rho(K,u),$$

for all $x \in \mathbb{R}^n$ and $r > 0$. Let $\tilde{\delta}$ denote the radial Hausdorff metric, as follows, if $K, L \in \mathcal{S}^n$, then (see, e.g., [40])

$$\tilde{\delta}(K, L) = |\rho(K, u) - \rho(L, u)|_{\infty}.$$

From the definition of the radial function, it follows immediately that for $A \in GL(n)$ the radial function of the image $AK = \{Ay : y \in K\}$ of K is given by

$$\rho(AK, x) = \rho(K, A^{-1}x),$$

for all $x \in \mathbb{R}^n$.

For $K_i \in \mathcal{S}^n, i = 1, \ldots, m$, define the real numbers R_{K_i} and r_{K_i} by

$$R_{K_i} = \max_{u \in S^{n-1}} \rho(K_i, u), \quad \text{and } r_{K_i} = \min_{u \in S^{n-1}} \rho(K_i, u),$$

obviously, $0 < r_{K_i} < R_{K_i}$, for all $K_i \in \mathcal{S}^n$, and writing $R = \max\{R_{K_i}\}$ and $r = \min\{r_{K_i}\}$, where $i = 1, \ldots, m$.

Mixed p-Quermassintegrals of Convex Bodies

The mixed p-quermassintegrals $W_{p,i}(K, L)$, for $K, L \in \mathcal{K}^n$, $0 \leq i < n$ and $p \geq 1$, defined by Lutwak (see [31])

$$W_{p,i}(K, L) = \frac{1}{n} \int_{S^{n-1}} h(L, u)^p dS_{p,i}(K, u),$$

where $S_{p,i}(K, \cdot)$ denotes the Boel measure on S^{n-1}. The measure $S_{p,i}(K, \cdot)$ are absolutely continuous with respect to $S_i(K, \cdot)$ and have Radon-Nikodym derivative

$$\frac{dS_{p,i}(K, \cdot)}{dS_i(K, \cdot)} = h(K, \cdot)^{1-p},$$

where $S_i(K, \cdot)$ is a regular Boel measure of convex body K on S^{n-1}. The Minkowski inequality for the mixed p-quermassintegrals was established: If $K, L \in \mathcal{K}^n$ and $0 \leq i < n - 1$ and $p > 1$, then

$$W_{p,i}(K, L)^{n-i} \geq W_i(K, L)^{n-i-p} W_i(L)^p,$$

with equality if and only if K and L are homothetic. The Brunn-Minkowski inequality for the mixed p-quermassintegrals was also established: If $K, L \in \mathcal{K}^n$ and $0 \leq i < n - 1$ and $p > 1$, then

$$W_i(K +_p L)^{p/(n-i)} \geq W_i(K)^{p/(n-i)} + W_i(L)^{p/(n-i)},$$

with equality if and only if K and L are homothetic (see [31]).

Dual Mixed Quermassintegrals

If $K_1, \ldots, K_n \in \mathscr{S}^n$, the dual mixed volume $\widetilde{V}(K_1, \ldots, K_n)$ defined by (see [28])

$$\widetilde{V}(K_1, \ldots, K_n) = \frac{1}{n} \int_{S^{n-1}} \rho(K_1, u) \cdots \rho(K_n, u) dS(u).$$

If $K_1 = \cdots = K_{n-i} = K, K_{n-i+1} = \cdots = K_n = L$, the dual mixed volume $\widetilde{V}(K_1, \ldots, K_n)$ is written as $\widetilde{V}_i(K, L)$. If $L = B$, the dual mixed volume $\widetilde{V}_i(K, L) = \widetilde{V}_i(K, B)$ is written as $\widetilde{W}_i(K)$ and call dual quermassintegral of K. Obviously, For $K \in \mathscr{S}^n$ and $0 \leq i < n$, we have

$$\widetilde{W}_i(K) = \frac{1}{n} \int_{S^{n-1}} \rho(K, u)^{n-i} dS(u). \tag{23}$$

If $K_1 = \cdots = K_{n-i-1} = K, K_{n-i} = \cdots = K_{n-1} = B$ and $K_n = L$, the dual mixed volume $\widetilde{V}(\underbrace{K, \ldots, K}_{n-i-1}, \underbrace{B, \ldots, B}_{i}, L)$ is written as $\widetilde{W}_i(K, L)$ and call dual mixed quermassintegral of K and L. For $K, L \in \mathscr{S}^n$ and $0 \leq i < n$, it is easy that [44]

$$\widetilde{W}_i(K, L) = \lim_{\varepsilon \to 0^+} \frac{\widetilde{W}_i(K \widetilde{+} \varepsilon \cdot L) - \widetilde{W}_i(K)}{\varepsilon} = \frac{1}{n} \int_{S^{n-1}} \rho(K, u)^{n-i-1} \rho(L, u) dS(u). \tag{24}$$

The fundamental inequality for dual mixed quermassintegral stated that if $K, L \in \mathscr{S}^n$ and $0 \leq i < n$, then

$$\widetilde{W}_i(K, L)^{n-i} \leq \widetilde{W}_i(K)^{n-1-i} \widetilde{W}_i(L), \tag{25}$$

with equality if and only if K and L are dilates. The Brunn-Minkowski inequality for dual quermassintegral is the following: If $K, L \in \mathscr{S}^n$ and $0 \leq i < n$, then

$$\widetilde{W}_i(K \widetilde{+} L)^{1/(n-i)} \leq \widetilde{W}_i(K)^{1/(n-i)} + \widetilde{W}_i(L)^{1/(n-i)}, \tag{26}$$

with equality if and only if K and L are dilates.

Dual Mixed p-Quermassintegrals

The following result follows immediately form (3) with $p < 0$ (or $p > 0$).

$$\frac{p}{n-i} \lim_{\varepsilon \to 0^+} \frac{\widetilde{W}_i(K \widetilde{+}_p \varepsilon \cdot L) - \widetilde{W}_i(L)}{\varepsilon} = \frac{1}{n} \int_{S^{n-1}} \rho(K \cdot u)^{n-i-p} \rho(L \cdot u)^p dS(u).$$

Definition 1 Let $K, L \in \mathscr{S}^n$, $0 \leq i < n$ and $p < 0$, define dual mixed p-quermassintegral of star K and L, $\widetilde{W}_{p,i}(K, L)$, by

$$\widetilde{W}_{p,i}(K, L) = \frac{1}{n} \int_{S^{n-1}} \rho(K \cdot u)^{n-i-p} \rho(L \cdot u)^p dS(u). \tag{27}$$

Obviously, when $K = L$, the dual mixed p-quermassintegral $\widetilde{W}_{p,i}(K, L)$ becomes the dual quermassintegral $\widetilde{W}_i(K)$. This integral representation (27), together with the Hölder inequality, immediately gives:

Proposition 2 *If* $K, L \in \mathscr{S}^n$, $0 \leq i < n$ *and* $p < 0$, *then*

$$\widetilde{W}_{p,i}(K, L)^{n-i} \geq \widetilde{W}_i(K)^{n-i-p} \widetilde{W}_i(L)^p, \tag{28}$$

with equality if and only if K *and* L *are dilates.*

Taking $i = 0$ in (28), (28) becomes (18).

Proposition 3 *If* $K, L \in \mathscr{S}^n$, $0 \leq i < n$ *and* $p < 0$, *then*

$$\widetilde{W}_i(K \widetilde{+}_p L)^{p/(n-i)} \geq \widetilde{W}_i(K)^{p/(n-i)} + \widetilde{W}_i(L)^{p/(n-i)}, \tag{29}$$

with equality if and only if K *and* L *are dilates.*

Proof From (27) and (3), it is easily seen that the dual mixed p-quermassintegral is linear with respect to the p-radial addition and together with inequality (28) show that

$$\begin{aligned}\widetilde{W}_{p,i}(Q, K \widetilde{+}_p L) &= \widetilde{W}_{p,i}(Q, K) + \widetilde{W}_{p,i}(Q, L)\\ &\geq \widetilde{W}_i(Q)^{(n-i-p)/(n-i)} (\widetilde{W}_i(K)^{p/(n-i)} + \widetilde{W}_i(L)^{p/(n-i)}),\end{aligned}$$

with equality if and only if K and L are dilates of Q.

Take $K \widetilde{+}_p L$ for Q, recall that $\widetilde{W}_{p,i}(Q, Q) = \widetilde{W}_i(Q)$, inequality (29) follows easy.

Taking $i = 0$ in (29), (29) becomes (8).

Dual Mixed Harmonic p-Quermassintegrals

The following result follows immediately form (4) with $p \geq 1$.

$$-\frac{p}{n-i} \lim_{\varepsilon \to 0^+} \frac{\widetilde{W}_i(K \widehat{+}_p \varepsilon \cdot L) - \widetilde{W}_i(L)}{\varepsilon} = \frac{1}{n} \int_{S^{n-1}} \rho(K \cdot u)^{n-i+p} \rho(L \cdot u)^{-p} dS(u).$$

Definition 4 Let $K, L \in \mathscr{S}^n$, $0 \leq i < n$ and $p \geq 1$, define dual mixed harmonic p-quermassintegral of star K and L, $\widetilde{W}_{-p,i}(K, L)$, by

$$\widetilde{W}_{-p,i}(K, L) = \frac{1}{n} \int_{S^{n-1}} \rho(K \cdot u)^{n-i+p} \rho(L \cdot u)^{-p} dS(u). \tag{30}$$

Obviously, when $K = L$, the dual mixed harmonic p-quermassintegral $\widetilde{W}_{-p,i}(K, L)$ becomes the dual quermassintegral $\widetilde{W}_i(K)$. This integral representation (30), together with the Hölder inequality, immediately gives:

Proposition 5 *If $K, L \in \mathscr{S}^n$, $0 \leq i < n$ and $p \geq 1$, then*

$$\widetilde{W}_{-p,i}(K, L)^{n-i} \geq \widetilde{W}_i(K)^{n-i+p} \widetilde{W}_i(L)^{-p}, \tag{31}$$

with equality if and only if K and L are dilates.

Taking $i = 0$ in (31), (31) becomes (20).

Proposition 6 *If $K, L \in \mathscr{S}^n$, $0 \leq i < n$ and $p \geq 1$, then*

$$\widetilde{W}_i(K \widehat{+}_p L)^{-p/(n-i)} \geq \widetilde{W}_i(K)^{-p/(n-i)} + \widetilde{W}_i(L)^{-p/(n-i)}, \tag{32}$$

with equality if and only if K and L are dilates.

Proof From (30) and (4), it is easily seen that the dual mixed harmonic p-quermassintegral is linear with respect to the p-harmonic addition and together with inequality (31) show that

$$\begin{aligned} \widetilde{W}_{-p,i}(Q, K \widehat{+}_p L) &= \widetilde{W}_{-p,i}(Q, K) + \widetilde{W}_{-p,i}(Q, L) \\ &\geq \widetilde{W}_i(Q)^{(n-i+p)/(n-i)} (\widetilde{W}_i(K)^{-p/(n-i)} + \widetilde{W}_i(L)^{-p/(n-i)}), \end{aligned}$$

with equality if and only if K and L are dilates of Q.

Take $K \widehat{+}_p L$ for Q, recall that $\widetilde{W}_{-p,i}(Q, Q) = \widetilde{W}_i(Q)$, inequality (32) follows.

Taking $i = 0$ in (32), (32) becomes (10).

For a systematic investigation of fundamental characteristics of additions of convex and star bodies, we refer to [12, 13] and [14]. For different variants of the classical Brunn-Minkowski inequalities, we refer to [1, 2, 4, 42], and [46] and the references therein.

Throughout the paper, the standard orthonormal basis for $\mathbb{R}^n$ will be $\{e_1, \ldots, e_n\}$. Let Φ_n, $n \in \mathbb{N}$, denote the set of convex function $\varphi : [0, \infty)^n \to (0, \infty)$ that are strictly decreasing in each variable and satisfy $\varphi(0) = \infty$ and $\varphi(e_j) = 1$, $j = 1, \ldots, n$. When $n = 1$, we shall write Φ instead of Φ_1. The left derivative and right derivative of a real-valued function f are denoted by $(f)'_l$ and $(f)'_r$, respectively.

Orlicz Radial Harmonic Addition

We first define the Orlicz radial harmonic addition.

Definition 7 Let $m \geq 2$, $\varphi \in \Phi_m$, $K_j \in \mathscr{S}^n$ and $j = 1, \ldots, m$ define the Orlicz radial harmonic addition of $K_1, \ldots, K_m$, denoted by $\widetilde{\widetilde{+}}_{\varphi}(K_1, \ldots, K_m)$, defined by

$$\rho(\widetilde{\widetilde{+}}_{\varphi}(K_1, \ldots, K_m), x) = \sup\left\{\lambda > 0 : \varphi\left(\frac{\rho(K_1, x)}{\lambda}, \ldots, \frac{\rho(K_m, x)}{\lambda}\right) \leq 1\right\}, \tag{33}$$

for $x \in \mathbb{R}^n$.

Equivalently, the Orlicz radial harmonic addition $\widetilde{\widetilde{+}}_{\varphi}(K_1, \ldots, K_m)$ can be defined implicitly by

$$\varphi\left(\frac{\rho(K_1, x)}{\rho(\widetilde{\widetilde{+}}_{\varphi}(K_1, \ldots, K_m), x)}, \ldots, \frac{\rho(K_m, x)}{\rho(\widetilde{\widetilde{+}}_{\varphi}(K_1, \ldots, K_m), x)}\right) = 1, \tag{34}$$

for all $x \in \mathbb{R}^n$.

An important special case is obtained when

$$\varphi(x_1, \ldots, x_m) = \sum_{j=1}^{m} \varphi_j(x_j),$$

for some fixed $\varphi_j \in \Phi$ such that $\varphi_1(1) = \cdots = \varphi_m(1) = 1$. We then write $\widetilde{\widetilde{+}}_{\varphi}(K_1, \ldots, K_m) = K_1 \widetilde{\widetilde{+}}_{\varphi} \cdots \widetilde{\widetilde{+}}_{\varphi} K_m$. This means that $K_1 \widetilde{\widetilde{+}}_{\varphi} \cdots \widetilde{\widetilde{+}}_{\varphi} K_m$ is defined either by

$$\rho(K_1 \widetilde{\widetilde{+}}_{\varphi} \cdots \widetilde{\widetilde{+}}_{\varphi} K_m, u) = \sup\left\{\lambda > 0 : \sum_{j=1}^{m} \varphi_j\left(\frac{\rho(K_j, x)}{\lambda}\right) \leq 1\right\}, \tag{35}$$

for all $x \in \mathbb{R}^n$ or by the corresponding special case of (34).

Lemma 8 *The Orlicz radial harmonic addition* $\widetilde{\widetilde{+}}_{\varphi} : (\mathscr{S}^n)^m \to \mathscr{S}^n$ *is monotonic and has the identity property.*

Proof Suppose $K_j \subset L_j$, where $K_j, L_j \in \mathscr{S}^n$, $j = 1, \ldots, m$. By using (33), $K_1 \subset L_1$ and the fact that φ is decreasing in the first variable, we obtain

$$\rho(\widetilde{\widetilde{+}}_{\varphi}(L_1, K_2 \ldots, K_m), x)$$
$$= \sup\left\{\lambda > 0 : \varphi\left(\frac{\rho(L_1, x)}{\lambda}, \frac{\rho(K_2, x)}{\lambda}, \ldots, \frac{\rho(K_m, x)}{\lambda}\right) \leq 1\right\}$$

$$\geq \sup\left\{\lambda > 0 : \varphi\left(\frac{\rho(K_1, x)}{\lambda}, \frac{\rho(K_2, x)}{\lambda}, \ldots, \frac{\rho(K_m, x)}{\lambda}\right) \leq 1\right\}$$
$$= \rho(\widetilde{\widetilde{+}}_\varphi(K_1, K_2, \ldots, K_m), x).$$

By repeating this argument for each of the other $(m-1)$ variables, we have $\rho(\widetilde{\widetilde{+}}_\varphi(K_1, \ldots, K_m), x) \leq \rho(\widetilde{\widetilde{+}}_\varphi(L_1, \ldots, L_m), x)$.

The identity property is obvious from (34).

Lemma 9 *The Orlicz radial harmonic addition* $\widetilde{\widetilde{+}}_\varphi : (\mathscr{S}^n)^m \to \mathscr{S}^n$ *is* $GL(n)$ *covariant.*

Proof For $A \in GL(n)$, we obtain

$$\rho(\widetilde{\widetilde{+}}_\varphi(AK_1, AK_2 \ldots, AK_m), x)$$
$$= \sup\left\{\lambda > 0 : \varphi\left(\frac{\rho(AK_1, x)}{\lambda}, \frac{\rho(AK_2, x)}{\lambda}, \ldots, \frac{\rho(AK_m, x)}{\lambda}\right) \leq 1\right\}$$
$$= \sup\left\{\lambda > 0 : \varphi\left(\frac{\rho(K_1, A^{-1}x)}{\lambda}, \frac{\rho(K_2, A^{-1}x)}{\lambda}, \ldots, \frac{\rho(K_m, A^{-1}x)}{\lambda}\right) \leq 1\right\}$$
$$= \rho(\widetilde{\widetilde{+}}_\varphi(K_1, \ldots, K_m), A^{-1}x)$$
$$= \rho(\widetilde{\widetilde{+}}_\varphi(K_1, \ldots, K_m), x).$$

This shows Orlicz radial harmonic addition $\widetilde{\widetilde{+}}_\varphi$ is $GL(n)$ covariant.

Lemma 10 *Suppose* $K, \ldots, K_m \in \mathscr{S}^n$. *If* $\varphi \in \Phi$, *then*

$$\varphi\left(\frac{\rho(K_1, x)}{t}\right) + \cdots + \varphi\left(\frac{\rho(K_m, x)}{t}\right) = 1$$

if and only if

$$\rho(\widetilde{\widetilde{+}}_\varphi(K_1, \ldots, K_m), x) = t$$

Proof This follows immediately from Definition 7.

Lemma 11 *Suppose* $K_m, \ldots, K_m \in \mathscr{S}^n$. *If* $\varphi \in \Phi$, *then*

$$\frac{r}{\varphi^{-1}(\frac{1}{m})} \leq \rho(\widetilde{\widetilde{+}}_\varphi(K_1, \ldots, K_m), x) \leq \frac{R}{\varphi^{-1}(\frac{1}{m})}.$$

Proof Suppose $\rho(\widetilde{\widetilde{+}}_\varphi(K_1, \ldots, K_m), x) = t$. From Lemma 10 and noting that φ is strictly deceasing on $(0, \infty)$, we have

$$\begin{aligned}1 &= \varphi\left(\frac{\rho(K_1,x)}{t}\right)+\cdots+\varphi\left(\frac{\rho(K_m,x)}{t}\right)\\ &\le \varphi\left(\frac{r_{K_1}}{t}\right)+\cdots+\varphi\left(\frac{r_{K_m}}{t}\right)\\ &= m\varphi\left(\frac{r}{t}\right).\end{aligned}$$

Noting that the inverse φ^{-1} of φ is strictly deceasing on $(0,\infty)$, we obtain the lower bound for $\rho(\widetilde{\widetilde{+}}_\varphi(K_1,\ldots,K_m),x)$:

$$t \ge \frac{r}{\varphi^{-1}(\frac{1}{m})}.$$

To obtain the upper estimate, observe that from Lemma 10, together with the convexity and the fact φ is strictly deceasing on $(0,\infty)$, we have

$$\begin{aligned}1 &= \varphi\left(\frac{\rho(K_1,x)}{t}\right)+\cdots+\varphi\left(\frac{\rho(K_m,x)}{t}\right)\\ &\ge m\varphi\left(\frac{\rho(K_1,x)+\cdots+\rho(K_m,x)}{mt}\right)\\ &\ge m\varphi\left(\frac{R}{t}\right).\end{aligned}$$

Then we obtain the upper estimate:

$$t \le \frac{R}{\varphi^{-1}(\frac{1}{m})}.$$

Lemma 12 *The Orlicz radial harmonic addition* $\widetilde{\widetilde{+}}_\varphi : (\mathscr{S}^n)^m \to \mathscr{S}^n$ *is continuous.*

Proof To see this, indeed, let $K_{ij}\in\mathscr{S}^n, i\in\mathbb{N}\cup\{0\}, j=1,\ldots,m$, be such that $K_{ij}\to K_{0j}$ as $i\to\infty$. Let

$$\rho(\widetilde{\widetilde{+}}_\varphi(K_{i1},\ldots,K_{im}),x)=t_i.$$

Then Lemma 11 shows

$$\frac{r_{ij}}{\varphi^{-1}(\frac{1}{m})}\le t_i\le\frac{R_{ij}}{\varphi^{-1}(\frac{1}{m})},$$

where $r_{ij}=\min\{r_{K_{ij}}\}$ and $R_{ij}=\max\{R_{K_{ij}}\}$.

Since $K_{ij}\to K_{0j}$, we have $R_{K_{ij}}\to R_{K_{0j}}<\infty$ and $r_{K_{ij}}\to r_{K_{0j}}>0$, and thus there exist a,b such that $0<a\le t_i\le b<\infty$ for all i. To show that the bounded sequence $\{t_i\}$ converges to $\rho(\widetilde{\widetilde{+}}_\varphi(K_{01},\ldots,K_{0m}),u)$, we show that every convergent subsequence of $\{t_i\}$ converges to $\rho(\widetilde{\widetilde{+}}_\varphi(K_{01}, \ldots, K_{0m}), x)$. Denote any subsequence of $\{t_i\}$ by $\{t_i\}$ as well, and suppose that for this subsequence, we have

$$t_i \to t_*.$$

Obviously $a \le t_* \le b$. Noting that φ is continuous function, we obtain

$$\begin{aligned} t_* \to \sup\left\{t_* > 0 : \varphi\left(\frac{\rho(K_{01}, x)}{t_*}, \ldots, \frac{\rho(K_{0m}, x)}{t_*}\right) \le 1\right\} \\ = \rho(\widetilde{\widetilde{+}}_\varphi(K_{01}, \ldots, K_{0m}), x). \end{aligned}$$

Hence

$$\rho(\widetilde{\widetilde{+}}_\varphi(K_{i1}, \ldots, K_{im}), x) \to \rho(\widetilde{\widetilde{+}}_\varphi(K_{01}, \ldots, K_{0m}), x)$$

as $i \to \infty$.

This shows that the Orlicz radial harmonic addition $\widetilde{\widetilde{+}}_\varphi : (\mathscr{S}^n)^m \to \mathscr{S}^n$ is continuous.

Next, we define the Orlicz radial harmonic linear combination on the case $m = 2$.

Definition 13 Orlicz radial harmonic linear combination $\widetilde{\widetilde{+}}_\varphi(K, L, \alpha, \beta)$ for $K, L \in \mathscr{S}^n$, and $\alpha, \beta \ge 0$ (not both zero), defined by

$$\alpha\varphi_1\left(\frac{\rho(K, x)}{\rho(\widetilde{\widetilde{+}}_\varphi(K, L, \alpha, \beta), x)}\right) + \beta\varphi_2\left(\frac{\rho(L, x)}{\rho(\widetilde{\widetilde{+}}_\varphi(K, L, \alpha, \beta), x)}\right) = 1, \tag{36}$$

for all $x \in \mathbb{R}^n$.

When $\varphi_1(t) = \varphi_2(t) = t^p$ and $p < 0$, then Orlicz radial harmonic linear combination $\widetilde{\widetilde{+}}_\varphi(K, L, \alpha, \beta)$ changes to the p-radial linear combination $\alpha \cdot K\widetilde{+}_p\beta \cdot L$. When $\varphi_1(t) = \varphi_2(t) = t^{-p}$ and $p \ge 1$, then Orlicz radial harmonic linear combination $\widetilde{\widetilde{+}}_\varphi(K, L, \alpha, \beta)$ changes to the harmonic p-linear combination $\alpha\Diamond K\widehat{+}_p\beta\Diamond L$. Moreover, we shall write $K\widetilde{\widetilde{+}}_{\varphi,\varepsilon}L$ instead of $\widetilde{\widetilde{+}}_\varphi(K, L, 1, \varepsilon)$, for $\varepsilon \ge 0$ and assume throughout that this is defined by (33), where $\alpha = 1$, $\beta = \varepsilon$ and $\varphi \in \Phi$. It is easy that $\widetilde{\widetilde{+}}_\varphi(K, L, 1, 1) = K\widetilde{\widetilde{+}}_\varphi L$. Obviously, $K\widetilde{\widetilde{+}}_{\varphi,\varepsilon}L$ and $K\widetilde{\widetilde{+}}_\varphi\varepsilon \cdot L$ are not the same meaning.

Orlicz Dual Mixed Quermassintegral

In order to define Orlicz dual mixed quermassintegral, we need the following Lemmas 14–17.

Lemma 14 *Let $\varphi \in \Phi$ and $\varepsilon > 0$. If $K, L \in \mathscr{S}^n$, then $K\widetilde{\widetilde{+}}_{\varphi,\varepsilon}L \in \mathscr{S}^n$.*

Proof Let $u_0 \in S^{n-1}$, for any subsequence $\{u_i\} \subset S^{n-1}$ such that $u_i \to u_0$ as $i \to \infty$.

Let

$$\rho(K\tilde{\tilde{+}}_{\varphi}L, u_i) = \lambda_i.$$

Then Lemma 11 shows

$$\frac{r}{\varphi^{-1}\left(\frac{1}{2}\right)} \leq \lambda_i \leq \frac{R}{\varphi^{-1}\left(\frac{1}{2}\right)},$$

where $R = \max\{R_K, R_L\}$ and $r = \min\{r_K, r_L\}$.

Since $K, L \in \mathscr{S}^n$, we have $0 < r_K \leq R_K < \infty$ and $0 < r_L \leq R_L < \infty$, and thus there exist a, b such that $0 < a \leq \lambda_i \leq b < \infty$ for all i. To show that the bounded sequence $\{\lambda_i\}$ converges to $\rho(K\tilde{\tilde{+}}L, u_0)$, we show that every convergent subsequence of $\{\lambda_i\}$ converges to $\rho(K\tilde{\tilde{+}}L, u_0)$. Denote any subsequence of $\{\lambda_i\}$ by $\{\lambda_i\}$ as well, and suppose that for this subsequence, we have

$$\lambda_i \to \lambda_0.$$

Obviously $a \leq \lambda_0 \leq b$. From (36) and note that φ_1, φ_2 are continuous functions, so φ_1^{-1} is continuous, we obtain

$$\lambda_i \to \frac{\rho(K, u_0)}{\varphi_1^{-1}\left(1 - \varepsilon\varphi_2\left(\frac{\rho(L, u_0)}{\lambda_0}\right)\right)}$$

as $i \to \infty$.

Hence

$$\varphi_1\left(\frac{\rho(K, u_0)}{\lambda_0}\right) + \varepsilon\varphi_2\left(\frac{\rho(L, u_0)}{\lambda_0}\right) = 1.$$

Therefore

$$\lambda_0 = \rho(K\tilde{\tilde{+}}_{\varphi,\varepsilon}L, u_0).$$

Namely,

$$\rho(K\tilde{\tilde{+}}_{\varphi,\varepsilon}L, u_i) \to \rho(K\tilde{\tilde{+}}_{\varphi,\varepsilon}L, u_0).$$

as $i \to \infty$.

This shows that $K\tilde{\tilde{+}}_{\varphi,\varepsilon}L \in \mathscr{S}^n$.

Lemma 15 *If $K, L \in \mathscr{S}^n$, $\varepsilon > 0$ and $\varphi \in \Phi$, then*

$$K\tilde{\tilde{+}}_{\varphi,\varepsilon}L \to K \tag{37}$$

in the radial Hausdorff metric as $\varepsilon \to 0^+$.

This is a useful result. In the following, we give two different proofs.

First Proof From (36) with $\alpha = 1$ and $\beta = \varepsilon$, we have

$$\rho(K, u) \leq \rho(K \widetilde{\widetilde{+}}_{\varphi,\varepsilon} L, u) \leq \rho(K \widetilde{\widetilde{+}}_{\varphi,1} L, u),$$

for all $\varepsilon \in (0, 1]$.

Notes that $\rho(K, u) > 0$, we conclude from

$$\varphi_1 \left(\frac{\rho(K, u)}{\rho(K \widetilde{\widetilde{+}}_{\varphi,\varepsilon} L, u)} \right) + \varepsilon \varphi_2 \left(\frac{\rho(L, u)}{\rho(K \widetilde{\widetilde{+}}_{\varphi,\varepsilon} L, u)} \right) = 1$$

that if $\varepsilon \to 0^+$ and if a subsequence of $\{\rho(K \widetilde{\widetilde{+}}_{\varphi,\varepsilon} L, u) : \varepsilon \in (0, 1]\}$ converges to a constant $0 < \mu \leq \rho(K, u)$, therefore

$$\varphi_1 \left(\frac{\rho(K, u)}{\mu} \right) = \varphi_1 \left(\frac{\rho(K, u)}{\mu} \right) + 0 \cdot \varphi_2 \left(\frac{\rho(L, u)}{\mu} \right) = 1 = \varphi_1(1).$$

This shows $\mu = \rho(K, u)$.

Hence yields that

$$\rho(K \widetilde{\widetilde{+}}_{\varphi,\varepsilon} L, u) \to \rho(K, u)$$

as $\varepsilon \to 0^+$.

Second Proof From (36), we obtain

$$\rho(K \widetilde{\widetilde{+}}_{\varphi,\varepsilon} L, u) \to \frac{\rho(K, u)}{\varphi_1^{-1} \left(1 - \varepsilon \varphi_2 \left(\frac{\rho(L, u)}{\rho(K \widetilde{\widetilde{+}}_{\varphi,\varepsilon} L, u)} \right) \right)}$$

as $\varepsilon \to 0^+$.

Since φ_1^{-1} is continuous, φ_2 is bounded and in view of $\varphi_1^{-1}(1) = 1$, we have

$$\varphi_1^{-1} \left(1 - \varepsilon \varphi_2 \left(\frac{\rho(L, u)}{\rho(K \widetilde{\widetilde{+}}_{\varphi,\varepsilon} L, u)} \right) \right) \to 1 \tag{38}$$

as $\varepsilon \to 0^+$.

This yields

$$\rho(K \widetilde{\widetilde{+}}_{\varphi,\varepsilon} L, u) \to \rho(K, u)$$

as $\varepsilon \to 0^+$.

Lemma 16 *If $K, L \in \mathscr{S}^n$, $0 \le i < n$, $\varphi \in \Phi_2$ and $\varphi_1, \varphi_2 \in \Phi$, then*

$$\lim_{\varepsilon\to 0^+} \frac{\rho(K\widetilde{+}_{\varphi,\varepsilon}L, u)^{n-i} - \rho(K,u)^{n-i}}{\varepsilon} = \frac{n-i}{((\varphi_1)'_l(1)} \cdot \varphi_2\left(\frac{\rho(L,u)}{\rho(K,u)}\right)\rho(K,u)^{n-i} \tag{39}$$

uniformly for $u \in S^{n-1}$.

This lemma plays a central role in our deriving new concept of Orlicz dual mixed quermassintegral. It would be possible to apply the argument of [12, Lemma 8.4] (cf. also the argument in [53, Theorem 4.1], which is attributed to Gardner). Here, we provide an alternative proof

Proof Form (36), (38), Lemma 15 and notice that φ_2 is continuous function, we obtain for $0 \le i < n$

$$\begin{aligned}
&\lim_{\varepsilon\to 0^+} \frac{\rho(K\widetilde{+}_{\varphi,\varepsilon}L, u)^{n-i} - \rho(K,u)^{n-i}}{\varepsilon}\\
&= \lim_{\varepsilon\to 0^+} \frac{\left(\dfrac{\rho(K,u)}{\varphi_1^{-1}\left(1-\varepsilon\varphi_2\left(\dfrac{\rho(L,u)}{\rho(K\widetilde{+}_{\varphi,\varepsilon}L,u)}\right)\right)}\right)^{n-i} - \rho(K,u)^{n-i}}{\varepsilon}\\
&= \lim_{\varepsilon\to 0^+} (n-i)\rho(K,u)^{n-i-1}\left(\rho(K,u)\varphi_2\left(\frac{\rho(L,u)}{\rho(K\widetilde{+}_{\varphi,\varepsilon}L,u)}\right)\right)\\
&\quad \lim_{y\to 1^-} \frac{\varphi_1^{-1}(y) - \varphi_1^{-1}(1)}{y-1}\\
&= \frac{n-i}{((\varphi_1)'_l(1)} \cdot \varphi_2\left(\frac{\rho(L,u)}{\rho(K,u)}\right)\rho(K,u)^{n-i},
\end{aligned}$$

where

$$y = \varphi_1^{-1}\left(1-\varepsilon\varphi_2\left(\frac{\rho(L,u)}{\rho(K\widetilde{+}_{\varphi,\varepsilon}L,u)}\right)\right),$$

and note that $y \to 1^-$ as $i \to 0^+$.

Lemma 17 *If $\varphi \in \Phi_2$, $0 \le i < n$ and $K, L \in \mathscr{S}^n$, then*

$$\frac{(\varphi_1)'_l(1)}{n-i} \lim_{\varepsilon\to 0^+} \frac{\widetilde{W}_i(K\widetilde{+}_{\varphi,\varepsilon}L) - \widetilde{W}_i(K)}{\varepsilon} = \frac{1}{n}\int_{S^{n-1}} \varphi_2\left(\frac{\rho(L,u)}{\rho(K,u)}\right)\rho(K,u)^{n-i}dS(u). \tag{40}$$

Proof This follows immediately from (23) and Lemma 15.

Denoting by $\widetilde{W}_{\varphi,i}(K,L)$, for any $\varphi \in \Phi$ and $0 \le i < n$, the integral on the right-hand side of (40) with φ_2 replaced by φ, we see that either side of Eq. (40) is equal to $\widetilde{W}_{\varphi_2,i}(K,L)$, and hence this new Orlicz dual mixed volume $\widetilde{W}_{\varphi,i}(K,L)$ has been born.

Definition 18 For $\varphi \in \Phi$ and $0 \le i < n$, Orlicz dual mixed quermassintegral of star bodies K and L, $\widetilde{W}_{\varphi,i}(K,L)$, defined by

$$\widetilde{W}_{\varphi,i}(K,L) =: \frac{1}{n}\int_{S^{n-1}} \varphi\left(\frac{\rho(L,u)}{\rho(K,u)}\right)\rho(K,u)^{n-i}dS(u). \tag{41}$$

Lemma 19 *If $\varphi_1, \varphi_2 \in \Phi$, $\varphi \in \Phi_2$, $0 \le i < n$ and $K, L \in \mathscr{S}^n$, then*

$$\widetilde{W}_{\varphi_2,i}(K,L) = \frac{(\varphi_1)'_l(1)}{n-i}\lim_{\varepsilon\to 0^+}\frac{\widetilde{W}_i(K\widetilde{\widetilde{+}}_{\varphi,\varepsilon}L)-\widetilde{W}_i(K)}{\varepsilon}. \tag{42}$$

Proof This follows immediately from Lemma 17 and (41).

From Lemma 10 and the definition of the Orlicz dual mixed quermassintegral, we easy find that $\widetilde{W}_{\varphi,i}(K,L)$ is invariant under simultaneous unimodular centro-affine transformation.

Lemma 20 *If $\varphi \in \Phi$, $0 \le i < n$ and $K, L \in \mathscr{S}^n$, then for $A \in SL(n)$,*

$$\widetilde{W}_{\varphi,i}(AK,AL) = \widetilde{W}_{\varphi,i}(K,L). \tag{43}$$

Orlicz Dual Minkowski Inequality

In this section, we need to define a Borel measure in S^{n-1}, $\widetilde{W}_{n,i}(K,\upsilon)$, called i-dual normalized cone measure of star body K.

Definition 21 Let $K \in \mathscr{S}^n$ and $0 \le i < n$, i-dual normalized cone measure, $\widetilde{W}_{n,i}(K,\upsilon)$, defined by

$$d\widetilde{W}_{n,i}(K,\upsilon) = \frac{\rho(K,\upsilon)^{n-i}}{n\widetilde{W}_i(K)}dS(\upsilon). \tag{44}$$

When $i = 0$, i-dual normalized cone measure $d\widetilde{W}_{n,i}(K,\upsilon)$ becomes the dual normalized cone measure $d\widetilde{V}_n(K,\upsilon)$.

Lemma 22 (Jensen's Inequality) *Let μ be a probability measure on a space X and $g : X \to I \subset \mathbb{R}$ is a μ-integrable function, where I is a possibly infinite interval. If $\varphi : I \to \mathbb{R}$ is a convex function, then*

$$\int_X \varphi(g(x))d\mu(x) \geq \varphi\left(\int_X g(x)d\mu(x)\right). \tag{45}$$

If φ is strictly convex, equality holds if and only if $g(x)$ is constant for μ-almost all $x \in X$ (see [22, p.165]).

Lemma 23 *Suppose that $\varphi : (0, \infty) \to (0, \infty)$ is decreasing and convex with $\varphi(0) = \infty$. If $K, L \in \mathscr{S}^n$ and $0 \leq i < n$, then*

$$\frac{1}{n\widetilde{W}_i(K)} \int_{S^{n-1}} \varphi\left(\frac{\rho(L,u)}{\rho(K,u)}\right) \rho(K,u)^{n-i} dS(u) \geq \varphi\left(\left(\frac{\widetilde{W}_i(L)}{\widetilde{W}_i(K)}\right)^{1/(n-i)}\right). \tag{46}$$

If φ is strictly convex, equality holds if and only if K and L are dilates.

Proof For $K \in \mathscr{S}^{n-1}$, $0 \leq i < n$ and any $u \in S^{n-1}$, since

$$\frac{1}{n} \int_{S^{n-1}} \rho(K,u)^{n-i} dS(u) = \tilde{W}_i(K),$$

so the i-dual normalized cone measure $\dfrac{\rho(K,u)^{n-i}}{n\tilde{W}_i(K)} dS(u)$ is a probability measure on S^{n-1}.

Hence, by using Jensen's inequality (45), (25) and in view of φ is decreasing, we obtain

$$\begin{aligned}
&\frac{1}{n\widetilde{W}_i(K)} \int_{S^{n-1}} \varphi\left(\frac{\rho(L,u)}{\rho(K,u)}\right) \rho(K,u)^{n-i} dS(u) \\
&= \int_{S^{n-1}} \varphi\left(\frac{\rho(L,u)}{\rho(K,u)}\right) d\widetilde{W}_{n,i}(K,u) \\
&\geq \varphi\left(\frac{\widetilde{W}_i(K,L)}{\widetilde{W}_i(K)}\right) \\
&\geq \varphi\left(\left(\frac{\widetilde{W}_i(L)}{\widetilde{W}_i(K)}\right)^{1/(n-i)}\right).
\end{aligned}$$

Next, we discuss the equal condition of (46). Suppose the equality holds in (46) and φ is strictly convex, form the equality condition of (25), so there exist $r > 0$ such that $L = rK$ and hence

$$\rho(L,u) = r\rho(K,u),$$

for all $u \in S^{n-1}$. On the other hand, suppose that K and L are dilates, i.e., there exist $\lambda > 0$ such that $\rho(L,u) = \lambda\rho(K,u)$ for all $u \in S^{n-1}$. Hence

$$\frac{1}{n\widetilde{W}_i(K)}\int_{S^{n-1}}\varphi\left(\frac{\rho(L,u)}{\rho(K,u)}\right)\rho(K,u)^{n-i}dS(u)$$
$$=\frac{1}{n\widetilde{W}_i(K)}\int_{S^{n-1}}\varphi\left(\left(\frac{\widetilde{W}_i(L)}{\widetilde{W}_i(K)}\right)^{1/(n-i)}\right)\rho(K,u)^{n-i}dS(u)$$
$$=\varphi\left(\left(\frac{\widetilde{W}_i(L)}{\widetilde{W}_i(K)}\right)^{1/(n-i)}\right).$$

This implies the equality in (46) holds.

Theorem 24 (Orlicz Dual Minkowski Inequality for Quermassintegrals) *If $K, L \in \mathscr{S}^n$, $0 \le i < n$ and $\varphi \in \Phi$, then*

$$\widetilde{W}_{\varphi,i}(K,L) \ge \widetilde{W}_i(K)\varphi\left(\left(\frac{\widetilde{W}_i(L)}{\widetilde{W}_i(K)}\right)^{1/(n-i)}\right). \tag{47}$$

If φ is strictly convex, equality holds if and only if K and L are dilates.

Proof This follows immediately from (41) and Lemma 23.

Corollary 25 *If $K, L \in \mathscr{S}^n$, $0 \le i < n$ and $p < 0$, then*

$$\widetilde{W}_{p,i}(K,L)^{n-i} \ge \widetilde{W}_i(K)^{n-i-p}\widetilde{W}_i(L)^p, \tag{48}$$

with equality if and only if K and L are dilates.

Proof This follows immediately from (47) with $\varphi_1(t) = \varphi_2(t) = t^p$ and $p < 0$.

Corollary 26 *If $K, L \in \mathscr{S}^n$, $0 \le i < n$ and $p \ge 1$, then*

$$\widetilde{W}_{-p,i}(K,L)^{n-i} \ge \widetilde{W}_i(K)^{n-i+p}\widetilde{W}_i(L)^{-p}, \tag{49}$$

with equality if and only if K and L are dilates.

Proof This follows immediately from (47) with $\varphi_1(t) = \varphi_2(t) = t^{-p}$ and $p \ge 1$.

One obvious immediate consequence of Theorem 24 is the following:

Corollary 27 *Let $K, L \in \mathscr{M} \subset \mathscr{S}^n$, $0 \le i < n$ and $\varphi \in \Phi$, and if either*

$$\widetilde{W}_{\varphi,i}(K,Q) = \widetilde{W}_{\varphi,i}(L,Q),\ \ for\ all\ Q \in \mathscr{M}$$

or

$$\frac{\widetilde{W}_{\varphi,i}(K,Q)}{\widetilde{W}_i(K)} = \frac{\widetilde{W}_{\varphi,i}(L,Q)}{\widetilde{W}_i(L)},\ \ for\ all\ Q \in \mathscr{M},$$

then $K = L$.

Remark 28 When $\varphi_1(t) = \varphi_2(t) = t^{-p}$, $p \geq 1$ and $i = 0$, Corollary 27 becomes the following result proved by Lutwak [32].

Let $K, L \in \mathscr{M} \subset \mathscr{S}^n$ and $p \geq 1$, and if either

$$\widetilde{V}_{-p}(K, Q) = \widetilde{V}_{-p}(L, Q), \text{ for all } Q \in \mathscr{M}$$

or

$$\frac{\widetilde{V}_{-p}(K, Q)}{V(K)} = \frac{\widetilde{V}_{-p}(L, Q)}{V(L)}, \text{ for all } Q \in \mathscr{M},$$

then $K = L$.

When $\varphi_1(t) = \varphi_2(t) = t^p$, $p < 0$ and $i = 0$, Corollary 27 becomes the following result.

Let $K, L \in \mathscr{M} \subset \mathscr{S}^n$ and $p < 0$, and if either

$$\widetilde{V}_p(K, Q) = \widetilde{V}_p(L, Q), \text{ for all } Q \in \mathscr{M}$$

or

$$\frac{\widetilde{V}_p(K, Q)}{V(K)} = \frac{\widetilde{V}_p(L, Q)}{V(L)}, \text{ for all } Q \in \mathscr{M},$$

then $K = L$.

Corollary 29 ([56] (Orlicz Dual Minkowski Inequality for Dual Mixed Volumes)) *If $K, L \in \mathscr{S}^n$ and $\varphi \in \Phi$, then*

$$\widetilde{V}_\varphi(K, L) \geq V(K)\varphi\left(\left(\frac{V(L)}{V(K)}\right)^{1/n}\right).$$

If φ is strictly convex, equality holds if and only if K and L are dilates.

Proof This follows immediately from (47) and with $i = 0$.

Orlicz Dual Brunn-Minkowski Inequality

Lemma 30 *If $K, L \in \mathscr{S}^n$, $0 \leq i < n$ and $\varphi \in \Phi_2$, and $\varphi_1, \varphi_2 \in \Phi$, then*

$$\widetilde{W}_i(K \widetilde{\widetilde{+}}_\varphi L) = \widetilde{W}_{\varphi_1, i}(K \widetilde{\widetilde{+}}_\varphi L, K) + \widetilde{W}_{\varphi_2, i}(K \widetilde{\widetilde{+}}_\varphi L, L). \tag{50}$$

Proof From (41), we have for any $Q \in \mathscr{S}^n$

$$\begin{aligned}&\widetilde{W}_{\varphi_1,i}(Q,K)+\widetilde{W}_{\varphi_2,i}(Q,L)\\&\quad=\frac{1}{n}\int_{S^{n-1}}\left(\varphi_1\left(\frac{\rho(K,u)}{\rho(Q,u)}\right)+\varphi_2\left(\frac{\rho(L,u)}{\rho(Q,u)}\right)\right)\rho(Q,u)^{n-i}dS(u)\\&\quad=\frac{1}{n}\int_{S^{n-1}}\varphi\left(\frac{\rho(K,u)}{\rho(Q,u)},\frac{\rho(L,u)}{\rho(Q,u)}\right)\rho(Q,u)^{n-i}dS(u).\end{aligned}\tag{51}$$

Putting $Q=K\widetilde{\widetilde{+}}_\varphi L$ in (51), (51) changes (50).

Theorem 31 (Orlicz Dual Brunn-Minkowski Inequality for Dual Quermassintegrals) *If $K,L\in\mathscr{S}^n$, $0\le i<n$ and $\varphi\in\Phi_2$, then*

$$1\geq\varphi\left(\left(\frac{\widetilde{W}_i(K)}{\widetilde{W}_i(K\widetilde{\widetilde{+}}_\varphi L)}\right)^{1/(n-i)},\left(\frac{\widetilde{W}_i(L)}{\widetilde{W}_i(K\widetilde{\widetilde{+}}_\varphi L)}\right)^{1/(n-i)}\right).\tag{52}$$

If φ is strictly convex, equality holds if and only if K and L are dilates.

Proof From Lemma 30 and (47), we have

$$\begin{aligned}\widetilde{W}_i(K\widetilde{\widetilde{+}}_\varphi L)&=\widetilde{W}_{\varphi_1,i}(K\widetilde{\widetilde{+}}_\varphi L,K)+\widetilde{W}_{\varphi_2,i}(K\widetilde{\widetilde{+}}_\varphi L,L)\\&\geq\widetilde{W}_i(K\widetilde{\widetilde{+}}_\varphi L)\left(\varphi_1\left(\left(\frac{\widetilde{W}_i(K)}{\widetilde{W}_i(K\widetilde{\widetilde{+}}_\varphi L)}\right)^{1/(n-i)}\right)\right.\\&\quad\left.+\varphi_2\left(\left(\frac{\widetilde{W}_i(L)}{k\widetilde{W}_i(K\widetilde{\widetilde{+}}_\varphi L)}\right)^{1/(n-i)}\right)\right)\\&=\widetilde{W}_i(K\widetilde{\widetilde{+}}_\varphi L)\varphi\left(\left(\frac{\widetilde{W}_i(K)}{\widetilde{W}_i(K\widetilde{\widetilde{+}}_\varphi L)}\right)^{1/(n-i)},\left(\frac{\widetilde{W}_i(L)}{\widetilde{W}_i(K\widetilde{\widetilde{+}}_\varphi L)}\right)^{1/(n-i)}\right).\end{aligned}$$

This is just inequality (52).

From the equality condition of (47), if follows that if φ is strictly convex equality in (52) holds if and only if K and L are dilates.

Corollary 32 *If $K,L\in\mathscr{S}^n$, $0\le i<n$ and $p<0$, then*

$$\widetilde{W}_i(K\widetilde{+}_pL)^{p/(n-i)}\geq\widetilde{W}_i(K)^{p/(n-i)}+\widetilde{W}_i(L)^{p/(n-i)},$$

with equality if and only if K and L are dilates.

Proof The result follows immediately from Theorem 31 with $\varphi(x_1,x_2)=x_1^p+x_2^p$ and $p<0$.

Corollary 33 *If $K, L \in \mathscr{S}^n$, $0 \le i < n$ and $p \ge 1$, then*

$$\widetilde{W}_i(K\widehat{+}_pL)^{-p/(n-i)} \ge \widetilde{W}_i(K)^{-p/(n-i)} + \widetilde{W}_i(L)^{-p/(n-i)},$$

with equality if and only if K and L are dilates.

Proof The result follows immediately from Theorem 31 with $\varphi(x_1, x_2) = x_1^{-p} + x_2^{-p}$ and $p \ge 1$.

Corollary 34 (Orlicz Dual Brunn-Minkowski Inequality for Volumes) *If $K, L \in \mathscr{S}^n$ and $\varphi \in \Phi_2$, then*

$$1 \ge \varphi\left(\left(\frac{V(K)}{V(K\widetilde{\widetilde{+}}_\varphi L)}\right)^{1/n}, \left(\frac{V(L)}{V(K\widetilde{\widetilde{+}}_\varphi L)}\right)^{1/n}\right).$$

If φ is strictly convex, equality holds if and only if K and L are dilates.

Proof This follows immediately from (52) and with $i = 0$.

Dual Log-Minkowski Inequality

Assume that K, L is the nonempty compact convex subsets of $\mathbb{R}^n$ containing the origin in their interiors, then the log Minkowski combination, $(1-\lambda)\cdot K +_o \lambda \cdot L$, defined by

$$(1-\lambda)\cdot K +_o \lambda\cdot L = \bigcap_{u\in S^{n-1}} \{x \in \mathbb{R}^n : x\cdot u \le h(K,u)^{1-\lambda} h(L,u)^{\lambda}\},$$

for all real $\lambda \in [0, 1]$. Böröczky et al. [4] conjecture that for origin-symmetric convex bodies K and L in $\mathbb{R}^n$ and $0 \le \lambda \le 1$, is it true that

$$V((1-\lambda)\cdot K +_o \lambda \cdot L) \ge V(K)^{1-\lambda} V(L)^{\lambda}\ ? \tag{53}$$

They proved (53) only when $n = 2$ and K, L are origin-symmetric convex bodies and note that while it is not true for general convex bodies. Moreover, they also show that (53), for all n, is equivalent to the following log-Minkowski inequality (guess). For origin-symmetric convex bodies K and L, is it true that

$$\int_{S^{n-1}} \log\left(\frac{h(L,u)}{h(K,u)}\right) d\bar{V}_n(K,u) \ge \frac{1}{n}\log\left(\frac{V(L)}{V(K)}\right)\ ? \tag{54}$$

In this section, we establish a dual log-Minkowski inequality. First, we derive a lemma.

Lemma 35 *Let $\varphi : (0, 1] \to [0, \infty)$ is decreasing and convex with $\varphi(0) = \infty$. If $K, L \in \mathscr{S}^n$ be such that $L \subset \operatorname{int}(K)$, then*

$$\int_{S^{n-1}} \varphi\left(\frac{\rho(L, u)}{\rho(K, u)}\right) d\widetilde{V}_n(K, u) \geq \varphi\left(\frac{V(L)^{1/n}}{V(K)^{1/n}}\right). \tag{55}$$

If φ is strictly convex, equality holds if and only if K and L are dilates.

Proof Note that if $L \subset \operatorname{int}(K)$, we have $\rho(L, u)/\rho(K, u) \in (0, 1]$ for all $u \in S^{n-1}$. Obviously, the dual normalized cone measure $d\tilde{V}_n(K, u)$ is a probability measure on S^{n-1}. By Jensen's inequality and the classical dual Minkowski inequality, we obtain

$$\int_{S^{n-1}} \varphi\left(\frac{\rho(L, u)}{\rho(K, u)}\right) d\widetilde{V}_n(K, u) \geq \varphi\left(\frac{V_1(K, L)}{V(K)}\right) \geq \varphi\left(\left(\frac{V(L)}{V(K)}\right)^{1/n}\right).$$

If φ is strictly convex, from equality condition of the dual Minkowski inequality, it follows that if φ is strictly convex equality in (55) holds if and only if K and L are dilates.

Theorem 36 (Dual Log-Minkowski Inequality) *If $K, L \in \mathscr{S}^n$ such that $L \subset \operatorname{int} K$, then*

$$\int_{S^{n-1}} \log\left(\frac{\rho(L, u)}{\rho(K, u)}\right) d\widetilde{V}_n(K, u) \leq \frac{1}{n} \log\left(\frac{V(L)}{V(K)}\right), \tag{56}$$

with equality if and only if K and L are dilates.

Proof Let $\varphi(t) = -\log t$, obviously φ is strictly decreasing and convex on $(0, 1]$ with $\varphi(t) \to \infty$ as $t \to 0^+$ and $\varphi(t) \in [0, \infty)$. If $K, L \in \mathscr{S}^n$ are such that $L \subset \operatorname{int} K$, then $\rho(L, u)/\rho(K, u) \in (0, 1]$. Hence (56) is a direct consequence of Lemma 35.

Acknowledgment This research was supported by the National Natural Sciences Foundation of China (11371334) and a HKU Seed Grant for Basic Research.

References

1. Abardia, J., Bernig A.: Projection bodies in complex vector spaces. Adv. Math. **227**, 830–846 (2011)
2. Alesker, S., Bernig, A., Schuster, F.E.: Harmonic analysis of translation invariant valuations. Geom. Funct. Anal. **21**, 751–773 (2011)
3. Berck, G.: Convexity of L_p-intersection bodies. Adv. Math. **222**, 920–936 (2009)
4. Böröczky, K.J., Lutwak, E., Yang, D., Zhang, G.: The log-Brunn-Minkowski inequality. Adv. Math. **231**, 1974–1997 (2012)

5. Feng, Y.B., Wang, W.D.: Shephard type probems for L_p-centroid bodies. Math. Inqu. Appl. **17**(3), 865–877 (2014)
6. Firey, W.J.: Polar means of convex bodies and a dual to the Brunn-Minkowski theorem. Can. J. Math. **13**, 444–453 (1961)
7. Firey, W.J.: p-means of convex bodies. Math. Scand. **10**, 17–24 (1962)
8. Gardner, R.J.: A positive answer to the Busemann-Petty problem in three dimensions. Ann. Math. **140**(2), 435–447 (1994)
9. Gardner, R.J.: Geometric Tomography. Cambridge University Press, New York (1996)
10. Gardner, R.J.: The Brunn-Minkowski inequality. Bull. Am. Math. Soc. **39**, 355–405 (2002)
11. Gardner, R.J., Koldobsky, A., Schlumprecht, T.: An analytic solution to the Busemann-Petty problem on sections of convex bodies. Ann. Math. **149**, 691–703 (1999)
12. Gardner, R.J., Hug, D., Weil, W.: The Orlicz-Brunn-Minkowski theory: a general framework, additions, and inequalities. J. Diff. Geom. **97**(3), 427–476 (2014)
13. Gardner, R.J., Parapatits, L., Schuster, F.E.: A Characterization of Blaschke Addition. Advances in Math. **254**, 396–418 (2014)
14. Gardner, R.J., Hug, D., Weil, W.: Operations between sets in geometry. J. Eur. Math. Soc. **15**(6), 2297–2352 (2013)
15. Haberl, C.: L_p intersection bodies. Adv. Math. **217**, 2599–2624 (2008)
16. Haberl, C., Ludwig, M.: A characterization of L_p intersection bodies. Int. Math. Res. Not. 2006, Art. ID 10548, 29 pp. (2006)
17. Haberl, C., Parapatits, L.: The centro-affine Hadwiger theorem. J. Am. Math. Soc. **27**, 685–705 (2014)
18. Haberl, C., Schuster, F.E.: Asymmetric affine L_p Sobolev inequalities. J. Funct. Anal. **257**, 641–658 (2009)
19. Haberl, C., Schuster, F.E.: General L_p affine isoperimetric inequalities. J. Differ. Geom. **83**, 1–26 (2009)
20. Haberl, C., Lutwak E., Yang, D., Zhang G.: The even Orlicz Minkowski problem. Adv. Math. **224**, 2485–2510 (2010)
21. Haberl, C., Schuster, F.E., Xiao, J.: An asymmetric affine Pólya-Szegö principle. Math. Ann. **352**, 517–542 (2012)
22. Hoffmann-Jϕgensen, J.: Probability With a View Toward Statistics, vol. 1, pp. 165–243. Chapman and Hall, New York (1994)
23. Huang, Q., He, B.: On the Orlicz Minkowski problem for polytopes. Discr. Comput. Geom. **48**, 281–297 (2012)
24. Koldobsky, A.: Fourier Analysis in Convex Geometry. Mathematical Surveys and Monographs, vol. 116, American Mathematical Society, Providence (2005)
25. Krasnosel'skii, M.A., Rutickii, Y.B.: Convex Functions and Orlicz Spaces. P. Noordhoff Ltd., Groningen (1961)
26. Ludwig, M.: Intersection bodies and valuations. Am. J. Math. **128**, 1409–1428 (2006)
27. Ludwig, M., Reitzner, M.: A classification of $SL(n)$ invariant valuations. Ann. Math. **172**, 1223–1271 (2010)
28. Lutwak, E.: Dual mixed volumes. Pac. J. Math. **58**, 531–538 (1975)
29. Lutwak, E.: Intersection bodies and dual mixed volumes. Adv. Math. **71**, 232–261 (1988)
30. Lutwak, E.: Centroid bodies and dual mixed volumes. Proc. Lond. Math. Soc. **60**, 365–391 (1990)
31. Lutwak, E.: The Brunn-Minkowski-Firey theory I: mixed volumes and the Minkowski problem. J. Differ. Geom. **38**, 131–150 (1993).
32. Lutwak, E.: The Brunn-Minkowski-Firey theory. II. Affine and geominimal surface areas. Adv. Math. **118**, 244–294 (1996)
33. Lutwak, E., Yang, D., Zhang, G.: L_p affine isoperimetric inequalities. J. Differ. Geom. **56**, 111–132 (2000)
34. Lutwak, E., Yang, D., Zhang, G.: Sharp affine L_p Sobolev inequalities. J. Differ. Geom. **62**, 17–38 (2002)

35. Lutwak, E., Yang, D., Zhang, G.: On the L_p-Minkowski problem. Trans. Am. Math. Soc. **356**, 4359–4370 (2004)
36. Lutwak, E., Yang, D., Zhang, G.: L_p John ellipsoids. Proc. Lond. Math. Soc. **90**, 497–520 (2005)
37. Lutwak, E., Yang, D., Zhang, G.: Orlicz projection bodies. Adv. Math. **223**(2), 220–24 (2010)
38. Lutwak, E., Yang, D., Zhang, G.: Orlicz centroid bodies. J. Differ. Geom. **84**, 365–387 (2010)
39. Lutwak, E., Yang, D., Zhang, G.: The Brunn-Minkowski-Firey inequality for nonconvex sets. Adv. Appl. Math. **48**, 407–413 (2012)
40. Parapatits, L.: $SL(n)$-covariant L_p-Minkowski valuations. J. Lond. Math. Soc. **89**(2), 397–414 (2014)
41. Parapatits, L.: $SL(n)$-contravariant L_p-Minkowski valuations. Trans. Am. Math. Soc. **366**, 1195–1211 (2014)
42. Parapatits, L., Schuster, F.E.: The Steiner formula for Minkowski valuations. Adv. Math. **230**, 978–994 (2012)
43. Rao, M.M., Ren, Z.D.: Theory of Orlicz Spaces. Marcel Dekker, New York (1991)
44. Schneider, R.: Convex Bodies: The Brunn-Minkowski Theory. Cambridge University Press, Cambridge (1993)
45. Schuster, F.E.: Valuations and Busemann-Petty type problems. Adv. Math. **219**, 344–368 (2008)
46. Schuster, F.E.: Crofton measures and Minkowski valuations. Duke Math. J. **154**, 1–30 (2010)
47. Schütt, C., Werner, E.: Surface bodies and p-affine surface area. Adv. Math. **187**, 98–145 (2004)
48. Werner, E.M.: Rényi divergence and L_p-affine surface area for convex bodies. Adv. Math. **230**, 1040–1059 (2012)
49. Werner, E., Ye, D.P.: New L_p affine isoperimetric inequalities. Adv. Math. **218**, 762–780 (2008)
50. Xi, D., Jin, H., Leng, G.: The Brunn-Minkwski inequality. Adv. Math. **260**, 350–374 (2014)
51. Zhang, G.: A positive answer to the Busemann-Petty problem in four dimensions. Ann. Math. **149**(2), 535–543 (1999)
52. Zhang, G., Grinberg, E.: Convolutions, transforms, and convex bodies. Proc. Lond. Math. Soc. **78**, 77–115 (1999)
53. Zhao, C.J.: Orlicz dual mixed volumes. Results Math. **68**, 93–104 (2015)
54. Zhao, C.J.: On the Orlicz-Brunn-Minkowski theory. Balkan J. Geom. Appl. **22**, 98–121 (2017)
55. Zhu, G.: The Orlicz centroid inequality for star bodies. Adv. Appl. Math. **48**, 432–445 (2012)
56. Zhu, B., Zhou, J., Xu, W.: Dual Orlicz-Brunn-Minkwski theory. Adv. Math. **264**, 700–725 (2014)

Modeling Cyber-Security

Nicholas J. Daras and Argyrios Alexopoulos

Abstract This paper documents a holistic mathematical modeling theory to provide a rigorous description of cyber-attacks and cyber-security. After determining valuations and vulnerabilities of parts of a node constituent, we recall the definitions of cyber-effect and cyber-interaction. Based on these concepts, we give the mathematical definitions of cyber navigation and infected node and we explain what is meant by dangerous cyber navigation and protection of cyber nodes from unplanned attacks. Our discussion proceeds to a rigorous description of passive and active cyber-attacks, as well as the relevant protections.

Keywords Mathematical modeling (models of systems) · Internet topics · Measure theory · Complex spaces · Valuation of a part of node constituent · Vulnerability of a part of node constituent · Node supervision · Cyber-effect · Cyber-interaction · Germ of cyber-attack · Cyber defense · Proactive cyber protection

Introduction

In many modern scientific studies, quantifying assumptions, data, and variables can contribute to the accurate description of the phenomena through appropriate mathematical models. So, in many disciplines, the analysts resort to a mathematical foundation of the concepts, in order to create a solid base for the theoretical formulation and solving all relevant problems. In this direction, there have been numerous significant contributions on the mathematical modeling of several branches of

N. J. Daras (✉)
Department of Mathematics, Hellenic Military Academy, Vari Attikis, Greece
e-mail: ndaras@sse.gr

A. Alexopoulos
Croisy Sur Seine, France
e-mail: a.d.alexopoulos@army.gr

N. J. Daras, Th. M. Rassias (eds.), *Modern Discrete Mathematics and Analysis*, Springer Optimization and Its Applications 131,
https://doi.org/10.1007/978-3-319-74325-7_3

theoretical engineering disciplines, such as theoretical computer science, network security, electronics, and artificial intelligence. Especially, in the case of cyber-security, we may mention several remarkable contributions [1, 6–8] and [10–15]. Indicative of the great interest shown for the mathematization of cyber-security is the regular organization of international conferences and workshops of major interest. However, although these presentations are innovative and promising, it seems that they lack a holistic view of the cyber environment. Moreover, there is no predictability of cyber-attacks, nor any opportunity to have given a strict definition of defensive protection so that we can look for an optimal design and organization of cyber defense. As a consequence thereof, one cannot build a solid foundation for a complete theory containing assumptions, definitions, theorems, and conclusions. But, this prevents the researcher to understand deeper behaviors, and requires limiting ourselves solely to practical techniques [9, 16].

The aim of the present chapter is to document a holistic mathematical theory to provide a rigorous description of cyber-attacks and cyber-security. To this end, section "Mathematical Definition of Cyberspace" recalls in brief the mathematical definition of cyberspace given in [4]. Next, in section "Mathematical Description of Cyber-Attacks", we first remind the concepts of *valuations* and *vulnerabilities* of the parts of a node constituent, and then, based on these two concepts, we give the definitions of *node supervision*, *cyber-effect*, and *cyber-interaction*. With this background, in section "Description of Cyber Navigations and Protection from Unplanned Attacks", we provide a mathematical definition of cyber navigation and, after giving the rigorous meaning of infected nodes, we determine what is meant by dangerous cyber navigation and protection of cyber nodes from unplanned attacks. Finally, in section "Description of Various Types of Cyber-Attacks and Protection", we describe a rigorous outline of passive and active cyber-attacks as well as an identification of the relevant proactive defense from such attacks. For a first study of proactive defense, one can consult [3]. Concrete examples to special types of germs of cyber-attacks are given in [2] and [5].

Mathematical Definition of Cyberspace

A multilayered weighted (finite or infinite) graph $\mathcal{X}$ with **N** interconnected layers is said to be an **N-cyber-archetype germ**. An **e-manifestation** gives a geographical qualifier at each node of $\mathcal{X}$. It is an embedding of $\mathcal{X}$ into a Cartesian product of **N** complex projective spaces $\mathbb{C}\mathbf{P}^{\mathbf{n_k}} \equiv \mathbf{P}\left(\mathbb{C}^{\mathbf{n_k}+1}\right)$, such that all nodes of $\mathcal{X}$ in the **k**-layer, called **e-node manifestations**, are illustrated at weighted points of the set $\mathbb{C}\mathbf{P}^{\mathbf{n_k}}$ and all directed edges (flows) of $\mathcal{X}$ in the **k**-layer, called **e-edge manifestations**, are given by simple weighted edges, i.e. by weighted homeomorphic images of the closed interval $[\mathbf{0}, \mathbf{1}]$ on $\mathbb{C}\mathbf{P}^{\mathbf{n_k}}$, so that, for any $\mathbf{k}=\mathbf{1}, \mathbf{2}, \ldots, \mathbf{N}$,

- the end points of each *e*-edge manifestation on $\mathbb{C}\mathbf{P}^{\mathbf{n_k}}$ must be images of end points of a corresponding original directed edge of $\mathcal{X}$ in the **k**-layer

– there should not be any *e*-edge manifestation on $\mathbb{C}\mathbf{P}^{\mathbf{n_k}}$ derived from directed *e*-edge of $\mathscr{X}$ in the **k**-layer into which belong points of *e*-edge manifestations that are defined by other nodes of $\mathscr{X}$ in the same layer.

The set $\mathscr{S}_e = \mathscr{S}_e\,(\mathbb{C}\mathbf{P}^{\mathbf{n_1}} \times \cdots \times \mathbb{C}\mathbf{P}^{\mathbf{n_N}})$ of *e*-manifestations of **N**-cyber archetype germs is the **e-superclass** in $\mathbb{C}\mathbf{P}^{\mathbf{n_1}} \times \cdots \times \mathbb{C}\mathbf{P}^{\mathbf{n_N}}$. An **e-graph category** $\mathscr{E}_{\mathscr{C}} = \mathscr{E}_{\mathscr{C}}\,(\mathbb{C}\mathbf{P}^{\mathbf{n_1}} \times \cdots \times \mathbb{C}\mathbf{P}^{\mathbf{n_N}})$ is a category consisting of the class $\mathbf{ob}\,(\mathscr{E}_{\mathscr{C}})$, whose elements, called **e-objects**, are the pairs $\mathscr{X} = (\mathbf{V}, \mathbf{E}) \in \mathscr{S}_e$, endowed with a class $\hom\,(\mathscr{E}_{\mathscr{C}})$ of **e-morphisms** on $\mathrm{ob}\,(\mathscr{E}_{\mathscr{C}})$ and an associative binary operation $\circ$ with identity.

Generalizing, one may consider additionally the following other four basic **e-categories**: The **e-set category** $\mathbf{e}_{\mathrm{Set}} = \mathbf{e}_{\mathrm{Set}}\,(\mathbb{C}\mathbf{P}^{\mathbf{n_1}} \times \cdots \times \mathbb{C}\mathbf{P}^{\mathbf{n_N}})$ where the objects are subsets of $\mathscr{E}_{\mathscr{C}}$, the **e-homomorphism category** $\mathbf{e}_{\mathrm{Hom}} = \mathbf{e}_{\mathrm{Hom}}\,(\mathbb{C}\mathbf{P}^{\mathbf{n_1}} \times \cdots \times \mathbb{C}\mathbf{P}^{\mathbf{n_N}})$ where the objects are sets of homomorphisms between subsets of $\mathbf{e}_{\mathrm{Set}}$, the **e-group category** $\mathbf{e}_{\mathrm{Grp}} = \mathbf{e}_{\mathrm{Gpr}}\,(\mathbb{C}\mathbf{P}^{\mathbf{n_1}} \times \cdots \times \mathbb{C}\mathbf{P}^{\mathbf{n_N}})$ where the objects are the groups of $\mathscr{E}_{\mathscr{C}}$, and the **e-topological category** $\mathbf{e}_{\mathrm{Top}} = \mathbf{e}_{\mathrm{Top}}\,(\mathbb{C}\mathbf{P}^{\mathbf{n_1}} \times \cdots \times \mathbb{C}\mathbf{P}^{\mathbf{n_N}})$ where the objects are topological subcategories of $\mathscr{E}_{\mathscr{C}}$. For reasons of homogenization of symbolism, we will adopt the following common notation $\mathscr{W}_{\mathbf{e}} = \left\{\mathscr{E}_{\mathscr{C}}, \mathbf{e}_{\mathrm{Set}}, \mathbf{e}_{\mathrm{Hom}}, \mathbf{e}_{\mathrm{Grp}}, \mathbf{e}_{\mathrm{Top}}\right\}$. The objects of each *e*-category $\mathbf{W_e} = \mathbf{W_e}\,(\mathbb{C}\mathbf{P}^{\mathbf{n_1}} \times \cdots \times \mathbb{C}\mathbf{P}^{\mathbf{n_N}}) \in \mathscr{W}_{\mathbf{e}}$ will be called **e-manifestations**. An easy **algebraic** structure in the (infinite) set of all these *e*-manifestations $(\mathbf{V}, \mathbf{E})$ and simultaneously, a compatible **topological** structure to allow for a detailed analytic study of $\mathscr{S}_e$ is given in [4]. Further, in [4], we investigate the possibility of allocating suitable vector weights to all the objects and morphisms of any *e*-category $\mathbf{W_e} \in \mathscr{W}_{\mathbf{e}} = \left\{\mathscr{E}_{\mathscr{C}}, \mathbf{e}_{\mathrm{Set}}, \mathbf{e}_{\mathrm{Grp}}, \mathbf{e}_{\mathrm{Top}}\right\}$. Towards this end, we consider two types of vector weights that can be attached to any object and/or morphism of such an *e*-category: the maximum weight and the square weight. Any such weight will be a point in the positive quadrant of the plane. Taking this into account, any *e*-category $\mathbf{W_e} \in \mathscr{W}_{\mathbf{e}} = \left\{\mathscr{E}_{\mathscr{C}}, \mathbf{e}_{\mathrm{Set}}, \mathbf{e}_{\mathrm{Hom}}, \mathbf{e}_{\mathrm{Gpr}}, \mathbf{e}_{\mathrm{Top}}\right\}$ can be viewed as an **infinite** *e-graph* $(\mathbb{V}, \mathbb{E})$ with *vector weights*, in such a way that the *e*-nodes in $\mathbb{V}$ are the *e*-objects $\mathbf{X} \in \mathbf{ob}\,(\mathbf{W_e})$, while the *e*-edges in $\mathbb{E}$ are the *e*-morphisms $\mathbf{h} \in \hom\,(\mathbf{W_e})$. For such an *e*-graph $\mathfrak{G}_{\mathbf{W_e}}$ corresponding to an *e*-category $\mathbf{W_e} \in \mathscr{W}_{\mathbf{e}}$, the vector weight of the *e*-node associated to the *e*-manifestation $\mathscr{X} = (\mathbf{V}, \mathbf{E}) \in \mathbb{V} = \mathbf{ob}\,(\mathbf{W_e})$ is equal to a weight of $\mathscr{X}$. Bearing all this in mind, in [4], we introduced a suitable intrinsic metric $\mathbf{d}_{\mathbf{W_e}}$ in the set $\mathbf{ob}\,(\mathbf{W_e})$ of objects of an *e*-category $\mathbf{W_e}$. The most significant benefits coming from such a consideration can be derived from the definitions of *cyber-evolution* and *cyber-domain*. To do this, we first defined the concept of *e-dynamics*, as a mapping of the form $cy : [0, 1] \to (\mathbf{ob}\,(\mathbf{W_e}), \mathbf{d}_{\mathbf{W_e}})$; its image is an *e-arrangement*. Each point $cy\,(\mathbf{t}) \in cy([0, 1])$ is an (instantaneous) local *e*-node manifestation with an interrelated *e*-edge manifestation. An *e*-arrangement together with all of its (instantaneous) *e*-morphisms is an *e-regularization*. The elements of the completion $\overline{\mathbf{ob}\,(\mathbf{W_e})}$ of $\mathbf{ob}\,(\mathbf{W_e})$ in $\overline{\mathbb{C}\mathbf{P}^{\mathbf{n_1}} \times \cdots \times \mathbb{C}\mathbf{P}^{\mathbf{n_N}}}$ are the *cyber-elements*, while the topological space $\left(\overline{\mathbf{ob}\,(\mathbf{W_e})}, \mathbf{d}_{\mathbf{W_e}}\right)$ is a *cyber-domain*. With this notation, a continuous *e*-dynamics $cy : [0, 1] \to \left(\overline{\mathbf{ob}\,(\mathbf{W_e})}, \mathbf{d}_{\mathbf{W_e}}\right)$ is said to be a *cyber-*

evolutionary path or simply *cyber-evolution* in the cyber-domain $\left(\overline{\mathbf{ob}\,(\mathbf{W_e})}, \mathbf{d_{W_e}}\right)$. Its image is said to be a *cyber-arrangement.* A cyber-arrangement together with all of its (instantaneous) cyber-morphisms is called a *cyberspace*.

In view of the above concepts, in [4], we investigated conditions under which an *e*-regularization may be susceptible of a projective *e*-limit. It is important to know if an *e-sub*-regularization is projective *e*-system. Subsequently, we defined and discussed the concept of the *length* in a cyber-domain. For the intrinsic cyber-metric $\mathbf{d_{W_e}}$, the distance between two cyber-elements is the length of the "shortest cyber-track" between these cyber-elements. The term shortest cyber-track is defined and is crucial for understanding the concept of *cyber-geodesic*. Although every shortest cyber track on a cyber-length space is a cyber-geodesic, the reverse argument is not valid. In fact, *some cyber-geodesics may fail to be shortest cyber-tracks on large scales*. However, since each cyber-domain $\left(\overline{\mathbf{ob}\,(\mathbf{W_e})}, \mathbf{d_{W_e}}\right)$ is a compact, complete metric space, and since for any pair of cyber-elements in $\overline{\mathbf{ob}\,(\mathbf{W_e})}$ there is a cyber-evolutionary path of finite length joining them, one can easily ascertain the following converse result: any *pair of two cyber-elements in each cyber-domain* $\left(\overline{\mathbf{ob}\,(\mathbf{W_e})}, \mathbf{d_{W_e}}\right)$ *has a shortest cyber track joining them*. Finally, in [4], we gave a discussion about the *speed* (: *cyber-speed*) of a cyber-evolution and the *convergence* of a sequence of cyber-evolutions.

Mathematical Description of Cyber-Attacks

At any moment **t**, a **node V** in the cyber-domain $\left(\overline{\mathbf{ob}\,(\mathbf{W_e})}, \mathbf{d_{W_e}}\right)$ is composed of cyber constituents consisting in devices $\mathbf{D_j^{(V)}}$ (:sensors, regulators of information flow, etc.) and resources $\mathbf{R_k^{(V)}}$ (:services, data, messages, etc.), the number of which depend potentially from the three geographical coordinates $\mathbf{x_1}, \mathbf{x_2}, \mathbf{x_3}$ and the time **t**. The order of any used quote of devices $\mathbf{D_1^{(V)}}$, $\mathbf{D_2^{(V)}}, \ldots$ and resources $\mathbf{R_1^{(V)}}$, $\mathbf{R_2^{(V)}}, \ldots$ is assumed to be given, pre-assigned, and well defined. We will assume uninterruptedly that:

1. the potential number of all ***possible*** devices and resources of **V** is equal to $\mathcal{M}_\mathbf{V} \gg \mathbf{0}$ and $\mathcal{L}_\mathbf{V} \gg \mathbf{0}$, respectively, and
2. the number of **V**'s ***available*** devices and resources is only $m_\mathbf{V}=m_\mathbf{V}(t)$ and $\ell_\mathbf{V}=\ell_\mathbf{V}\,(t)$, respectively, with $m_\mathbf{V}<\mathcal{M}_\mathbf{V}$ and $\ell_\mathbf{V}<\mathcal{L}_\mathbf{V}$.

Valuations and Vulnerabilities of Parts of a Node Constituent

Let **U**, **V** be two nodes in the cyber-domain $\left(\overline{\mathbf{ob}\,(\mathbf{W_e})}, \mathbf{d}_{\mathbf{W_e}}\right)$ and let $\mathscr{K}^{(\mathbf{V})}$ be an available constituent in **V**:

$$\mathscr{K} = \begin{cases} \mathbf{D}, \text{ if the constituent is a device,} \\ \mathbf{R}, \text{ if the constituent is a resource element.} \end{cases}$$

Obviously, $\mathscr{K}^{(\mathbf{V})}$ may also be viewed as a nonempty collection of a number of elements. It is easy to see that one can make as much finite $\boldsymbol{\sigma}$-algebras as partitions on $\mathscr{K}^{(\mathbf{V})}$.

Definition 1 For every partition $\mathscr{P}$ of $\mathscr{K}^{(\mathbf{V})}$, let us consider a corresponding $\boldsymbol{\sigma}$-algebra $\mathfrak{U}_{\mathscr{P}}$ of subsets of $\mathscr{K}^{(\mathbf{V})}$ as well as a monotonic measure $\boldsymbol{\mu}$ defined on $\mathfrak{U}_{\mathscr{P}}$. Let also $\mathrm{Cr}_1, \mathrm{Cr}_2, \ldots, \mathrm{Cr}_{\mathfrak{N}}$ be $\mathfrak{N}=\mathfrak{N}\left(\mathscr{K}^{(\mathbf{V})}, \mathscr{P}\right)$ objective quantifiable criteria for the assessment of the points of $\mathscr{K}^{(\mathbf{V})}$. Denoting by $\mathrm{Cr}_\mathbf{j}\,(\mathbf{p}) \in \mathbb{R}$ the value of $\mathrm{Cr}_\mathbf{j}$ on $\mathbf{p} \in \mathscr{K}^{(\mathbf{V})}$ at a point $(\mathbf{x}_1, \mathbf{x}_2, \mathbf{x}_3, \mathbf{t}) \in \mathbb{R}^3 \times [\mathbf{0}, \mathbf{1}]$, suppose

1. the functions $\mathrm{Cr}_\mathbf{j}\,(\mathbf{p})$ are measurable with respect to $\boldsymbol{\mu}$ and
2. a valuation weight $\mathbf{u}_\mathbf{j}\,(\mathbf{p})$ is attributed by (the user(s) of) **U** to the Criterion $\mathrm{Cr}_\mathbf{j}$ on $\mathbf{p} \in \mathscr{K}^{(\mathbf{V})}$ at $(\mathbf{x}_1, \mathbf{x}_2, \mathbf{x}_3, \mathbf{t}) \in \mathbb{R}^4$.

If $\mathbf{E} \in \mathfrak{U}_{\mathscr{P}}$ is a part of $\mathscr{K}^{(\mathbf{V})}$and $\mathfrak{n} \leq \mathfrak{N}$, then a **relative valuation of E from the viewpoint (of user(s)) of node U** with respect to the $\mathfrak{n}$ criteria $\mathrm{Cr}_1, \mathrm{Cr}_2, \ldots, \mathrm{Cr}_\mathfrak{n}$ at the spatiotemporal point $(\mathbf{x}_1, \mathbf{x}_2, \mathbf{x}_3, \mathbf{t}) \in \mathbb{R}^4$ is any vector

$$\mathbf{A}^{(\mathbf{U} \rightsquigarrow \mathbf{V})}\,(\mathbf{E}) = \left(\mathbf{a}_1^{(\mathbf{U} \rightsquigarrow \mathbf{V})}\,(\mathbf{E})\,, \mathbf{a}_2^{(\mathbf{U} \rightsquigarrow \mathbf{V})}\,(\mathbf{E})\,, \ldots, \mathbf{a}_\mathfrak{n}^{(\mathbf{U} \rightsquigarrow \mathbf{V})}\,(\mathbf{E})\right)^\mathbf{T} \in \mathbb{R}^\mathfrak{n}$$

where each definite integral

$$\mathbf{a}_\mathbf{j}^{(\mathbf{U} \rightsquigarrow \mathbf{V})}\,(\mathbf{E}) := \int_\mathbf{E} \mathrm{Cr}_\mathbf{j}\,(\mathbf{p})\,\mathbf{u}_\mathbf{j}\,(\mathbf{p})\;\mathbf{d}\boldsymbol{\mu}\,(\mathbf{p})$$

is the **component valuation of E from the viewpoint (of user(s)) of the node U into the constituent** $\mathscr{K}^{(\mathbf{V})}$ at $(\mathbf{x}_1, \mathbf{x}_2, \mathbf{x}_3, \mathbf{t})$. The number $\mathfrak{n}$ is the **dimension of the valuation**.

There is a special category of valuations of particular interest, determined in regard to the low degree of "security" of the constituents of the node. The low degree of security is described completely by the concept of vulnerability.

Definition 2 For every partition $\mathscr{P}$ of $\mathscr{K}^{(\mathbf{V})}$, let us consider a corresponding $\boldsymbol{\sigma}$-algebra $\mathfrak{U}_{\mathscr{P}}$ of subsets of $\mathscr{K}^{(\mathbf{V})}$ as well as a monotonic measure $\boldsymbol{\lambda}$ defined on $\mathfrak{U}_{\mathscr{P}}$. Let also $\mathrm{SCr}_1, \mathrm{SCr}_2, \ldots, \mathrm{SCr}_{\mathfrak{M}}$ be $\mathfrak{M}=\mathfrak{M}\left(\mathscr{K}^{(\mathbf{V})}, \mathscr{P}\right)$ objective quantifiable criteria for the security assessment of the points of $\mathcal{K}^{(\mathbf{V})}$. Denoting by $\mathrm{SCr_j}\,(\mathbf{p}) \in \mathbb{R}$ the value of $\mathrm{SeCr_j}$ on $\mathbf{p} \in \mathcal{K}^{(\mathbf{V})}$ at a spatiotemporal point $(\mathbf{x_1}, \mathbf{x_2}, \mathbf{x_3}, \mathbf{t}) \in \mathbb{R}^3 \times [\mathbf{0}, \mathbf{1}]$, suppose

1. the functions $\mathrm{SCr_j}\,(\mathbf{p})$ are measurable with respect to $\boldsymbol{\lambda}$ and
2. a **vulnerability weight** $u_\mathrm{j}\,(\mathbf{p})$ is attributed by (the (user(s) of) node **U** to the security criterion $\mathrm{SCr_j}$ on $\mathbf{p} \in \mathcal{K}^{(\mathbf{V})}$ at $(\mathbf{x_1}, \mathbf{x_2}, \mathbf{x_3}, \mathbf{t}) \in \mathbb{R}^4$.

If $\mathbf{E} \in \mathfrak{U}_{\mathscr{P}}$ is a part of $\mathcal{K}^{(\mathbf{V})}$and $\mathfrak{m} \le \mathfrak{M}$, then a **relative vulnerability of E from the viewpoint (of the user(s)) of node U** with respect to the $\mathfrak{m}$ security criteria $\mathrm{SCr_1}, \mathrm{SCr_2}, \ldots, \mathrm{SCr_m}$ at $(\mathbf{x_1}, \mathbf{x_2}, \mathbf{x_3}, \mathbf{t}) \in \mathbb{R}^4$ is any vector

$$\mathbf{B}^{(\mathbf{U} \rightsquigarrow \mathbf{V})}\,(\mathbf{E}) = \left(\mathbf{b}_1^{(\mathbf{U} \rightsquigarrow \mathbf{V})}\,(\mathbf{E}), \mathbf{b}_2^{(\mathbf{U} \rightsquigarrow \mathbf{V})}\,(\mathbf{E}), \ldots, \mathbf{b}_\mathrm{m}^{(\mathbf{U} \rightsquigarrow \mathbf{V})}\,(\mathbf{E})\right)^\mathbf{T} \in \mathbb{R}^\mathbf{m}$$

where each definite integral

$$\mathbf{b}_\mathbf{j}^{(\mathbf{U} \rightsquigarrow \mathbf{V})}\,(\mathbf{E}) = \int_\mathbf{E} \mathrm{SCr_j}\,(\mathbf{p})\, u_\mathrm{j}\,(\mathbf{p})\ \mathbf{d}\boldsymbol{\lambda}\,(\mathbf{p})$$

is the **component vulnerability of E from the viewpoint (of the (user(s)) of the node U into the constituent** $\mathcal{K}^{(\mathbf{V})}$ at $(\mathbf{x_1}, \mathbf{x_2}, \mathbf{x_3}, \mathbf{t})$. The number $\mathfrak{m}$ is the **dimension of the vulnerability**.

In what follows, a part **E** of a ***possible*** device $\mathbf{D}_\kappa^{(\mathbf{V})}$ or/and resource $\mathbf{R}_\xi^{(\mathbf{V})}$ of **V** that is evaluated from the viewpoint (of the user(s)) of node **U** may be denoted by $\mathrm{fr}\left(\mathbf{D}_\kappa^{(\mathbf{V})}\right)$ or/and $\mathrm{fr}\left(\mathbf{R}_\xi^{(\mathbf{V})}\right)$, respectively ($\kappa$=**1**, **2**, ..., $\mathcal{M}_\mathbf{V}$, ξ=**1**, **2**, ..., $\mathcal{L}_\mathbf{V}$). However, to denote both $\mathbf{A}^{(\mathbf{U} \rightsquigarrow \mathbf{V})}\left(\mathrm{fr}\left(\mathbf{D}_\kappa^{(\mathbf{V})}\right)\right)$ and $\mathbf{A}^{(\mathbf{U} \rightsquigarrow \mathbf{V})}\left(\mathrm{fr}\left(\mathbf{R}_\xi^{(\mathbf{V})}\right)\right)$ we will prefer to use the common notation $\mathbf{A}_\mathbf{s}^{(\mathbf{U} \rightsquigarrow \mathbf{V})}$:

$$\begin{aligned}\mathbf{A}_\mathbf{s}^{(\mathbf{U} \rightsquigarrow \mathbf{V})} &= \left(\mathbf{a}_{1,\mathbf{s}}^{(\mathbf{U} \rightsquigarrow \mathbf{V})}, \ldots, \mathbf{a}_{\mathrm{n},\mathbf{s}}^{(\mathbf{U} \rightsquigarrow \mathbf{V})}\right)^\mathbf{T} \\ &= \begin{cases} \mathbf{A}_\mathbf{U}\left(\mathrm{fr}\left(\mathbf{D}_\mathbf{s}^{(\mathbf{V})}\right)\right), \ \textbf{if s=1, 2, \ldots, } \mathcal{M}_\mathbf{V} \\ \mathbf{A}_\mathbf{U}\left(\mathrm{fr}\left(\mathbf{R}_{\mathbf{s}-\mathcal{M}_\mathbf{V}}^{(\mathbf{V})}\right)\right) \ \textbf{if s=}\mathcal{M}_\mathbf{V}\textbf{+1, }\mathcal{M}_\mathbf{V}\textbf{+2, \ldots, }\mathcal{M}_\mathbf{V}\textbf{+}\mathcal{L}_\mathbf{V}. \end{cases}\end{aligned}$$

Similarly, to denote both $\mathbf{B}^{(\mathbf{U} \rightsquigarrow \mathbf{V})}\left(\mathrm{fr}\left(\mathbf{D}_\kappa^{(\mathbf{V})}\right)\right)$, κ=**1**, **2**, ..., $\mathcal{M}_\mathbf{V}$ and $\mathbf{B}^{(\mathbf{U} \rightsquigarrow \mathbf{V})}\left(\mathrm{fr}\left(\mathbf{R}_\xi^{(\mathbf{V})}\right)\right)$, ξ=**1**, **2**, ..., $\mathcal{L}_\mathbf{V}$, we will prefer to adopt the notation

$$\mathbf{B}_\mathbf{s}^{(\mathbf{U} \rightsquigarrow \mathbf{V})} = \left(\mathbf{b}_{1,\mathbf{s}}^{(\mathbf{U} \rightsquigarrow \mathbf{V})}, \ldots, \mathbf{b}_{\mathrm{m},\mathbf{s}}^{(\mathbf{U} \rightsquigarrow \mathbf{V})}\right)^\mathbf{T}$$

$$= \begin{cases} \mathbf{B_U}\left(\mathrm{fr}\left(\mathbf{D_s^{(V)}}\right)\right), \ \mathbf{if\ s{=}1, 2, \ldots, \mathscr{M}_V} \\ \mathbf{B_U}\left(\mathrm{fr}\left(\mathbf{R_{s-\mathscr{M}_V}^{(V)}}\right)\right) \ \mathbf{if\ s{=}\mathscr{M}_V{+}1, \mathscr{M}_V{+}2, \ldots, \mathscr{M}_V{+}\mathscr{L}_V}. \end{cases}$$

Cyber-Effects and Cyber-Interactions

We are now in position to proceed towards a description of homomorphisms between cyber nodes. Let $\mathbf{U}$, $\mathbf{V}$ be two nodes in the cyber-domain $\left(\overline{\mathbf{ob}\,(\mathbf{W_e})}, \mathbf{d_{W_e}}\right)$. Without loss of generality, we may suppose the numbers $\mathscr{M}_\mathrm{V}+\mathscr{L}_\mathrm{V}$ and $\mathscr{M}_\mathrm{U}+\mathscr{L}_\mathrm{U}$ are both enough large, so that $k := \mathscr{M}_\mathrm{V}+\mathscr{L}_\mathrm{V}=\mathscr{M}_\mathrm{U}+\mathscr{L}_\mathrm{U}$. We consider the following sets.

1.

$$\mathfrak{C}^{(\text{fraction})}\,(\mathbf{V}) = \left\{\left(\mathrm{fr}\left(\mathbf{D_1^{(V)}}\right), \ldots, \mathrm{fr}\left(\mathbf{D_{\mathscr{M}_V}^{(V)}}\right), \mathrm{fr}\left(\mathbf{R_1^{(V)}}\right), \ldots, \mathrm{fr}\left(\mathbf{R_{\mathscr{L}_V}^{(V)}}\right)\right) : \right.$$

$$\left. \mathrm{fr}\left(\mathbf{D}_\kappa^{(\mathbf{V})}\right), \mathrm{fr}\left(\mathbf{R}_\xi^{(\mathbf{V})}\right) \in \mathfrak{U}_{\mathscr{P}}, \kappa \le \mathscr{M}_\mathrm{V},\ \xi \le \mathscr{L}_\mathrm{V}\right\}$$

the set of ordered columns of possible parts of constituents of $\mathbf{V}$;

2.

$$\mathscr{A}_\mathrm{U}\mathfrak{C}^{(\text{fraction})}\,(\mathbf{V}) = \left\{\left(\mathbf{A}_1^{(\mathbf{U}\rightsquigarrow\mathbf{V})}, \ldots, \mathbf{A}_k^{(\mathbf{U}\rightsquigarrow\mathbf{V})}\right) : \right.$$

$$\left. \mathbf{A}_\mathbf{s}^{(\mathbf{U}\rightsquigarrow\mathbf{V})} \in \mathbb{R}^\mathrm{n}, \mathbf{s{=}1, 2, \ldots}, k\right\} \equiv \mathbb{R}^{\mathrm{n}\times k} :$$

the set of ordered columns of relative *valuations* of parts of possible constituents of $\mathbf{V}$, from the viewpoint of $\mathbf{U}$, over the space time $\mathbb{R}^3\times[\mathbf{0}, \mathbf{1}]$;

3.

$$\mathscr{B}_\mathrm{U}\mathfrak{C}^{(\text{fraction})}\,(\mathbf{V}) = \left\{\left(\mathbf{B}_1^{(\mathbf{U}\rightsquigarrow\mathbf{V})}, \ldots, \mathbf{B}_k^{(\mathbf{U}\rightsquigarrow\mathbf{V})}\right) : \right.$$

$$\left. \mathbf{B}_\mathbf{s}^{(\mathbf{U}\rightsquigarrow\mathbf{V})} \in \mathbb{R}^\mathrm{n}, \mathbf{s{=}1, 2, \ldots}, k\right\} \equiv \mathbb{R}^{\mathrm{n}\times k} :$$

the set of all ordered columns of relative *vulnerabilities* of parts of possible constituents in $\mathbf{V}$, from the viewpoint of $\mathbf{U}$, over $\mathbb{R}^3\times[\mathbf{0}, \mathbf{1}]$.

Definition 3 The triplet

$$\mathscr{P}{=}\mathscr{P}\,(\mathbf{V}) = \left(\mathfrak{C}^{(\text{fraction})}\,(\mathbf{V})\,,\, \mathscr{A}_\mathrm{U}\mathfrak{C}^{(\text{fraction})}\,(\mathbf{V})\,,\, \mathscr{B}_\mathrm{U}\mathfrak{C}^{(\text{fraction})}\,(\mathbf{V})\right)$$

is called the **cyber-range of V from the viewpoint of (the users of) U**. Its elements p are the **(threefold) cyber situations**. Especially, if **U**=**V**, the cyber-field $\mathscr{P}=\mathscr{P}$ (**V**) is the **cyber-purview** of **V** and is denoted $\mathscr{P}^{(\mathrm{self})}=\mathscr{P}^{(\mathrm{self})}$ (**V**). Its elements are represented by $\hat{p}$.

Given an ordered set

$$\mathbf{FR}^{(\mathbf{V})} := \left(\mathrm{fr}\left(\mathbf{D}_{\mathbf{1}}^{(\mathbf{V})}\right), \ldots, \mathrm{fr}\left(\mathbf{D}_{\mathscr{M}_{\mathbf{V}}}^{(\mathbf{V})}\right), \mathrm{fr}\left(\mathbf{R}_{\mathbf{1}}^{(\mathbf{V})}\right), \ldots, \mathrm{fr}\left(\mathbf{R}_{\mathscr{L}_{\mathbf{V}}}^{(\mathbf{V})}\right)\right)$$

of ordered columns of parts of constituents of **V**, a cyber situation p on **V** can be viewed as an ordered pair of matrices

$$p=\left(\mathbb{A}^{(\mathbf{U}\rightsquigarrow\mathbf{V})}, \mathbb{B}^{(\mathbf{U}\rightsquigarrow\mathbf{V})}\right) = \left(\left(\mathbf{a}_{\mathrm{i,j}}\right), \left(\mathbf{b}_{\mathrm{i,j}}\right)\right) \in \mathbb{R}^{\mathrm{n}\times k}\times\mathbb{R}^{\mathrm{m}\times k}$$

where

$$\mathbb{A}^{(\mathbf{U}\rightsquigarrow\mathbf{V})}=\left(\mathbf{A}_{1}^{(\mathbf{U}\rightsquigarrow\mathbf{V})}, \ldots, \mathbf{A}_{k}^{(\mathbf{U}\rightsquigarrow\mathbf{V})}\right) = \left(\mathbf{a}_{\mathrm{i,j}}\right) = \begin{pmatrix} \mathbf{a}_{1,1}^{(\mathbf{U}\rightsquigarrow\mathbf{V})} & \cdots & \mathbf{a}_{1,k}^{(\mathbf{U}\rightsquigarrow\mathbf{V})} \\ \vdots & \vdots & \vdots \\ \mathbf{a}_{\mathrm{n},1}^{(\mathbf{U}\rightsquigarrow\mathbf{V})} & \cdots & \mathbf{a}_{\mathrm{n},k}^{(\mathbf{U}\rightsquigarrow\mathbf{V})} \end{pmatrix} \text{ and}$$

$$\mathbb{B}^{(\mathbf{U}\rightsquigarrow\mathbf{V})}=\left(\mathbf{B}_{\mathbf{1}}^{(\mathbf{U}\rightsquigarrow\mathbf{V})}, \ldots, \mathbf{B}_{k}^{(\mathbf{U}\rightsquigarrow\mathbf{V})}\right) = \left(\mathbf{b}_{\mathrm{i,j}}\right) = \begin{pmatrix} \mathbf{b}_{1,1}^{(\mathbf{U}\rightsquigarrow\mathbf{V})} & \cdots & \mathbf{b}_{1,k}^{(\mathbf{U}\rightsquigarrow\mathbf{V})} \\ \vdots & \vdots & \vdots \\ \mathbf{b}_{\mathrm{m},1}^{(\mathbf{U}\rightsquigarrow\mathbf{V})} & \cdots & \mathbf{b}_{\mathrm{m},k}^{(\mathbf{U}\rightsquigarrow\mathbf{V})} \end{pmatrix}.$$

In particular, any purview $\hat{p}$ on **V** can simply be viewed as an ordered pair

$$\hat{p}=\left(\widehat{\mathbb{A}}^{(\mathbf{V}\rightsquigarrow\mathbf{V})}, \widehat{\mathbb{B}}^{(\mathbf{V}\rightsquigarrow\mathbf{V})}\right) = \left(\left(\widehat{\mathbf{a}}_{\mathrm{i,j}}\right), \left(\widehat{\mathbf{b}}_{\mathrm{i,j}}\right)\right) \in \mathbb{R}^{\mathrm{n}\times k}\times\mathbb{R}^{\mathrm{m}\times k}$$

with

$$\widehat{\mathbb{A}}^{(\mathbf{V}\rightsquigarrow\mathbf{V})}= \left(\widehat{\mathbf{a}}_{\mathrm{i,j}}\right) = \begin{pmatrix} \mathbf{a}_{1,1}^{(\mathbf{V}\rightsquigarrow\mathbf{V})} & \cdots & \mathbf{a}_{1,k}^{(\mathbf{V}\rightsquigarrow\mathbf{V})} \\ \vdots & \vdots & \vdots \\ \mathbf{a}_{\mathrm{n},1}^{(\mathbf{V}\rightsquigarrow\mathbf{V})} & \cdots & \mathbf{a}_{\mathrm{n},k}^{(\mathbf{V}\rightsquigarrow\mathbf{V})} \end{pmatrix}$$

and

$$\widehat{\mathbb{B}}^{(\mathbf{V}\rightsquigarrow\mathbf{V})}= \left(\widehat{\mathbf{b}}_{\mathrm{i,j}}\right) = \begin{pmatrix} \mathbf{b}_{1,1}^{(\mathbf{V}\rightsquigarrow\mathbf{V})} & \cdots & \mathbf{b}_{1,k}^{(\mathbf{V}\rightsquigarrow\mathbf{V})} \\ \vdots & \vdots & \vdots \\ \mathbf{b}_{\mathrm{m},1}^{(\mathbf{V}\rightsquigarrow\mathbf{V})} & \cdots & \mathbf{b}_{\mathrm{m},k}^{(\mathbf{V}\rightsquigarrow\mathbf{V})} \end{pmatrix}.$$

To simplify our approach, in what follows we will assume that ***the location*** $(\mathbf{x}_1, \mathbf{x}_2, \mathbf{x}_3) \in \mathbb{R}^3$ ***of* V *remains constantly fixed***.

Definition 4 The **supervision vector of V in the node system** (**V**, **U**) at a given time moment **t** ∈ [0, 1] is defined to be the pair

$$(z, w)(\mathbf{t}) = \left(\mathbb{A}_{\mathbf{U}\to\mathbf{V}} + \mathbf{i}\widehat{\mathbb{A}}_{\mathbf{V}\to\mathbf{V}},\ \mathbb{B}_{\mathbf{U}\to\mathbf{V}} + \mathbf{i}\widehat{\mathbb{B}}_{\mathbf{V}\to\mathbf{V}}\right)(\mathbf{t}) \in \mathbb{C}^{\mathrm{n}\times k}\times\mathbb{C}^{\mathrm{m}\times k}$$

with $\mathbf{i}{:=}\sqrt{-\mathbf{1}} \in \mathbb{C}$. Especially, the complex matrices z and w are called **supervisory perceptions of V in the node system** (**V**, **U**) at moment **t**. The mapping defined by

$$\gamma_{\mathbf{V}}\colon [0, 1] \to \mathbb{C}^{\mathrm{n}\times k}\times\mathbb{C}^{\mathrm{m}\times k}{:}\mathbf{t} \mapsto \gamma_{\mathbf{V}}(\mathbf{t}) = (z, w)(\mathbf{t})$$

is the **supervisory perception curve of V in the node system** (**V**, **U**) **during the whole of time interval** [**0**, **1**]. The **supervisory perception domain of V in the node system** (**V**, **U**) is the range $\gamma_{\mathbf{V}}$ ([**0**, **1**]) of $\Gamma_{\mathbf{V}}$, denoted by $\gamma_{\mathbf{V}}^{*}$.

Theoretically, each point in the space $\mathbb{C}^{\mathrm{n}\times k}\times\mathbb{C}^{\mathrm{m}\times k}$ can be viewed as a supervision vector of **V** in the system of nodes **V** and **U**. Since in many cases, it suffices (or is preferable) to use only specific supervisions from the viewpoint of **U** or **V**:

$$(\mathbb{A}_{\mathbf{U}\to\mathbf{V}}, \mathbb{B}_{\mathbf{U}\to\mathbf{V}})(\mathbf{t}) \text{ or } \left(\widehat{\mathbb{A}}_{\mathbf{V}\to\mathbf{V}}, \widehat{\mathbb{B}}_{\mathbf{V}\to\mathbf{V}}\right)(\mathbf{t})$$

$$\text{or } \left(\mathbb{A}_{\mathbf{U}\to\mathbf{V}}, \mathbf{i}\widehat{\mathbb{B}}_{\mathbf{V}\to\mathbf{V}}\right)(\mathbf{t}) \text{ or } \left(\mathbf{i}\widehat{\mathbb{A}}_{\mathbf{V}\to\mathbf{V}}, \mathbb{B}_{\mathbf{U}\to\mathbf{V}}\right)(\mathbf{t})$$

it is natural and imperative to consider two main vector fields **X1** and **X2** defined on $\gamma_{\mathbf{V}}^{*}$, as follows.

- The vector field **X1** which assigns to each point

$$(z, w)(\mathbf{t}) = \left(\mathbb{A}_{\mathbf{U}\to\mathbf{V}} + \mathbf{i}\widehat{\mathbb{A}}_{\mathbf{V}\to\mathbf{V}},\ \mathbb{B}_{\mathbf{U}\to\mathbf{V}} + \mathbf{i}\widehat{\mathbb{B}}_{\mathbf{V}\to\mathbf{V}}\right)(\mathbf{t})$$

  of $\gamma_{\mathbf{V}}^{*}$ the vector

$$(\mathbf{Re}z, \mathbf{Re}w)(\mathbf{t}) \equiv (\mathbb{A}_{\mathbf{U}\to\mathbf{V}} + \mathbf{i0}, \mathbb{B}_{\mathbf{U}\to\mathbf{V}} + \mathbf{i0})(\mathbf{t}) \in \mathbb{R}^{\mathrm{n}\times k}\times\mathbb{R}^{\mathrm{m}\times k},$$

  i.e., the vector of the relative *valuations* and *vulnerabilities* of $\mathbf{FR}^{(\mathbf{V})}$ at **t**, considered from the viewpoint of **U**; in particular, we may also define the vector fields **Y1** and **Z1** assigning to each point

$$(z, w)(\mathbf{t}) = \left(\mathbb{A}_{\mathbf{U}\to\mathbf{V}} + \mathbf{i}\widehat{\mathbb{A}}_{\mathbf{V}\to\mathbf{V}},\ \mathbb{B}_{\mathbf{U}\to\mathbf{V}} + \mathbf{i}\widehat{\mathbb{B}}_{\mathbf{V}\to\mathbf{V}}\right)(\mathbf{t})$$

  of $\gamma_{\mathbf{V}}^{*}$ the relative *valuations* and relative *vulnerabilities* of $\mathbf{FR}^{(\mathbf{V})}$ at **t**, considered from the viewpoint of **U**:

$$\mathbf{Re}z(\mathbf{t}) \equiv \mathbb{A}_{\mathbf{U}\to\mathbf{V}}(\mathbf{t}) \in \mathbb{R}^{\mathrm{n}\times k} \text{ and } \mathbf{Re}w(\mathbf{t}) \equiv \mathbb{B}_{\mathbf{U}\to\mathbf{V}}(\mathbf{t}) \in \mathbb{R}^{\mathrm{m}\times k}.$$

– The vector field **X2** which assigns to each point

$$(z,\,w)\,(\mathbf{t}) = \left(\mathbb{A}_{\mathbf{U}\to\mathbf{V}} + \mathbf{i}\widehat{\mathbb{A}}_{\mathbf{V}\to\mathbf{V}},\, \mathbb{B}_{\mathbf{U}\to\mathbf{V}} + \mathbf{i}\widehat{\mathbb{B}}_{\mathbf{V}\to\mathbf{V}}\right)(\mathbf{t})$$

of $\gamma^*_{\mathbf{V}}$ the vector

$$(\mathbf{Im}z,\,\mathbf{Im}w)\,(\mathbf{t}) \equiv \left(\mathbf{0} + \mathbf{i}\widehat{\mathbb{A}}_{\mathbf{V}\to\mathbf{V}},\, \mathbf{0} + \mathbf{i}\widehat{\mathbb{B}}_{\mathbf{V}\to\mathbf{V}}\right)(\mathbf{t}) \in \mathbb{R}^{n\times k}\times\mathbb{R}^{m\times k},$$

i.e., the vector of the *valuations* and *vulnerabilities* of $\mathbf{FR}^{(\mathbf{V})}$ at **t**, considered from the viewpoint of **V** itself; subsequently, we may define the vector fields **Y2** and **Z2** assigning to each point $(z,\,w)\,(\mathbf{t}) = \left(\mathbb{A}_{\mathbf{U}\to\mathbf{V}} + \mathbf{i}\widehat{\mathbb{A}}_{\mathbf{V}\to\mathbf{V}},\, \mathbb{B}_{\mathbf{U}\to\mathbf{V}} + \mathbf{i}\widehat{\mathbb{B}}_{\mathbf{V}\to\mathbf{V}}\right)(\mathbf{t})$ of $\gamma^*_{\mathbf{V}}$ the vectors of *valuations* and *vulnerabilities* of $\mathbf{FR}^{(\mathbf{V})}$ at **t**, considered from the viewpoint of **V** itself:

$$\mathbf{Im}z\,(\mathbf{t}) \equiv \widehat{\mathbb{A}}_{\mathbf{V}\to\mathbf{V}}\,(\mathbf{t}) \in \mathbb{R}^{n\times k} \text{ and } \mathbf{Im}w\,(\mathbf{t}) \equiv \widehat{\mathbb{B}}_{\mathbf{V}\to\mathbf{V}}\,(\mathbf{t}) \in \mathbb{R}^{m\times k}.$$

Of course, we may consider combinatorial vector fields, for instance the vector field **X3** which assigns to each point

$$(z,\,w)\,(\mathbf{t}) = \left(\mathbb{A}_{\mathbf{U}\to\mathbf{V}} + \mathbf{i}\widehat{\mathbb{A}}_{\mathbf{V}\to\mathbf{V}},\, \mathbb{B}_{\mathbf{U}\to\mathbf{V}} + \mathbf{i}\widehat{\mathbb{B}}_{\mathbf{V}\to\mathbf{V}}\right)(\mathbf{t})$$

of $\gamma^*_{\mathbf{V}}$ the vector

$$(\mathbf{Re}z,\,\mathbf{Im}w)\,(\mathbf{t}) \equiv \left(\mathbb{A}_{\mathbf{U}\to\mathbf{V}} + \mathbf{i0},\, \mathbf{0} + \mathbf{i}\widehat{\mathbb{B}}_{\mathbf{V}\to\mathbf{V}}\right)(\mathbf{t}) \in \mathbb{R}^{n\times k}\times\mathbb{R}^{m\times k},$$

i.e., the vector containing relative *valuations* of $\mathbf{FR}^{(\mathbf{V})}$ at **t** considered from the viewpoint of **U** and *vulnerabilities* of $\mathbf{FR}^{(\mathbf{V})}$ at **t** considered from the viewpoint of **V** itself, or the vector field **X4** which assigns to each point

$$(z,\,w)\,(\mathbf{t}) = \left(\mathbb{A}_{\mathbf{U}\to\mathbf{V}} + \mathbf{i}\widehat{\mathbb{A}}_{\mathbf{V}\to\mathbf{V}},\, \mathbb{B}_{\mathbf{U}\to\mathbf{V}} + \mathbf{i}\widehat{\mathbb{B}}_{\mathbf{V}\to\mathbf{V}}\right)(\mathbf{t})$$

of $\gamma^*_{\mathbf{V}}$ the vector

$$(\mathbf{Im}z,\,\mathbf{Re}w)\,(\mathbf{t}) \equiv \left(\mathbf{0} + \mathbf{i}\widehat{\mathbb{A}}_{\mathbf{V}\to\mathbf{V}},\, \mathbb{B}_{\mathbf{U}\to\mathbf{V}} + \mathbf{i0}\right)(\mathbf{t})\,.$$

i.e., the vector containing *valuations* of $\mathbf{FR}^{(\mathbf{V})}$ at **t** considered from the viewpoint of **V** itself and relative *vulnerabilities* of $\mathbf{FR}^{(\mathbf{V})}$ at **t** considered from the viewpoint of **U** itself.

With these definitions and notations, we can go further. The concept of supervisory perception curve is a concept that provides a clear overall relative evaluation of a node along time and particularly contains the changes of the quantitative overview on the node. In this sense, the supervisory perception curve could be considered as a concept that provides for the appearance of an action which could lead to changes. However, such a concept can not describe any changes that may occur in a node.

For this purpose, it should be noted that the concept of such an action is clearly local. Having regard to the above, it is therefore necessary to seek for a momentary consideration of the above defined supervisory concepts and, particularly, to proceed to a local study of the relevant curves. To this end, we fix a time moment $\mathbf{t_0} \in]\mathbf{0}, \mathbf{1}[$. A **supervision element** $(\gamma_{\mathbf{V}}, \mathscr{N})$ at $\mathbf{t_0}$ consists of a supervisory perception curve of $\mathbf{V}$ (in the system of nodes $\mathbf{V}$ and $\mathbf{U}$) defined on an open neighborhood $\mathscr{N}=]\mathbf{t_0}-\boldsymbol{\varepsilon}, \mathbf{t_0}+\boldsymbol{\varepsilon}[$ of $\mathbf{t_0}$. Two supervision elements $(\gamma_{\mathbf{V}}, \mathscr{N})$ and $(\boldsymbol{\delta}_{\mathbf{V}}, \mathscr{M})$ at $\mathbf{t_0}$ are equivalent (at $\mathbf{t_0}$) if there is an open neighborhood $\mathscr{L} \subset \mathscr{N} \bigcap \mathscr{M}$ of $\mathbf{t_0}$, such that $\gamma_{\mathbf{V}}|_{\mathscr{L}} = \boldsymbol{\delta}_{\mathbf{V}}|_{\mathscr{L}}$. The set of equivalence classes of supervision elements at $\mathbf{t_0}$ is called **the set of germs of supervisory perceptions** of $\mathbf{V}$ (in the system of nodes $\mathbf{V}$ and $\mathbf{U}$) at a given time moment $\mathbf{t_0}$ and is denoted by $\mathbb{G}_{\mathbf{t_0}}^{(\mathbf{V})}$:

$$\mathbb{G}_{\mathbf{t_0}}^{(\mathbf{V})} := \left\{ \gamma_{\mathbf{V}}(\mathbf{t}) = (z, w)(\mathbf{t}) \equiv \left(\mathbb{A}_{\mathbf{U}\to\mathbf{V}} + \mathbf{i}\widehat{\mathbb{A}}_{\mathbf{V}\to\mathbf{V}}, \mathbb{B}_{\mathbf{U}\to\mathbf{V}} + \mathbf{i}\widehat{\mathbb{B}}_{\mathbf{V}\to\mathbf{V}} \right)(\mathbf{t}) : \right.$$
$$\left. \mathbf{t} \in]\mathbf{t_0}-\boldsymbol{\varepsilon}, \mathbf{t_0}+\boldsymbol{\varepsilon}[, \boldsymbol{\varepsilon} > 0 \right\}.$$

The equivalence class of $(\gamma_{\mathbf{V}}, \mathscr{N})$ at $\mathbf{t_0}$ is denoted by $\gamma_{\mathbf{V}}^{\mathbf{t_0}}$, and $(\Gamma_{\mathbf{V}}, \mathscr{N})$ is called a **representative** of the germ $\Gamma_{\mathbf{V}}^{\mathbf{t_0}}$. A supervision element $(\gamma_{\mathbf{V}}, \mathscr{N})$ defines germs $\Gamma_{\mathbf{V}}^{\mathbf{t_0}}$ of supervisory perceptions for each $\mathbf{t_0} \in \mathscr{N}$. Since $\mathbb{G}_{\mathbf{t_0}}^{(\mathbf{V})} \bigcap \mathbb{G}_{\mathbf{s_0}}^{(\mathbf{V})} = \emptyset$ for $\mathbf{t_0} \neq \mathbf{s_0}$, we have $\Gamma_{\mathbf{V}}^{\mathbf{t_0}} \neq \Gamma_{\mathbf{V}}^{\mathbf{s_0}}$ for $\mathbf{t_0} \neq \mathbf{s_0}$.

Definition 5 A **cyber-effect** of $\mathbf{U}$ on $\mathbf{V}$ in the cyber-domain $\left(\overline{\mathbf{ob}\,(\mathbf{W_e})}, \mathbf{d}_{\mathbf{W_e}}\right)$ is a collection of mappings from the set $\mathbb{G}_{\mathbf{t}}^{(\mathbf{U})}$ of germs of supervisory perceptions of $\mathbf{U}$ at time $\mathbf{t} \in]\sigma, \tau[\subset\subset]\mathbf{0}, \mathbf{1}[$ into the set $\mathbb{G}_{\mathbf{t}+\Delta\mathbf{t}}^{(\mathbf{V})}$ of germs of supervisory perceptions of $\mathbf{V}$ into a cyber field $\mathscr{P}$ of $\mathbf{V}$ at another time $\mathbf{t}' := \mathbf{t}+\Delta\mathbf{t} \in]\sigma, \tau[$:

$$\left(g_{\mathbf{t}} : \mathbb{G}_{\mathbf{t}}^{(\mathbf{U})} \to \mathbb{G}_{\mathbf{t}+\Delta\mathbf{t}}^{(\mathbf{V})} : \boldsymbol{\delta}_{\mathbf{U}}(\mathbf{t}) \longmapsto \gamma_{\mathbf{V}}\left(\mathbf{t}'\right) \right)_{\mathbf{t}\in]\sigma,\tau[} (\mathbf{t}' := \mathbf{t}+\Delta\mathbf{t}).$$

Notice that the case $\Delta\mathbf{t}=\mathbf{0}$ is not excluded.

Although the concept of cyber-effect at a time moment $\mathbf{t}$ seems to be rather sufficient, we care to describe the **interaction** that has one cyber-node on each other, as well as the mutual effects resulting at a later time. In this case, the putative mutuality is influenced directly by the users' **subjectivity** of the cyber nodes. So, frequently, instead of the concept of a momentary cyber-effect, we are forced to consider mappings describing mutual influences.

Definition 6 A **cyber-activity** of $\mathbf{U}$ on $\mathbf{V}$ over the time interval $]\sigma, \tau[\subset\subset]\mathbf{0}, \mathbf{1}[$ is a collection of correspondences from the product $\mathbb{G}_{\mathbf{t}}^{(\mathbf{U})} \times \mathbb{G}_{\mathbf{t}}^{(\mathbf{V})}$ into the set

$$\mathbb{G}_{\mathbf{t}+\Delta\mathbf{t}}^{(\mathbf{U})} \times \mathbb{G}_{\mathbf{t}+\Delta\mathbf{t}}^{(\mathbf{V})} :$$

$$\left(\mathscr{G}_t : \mathbb{G}_t^{(U)} \times \mathbb{G}_t^{(V)} \to \mathbb{G}_{t+\Delta t}^{(U)} \times \mathbb{G}_{t+\Delta t}^{(V)} : \right.$$

$$\left.(\delta_U(t), \gamma_V(t)) \longmapsto \left(\delta_U\left(t'\right), \gamma_V\left(t'\right)\right)\right)_{t\in]\sigma,\tau[} (t' := t+\Delta t \in]\sigma, \tau[).$$

Notice that the case $\Delta t=0$ is not excluded. A **cyber-interplay** of the ordered cyber pair (**V**, **U**) over the time interval $]\sigma, \tau[\subset\subset [0, \infty[$ is an open shift curve

$$\mathscr{G}:]\sigma, \tau[\to \mathbb{G}_t^{(U)} \times \mathbb{G}_t^{(V)} \times \mathbb{G}_{t+\Delta t}^{(U)} \times \mathbb{G}_{t+\Delta t}^{(V)} : t \mapsto \mathscr{G}(t) :=$$

$$(\delta_U(t), \gamma_V(t), \; \delta_U(t+\Delta t), \; \gamma_V(t+\Delta t)) \; (t+\Delta t \in]\sigma, \tau[).$$

If the cyber-interplay $\mathscr{G}$ is composition of several separate interplays, we say that $\mathscr{G}$ is **sequential**; otherwise is called **elementary**.

In that regard to the concept of cyber-activity, we have the concept of cyber-interaction.

Definition 7 A **cyber-interaction** between **U** and **V** at a given time moment $t_0 \in]\sigma, \tau[$ is a tetrad

$$Z = Z_{(U,V)}(t_0) = ((z_1, w_1), (z_2, w_2), (z_3, w_3), (z_4, w_4)) \in \left(\mathbb{C}^{n\times k} \times \mathbb{C}^{m\times k}\right)^4$$

for which there is an associated cyber-activity of **U** on **V**:

$$\left(g_t = g_t^{(Z)} : \mathbb{G}_t^{(U)} \times \mathbb{G}_t^{(V)} \to \mathbb{G}_{t+\Delta t}^{(U)} \times \mathbb{G}_{t+\Delta t}^{(V)} : \right.$$

$$\left.(\delta_U(t), \gamma_V(t)) \longmapsto \left(\delta_U\left(t'\right), \gamma_V\left(t'\right)\right)\right)_{t\in]\sigma,\tau[} (t' := t+\Delta t \in]\sigma, \tau[),$$

such that

$$(z_1, w_1) = \delta_U(t_0) = \left(\mathbb{A}_{V\to U} + i\widehat{\mathbb{A}}_{U\to U}, \mathbb{B}_{V\to U} + i\widehat{\mathbb{B}}_{U\to U}\right) \in \mathbb{C}^{n\times k} \times \mathbb{C}^{m\times k},$$

$$(z_2, w_2) = \gamma_V(t_0) = \left(\mathbb{A}_{U\to V} + i\widehat{\mathbb{A}}_{V\to V}, \mathbb{B}_{U\to V} + i\widehat{\mathbb{B}}_{V\to V}\right) \in \mathbb{C}^{n\times k} \times \mathbb{C}^{m\times k},$$

$$(z_3, w_3) = \delta_U\left(t_0'\right) = \left(\mathbb{A}'_{V\to U} + i\widehat{\mathbb{A}}'_{U\to U}, \mathbb{B}'_{V\to U} + i\widehat{\mathbb{B}}'_{U\to U}\right) \in \mathbb{C}^{n\times k} \times \mathbb{C}^{m\times k},$$

$$(z_4, w_4) = \gamma_V\left(t_0'\right) = \left(\mathbb{A}'_{U\to V} + i\widehat{\mathbb{A}}'_{V\to V}, \mathbb{B}'_{U\to V} + i\widehat{\mathbb{B}}'_{V\to V}\right) \in \mathbb{C}^{n\times k} \times \mathbb{C}^{m\times k}.$$

Obviously, keeping a fixed supervisory perception $\gamma_V(t_0)$ in the archetype germ $\mathbb{G}_t^{(V)}$ and a fixed supervisory perception $\gamma_U(t+\Delta t)$ in the component image germ $\mathbb{G}_{t+\Delta t}^{(U)}$, the corresponding cyber-interaction becomes a cyber-effect. And, as we shall see below, proper management of cyber-effects is enough to study cyber navigations. However, in most cases, as in the case of cyber-attacks, it is necessary to consider cyber-interactions. So, because cyber-effects are a partial case of cyber-interactions, we will give a slight priority in the most general context of cyber-interactions.

It is easily verified that the general form of a cyber-interaction is as follows.

$$\begin{aligned}\mathbf{Z} &= ((z_1, w_1), (z_2, w_2), (z_3, w_3), (z_4, w_4)) \\ &= ((z_1, w_1), (z_2, w_2), (z_3, w_3), (z_4, w_4))(\mathbf{t_0}) \\ &= \Bigg(\underbrace{\mathbb{A}_{\mathbf{V}\to\mathbf{U}}+\mathbf{i}\widehat{\mathbb{A}}_{\mathbf{U}\to\mathbf{U}}, \mathbb{B}_{\mathbf{V}\to\mathbf{U}}+\mathbf{i}\widehat{\mathbb{B}}_{\mathbf{U}\to\mathbf{U}}}_{\delta_{\mathbf{U}}(\mathbf{t_0})}, \underbrace{\mathbb{A}_{\mathbf{U}\to\mathbf{V}}+\mathbf{i}\widehat{\mathbb{A}}_{\mathbf{V}\to\mathbf{V}}, \mathbb{B}_{\mathbf{U}\to\mathbf{V}}+\mathbf{i}\widehat{\mathbb{B}}_{\mathbf{V}\to\mathbf{V}}}_{\gamma_{\mathbf{V}}(\mathbf{t_0})}, \\ &\quad \underbrace{\mathbb{A}'_{\mathbf{V}\to\mathbf{U}}+\mathbf{i}\widehat{\mathbb{A}}'_{\mathbf{U}\to\mathbf{U}}, \mathbb{B}'_{\mathbf{V}\to\mathbf{U}}+\mathbf{i}\widehat{\mathbb{B}}'_{\mathbf{U}\to\mathbf{U}}}_{\delta_{\mathbf{U}}(\mathbf{t}'_0)}, \underbrace{\mathbb{A}'_{\mathbf{U}\to\mathbf{V}}+\mathbf{i}\widehat{\mathbb{A}}'_{\mathbf{V}\to\mathbf{V}}, \mathbb{B}'_{\mathbf{U}\to\mathbf{V}}+\mathbf{i}\widehat{\mathbb{B}}'_{\mathbf{V}\to\mathbf{V}}}_{\gamma_{\mathbf{V}}(\mathbf{t}'_0)} \Bigg).\end{aligned}$$

Suppose now

$$\begin{aligned}\mathbf{Z}=\mathbf{Z}_{(\mathbf{U},\mathbf{V})}(\mathbf{t_0}) &= ((z_1, w_1), (z_2, w_2), (z_3, w_3), (z_4, w_4)) \\ &\in \mathbb{C}^{\mathrm{n}\times k}\times\mathbb{C}^{\mathrm{m}\times k}\times\mathbb{C}^{\mathrm{n}\times k}\times\mathbb{C}^{\mathrm{m}\times k}\end{aligned}$$

is a cyber interaction between $\mathbf{U}, \mathbf{V}$ at a fixed time moment $\mathbf{t_0} \in]\sigma, \tau[\subset\subset]\mathbf{0}, \mathbf{1}[$, with corresponding cyber-interplay

$$\begin{aligned}g:]\sigma, \tau[\to \mathbb{G}_{\mathbf{t}}^{(\mathbf{U})}\times\mathbb{G}_{\mathbf{t}}^{(\mathbf{V})}\times\mathbb{G}_{\mathbf{t}+\Delta\mathbf{t}}^{(\mathbf{U})}\times\mathbb{G}_{\mathbf{t}+\Delta\mathbf{t}}^{(\mathbf{V})}&:\mathbf{t}\mapsto g(\mathbf{t}) \\ := \left(\delta_{\mathbf{U}}(\mathbf{t}), \gamma_{\mathbf{V}}(\mathbf{t}), \delta_{\mathbf{U}}(\mathbf{t}+\Delta\mathbf{t}), \gamma_{\mathbf{V}}(\mathbf{t}+\Delta\mathbf{t})\right)&(\mathbf{t}+\Delta\mathbf{t} \in]\sigma, \tau[)\end{aligned}$$

and cyber-activity

$$\begin{aligned}&\Big(g_{\mathbf{t}} : \mathbb{G}_{\mathbf{t}}^{(\mathbf{U})}\times\mathbb{G}_{\mathbf{t}}^{(\mathbf{V})} \to \mathbb{G}_{\mathbf{t}+\Delta\mathbf{t}}^{(\mathbf{U})}\times\mathbb{G}_{\mathbf{t}+\Delta\mathbf{t}}^{(\mathbf{V})} : \\ &(\delta_{\mathbf{U}}(\mathbf{t}), \gamma_{\mathbf{V}}(\mathbf{t})) \longmapsto \left(\delta_{\mathbf{U}}\left(\mathbf{t}'\right), \gamma_{\mathbf{V}}\left(\mathbf{t}'\right)\right)\Big)_{\mathbf{t}\in]\sigma,\tau[} (\mathbf{t}':=\mathbf{t}+\Delta\mathbf{t} \in]\sigma, \tau[).\end{aligned}$$

Definition 8 A **forced cyber-reflection** of $\mathbf{Z}$ at a time moment $\mathbf{t_0} \in]\sigma, \tau[$ is another cyber-interaction

$$\begin{aligned}\mathbf{Z}'=\mathbf{Z}'_{(\mathbf{U},\mathbf{V})}(\mathbf{t_0}) &= \left(\left(z'_1, w'_1\right), \left(z'_2, w'_2\right), \left(z'_3, w'_3\right), \left(z'_4, w'_4\right)\right) \\ &= \left(\left(z'_1, w'_1\right), \left(z'_2, w'_2\right), \left(z'_3, w'_3\right), \left(z'_4, w'_4\right)\right)\left(\mathbf{t}'_0\right) \in \left(\mathbb{C}^{\mathrm{n}\times k}\times\mathbb{C}^{\mathrm{m}\times k}\right)^4\end{aligned}$$

between **U** and **V** at a next time moment $\mathbf{t}_0'=\mathbf{t}_0+\Delta\mathbf{t}_0 \in \,]\sigma,\tau[$, with corresponding **forced cyber-interplay**

$$g':]\sigma,\tau[\to \mathbb{G}_{\mathbf{t}}^{(\mathbf{U})}\times\mathbb{G}_{\mathbf{t}}^{(\mathbf{V})}\times\mathbb{G}_{\mathbf{t}+\Delta\mathbf{t}}^{(\mathbf{U})}\times\mathbb{G}_{\mathbf{t}+\Delta\mathbf{t}}^{(\mathbf{V})}:\mathbf{t}\mapsto g'(\mathbf{t})$$
$$:=\left(\delta_{\mathbf{U}}\left(\mathbf{t}'\right),\gamma_{\mathbf{V}}\left(\mathbf{t}'\right),\ \delta_{\mathbf{U}}\left(\mathbf{t}''\right),\ \gamma_{\mathbf{V}}\left(\mathbf{t}''\right)\right)(\mathbf{t}'':=\mathbf{t}'+\Delta\mathbf{t}' \in \,]\sigma,\tau[)$$

and associated **forced cyber-activity**:

$$\left(g'_{\mathbf{t}'}:\mathbb{G}_{\mathbf{t}+\Delta\mathbf{t}}^{(\mathbf{U})}\times\mathbb{G}_{\mathbf{t}+\Delta\mathbf{t}}^{(\mathbf{V})}\to\mathbb{G}_{\mathbf{t}'+\Delta\mathbf{t}'}^{(\mathbf{U})}\times\mathbb{G}_{\mathbf{t}'+\Delta\mathbf{t}'}^{(\mathbf{V})}\right.$$
$$\left.:\left(\delta_{\mathbf{U}}\left(\mathbf{t}'\right),\gamma_{\mathbf{V}}\left(\mathbf{t}'\right)\right)\longmapsto\left(\delta_{\mathbf{U}}\left(\mathbf{t}''\right),\gamma_{\mathbf{V}}\left(\mathbf{t}''\right)\right)\right)_{\mathbf{t}\in]\sigma,\tau[}(\mathbf{t}'':=\mathbf{t}'+\Delta\mathbf{t}' \in \,]\sigma,\tau[),$$

that satisfies the following property: *there is an open neighborhood* $]\mathbf{t}_0-\boldsymbol{\varepsilon},\mathbf{t}_0+\boldsymbol{\varepsilon}[\,\subset\,]\sigma,\tau[$ *of* $\mathbf{t}_0$, *into which presence of cyber-reflection* of **Z** *forces application of* $\mathbf{Z}'$, *in the sense that activity* $g_{\mathbf{t}}$ *is obliged to push forward its composition with activity* $g'_{\mathbf{t}'}$ *in such a way that occurrence of* g *guarantees appearance of the composition* $g'_{\mathbf{t}'}\circ g_{\mathbf{t}}$. In such a case, the cyber-activity $g_{\mathbf{t}}$ together with its forced cyber-activity $g'_{\mathbf{t}'}$ is a **reflexive cyber-activity** between **U** and **V** during the period $]\mathbf{t}_0-\boldsymbol{\varepsilon},\mathbf{t}_0+\boldsymbol{\varepsilon}[$. Their composition

$$g'\circ g:\mathbf{t}\longmapsto g'_{\mathbf{t}'}\circ g_{\mathbf{t}}$$

is a **self-inflicted cyber-activity** between **U** and **V** during the period $]\mathbf{t}_0-\boldsymbol{\varepsilon},\mathbf{t}_0+\boldsymbol{\varepsilon}[$. In particular, the interaction $\mathbf{Z}'=\mathbf{Z}'_{(\mathbf{U},\mathbf{V})}\left(\mathbf{t}_0'\right)$ is called **forced cyber-reflection** of $\mathbf{Z}=\mathbf{Z}_{(\mathbf{U},\mathbf{V})}\,(\mathbf{t}_0)$ at $\mathbf{t}_0$. A mapping

$$\Phi:\left(\mathbb{C}^{\mathrm{n}\times k}\times\mathbb{C}^{\mathrm{m}\times k}\right)^2\to\left(\mathbb{C}^{\mathrm{n}\times k}\times\mathbb{C}^{\mathrm{m}\times k}\right)^2$$

which maps the cyber-interaction $\mathbf{Z}=\mathbf{Z}_{(\mathbf{U},\mathbf{V})}\,(\mathbf{t}_0)$ to its forced cyber-reflection $\mathbf{Z}'=\mathbf{Z}'_{(\mathbf{U},\mathbf{V})}\left(\mathbf{t}_0'\right)$ is called **reflexive cyber-interaction mapping** at $\mathbf{t}_0$.

It is frequent that, under a self-inflicted cyber-activity

$$\left(g'_{\mathbf{t}'}\circ g_{\mathbf{t}}:\mathbb{G}_{\mathbf{t}}^{(\mathbf{U})}\times\mathbb{G}_{\mathbf{t}}^{(\mathbf{V})}\to\mathbb{G}_{\mathbf{t}'+\Delta\mathbf{t}'}^{(\mathbf{U})}\times\mathbb{G}_{\mathbf{t}'+\Delta\mathbf{t}'}^{(\mathbf{V})}:\right.$$
$$\left.(\delta_{\mathbf{U}}\,(\mathbf{t}),\gamma_{\mathbf{V}}\,(\mathbf{t}))\longmapsto\left(\delta_{\mathbf{U}}\left(\mathbf{t}'+\Delta\mathbf{t}'\right),\gamma_{\mathbf{V}}\left(\mathbf{t}'+\Delta\mathbf{t}'\right)\right)\right)_{\mathbf{t}\in]\sigma,\tau[}(\mathbf{t}':=\mathbf{t}+\Delta\mathbf{t} \in \,]\sigma,\tau[),$$

between **U** and **V** during the period $]\mathbf{t}_0-\boldsymbol{\varepsilon},\mathbf{t}_0+\boldsymbol{\varepsilon}[\,\subset\,]\sigma,\tau[$, some valuations and vulnerabilities of the initial node **U** change at the moment $\mathbf{t}_0$. *For emphasis*, this "new" node is called **variant node** of **U** and is denoted by $\mathbf{U}'$, or sometimes, without any risk of confusion, again by **U**. In such a case, the forced cyber-reflection $\mathbf{Z}'=\mathbf{Z}'_{(\mathbf{U},\mathbf{V})}\left(\mathbf{t}'_0\right)$ is a **cyber parallax** of the cyber-interaction $\mathbf{Z}=\mathbf{Z}_{(\mathbf{U},\mathbf{V})}\,(\mathbf{t}_0)$ at $\mathbf{t}_0$ and the forced cyber-activity $g'_{\mathbf{t}'}$ is a **parallactic cyber-activity** which gives rise to a **parallactic cyber-interaction** at $\mathbf{t}_0$.

Definition 9 Let $\mathbf{E} \in \mathfrak{U}_{\mathscr{P}}$ be a part of $\mathscr{K}^{(\mathbf{U})}$ where:

$$\mathscr{K}=\begin{cases}\mathbf{D}, \text{ if the constituent is a device,}\\ \mathbf{R}, \text{ if the constituent is a resource element}\end{cases}$$

i. A **node shield** containing **E** in the node **U** at **t** is an intermediate fixed node $\overline{\overline{\mathbf{U}}}=\overline{\overline{\mathbf{U}}}_\mathbf{t}$ which, at this time, is interposed in each cyber parallax $\mathbf{Z}'$ that aims at **E** in the node **U**, so that the self-inflicted parallactic cyber-activity $g' \circ g$ between **U** and **V** at moment time **t** ends up in the intermediate node $\overline{\overline{\mathbf{U}}}$, and never can reach part **E** of the initial target **U**.
ii. A **node filter** in part **E** of the constituent $\mathscr{K}^{(\mathbf{U})}$ in **U** at **t** is an intermediate fixed node $\overline{\mathbf{U}}^{(\mathbf{E})}$ which, at this time moment, is interposed in each parallactic cyber-activity g' that aims at part **E** of node **U**, so that the filter $\overline{\mathbf{U}}^{(\mathbf{E})}$ allows the self-inflicted parallactic cyber-activity $g' \circ g$ at **t** to reach only constituent parts of the initial target **U** that differ from **E**.

Description of Cyber Navigations and Protection from Unplanned Attacks

Cyber Navigations

Cyber navigation refers to the process of navigating a network of information resources in cyberspace, which is organized as hypertext or hypermedia. The mathematical modeling of cyber-navigation and its risks, as well as protection against such risks will be the main theme of this session. To this direction, let us begin with the following definition.

Definition 10 Suppose $\mathbf{t}=\mathbf{t}_0<\mathbf{t}_1<\cdots<\mathbf{t}_\mathbf{k}=\mathbf{t}'$ is a partition of the interval $\left[\mathbf{t},\mathbf{t}'\right] \subset\ ]\mathbf{0},\mathbf{1}[$.

i. The corresponding **cyber walk** with start node $\mathbf{V}_{(\mathbf{x}_1,\mathbf{x}_2,\mathbf{x}_3,\mathbf{t}_0)}$ in the source $\mathbf{ob}\,(cy\,(\mathbf{t}_0))$ and final node $\mathbf{V}_{(\mathbf{x}_1,\mathbf{x}_2,\mathbf{x}_3,\mathbf{t}_\mathbf{k})}$ in the ending $\mathbf{ob}\,(cy\,(\mathbf{t}_\mathbf{k}))$ is an ordered node quote

$$\mathbf{V_0 V_1 \ldots V_k} = \underbrace{\mathbf{V}_{(\mathbf{x_1},\mathbf{x_2},\mathbf{x_3},\mathbf{t_0})}}_{\in \mathbf{ob}(cy(\mathbf{t_0}))} \underbrace{\mathbf{V}_{(\mathbf{x_1},\mathbf{x_2},\mathbf{x_3},\mathbf{t_1})}}_{\in \mathbf{ob}(\mathbf{F_1}[cy(\mathbf{t_0})])} \cdots \underbrace{\mathbf{V}_{(\mathbf{x_1},\mathbf{x_2},\mathbf{x_3},\mathbf{t_k})}}_{\in \mathbf{ob}([\mathbf{F_k} \circ \ldots \circ \mathbf{F_1}][cy(\mathbf{t_0})])},$$

defined by given mappings

$$\mathbf{F_i}: \underbrace{\{cy:\mathbb{I} \to ([\mathbf{ob}\,(\mathbf{W_e})]\,, \mathbf{d_{W_e}})\}}_{\mathbf{T}} \to \underbrace{\{cy:\mathbb{I} \to ([\mathbf{ob}\,(\mathbf{W_e})]\,, \mathbf{d_{W_e}})\}}_{\mathbf{T}}, \mathbf{i=1, 2, \ldots, k}$$

with the following three properties

1. $cy\,(\mathbf{t}_\nu) = [\mathbf{F}_\nu \circ \ldots \circ \mathbf{F_1}]\,[cy\,(\mathbf{t_0})]$, $\nu\mathbf{=1, 2, \ldots, k}$
2. $\mathbf{V_0}, \mathbf{V_1} \in \mathbf{ob}\,(\mathbf{F_1}\,[cy\,(\mathbf{t_0})])$, $\mathbf{V_1}, \mathbf{V_2} \in \mathbf{ob}\,([\mathbf{F_2} \circ \mathbf{F_1}]\,[cy\,(\mathbf{t_0})])$,..., $\mathbf{V_{k-1}}, \mathbf{V_k} \in \mathbf{ob}\,([\mathbf{F_k} \circ \ldots \circ \mathbf{F_1}]\,[cy\,(\mathbf{t_0})])$
3. $\mathbf{h_1} = [\mathbf{V_0}, \mathbf{V_1}] \in \hom\,(\mathbf{F_1}\,[cy\,(\mathbf{t_0})])$,..., $\mathbf{h_k} = [\mathbf{V_{k-1}}, \mathbf{V_k}] \in \hom\,([\mathbf{F_k} \circ \ldots \circ \mathbf{F_1}]\,[cy\,(\mathbf{t_0})])$.

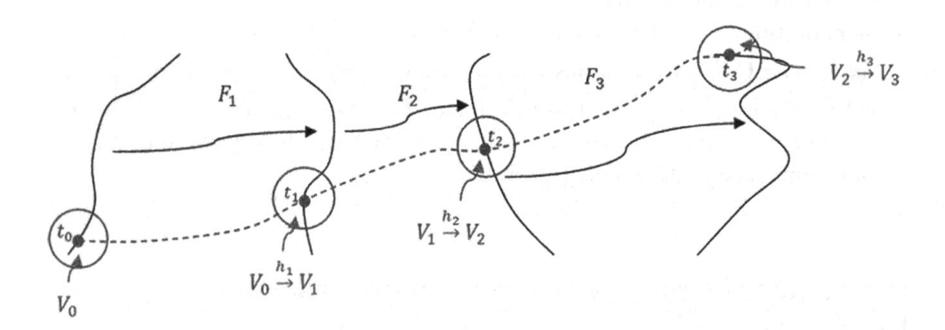

ii. A **cyber navigation** of the cyber node $\mathbf{W} = \mathbf{W}_{(\mathbf{x_1},\mathbf{x_2},\mathbf{x_3},\mathbf{t})} \in \bigcap_{\alpha=1}^{\mathbf{k}} \mathbf{ob}\,(cy\,(\mathbf{t}_\alpha))$ (over a cyber walk from the node $\mathbf{V_0}$ up to the node $\mathbf{V_k}$) is a finite sequence of reflexive cyber-effects

$$\mathfrak{N} = (g_0, g_1, \ldots, g_{\mathbf{k-1}}, g_{\mathbf{k}})$$

such that the ordered node quote $\mathbf{V_0 V_1 \ldots V_k}$ is a cyber walk and the diagrams below commute

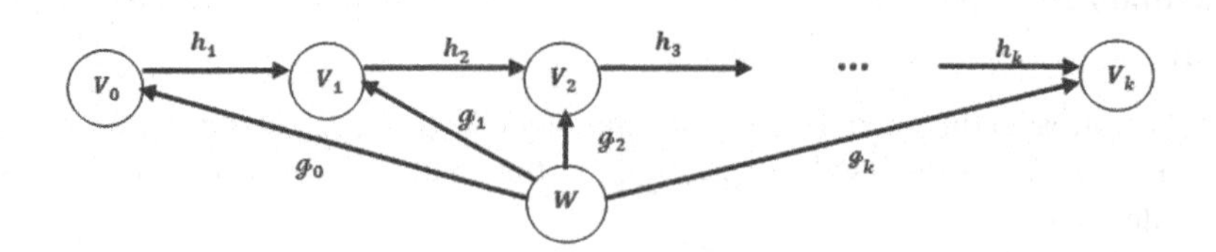

in the sense that $g_1 = \mathbf{h_1} \circ g_0$, $g_2 = \mathbf{h_2} \circ g_1, \ldots, g_{\mathbf{k}} = \mathbf{h_k} \circ g_{\mathbf{k-1}}$ It is clear that $g_{\mathbf{k}} = \mathbf{h_k} \circ \mathbf{h_{k-1}} \circ \ldots \circ \mathbf{h_2} \circ \mathbf{h_1} \circ g_0 = \mathbf{h} \circ g_0$ where $\mathbf{h} := \mathbf{h_k} \circ \ldots \circ \mathbf{h_1}$.

Inadequacy of Cyber Nodes

Suppose $\mathbf{t}=\mathbf{t_0}<\mathbf{t_1}<\cdots<\mathbf{t_k}=\mathbf{t}'$ is a partition of the interval $\left[\mathbf{t},\mathbf{t}'\right]\subset]\mathbf{0},\mathbf{1}[$. Let

$$\mathbf{V_0V_1\ldots V_k}=\underbrace{\mathbf{V}_{(\mathbf{x_1},\mathbf{x_2},\mathbf{x_3})(\mathbf{t_0})}}_{\in\mathbf{ob}(cy(\mathbf{t_0}))}\underbrace{\mathbf{V}_{(\mathbf{x_1},\mathbf{x_2},\mathbf{x_3})(\mathbf{t_1})}}_{\in\mathbf{ob}(\mathbf{F_1}[cy(\mathbf{t_0})])}\cdots\underbrace{\mathbf{V}_{(\mathbf{x_1},\mathbf{x_2},\mathbf{x_3})(\mathbf{t_n})}}_{\in\mathbf{ob}([\mathbf{F_k}\circ\ldots\circ\mathbf{F_1}][cy(\mathbf{t_0})])}$$

be corresponding walk with starting node $\mathbf{V_0}=\mathbf{V}_{(\mathbf{x_1},\mathbf{x_2},\mathbf{x_3})(\mathbf{t_0})}$ in the source $\mathbf{ob}\,(cy\,(\mathbf{t_0}))$ and defined by the mappings

$$\mathbf{F_i}:\underbrace{\left\{cy{:}\mathbb{I}\to\left(\lceil\mathbf{ob}\,(\mathbf{W_e})\rceil,\mathbf{d}_{\mathbf{W_e}}\right)\right\}}_{\mathbf{T}}\to\underbrace{\left\{cy{:}\mathbb{I}\to\left(\lceil\mathbf{ob}\,(\mathbf{W_e})\rceil,\mathbf{d}_{\mathbf{W_e}}\right)\right\}}_{\mathbf{T}},\ \mathbf{i}=\mathbf{1},\mathbf{2},\ldots,\mathbf{k}.$$

Let also a cyber navigation

$$\mathfrak{N}=(g_0,g_1,\ldots,g_{k-1},g_k)$$

of a cyber node $\mathbf{W}=\mathbf{W}_{(\mathbf{x_1},\mathbf{x_2},\mathbf{x_3},\mathbf{t})}\in\bigcap_{\alpha=1}^{\mathbf{k}}\mathbf{ob}\,(cy\,(\mathbf{t}_\alpha))$ over a cyber walk from the node $\mathbf{V_0}$ up to the node $\mathbf{V_k}$.

Definition 11 To each part $\mathbf{E}=\mathbf{fr}\left(\mathscr{K}^{(\mathbf{W})}\right)$ in the $\sigma-$algebra $\mathfrak{U}_{\mathscr{P}}$ of subsets of available (or not) constituents in the node $\mathbf{W}$:

$$\mathscr{K}=\begin{cases}\mathbf{dev}, \text{ if the constituent is a device,}\\ \mathbf{res}, \text{ if the constituent is a resource element}\end{cases}$$

the user(s) of a cyber-node $\mathbf{Z}$ (possibly identical to $\mathbf{W}$) associate an **efficiency threshold vector**

$$\mathbf{B}\,(\mathbf{E})=(\mathbf{B_1}\,(\mathbf{E}),\ldots,\mathbf{B_n}\,(\mathbf{E}))\in]\mathbf{0},+\infty[^{\mathbf{n}}.$$

i. The cyber node $\mathbf{W}$ is said to be **partially inadequate** in its part $\mathbf{E}=\mathbf{fr}\left(\mathscr{K}^{(\mathbf{W})}\right)$ over the cyber walk $\mathbf{V_0V_1\ldots V_k}$ from the viewpoint of the user(s) of $\mathbf{Z}$, if there is a variant node $\mathbf{W}'=\mathbf{W}_{\mathbf{t}_\lambda}$ and a valuation

$$\begin{aligned}\mathbf{b}^{(\mathbf{Z})}_{\mathbf{W}',\mathbf{FR}(\mathscr{K}^{(\mathbf{W})})}&=\mathbf{S}^{(\mathbf{Z})}_{\mathbf{W}'}\left(\mathbf{fr}\left(\mathscr{K}^{(\mathbf{W})}\right)\right)=\mathbf{S}^{(\mathbf{Z})}_{\mathbf{W}'}\,[\mathbf{x_1},\mathbf{x_2},\mathbf{x_3},\mathbf{t_k}]\left(\mathbf{fr}\left(\mathscr{K}^{(\mathbf{W})}\right)\right)\\&=\left(\mathbf{s}^{(\mathbf{Z})}_{\mathbf{W}',\mathbf{1}}\left(\mathbf{fr}\left(\mathscr{K}^{(\mathbf{W})}\right)\right),\mathbf{s}^{(\mathbf{Z})}_{\mathbf{W}',\mathbf{2}}\left(\mathbf{fr}\left(\mathscr{K}^{(\mathbf{W})}\right)\right),\right.\\&\qquad\left.\ldots,\mathbf{s}^{(\mathbf{Z})}_{\mathbf{W}',\mathbf{n}}\left(\mathbf{fr}\left(\mathscr{K}^{(\mathbf{W})}\right)\right)\right)\end{aligned}$$

of $\mathscr{K}^{(\mathbf{W})}$ in $\mathbf{W}'$ from the viewpoint of the user(s) of $\mathbf{Z}$, with some coordinates **less** than the corresponding coordinates of the efficiency threshold vector:

$$\mathbf{s}^{(\mathbf{Z})}_{\mathbf{W}',\mathbf{i_j}}\left(\mathbf{fr}\left(\mathscr{K}^{(\mathbf{W})}\right)\right) < \mathbf{B}_{\mathbf{i_j}}(\mathbf{E}), \mathbf{1} \le \mathbf{j} \le \mathrm{n}.$$

The number

$$\rho := \mathbf{max}_{\mathbf{1} \le \mathbf{j} \le \mathrm{n}} \left| \mathbf{B}_{\mathbf{i_j}}(\mathbf{E}) - \mathbf{s}^{(\mathbf{Z})}_{\mathbf{W}',\mathbf{i_j}}\left(\mathbf{fr}\left(\mathscr{K}^{(\mathbf{W})}\right)\right) \right|$$

is called the **degree of partial inadequacy** of part $\mathbf{E}=\mathbf{fr}\left(\mathscr{K}^{(\mathbf{W})}\right)$ in the cyber node $\mathbf{W}$ over the cyber walk $\mathbf{V_0 V_1 \ldots V_k}$ from the viewpoint of the user(s) of $\mathbf{Z}$. In the particular case where $\mathbf{s}^{(\mathbf{Z})}_{\mathbf{W}',\mathbf{i}}\left(\mathbf{fr}\left(\mathscr{K}^{(\mathbf{W})}\right)\right) < \mathbf{B_i}(\mathbf{E})$ whenever $\mathbf{i}=\mathbf{1}, \mathbf{2}, \ldots, \mathrm{n}$, we say that $\mathbf{W}$ is **completely inadequate** in its part $\mathbf{E}=\mathbf{fr}\left(\mathscr{K}^{(\mathbf{W})}\right)$ over the cyber walk $\mathbf{V_0 V_1 \ldots V_k}$ from the viewpoint of the user(s) of $\mathbf{Z}$.

ii. The cyber node $\mathbf{W}$ is said to be **totally inadequate** in its part $\mathbf{E}=\mathbf{fr}\left(\mathscr{K}^{(\mathbf{W})}\right)$ over the cyber walk $\mathbf{V_0 V_1 \ldots V_k}$ from the viewpoint of the user(s) of $\mathbf{Z}$, if there is a variant node $\mathbf{W}'=\mathbf{W}_{\mathbf{t}_\lambda}$ and a valuation

$$\begin{aligned}\mathbf{b}^{(\mathbf{Z})}_{\mathbf{W}',\mathbf{FR}(\mathscr{K}^{(\mathbf{W})})} &= \mathbf{S}^{(\mathbf{Z})}_{\mathbf{W}'}\left(\mathbf{fr}\left(\mathscr{K}^{(\mathbf{W})}\right)\right) = \mathbf{S}^{(\mathbf{Z})}_{\mathbf{W}'}[\mathbf{x_1}, \mathbf{x_2}, \mathbf{x_3}, \mathbf{t_k}]\left(\mathbf{fr}\left(\mathscr{K}^{(\mathbf{W})}\right)\right)\\ &= \left(\mathbf{s}^{(\mathbf{Z})}_{\mathbf{W}',\mathbf{1}}\left(\mathbf{fr}\left(\mathscr{K}^{(\mathbf{W})}\right)\right), \mathbf{s}^{(\mathbf{Z})}_{\mathbf{W}',\mathbf{2}}\left(\mathbf{fr}\left(\mathscr{K}^{(\mathbf{W})}\right)\right),\right.\\ &\qquad \left.\ldots, \mathbf{s}^{(\mathbf{Z})}_{\mathbf{W}',\mathrm{n}}\left(\mathbf{fr}\left(\mathscr{K}^{(\mathbf{W})}\right)\right)\right)\end{aligned}$$

of $\mathscr{K}^{(\mathbf{W})}$ in $\mathbf{W}'$ from the viewpoint of the user(s) of $\mathbf{Z}$, with (Euclidean or not) norm **less** than the (corresponding Euclidean or not) norm of the efficiency threshold vector:

$$\left\|\mathbf{S}^{(\mathbf{Z})}_{\mathbf{W}'}\left(\mathbf{fr}\left(\mathscr{K}^{(\mathbf{W})}\right)\right)\right\| < \|\mathbf{B}(\mathbf{E})\|.$$

The number

$$\rho^{(\infty)} := \|\mathbf{B}(\mathbf{E})\| - \left\|\mathbf{S}^{(\mathbf{Z})}_{\mathbf{W}'}\left(\mathbf{fr}\left(\mathscr{K}^{(\mathbf{W})}\right)\right)\right\|$$

is the **degree of total inadequacy** of part $\mathbf{E}=\mathbf{fr}\left(\mathscr{K}^{(\mathbf{W})}\right)$ in the cyber node $\mathbf{W}$ over the cyber walk $\mathbf{V_0 V_1 \ldots V_k}$ from the viewpoint of the user(s) of $\mathbf{Z}$. In the contrary case, where $\mathbf{W}$ is not **partially inadequate** and not **totally inadequate** in its part $\mathbf{E}=\mathbf{fr}\left(\mathscr{K}^{(\mathbf{W})}\right)$ over the cyber walk $\mathbf{V_0 V_1 \ldots V_k}$ from the viewpoint of the user(s) of $\mathbf{Z}$, the node $\mathbf{W}$ is said to be **adequate** in its part $\mathbf{E}=\mathbf{fr}\left(\mathscr{K}^{(\mathbf{W})}\right)$ over the cyber walk $\mathbf{V_0 V_1 \ldots V_k}$ from the viewpoint of the user(s) of $\mathbf{Z}$.

Infected Cyber Nodes

Suppose $\mathbf{t}=\mathbf{t_0}<\mathbf{t_1}<\cdots<\mathbf{t_k}=\mathbf{t'}$ is a partition of the interval $\left[\mathbf{t},\mathbf{t'}\right]\subset]\mathbf{0},\mathbf{1}[$. Let

$$\mathbf{V_0V_1\ldots V_k}=\underbrace{\mathbf{V}_{(\mathbf{x_1},\mathbf{x_2},\mathbf{x_3})(\mathbf{t_0})}}_{\in\mathbf{ob}(cy(\mathbf{t_0}))}\underbrace{\mathbf{V}_{(\mathbf{x_1},\mathbf{x_2},\mathbf{x_3})(\mathbf{t_1})}}_{\in\mathbf{ob}(\mathbf{F_1}[cy(\mathbf{t_0})])}\cdots\underbrace{\mathbf{V}_{(\mathbf{x_1},\mathbf{x_2},\mathbf{x_3})(\mathbf{t_n})}}_{\in\mathbf{ob}([\mathbf{F_k}\circ\ldots\circ\mathbf{F_1}][cy(\mathbf{t_0})])}$$

be corresponding walk with starting node $\mathbf{V_0}=\mathbf{V}_{(\mathbf{x_1},\mathbf{x_2},\mathbf{x_3})(\mathbf{t_0})}$ in the source $\mathbf{ob}\,(cy\,(\mathbf{t_0}))$ and defined by the mappings

$$\mathbf{F_i}:\underbrace{\left\{cy{:}\mathbb{I}\rightarrow\left(\lceil\mathbf{ob}\,(\mathbf{W_e})\rceil\,,\mathbf{d}_{\mathbf{W_e}}\right)\right\}}_{\mathbf{T}}\rightarrow\underbrace{\left\{cy{:}\mathbb{I}\rightarrow\left(\lceil\mathbf{ob}\,(\mathbf{W_e})\rceil\,,\mathbf{d}_{\mathbf{W_e}}\right)\right\}}_{\mathbf{T}},\quad\mathbf{i}=\mathbf{1},\mathbf{2},\ldots,\mathbf{k}.$$

Let also a cyber navigation

$$\mathfrak{N}=(g_0,g_1,\ldots,g_{\mathbf{k-1}},g_{\mathbf{k}})$$

of a cyber node $\mathbf{W}=\mathbf{W}_{(\mathbf{x_1},\mathbf{x_2},\mathbf{x_3},\mathbf{t})}\in\bigcap_{\alpha=1}^{\mathbf{k}}\mathbf{ob}\,(cy\,(\mathbf{t}_\alpha))$ over a cyber walk from the node $\mathbf{V_0}$ up to the node $\mathbf{V_k}$.

To each part $\mathbf{E}=\mathbf{fr}\left(\mathscr{K}^{(\mathbf{W})}\right)$ in the $\boldsymbol{\sigma}$−algebra $\mathfrak{U}_{\mathscr{P}}$ of subsets of available (or not) constituents of the node $\mathbf{W}$:

$$\mathscr{K}=\begin{cases}\mathbf{dev}, \text{ if the constituent is a device,}\\ \mathbf{res}, \text{ if the constituent is a resource element}\end{cases}$$

the user(s) of a cyber-node $\mathbf{Z}$ (possibly identical to $\mathbf{W}$) associate a **health tolerance vector**

$$\Gamma\,(\mathbf{E})=(\Gamma_1\,(\mathbf{E})\,,\ldots,\Gamma_{\mathrm{m}}\,(\mathbf{E}))\in]\mathbf{0},+\infty[^{\mathrm{m}}.$$

Definition 12

i. The cyber node $\mathbf{W}$ is said to be **partially infected** in its part $\mathbf{E}=\mathbf{fr}\left(\mathscr{K}^{(\mathbf{W})}\right)$ over the cyber walk $\mathbf{V_0V_1\ldots V_k}$ from the viewpoint of the user(s) of $\mathbf{Z}$, if there is a variant node $\mathbf{W'}=\mathbf{W}_{\mathbf{t}_\lambda}$ and a vulnerability

$$\begin{aligned}\mathbf{c}^{(\mathbf{Z})}_{\mathbf{W'},\mathbf{fr}(\mathscr{K}^{(\mathbf{W})})}&=\mathbf{U}^{(\mathbf{Z})}_{\mathbf{W'}}\left(\mathbf{fr}\left(\mathscr{K}^{(\mathbf{W})}\right)\right)=\mathbf{U}^{(\mathbf{Z})}_{\mathbf{W'}}\,[\mathbf{x_1},\mathbf{x_2},\mathbf{x_3},\mathbf{t_k}]\left(\mathbf{FR}\left(\mathscr{K}^{(\mathbf{W})}\right)\right)\\&=\left(\mathbf{u}^{(\mathbf{Z})}_{\mathbf{W'},1}\left(\mathbf{fr}\left(\mathscr{K}^{(\mathbf{W})}\right)\right),\mathbf{u}^{(\mathbf{Z})}_{\mathbf{W'},2}\left(\mathbf{fr}\left(\mathscr{K}^{(\mathbf{W})}\right)\right),\right.\\&\qquad\left.\ldots,\mathbf{u}^{(\mathbf{Z})}_{\mathbf{W'},\mathrm{m}}\left(\mathbf{fr}\left(\mathscr{K}^{(\mathbf{W})}\right)\right)\right)\end{aligned}$$

of $\mathscr{K}^{(W)}$ in $\mathbf{W}'$ from the viewpoint of the user(s) of **Z**, with some coordinates **greater** than the corresponding coordinates of the **health tolerance vector**:

$$\mathbf{u}^{(\mathbf{Z})}_{\mathbf{W}',\mathbf{i_j}}\left(\mathbf{fr}\left(\mathscr{K}^{(\mathbf{W})}\right)\right) > \Gamma_{\mathbf{i_j}}(\mathbf{E}), 1 \leq \mathbf{j} \leq \mathrm{m}.$$

The number

$$\delta := \mathbf{min}_{1\leq \mathbf{j}\leq \mathrm{m}}\left|\Gamma_{\mathbf{i_j}}(\mathbf{E}) - \mathbf{u}^{(\mathbf{Z})}_{\mathbf{W}',\mathbf{i_j}}\left(\mathbf{fr}\left(\mathscr{K}^{(\mathbf{W})}\right)\right)\right|$$

is the **degree of partial infection** of part $\mathbf{E}=\mathbf{fr}\left(\mathscr{K}^{(\mathbf{W})}\right)$ in the cyber node **W** over the cyber walk $\mathbf{V_0V_1\ldots V_k}$ from the viewpoint of the user(s) of **Z**. In the particular case where $\mathbf{u}^{(\mathbf{Z})}_{\mathbf{W}',\mathbf{i}}\left(\mathbf{fr}\left(\mathscr{K}^{(\mathbf{W})}\right)\right) > \Gamma_{\mathbf{i}}(\mathbf{E})$ whenever $\mathbf{i}=\mathbf{1},\mathbf{2},\ldots,\mathrm{m}$, we say that **W** is **completely infected** in its part $\mathbf{E}=\mathbf{fr}\left(\mathscr{K}^{(\mathbf{W})}\right)$ over the cyber walk $\mathbf{V_0V_1\ldots V_k}$ from the viewpoint of the user(s) of **Z**.

ii. The cyber node **W** is said to be totally infected (or totally compromised) in its part $\mathbf{E}=\mathbf{fr}\left(\mathscr{K}^{(\mathbf{W})}\right)$ over the cyber walk $\mathbf{V_0V_1\ldots V_k}$ from the viewpoint of the user(s) of **Z**, if there is a variant node $\mathbf{W}'=\mathbf{W}_{\mathbf{t}_\lambda}$ and a valuation

$$\begin{aligned}\mathbf{c}^{(\mathbf{Z})}_{\mathbf{W}',\mathbf{fr}(\mathscr{K}^{(\mathbf{W})})} &= \mathbf{U}^{(\mathbf{Z})}_{\mathbf{W}'}\left(\mathbf{fr}\left(\mathscr{K}^{(\mathbf{W})}\right)\right) = \mathbf{U}^{(\mathbf{Z})}_{\mathbf{W}'}[\mathbf{x_1},\mathbf{x_2},\mathbf{x_3},\mathbf{t_k}]\left(\mathbf{fr}\left(\mathscr{K}^{(\mathbf{W})}\right)\right)\\ &= \left(\mathbf{u}^{(\mathbf{Z})}_{\mathbf{W}',1}\left(\mathbf{fr}\left(\mathscr{K}^{(\mathbf{W})}\right)\right), \mathbf{u}^{(\mathbf{Z})}_{\mathbf{W}',2}\left(\mathbf{fr}\left(\mathscr{K}^{(\mathbf{W})}\right)\right),\right.\\ &\qquad \left.\ldots, \mathbf{u}^{(\mathbf{Z})}_{\mathbf{W}',\mathrm{m}}\left(\mathbf{fr}\left(\mathscr{K}^{(\mathbf{W})}\right)\right)\right)\end{aligned}$$

of $\mathscr{K}^{(W)}$ in $\mathbf{W}'$ from the viewpoint of the user(s) of **Z**, with (Euclidean or not) norm **greater** than the (corresponding Euclidean or not) norm of the health tolerance vector:

$$\left\|\mathbf{c}^{(\mathbf{Z})}_{\mathbf{W}',\mathbf{fr}(\mathscr{K}^{(\mathbf{W})})}\right\| > \|\Gamma(\mathbf{E})\|.$$

The number

$$\delta^{(\infty)} := \left\|\mathbf{c}^{(\mathbf{Z})}_{\mathbf{W}',\mathbf{fr}(\mathscr{K}^{(\mathbf{W})})}\right\| - \|\Gamma(\mathbf{E})\|$$

is the degree of the total infection of part $\mathbf{E}=\mathbf{fr}\left(\mathscr{K}^{(\mathbf{W})}\right)$ in the cyber node **W** over the cyber walk $\mathbf{V_0V_1\ldots V_k}$ from the viewpoint of the user(s) of **Z**. In the contrary case, where **W** is not partially infected and not totally infected in its part $\mathbf{E}=\mathbf{fr}\left(\mathscr{K}^{(\mathbf{W})}\right)$ over the cyber walk $\mathbf{V_0V_1\ldots V_k}$ from the viewpoint of the user(s) of **Z**, the node **W** is said to be healthy in its part $\mathbf{E}=\mathbf{fr}\left(\mathscr{K}^{(\mathbf{W})}\right)$ over the cyber walk $\mathbf{V_0V_1\ldots V_k}$ from the viewpoint of the user(s) of **Z**.

Dangerous Navigations

Let again $\mathbf{E}=\mathbf{fr}\left(\mathscr{K}^{(\mathbf{W})}\right)$ be a set in the σ-algebra $\mathfrak{U}_{\mathscr{P}}$ of subsets of available (or not) constituents of the cyber node **W**:

$$\mathscr{K}=\begin{cases}\mathbf{dev}, \text{ if the constituent is a device,}\\ \mathbf{res}, \text{ if the constituent is a resource element}\end{cases}$$

Suppose the user(s) of a cyber-node **Z** (possibly identical to **W**) associate an efficiency threshold vector

$$\mathbf{B}\left(\mathbf{E}\right)=\left(\mathbf{B}_{1}\left(\mathbf{E}\right),\ldots,\mathbf{B}_{n}\left(\mathbf{E}\right)\right)\in\left]\mathbf{0},+\infty\right[^{n},$$

as well as a health tolerance vector

$$\Gamma\left(\mathbf{E}\right)=\left(\Gamma_{1}\left(\mathbf{E}\right),\ldots,\Gamma_{m}\left(\mathbf{E}\right)\right)\in\left]\mathbf{0},+\infty\right[^{m}.$$

Definition 13 The navigation

$$\mathfrak{N}=\left(g_{0},g_{1},\ldots,g_{\mathbf{k}-1},g_{\mathbf{k}}\right)$$

of an adequate and healthy cyber node $\mathbf{W}=\mathbf{W}_{(\mathbf{x}_1,\mathbf{x}_2,\mathbf{x}_3,\mathbf{t})}\in\bigcap_{\alpha=1}^{\mathbf{k}}\mathbf{ob}\left(cy\left(\mathbf{t}_{\alpha}\right)\right)$ (over a cyber node homomorphism from a node $\mathbf{V}_0$ up to an infected node $\mathbf{V}_{\mathbf{k}}$) is said to be a **dangerous navigation** or an **unplanned attack with degree of danger** $\mathbf{d}:=\mathbf{max}\left\{\rho,\rho^{(\infty)}\right\}+\mathbf{max}\left\{\delta,\delta^{(\infty)}\right\}$ in its part **E** over the cyber walk $\mathbf{V}_0\mathbf{V}_1\ldots\mathbf{V}_{\mathbf{k}}$ from the viewpoint of the user(s) of **Z**, if the node **W** becomes

1. inadequate in its part $\mathbf{E}=\mathbf{fr}\left(\mathscr{K}^{(\mathbf{W})}\right)$ over the cyber walk $\mathbf{V}_0\mathbf{V}_1\ldots\mathbf{V}_{\mathbf{k}}$ from the viewpoint of the user(s) of **Z**, with degree of partial inadequacy equal to ρ and degree of total inadequacy equal to $\rho^{(\infty)}$ and
2. infected in its part $\mathbf{E}=\mathbf{fr}\left(\mathscr{K}^{(\mathbf{W})}\right)$ over the cyber walk $\mathbf{V}_0\mathbf{V}_1\ldots\mathbf{V}_{\mathbf{k}}$ from the viewpoint of the user(s) of **Z**, with degree of partial infection equal to δ and degree of total infection equal to $\delta^{(\infty)}$.

Protection of Cyber Nodes from Unplanned Attacks

Let again $\mathbf{E}=\mathbf{fr}\left(\mathscr{K}^{(\mathbf{W})}\right)$ in the σ-algebra $\mathfrak{U}_{\mathscr{P}}$ of subsets of an available or not constituent $\mathscr{K}^{(\mathbf{W})}$ in node **W**:

$$\mathscr{K}=\begin{cases}\mathbf{dev}, \text{ if the constituent is a device,}\\ \mathbf{res}, \text{ if the constituent is a resource element}\end{cases}$$

Suppose the user(s) of a cyber-node **Z** (possibly identical to **W**) associate an efficiency threshold vector

$$\mathbf{B}(\mathbf{E}) = (\mathbf{B}_1(\mathbf{E}), \ldots, \mathbf{B}_n(\mathbf{E})) \in]\mathbf{0}, +\infty[^n,$$

as well as a **health tolerance vector**

$$\Theta(\mathbf{E}) = (\Theta_1(\mathbf{E}), \ldots, \Theta_m(\mathbf{E})) \in]\mathbf{0}, +\infty[^m.$$

Definition 14

i. At a given time, the constituent part **E** of node **W** is said to be **protected from unplanned attacks, with degree of protection p** $\in$]**0**, **1**], if, at this time, there is a nodal fixed filter system $\overline{\mathbf{W}}^{(\mathbf{E})}$ in part **E** that allows every **self-inflicted parallactic cyber-effect** $g_j^{'} \circ g_j$ in any cyber-navigation of degree of danger

$$\mathbf{d} = -\log \mathbf{p}$$

to reach only constituent parts of the initial target **U** that are different from part **E** of $\mathscr{K}^{(\mathbf{U})}$.

ii. At a given time, the node **U** is said to be **completely protected from unplanned attacks of danger degree d**, if, at this time, any part of every constituent of **U** is protected from unplanned attacks with degree of protection

$$\mathbf{p} = \mathbf{e}^{-\mathbf{d}}.$$

The node **U** is said to be **completely protected from unplanned attacks** at a given time, if, at this time, any constituent part of **U** is protected from unplanned attacks with degree of protection **p=1**.

Description of Various Types of Cyber-Attacks and Protection

Passive Cyber-Attacks

A passive attack is a network attack in which a system is monitored and sometimes scanned for open ports and vulnerabilities. The purpose is solely to gain information about the target and no data is changed on the target. So, a passive attack contrasts with an active attack, in which an intruder attempts to alter data on the target system or data en route for the target system.

Let $\mathbf{U}, \mathbf{V} \in \mathbf{ob}\,(cy\,(\mathbf{t}))$, whenever $\mathbf{t}$ is in an arbitrary subset $\mathbb{I}$ of $]\boldsymbol{\alpha}, \boldsymbol{\beta}[\subset\subset [\mathbf{0}, \mathbf{1}]$. Let also $\gamma_{\mathbf{V}}$ and $\gamma_{\mathbf{U}}$ be two **supervisory perception curves** of $\mathbf{V}$ and $\mathbf{U}$ in the node system $(\mathbf{V}, \mathbf{U})$. Suppose $\mathbf{X}, \mathbf{Y} \in \{\mathbf{V}, \mathbf{U}\}$ and $r > \mathbf{0}$. A family of interactions

$$\mathscr{F} = \Big\{\mathbf{Z}{=}\mathbf{Z}_{(\mathbf{Y},\mathbf{X})}\,(\mathbf{t}) = ((z_1, w_1),\ (z_2, w_2), (z_3, w_3), (z_4, w_4)) \in \left(\mathbb{C}^{n\times k}{\times}\mathbb{C}^{m\times k}\right)^{\mathbf{4}}, \mathbf{t} \in \mathbb{I}\Big\},$$

with associated family of cyber-interplays of the ordered cyber pair $(\mathbf{Y}, \mathbf{X})$ over the time $\mathbf{t} \in\ ]\boldsymbol{\alpha}, \boldsymbol{\beta}[$

$$\mathscr{D}_{\mathscr{F}} = \Big\{g{=}g^{(\mathbf{Z})}{:}\mathbb{I} \to \mathbb{G}_t^{(Y)}{\times}\mathbb{G}_t^{(X)}{\times}\mathbb{G}_t^{(Y)}{\times}\mathbb{G}_t^{(X)}{:}\mathbf{t} \mapsto g^{(\mathbf{Z})}\,(\mathbf{t}) := \left(\gamma_{\mathbf{Y}}^{(\mathbf{Z})}\,(\mathbf{t}),\ \gamma_{\mathbf{X}}^{(\mathbf{Z})}\,(\mathbf{t}),\ \gamma_{\mathbf{Y}}^{(\mathbf{Z})}\,(\mathbf{t}{+}\Delta\mathbf{t}),\ \gamma_{\mathbf{X}}^{(\mathbf{Z})}\,(\mathbf{t}{+}\Delta\mathbf{t})\right) : \mathbf{Z} \in \mathscr{F}\Big\},$$

is called **coherent interactive family** in $\mathbb{I}$, if there is a homotopy

$$\mathbf{H}{:}\mathbb{I}{\times}\,[\mathbf{0}, \mathbf{1}] \to \mathbb{G}_t^{(Y)}{\times}\mathbb{G}_t^{(X)}{\times}\mathbb{G}_t^{(Y)}{\times}\mathbb{G}_t^{(X)}$$

such that, for each cyber-interplay $g{=}g^{(\mathbf{Z})} \in \mathscr{D}_{\mathscr{F}}$ there is a $\mathbf{p} \in [\mathbf{0}, \mathbf{1}]$ satisfying $\mathbf{H}\,(\mathbf{t}, \mathbf{p}) = g\,(\mathbf{t})$ at any moment time $\mathbf{t} \in \mathbb{I}$ on which the cyber-interplay $g{=}g^{(\mathbf{Z})}$ implements the interaction $\mathbf{Z}$.

Definition 15 A family of coherent interactions

$$\mathscr{F} = \Big\{\mathbf{Z}{=}\mathbf{Z}_{(\mathbf{U},\mathbf{V})}\,(\mathbf{t}) = ((z_1, w_1),\ (z_2, w_2), (z_3, w_3), (z_4, w_4)) = ((z_1, w_1),\ (z_2, w_2), (z_3, w_3), (z_4, w_4))\,(\mathbf{t}) \in \left(\mathbb{C}^{n\times k}{\times}\mathbb{C}^{m\times k}\right)^{\mathbf{4}}, \mathbf{t} \in \mathbb{I}\Big\},$$

lying in the partial danger sector $\mathscr{E}{=}\mathscr{E}_{\mathbf{U}\to\mathbf{V}}$ to the node $\mathbf{V}$ from the node $\mathbf{U}$ during the entire time set $\mathbb{I}$, is a **germ of (partial) passive attack from U against the** $(\kappa_1, \ldots, \kappa_\lambda)$**-resource parts** $\mathbf{fr}\left(\mathbf{res}_{\kappa_1}^{(\mathbf{V})}\right), \mathbf{fr}\left(\mathbf{res}_{\kappa_2}^{(\mathbf{V})}\right), \ldots, \mathbf{fr}\left(\mathbf{res}_{\kappa_\lambda}^{(\mathbf{V})}\right)$ of $\mathbf{V}$, during a given time subset $\mathbb{I}$ of a subinterval $[\boldsymbol{\alpha}, \boldsymbol{\beta}] \subset\subset [\mathbf{0}, \mathbf{1}]$, if, whenever $\mathbf{t} \in \mathbb{I}$, the pair $((z_1, w_1),\ (z_2, w_2)) \in \delta_{\mathbf{U}}{\times}\delta_{\mathbf{V}}$ of supervisory resource perceptions of $\mathbf{U}$ and $\mathbf{V}$ in the system of nodes $\mathbf{U}$ and $\mathbf{V}$ has the form

$$((z_1, w_1), (z_2, w_2)) = \left(\left(\begin{pmatrix} \mathbf{0} & \cdots & \mathbf{0} \\ \cdots & & \cdots \\ \mathbf{0} & \cdots & \mathbf{0} \\ \beta^{(\mathbf{U}\rightsquigarrow\mathbf{V})}_{\mathscr{M}_{\mathbf{V}+1,1}} + \mathbf{i}\, \widehat{\beta}^{(\mathbf{V}\rightsquigarrow\mathbf{V})}_{\mathscr{M}_{\mathbf{V}+1,1}} & & \beta^{(\mathbf{U}\rightsquigarrow\mathbf{V})}_{\mathscr{M}_{\mathbf{V}+1,n}} + \mathbf{i}\, \widehat{\beta}^{(\mathbf{V}\rightsquigarrow\mathbf{V})}_{\mathscr{M}_{\mathbf{V}+1,n}} \\ \cdots & & \cdots \\ \beta^{(\mathbf{U}\rightsquigarrow\mathbf{V})}_{\mathscr{M}_{\mathbf{V}+\ell_{\mathbf{V}},1}} + \mathbf{i}\, \widehat{\beta}^{(\mathbf{V}\rightsquigarrow\mathbf{V})}_{\mathscr{M}_{\mathbf{V}+\ell_{\mathbf{V}},1}} & & \beta^{(\mathbf{U}\rightsquigarrow\mathbf{V})}_{\mathscr{M}_{\mathbf{V}+\ell_{\mathbf{V}},n}} + \mathbf{i}\, \widehat{\beta}^{(\mathbf{V}\rightsquigarrow\mathbf{V})}_{\mathscr{M}_{\mathbf{V}+\ell_{\mathbf{V}},n}} \\ \mathbf{0} & \cdots & \mathbf{0} \\ \cdots & & \cdots \\ \mathbf{0} & & \mathbf{0} \end{pmatrix}, \begin{pmatrix} \mathbf{0} & \cdots & \mathbf{0} \\ \cdots & & \cdots \\ \mathbf{0} & \cdots & \mathbf{0} \\ \varepsilon^{(\mathbf{U}\rightsquigarrow\mathbf{V})}_{\mathscr{M}_{\mathbf{V}+1,1}} + \mathbf{i}\, \widehat{\varepsilon}^{(\mathbf{V}\rightsquigarrow\mathbf{V})}_{\mathscr{M}_{\mathbf{V}+1,1}} & & \varepsilon^{(\mathbf{U}\rightsquigarrow\mathbf{V})}_{\mathscr{M}_{\mathbf{V}+1,m}} + \mathbf{i}\, \widehat{\varepsilon}^{(\mathbf{V}\rightsquigarrow\mathbf{V})}_{\mathscr{M}_{\mathbf{V}+1,m}} \\ \cdots & & \cdots \\ \varepsilon^{(\mathbf{U}\rightsquigarrow\mathbf{V})}_{\mathscr{M}_{\mathbf{V}+\ell_{\mathbf{V}},1}} + \mathbf{i}\, \widehat{\varepsilon}^{(\mathbf{V}\rightsquigarrow\mathbf{V})}_{\mathscr{M}_{\mathbf{V}+\ell_{\mathbf{V}},1}} & & \varepsilon^{(\mathbf{U}\rightsquigarrow\mathbf{V})}_{\mathscr{M}_{\mathbf{V}+\ell_{\mathbf{V}},m}} + \mathbf{i}\, \widehat{\varepsilon}^{(\mathbf{V}\rightsquigarrow\mathbf{V})}_{\mathscr{M}_{\mathbf{V}+\ell_{\mathbf{V}},m}} \\ \mathbf{0} & \cdots & \mathbf{0} \\ \cdots & & \cdots \\ \mathbf{0} & & \mathbf{0} \end{pmatrix} \right), \left(\begin{pmatrix} \mathbf{0} & \cdots & \mathbf{0} \\ \cdots & & \cdots \\ \mathbf{0} & \cdots & \mathbf{0} \\ \beta^{(\mathbf{V}\rightsquigarrow\mathbf{U})}_{\mathscr{M}_{\mathbf{U}+1,1}} + \mathbf{i}\, \widehat{\beta}^{(\mathbf{U}\rightsquigarrow\mathbf{U})}_{\mathscr{M}_{\mathbf{U}+1,1}} & & \beta^{(\mathbf{V}\rightsquigarrow\mathbf{U})}_{\mathscr{M}_{\mathbf{U}+1,n}} + \mathbf{i}\, \widehat{\beta}^{(\mathbf{U}\rightsquigarrow\mathbf{U})}_{\mathscr{M}_{\mathbf{U}+1,n}} \\ \cdots & & \cdots \\ \beta^{(\mathbf{V}\rightsquigarrow\mathbf{U})}_{\mathscr{M}_{\mathbf{U}+\ell_{\mathbf{U}},1}} + \mathbf{i}\, \widehat{\beta}^{(\mathbf{U}\rightsquigarrow\mathbf{U})}_{\mathscr{M}_{\mathbf{U}+\ell_{\mathbf{U}},1}} & & \beta^{(\mathbf{V}\rightsquigarrow\mathbf{U})}_{\mathscr{M}_{\mathbf{U}+\ell_{\mathbf{U}},n}} + \mathbf{i}\, \widehat{\beta}^{(\mathbf{U}\rightsquigarrow\mathbf{U})}_{\mathscr{M}_{\mathbf{U}+\ell_{\mathbf{U}},n}} \\ \mathbf{0} & \cdots & \mathbf{0} \\ \cdots & & \cdots \\ \mathbf{0} & & \mathbf{0} \end{pmatrix}, \begin{pmatrix} \mathbf{0} & \cdots & \mathbf{0} \\ \cdots & & \cdots \\ \mathbf{0} & \cdots & \mathbf{0} \\ \varepsilon^{(\mathbf{V}\rightsquigarrow\mathbf{U})}_{\mathscr{M}_{\mathbf{U}+1,1}} + \mathbf{i}\, \widehat{\varepsilon}^{(\mathbf{U}\rightsquigarrow\mathbf{U})}_{\mathscr{M}_{\mathbf{U}+1,1}} & & \varepsilon^{(\mathbf{V}\rightsquigarrow\mathbf{U})}_{\mathscr{M}_{\mathbf{U}+1,m}} + \mathbf{i}\, \widehat{\varepsilon}^{(\mathbf{U}\rightsquigarrow\mathbf{U})}_{\mathscr{M}_{\mathbf{U}+1,m}} \\ \cdots & & \cdots \\ \varepsilon^{(\mathbf{V}\rightsquigarrow\mathbf{U})}_{\mathscr{M}_{\mathbf{U}+\ell_{\mathbf{U}},1}} + \mathbf{i}\, \widehat{\varepsilon}^{(\mathbf{U}\rightsquigarrow\mathbf{U})}_{\mathscr{M}_{\mathbf{U}+\ell_{\mathbf{U}},1}} & & \varepsilon^{(\mathbf{V}\rightsquigarrow\mathbf{U})}_{\mathscr{M}_{\mathbf{U}+\ell_{\mathbf{U}},m}} + \mathbf{i}\, \widehat{\varepsilon}^{(\mathbf{U}\rightsquigarrow\mathbf{U})}_{\mathscr{M}_{\mathbf{U}+\ell_{\mathbf{U}},m}} \\ \mathbf{0} & \cdots & \mathbf{0} \\ \cdots & & \cdots \\ \mathbf{0} & & \mathbf{0} \end{pmatrix} \right) \right)$$

and is depicted, at a next moment $\mathbf{t}^{'}=\mathbf{t}+\Delta\mathbf{t}$, via the associated family of cyber-activities

$$\mathscr{D}_{\mathscr{F}}=\left(g_{\mathbf{t}}=g_{\mathbf{t}}^{(\mathbf{Z})}\colon \mathbb{G}_{t}^{(U)}\times\mathbb{G}_{t}^{(V)}\rightarrow\mathbb{G}_{t}^{(U)}\times\mathbb{G}_{t}^{(V)}\colon(\delta_{\mathbf{U}}\left(\mathbf{t}\right),\gamma_{\mathbf{V}}\left(\mathbf{t}\right))\right.$$
$$\left.\longmapsto\left(\delta_{\mathbf{U}}^{'}\left(\mathbf{t}^{'}\right),\gamma_{\mathbf{V}}^{'}\left(\mathbf{t}^{'}\right)\right)\right)_{\mathbf{t}\in]\alpha,\beta[}$$

over the time $\mathbf{t}\in\,]\boldsymbol{\alpha},\boldsymbol{\beta}[$ at $((z_{3},w_{3}),(z_{4},w_{4}))\in\delta_{\mathbf{U}}\times\gamma_{\mathbf{V}}$ of supervisory resource perceptions of **U** and **V** having the form

$$((z_3, w_3), (z_4, w_4)) =$$

$$\left(\left(\left(\begin{array}{ccc} \mathbf{0} & \cdots & \mathbf{0} \\ \cdots & & \cdots \\ \mathbf{0} & \cdots & \mathbf{0} \\ \beta'^{(\mathbf{U}\rightsquigarrow\mathbf{V})}_{\mathscr{M}_{\mathbf{V}+1,1}} + \mathbf{i}\ \widehat{\beta'}^{(\mathbf{V}\rightsquigarrow\mathbf{V})}_{\mathscr{M}_{\mathbf{V}+1,1}} & & \beta'^{(\mathbf{U}\rightsquigarrow\mathbf{V})}_{\mathscr{M}_{\mathbf{V}+1,\mathrm{n}}} + \mathbf{i}\ \widehat{\beta'}^{(\mathbf{V}\rightsquigarrow\mathbf{V})}_{\mathscr{M}_{\mathbf{V}+1,\mathrm{n}}} \\ \cdots & & \cdots \\ \beta'^{(\mathbf{U}\rightsquigarrow\mathbf{V})}_{\mathscr{M}_{\mathbf{V}+\ell_{\mathbf{V}},1}} + \mathbf{i}\ \widehat{\beta'}^{(\mathbf{V}\rightsquigarrow\mathbf{V})}_{\mathscr{M}_{\mathbf{V}+\ell_{\mathbf{V}},1}} & & \beta'^{(\mathbf{U}\rightsquigarrow\mathbf{V})}_{\mathscr{M}_{\mathbf{V}+\ell_{\mathbf{V}},\mathrm{n}}} + \mathbf{i}\ \widehat{\beta'}^{(\mathbf{V}\rightsquigarrow\mathbf{V})}_{\mathscr{M}_{\mathbf{V}+\ell_{\mathbf{V}},\mathrm{n}}} \\ \mathbf{0} & \cdots & \mathbf{0} \\ \cdots & & \cdots \\ \mathbf{0} & & \mathbf{0} \end{array}\right), \left(\begin{array}{ccc} \mathbf{0} & \cdots & \mathbf{0} \\ \cdots & & \cdots \\ \mathbf{0} & \cdots & \mathbf{0} \\ \varepsilon'^{(\mathbf{U}\rightsquigarrow\mathbf{V})}_{\mathscr{M}_{\mathbf{V}+1,1}} + \mathbf{i}\ \widehat{\varepsilon'}^{(\mathbf{V}\rightsquigarrow\mathbf{V})}_{\mathscr{M}_{\mathbf{V}+1,1}} & & \varepsilon'^{(\mathbf{U}\rightsquigarrow\mathbf{V})}_{\mathscr{M}_{\mathbf{V}+1,\mathrm{m}}} + \mathbf{i}\ \widehat{\varepsilon'}^{(\mathbf{V}\rightsquigarrow\mathbf{V})}_{\mathscr{M}_{\mathbf{V}+1,\mathrm{m}}} \\ \cdots & & \cdots \\ \varepsilon'^{(\mathbf{U}\rightsquigarrow\mathbf{V})}_{\mathscr{M}_{\mathbf{V}+\ell_{\mathbf{V}},1}} + \mathbf{i}\ \widehat{\varepsilon'}^{(\mathbf{V}\rightsquigarrow\mathbf{V})}_{\mathscr{M}_{\mathbf{V}+\ell_{\mathbf{V}},1}} & & \varepsilon'^{(\mathbf{U}\rightsquigarrow\mathbf{V})}_{\mathscr{M}_{\mathbf{V}+\ell_{\mathbf{V}},\mathrm{m}}} + \mathbf{i}\ \widehat{\varepsilon'}^{(\mathbf{V}\rightsquigarrow\mathbf{V})}_{\mathscr{M}_{\mathbf{V}+\ell_{\mathbf{V}},\mathrm{m}}} \\ \mathbf{0} & \cdots & \mathbf{0} \\ \cdots & & \cdots \\ \mathbf{0} & & \mathbf{0} \end{array}\right)\right),\right.$$

$$\left.\left(\left(\begin{array}{ccc} \mathbf{0} & \cdots & \mathbf{0} \\ \cdots & & \cdots \\ \mathbf{0} & \cdots & \mathbf{0} \\ \beta'^{(\mathbf{V}\rightsquigarrow\mathbf{U})}_{\mathscr{M}_{\mathbf{U}+1,1}} + \mathbf{i}\ \widehat{\beta'}^{(\mathbf{U}\rightsquigarrow\mathbf{U})}_{\mathscr{M}_{\mathbf{U}+1,1}} & & \beta'^{(\mathbf{V}\rightsquigarrow\mathbf{U})}_{\mathscr{M}_{\mathbf{U}+1,\mathrm{n}}} + \mathbf{i}\ \widehat{\beta'}^{(\mathbf{U}\rightsquigarrow\mathbf{U})}_{\mathscr{M}_{\mathbf{U}+1,\mathrm{n}}} \\ \cdots & & \cdots \\ \beta'^{(\mathbf{V}\rightsquigarrow\mathbf{U})}_{\mathscr{M}_{\mathbf{U}+\ell_{\mathbf{U}},1}} + \mathbf{i}\ \widehat{\beta'}^{(\mathbf{U}\rightsquigarrow\mathbf{U})}_{\mathscr{M}_{\mathbf{U}+\ell_{\mathbf{U}},1}} & & \beta'^{(\mathbf{V}\rightsquigarrow\mathbf{U})}_{\mathscr{M}_{\mathbf{U}+\ell_{\mathbf{U}},\mathrm{n}}} + \mathbf{i}\ \widehat{\beta'}^{(\mathbf{U}\rightsquigarrow\mathbf{U})}_{\mathscr{M}_{\mathbf{U}+\ell_{\mathbf{U}},\mathrm{n}}} \\ \beta'^{(\mathbf{V}\rightsquigarrow\mathbf{U})}_{\mathscr{M}_{\mathbf{U}+\ell_{\mathbf{U}}+1,1}} + \mathbf{i}\ \widehat{\beta'}^{(\mathbf{U}\rightsquigarrow\mathbf{U})}_{\mathscr{M}_{\mathbf{U}+\ell_{\mathbf{U}}+1,1}} & & \beta'^{(\mathbf{V}\rightsquigarrow\mathbf{U})}_{\mathscr{M}_{\mathbf{U}+\ell_{\mathbf{U}}+1,\mathrm{n}}} + \mathbf{i}\ \widehat{\beta'}^{(\mathbf{U}\rightsquigarrow\mathbf{U})}_{\mathscr{M}_{\mathbf{U}+\ell_{\mathbf{U}}+1,\mathrm{n}}} \\ \cdots & \cdots & \cdots \\ \beta'^{(\mathbf{V}\rightsquigarrow\mathbf{U})}_{\mathscr{M}_{\mathbf{U}+\ell_{\mathbf{U}}+\nu,1}} + \mathbf{i}\ \widehat{\beta'}^{(\mathbf{U}\rightsquigarrow\mathbf{U})}_{\mathscr{M}_{\mathbf{U}+\ell_{\mathbf{U}}+\nu,1}} & & \beta'^{(\mathbf{V}\rightsquigarrow\mathbf{U})}_{\mathscr{M}_{\mathbf{U}+\ell_{\mathbf{U}}+\nu,\mathrm{n}}} + \mathbf{i}\ \widehat{\beta'}^{(\mathbf{U}\rightsquigarrow\mathbf{U})}_{\mathscr{M}_{\mathbf{U}+\ell_{\mathbf{U}}+,\mathrm{n}}} \\ \mathbf{0} & & \mathbf{0} \\ \cdots & & \cdots \\ \mathbf{0} & & \mathbf{0} \end{array}\right), \left(\begin{array}{ccc} \mathbf{0} & \cdots & \mathbf{0} \\ \cdots & & \cdots \\ \mathbf{0} & \cdots & \mathbf{0} \\ \varepsilon'^{(\mathbf{V}\rightsquigarrow\mathbf{U})}_{\mathscr{M}_{\mathbf{U}+1,1}} + \mathbf{i}\ \widehat{\varepsilon'}^{(\mathbf{U}\rightsquigarrow\mathbf{U})}_{\mathscr{M}_{\mathbf{U}+1,1}} & & \varepsilon'^{(\mathbf{V}\rightsquigarrow\mathbf{U})}_{\mathscr{M}_{\mathbf{U}+1,\mathrm{m}}} + \mathbf{i}\ \widehat{\varepsilon'}^{(\mathbf{U}\rightsquigarrow\mathbf{U})}_{\mathscr{M}_{\mathbf{U}+1,\mathrm{m}}} \\ \cdots & & \cdots \\ \varepsilon'^{(\mathbf{V}\rightsquigarrow\mathbf{U})}_{\mathscr{M}_{\mathbf{U}+\ell_{\mathbf{U}},1}} + \mathbf{i}\ \widehat{\varepsilon'}^{(\mathbf{U}\rightsquigarrow\mathbf{U})}_{\mathscr{M}_{\mathbf{U}+\ell_{\mathbf{U}},1}} & & \varepsilon'^{(\mathbf{V}\rightsquigarrow\mathbf{U})}_{\mathscr{M}_{\mathbf{U}+\ell_{\mathbf{U}},\mathrm{m}}} + \mathbf{i}\ \widehat{\varepsilon'}^{(\mathbf{U}\rightsquigarrow\mathbf{U})}_{\mathscr{M}_{\mathbf{U}+\ell_{\mathbf{U}},\mathrm{m}}} \\ \varepsilon'^{(\mathbf{V}\rightsquigarrow\mathbf{U})}_{\mathscr{M}_{\mathbf{U}+\ell_{\mathbf{U}}+1,1}} + \mathbf{i}\ \widehat{\varepsilon'}^{(\mathbf{U}\rightsquigarrow\mathbf{U})}_{\mathscr{M}_{\mathbf{U}+\ell_{\mathbf{U}}+1,1}} & & \varepsilon'^{(\mathbf{V}\rightsquigarrow\mathbf{U})}_{\mathscr{M}_{\mathbf{U}+\ell_{\mathbf{U}}+1,\mathrm{m}}} + \mathbf{i}\ \widehat{\varepsilon'}^{(\mathbf{U}\rightsquigarrow\mathbf{U})}_{\mathscr{M}_{\mathbf{U}+\ell_{\mathbf{U}}+1,\mathrm{m}}} \\ \cdots & \cdots & \cdots \\ \varepsilon'^{(\mathbf{V}\rightsquigarrow\mathbf{U})}_{\mathscr{M}_{\mathbf{U}+\ell_{\mathbf{U}}+\nu,1}} + \mathbf{i}\ \widehat{\varepsilon'}^{(\mathbf{U}\rightsquigarrow\mathbf{U})}_{\mathscr{M}_{\mathbf{U}+\ell_{\mathbf{U}}+\nu,1}} & & \varepsilon'^{(\mathbf{V}\rightsquigarrow\mathbf{U})}_{\mathscr{M}_{\mathbf{U}+\ell_{\mathbf{U}}+\nu,\mathrm{m}}} + \mathbf{i}\ \widehat{\varepsilon'}^{(\mathbf{U}\rightsquigarrow\mathbf{U})}_{\mathscr{M}_{\mathbf{U}+\ell_{\mathbf{U}}+\nu,\mathrm{m}}} \\ \mathbf{0} & & \mathbf{0} \\ \cdots & & \cdots \\ \mathbf{0} & & \mathbf{0} \end{array}\right)\right)\right).$$

It is easy to prove and/or verify the next two results.

Proposition 1 *In a passive attack* $\mathscr{F}$ *from* **U** *against* **V**, *the number of resource parts in* **U** *at a moment* $\mathbf{t}^{'}=\mathbf{t}+\Delta\mathbf{t}$ *has increased by at least* $\boldsymbol{\lambda}$ *new resource parts, say* $\mathbf{fr}\left(\mathbf{res}^{(\mathbf{U})}_{\mathscr{M}_{\mathrm{U}}+\ell_{\mathrm{U}}+1}\right)$, $\mathbf{fr}\left(\mathbf{res}^{(\mathbf{U})}_{\mathscr{M}_{\mathrm{U}}+\ell_{\mathrm{U}}+2}\right)$,..., $\mathbf{fr}\left(\mathbf{res}^{(\mathbf{U})}_{\mathscr{M}_{\mathrm{U}}+\ell_{\mathrm{U}}+\lambda}\right)$, *derived from the resource parts* $\mathbf{fr}\left(\mathbf{res}^{(\mathbf{V})}_{\kappa_1}\right)$, $\mathbf{fr}\left(\mathbf{res}^{(\mathbf{V})}_{\kappa_2}\right)$,..., $\mathbf{fr}\left(\mathbf{res}^{(\mathbf{V})}_{\kappa_\lambda}\right)$ *that existed in the node* **V** *the previous moment* **t**, *in such a way that the following elementary properties hold.*

i. *If the relative valuations of*

$$\mathbf{fr}\left(\mathbf{res}^{(\mathbf{U})}_{\mathscr{M}_{\mathrm{V}}+\ell_{\mathrm{V}}+1}\right),\ \mathbf{fr}\left(\mathbf{res}^{(\mathbf{U})}_{\mathscr{M}_{\mathrm{V}}+\ell_{\mathrm{V}}+2}\right),\ldots,\mathbf{fr}\left(\mathbf{res}^{(\mathbf{U})}_{\mathscr{M}_{\mathrm{V}}+\ell_{\mathrm{V}}+\lambda}\right)$$

from the viewpoint of the (user(s) of) node **U** *at the previous moment* **t** *are*

$$\left(\boldsymbol{\beta}^{(\mathbf{U}\rightsquigarrow\mathbf{V})}_{\mathscr{M}_{\mathrm{V}}+\mu_1,1},\ldots,\boldsymbol{\beta}^{(\mathbf{U}\rightsquigarrow\mathbf{V})}_{\mathscr{M}_{\mathrm{V}}+\mu_1,\mathrm{n}}\right),\left(\boldsymbol{\beta}^{(\mathbf{U}\rightsquigarrow\mathbf{V})}_{\mathscr{M}_{\mathrm{V}}+\mu_2,1},\ldots,\boldsymbol{\beta}^{(\mathbf{U}\rightsquigarrow\mathbf{V})}_{\mathscr{M}_{\mathrm{V}}+\mu_2,\mathrm{n}}\right),$$
$$\ldots,\left(\boldsymbol{\beta}^{(\mathbf{U}\rightsquigarrow\mathbf{V})}_{\mathscr{M}_{\mathrm{V}}+\mu_\lambda,1},\ldots,\boldsymbol{\beta}^{(\mathbf{U}\rightsquigarrow\mathbf{V})}_{\mathscr{M}_{\mathrm{V}}+\mu_\lambda,\mathrm{n}}\right)$$

respectively, with $\boldsymbol{\mu}_1,\ldots,\boldsymbol{\mu}_\lambda \in \{\mathbf{1},\mathbf{2},\ldots,\ell_{\mathrm{V}}\}$, *then the resulting valuation vectors*

$$\left(\widehat{\boldsymbol{\beta}'}^{(\mathbf{U}\rightsquigarrow\mathbf{U})}_{\mathscr{M}_{\mathrm{U}}+\ell_{\mathrm{U}}+1,1},\ldots,\widehat{\boldsymbol{\beta}'}^{(\mathbf{U}\rightsquigarrow\mathbf{U})}_{\mathscr{M}_{\mathrm{U}}+\ell_{\mathrm{U}}+1,\mathrm{n}}\right),\ldots,\left(\widehat{\boldsymbol{\beta}'}^{(\mathbf{U}\rightsquigarrow\mathbf{U})}_{\mathscr{M}_{\mathrm{U}}+\ell_{\mathrm{U}}+\lambda,1},\ldots,\widehat{\boldsymbol{\beta}'}^{(\mathbf{U}\rightsquigarrow\mathbf{U})}_{\mathscr{M}_{\mathrm{U}}+\ell_{\mathrm{U}}+\lambda,\mathrm{n}}\right)$$

of the new resource parts

$$\mathbf{fr}\left(\mathbf{res}^{(\mathbf{U})}_{\mathscr{M}_{\mathrm{U}}+\ell_{\mathrm{U}}+1}\right),\mathbf{fr}\left(\mathbf{res}^{(\mathbf{U})}_{\mathscr{M}_{\mathrm{U}}+\ell_{\mathrm{U}}+2}\right),\ldots,\mathbf{fr}\left(\mathbf{res}^{(\mathbf{U})}_{\mathscr{M}_{\mathrm{U}}+\ell_{\mathrm{U}}+\lambda}\right)$$

in **U**, *as evaluated from the viewpoint of the user(s) of* **U** *at a next moment* $\mathbf{t}^{'}=\mathbf{t}+\Delta\mathbf{t}$ *are equal to*

$$\left(\boldsymbol{\beta}^{(\mathbf{U}\rightsquigarrow\mathbf{V})}_{\mathscr{M}_{\mathrm{V}}+\mu_1,1},\ldots,\boldsymbol{\beta}^{(\mathbf{U}\rightsquigarrow\mathbf{V})}_{\mathscr{M}_{\mathrm{V}}+\mu_1,\mathrm{n}}\right),\ldots,\left(\boldsymbol{\beta}^{(\mathbf{U}\rightsquigarrow\mathbf{V})}_{\mathscr{M}_{\mathrm{V}}+\mu_\lambda,1},\ldots,\boldsymbol{\beta}^{(\mathbf{U}\rightsquigarrow\mathbf{V})}_{\mathscr{M}_{\mathrm{V}}+\mu_\lambda,\mathrm{n}}\right):$$
$$\left(\widehat{\boldsymbol{\beta}'}^{(\mathbf{U}\rightsquigarrow\mathbf{U})}_{\mathscr{M}_{\mathrm{U}}+\ell_{\mathrm{U}}+\alpha,1},\ldots,\widehat{\boldsymbol{\beta}'}^{(\mathbf{U}\rightsquigarrow\mathbf{U})}_{\mathscr{M}_{\mathrm{U}}+\ell_{\mathrm{U}}+\alpha,\mathrm{n}}\right)=\left(\boldsymbol{\beta}^{(\mathbf{U}\rightsquigarrow\mathbf{V})}_{\mathscr{M}_{\mathrm{V}}+\mu_\alpha,1},\ldots,\boldsymbol{\beta}^{(\mathbf{U}\rightsquigarrow\mathbf{V})}_{\mathscr{M}_{\mathrm{V}}+\mu_\alpha,\mathrm{n}}\right),$$
$$\forall\alpha\in\{\mathbf{1},\mathbf{2},\ldots,\boldsymbol{\lambda}\}.$$

ii. *All resulting valuations and vulnerabilities of new resource parts*

$$\mathbf{fr}\left(\mathbf{res}^{(\mathbf{U})}_{\mathscr{M}_{\mathrm{U}}+\ell_{\mathrm{U}}+1}\right),\ldots,\mathbf{fr}\left(\mathbf{res}^{(\mathbf{U})}_{\mathscr{M}_{\mathrm{U}}+\ell_{\mathrm{U}}+\lambda}\right)$$

in **U** *from the viewpoint of the user(s) of* **V** *remain equal to* **0**:

$$\forall \mathbf{j} \in \{1, 2, \ldots, \mathrm{n}\} \text{ and } \forall \alpha \in \{1, 2, \ldots, \boldsymbol{\lambda}\} \Longrightarrow \boldsymbol{\beta}'^{(\mathbf{V} \rightsquigarrow \mathbf{U})}_{\mathscr{M}_{\mathbf{U}}+\ell_{\mathbf{U}}+\alpha,\mathbf{j}} = \mathbf{0},$$

$$\forall \mathbf{k} \in \{1, 2, \ldots, \mathrm{m}\} \text{ and } \forall \alpha \in \{1, 2, \ldots, \boldsymbol{\lambda}\} \Longrightarrow \varepsilon'^{(\mathbf{V} \rightsquigarrow \mathbf{U})}_{\mathscr{M}_{\mathbf{U}}+\ell_{\mathbf{U}}+\alpha,\mathbf{k}} = \mathbf{0}.$$

iii. *There is at least one resulting valuation* $\boldsymbol{\beta}'^{(\mathbf{U} \rightsquigarrow \mathbf{V})}_{\mathscr{M}_{\mathbf{V}}+\boldsymbol{\lambda}_{\alpha},\mathbf{j}}$ *of a part* $\mathbf{fr}\left(\mathbf{res}^{(\mathbf{V})}_{\kappa_{\alpha}}\right)$ *in* $\mathbf{V}$ *from the viewpoint of the user(s) of* $\mathbf{U}$ *which decreases:*

$$\exists \mathbf{j} \in \{1, 2, \ldots, \mathrm{n}\} \text{ and } \exists \boldsymbol{\lambda}_{\alpha} \in \{\mathscr{M}_{\mathbf{V}}+1, \ldots, \mathscr{M}_{\mathbf{V}}+\ell_{\mathbf{V}}\} : \boldsymbol{\beta}'^{(\mathbf{U} \rightsquigarrow \mathbf{V})}_{\mathscr{M}_{\mathbf{V}}+\boldsymbol{\lambda}_{\alpha},\mathbf{j}} < \boldsymbol{\beta}^{(\mathbf{U} \rightsquigarrow \mathbf{V})}_{\mathscr{M}_{\mathbf{V}}+\boldsymbol{\lambda}_{\alpha},\mathbf{j}};$$

similarly, there is at least one vulnerability $\varepsilon'^{(\mathbf{U} \rightsquigarrow \mathbf{V})}_{\mathscr{M}_{\mathbf{V}}+\rho_{\alpha},\mathbf{k}}$ *of part* $\mathbf{fr}\left(\mathbf{res}^{(\mathbf{V})}_{\kappa_{\alpha}}\right)$ *in* $\mathbf{V}$ *from the viewpoint of the user(s) of* $\mathbf{U}$ *which increases*

$$\exists \mathbf{k} \in \{1, 2, \ldots, \mathrm{m}\} \text{ and } \exists \rho_{\alpha} \in \{\mathscr{M}_{\mathbf{V}}+1, \ldots, \mathscr{M}_{\mathbf{V}}+\ell_{\mathbf{V}}\} : \varepsilon'^{(\mathbf{U} \rightsquigarrow \mathbf{V})}_{\mathscr{M}_{\mathbf{V}}+\rho_{\alpha},\mathbf{k}} > \varepsilon^{(\mathbf{U} \rightsquigarrow \mathbf{V})}_{\mathscr{M}_{\mathbf{V}}+\rho_{\alpha},\mathbf{k}}.$$

iv. *The valuations and vulnerabilities of each part* $\mathbf{fr}\left(\mathbf{res}^{(\mathbf{V})}_{\kappa_{\alpha}}\right)$ *in* $\mathbf{V}$ *from the viewpoint of the user(s) of* $\mathbf{V}$ *remain unchanged:*

$$\forall \mathbf{j} \in \{1, 2, \ldots, \mathrm{n}\} \text{ and } \forall \boldsymbol{\lambda}_{\alpha} \in \{\mathscr{M}_{\mathbf{V}}+1, \ldots, \mathscr{M}_{\mathbf{V}}+\ell_{\mathbf{V}}\} \Rightarrow \widehat{\boldsymbol{\beta}'}^{(\mathbf{V} \rightsquigarrow \mathbf{V})}_{\mathscr{M}_{\mathbf{V}}+\boldsymbol{\lambda}_{\alpha},\mathbf{j}} = \widehat{\boldsymbol{\beta}}^{(\mathbf{V} \rightsquigarrow \mathbf{V})}_{\mathscr{M}_{\mathbf{V}}+\boldsymbol{\lambda}_{\alpha},\mathbf{j}},$$

$$\forall \mathbf{k} \in \{1, 2, \ldots, \mathrm{m}\} \text{ and } \forall \mu_{\alpha} \in \{\mathscr{M}_{\mathbf{V}}+1, \ldots, \mathscr{M}_{\mathbf{V}}+\ell_{\mathbf{V}}\} \Rightarrow \widehat{\varepsilon'}^{(\mathbf{V} \rightsquigarrow \mathbf{V})}_{\mathscr{M}_{\mathbf{V}}+\ell_{\mathbf{V}}+\mu_{\alpha},\mathbf{k}} = \widehat{\varepsilon}^{(\mathbf{V} \rightsquigarrow \mathbf{V})}_{\mathscr{M}_{\mathbf{V}}+\ell_{\mathbf{V}}+\mu_{\alpha},\mathbf{k}}.$$

Proposition 2 *In a passive attack* $\mathscr{F}$ *from* $\mathbf{U}$ *against* $\mathbf{V}$, *the number of resource parts in* $\mathbf{U}$ *at a moment* $\mathbf{t}' = \mathbf{t} + \Delta\mathbf{t}$ *has increased by at least* $\boldsymbol{\lambda}$ *new resource parts, say* $\mathbf{fr}\left(\mathbf{res}^{(\mathbf{U})}_{\mathscr{M}_{\mathbf{U}}+\ell_{\mathbf{U}}+1}\right)$, $\mathbf{fr}\left(\mathbf{res}^{(\mathbf{U})}_{\mathscr{M}_{\mathbf{U}}+\ell_{\mathbf{U}}+2}\right), \ldots,$ $\mathbf{fr}\left(\mathbf{res}^{(\mathbf{U})}_{\mathscr{M}_{\mathbf{U}}+\ell_{\mathbf{U}}+\boldsymbol{\lambda}}\right)$, *derived from the resource parts* $\mathbf{fr}\left(\mathbf{res}^{(\mathbf{V})}_{\kappa_1}\right)$, $\mathbf{fr}\left(\mathbf{res}^{(\mathbf{V})}_{\kappa_2}\right), \ldots,$ $\mathbf{fr}\left(\mathbf{res}^{(\mathbf{V})}_{\kappa_{\boldsymbol{\lambda}}}\right)$ *that existed in the node* $\mathbf{V}$ *the previous moment* $\mathbf{t}$, *in such a way that the following elementary properties hold.*

i. *The (Euclidean) norm* $\left\|\widehat{\boldsymbol{\beta}'}^{\left(\mathbf{U}' \rightsquigarrow \mathbf{U}'\right)}\right\| := \left(\sum_{j=1}^{\mathrm{n}} \sum_{\nu=1}^{\ell_{\mathbf{U}}+\boldsymbol{\lambda}} \left|\widehat{\boldsymbol{\beta}'}^{(\mathbf{U} \rightsquigarrow \mathbf{U})}_{\mathscr{M}_{\mathbf{U}}+\boldsymbol{\nu},\mathbf{j}}\right|^2\right)^{1/2}$ *of the resulting overall valuation in the variant node* $\mathbf{U}'$ *as evaluated from the viewpoint of the user(s) of* $\mathbf{U}$ *at the next moment* $\mathbf{t}'$ *is* ***greater*** *than the (Euclidean) norms*

$$\left\|\widehat{\boldsymbol{\beta}}^{(\mathbf{U} \rightsquigarrow \mathbf{U})}\right\| := \left(\sum_{j=1}^{\mathrm{n}} \sum_{\nu=1}^{\ell_{\mathbf{U}}} \left|\widehat{\boldsymbol{\beta}}^{(\mathbf{U} \rightsquigarrow \mathbf{U})}_{\mathscr{M}_{\mathbf{U}}+\boldsymbol{\nu},\mathbf{j}}\right|^2\right)^{1/2}$$

and

$$\left\|\boldsymbol{\beta}^{(\mathbf{U}\rightsquigarrow\mathbf{V})}\right\| := \left(\sum_{j=1}^{n}\sum_{\nu=1}^{\ell_{\mathbf{V}}}\left|\boldsymbol{\beta}^{(\mathbf{U}\rightsquigarrow\mathbf{V})}_{\mathscr{M}_{\mathbf{V}}+\boldsymbol{\nu},\mathbf{j}}\right|^{2}\right)^{1/2}$$

of the initial overall valuations in the nodes **U** *and* **V** *as evaluated from the viewpoint of the users of* **U** *at the preceding moment* **t***:*

$$\left\|\widehat{\boldsymbol{\beta}'}^{\left(\mathbf{U}'\rightsquigarrow\mathbf{U}'\right)}\right\| > max\left\{\left\|\widehat{\boldsymbol{\beta}}^{(\mathbf{U}\rightsquigarrow\mathbf{U})}\right\|, \left\|\boldsymbol{\beta}^{(\mathbf{U}\rightsquigarrow\mathbf{V})}\right\|\right\}.$$

ii. *The norm* $\left\|\boldsymbol{\beta}'^{\left(\mathbf{U}'\rightsquigarrow\mathbf{V}'\right)}\right\| := \left(\sum_{j=1}^{n}\sum_{\nu=1}^{\ell_{\mathbf{V}}}\left|\boldsymbol{\beta}'^{(\mathbf{U}\rightsquigarrow\mathbf{V})}_{\mathscr{M}_{\mathbf{V}}+\boldsymbol{\nu},\mathbf{j}}\right|^{2}\right)^{1/2}$ *of the resulting overall valuation in the node* $\mathbf{V}'$ *as evaluated from the viewpoint of the user(s) of* **U** *at the next moment* $\mathbf{t}'$ *is* ***less*** *than the norm*

$$\left\|\boldsymbol{\beta}^{(\mathbf{U}\rightsquigarrow\mathbf{V})}\right\| := \left(\sum_{j=1}^{n}\sum_{\nu=1}^{\ell_{\mathbf{V}}}\left|\boldsymbol{\beta}^{(\mathbf{U}\rightsquigarrow\mathbf{V})}_{\mathscr{M}_{\mathbf{V}}+\boldsymbol{\nu},\mathbf{j}}\right|^{2}\right)^{1/2}$$

of the initial overall valuation in the node **V** *as evaluated from the viewpoint of the users of* **U** *at the preceding moment* **t***:*

$$\left\|\boldsymbol{\beta}'^{\left(\mathbf{U}'\rightsquigarrow\mathbf{V}'\right)}\right\| < \left\|\boldsymbol{\beta}^{(\mathbf{U}\rightsquigarrow\mathbf{V})}\right\|.$$

iii. *The norm* $\left\|\widehat{\varepsilon'}^{\left(\mathbf{U}'\rightsquigarrow\mathbf{U}'\right)}\right\| := \left(\sum_{j=1}^{m}\sum_{\lambda=1}^{\ell_{\mathbf{U}}+\boldsymbol{\nu}}\left|\widehat{\varepsilon'}^{(\mathbf{U}\rightsquigarrow\mathbf{U})}_{\mathscr{M}_{\mathbf{U}}+\boldsymbol{\lambda},\mathbf{j}}\right|^{2}\right)^{1/2}$ *of the resulting overall vulnerability in the variant node* $\mathbf{U}'$ *as evaluated from the viewpoint of the user(s) of* **U** *at the next moment* $\mathbf{t}'$ *is* ***less or equal*** *than the norms*

$$\left\|\widehat{\varepsilon}^{(\mathbf{U}\rightsquigarrow\mathbf{U})}\right\| := \left(\sum_{j=1}^{m}\sum_{\nu=1}^{\ell_{\mathbf{U}}}\left|\widehat{\varepsilon}^{(\mathbf{U}\rightsquigarrow\mathbf{U})}_{\mathscr{M}_{\mathbf{U}}+\boldsymbol{\nu},\mathbf{j}}\right|^{2}\right)^{1/2}$$

and

$$\left\|\varepsilon^{(\mathbf{U}\rightsquigarrow\mathbf{V})}\right\| := \left(\sum_{j=1}^{m}\sum_{\nu=1}^{\ell_{\mathbf{V}}}\left|\varepsilon^{(\mathbf{U}\rightsquigarrow\mathbf{V})}_{\mathscr{M}_{\mathbf{V}}+\boldsymbol{\nu},\mathbf{j}}\right|^{2}\right)^{1/2}$$

of the initial overall vulnerabilities in the nodes **U** *and* **V** *as evaluated from the viewpoint of the users of* **U** *at the preceding moment* **t***:*

$$\left\| \widehat{\gamma'}^{\left(\mathbf{U}' \rightsquigarrow \mathbf{U}'\right)} \right\| = \min \left\{ \left\| \widehat{\gamma}^{(\mathbf{U} \rightsquigarrow \mathbf{U})} \right\|, \left\| \gamma^{(\mathbf{U} \rightsquigarrow \mathbf{V})} \right\| \right\}.$$

iv. The norm $\left\| \varepsilon'^{\left(\mathbf{U}' \rightsquigarrow \mathbf{V}'\right)} \right\| := \left(\sum_{j=1}^{m} \sum_{\nu=1}^{\ell_{\mathbf{V}}} \left| \varepsilon'^{(\mathbf{U} \rightsquigarrow \mathbf{V})}_{\mathscr{M}_{\mathbf{V}}+\nu,\mathbf{j}} \right|^2 \right)^{1/2}$ *of the resulting overall vulnerability in the node* $\mathbf{V}'$ *as evaluated from the viewpoint of the users of* $\mathbf{U}$ *at the next moment* $\mathbf{t}'$ *is* ***greater*** *than the norm*

$$\left\| \varepsilon^{(\mathbf{U} \rightsquigarrow \mathbf{V})} \right\| := \left(\sum_{j=1}^{m} \sum_{\nu=1}^{\ell_{\mathbf{V}}} \left| \varepsilon^{(\mathbf{U} \rightsquigarrow \mathbf{V})}_{\mathscr{M}_{\mathbf{U}}+\nu,\mathbf{j}} \right|^2 \right)^{1/2}$$

of the initial overall vulnerability in the node $\mathbf{V}$ *as evaluated from the viewpoint of the user(s) of* $\mathbf{U}$ *at the preceding moment* $\mathbf{t}$*:* $\left\| \varepsilon'^{\left(\mathbf{U}' \rightsquigarrow \mathbf{V}'\right)} \right\| > \left\| \varepsilon^{(\mathbf{U} \rightsquigarrow \mathbf{V})} \right\|$.

The **degree** $\mathbf{d}=\mathbf{d}_{\kappa_1,\ldots,\kappa_\nu}$ **of the passive attack f against the resource parts** $\mathbf{fr}\left(\mathbf{res}^{(\mathbf{V})}_{\kappa_1}\right)$, $\mathbf{fr}\left(\mathbf{res}^{(\mathbf{V})}_{\kappa_2}\right), \ldots, \mathbf{fr}\left(\mathbf{res}^{(\mathbf{V})}_{\kappa_\lambda}\right)$ of node $\mathbf{V}$ from the offensive node $\mathbf{U}$ at time moment $\mathbf{t} \in \mathbb{I}$ is the maximum of the two quotients
$\mathbf{d}_1 := \left\| \widehat{\beta'}^{\left(\mathbf{U}' \rightsquigarrow \mathbf{U}'\right)} \right\| / \left\| \beta'^{\left(\mathbf{U}' \rightsquigarrow \mathbf{V}'\right)} \right\|$ and $\mathbf{d}_2 := \left(\left\| \widehat{\varepsilon'}^{\left(\mathbf{U}' \rightsquigarrow \mathbf{U}'\right)} \right\| / \left\| \varepsilon'^{\left(\mathbf{U}' \rightsquigarrow \mathbf{V}'\right)} \right\| \right)^{-1}$.
Thus

$$\mathbf{d}=\mathbf{d}_{\kappa_1,\ldots,\kappa_\lambda} := \mathbf{max}\left\{\mathbf{d}_1, \mathbf{d}_2\right\}.$$

If the degree $\mathbf{d}$ surpasses a given threshold $\mathscr{S}^{(\mathbf{U},\mathbf{V})}_{\kappa_1,\ldots,\kappa_\lambda} \in [\mathbf{0}, \infty[$, called the **passive attack threshold in the resource parts** $\mathbf{fr}\left(\mathbf{res}^{(\mathbf{V})}_{\kappa_1}\right)$, $\mathbf{fr}\left(\mathbf{res}^{(\mathbf{V})}_{\kappa_2}\right), \ldots, \mathbf{fr}\left(\mathbf{res}^{(\mathbf{V})}_{\kappa_\lambda}\right)$ of $\mathbf{V}$ at time moment $\mathbf{t} \in \mathbb{I}$, we say that the passive attack $\mathbf{f}$ is **dangerous with degree of danger d** in the resource parts $\mathbf{fr}\left(\mathbf{res}^{(\mathbf{V})}_{\kappa_1}\right)$, $\mathbf{fr}\left(\mathbf{res}^{(\mathbf{V})}_{\kappa_2}\right), \ldots, \mathbf{fr}\left(\mathbf{res}^{(\mathbf{V})}_{\kappa_\lambda}\right)$ of $\mathbf{V}$.

Protected Cyber Nodes from Passive Attacks

Definition 16

i. The node $\mathbf{V}$ is said to be **protected from passive attacks, with degree of protection** $\mathbf{p} \in]\mathbf{0}, \mathbf{1}]$ over the resource parts

$$\mathbf{fr}\left(\mathbf{res}^{(\mathbf{V})}_{\kappa_1}\right), \mathbf{fr}\left(\mathbf{res}^{(\mathbf{V})}_{\kappa_2}\right), \ldots, \mathbf{fr}\left(\mathbf{res}^{(\mathbf{V})}_{\kappa_\nu}\right)$$

of $\mathbf{V}$ over a time period $\mathbb{I}$, if, during this time period, there is a nodal fixed filter system $\overline{\mathbf{V}}^{(\kappa_1,\ldots,\kappa_\nu)}$ in the union $\mathbf{E}=\mathbf{fr}\left(\mathbf{res}^{(\mathbf{V})}_{\kappa_1}\right) \bigcup \mathbf{fr}\left(\mathbf{res}^{(\mathbf{V})}_{\kappa_2}\right) \bigcup \ldots \bigcup \mathbf{fr}\left(\mathbf{res}^{(\mathbf{V})}_{\kappa_\nu}\right)$

that allow every parallactic cyber passive attack against the resource parts (from any offensive node **U**) with degree of danger

$$\mathbf{d} = -\log \mathbf{p}$$

to reach only resource parts **K** of the initial target **V** that are disjoint from **E**.

ii. During the time period $\mathbb{I}$, the node **V** is said to be **completely protected from passive attacks of danger degree d**, if, at this time period, any resource part in **V** is protected from passive attacks against **V**, with degree of protection

$$\mathbf{p} = \mathbf{e}^{-\mathbf{d}}.$$

The node **V** is said to be completely protected from passive attacks at a given time period, if, during this time period, any resource part of **V** is protected from active attacks against **V** with degree of protection

$$\mathbf{p=1}.$$

Active Cyber-Attacks

An attack is active if it is an attack with data transmission to all parties thereby acting as a liaison enabling severe compromise. The purpose is to alter system resources or affect their operation. So, in an active attack, an intruder attempts to alter data on the target system or data en route for the target system. Let $\mathbf{U}, \mathbf{V} \in \mathbf{ob}\,(cy\,(\mathbf{t}))$, whenever $\mathbf{t}$ is in an arbitrary subset $\mathbb{I}$ of $]\boldsymbol{\alpha}, \boldsymbol{\beta}[\subset\subset [\mathbf{0}, \mathbf{1}]$. Let also $\varepsilon_{\mathbf{V}}$ and $\varepsilon_{\mathbf{U}}$ be two supervisory perception curves of **V** and **U** in the node system (**V**, **U**).

Definition 17 A family of coherent interactions

$$\mathscr{F}=\{\mathbf{Z}{=}\mathbf{Z}_{(\mathbf{U},\mathbf{V})}\,(\mathbf{t}) = ((z_1, w_1), (z_2, w_2), (z_3, w_3), (z_4, w_4)) = \\ ((z_1, w_1), (z_2, w_2), (z_3, w_3), (z_4, w_4))\,(\mathbf{t}) \in \left(\mathbb{C}^{n\times k}\times\mathbb{C}^{m\times k}\right)^{\mathbf{4}}, \mathbf{t} \in \mathbb{I}\right\},$$

lying in the partial danger sector $\mathscr{E}{=}\mathscr{E}_{\mathbf{U}\to\mathbf{V}}$ to the node **V** from the node **U** during the entire time set $\mathbb{I}$, is a **germ of (partial) active attack** against the $(\mu_1, \ldots, \mu_\nu)$-device parts $\mathbf{fr}\left(\mathbf{dev}_{\mu_1}^{(\mathbf{V})}\right)$, $\mathbf{fr}\left(\mathbf{dev}_{\mu_2}^{(\mathbf{V})}\right), \ldots, \mathbf{fr}\left(\mathbf{dev}_{\mu_\nu}^{(\mathbf{V})}\right)$ of **V** and the $(\kappa_1, \ldots, \kappa_\lambda)$-resource parts $\mathbf{fr}\left(\mathbf{res}_{\kappa_1}^{(\mathbf{V})}\right)$, $\mathbf{fr}\left(\mathbf{res}_{\kappa_2}^{(\mathbf{V})}\right), \ldots, \mathbf{fr}\left(\mathbf{res}_{\kappa_\lambda}^{(\mathbf{V})}\right)$ of **V**, during a given time subset $\mathbb{I}$ of a subinterval $[\boldsymbol{\alpha}, \boldsymbol{\beta}] \subset\subset [\mathbf{0}, \mathbf{1}]$, if, whenever $\mathbf{t} \in \mathbb{I}$, the pair $((z_1, w_1), (z_2, w_2)) \in \gamma_{\mathbf{U}}\times\gamma_{\mathbf{V}}$ of supervisory resource perceptions of **U** and **V** in the system of nodes **U** and **V** has the form

$$((z_1, w_1), (z_2, w_2)) =$$

$$\left(\left(\begin{pmatrix} \beta^{(U\rightsquigarrow V)}_{1,1} + i\,\widehat{\beta}^{(V\rightsquigarrow V)}_{1,1} & \cdots & \beta^{(U\rightsquigarrow V)}_{1,n} + i\,\widehat{\beta}^{(V\rightsquigarrow V)}_{1,n} \\ \cdots & \cdots & \cdots \\ \beta^{(U\rightsquigarrow V)}_{m_V,1} + i\,\widehat{\beta}^{(V\rightsquigarrow V)}_{m_V,1} & \cdots & \beta^{(U\rightsquigarrow V)}_{m_V,n} + i\,\widehat{\beta}^{(V\rightsquigarrow V)}_{m_V,n} \\ 0 & \cdots & 0 \\ \cdots & \cdots & \cdots \\ 0 & \cdots & 0 \\ \beta^{(U\rightsquigarrow V)}_{\mathscr{M}_V+1,1} + i\,\widehat{\beta}^{(V\rightsquigarrow V)}_{\mathscr{M}_V+1,1} & \cdots & \beta^{(U\rightsquigarrow V)}_{\mathscr{M}_V+1,n} + i\,\widehat{\beta}^{(V\rightsquigarrow V)}_{\mathscr{M}_V+1,n} \\ \cdots & \cdots & \cdots \\ \beta^{(U\rightsquigarrow V)}_{\mathscr{M}_V+\ell_V,1} + i\,\widehat{\beta}^{(V\rightsquigarrow V)}_{\mathscr{M}_V+\ell_V,1} & \cdots & \beta^{(U\rightsquigarrow V)}_{\mathscr{M}_V+\ell_V,n} + i\,\widehat{\beta}^{(V\rightsquigarrow V)}_{\mathscr{M}_V+\ell_V,n} \\ 0 & \cdots & 0 \\ \cdots & \cdots & \cdots \\ 0 & \cdots & 0 \end{pmatrix}, \begin{pmatrix} \varepsilon^{(U\rightsquigarrow V)}_{1,1} + i\,\widehat{\varepsilon}^{(V\rightsquigarrow V)}_{1,1} & \cdots & \varepsilon^{(U\rightsquigarrow V)}_{1,m} + i\,\widehat{\varepsilon}^{(V\rightsquigarrow V)}_{1,m} \\ \cdots & \cdots & \cdots \\ \varepsilon^{(U\rightsquigarrow V)}_{m_V,1} + i\,\widehat{\varepsilon}^{(V\rightsquigarrow V)}_{m_V,1} & \cdots & \varepsilon^{(U\rightsquigarrow V)}_{m_V,m} + i\,\widehat{\varepsilon}^{(V\rightsquigarrow V)}_{m_V,m} \\ 0 & \cdots & 0 \\ \cdots & \cdots & \cdots \\ 0 & \cdots & 0 \\ \varepsilon^{(U\rightsquigarrow V)}_{\mathscr{M}_V+1,1} + i\,\widehat{\varepsilon}^{(V\rightsquigarrow V)}_{\mathscr{M}_V+1,1} & \cdots & \varepsilon^{(U\rightsquigarrow V)}_{\mathscr{M}_V+1,m} + i\,\widehat{\varepsilon}^{(V\rightsquigarrow V)}_{\mathscr{M}_V+1,m} \\ \cdots & \cdots & \cdots \\ \varepsilon^{(U\rightsquigarrow V)}_{\mathscr{M}_V+\ell_V,1} + i\,\widehat{\varepsilon}^{(V\rightsquigarrow V)}_{\mathscr{M}_V+\ell_V,1} & \cdots & \varepsilon^{(U\rightsquigarrow V)}_{\mathscr{M}_V+\ell_V,m} + i\,\widehat{\varepsilon}^{(V\rightsquigarrow V)}_{\mathscr{M}_V+\ell_V,m} \\ 0 & \cdots & 0 \\ \cdots & \cdots & \cdots \\ 0 & \cdots & 0 \end{pmatrix}\right),\right.$$

$$\left.\left(\begin{pmatrix} \beta^{(V\rightsquigarrow U)}_{1,1} + i\,\widehat{\beta}^{(U\rightsquigarrow U)}_{1,1} & \cdots & \beta^{(V\rightsquigarrow U)}_{1,n} + i\,\widehat{\beta}^{(U\rightsquigarrow U)}_{1,n} \\ \cdots & \cdots & \cdots \\ \beta^{(V\rightsquigarrow U)}_{m_U,1} + i\,\widehat{\beta}^{(U\rightsquigarrow U)}_{m_U,1} & \cdots & \beta^{(V\rightsquigarrow U)}_{m_U,n} + i\,\widehat{\beta}^{(U\rightsquigarrow U)}_{m_U,n} \\ 0 & \cdots & 0 \\ \cdots & \cdots & \cdots \\ 0 & \cdots & 0 \\ \beta^{(V\rightsquigarrow U)}_{\mathscr{M}_U+1,1} + i\,\widehat{\beta}^{(U\rightsquigarrow U)}_{\mathscr{M}_U+1,1} & \cdots & \beta^{(V\rightsquigarrow U)}_{\mathscr{M}_U+1,n} + i\,\widehat{\beta}^{(U\rightsquigarrow U)}_{\mathscr{M}_U+1,n} \\ \cdots & \cdots & \cdots \\ \beta^{(V\rightsquigarrow U)}_{\mathscr{M}_U+\ell_U,1} + i\,\widehat{\beta}^{(U\rightsquigarrow U)}_{\mathscr{M}_U+\ell_U,1} & \cdots & \beta^{(V\rightsquigarrow U)}_{\mathscr{M}_U+\ell_U,n} + i\,\widehat{\beta}^{(U\rightsquigarrow U)}_{\mathscr{M}_U+\ell_U,n} \\ 0 & \cdots & 0 \\ \cdots & \cdots & \cdots \\ 0 & \cdots & 0 \end{pmatrix}, \begin{pmatrix} \varepsilon^{(V\rightsquigarrow U)}_{1,1} + i\,\widehat{\varepsilon}^{(U\rightsquigarrow U)}_{1,1} & \cdots & \varepsilon^{(V\rightsquigarrow U)}_{1,m} + i\,\widehat{\varepsilon}^{(U\rightsquigarrow U)}_{1,m} \\ \cdots & \cdots & \cdots \\ \varepsilon^{(V\rightsquigarrow U)}_{m_V,1} + i\,\widehat{\varepsilon}^{(U\rightsquigarrow U)}_{m_V,1} & \cdots & \varepsilon^{(V\rightsquigarrow U)}_{m_V,m} + i\,\widehat{\varepsilon}^{(U\rightsquigarrow U)}_{m_V,m} \\ 0 & \cdots & 0 \\ \cdots & \cdots & \cdots \\ 0 & \cdots & 0 \\ \varepsilon^{(V\rightsquigarrow U)}_{\mathscr{M}_V+1,1} + i\,\widehat{\varepsilon}^{(U\rightsquigarrow U)}_{\mathscr{M}_V+1,1} & \cdots & \varepsilon^{(V\rightsquigarrow U)}_{\mathscr{M}_V+1,m} + i\,\widehat{\varepsilon}^{(U\rightsquigarrow U)}_{\mathscr{M}_V+1,m} \\ \cdots & \cdots & \cdots \\ \varepsilon^{(V\rightsquigarrow U)}_{\mathscr{M}_V+\ell_V,1} + i\,\widehat{\varepsilon}^{(U\rightsquigarrow U)}_{\mathscr{M}_V+\ell_V,1} & \cdots & \varepsilon^{(V\rightsquigarrow U)}_{\mathscr{M}_V+\ell_V,m} + i\,\widehat{\varepsilon}^{(U\rightsquigarrow U)}_{\mathscr{M}_V+\ell_V,m} \\ 0 & \cdots & 0 \\ \cdots & \cdots & \cdots \\ 0 & \cdots & 0 \end{pmatrix}\right)\right)$$

and is depicted, at a next moment $\mathbf{t}'=\mathbf{t}+\Delta\mathbf{t}$, via the associated family of cyber-activities

$$\mathcal{D}_{\mathcal{F}}=\left(g_{\mathbf{t}}=g_{\mathbf{t}}^{(\mathbf{Z})}: \mathbb{G}_t^{(U)}\times\mathbb{G}_t^{(V)} \rightarrow \mathbb{G}_t^{(U)}\times\mathbb{G}_t^{(V)}: (\delta_{\mathbf{U}}(\mathbf{t}), \gamma_{\mathbf{V}}(\mathbf{t})) \longmapsto \left(\delta'_{\mathbf{U}}\left(\mathbf{t}'\right), \gamma'_{\mathbf{V}}\left(\mathbf{t}'\right)\right)\right)_{\mathbf{t}\in]\boldsymbol{\alpha},\boldsymbol{\beta}[}$$

over the time $\mathbf{t} \in]\boldsymbol{\alpha}, \boldsymbol{\beta}[$ at the pair $((z_3, w_3), (z_4, w_4)) \in \mathbb{G}_t^{(U)}\times\mathbb{G}_t^{(V)}$ of supervisory resource perceptions of **U** and **V** having the form

$$((z_3, w_3), (z_4, w_4)) =$$

$$\left(\left(\begin{pmatrix}
\beta'^{(U\rightsquigarrow V)}_{1,1} + \mathbf{i}\, \widehat{\beta'}^{(V\rightsquigarrow V)}_{1,1} & \cdots & \beta'^{(U\rightsquigarrow V)}_{1,n} + \mathbf{i}\, \widehat{\beta'}^{(V\rightsquigarrow V)}_{1,n} \\
\cdots & \cdots & \cdots \\
\beta'^{(U\rightsquigarrow V)}_{m_V,1} + \mathbf{i}\, \widehat{\beta'}^{(V\rightsquigarrow V)}_{m_V,1} & \cdots & \beta'^{(U\rightsquigarrow V)}_{m_V,n} + \mathbf{i}\, \widehat{\beta'}^{(V\rightsquigarrow V)}_{m_V,n} \\
\mathbf{0} & \cdots & \mathbf{0} \\
\cdots & \cdots & \cdots \\
\mathbf{0} & \cdots & \mathbf{0} \\
\beta'^{(U\rightsquigarrow V)}_{\mathscr{M}_V+1,1} + \mathbf{i}\, \widehat{\beta'}^{(V\rightsquigarrow V)}_{\mathscr{M}_V+1,1} & \cdots & \beta'^{(U\rightsquigarrow V)}_{\mathscr{M}_V+1,n} + \mathbf{i}\, \widehat{\beta'}^{(V\rightsquigarrow V)}_{\mathscr{M}_V+1,n} \\
\cdots & \cdots & \cdots \\
\beta'^{(U\rightsquigarrow V)}_{\mathscr{M}_V+\ell_V,1} + \mathbf{i}\, \widehat{\beta'}^{(V\rightsquigarrow V)}_{\mathscr{M}_V+\ell_V,1} & \cdots & \beta'^{(U\rightsquigarrow V)}_{\mathscr{M}_V+\ell_V,n} + \mathbf{i}\, \widehat{\beta'}^{(V\rightsquigarrow V)}_{\mathscr{M}_V+\ell_V,n} \\
\mathbf{0} & \cdots & \mathbf{0} \\
\cdots & \cdots & \cdots \\
\mathbf{0} & & \mathbf{0}
\end{pmatrix},
\begin{pmatrix}
\varepsilon'^{(U\rightsquigarrow V)}_{1,1} + \mathbf{i}\, \widehat{\varepsilon'}^{(V\rightsquigarrow V)}_{1,1} & \cdots & \varepsilon'^{(U\rightsquigarrow V)}_{1,m} + \mathbf{i}\, \widehat{\varepsilon'}^{(V\rightsquigarrow V)}_{1,m} \\
\cdots & \cdots & \cdots \\
\varepsilon'^{(U\rightsquigarrow V)}_{m_V,1} + \mathbf{i}\, \widehat{\varepsilon'}^{(V\rightsquigarrow V)}_{m_V,1} & \cdots & \varepsilon'^{(U\rightsquigarrow V)}_{m_V,m} + \mathbf{i}\, \widehat{\varepsilon'}^{(V\rightsquigarrow V)}_{m_V,m} \\
\mathbf{0} & & \mathbf{0} \\
\cdots & \cdots & \cdots \\
\mathbf{0} & \cdots & \mathbf{0} \\
\varepsilon'^{(U\rightsquigarrow V)}_{\mathscr{M}_V+1,1} + \mathbf{i}\, \widehat{\varepsilon'}^{(V\rightsquigarrow V)}_{\mathscr{M}_V+1,1} & \cdots & \varepsilon'^{(U\rightsquigarrow V)}_{\mathscr{M}_V+1,m} + \mathbf{i}\, \widehat{\varepsilon'}^{(V\rightsquigarrow V)}_{\mathscr{M}_V+1,m} \\
\cdots & & \cdots \\
\varepsilon'^{(U\rightsquigarrow V)}_{\mathscr{M}_V+\ell_V,1} + \mathbf{i}\, \widehat{\varepsilon'}^{(V\rightsquigarrow V)}_{\mathscr{M}_V+\ell_V,1} & \cdots & \varepsilon'^{(U\rightsquigarrow V)}_{\mathscr{M}_V+\ell_V,m} + \mathbf{i}\, \widehat{\varepsilon'}^{(V\rightsquigarrow V)}_{\mathscr{M}_V+\ell_V,m} \\
\mathbf{0} & \cdots & \mathbf{0} \\
\cdots & \cdots & \cdots \\
\mathbf{0} & & \mathbf{0}
\end{pmatrix}\right),\right.$$

$$\left.\left(\begin{pmatrix}
\beta'^{(V\rightsquigarrow U)}_{1,1} + \mathbf{i}\, \widehat{\beta'}^{(U\rightsquigarrow U)}_{1,1} & \cdots & \beta'^{(V\rightsquigarrow U)}_{1,n} + \mathbf{i}\, \widehat{\beta'}^{(U\rightsquigarrow U)}_{1,n} \\
\cdots & \cdots & \cdots \\
\beta'^{(V\rightsquigarrow U)}_{m_U,1} + \mathbf{i}\, \widehat{\beta'}^{(U\rightsquigarrow U)}_{m_U,1} & \cdots & \beta'^{(V\rightsquigarrow U)}_{m_U,n} + \mathbf{i}\, \widehat{\beta'}^{(U\rightsquigarrow U)}_{m_U,n} \\
\mathbf{0} & & \mathbf{0} \\
\cdots & \cdots & \cdots \\
\mathbf{0} & \cdots & \mathbf{0} \\
\beta'^{(V\rightsquigarrow U)}_{\mathscr{M}_U+1,1} + \mathbf{i}\, \widehat{\beta'}^{(U\rightsquigarrow U)}_{\mathscr{M}_U+1,1} & \cdots & \beta'^{(V\rightsquigarrow U)}_{\mathscr{M}_U+1,n} + \mathbf{i}\, \widehat{\beta'}^{(U\rightsquigarrow U)}_{\mathscr{M}_U+1,n} \\
\cdots & & \cdots \\
\beta'^{(V\rightsquigarrow U)}_{\mathscr{M}_U+\ell_U,1} + \mathbf{i}\, \widehat{\beta'}^{(U\rightsquigarrow U)}_{\mathscr{M}_U+\ell_U,1} & & \beta'^{(V\rightsquigarrow U)}_{\mathscr{M}_U+\ell_U,n} + \mathbf{i}\, \widehat{\beta'}^{(U\rightsquigarrow U)}_{\mathscr{M}_U+\ell_U,n} \\
\beta'^{(V\rightsquigarrow U)}_{\mathscr{M}_U+\ell_U+1,1} + \mathbf{i}\, \widehat{\beta'}^{(U\rightsquigarrow U)}_{\mathscr{M}_U+\ell_U+1,1} & \cdots & \beta'^{(V\rightsquigarrow U)}_{\mathscr{M}_U+\ell_U+1,n} + \mathbf{i}\, \widehat{\beta'}^{(U\rightsquigarrow U)}_{\mathscr{M}_U+\ell_U+1,n} \\
\cdots & \cdots & \cdots \\
\beta'^{(V\rightsquigarrow U)}_{\mathscr{M}_U+\ell_U+N,1} + \mathbf{i}\, \widehat{\beta'}^{(U\rightsquigarrow U)}_{\mathscr{M}_U+\ell_U+N,1} & \cdots & \beta'^{(V\rightsquigarrow U)}_{\mathscr{M}_U+\ell_U+N,n} + \mathbf{i}\, \widehat{\beta'}^{(U\rightsquigarrow U)}_{\mathscr{M}_U+\ell_U+N,n} \\
\mathbf{0} & & \mathbf{0} \\
\cdots & & \cdots \\
\mathbf{0} & & \mathbf{0}
\end{pmatrix},
\begin{pmatrix}
\varepsilon'^{(V\rightsquigarrow U)}_{1,1} + \mathbf{i}\, \widehat{\varepsilon'}^{(U\rightsquigarrow U)}_{1,1} & \cdots & \varepsilon'^{(V\rightsquigarrow U)}_{1,m} + \mathbf{i}\, \widehat{\varepsilon'}^{(U\rightsquigarrow U)}_{1,m} \\
\cdots & \cdots & \cdots \\
\varepsilon'^{(V\rightsquigarrow U)}_{m_V,1} + \mathbf{i}\, \widehat{\varepsilon'}^{(U\rightsquigarrow U)}_{m_V,1} & \cdots & \varepsilon'^{(V\rightsquigarrow U)}_{m_V,m} + \mathbf{i}\, \widehat{\varepsilon'}^{(U\rightsquigarrow U)}_{m_V,m} \\
\mathbf{0} & & \mathbf{0} \\
\cdots & \cdots & \cdots \\
\mathbf{0} & \cdots & \mathbf{0} \\
\varepsilon'^{(V\rightsquigarrow U)}_{\mathscr{M}_V+1,1} + \mathbf{i}\, \widehat{\varepsilon'}^{(U\rightsquigarrow U)}_{\mathscr{M}_V+1,1} & \cdots & \varepsilon'^{(V\rightsquigarrow U)}_{\mathscr{M}_V+1,m} + \mathbf{i}\, \widehat{\varepsilon'}^{(U\rightsquigarrow U)}_{\mathscr{M}_V+1,m} \\
\cdots & & \cdots \\
\varepsilon'^{(V\rightsquigarrow U)}_{\mathscr{M}_V+\ell_V,1} + \mathbf{i}\, \widehat{\varepsilon'}^{(U\rightsquigarrow U)}_{\mathscr{M}_V+\ell_V,1} & & \varepsilon'^{(V\rightsquigarrow U)}_{\mathscr{M}_V+\ell_V,m} + \mathbf{i}\, \widehat{\varepsilon'}^{(U\rightsquigarrow U)}_{\mathscr{M}_V+\ell_V,m} \\
\varepsilon'^{(V\rightsquigarrow U)}_{\mathscr{M}_U+\ell_U+1,1} + \mathbf{i}\, \widehat{\varepsilon'}^{(U\rightsquigarrow U)}_{\mathscr{M}_U+\ell_U+1,1} & \cdots & \varepsilon'^{(V\rightsquigarrow U)}_{\mathscr{M}_U+\ell_U+1,m} + \mathbf{i}\, \widehat{\varepsilon'}^{(U\rightsquigarrow U)}_{\mathscr{M}_U+\ell_U+1,m} \\
\cdots & \cdots & \cdots \\
\varepsilon'^{(V\rightsquigarrow U)}_{\mathscr{M}_U+\ell_U+N,1} + \mathbf{i}\, \widehat{\varepsilon'}^{(U\rightsquigarrow U)}_{\mathscr{M}_U+\ell_U+N,1} & \cdots & \varepsilon'^{(V\rightsquigarrow U)}_{\mathscr{M}_U+\ell_U+N,m} + \mathbf{i}\, \widehat{\varepsilon'}^{(U\rightsquigarrow U)}_{\mathscr{M}_U+\ell_U+N,m} \\
\mathbf{0} & & \mathbf{0} \\
\cdots & & \cdots \\
\mathbf{0} & & \mathbf{0}
\end{pmatrix}\right)\right).$$

It is easy to prove and/or verify the next two results.

Proposition 3 *In an active attack* $\mathscr{F}$ *from* **U** *against the* $(\mu_1, \ldots, \mu_\nu)$*-device parts* $\mathbf{fr}\left(\mathbf{dev}_{\mu_1}^{(\mathbf{V})}\right), \ldots, \mathbf{fr}\left(\mathbf{dev}_{\mu_\nu}^{(\mathbf{V})}\right)$ *of* **V** *and the* $(\kappa_1, \ldots, \kappa_\lambda)$*-resource parts* $\mathbf{fr}\left(\mathbf{res}_{\kappa_1}^{(\mathbf{V})}\right), \ldots, \mathbf{fr}\left(\mathbf{res}_{\kappa_\lambda}^{(\mathbf{V})}\right)$ *of* **V***, the following elementary properties hold.*

i. *All new resource valuations of the offensive node* **U** *are derived from the set of all initial resource valuations of* **V***, i.e., for any*

$$\mathbf{j} \in \{\mathscr{M}_\mathbf{U}+\ell_\mathbf{U}+\mathbf{1}, \ldots, \mathscr{M}_\mathbf{U}+\ell_\mathbf{U}+\mathbf{N}\}$$

and any

$$\mathbf{k} \in \{\mathbf{1}, \mathbf{2}, \ldots, \mathfrak{n}\},$$

the new valuations

$$\boldsymbol{\beta}'^{(\mathbf{V}\rightsquigarrow\mathbf{U})}_{\mathbf{j,k}} + \mathbf{i}\ \widehat{\boldsymbol{\beta}}'^{(\mathbf{U}\rightsquigarrow\mathbf{U})}_{\mathbf{j,k}}$$

are obtained as functions of the initial valuations

$$\boldsymbol{\beta}^{(\mathbf{U}\rightsquigarrow\mathbf{V})}_{\mathbf{p,l}} + \mathbf{i}\ \widehat{\boldsymbol{\beta}}^{(\mathbf{V}\rightsquigarrow\mathbf{V})}_{\mathbf{p,l}},$$

$$\mathbf{p} \in \{\mathbf{1}, \mathbf{2}, \ldots, m_\mathbf{V}, \mathscr{M}_\mathbf{V}+\mathbf{1}, \ldots, \mathscr{M}_\mathbf{V}+\ell_\mathbf{V}\}, \mathbf{l} \in \{\mathbf{1}, \mathbf{2}, \ldots, \mathfrak{n}\}.$$

ii. *Similarly, all new resource vulnerabilities of the offensive node* **U** *are derived from the set of all initial resource vulnerabilities of* **V***, i.e., for any* $\mathbf{j} \in \{\mathscr{M}_\mathbf{U}+\ell_\mathbf{U}+\mathbf{1}, \ldots, \mathscr{M}_\mathbf{U}+\ell_\mathbf{U}+\mathbf{N}\}$ *and any* $\mathbf{k} \in \{\mathbf{1}, \mathbf{2}, \ldots, \mathfrak{n}\}$*, the new vulnerabilities*

$$\varepsilon'^{(\mathbf{V}\rightsquigarrow\mathbf{U})}_{\mathbf{j,k}} + \mathbf{i}\ \widehat{\varepsilon}'^{(\mathbf{U}\rightsquigarrow\mathbf{U})}_{\mathbf{j,k}}$$

are obtained as functions of the initial vulnerabilities

$$\varepsilon^{(\mathbf{U}\rightsquigarrow\mathbf{V})}_{\mathbf{p,l}} + \mathbf{i}\ \widehat{\varepsilon}^{(\mathbf{V}\rightsquigarrow\mathbf{V})}_{\mathbf{p,l}}, \mathbf{p} \in \{\mathbf{1}, \mathbf{2}, \ldots, m_\mathbf{V}, \mathscr{M}_\mathbf{V}+\mathbf{1}, \ldots, \mathscr{M}_\mathbf{V}+\ell_\mathbf{V}\},$$

$$\mathbf{k} \in \{\mathbf{1}, \mathbf{2}, \ldots, \mathfrak{m}\}.$$

iii. *Finally, from the viewpoint of the (user(s) of) node* **V***, all valuations of* **U** *remain unchanged, i.e., if*

$$\mathbf{j} \in \{\mathbf{1}, \mathbf{2}, \ldots, m_\mathbf{U}, \mathscr{M}_\mathbf{U}+\mathbf{1}, \ldots, \mathscr{M}_\mathbf{U}+\ell_\mathbf{U}\},$$

then

$$\boldsymbol{\beta}_{\mathbf{j,k}}^{(\mathbf{V}\rightsquigarrow\mathbf{U})}=\boldsymbol{\beta}'^{(\mathbf{V}\rightsquigarrow\mathbf{U})}_{\mathbf{j,k}}$$

for any $\mathbf{k} \in \{\mathbf{1}, \mathbf{2}, \ldots, \mathfrak{n}\}$ *and*

$$\gamma_{\mathbf{j,k}}^{(\mathbf{V}\rightsquigarrow\mathbf{U})}=\gamma'^{(\mathbf{V}\rightsquigarrow\mathbf{U})}_{\mathbf{j,k}} \text{ for any } \mathbf{k} \in \{\mathbf{1}, \mathbf{2}, \ldots, \mathfrak{m}\}.$$

Proposition 4 *In an active attack* $\mathscr{F}$ *from* $\mathbf{U}$ *against the* $(\boldsymbol{\mu}_1, \ldots, \boldsymbol{\mu}_\nu)$*-device parts* $\mathbf{fr}\left(\mathbf{dev}_{\mu_1}^{(\mathbf{V})}\right), \ldots, \mathbf{fr}\left(\mathbf{dev}_{\mu_\nu}^{(\mathbf{V})}\right)$ *of* $\mathbf{V}$ *and the* $(\boldsymbol{\kappa}_1, \ldots, \boldsymbol{\kappa}_\lambda)$*-resource parts* $\mathbf{fr}\left(\mathbf{res}_{\kappa_1}^{(\mathbf{V})}\right), \ldots, \mathbf{fr}\left(\mathbf{res}_{\kappa_\lambda}^{(\mathbf{V})}\right)$ *of* $\mathbf{V}$*, the following elementary properties hold.*

i. *The (Euclidean) norm* $\left\| \boldsymbol{\beta}'^{\left(\mathbf{U}'\rightsquigarrow\mathbf{V}'\right)} \right\| := \left(\sum_{j=1}^{\mathfrak{n}} \sum_{\lambda=1}^{\ell_{\mathbf{V}}} \left|\boldsymbol{\beta}'^{(\mathbf{U}\rightsquigarrow\mathbf{V})}_{\mathscr{M}_{\mathbf{U}}+\boldsymbol{\lambda},\mathbf{j}}\right|^2\right)^{1/2}$ *of the resulting overall valuation in node* $\mathbf{V}'$ *as evaluated from the viewpoint of the user(s) of* $\mathbf{U}$ *at the next moment* $\mathbf{t}'$ *is less than the (Euclidean) norm*

$$\left\|\boldsymbol{\beta}^{(\mathbf{U}\rightsquigarrow\mathbf{V})}\right\| := \left(\sum_{j=1}^{\mathfrak{n}} \sum_{\lambda=1}^{\ell_{\mathbf{V}}} \left|\boldsymbol{\beta}^{(\mathbf{U}\rightsquigarrow\mathbf{V})}_{\mathscr{M}_{\mathbf{U}}+\boldsymbol{\lambda},\mathbf{j}}\right|^2\right)^{\mathbf{1/2}}$$

of the initial overall valuation in $\mathbf{V}$ *as evaluated from the viewpoint of the user(s) of* $\mathbf{U}$ *at the preceding moment* $\mathbf{t}$*:*

$$\left\| \boldsymbol{\beta}'^{\left(\mathbf{U}'\rightsquigarrow\mathbf{V}'\right)} \right\| < \left\|\boldsymbol{\beta}^{(\mathbf{U}\rightsquigarrow\mathbf{V})}\right\|.$$

ii. *The (Euclidean) norm* $\left\| \varepsilon'^{\left(\mathbf{U}'\rightsquigarrow\mathbf{V}'\right)} \right\| := \left(\sum_{j=1}^{\mathfrak{m}} \sum_{\lambda=1}^{\ell_{\mathbf{V}}} \left|\varepsilon'^{(\mathbf{U}\rightsquigarrow\mathbf{V})}_{\mathscr{M}_{\mathbf{U}}+\boldsymbol{\lambda},\mathbf{j}}\right|^2\right)^{1/2}$ *of the resulting overall vulnerability in the node* $\mathbf{V}'$ *as evaluated from the viewpoint of the user(s) of* $\mathbf{U}$ *at the next moment* $\mathbf{t}'$ *is greater than the (Euclidean) norm* $\left\|\varepsilon^{(\mathbf{U}\rightsquigarrow\mathbf{V})}\right\| := \left(\sum_{j=1}^{\mathfrak{m}} \sum_{\lambda=1}^{\ell_{\mathbf{V}}} \left|\varepsilon^{(\mathbf{U}\rightsquigarrow\mathbf{V})}_{\mathscr{M}_{\mathbf{U}}+\boldsymbol{\lambda},\mathbf{j}}\right|^2\right)^{\mathbf{1/2}}$ *of the initial overall vulnerability in the node* $\mathbf{V}$ *as evaluated from the viewpoint of the user(s) of* $\mathbf{U}$ *at the preceding moment* $\mathbf{t}$*:*

$$\left\| \varepsilon'^{\left(\mathbf{U}'\rightsquigarrow\mathbf{V}'\right)} \right\| > \left\|\varepsilon^{(\mathbf{U}\rightsquigarrow\mathbf{V})}\right\|.$$

iii. *The (Euclidean) norm*

$$\left\| \widehat{\boldsymbol{\beta}'}^{\left(\mathbf{U}'\rightsquigarrow\mathbf{U}'\right)} \right\| := \left(\sum_{j=1}^{\mathfrak{n}} \left\{ \sum_{\lambda=1}^{m_{\mathbf{U}}} \left|\widehat{\boldsymbol{\beta}'}^{(\mathbf{U}\rightsquigarrow\mathbf{U})}_{\boldsymbol{\lambda},\mathbf{j}}\right|^{\mathbf{2}} + \sum_{\lambda=1}^{\ell_{\mathbf{U}}+\mathbf{N}} \left|\widehat{\boldsymbol{\beta}'}^{(\mathbf{U}\rightsquigarrow\mathbf{U})}_{\mathscr{M}_{\mathbf{U}}+\boldsymbol{\lambda},\mathbf{j}}\right|^{\mathbf{2}} \right\}\right)^{1/2}$$

of the resulting overall valuation in the variant node $\mathbf{U}'$ *as evaluated from the viewpoint of the user(s) of* $\mathbf{U}$ *at the next moment* $\mathbf{t}'$ *is greater than the (Euclidean) norms* $\left\|\widehat{\boldsymbol{\beta}}^{(\mathbf{U}\rightsquigarrow\mathbf{U})}\right\|:=\left(\sum_{j=1}^{n}\left\{\sum_{\lambda=1}^{m_{\mathbf{U}}}\left|\widehat{\boldsymbol{\beta}}_{\boldsymbol{\lambda},\mathbf{j}}^{(\mathbf{U}\rightsquigarrow\mathbf{U})}\right|^{2}+\sum_{\lambda=1}^{\ell_{\mathbf{U}}}\left|\widehat{\boldsymbol{\beta}}_{\mathscr{M}_{\mathbf{U}}+\boldsymbol{\lambda},\mathbf{j}}^{(\mathbf{U}\rightsquigarrow\mathbf{U})}\right|^{2}\right\}\right)^{1/2}$ *and*

$$\left\|\boldsymbol{\beta}^{(\mathbf{U}\rightsquigarrow\mathbf{V})}\right\|:=\left(\sum_{j=1}^{n}\left\{\sum_{\lambda=1}^{m_{\mathbf{V}}}\left|\boldsymbol{\beta}_{\boldsymbol{\lambda},\mathbf{j}}^{(\mathbf{U}\rightsquigarrow\mathbf{V})}\right|^{2}+\sum_{\lambda=1}^{\ell_{\mathbf{V}}}\left|\boldsymbol{\beta}_{\mathscr{M}_{\mathbf{V}}+\boldsymbol{\lambda},\mathbf{j}}^{(\mathbf{U}\rightsquigarrow\mathbf{V})}\right|^{2}\right\}\right)^{1/2}$$

of the initial overall valuations in the nodes $\mathbf{U}$ *and* $\mathbf{V}$ *as evaluated from the viewpoint of the user(s) of* $\mathbf{U}$ *at the preceding moment* $\mathbf{t}$*:*

$$\left\|\widehat{\boldsymbol{\beta}'}^{\left(\mathbf{U}'\rightsquigarrow\mathbf{U}'\right)}\right\|>max\left\{\left\|\widehat{\boldsymbol{\beta}}^{(\mathbf{U}\rightsquigarrow\mathbf{U})}\right\|,\left\|\boldsymbol{\beta}^{(\mathbf{U}\rightsquigarrow\mathbf{V})}\right\|\right\}.$$

iv. *The (Euclidean) norm* $\left\|\widehat{\varepsilon'}^{\left(\mathbf{U}'\rightsquigarrow\mathbf{U}'\right)}\right\|:=\left(\sum_{j=1}^{\mathrm{m}}\sum_{\lambda=1}^{\ell_{\mathbf{U}}+\nu}\left|\widehat{\varepsilon'}_{\mathscr{M}_{\mathbf{U}}+\boldsymbol{\lambda},\mathbf{j}}^{(\mathbf{U}\rightsquigarrow\mathbf{U})}\right|^{2}\right)^{1/2}$ *of the resulting overall vulnerability in the variant node* $\mathbf{U}'$ *as evaluated from the viewpoint of the user(s) of* $\mathbf{U}$ *at the next moment* $\mathbf{t}'$ *is less or equal than the (Euclidean) norms*

$$\left\|\widehat{\varepsilon}^{(\mathbf{U}\rightsquigarrow\mathbf{U})}\right\|:=\left(\sum_{j=1}^{\mathrm{m}}\sum_{\lambda=1}^{\ell_{\mathbf{U}}}\left|\varepsilon_{\mathscr{M}_{\mathbf{U}}+\boldsymbol{\lambda},\mathbf{j}}^{(\mathbf{U}\rightsquigarrow\mathbf{U})}\right|^{2}\right)^{1/2}$$

and

$$\left\|\varepsilon^{(\mathbf{U}\rightsquigarrow\mathbf{V})}\right\|:=\left(\sum_{j=1}^{\mathrm{m}}\sum_{\lambda=1}^{\ell_{\mathbf{V}}}\left|\varepsilon_{\mathscr{M}_{\mathbf{U}}+\boldsymbol{\lambda},\mathbf{j}}^{(\mathbf{U}\rightsquigarrow\mathbf{V})}\right|^{2}\right)^{1/2}$$

of the initial overall vulnerabilities in the nodes $\mathbf{U}$ *and* $\mathbf{V}$ *as evaluated from the viewpoint of the user(s) of* $\mathbf{U}$ *at the preceding moment* $\mathbf{t}$*:*

$$\left\|\widehat{\varepsilon'}^{\left(\mathbf{U}'\rightsquigarrow\mathbf{U}'\right)}\right\|=\min\left\{\left\|\widehat{\varepsilon}^{(\mathbf{U}\rightsquigarrow\mathbf{U})}\right\|,\left\|\varepsilon^{(\mathbf{U}\rightsquigarrow\mathbf{V})}\right\|\right\}.$$

The **degree** $\mathbf{d}=\mathbf{d}_{\{\mu_1,\ldots,\mu_\nu\}\bigcup\{\kappa_1,\ldots,\kappa_\lambda\}}$ **of the active attack f against the** $(\mu_1,\ldots,\mu_\nu)$**-device parts**

$$\mathbf{fr}\left(\mathbf{dev}_{\mu_1}^{(\mathbf{V})}\right),\mathbf{fr}\left(\mathbf{dev}_{\mu_2}^{(\mathbf{V})}\right),\ldots,\ \mathbf{fr}\left(\mathbf{dev}_{\mu_\nu}^{(\mathbf{V})}\right)$$

of $\mathbf{V}$ and the $(\kappa_1, \ldots, \kappa_\lambda)$-**resource parts** $\mathbf{fr}\left(\mathbf{res}_{\kappa_1}^{(\mathbf{V})}\right), \mathbf{fr}\left(\mathbf{res}_{\kappa_2}^{(\mathbf{V})}\right), \ldots, \mathbf{fr}\left(\mathbf{res}_{\kappa_\lambda}^{(\mathbf{V})}\right)$ of $\mathbf{V}$ from the offensive node $\mathbf{U}$ at time moment $\mathbf{t} \in \mathbb{I}$ is defined to be the maximum of the two quotients

$\mathbf{d_1}{:=}\left\|\widehat{\beta'}^{(\mathbf{U}' \rightsquigarrow \mathbf{U}')}\right\| / \left\|\beta'^{(\mathbf{U}' \rightsquigarrow \mathbf{V}')}\right\|$ and $\mathbf{d_2}{:=}\left(\left\|\widehat{\gamma'}^{(\mathbf{U}' \rightsquigarrow \mathbf{U}')}\right\| / \left\|\gamma'^{(\mathbf{U}' \rightsquigarrow \mathbf{V}')}\right\|\right)^{-1}$.

Thus

$$\mathbf{d}{=}\mathbf{d}_{\{\mu_1,\ldots,\mu_\nu\}\bigcup\{\kappa_1,\ldots,\kappa_\lambda\}}{:=}\mathbf{max}\,\{\mathbf{d_1}, \mathbf{d_2}\}\,.$$

If the degree $\mathbf{d}$ surpasses a given threshold $\mathscr{T}^{(\mathbf{U},\mathbf{V})}_{\{\mu_1,\ldots,\mu_\nu\}\bigcup\{\kappa_1,\ldots,\kappa_\lambda\}} \in [\mathbf{0}, \infty[$, called **threshold of active attack** from $\mathbf{U}$ against the $(\mu_1, \ldots, \mu_\nu)$-device parts

$$\mathbf{fr}\left(\mathbf{dev}_{\mu_1}^{(\mathbf{V})}\right), \ldots, \mathbf{fr}\left(\mathbf{dev}_{\mu_\nu}^{(\mathbf{V})}\right)$$

of $\mathbf{V}$ and the $(\kappa_1, \ldots, \kappa_\lambda)$-resource parts

$$\mathbf{fr}\left(\mathbf{res}_{\kappa_1}^{(\mathbf{V})}\right), \ldots, \mathbf{fr}\left(\mathbf{res}_{\kappa_\lambda}^{(\mathbf{V})}\right)$$

of $\mathbf{V}$ at time moment $\mathbf{t} \in \mathbb{I}$, we say that **the passive attack f is dangerous with degree of danger d in the** $(\mu_1, \ldots, \mu_\nu)$**-device parts** $\mathbf{fr}\left(\mathbf{dev}_{\mu_1}^{(\mathbf{V})}\right), \ldots, \mathbf{fr}\left(\mathbf{dev}_{\mu_\nu}^{(\mathbf{V})}\right)$ **of V and the** $(\kappa_1, \ldots, \kappa_\lambda)$**-resource parts** $\mathbf{fr}\left(\mathbf{res}_{\kappa_1}^{(\mathbf{V})}\right), \mathbf{fr}\left(\mathbf{res}_{\kappa_2}^{(\mathbf{V})}\right), \ldots, \mathbf{fr}\left(\mathbf{res}_{\kappa_\lambda}^{(\mathbf{V})}\right)$ **of V**.

Remark 1 It is easy to verify that the following conditions 1–4 can be considered as stronger forms of the corresponding conditions in Proposition 4.

i. *1st Condition*: From the point of view of users of nodes $\mathbf{U}$ and $\mathbf{V}$, every attacked device part, as well as any attacked resource part, acquires new valuation measures that are smaller than the original corresponding valuations in node $\mathbf{V}$, with (at least) one such a valuation measure very reduced, i.e., for any $\mathbf{j} \in \{\mu_1, \ldots, \nu\}\bigcup\{\kappa_1, \ldots, \kappa_\lambda\}$, it holds

$$\sum_{\mathbf{k=1}}^{\mathrm{n}} \left|\beta_{\mathbf{j,k}}^{(\mathbf{X} \rightsquigarrow \mathbf{V})} + \mathbf{i}\ \widehat{\beta}_{\mathbf{j,k}}^{(\mathbf{X} \rightsquigarrow \mathbf{V})}\right|^2 > \sum_{\mathbf{k=1}}^{\mathrm{n}} \left|\beta'^{(\mathbf{X} \rightsquigarrow \mathbf{V})}_{\mathbf{j,k}} + \mathbf{i}\ \widehat{\beta'}^{(\mathbf{X} \rightsquigarrow \mathbf{V})}_{\mathbf{j,k}}\right|^2$$

with at least one index $\mathbf{k} \in \{\mathbf{1}, \mathbf{2}, \ldots, \mathrm{n}\}$ being such that

$$\left|\beta_{\mathbf{j,k}}^{(\mathbf{X} \rightsquigarrow \mathbf{V})} + \mathbf{i}\ \widehat{\beta}_{\mathbf{j,k}}^{(\mathbf{X} \rightsquigarrow \mathbf{V})}\right| \gg \left|\beta'^{(\mathbf{X} \rightsquigarrow \mathbf{V})}_{\mathbf{j,k}} + \mathbf{i}\ \widehat{\beta'}^{(\mathbf{X} \rightsquigarrow \mathbf{V})}_{\mathbf{j,k}}\right|$$

whenever $\mathbf{X}{=}\mathbf{V}, \mathbf{U}$.

ii. *2nd Condition*: Similarly, from the point of view of users of nodes $\mathbf{U}$ and $\mathbf{V}$, every attacked device part, as well as any attacked resource part, acquires new vulnerability measures that are smaller than the original corresponding vulnerabilities in node $\mathbf{V}$, with (at least) one such a vulnerability measure very reduced, i.e., for any $\mathbf{j} \in \{\mu_1, \ldots, \mu_\nu\}\bigcup\{\kappa_1, \ldots, \kappa_\lambda\}$, it holds

$$\sum_{k=1}^{\mathfrak{m}} \left| \varepsilon_{\mathbf{j,k}}^{(\mathbf{X}\rightsquigarrow\mathbf{V})} + \mathbf{i}\ \widehat{\varepsilon}_{\mathbf{j,k}}^{(\mathbf{X}\rightsquigarrow\mathbf{V})} \right|^2 = \sum_{k=1}^{\mathfrak{m}} \left| \varepsilon_{\mathbf{j,k}}^{(\mathbf{X}\rightsquigarrow\mathbf{V})} + \mathbf{i}\ \widehat{\varepsilon}_{\mathbf{j,k}}^{(\mathbf{X}\rightsquigarrow\mathbf{V})} \right|^2$$

with at least one index $\mathbf{k} \in \{\mathbf{1}, \mathbf{2}, \ldots, \mathfrak{m}\}$ being such that

$$\left| \varepsilon_{\mathbf{j,k}}^{(\mathbf{X}\rightsquigarrow\mathbf{V})} + \mathbf{i}\ \widehat{\varepsilon}_{\mathbf{j,k}}^{(\mathbf{X}\rightsquigarrow\mathbf{V})} \right| < \left| \varepsilon'^{(\mathbf{X}\rightsquigarrow\mathbf{V})}_{\mathbf{j,k}} + \mathbf{i}\ \widehat{\varepsilon'}^{(\mathbf{X}\rightsquigarrow\mathbf{V})}_{\mathbf{j,k}} \right|$$

whenever $\mathbf{X}=\mathbf{V}, \mathbf{U}$.

iii. *3rd Condition*: From the viewpoint of the (user(s) of) node **U**, in the offensive node **U** there are strongly growing valuations, i.e., there are

$$\mathbf{j} \in \{\mathbf{1}, \mathbf{2}, \ldots, m_{\mathrm{U}}, \mathscr{M}_{\mathrm{U}}+\mathbf{1}, \ldots, \mathscr{M}_{\mathrm{U}}+\ell_{\mathrm{U}}\}$$

and $\mathbf{k} \in \{\mathbf{1}, \mathbf{2}, \ldots, \mathfrak{n}\}$, such that

$$\left| \widehat{\boldsymbol{\beta}}_{\mathbf{j,k}}^{(\mathbf{U}\rightsquigarrow\mathbf{U})} \right| \ll \left| \widehat{\boldsymbol{\beta}'}^{(\mathbf{U}\rightsquigarrow\mathbf{U})}_{\mathbf{j,k}} \right|.$$

iv. *4th Condition*: From the viewpoint of the (user(s) of) node **U**, in the offensive node **U** there is no growing vulnerability, i.e., for any

$$\mathbf{j} \in \{\mathbf{1}, \mathbf{2}, \ldots, m_{\mathrm{U}}, \mathscr{M}_{\mathrm{U}}+\mathbf{1}, \ldots, \mathscr{M}_{\mathrm{U}}+\ell_{\mathrm{U}}\}$$

and any

$$\mathbf{k} \in \{\mathbf{1}, \mathbf{2}, \ldots, \mathfrak{m}\},$$

it holds

$$\left| \widehat{\varepsilon}_{\mathbf{j,k}}^{(\mathbf{U}\rightsquigarrow\mathbf{U})} \right| = \left| \widehat{\varepsilon'}^{(\mathbf{U}\rightsquigarrow\mathbf{U})}_{\mathbf{j,k}} \right|.$$

Protected Cyber Nodes from Active Attacks

Finally, let's see how we could define the concept of protection from active cyber-attacks.

Definition 18

i. The node **V** is said to be **protected from active attacks, with degree of protection** $\mathbf{p} \in]\mathbf{0}, \mathbf{1}]$ over the $(\mu_1, \ldots, \mu_\nu)$-device parts

$$\mathbf{fr}\left(\mathbf{dev}_{\mu_1}^{(\mathbf{V})}\right), \ldots, \mathbf{fr}\left(\mathbf{dev}_{\mu_\nu}^{(\mathbf{V})}\right)$$

of **V** and the $(\kappa_1, \ldots, \kappa_\lambda)$-resource parts $\mathbf{fr}\left(\mathbf{res}_{\kappa_1}^{(\mathbf{V})}\right), \ldots, \mathbf{fr}\left(\mathbf{res}_{\kappa_\lambda}^{(\mathbf{V})}\right)$ of **V** over a time period $\mathbb{I}$, if, during this time period, there is a nodal fixed filter system $\overline{\mathbf{V}}^{\{\mu_1,\ldots,\mu_\nu\}\bigcup\{\kappa_1,\ldots,\kappa_\lambda\}}$ in the union

$$\mathbf{E}{=}\mathbf{fr}\left(\mathbf{dev}_{\mu_1}^{(\mathbf{V})}\right)\bigcup\cdots\bigcup\mathbf{fr}\left(\mathbf{dev}_{\mu_\nu}^{(\mathbf{V})}\right)\bigcup\mathbf{fr}\left(\mathbf{res}_{\kappa_1}^{(\mathbf{V})}\right)\bigcup\ldots\bigcup\mathbf{fr}\left(\mathbf{res}_{\kappa_\nu}^{(\mathbf{V})}\right)$$

that allow every parallactic cyber active attack against the $(\mu_1, \ldots, \mu_\nu)$-device parts

$$\mathbf{fr}\left(\mathbf{dev}_{\mu_1}^{(\mathbf{V})}\right), \ldots, \mathbf{fr}\left(\mathbf{dev}_{\mu_\nu}^{(\mathbf{V})}\right)$$

of **V** and the $(\kappa_1, \ldots, \kappa_\lambda)$-resource parts $\mathbf{fr}\left(\mathbf{res}_{\kappa_1}^{(\mathbf{V})}\right), \ldots, \mathbf{fr}\left(\mathbf{res}_{\kappa_\lambda}^{(\mathbf{V})}\right)$ of node **V** (from any offensive node **U**) with degree of danger

$$\mathbf{d} = -\mathbf{log}\,\mathbf{p}$$

to reach only resource parts **K** of the initial target **V** that are disjoint from **E**.

ii. During the time period $\mathbb{I}$, the node **V** is said to be **completely protected from active attacks of danger degree d**, if, at this time period, any resource part in **V** is protected from active attacks against **V**, with degree of protection

$$\mathbf{p} = \mathbf{e}^{-\mathbf{d}}.$$

The node **V** is said to be **completely protected from active attacks at a given time period**, if, during this time period, any resource part of **V** is protected from active attacks against **V** with degree of protection

$$\mathbf{p}{=}\mathbf{1}.$$

References

1. Adams, N., Heard, N. (eds.): Data Analysis for Network Cyber-Security, 200 pp. World Scientific, Singapore (2014). ISBN: 978-1-78326-374-5 (hardcover)/ISBN: 978-1-78326-376-9 (ebook)
2. Alexopoulos, A., Daras, N.: Case study of proactive defense against various types of cyber attacks, submitted
3. Colbaugh, R., Glass, K.: Proactive Cyber Defense. Springer Integrated Series on Intelligent Systems. Springer, Berlin (2012). Document No. 5299122, SAND 2011-8794P
4. Daras, N.J.: On the mathematical definition of cyberspace. 4th International Conference on Operational Planning, Technological Innovations and Mathematical Applications (OPTIMA), Hellenic Army Academy, Vari Attikis, Greece (2017)

5. Daras, N., Alexopoulos, A.: Mathematical description of cyber-attacks and proactive defenses. Journal of Applied Mathematics & Bioinformatics, **7**(1), 71–142 (2017)
6. Dunlavy, D.M., Hendrickson, B., Kolda, T.G.: Mathematical challenges in cyber-security. Sandia National Laboratories (2009). www.sandia.gov/~dmdunla/publications/SAND2009-0805.pdf
7. Luccio, F., Pagli, L., Stee, G.: Mathematical and Algorithmic Foundations of the Internet, pp. 221. CRC, Boca Raton (2011)
8. Malkevitch, J.: Mathematics and internet security (2015). http://www.ams.org/samplings/feature-column/fcarc-internet
9. Meza, J., Campbell, S., Bailey, D.: Mathematical and statistical opportunities in cyber security (2009). http://arxiv.org/pdf/0904.1616v1.pdf. http://www.davidhbailey.com/dhbpapers/CyberMath.pdf
10. Ramsey, D.: Mathematical problems in cyber security. In: DOE Workshop on Mathematical Research Challenges in Optimization of Complex Systems, 7–8 December 2006. http://www.courant.nyu.edu/ComplexSystems/
11. Rowe, N.C., Custy E.J., Duong B.T.: Defending cyberspace with fake honeypots. J. Comput. **2**(2), 25–36 (2007). http://www.academypublisher.com/jcp/vol02/no02/jcp02022536.pdf
12. Ryan, P.Y.A.: Mathematical models of computer security. In: Focardi, R., Gorrieri, R. (eds.) Foundations of Security Analysis and Design: Tutorial Lectures (FOSAD 2000). Lecture Notes in Computer Science, vol. 2171, pp. 1–62. Springer, Berlin, Heidelberg (2001). http://www.ece.cmu.edu/~ece732/readings/mathematical-models-of-computer-security.pdf
13. Saini, D.K.: Cyber defense: mathematical modeling and simulation. Int. J. Appl. Phys. Math. **2**(5), 312–315 (2012)
14. Shen, D., Chen, G., Cruz, J.B. Jr., Blasch, E., Pham, K.: An adaptive Markov game model for cyber threat intent inference. In: Theory and Novel Applications of Machine Learning, pp. 376. I-Tech, Vienna (2009). ISBN 978-3-902613-55-4. http://cdn.intechweb.org/pdfs/6200.pdf
15. Subil, A., Suku, N.: Cyber security analytics: a stochastic model for security quantification using absorbing Markov chains. J. Commun. **9**(12), 899–907 (2014)
16. Wendelberger, J., Griffin, C., Wilder, L., Jiao, Y., Kolda T.: A mathematical basis for science-based cyber-security (with alternate title "Mathematical Underpinnings for Science-Based Cyber-Security"), February 2008, Sandia National Laboratories, Chad Scherrer, Pacific Northwest National Laboratory. https://wiki.cac.washington.edu/download/attachments/7479040/doecybermath25feb08.doc

Solutions of Hard Knapsack Problems Using Extreme Pruning

E. Daravigkas, K. A. Draziotis, and A. Papadopoulou

Abstract In the present study we provide a review for the state-of-the-art attacks to the knapsack problem. We implemented the Schnorr-Shevchenko lattice attack, and we applied the new reduction strategy, BKZ 2.0. Finally, we compared the two implementations.

Keywords Knapsack problem · Subset sum problem · Lattice · LLL reduction · BKZ reduction · Extreme pruning

Introduction

In this work we address the class of knapsack problems with density close to 1. This problem except its cryptographic applications [18] is also very interesting in computational number theory. The last years, there was a better understanding of lattice reductions, and new methods were discovered [3]. This had a dramatic effect in the cryptanalysis of lattice-based cryptosystems, changing the security parameters to some of them [23]. We shall apply these new implementations to a lattice attack presented in [22]. As far as we know, these are the first experiments in this direction.

We start by describing the problem. Given a list of n positive integers $\{a_1, \ldots, a_n\}$ and an integer s such that

$$\max\{a_i\}_i \leq s \leq \sum_{i=1}^{n} a_i,$$

E. Daravigkas · K. A. Draziotis (✉)
Department of Informatics, Aristotle University of Thessaloniki, Thessaloniki, Greece
e-mail: drazioti@csd.auth.gr

A. Papadopoulou
Department of Mathematics, Aristotle University of Thessaloniki, Thessaloniki, Greece

N. J. Daras, Th. M. Rassias (eds.), *Modern Discrete Mathematics and Analysis*,
Springer Optimization and Its Applications 131,
https://doi.org/10.1007/978-3-319-74325-7_4

find a binary vector

$$\mathbf{x} = (x_i)_i \text{ with, } \sum_{i=1}^{n} x_i a_i = s. \tag{1}$$

We define the density of the knapsack (or subset sum problem) to be

$$d = \frac{n}{\log_2 \max_i \{a_i\}_i}.$$

The decision version of the problem is known to be NP-complete [10]. In cryptography we are interested in values of d less than one. Since if $d > 1$, then there are many solutions of the knapsack problem, and so we cannot transmit efficient information. Also, because of the low-density attacks [2, 13], we consider the density close to 1.

Roadmap In section "Lattices" we provide some basic facts about lattices and the LLL-reduction. In the next section, we describe the possible attacks to the knapsack problem, and we present the BKZ-reduction which is a generalization of LLL reduction. Finally, in section "Attacks to Subset Sum Problem," we present our experiments. In section "Experiments" we provide some concluding remarks.

Lattices

For an introduction to lattices see [6, 12, 15].

Definition 1 A lattice $\mathcal{L}$ is a discrete subgroup of $\mathbb{R}^m$.

So a lattice is a subgroup of the Abelian group $\langle \mathbb{R}^m, + \rangle$ thus $\mathbf{0} \in \mathcal{L}$, and is discrete as a subset of $\mathbb{R}^m$. That is, $\mathbf{0}$ is not a limit point. This means that the intersection of $\mathcal{L}$ and any bounded set of $\mathbb{R}^m$ is a finite set of points. For instance, $\{\mathbf{0}\}$ and every subgroup of the additive group $\langle \mathbb{Z}^m, + \rangle$ are lattices of $\mathbb{R}^m$. A set of vectors $\mathcal{B} = \{\mathbf{b}_1, \ldots, \mathbf{b}_n\} \subset \mathbb{R}^m$ (where $n \leq m$) is called generator of $\mathcal{L}$ if each vector of $\mathcal{L}$ is a linear combination of vectors in $\mathcal{B}$ with integer coefficients. If the vectors of this set are also independent, we call $\mathcal{B}$ a basis of the lattice. In this case, the number n is the same for all bases, and we call it *rank* of the lattice, i.e., $n = \dim(span(\{\mathbf{b}_1, \ldots, \mathbf{b}_n\}))$. The integer m is called *dimension* of the lattice. If $n = m$, then we call the lattice, *full-rank lattice*. Some authors call the rank as the dimension of the lattice. Usually we consider lattices which contain vectors only with integer coordinates. Such lattices are called *integer lattices*. A basis $\mathcal{B}$ is given by an $n \times m$ matrix, with rows the vectors of a basis of a lattice. If we have two distinct basis-matrices B and B', then there is a unimodular integer (square) matrix U with dimension n, such that $B = UB'$. It turns out that the Gram determinant $\det(BB^T)$ does not depend on the basis B or B'. This number is always positive, and its square root is called determinant or volume of the lattice $\mathcal{L}$.

Definition 2 We call determinant or volume of the lattice $\mathcal{L}$ the positive real number

$$\det(\mathcal{L}) = vol(\mathcal{L}) = \sqrt{\det(BB^T)}.$$

The volume is in fact the (geometric) volume of the parallelepiped spanned by any basis of the lattice.

If $\mathcal{L}$ is a full-rank lattice, then $vol(\mathcal{L}) = \det(B)$, where B is the square matrix representing a basis of $\mathcal{L}$. Having a matrix B, we can write the lattice generated by the basis $\mathcal{B}$ as:

$$\mathcal{L}(\mathcal{B}) = \Big\{\sum_{j=1}^{n} k_j \mathbf{b}_j : k_j \in \mathbb{Z}\Big\} \text{ or } \{\mathbf{x}B : \mathbf{x} \in \mathbb{Z}^n\}.$$

Since a lattice is a discrete subset of $\mathbb{R}^m$, then there is a nonzero vector $\mathbf{x} \in \mathcal{L}$ with the lowest Euclidean norm. We denote this norm by $\lambda_1(\mathcal{L})$, and we call it *first successive minima*. In general, not only one vector $\mathbf{v} \in \mathcal{L} - \{\mathbf{0}\}$ has $||\mathbf{v}|| = \lambda_1$. The number $|\{\mathbf{v} \in \mathcal{L} - \{\mathbf{0}\} : ||\mathbf{v}|| = \lambda_1\}|$ is called *kissing number* of the lattice. This number is at least 2 since, also $-\mathbf{v}$ has length λ_1. Further, Minkowski proved that $\lambda_1(L) \leq \sqrt{n} vol(L)^{1/n}$.

We set $\overline{B}_m(r) = \{\mathbf{x} \in \mathbb{R}^m : ||\mathbf{x}|| \leq r\}$. Then, $\lambda_1(\mathcal{L})$ is the smallest real number r such that, the real vector space $M_r(\mathcal{L}) = span(\mathcal{L} \cap \overline{B}_m(r))$, has at least one independent vector. Equivalently $\dim(M_r(\mathcal{L})) \geq 1$. In general we define $\lambda_j(\mathcal{L})$ to be the positive number $\inf\{r : \dim(M_r(\mathcal{L})) \geq j\}$. Always we have

$$\lambda_1(\mathcal{L}) \leq \lambda_2(\mathcal{L}) \leq \cdots \leq \lambda_n(\mathcal{L}).$$

An upper bound for the λ_i's was given by Minkowski.

Proposition 1 (Second Theorem of Minkowski) *For every integer $n > 1$, there exists a positive constant γ_n such that for any lattice with rank n and every $j \in \{1, 2, .., n\}$, we have the following inequality*

$$\Big(\prod_{i=1}^{n} \lambda_i\Big)^{1/n} \leq \sqrt{\gamma_n} vol(\mathcal{L})^{1/n}.$$

The problem of finding a shortest vector in a lattice is called shortest vector problem (SVP) and is one of the most important and difficult problems in lattices. Since this problem is hard (is proved that is NP-hard under randomized reductions by Ajtai [1]), we try to attack the approximation problem. For instance, we are looking for lattice vectors $\mathbf{v}$ such that $||\mathbf{v}|| < \gamma(n)\lambda_1(\mathcal{L})$, where $n = rank(\mathcal{L})$. In fact LLL, as we shall see in the next section, solves SVP for approximating factors $\gamma(n) = 2^{(n-1)/2}$.

Since lattice bases always exist, we are interested in bases with short and nearly orthogonal vectors. The procedure of finding such a basis (given one) is called lattice reduction.

Lenstra-Lenstra-Lovász Algorithm (LLL)

Before we discuss LLL algorithm which is a reduction algorithm for lattices in $\mathbb{R}^m$, we shall remind the reader the Gram-Schmidt procedure, which is a basic subroutine in the LLL algorithm.

Gram-Schmidt Orthogonalization (GSO)

Definition 3 Let $\mathcal{B} = \{\mathbf{b}_1, \ldots, \mathbf{b}_n\}$ be an independent set of vectors in $\mathbb{R}^m$. Then, their GSO is a set of orthogonal vectors $\{\mathbf{b}_1^*, \ldots, \mathbf{b}_n^*\}$ defined by the following relations:

$$\mathbf{b}_1^* = \mathbf{b}_1,\ \mathbf{b}_i^* = proj_{M_{i-1}}(\mathbf{b}_i),\ 2 \le i \le n,$$

where $M_{i-1} = span(\{\mathbf{b}_1, \ldots, \mathbf{b}_{i-1}\})^{\perp}$, is the orthogonal span of $\{\mathbf{b}_1, \ldots, \mathbf{b}_{i-1}\}$.

We rewrite the previous relations as

$$\mathbf{b}_1 = \mathbf{b}_1^*,\ \mathbf{b}_i^* = \mathbf{b}_i - \sum_{j=1}^{i-1} \mu_{i,j}\mathbf{b}_j^*,\ \mu_{i,j} = \frac{\mathbf{b}_i \cdot \mathbf{b}_j^*}{||\mathbf{b}_j^*||^2},\ 2 \le j < i \le n.$$

If we consider the orthonormal basis of $span(\mathcal{B})$,

$$\mathcal{B}' = \left\{\mathbf{b}_1' = \frac{\mathbf{b}_1^*}{||\mathbf{b}_1^*||}, \mathbf{b}_2' = \frac{\mathbf{b}_2^*}{||\mathbf{b}_2^*||}, \ldots, \mathbf{b}_n' = \frac{\mathbf{b}_n^*}{||\mathbf{b}_n^*||}\right\},$$

then the transition matrix $T = T_{\mathcal{B}' \leftarrow \mathcal{B}}$ is the upper triangular matrix $(n \times n)$

$$T = \begin{pmatrix} ||\mathbf{b}_1^*|| & \mu_{2,1}||\mathbf{b}_1^*|| & \cdots & \mu_{n,1}||\mathbf{b}_1^*|| \\ 0 & ||\mathbf{b}_2^*|| & \cdots & \mu_{n,2}||\mathbf{b}_2^*|| \\ \vdots & \vdots & \ddots & \vdots \\ 0 & 0 & \cdots & ||\mathbf{b}_n^*|| \end{pmatrix} \tag{2}$$

and $(\mathbf{b}_1, \ldots, \mathbf{b}_n) = (\mathbf{b}_1', \ldots, \mathbf{b}_n')T$. If we consider the $n \times n$ lower-triangular matrix

$$\mu = \begin{pmatrix} 1 & 0 & \cdots & 0 \\ \mu_{2,1} & 1 & \cdots & 0 \\ \vdots & \vdots & \ddots & \vdots \\ \mu_{n,1} & \mu_{n,2} & \cdots & 1 \end{pmatrix}$$

then $(\mathbf{b}_1, \ldots, \mathbf{b}_n)^T = \mu(\mathbf{b}_1^*, \ldots, \mathbf{b}_n^*)^T$. So the GSO of $\mathcal{B}$ is given by the vectors $(||\mathbf{b}_i^*||)_i$ and the triangular matrix μ.

GSO Pseudocode :

INPUT: A basis $\mathcal{B} = \{\mathbf{b}_1, \mathbf{b}_2, \ldots, \mathbf{b}_n\}$ of $V = span(\mathcal{L})$
OUTPUT: An orthogonal basis of V and the Gram-Schmidt matrix μ

01. $\mathbf{R}_1 \leftarrow \mathbf{b}_1$
02. **for** $i = 1$ to n
03. $\quad \mathbf{t}_1 \leftarrow \mathbf{b}_i$
04. $\quad \mathbf{t}_3 \leftarrow \mathbf{t}_1$
05. $\quad$ **for** $j = i - 1$ to 1
06. $\quad\quad \mu_{i,j} \leftarrow \frac{\mathbf{t}_1 \cdot \mathbf{R}_j}{||\mathbf{R}_j||^2}$
07. $\quad\quad \mathbf{t}_2 \leftarrow \mathbf{t}_3 - proj_{\mathbf{R}_j}(\mathbf{t}_1)$
08. $\quad\quad \mathbf{t}_3 \leftarrow \mathbf{t}_2$
09. $\quad$ **end for**
10. $\quad \mathbf{R}_j \leftarrow \mathbf{t}_2$
11. **end for**
12. Return $R = (\mathbf{R}_1, \ldots, \mathbf{R}_n)$, $\mu = ([\mu_{i,j}])_{1 \leq i,j \leq n}$

The running time of GSO is polynomial with respect to $\max\{n, \log \max_i ||\mathbf{b}_i||\}$.

LLL-Pseudocode

Definition 4 A lattice basis $\mathcal{B} = \{\mathbf{b}_1, \ldots, \mathbf{b}_n\}$ is called *size-reduced* if and only if

$$|\mu_{i,j}| = \frac{|\mathbf{b}_i \cdot \mathbf{b}_j^*|}{||\mathbf{b}_j^*||^2} \leq \frac{1}{2}, \; 1 \leq j < i \leq n.$$

$\mathcal{B}$ is called $\delta-$LLL reduced if in advance

$$\text{(Lovász condition) } \delta||\mathbf{b}_{i-1}^*||^2 \leq ||\mu_{i,i-1}\mathbf{b}_{i-1}^* + \mathbf{b}_i^*||^2, \; 2 \leq i \leq n.$$

In LLL algorithm we first apply size reduction. That is, we consider vectors of the lattice near the Gram-Schmidt basis. This property is introduced by Lagrange. Then, we continue by checking if Lovász condition does not hold, in two successive vectors, and in this case we swap the vectors.

LLL Pseudocode

INPUT: A basis $\mathcal{B} = \{\mathbf{b}_1, \ldots, \mathbf{b}_n\} \subset \mathbb{Z}^m$
of the lattice $\mathcal{L}(\mathcal{B})$ and a real number $\delta \in (1/4, 1)$
We use the function $gso = GSO[\mathcal{B}]$ which returns the orthogonal basis gso[0] and the Gram-Schmidt matrix gso[1]

```
OUTPUT: δ-LLL-reduced basis for 𝓛

--   Initialization:
01. i = 2
02. gso ← GSO(𝓑)
03. (b_i)_i ← gso[1]
04. (μ_{i,j})_{i,j} ← gso[2]
--   Size reduction step:
05. While i ≤ n do
06.    for j = i − 1 to 1 do
07.       c_{i,j} ← ⌊μ_{i,j}⌉         #⌊x⌉ = ⌊x + 0.5⌋
08.       b_i ← b_i − c_{i,j} b_j
09.       update 𝓑 and gso ← GSO(𝓑)
10.       (b_i)_i ← gso[1]
11.       (μ_{i,j})_{i,j} ← gso[2]
12.    end for
--    Swap step:
13.    if δ||b*_i||^2 > ||μ_{i+1,i} b*_i + b*_{i+1}||^2 then
14.         b_i ↔ b_{i+1}
15.        i = max(2, i − 1)
16.        update 𝓑 and gso ← GSO(𝓑)
17.        (b_i)_i ← gso[1]
18.        (μ_{i,j})_{i,j} ← gso[2]
19.    else
20.        i ← i + 1
21.    end if
22.   end while
23. Return 𝓑
```

There are two basic efficient (and stable) implementations, using floating-point arithmetic of suitable precision: NTL [23] and fpLLL [25]. A version of LLL using floating-point arithmetic was first presented by Schnorr in [20] and later by Nguyen and Stehlé [17].

Analysis of LLL

LLL-reduction algorithm applies two basic operations to the initial basis. Size reduction and swap of two vectors, if Lovász condition fails. After the first operation, lines 06–12, we get $|\mu_{i,j}| < 1/2$. Then we check the Lovász condition, lines 13–20. Accordingly if the previous check is true or false, we swap or not the vectors. This schematically can be explained in terms of matrices. We consider the 2×2 block of matrix 2, indexed at position (i, i). That is,

$$\begin{pmatrix} ||\mathbf{b}_i^*|| & \mu_{i+1,i}||\mathbf{b}_i^*|| \\ 0 & ||\mathbf{b}_{i+1}^*|| \end{pmatrix}.$$

Then Lovász condition holds, if the second column has length at least $\sqrt{\delta}$ times the length of the first column. The following theorem was proved in [14].

Theorem 1 *Let $\mathcal{L}(\mathcal{B})$ with $\mathcal{B} = \{\mathbf{b}_1, \ldots, \mathbf{b}_n\}$, $\mathbf{b}_i \in \mathbb{Z}^m$, $\Lambda = \max_i(||\mathbf{b}_i||^2)$ and $\delta = 3/4$. Then the previous algorithm terminates after $O(n^4 \log \Lambda)$ arithmetic operations, and the integers on which these operations are performed have bit length $O(n \log \Lambda)$.*

Proof Proposition 1.26 in [14]. □

The first proof about the complexity of LLL concerns only integer lattices and $\delta = 3/4$ later was proved for every $\delta \in (1/4, 1)$. A basic consequence is the following proposition.

Proposition 2 *If $\{\mathbf{v}_1, \ldots, \mathbf{v}_n\}$ is a δ-LLL-reduced basis (with $\delta = 3/4$) of the lattice $\mathcal{L}(\mathcal{B})$ with $\mathcal{B} = \{\mathbf{b}_1, \ldots, \mathbf{b}_n\}$, $\mathbf{b}_i \in \mathbb{R}^m$. Then,*

$$||\mathbf{v}_1|| \leq 2^{(n-1)/2}\lambda_1(\mathcal{L}).$$

Proof [6, Lemma 17.2.12] □

So, LLL provides a solution to the approximate SVP_γ, with factor $\gamma = 2^{(n-1)/2}$.

Blockwise Korkine-Zolotarev (BKZ)

In LLL algorithm we check Lovász condition in a matrix of blocksize 2. In BKZ algorithm Lovász condition is generalized to larger blocksizes. This has the positive effect of providing better quality of the output basis than in LLL but with a cost in the running time. BKZ does not run in polynomial time; in fact Gama and Nguyen [8] experimentally provide strong evidences that BKZ runs exponentially with the rank of the lattice. The running time of the reduction increases significantly with higher blocksizes. However, increasing blocksizes also means an improvement in the quality of the output basis, i.e., basis with smaller length. If the blocksize is equal to the rank of the lattice, then we get a Hermite-Korkine-Zolotarev (HKZ)-reduced basis.

BKZ is an approximation algorithm, which takes as input the blocksize β and a parameter δ (the same as in LLL), and then it calls an enumeration subroutine many times looking for a shortest vector in projected lattices of dimension $\leq \beta$. The cost of the enumeration subroutine is $2^{O(\beta^2)}$ operations [3].

We provide the description of BKZ and then its pseudocode. As we mentioned above, the input basis provides some local blocks with dimension β. The first block consists of the first basis vector and the $\beta - 1$ succeeding vectors. For the second block, we take the projections of the second basis vector and the succeeding vectors onto the orthogonal complement of the vector. The remaining blocks are constructed in the same way. The index l of the last vector in a block is computed as $l = \min(i + \beta - 1, n)$, where n is the rank of the lattice. We start with two definitions.

Definition 5 Let $\mathcal{L} = \mathcal{L}(\mathcal{B})$ be a lattice with basis $\mathcal{B} = \{\mathbf{b}_1, \ldots, \mathbf{b}_n\}$. We define,

$$\pi_i : \mathcal{L} \rightarrow M_{i-1} = span(\mathbf{b}_1, \ldots, \mathbf{b}_{i-1})^{\perp}$$

$$\pi_i(\mathbf{x}) = proj_{M_{i-1}}(\mathbf{x}).$$

With $\mathcal{L}_{[i,k]}$ we denote the lattice generated by the vectors

$$\mathcal{B}_{[i,k]} = \{\pi_i(\mathbf{b}_i), \pi_i(\mathbf{b}_{i+1}), \ldots, \pi_i(\mathbf{b}_k)\},$$

for $1 \leq i \leq k \leq n$. Remark that $\pi_i(\mathcal{L}) = \mathcal{L}_{[i,n]}$ and $rank(\mathcal{L}_{[i,k]}) = k + 1 - i$.

For the given basis $\mathcal{B}$, we get $\pi_i(\mathbf{b}_j) = \mathbf{b}_j - \sum_{k=1}^{i-1} \mathbf{b}_k^*$.

Definition 6 (BKZ) We say that the basis $\mathcal{B} = \{\mathbf{b}_1, \ldots, \mathbf{b}_n\}$ is BKZ$-\beta$ reduced with blocksize β and parameter δ, if it is $\delta-$LLL reduced ($\delta \in (1/4, 1)$) and for each $i \in \{1, 2, \ldots, n\}$, we have $||\mathbf{b}_i^*|| = \lambda_1(\mathcal{L}_{[i,k]})$ for $k = \min(i + \beta - 1, n)$.

The actual improvement is achieved as follows: in every step of the algorithm, we ensure that the first vector of each block $\mathcal{B}_{[i,k]}$ is the shortest vector inside each lattice. If it is not, then the shortest vector of the lattice is inserted into the block, but this vector is linear dependent on the basis vectors. That means that the local block is not a basis for the local lattice any longer. For this reason, we use the LLL algorithm on the expanded block. The blocksize determines the dimension of most of the local projected lattices in which shortest vectors have to be found. It also determines the amount of basis vectors that are used as input for the LLL algorithm after insertion of a new vector into the local block.

The LLL algorithm reduces iteratively each local block $B_{[j,\min(j+\beta-1,n)]}$ for $j = 1, \ldots, n$, to ensure that the first vector is the shortest one. If $\beta = 2$, then BKZ coincides with LLL, since $B_{[j,\min(j+\beta-1,n)]} = B_{[j,j+1]}$. BKZ algorithm uses an enumeration algorithm to find a vector $\mathbf{v} = (v_1, \ldots, v_n)$ such that,

$$||\pi_j(\sum_{i=j}^{k} v_i \mathbf{b}_i)|| = \lambda_1(L_{[j,k]}).$$

In each iteration we check if $||\mathbf{b}_j^*|| > \lambda_1(L_{[j,k]})$. If it is true, then a new vector $\mathbf{b}' = \sum_{i=j}^{k} v_i \mathbf{b}_i$ is inserted in the basis, and the vectors are linear dependent from now. Then, LLL is called to solve this problem and provide a reduced basis. The procedure ends when all of the enumerations fail.

The Enumeration Subroutine

Enumeration constitutes a basic part of BKZ algorithm. It finds a shortest vector in a local projected lattice $L_{[j,k]}$. That means given as input two integers j, k such that $0 \leq j \leq k \leq n$ provides as output a vector $\mathbf{v} = (v_j, \dots, v_k)$ such that $||\pi_j(\sum_{i=j}^{k} v_i \mathbf{b}_i)|| = \lambda_1(L_{[j,k]})$. The subroutine enumerates the vectors that are within a cycle, with radius $R = ||\mathbf{b}_j^*||$: an initial bound of $\lambda_1(L_{[j,k]})$.

In order to find a shortest vector, enumeration algorithm goes through the *enumeration tree* which consists of the half vectors in the projected lattices $L_{[k,k]}$, $L_{[k-1,k]}, \dots, L_{[j,k]}$ of norm at most R. The reason why this tree formed only by the half number of vectors is the following: if $\mathbf{v} \in \mathcal{L}$, then $-\mathbf{v} \in \mathcal{L}$. Thus, we count only the *positive* nodes. The more reduced the basis is, the less nodes in the tree, and the enumeration is faster.

The tree has depth $k - j + 1$, and the root of the tree is the zero vector, while the leaves are all the vectors of L with norm $\leq R$. The parent of a node at depth k is at depth $k - 1$. Child nodes are ordered by increasing Euclidean norm. The Schnorr-Euchner algorithm performs a depth-first search of the tree to find a nonzero leaf of minimal norm.

The process of enumeration is quite expensive (especially for large β), and for this reason we attempt not to enumerate all the tree nodes and discard some nodes. Pruning can speed up the subroutine, but the output vector may not be a shortest one. Schnorr and Euchner first studied pruning enumeration, and the basic idea is replacing the radius R, with $R_k < R$ for $k = 1, \dots, n$.

The following algorithm is the block Korkin-Zolotarev (BKZ), as proposed by Chen and Nguyen [3].

BKZ pseudocode (without pruning):

INPUT: A basis $\mathcal{B} = \{\mathbf{b}_1, \dots, \mathbf{b}_n\}$ of $\mathcal{L}$,
blocksize β, the Gram-Schmidt triangular matrix μ and
the lengths $||\mathbf{b}_i^*||^2$, $i = 1, \dots, n$
OUTPUT: A BKZ-β reduced basis of $\mathcal{L}$

01. $z \leftarrow 0$
02. $j \leftarrow 0$
03. $LLL(\mathbf{b}_1, \dots, \mathbf{b}_n, \mu)$
04. **while** $z < n$
05. $\quad j \leftarrow (j \pmod{n-1} + 1)$
06. $\quad k \leftarrow \min(j + \beta - 1, n)$
07. $\quad h \leftarrow \min(k + 1, n)$
08. $\quad \mathbf{v} \leftarrow Enum(\mu_{[j,k]}, ||\mathbf{b}_j^*||^2, \dots, ||\mathbf{b}_k^*||^2)$
09. $\quad$ **if** $\mathbf{v} \neq (1, 0, \dots, 0)$ **then**
10. $\qquad z \leftarrow 0$
11. $\qquad LLL(\mathbf{b}_1, \dots, \sum_{i=j}^{k}, \dots, \mathbf{b}_h, \mu)$ at stage j

```
12.    else
13.       z ← z + 1
14.       LLL(b_1, ..., b_h, μ) at stage h − 1
15.    end if
16. end while
```

Extreme Pruning

As we mentioned above, enumeration routine is an exhaustive search trying to find a shortest basis vector, with norm $\leq R$. It runs in exponential time or worse. Pruned enumeration reduces the running time, by discarding the subtrees whose probability of finding the desired lattice point is too small. The extreme pruning, proposed by Gama, Nguyen, and Regev [9], is $\approx 1.414^n$ faster than the basic enumeration. Thus, it causes an exponential speedup and constitutes an important variant for BKZ. The main idea is to prune the branches using some bounding functions whose success probability is small enough. It takes as input the lattice basis and the following n numbers, $R_1^2 \leq R_2^2 \leq \cdots \leq R_n^2 = R^2$, where R_k stands for the pruning in depth k. The goal is to find a vector of length at most R. The algorithm is performed at two steps and repeated until a vector of length $\leq R$ is found:

1. Randomize the input basis and apply basis reduction
2. Run the enumeration on the tree with radii $R_1, R_2, \ldots, R_n$

It is important that there are a lot of methods of randomization which can affect the algorithm. The choice of the basis reduction has also a great effect on the overall running time.

Attacks to Subset Sum Problem

Birthday Attacks

The first algorithms that manage to solve the knapsack were based on birthday paradox.

Schroeppel-Shamir Algorithm This algorithm was the best for solving hard knapsacks until 2009, with time complexity $\tilde{O}(2^{n/2})$ and memory requirement $O(2^{n/4})$. The basic idea is decomposing the initial sum into two smaller sums. After constructing two lists that consist of the two separate solutions, then find a collision between them.

The Howgrave-Graham-Joux Algorithm In 2010, Howgrave-Graham and Joux [11] managed to solve hard knapsack problems with heuristic running time $\tilde{O}(2^{0.337n})$ and memory $\tilde{O}(2^{0.256n})$. This is an improvement of the previous algorithm, and the idea is allowing the solutions to overlap. That gives the problem more degrees of freedom and decreases the running time.

Becker, Coron, and Joux Improvement Another improvement of the previous algorithm that was presented in Eurocrypt 2011 [2] reduces the (heuristic) running time down to $\tilde{O}(2^{0.291n})$. The basic idea of Becker, Coron, and Joux algorithm is adding a bit more degrees of freedom. The solutions of the two sub-knapsacks consist of coefficients from $\{-1, 0, 1\}$. As a result, there are more solution's representations of the original knapsack.

Lattice Attacks

There is another class of attacks to knapsack problem, based on lattices and LLL or BKZ reductions.

Low-Density Attacks Lagarias and Odlyzko [13] solved knapsack problems with density $d < 0.6463$, assuming that there exists a SVP-oracle. The authors used the lattice generated by the rows of the following matrix:

$$B = \begin{bmatrix} 1 & 0 & \cdots & 0 & -a_1 \\ 0 & 1 & \cdots & 0 & -a_2 \\ \vdots & \vdots & \ddots & \vdots & \vdots \\ 0 & 0 & \cdots & 1 & -a_n \\ 0 & 0 & \cdots & 0 & s \end{bmatrix}.$$

They proved the following:

Theorem 2 *Assume that there is a SVP-oracle and the knapsack problem has a solution. Then with high probability, we can solve all knapsack problems with density* < 0.6463.

An SVP-oracle is a probabilistic polynomial algorithm which, given a lattice $\mathcal{L}$, provides a shortest vector of $\mathcal{L}$ with high probability. Unfortunately, in practice we do not have SVP-oracles. Experiments made by Nguyen and Gama [7] suggest that LLL behave as a SVP-oracle for dimensions ≤ 35 and BKZ-20 algorithm [21] and for dimensions ≤ 60. Further, two more simplified proofs of the previous theorem were also given in [4, 5]. Also in [19], the algorithm was tested experimentally providing some improvements.

The previous result was improved by Coster et al. [4]. The new density bound was improved to 0.9408. Their approach is in the same spirit of Lagarias and Odlyzko. So the assumption of the existence of a SVP-oracle remained. They applied LLL reduction algorithm to the lattice generated by the rows of the matrix:

$$B = \begin{bmatrix} 1 & 0 & \cdots & 0 & Na_1 \\ 0 & 1 & \cdots & 0 & Na_2 \\ \vdots & \vdots & \ddots & \vdots & \vdots \\ 0 & 0 & \cdots & 1 & Na_n \\ \frac{1}{2} & \frac{1}{2} & \cdots & \frac{1}{2} & Ns \end{bmatrix}, \tag{3}$$

where N is a positive integer $> \sqrt{n}/2$.

Finally in [16], they managed to solve $n-$dimensional hard knapsack problems making a single call to a CVP-oracle in a $(n-1)-$dimensional lattice (CVP: closest vector problem). Unfortunately, we do not have such oracles in practice, for lattice with large rank.

Schnorr and Shevchenko Algorithm [22] This algorithm uses BKZ-reduction into the rows of a matrix similar to (3). Schnorr and Shevchenko solve the knapsack problem faster, in practice, than the Becker, Coron, and Joux method.

The basis $\mathcal{B}$ that is used is presented by the rows of the matrix $B \in \mathbb{Z}^{(n+1)(n+3)}$:

$$B = \begin{bmatrix} 2 & 0 & \dots & 0 & Na_1 & 0 & N \\ 0 & 2 & \dots & 0 & Na_2 & 0 & N \\ \cdot & \cdot & \dots & \cdot & \cdot & \cdot & \cdot \\ \cdot & \cdot & \dots & \cdot & \cdot & \cdot & \cdot \\ \cdot & \cdot & \dots & \cdot & \cdot & \cdot & \cdot \\ 0 & 0 & \dots & 2 & Na_n & 0 & N \\ 1 & 1 & \dots & 1 & Ns & 1 & HN \end{bmatrix} \tag{4}$$

Let $\mathcal{L}(\mathcal{B}) \subset \mathbb{R}^{n+3}$ be the lattice generated by $\mathcal{B}$. The basis consists of these $n+1$ row vectors, with $n+3$ elements each one. The integer N must be larger than $\sqrt{n}$. Furthermore, we assume that n is even. At our examples $n \in \{80, 84\}$ and we choose $N = 16$. The parameter H is the hamming weight (the number of 1's in a solution). We set $H = n/2$ in order to take balanced solutions.

Let $\mathbf{b} = (b_1, b_2, \dots, b_{n+3})$ in $\mathcal{L}(\mathcal{B})$ that satisfies the following three equalities:

$$|b_i| = 1 \text{ for } i = 1, \dots, n$$

$$|b_{n+2}| = 1,$$

$$b_{n+1} = b_{n+3} = 0$$

then, the solution $\mathbf{x} = (x_1, \dots, x_n)$ consists of the following integers :

$$x_i = \frac{|b_i - b_{n+2}|}{2}, \quad \text{for } i = 1, 2, \dots, n,$$

with the property $\sum_{i=1}^{n} x_i = n/2$. The inverse fact is that every solution is written as previous. The integer $n/2$ can be replaced with an integer $H \in \{1, \ldots, n-1\}$.

Then we can apply the following algorithm, divided into two parts:

Schnorr-Shevchenko Algorithm

1. We apply BKZ-reduction to the rows of B without pruning. This step is repeated five times with blocksizes 2^k for k from 1 to 5. Before the reduction is suggested a permutation of the rows by Schnorr and Shevchenko. If the solution is found, the algorithm stops.
2. If the solution has not been found in the first part of the algorithm, then BKZ reduces the basis independently with blocksizes: $bs = 30, 31, 32, \ldots, 60$. The pruning parameter for each blocksize is $10, 11, 12, 10, 11, \ldots$ and so on. Always terminate if the solution has been found.

Experiments

We implemented Schnorr-Shevchenko (SS) algorithm using fpLLL 5.0 [25]. We used two implementations. The one proposed by the original Schnorr-Shevchenko and the other that uses extreme pruning instead of linear. We generated 73 random knapsack instances in Fig. 1 of dimension 80 (resp., in Fig. 2 with dim = 84) and density close to 1. The experiments were executed in a i7 3.8 GHz cpu with 8 GB ram. We got points $(x_i, y_i)_i$, where x_i is the (cpu) time in minutes for the usual SS attack (using BKZ with linear pruning) and y_i the (cpu) time for the same instance using extreme pruning. Finally, we sorted with respect to x_i. We summarized the results in Fig. 1 (resp., in Fig. 2). The average cpu time for the usual SS method was 13.3 m (resp., 67 m) and with extreme pruning was 6.8 m (resp., 35 m). For the

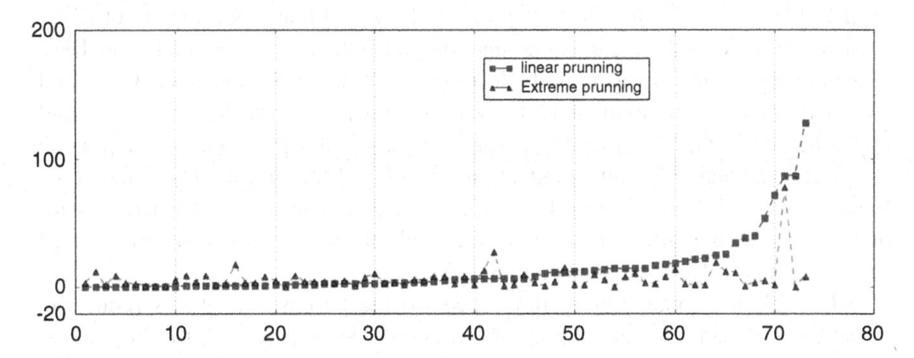

Fig. 1 We generated 73 random hard knapsack problems of dimension 80. We measured the cpu times produced by the original SS method and SS using extreme pruning

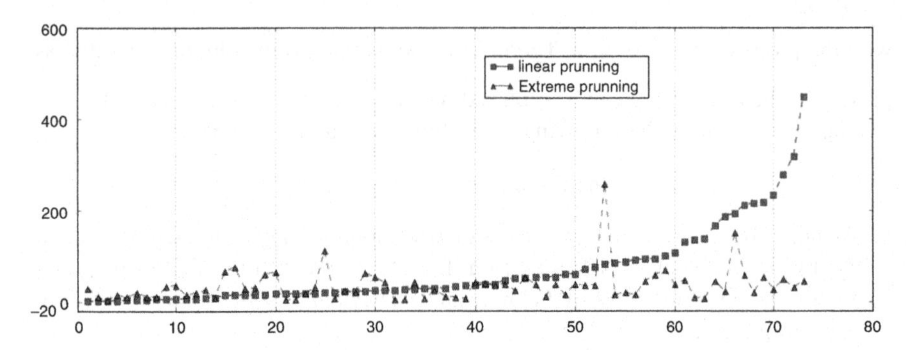

Fig. 2 We generated 73 random hard knapsack problems of dimension 84. We measured the cpu times produced by the original SS method and SS using extreme pruning

both experiments, almost half the examples had the same running times, and for the rest instances, the differences $y_i - x_i$ were positive and on average 18.8 m (resp., 76.4 m).

Conclusions

In this work we addressed the knapsack problem, where we apply the attack [22], but we used extreme pruning instead the usual linear pruning in BKZ. We executed experiments in dimension 80 and 84 using fplll, and our results suggest that extreme pruning is clearly faster than linear pruning using the Schnorr-Shevchenko method.

Appendix

fpLLL [25], standing for floating-point LLL, is a library developed for C++, implementing several fundamental functions on lattices, such as LLL and BKZ reduction algorithms. It was initially developed by Damien Stehlé, David Cadé, and Xavier Pujol, currently maintained by Martin Albrecht and Shi Bai and distributed under the GNU LGPL license. The purpose of the initial deployment was to provide practical benchmarks for lattice reduction algorithms for everyone. The name of the library derives from the floating-point arithmetic data type, as this is where it relies on for all the computations, in order to offer multiple ratios of speed and guaranteed success output. fpLLL is also used in SageMath [24].

NTL [23] is another library developed for the purpose of lattices reduction, initially developed by Victor Shoup. NTL is also developed for C++. In general, the library provides data structures and algorithms for manipulating arbitrary length integers, as well as other data types over integers and finite fields. Beyond others, it includes implementations of BKZ and LLL algorithms. The library can also be used in SageMath and can also be compiled in thread safe mode.

References

1. Ajtai, M.: The shortest vector problem in L_2 is NP-hard for randomized reduction. In: Proceedings of the 30th ACM Symposium on Theory of Computing (1998)
2. Becker, A., Coron, J.-S., Joux, A.: Improved generic algorithm for hard knapsacks. In: Eurocrypt 2011. Lecture Notes in Computer Science, vol. 6632 (2011)
3. Chen, Y., Nguyen, P.Q.: BKZ 2.0: better lattice security estimates. In: Asiacrypt. Lecture Notes in Computer Science, vol. 7073. Springer, Berlin (2011)
4. Coster, M.J., Joux, A., LaMacchia, B.A., Odlyzko, A.M., Schnorr, C.-P., Stern, J.: Improved low-density subset sum algorithms. Comput. Complex. **2**, 111–128 (1992)
5. Freize, A.M.: On the Lagarias-Odlyzko algorithm for the subset sum problem. SIAM J. Comput. **15**(2), 536–539 (1986)
6. Galbraith, S.: Mathematics of Public Key Cryptography. Cambridge University Press, Cambridge (2012)
7. Gama, N., Nguyen, P.Q.: Predicting Lattice Reduction. Lecture Notes in Computer Science, vol. 4965 (2008)
8. Gama, N., Nguyen, P.Q.: Predicting lattice reduction. In: Eurocrypt 2008. Advances in Cryptology. Springer, Berlin (2008)
9. Gama, N., Nguyen, P.Q., Regev, O.: Lattice enumeration using extreme pruning. In: Eurocrypt 2010. Advances in Cryptology. Springer, Berlin (2010)
10. Garey, M.R., Johnson, D.S.: Computers and Intractability: A Guide to the Theory of NP-Completeness. W. H. Freeman, San Francisco (1979)
11. Howgrave-Graham, N., Joux, A.: New generic algorithms for hard knapsacks. In: Eurocrypt 2010. Lecture Notes in Computer Science. vol. 6110 (2010)
12. Joux, A.: Algorithmic Cryptanalysis. Chapman & Hall/CRC Cryptography and Network Security. CRC Press, Boca Raton (2009)
13. Lagarias, J., Odlyzko, A.: Solving low-density attacks to low-weight knapsacks. In: Advance in Cryptology-ASIACRYPT 2005. Lecture Notes in Computer Science, vol. 3788 (2005)
14. Lenstra, A.K., Lenstra, H.W., Lovász, L.: Factoring polynomials with rational coefficients. Math. Ann. **261**(4), 515–534 (1982)
15. Micciancio, D., Goldwasser, S.: Complexity of Lattices : A Cryptographic Perspective. Springer, Berlin (2003)
16. Nguyen, P.Q., Stern, J.: Adapting density attacks to low-weight knapsacks. In: Advances in Cryptology - ASIACRYPT 2005. Lecture Notes in Computer Science, vol. 3788 (2005)
17. Nguyen, P.Q., Stehlé, D.: An LLL algorithm with quadratic complexity. SIAM J. Comput. **39**(3), 874–903 (2009)
18. Okamoto, T., Tanaka, K., Uchiyama, S.: Quantum public-key cryptosystems. In: On Proceedings of Crypto 2000 (2000)
19. Radziszowski, S., Kreher, D.: Solving subset sum problems with the L^3 algorithm. J. Comb. Math. Comb. Comput. **3**, 49–63 (1988)
20. Schnorr, C.-P.: A more efficient algorithm for lattice basis reduction algorithms. J. Algorithms **9**, 47–62 (1987)
21. Schnorr, C.-P., Euchner, M.: Lattice basis reduction: improved practical algorithms and solving subset sum problems. Math. Program. **66**(1–3), 181–199 (1994)
22. Schnorr, C.-P., Shevchenko, T.: Solving Subset Sum Problems of Density close to 1 by "randomized" BKZ-reduction. Cryptology. ePrint Archive: Report 2012/620
23. Shoup, V.: NTL: A library for doing number theory. http://www.shoup.net/ntl/
24. Stein, W.A., et al. Sage Mathematics Software. The Sage Development Team (2012). http://www.sagemath.org
25. The FPLLL development team, fplll, a lattice reduction library (2016). Available at https://github.com/fplll/fplll

References

A Computational Intelligence System Identifying Cyber-Attacks on Smart Energy Grids

Konstantinos Demertzis and Lazaros Iliadis

Abstract According to the latest projections of the International Energy Agency, smart grid technologies have become essential to handling the radical changes expected in international energy portfolios through 2030. A smart grid is an energy transmission and distribution network enhanced through digital control, monitoring, and telecommunication capabilities. It provides a real-time, two-way flow of energy and information to all stakeholders in the electricity chain, from the generation plant to the commercial, industrial, and residential end user. New digital equipment and devices can be strategically deployed to complement existing equipment. Using a combination of centralized IT and distributed intelligence within critical system control nodes ranging from thermal and renewable plant controls to grid and distribution utility servers to cities, commercial and industrial infrastructures, and homes a smart grid can bring unprecedented efficiency and stability to the energy system. Information and communication infrastructures will play an important role in connecting and optimizing the available grid layers. Grid operation depends on control systems called *Supervisory Control and Data Acquisition* (SCADA) that monitor and control the physical infrastructure. At the heart of these SCADA systems are specialized computers known as *Programmable Logic Controllers* (PLCs). There are destructive cyber-attacks against SCADA systems as *Advanced Persistent Threats* (APT) were able to take over the PLCs controlling the centrifuges, reprogramming them in order to speed up the centrifuges, leading to the destruction of many and yet displaying a normal operating speed in order to trick the centrifuge operators and finally can not only shut things down but can alter their function and permanently damage industrial equipment. This paper proposes a computational intelligence *System for Identification Cyber-Attacks on the Smart*

K. Demertzis (✉) · L. Iliadis
School of Engineering, Department of Civil Engineering, Faculty of Mathematics Programming and General Courses, Democritus University of Thrace, Xanthi, Greece
e-mail: kdemertz@fmenr.duth.gr; liliadis@civil.duth.gr

N. J. Daras, Th. M. Rassias (eds.), *Modern Discrete Mathematics and Analysis*, Springer Optimization and Its Applications 131,
https://doi.org/10.1007/978-3-319-74325-7_5

Energy Grids (SICASEG). It is a big data forensics tool which can capture, record, and analyze the smart energy grid network events to find the source of an attack to both prevent future attacks and perhaps for prosecution.

Keywords Smart energy grids · Cyber-attacks · Advanced persistent threats · Supervisory control and data acquisition—SCADA · Big data · Forensics tool

Introduction

Smart Energy Grids

It is a fact that the majority of research in electrical energy systems is related to Smart Energy Grids (SEG) [1]. There is a global effort on the way, aiming to overcome the problems of conventional systems and networks. The smart energy grid networks are using information and communication technologies (ICT) in order to offer optimal transfer and distribution of electrical energy from the providers to the customers [2]. On the other hand, SEG operate in electrical networks that use digital technology to monitor and transfer electricity from all sources, in order to cover the varying needs of the users. They also coordinate the needs and the potentials of the producers, managers, consumers, and all market entities in order to ensure that they function in the optimal way. Actually, they are minimizing the cost and the environmental consequences and at the same time they are enhancing reliability and stability [3].

Conceptual Framework

This new energy network which aims to cover a basic and crucial matter of common prosperity is integrated under the conceptual framework of heterogeneous infrastructure collective operation, in a status of innovation and financial investments of mid-long term payoff [4]. Under this point of view the SEG offer important contributions towards sustainable development.

The main advantages of this technology are briefly discussed below [5, 6]:

1. The SEG integrate distributed production of renewable energy sources
2. They offer reliability and quality of power, especially in areas with frequent voltage fluctuations.
3. They offer electricity with the use of distributed energy production in remote areas, e.g. antennas, small villages, oil oceanic platforms.
4. Demand forecasting based on statistical data is used to reduce distribution lines overloading and accidental interruptions of electrical supply. At the same time they incorporate instant restart potentials of *Black Start* type.

5. SEG respond directly and optimally in new power demands, by forecasting the actual needs under specific situations and time periods.
6. Microgrids offer energy sustainability and backup.
7. SEG automate the provided services of the system that records and financially evaluates the interruption and reconnection of electrical power.
8. They activate systems of energy, physical and logical security with mechanisms of multilevel control access plus cryptography.
9. They are using real-time controllers to offer management, correlation and warning of incidents, with technologies of Intrusion Prevention System (IPS) type.
10. They offer qualitative services of high added value in every phase of the energy cycle.

Conceptual Model

The standardization organizations have applied a division model of the energy cycle in partial primitive branches. This was achieved based on the general conceptual functional framework of the smart energy grids [6]. This division aims in classifying the involved entities based on their homogeneous elements of interest and on their specific functions. This conceptual model is based on the functional variation of each sector and it is not related to an architectural plan. It is a tool to provide the background for the description and analysis of the models' interfunctionality that also supports the development of emerging architectures in SEG technologies.

The conceptual model [7] comprises of seven basic sectors (domains) namely: *Bulk Generation, Transmission, Distribution, Customer, Service Provider, Operations*, and *Markets* (Fig. 1).

Cybersecurity for Smart Grid Systems

A New, Smart Era for the Energy Grids

Upgrading of the energy infrastructures by incorporating new technologies (especially the ones related to ICT and Internet) introduces risks and new threats for the security and the continuous function of the electrical energy network [8]. The exploitation of the vulnerable points of a cable or a wireless smart network can lead to the occupation of ε critical electronic devices and applications, the leak of top secret or personal information, the loss or block of necessary services, even to the total interruption of electricity with huge consequences [9].

Confronting the security issues combined with the application of a strong legal framework that would ensure integrity, security, and availability of the transferred energy information is a primitive target, a continuous challenge and a social demand for the transition to the new energy scheme.

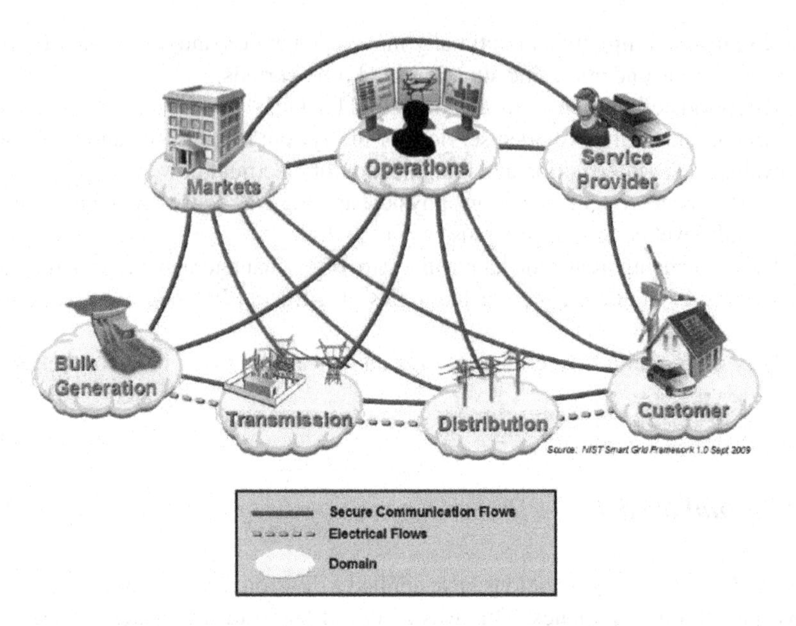

Fig. 1 Conceptual model of smart energy grids

Risks Involved

The smart network not only offers new functions but it introduces new risks in the electricity system as well. Given that the modern civilization is based on electricity and on the supporting infrastructure, this matter is of high importance. A potential extended interruption of the production or distribution services would have huge socio-economic consequences and it would lead to loss of human lives. The risks associated with the SEG application are mainly related to the telecommunications, automation systems, and data collection mechanisms [8–10].

Due to the fact that the basic core of a SEG net is the telecommunication network, the use of the most modern relative infrastructures such as *fiber optics, Broadband over Power Line* (BLP), and *wireless transmission* technologies is really crucial. However, this stratification and also this modeling approach increase the system's complexity and they create asymmetric threats [10].

Another problem is that the incorporation of SEG technology converts the previously isolated and closed network of power control systems to a public one, accessible to the general public. This fact combined with the rapid spread of the internet introduces new threats to the energy infrastructures. The advanced techniques undoubtedly offer significant advantages and possibilities, but also, they significantly increase the problems associated with the protection and availability of information [11]. Besides cyber threats [12], as malware, spyware, computer viruses, which currently threaten the ICT networks, the introduction of new technologies and services such as smart meters, sensors, and distributed access points may create vulnerabilities in SEG.

However, the smart energy grids are not only exposed to risks due to the vulnerabilities of the communications networks, but they also face risks inherent to the existing electrical networks, due to physical vulnerabilities of the existing old infrastructures [10].

The problems due to physical attacks are targeting to interrupt the production, transfer, and distribution of the electric power. However, the cyberattacks aim to gain remote access to users' data, endanger or control electronic devices and general infrastructure to their benefit [11].

Threats

A threat is a potential damage or an unpleasant development that can take place if nothing changes, or that can be caused by someone if his target will not comply with his demands [10–12]. The best known types of threats related to energy systems are presented below [10–12]:

1. *Physical threats*

   They require specific tools and natural presence. The lines can be undermined anywhere along the line or in the transmission tower. The distribution lines are positioned at a relatively low height and can be easily interrupted. Also, smart meters are extremely vulnerable to theft since they are installed at the customer premises [10–12].
2. *Cyber threats*

   They can be executed by any computer. Smart meters communicate and interface with other counters in the network and with smart home appliances and energy management systems. These interfaces increase the exposure of the SEG in remote threats, such as invasion of privacy through wiretapping and data traffic analysis, unauthorized access to stored data, attacks, interference or modification of communication networks [10–12].
3. *Cyber-Physical (combined threats)*

   They require combined knowledge, since the electronic attacks can have physical effects. On the other hand, physical attacks can affect the electronic infrastructure. For example, a disgruntled or complained employee with authorization to the computer network may enter the substation security system and disable the perimeter security, paving the way for any physical attack [10–12].

Types of Attacks

Attack is any attempt to breach the confidentiality, integrity, or availability of an energy system or network. It is also any unauthorized action that aims to prevent, bypass, or disable safety mechanisms and access control to a system or a network.

The goal of an attack varies depending on the capabilities and objectives of the intruder and on the degree of difficulty of the attempt regarding the measures and security mechanisms to be addressed.

There are four types of attacks [10–12]:

1. *Denial-of-Service* (DoS)
   The attacker (Bob) denies the source (Alice) access to the destination (Mary).
2. *Man-in-the-middle*
   Bob pretends to be Mary, so he receives all messages from Alice to Mary. Bob can change the messages and forward them changed to Mary.
3. *Spoofing*
   Bob impersonates as Alice so that he can create and send messages to Mary.
4. *Eavesdropping*
   Bob receives all messages sent from Alice to Mary, but both Alice and Mary do not know about it (Fig. 2).

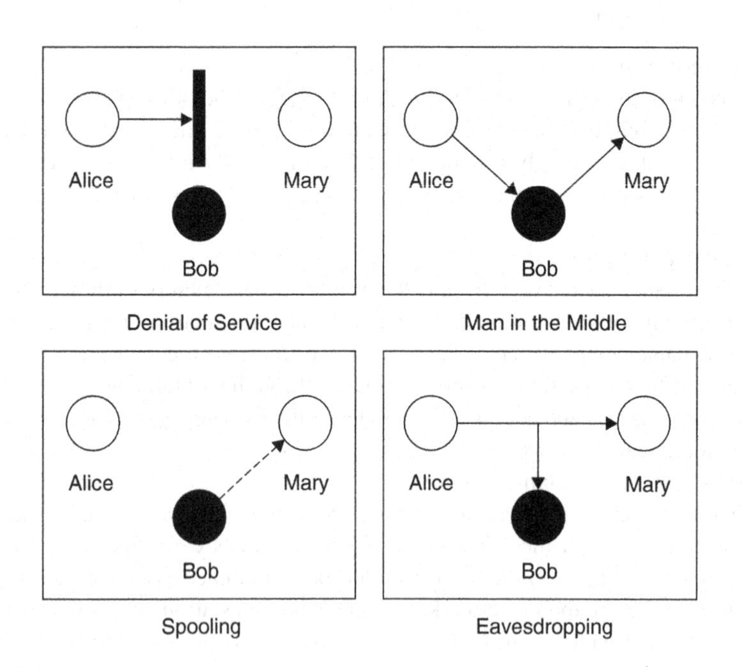

Fig. 2 Cyber attacks

Cyber Attacks on Smart Grid

Smart Energy Grids attacks can be classified based on the following [10–12]:

1. *On the motivation*
   The motive of the attackers can be categorized into five areas, namely: curiosity for information motivated attacks, immoral power theft, theft of power consumption information, economic benefits.
2. *On the number of attackers*
   They can be characterized as *single or individual*, aiming in collecting all the necessary information to commit a small scale blackout. Also they can be considered as *coordinated attacks* when they are organized by groups of attackers who cooperate to hit critical infrastructures.
3. *On the target*
   A hacking attempt can aim in any field of the electric power network, such as the production (targeting to interrupt the operation of generators) or the distribution and control.

The final target might be the change of the phase or other network status information, resulting in the sudden load change in critical locations of the electrical network. This could cause overload of the transmission lines and network collapse.

SCADA Systems

The term SCADA (supervisory control and data acquisition) [13] describes a class of industrial controllers and telemetry systems. The characteristic of the SCADA systems is that they comprise of local controllers, controlling individual components and units of an installation connected to a centralized Master Station. The central workstation can then communicate the data collected from the establishment in a number of workstations in the LAN or to transmit the plant data in remote locations via a telecommunication system, e.g. via the wired telephone network or via a wireless network or via the Internet [13].

It is also possible that each local controller is in a remote location and transmits the data to the master station via a single cable or via a wireless transceiver, always set by local controllers connected in a star topology to a master station.

The use of SCADA systems manages on-line monitoring (through PLCs) and continuous recording of all critical parameters of the electricity network, in order to achieve surveillance in real time.

The main functions of a SCADA system are the following [13]:

1. Data collection from the PLCs and the remote terminal unit (RTU). All of the desired signals are propagated towards the SCADA system through the network.

2. Data storage in the database and their representation through graphs. The selected information is represented either as such or after suitable processing.
3. Analyze data and alert personnel in fault cases. When data values get abnormal, the SCADA system notifies operators by using visual or audible signals in order to avoid unpleasant consequences.
4. Control of the closed loop processes. There exists the possibility of technical control application, automatically or manually.
5. Graphical representation of the process sections to mimic diagram and data presentations in active fields. The mimic diagrams depict realistic parts of the process in order to facilitate monitoring and understanding of the data from the system operators.
6. Recording of all events regular or not, for the creation of a historical archive of critical parameters in the form of a database. Support of a dual computer system with automatic switching if this is considered appropriate, based on the process under control. In high risk processes the occurrence of error due to failure of the equipment should be minimized as much as possible. For this reason, the SCADA systems support a second computer system that undertakes in case of error.
7. Transfer of data to other parts of the central management and information system.
8. Check the access of the operators to the various subsystems of the SCADA system.
9. Specific software applications such as C++ code execution or intelligent systems development.
10. Handling, managing, and processing of vast amounts of data.

Methods of Attack

The attack methodology might be followed by a specific hostile entity willing to cause more than a service interruption. Getting unauthorized access to a SCADA system is an extremely difficult task requiring skills and many hours of research.

Gaining control of the automation system of an electrical network requires three essential steps [14]:

1. *Access:* The first step of the attacker would be to gain access to the SCADA system. The attacker can gather as much information as possible (e.g., from the Internet) such as names and installed equipment. Then he targets specific elements of the system by using malware, he exploits weaknesses and gains access. The most common method for gaining unauthorized access is the external VPN access to the SCADA. The VPN access is mainly used by specialized personnel which logins from home or from work. Of course stealing the login details of such personnel is a huge problem.

2. *Discovery:* After intruding the SCADA the next step is to analyze and understand the specific network by discovering the processes running in it. The complexity of the network is a really good defense against the attacks, however an experienced intruder can cause serious problems and the collapse of services. First the attacker searches simple information sources such as web servers or workstations. The information traffic can be monitored for a long period of time and thus a vast volume of data can be discovered (e.g., FTP, Telnet, and HTTP certificates). The combination of all the above can offer a clear view of the network's function for the intruder.
3. *Control*: If the SCADA is analyzed, there are various methods to control the system. The engineers' workstations used to upgrade the software, the database systems and the application server (where various SCADA applications are saved providing control) are a potential target.

Additionally, another optional step which employs experienced invaders is hiding the attacks by deleting specific folders that can detect and report the presence of intruders in automation systems.

Literature Review

In an earlier research of our team we have made few hybrid computational intelligence systems [15–29]. Tao et al. described the network attack knowledge, based on the theory of the factor expression of knowledge, and studied the formal knowledge theory of SCADA network from the factor state space and equivalence partitioning. This approach utilizes the *factor neural network* (FNN) theory which contains high-level knowledge and quantitative reasoning described to establish a predictive model including analytic FNN and analogous FNN. This model abstracts and builds an equivalent and corresponding network attack and defense knowledge factors system. Also, the [30] introduces a new *European Framework-7* project *Cockpit CI* (*Critical Infrastructure*) and roles of intelligent machine learning methods to prevent SCADA systems from cyber-attacks. Qian and Sherif [31] apply autonomic computing technology to monitor SCADA system performance, and proactively estimate upcoming attacks for a given system model of a physical infrastructure. In addition, Soupionis et al. [32] propose a combinatorial method for automatic detection and classification of faults and cyber-attacks occurring on the power grid system when there is limited data from the power grid nodes due to cyber implications.

The efficiency of the proposed method is demonstrated via an extensive experimental phase measuring the false positive rate, false negative rate, and the delay of the detections. Finally, Qin et al. [33] put forward an analytic factor neuron model which combines reasoning machine based on the cloud generator with the FNN theory. The FNN model is realized based on mobile intelligent agent and malicious behavior perception technology.

The authors have acknowledged the potential of machine learning-based approaches in providing efficient and effective detection, but they have not provided a deeper insight on specific methods, neither the comparison of the approaches by detection performances and evaluation practices.

This research paper proposes the development of the *SICASEG*, a cyber-threat bio-inspired intelligence management system. Unlike other techniques that have been proposed from time to time and focus in single traffic analysis, SICASEG is an efficient SCADA supervision system which provides smart mechanisms for the supervision and categorization of networks. It provides intelligent approaches for the above task and it is capable of defending over sophisticated attacks and of exploiting effectively the hardware capabilities with minimum computational and resources cost. More specifically, this research proposes an innovative and very effective *Extreme Learning Machine* (ELM) model, which is optimized by the *Adaptive Elitist Differential Evolution* algorithm (AEDE). The AEDE is an improved version of the *Differential Evolution* (DE) algorithm and it is proper for big data resolution. This hybrid method combines two highly effective, biologically inspired, machine learning algorithms, for solving a multidimensional and complex cyber security problem.

Power System Attack Datasets

SCADA Power System Architecture

Figure 3 shows the power system framework configuration which is used in generating power event scenarios [33–37]. In the network diagram we have several components. G_1 and G_2 are power generators whereas R1 through R4 are Intelligent Electronic Devices (IEDs) that can switch the breakers on or off. These breakers are labeled BR_1 through BR_4. There are also two main lines. $Line_1$ spans from breaker one BR_1 to breaker two BR_2 and $Line_2$ spans from breaker three BR_3 to breaker four BR_4. Each IED automatically controls one breaker. R_1 controls BR_1, R_2 controls BR_2, and so on accordingly. The IEDs use a distance protection scheme which trips the breaker on detected faults whether actually valid or faked since they have no internal validation to detect the difference. Operators can also manually issue commands to the IEDs R_1 through R_4 to manually trip the breakers BR_1 through BR_4. The manual override is used when performing maintenance on the lines or other system components [33–37].

Types of Scenarios

There are five types of scenarios [33–37]:

1. *Short-circuit fault*—This is a short in a power line and can occur in various locations along the line, the location is indicated by the percentage range.

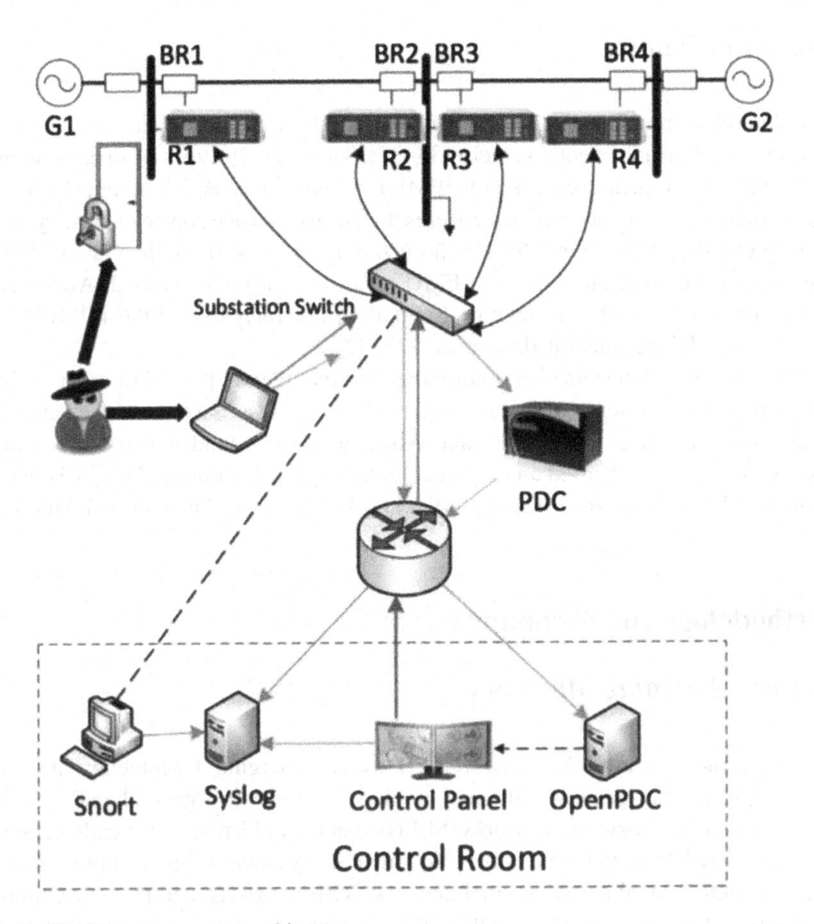

Fig. 3 SCADA power system architecture

2. *Line maintenance*—One or more relays are disabled on a specific line to do maintenance for that line.
3. *Remote tripping command injection* (Attack)—This is an attack that sends a command to a relay which causes a breaker to open. It can only be done once an attacker has penetrated outside defenses.
4. *Relay setting change* (Attack)—Relays are configured with a distance protection scheme and the attacker changes the setting to disable the relay function such that relay will not trip for a valid fault or a valid command.
5. *Data Injection* (Attack)—Here we imitate a valid fault by changing values to parameters such as current, voltage, and sequence components. This attack aims to blind the operator and causes a black out.

The Final Dataset

The dataset comprised of 128 independent variables and three classes—markers (No Events, Normal Events, Attack) [33–37]. There are 29 types of measurements from each phasor measurement units (PMU). A phasor measurement unit (PMU) or synchrophasor is a device which measures the electrical waves on an electricity grid, using a common time source for synchronization. In our system, there are 4 PMUs which measure 29 features for 116 PMU measurement columns total. Also, there are 12 features for control panel logs, Snort alerts and relay logs of the 4 PMU/relay (relay and PMU are integrated together) [33–37].

The dataset is determined and normalized to the interval [−1,1] in order to face the problem of prevalence of features with wider range over the ones with a narrower range, without being more important. Also, the outliers and the extreme values spotted were removed based on the Inter Quartile Range technique. The final dataset contains 159,045 patterns (48,455 No Events, 54,927 Natural and 55,663 Attack).

Methodology and Techniques

Extreme Learning Machines

The Extreme Learning Machine (ELM) as an emerging biologically inspired learning technique provides efficient unified solutions to "generalized" *Single-hidden Layer feed forward Networks* (SLFNs) but the hidden layer (or called feature mapping) in ELM need not be tuned [38]. Such SLFNs include but are not limited to support vector machine, polynomial network, RBF networks, and the conventional feed forward neural networks. All the hidden node parameters are independent from the target functions or the training datasets and the output weights of ELMs may be determined in different ways (with or without iterations, with or without incremental implementations). ELM has several advantages, ease of use, faster learning speed, higher generalization performance, suitable for many nonlinear activation function and kernel functions.

According to the ELM theory [38], the ELM with Gaussian Radial Basis Function kernel (GRBFk) $K(u, v) = \exp(-\gamma||u - v||^2)$ used in this approach. The hidden neurons are $k = 20$. Subsequently assigned random input weights w_i and biases b_i, $i = 1, \ldots, N$. To calculate the hidden layer output matrix H used the function (1):

$$H = \begin{bmatrix} h(x_1) \\ \vdots \\ h(x_N) \end{bmatrix} = \begin{bmatrix} h_1(x_1) & \cdots & h_L(x_1) \\ \vdots & & \vdots \\ h_1(x_N) & \cdots & h_L(x_N) \end{bmatrix} \tag{1}$$

$h(x) = [h_1(x), \ldots, h_L(x)]$ is the output (row) vector of the hidden layer with respect to the input x. *h(x)* actually maps the data from the d-dimensional input space to the L-dimensional hidden-layer feature space (ELM feature space) H, and thus, *h(x)* is indeed a feature mapping. ELM is to minimize the training error as well as the norm of the output weights:

$$\text{Minimize} : ||H\beta - T||^2 \text{ and } ||\beta||, \tag{2}$$

where H is the hidden-layer output matrix of the function (1). To minimize the norm of the output weights $||\beta||$ is actually to maximize the distance of the separating margins of the two different classes in the ELM feature space $2/||\beta||$.

The following function (3) is used to calculate the output weights β:

$$\beta = \left(\frac{I}{C} + H^T H\right)^{-1} H^T T, \tag{3}$$

where C is a positive constant is obtained and T resulting from the *Function Approximation of SLFNs with additive neurons* in which is an arbitrary distinct samples with $t_i = [t_{i1}, t_{i2}, \ldots, t_{im}]^T \in R^m$ and $T = \begin{bmatrix} t_1^T \\ \vdots \\ t_N^T \end{bmatrix}$ [38].

Adaptive Elitist Differential Evolution (AEDE)

In evolutionary computation, *Differential Evolution* (DE) [39] is a method that optimizes a problem by iteratively trying to improve a candidate solution with regard to a given measure of quality. In the DE, the parameters such as mutant factor F and crossover control parameter *CR*, and trial vector generation strategies have significant influence on its performance. To overcome the common limitations of optimization algorithms such as the use of a huge volume of resources (e.g., high computational cost) the Adaptive Elitist Differential Evolution algorithm (AEDE) [40] introduces two alternatives. The first one is applied in the mutation phase and the second one in the selection phase, in order to enhance the search capability as well as the convergence speed of the DE algorithm. The new adaptive mutation scheme of the DE uses two mutation operators. The first one is the "rand/1" which aims to ensure diversity of the population and prohibits the population from getting stuck in a local optimum. The second is the "current-to-best/1" which aims to accelerate convergence speed of the population by guiding the population toward the best individual. On the other hand, the new selection mechanism is performed as follows: Firstly, the children population C consisting of trial vectors is combined with the parent population P of target vectors to create a combined population Q. Then, *NP* best individuals are chosen from the Q to construct the population for

Algorithm 1 Elitist selection operator [40]

1: **Input**: Children population ***C*** and parent population ***P***
2: Assign $Q = C \cup P$
3: Select ***NP*** best individuals from ***Q*** and assign to ***P***
4: **Output: *P***

the next generation. In this way, the best individuals of the whole population are always stored for the next generation. This helps the algorithm to obtain a better convergence rate [40]. The elitist selection operator is presented in the following Algorithm 1.

The aeDE method is summarily shown as in Algorithm 2 below [40]:

where *tolerance* is the allowed error; *MaxIter* is the maximum number of iterations; and *randint*(1, D) is the function which returns a uniformly distributed random integer number between 1 and D.

Adaptive Elitist Differential Evolution ELM (AEDE-ELM)

Given that ELMs produce the initial weights (weights) and (bias) randomly, the process may not reach the optimal result, which may not imply as high classification accuracy as the desired one. The optimal choice of weights and bias creates the conditions for maximum potential accuracy and of course the best generalization performance of the ELMs [41]. To solve the above problem, we recommend the use of the AEDE optimization method for the optimal selection of weights and bias of the ELMs. Initially, each individual in the first generation is obtained randomly, and it is composed of the input weights and hidden biases: $x = [\omega_1, \omega_2, \ldots, \omega_l,\ b_1, b_2, \ldots, b_l]$.

Secondly, the corresponding output weights matrix for each individual is calculated in the manner of the ELM algorithm. Then, we apply AEDE to find the fitness for each individual in the population. Finally, when the evolution is over, we can use the optimal parameters of the ELM to perform the classification [41].

The procedure of AEDE-ELM algorithm is shown by Algorithm 3 [41].

Results and Comparative Analysis

It is extremely comforting and hopeful, the fact that the proposed system manages to solve a particularly complex cyber security problem with high accuracy. The performance of the proposed AEDE-ELM is evaluated by comparing it with RBFANN, GMDH, PANN, and FNNGA learning algorithms. Regarding the overall efficiency of the methods, the results show that the AEDE-ELM has much better generalization performance and more accurate classification output from the other

Algorithm 2 The adaptive elitist Differential Evolution (aeDE) algorithm [40]

1: Generate the initial population
2: Evaluate the fitness for each individual in the population
//Definition of searching criteria
3: **while** delta > tolerance or *MaxIter* is not reached **do**
//Find the best individuals
4: **for** i =1 to *NP* **do**
//Generate the initial mutation factor
5: F = rand[0.4, 1]
//Generate the initial crossover control parameter
6: CR = rand[0.7, 1]
//Select a random integer number between 1 and D
7: j_{rand} = randint(1, D)
//Find the optimal parameters
8: **for** j =1 **to** D **do**
//Check the crossover operation
9: **if** rand[0, 1] < CR or j == j_{rand} **then**
//Check the mutation
10: **if** delta > threshold **then**
//Select the optimal parameters
11: Select randomly r1 $\neq$ r2 $\neq$ r3 $\neq$ i;

$$\forall i \in \{1, \ldots, NP\}$$

12:

$$u_{ij} = x_{r1j} + F \times \left(x_{r2j} - x_{r3j}\right)$$

13: **else**
14: Select randomly r1 $\neq$ r2 $\neq$ best $\neq$ i;

$$\forall i \in \{1, \ldots, NP\}$$

15:

$$u_{ij} = x_{ij} + F \times \left(x_{bestj} - x_{ij}\right) + F \times \left(x_{r2j} - x_{r3j}\right)$$

16: **end if**
17: **else**
18:

$$u_{ij} = x_{ij}$$

19: **end if**
20: **end for**
21:Evaluate the trial vector $\mathbf{u}_i$
22: **end for**
23: Do selection phase based on **Algorithm 1**
24: Define f_{best}, f_{mean}
25: delta = $\left|\frac{f_{best}}{f_{mean}-1}\right|$
26: **end while**

Algorithm 3 aeDE-ELM algorithm [41]

Input:
Training set, testing set;
aeDE algorithm parameters, *NP*;
1: Create a random initial population;
2: Evaluate the fitness for each individual with training set;
3: **while** (stopping criteria not met) **do**
4: Randomly generate F_i and CR_i
5: **for** i=1 to NP **do**
6: Call the **Algorithm 2**;
7: Use the optimal parameters of ELM;
8: **end for**
9: **end while**
10: Evaluate the optimized model by testing set;
Output:
Classification result;

Table 1 Comparison between algorithms

	Classification accuracy & performance metrics						
Classifier	ACC (%)	RMSE	Precision (%)	Recall	F-Score (%)	ROC area	Validation
SaE-ELM	**96.55**	**0.1637**	**0.966**	**0.966**	**0.965**	**0.996**	**10-fcv**
RBF ANN	90.60	0.2463	0.909	0.907	0.907	0.905	10-fcv
GMDH	92.66	0.1828	0.927	0.927	0.927	0.980	10-fcv
PANN	91.34	0.2162	0.914	0.913	0.914	0.961	10-fcv
FNN-GA	94.71	0.2054	0.947	0.947	0.947	0.969	10-fcv

The best results from classification are marked with bold

compared algorithms. Table 1 presents the analytical values of the predictive power of the AEDE-ELM by using a 10-Fold Cross Validation approach (10-fcv) and the corresponding results when competitive algorithms were used.

The *Precision* measure shows what percentage of positive predictions were correct, whereas *Recall* measures the percentage of positive events that were correctly predicted. The *F-Score* can be interpreted as a weighted average of the precision and recall. Therefore, this score takes both false positives and false negatives into account. Intuitively it is not as easy to understand as accuracy, but F-Score is usually more useful than accuracy and it works best if false positives and false negatives have similar cost, in this case. Finally, the ROC curve is related in a direct and natural way to cost/benefit analysis of diagnostic decision making. This comparison generates encouraging expectations for the identification of the AEDE-ELM as a robust classification model suitable for difficult problems.

According to this comparative analysis, it appears that AEDE-ELM is a highly suitable method for applications with huge amounts of data such that traditional learning approaches that use the entire data set in aggregate are computationally infeasible. This algorithm successfully reduces the problem of entrapment in local minima in training process, with very fast convergence rates. These improvements are accompanied by high classification rates and low test errors as well. The performance of proposed model was evaluated in a high complex dataset and the real-world sophisticated scenarios. The experimental results showed that the AEDE-ELM has better generalization performance at a very fast learning speed and more accurate and reliable classification results. The final conclusion is that the proposed method has proven to be reliable and efficient and has outperformed at least for this security problem the other approaches.

Discussion: Conclusions

An innovative biologically inspired hybrid computational intelligence approach suitable for big data was presented in this research paper. It is a computational intelligence system for identification cyber-attacks on Smart Energy Grids. Specifically, the hybrid and innovative AEDE-ELM algorithm was suggested which uses the innovative and highly effective algorithm AEDE in order to optimize the operating parameters of an ELM. The classification performance and the accuracy of the proposed model were experimentally explored based on several scenarios and reported very promising results. Moreover, SICASEG is an effective cross-layer system of network supervision, with capabilities of automated control. This is done to enhance the energetic security and the mechanisms of reaction of the general system, without special requirements. In this way, it adds a higher degree of integrity to the rest of the security infrastructure of Smart Energy Grids. The most significant innovation of this methodology is that it offers high learning speed, ease of implementation, minimal human intervention, and minimum computational power and resources to properly classify SCADA attacks with high accuracy and generalization.

Future research could involve its model under a hybrid scheme, which will combine semi supervised methods and online learning for the trace and exploitation of hidden knowledge between the inhomogeneous data that might emerge. Also, SICASEG can be further improved with self-adaptive learning properties such as self-modified the number of hidden notes. Moreover, additional computational intelligence methods could be explored, tested, and compared on the same security task in an ensemble approach. Finally, the ultimate challenge would be the scalability of SICASEG with other bio-inspired optimization algorithms in parallel and distributed computing in a real-time system.

References

1. Blumsack, S., Fernandez, A.: Ready or not, here comes the smart grid!. Energy **37**, 61–68 (2012)
2. Coll-Mayora, D., Pagetb, M., Lightnerc, E.: Future intelligent power grids: analysis of the vision in the European Union and the United States. Energy Policy **35**, 2453–2465 (2007)
3. Gellings C.W.: The Smart Grid: Enabling Energy Efficiency and Demand Response. The Fairmont Press, Lilburn (2009)

4. Rohjans, S., Uslar, M., Bleiker, R., Gonzalez, J., Specht, M., Suding, T., Weidelt, T.: Survey of smart grid standardization studies and recommendations. In: First IEEE International Conference on Smart Grid Communications (2010). Print ISBN: 978-1-4244-6510-1
5. Smart Grid NIST: NIST Smart Grid Conceptual Model. IEEE, New York (2010). http://smartgrid.ieee.org
6. Wang, W., Tolk, A.: The levels of conceptual interoperability model: applying systems engineering principles to M&S. In: Proceeding, SpringSim '09 Proceedings of the 2009 Spring Simulation Multiconference, Article No. 168. Society for Computer Simulation International, San Diego (2009)
7. Widergren, S., Levinson, A., Mater, J., Drummond, R.: Smart grid interoperability maturity model. In: Power & Energy Society General Meeting, IEEE, New York (2010). E-ISBN: 978-1-4244-8357-0
8. Naruchitparames, J., Gunes, M.H., Evrenosoglu, C.Y.: Secure Communications in the Smart Grid. IEEE, New York (2012)
9. Massoud, S.A., Giacomoni, A.M.: Smart grid—safe, secure, self-healing. IEEE Power Energy Mag.—Keeping the Smart Grid Safe **10**(1), 33–40 (2012)
10. Liu, C.-C., Stefanov, A., Hong, J., Panciatici, P.: Intruders in the grid. IEEE Power Energy Mag.—Keeping the Smart Grid Safe **10**(1), 58–66 (2012)
11. Wei, D., Jafari, Y.L.M., Skare, P.M., Rohde, K.: Protecting smart grid automation systems against cyberattacks. IEEE Trans. Smart Grid **2**(4), 782–795 (2011)
12. Hahn A., Govindarasu, M.: Cyber attack exposure evaluation framework for the smart grid. IEEE Trans. Smart Grid **2**(4), 835–843 (2011)
13. Ahmed, M.M., Soo, W.L.: Supervisory Control and Data Acquisition System (SCADA) based customized Remote Terminal Unit (RTU) for distribution automation system. In: Power and Energy Conference, 2008. PECon 2008. IEEE 2nd International (2008). https://doi.org/10.1109/PECON.2008.4762744
14. Kalluri, E., Mahendra, L., Senthil Kumar, R.K., Ganga Prasad, G.L.: National Power Systems Conference (NPSC), pp. 1–5. IEEE Conference Publications (2016) https://doi.org/10.1109/NPSC.2016.7858908
15. Demertzis K., Iliadis, L.: Intelligent bio-inspired detection of food borne pathogen by DNA barcodes: the case of invasive fish species Lagocephalus Sceleratus. Eng. Appl. Neural Netw. **517**, 89–99 (2015). https://doi.org/10.1007/978-3-319-23983-5_9
16. Demertzis, K., Iliadis, L.: A hybrid network anomaly and intrusion detection approach based on evolving spiking neural network classification. In: E-Democracy, Security, Privacy and Trust in a Digital World. Communications in Computer and Information Science, vol. 441, pp. 11–23. Springer, Berlin (2014). https://doi.org/10.1007/978-3-319-11710-2_2
17. Demertzis, K., Iliadis, L.: Evolving computational intelligence system for malware detection. In: Advanced Information Systems Engineering Workshops. Lecture Notes in Business Information Processing, vol. 178, pp. 322–334. Springer, Berlin (2014). https://doi.org/10.1007/978-3-319-07869-4_30
18. Demertzis K., Iliadis L.: Bio-inspired hybrid artificial intelligence framework for cyber security. In: Springer Proceedings 2nd Conference on CryptAAF: Cryptography Network Security and Applications in the Armed Forces, Athens, pp. 161–193. Springer, Cham (2014). https://doi.org/10.1007/978-3-319-18275-9_7
19. Demertzis, K., Iliadis, L.: Bio-inspired hybrid intelligent method for detecting android malware. In: Proceedings of the 9th KICSS 2014, Knowledge Information and Creative Support Systems, Cyprus, pp. 231–243 (2014). ISBN: 978-9963-700-84-4
20. Demertzis, K., Iliadis, L.: Evolving smart URL filter in a zone-based policy firewall for detecting algorithmically generated malicious domains. In: Proceedings SLDS (Statistical Learning and Data Sciences) Conference. Lecture Notes in Artificial Intelligence, vol. 9047, pp. 223–233 (Springer, Royal Holloway University, London, 2015). https://doi.org/10.1007/978-3-319-17091-6_17

21. Demertzis K., Iliadis, L.: SAME: An intelligent anti-malware extension for android ART virtual machine. In: Proceedings of the 7th International Conference ICCCI 2015. Lecture Notes in Artificial Intelligence, vol. 9330, pp. 235–245. Springer, Cham (2015). https://doi.org/10.1007/978-3-319-24306-1_23
22. Demertzis, K., Iliadis, L.: Computational intelligence anti-malware framework for android OS. Vietnam J. Comput. Sci. (Special Issue) (2016). Springer, Berlin. https://doi.org/10.1007/s40595-017-0095-3
23. Demertzis, K., Iliadis, L.: Detecting invasive species with a bio-inspired semi supervised neurocomputing approach: the case of Lagocephalus Sceleratus. Neural Comput. Appl. (Special issues) (2016). Springer. https://doi.org/10.1007/s00521-016-2591-2
24. Demertzis, K., Iliadis, L.: SICASEG: a cyber threat bio-inspired intelligence management system. J. Appl. Math. Bioinform. **6**(3), 45–64 (2016). ISSN: 1792-6602 (print). 1792-6939 (online). Scienpress Ltd. (2016)
25. Bougoudis I., Demertzis K., Iliadis, L.: Fast and low cost prediction of extreme air pollution values with hybrid unsupervised learning. Integr. Comput. Aided Eng. **23**(2), 115–127 (2016). https://doi.org/10.3233/ICA-150505. IOS Press (2016)
26. Bougoudis, I., Demertzis, K., Iliadis, L.: HISYCOL a hybrid computational intelligence system for combined machine learning: the case of air pollution modeling in Athens. Neural Comput. Appl. 1–16 (2016). https://doi.org/10.1007/s00521-015-1927-7
27. Anezakis, V.D., Demertzis, K., Iliadis, L., Spartalis, S.: A hybrid soft computing approach producing robust forest fire risk indices. In: IFIP Advances in Information and Communication Technology, AIAI September 2016, Thessaloniki, vol. 475, pp. 191–203 (2016)
28. Anezakis, V.D., Dermetzis, K., Iliadis, L., Spartalis, S.: Fuzzy cognitive maps for long-term prognosis of the evolution of atmospheric pollution, based on climate change scenarios: the case of Athens. In: Lecture Notes in Computer Science. Lecture Notes in Artificial Intelligence and Lecture Notes in Bioinformatics, vol. 9875, pp. 175–186. Springer, Cham (2016). https://doi.org/10.1007/978-3-319-45243-2_16
29. Bougoudis, I., Demertzis, K., Iliadis, L., Anezakis, V.D., Papaleonidas. A.: Semi-supervised hybrid modeling of atmospheric pollution in urban centers. Commun. Comput. Inform. Sci. **629**, 51–63 (2016)
30. Yasakethu, S.L.P., Jiang, J.: Intrusion detection via machine learning for SCADA System Protection. In: Proceedings of the 1st International Symposium for ICS & SCADA Cyber Security Research 2013. Learning and Development Ltd. (2013)
31. Chen, Q., Abdelwahed, S.: A model-based approach to self-protection in computing system. In: Proceeding CAC '13 Proceedings of the 2013 ACM Cloud and Autonomic Computing Conference, Article No. 16 (2013)
32. Soupionis, Y., Ntalampiras, S., Giannopoulos, G.: Lecture Notes in Computer Science (2016) https://doi.org/10.1007/978-3-319-31664-2_29
33. Qin, Y., Cao, X., Liang, P.: Hu, Q.: Zhang, W.: Research on the analytic factor neuron model based on cloud generator and its application in oil&gas SCADA security defense. In: 2014 IEEE 3rd International Conference on Cloud Computing and Intelligence Systems (CCIS) (2014). https://doi.org/10.1109/CCIS.2014.7175721
34. Pan, S., Morris, T., Adhikari, U.: Developing a hybrid intrusion detection system using data mining for power systems. IEEE Trans. Smart Grid (2015). https://doi.org/10.1109/TSG.2015.2409775
35. Pan, S., Morris, T., Adhikari, U.: Classification of disturbances and cyber-attacks in power systems using heterogeneous time-synchronized data. IEEE Trans. Ind. Inform. (2015) https://doi.org/10.1109/TII.2015.2420951
36. Pan, S., Morris, T., Adhikari, U.: A specification-based intrusion detection framework for cyber-physical environment in electric power system. Int. J. Netw. Secur. **17**(2), 174–188 (2015)

37. Beaver, J., Borges, R., Buckner, M., Morris, T., Adhikari, U., Pan, S.: Machine learning for power system disturbance and cyber-attack discrimination. In: Proceedings of the 7th International Symposium on Resilient Control Systems, Denver, CO (2014)
38. Cambria, E., Guang-Bin, H.: Extreme learning machines. In: IEEE InTeLLIGenT SYSTemS. 541-1672/13 (2013)
39. Price, K., Storn, M., Lampinen, A.: Differential Evolution: A Practical Approach to Global Optimization. Springer, Berlin (2005). ISBN: 978-3-540-20950-8
40. Ho-Huu, V., Nguyen-Thoi, T., Vo-Duy, T., Nguyen-Trang, T.: An adaptive elitist differential evolution for optimization of truss structures with discrete design variables. Comput. Struct. **165**, 59–75 (2016)
41. Demertzis, K., Iliadis, L.: Adaptive elitist differential evolution extreme learning machines on big data: intelligent recognition of invasive species. In: International Neural Network Society Conference on Big Data (INNS Big Data 2016), Thessaloniki, Proceedings. Advances in Big Data, pp. 23–25. Advances in Intelligent Systems and Computing, vol. 529, pp. 333–345. Springer, Cham (2016) https://doi.org/10.1007/978-3-319-47898-2_34

Recent Developments of Discrete Inequalities for Convex Functions Defined on Linear Spaces with Applications

Silvestru Sever Dragomir

Abstract In this paper we survey some recent discrete inequalities for functions defined on convex subsets of general linear spaces. Various refinements and reverses of Jensen's discrete inequality are presented. The Slater inequality version for these functions is outlined. As applications, we establish several bounds for the *mean f-deviation* of an n-tuple of vectors as well as for the *f-divergence* of an n -tuple of vectors given a discrete probability distribution. Examples for the K. Pearson χ^2*-divergence,* the *Kullback-Leibler divergence,* the *Jeffreys divergence,* the *total variation distance* and other divergence measures are also provided.

Keywords Convex functions on linear spaces · Discrete Jensen's inequality · Reverse of Jensen's inequality · Discrete divergence measures · f-Divergence measures

Introduction

The Jensen inequality for convex functions plays a crucial role in the theory of inequalities due to the fact that other inequalities such as that arithmetic mean-geometric mean inequality, Hölder and Minkowski inequalities, Ky Fan's inequality, etc. can be obtained as particular cases of it. In order to state some recent reverses of Jensen's discrete inequality for functions of a real variable we need the following facts.

S. S. Dragomir (✉)
Mathematics, College of Engineering & Science, Victoria University, Melbourne City, MC, Australia

DST-NRF Centre of Excellence in the Mathematical and Statistical Sciences, School of Computer Science & Applied Mathematics, University of the Witwatersrand, Johannesburg, South Africa
e-mail: sever.dragomir@vu.edu.au
http://rgmia.org/dragomir

N. J. Daras, Th. M. Rassias (eds.), *Modern Discrete Mathematics and Analysis*, Springer Optimization and Its Applications 131,
https://doi.org/10.1007/978-3-319-74325-7_6

If $x_i, y_i \in \mathbb{R}$ and $w_i \geq 0\,(i = 1, \ldots, n)$ with $W_n := \sum_{i=1}^{n} w_i = 1$, then we may consider the *Čebyšev functional*:

$$T_w(x, y) := \sum_{i=1}^{n} w_i x_i y_i - \sum_{i=1}^{n} w_i x_i \sum_{i=1}^{n} w_i y_i. \tag{1}$$

The following result is known in the literature as the *Grüss inequality*:

$$|T_w(x, y)| \leq \frac{1}{4}(\Gamma - \gamma)(\Delta - \delta), \tag{2}$$

provided

$$-\infty < \gamma \leq x_i \leq \Gamma < \infty, \quad -\infty < \delta \leq y_i \leq \Delta < \infty \tag{3}$$

for $i = 1, \ldots, n$.

The constant $\frac{1}{4}$ is sharp in the sense that it cannot be replaced by a smaller constant.

If we assume that $-\infty < \gamma \leq x_i \leq \Gamma < \infty$ for $i = 1, \ldots, n$, then by the Grüss inequality for $y_i = x_i$ and by the Schwarz's discrete inequality, we have (see also [1])

$$\sum_{i=1}^{n} w_i \left| x_i - \sum_{j=1}^{n} w_j x_j \right| \leq \left[\sum_{i=1}^{n} w_i x_i^2 - \left(\sum_{j=1}^{n} w_j x_j \right)^2 \right]^{\frac{1}{2}} \leq \frac{1}{2}(\Gamma - \gamma). \tag{4}$$

In order to provide a reverse of the celebrated Jensen's inequality for convex functions of a real variable, S. S. Dragomir obtained in 2002 [14] the following result:

Theorem 1 *Let $f : [m, M] \to \mathbb{R}$ be a differentiable convex function on (m, M). If $x_i \in [m, M]$ and $w_i \geq 0$ $(i = 1, \ldots, n)$ with $W_n := \sum_{i=1}^{n} w_i = 1$, then one has the counterpart of Jensen's weighted discrete inequality:*

$$\begin{aligned} 0 &\leq \sum_{i=1}^{n} w_i f(x_i) - f\left(\sum_{i=1}^{n} w_i x_i \right) \\ &\leq \sum_{i=1}^{n} w_i f'(x_i) x_i - \sum_{i=1}^{n} w_i f'(x_i) \sum_{i=1}^{n} w_i x_i \\ &\leq \frac{1}{2}\left[f'(M) - f'(m) \right] \sum_{i=1}^{n} w_i \left| x_i - \sum_{j=1}^{n} w_j x_j \right|. \end{aligned} \tag{5}$$

Remark 1 We notice that the inequality between the first and the second term in (5) was proved in 1994 by Dragomir and Ionescu; see [28].

On making use of (4), we can state the following string of reverse inequalities:

$$\begin{aligned}0 &\le \sum_{i=1}^{n} w_i f(x_i) - f\left(\sum_{i=1}^{n} w_i x_i\right) \qquad (6)\\ &\le \sum_{i=1}^{n} w_i f'(x_i) x_i - \sum_{i=1}^{n} w_i f'(x_i) \sum_{i=1}^{n} w_i x_i \\ &\le \frac{1}{2}\left[f'(M) - f'(m)\right] \sum_{i=1}^{n} w_i \left| x_i - \sum_{j=1}^{n} w_j x_j \right| \\ &\le \frac{1}{2}\left[f'(M) - f'(m)\right] \left[\sum_{i=1}^{n} w_i x_i^2 - \left(\sum_{j=1}^{n} w_j x_j\right)^2\right]^{\frac{1}{2}} \\ &\le \frac{1}{4}\left[f'(M) - f'(m)\right](M-m),\end{aligned}$$

provided that $f : [m, M] \subset \mathbb{R} \to \mathbb{R}$ is a differentiable convex function on (m, M), $x_i \in [m, M]$ and $w_i \ge 0 \ (i = 1, \ldots, n)$ with $W_n := \sum_{i=1}^{n} w_i = 1$.

Remark 2 We notice that the inequality between the first, second and last term from (6) was proved in the general case of positive linear functionals in 2001 by S. S. Dragomir in [13].

The following reverse Jensen's inequality for convex functions of a real variable also holds:

Theorem 2 (Dragomir [24]) *Let $f : I \to \mathbb{R}$ be a continuous convex function on the interval of real numbers I and $m, M \in \mathbb{R}$, $m < M$ with $[m, M] \subset \mathring{I}$, $\mathring{I}$ is the interior of I. If $x_i \in [m, M]$ and $w_i \ge 0 \ (i = 1, \ldots, n)$ with $W_n := \sum_{i=1}^{n} w_i = 1$, then*

$$\begin{aligned}0 &\le \sum_{i=1}^{n} w_i f(x_i) - f\left(\sum_{i=1}^{n} w_i x_i\right) \qquad (7)\\ &\le \frac{\left(M - \sum_{i=1}^{n} w_i x_i\right)\left(\sum_{i=1}^{n} w_i x_i - m\right)}{M-m} \Psi_f\left(\sum_{i=1}^{n} w_i x_i; m, M\right) \\ &\le \frac{\left(M - \sum_{i=1}^{n} w_i x_i\right)\left(\sum_{i=1}^{n} w_i x_i - m\right)}{M-m} \sup_{t\in(m,M)} \Psi_f(t; m, M)\end{aligned}$$

$$\leq \left(M - \sum_{i=1}^{n} w_i x_i\right)\left(\sum_{i=1}^{n} w_i x_i - m\right) \frac{f'_-(M) - f'_+(m)}{M - m}$$

$$\leq \frac{1}{4}(M - m)\left[f'_-(M) - f'_+(m)\right],$$

where $\Psi_f(\cdot; m, M) : (m, M) \to \mathbb{R}$ *is defined by*

$$\Psi_f(t; m, M) = \frac{f(M) - f(t)}{M - t} - \frac{f(t) - f(m)}{t - m}.$$

We also have the inequality

$$0 \leq \sum_{i=1}^{n} w_i f(x_i) - f\left(\sum_{i=1}^{n} w_i x_i\right) \tag{8}$$

$$\leq \frac{\left(M - \sum_{i=1}^{n} w_i x_i\right)\left(\sum_{i=1}^{n} w_i x_i - m\right)}{M - m} \Psi_f\left(\sum_{i=1}^{n} w_i x_i; m, M\right)$$

$$\leq \frac{1}{4}(M - m)\Psi_f\left(\sum_{i=1}^{n} w_i x_i; m, M\right)$$

$$\leq \frac{1}{4}(M - m)\sup_{t\in(m,M)} \Psi_f(t; m, M)$$

$$\leq \frac{1}{4}(M - m)\left[f'_-(M) - f'_+(m)\right],$$

provided that $\sum_{i=1}^{n} w_i x_i \in (m, M)$.

The following result also holds:

Theorem 3 (Dragomir [24]) *With the assumptions of Theorem 2, we have the inequalities*

$$0 \leq \sum_{i=1}^{n} w_i f(x_i) - f\left(\sum_{i=1}^{n} w_i x_i\right) \tag{9}$$

$$\leq 2\max\left\{\frac{M - \sum_{i=1}^{n} w_i x_i}{M - m}, \frac{\sum_{i=1}^{n} w_i x_i - m}{M - m}\right\}$$

$$\times \left[\frac{f(m) + f(M)}{2} - f\left(\frac{m + M}{2}\right)\right]$$

$$\leq \frac{1}{2}\max\left\{M - \sum_{i=1}^{n} w_i x_i, \sum_{i=1}^{n} w_i x_i - m\right\}\left[f'_-(M) - f'_+(m)\right].$$

Remark 3 Since, obviously,

$$\frac{M-\sum_{i=1}^{n} w_i x_i}{M-m}, \frac{\sum_{i=1}^{n} w_i x_i - m}{M-m} \leq 1,$$

then we obtain from the first inequality in (9) the simpler, however, coarser inequality, namely

$$0 \leq \sum_{i=1}^{n} w_i f(x_i) - f\left(\sum_{i=1}^{n} w_i x_i\right) \leq 2\left[\frac{f(m)+f(M)}{2} - f\left(\frac{m+M}{2}\right)\right]. \tag{10}$$

This inequality was obtained in 2008 by S. Simić in [31].

The following result also holds:

Theorem 4 (Dragomir [25]) *Let $\Phi : I \to \mathbb{R}$ be a continuous convex function on the interval of real numbers I and $m, M \in \mathbb{R}$, $m < M$ with $[m, M] \subset \mathring{I}$; $\mathring{I}$ is the interior of I. If $x_i \in I$ and $w_i \geq 0$ for $i \in \{1, \ldots, n\}$ with $\sum_{i=1}^{n} w_i = 1$, denote $\bar{x}_w := \sum_{i=1}^{n} w_i x_i \in I$, then we have the inequality*

$$0 \leq \sum_{i=1}^{n} w_i \Phi(x_i) - \Phi(\bar{x}_w) \tag{11}$$

$$\leq \frac{(M-\bar{x}_w)\int_m^{\bar{x}_w} |\Phi'(t)|\, dt + (\bar{x}_w - m)\int_{\bar{x}_w}^{M} |\Phi'(t)|\, dt}{M-m} := \Theta_\Phi(\bar{x}_w; m, M),$$

where $\Theta_\Phi(\bar{x}_w; m, M)$ satisfies the bounds

$$\Theta_\Phi(\bar{x}_w; m, M) \tag{12}$$

$$\leq \begin{cases} \left[\frac{1}{2} + \frac{\left|\bar{x}_w - \frac{m+M}{2}\right|}{M-m}\right] \int_m^M |\Phi'(t)|\, dt \\ \\ \left[\frac{1}{2}\int_m^M |\Phi'(t)|\, dt + \frac{1}{2}\left|\int_{\bar{x}_w}^M |\Phi'(t)|\, dt - \int_m^{\bar{x}_w} |\Phi'(t)|\, dt\right|\right], \end{cases}$$

$$\Theta_\Phi(\bar{x}_w; m, M) \tag{13}$$

$$\leq \frac{(\bar{x}_w - m)(M-\bar{x}_w)}{M-m}\left[\|\Phi'\|_{[\bar{x}_w, M],\infty} + \|\Phi'\|_{[m,\bar{x}_w],\infty}\right]$$

$$\leq \frac{1}{2}(M-m)\frac{\|\Phi'\|_{[\bar{w}_p, M],\infty} + \|\Phi'\|_{[m,\bar{w}_p],\infty}}{2} \leq \frac{1}{2}(M-m)\|\Phi'\|_{[m,M],\infty}$$

and

$$\Theta_{\Phi}(\bar{x}_w; m, M) \tag{14}$$
$$\leq \frac{1}{M-m}\Big[(\bar{x}_w - m)(M-\bar{x}_w)^{1/q}\left\|\Phi'\right\|_{[\bar{x}_w, M], p}$$
$$+ (M-\bar{x}_w)(\bar{x}_w - m)^{1/q}\left\|\Phi'\right\|_{[m, \bar{x}_w], p}\Big]$$
$$\leq \frac{1}{M-m}\left[(\bar{x}_w - m)^q (M-\bar{x}_w) + (M-\bar{x}_w)^q (\bar{x}_w - m)\right]^{1/q}\left\|\Phi'\right\|_{[m, M], p}$$

where $p > 1, \frac{1}{p} + \frac{1}{q} = 1$.

For a real function $g : [m, M] \to \mathbb{R}$ and two distinct points $\alpha, \beta \in [m, M]$, we recall that the *divided difference* of g in these points is defined by

$$[\alpha, \beta; g] := \frac{g(\beta) - g(\alpha)}{\beta - \alpha}.$$

Theorem 5 (Dragomir [22]) *Let* $f : I \to \mathbb{R}$ *be a continuous convex function on the interval of real numbers* I *and* $m, M \in \mathbb{R}, m < M$ *with* $[m, M] \subset \mathring{I}$, $\mathring{I}$ *the interior of* I. *Let* $\overline{\mathbf{a}} = (a_1, \ldots, a_n), \overline{\mathbf{p}} = (p_1, \ldots, p_n)$ *be* n*-tuples of real numbers with* $p_i \geq 0\,(i \in \{1, \ldots, n\})$ *and* $\sum_{i=1}^{n} p_i = 1$. *If* $m \leq a_i \leq M, \quad i \in \{1, \ldots, n\}$, *with* $\sum_{i=1}^{n} p_i a_i \neq m, M$, *then*

$$\left|\sum_{i=1}^{n} p_i \left| f(a_i) - f\left(\sum_{j=1}^{n} p_j a_j\right)\right| sgn\left(a_i - \sum_{j=1}^{n} p_j a_j\right)\right| \tag{15}$$
$$\leq \sum_{i=1}^{n} p_i f(a_i) - f\left(\sum_{i=1}^{n} p_i a_i\right)$$
$$\leq \frac{1}{2}\left(\left[\sum_{i=1}^{n} p_i a_i, M; f\right] - \left[m, \sum_{i=1}^{n} p_i a_i; f\right]\right)\sum_{i=1}^{n} p_i \left|a_i - \sum_{j=1}^{n} p_j a_j\right|$$
$$\leq \frac{1}{2}\left(\left[\sum_{i=1}^{n} p_i a_i, M; f\right] - \left[m, \sum_{i=1}^{n} p_i a_i; f\right]\right)$$
$$\times \left[\sum_{i=1}^{n} p_i a_i^2 - \left(\sum_{j=1}^{n} p_j a_j\right)^2\right]^{1/2}.$$

If the lateral derivatives $f'_+(m)$ *and* $f'_-(M)$ *are finite, then we also have the inequalities*

$$
\begin{aligned}
0 &\leq \sum_{i=1}^{n} p_i f(a_i) - f\left(\sum_{i=1}^{n} p_i a_i\right) \qquad (16)\\
&\leq \frac{1}{2}\left(\left[\sum_{i=1}^{n} p_i a_i, M; f\right] - \left[m, \sum_{i=1}^{n} p_i a_i; f\right]\right) \sum_{i=1}^{n} p_i \left|a_i - \sum_{j=1}^{n} p_j a_j\right| \\
&\leq \frac{1}{2}\left[f'_-(M) - f'_+(m)\right] \sum_{i=1}^{n} p_i \left|a_i - \sum_{j=1}^{n} p_j a_j\right| \\
&\leq \frac{1}{2}\left[f'_-(M) - f'_+(m)\right] \left[\sum_{i=1}^{n} p_i a_i^2 - \left(\sum_{j=1}^{n} p_j a_j\right)^2\right]^{1/2}.
\end{aligned}
$$

In this paper we survey some recent discrete inequalities for functions defined on convex subsets of general linear spaces. Various refinements and reverses of Jensen's discrete inequality are presented. The Slater inequality version for these functions is outlined. As applications, we establish several bounds for the *mean f-deviation* of an n-tuple of vectors as well as for the *f-divergence* of an n-tuple of vectors given a discrete probability distribution. Examples for the K. Pearson *χ^2-divergence,* the *Kullback-Leibler divergence,* the *Jeffreys divergence,* the *total variation distance* and other divergence measures are also provided.

Refinements of Jensen's Inequality

Preliminary Facts

Let C be a convex subset of the linear space X and f a convex function on C. If $\mathbf{p} = (p_1, \ldots, p_n)$ is a probability sequence and $\mathbf{x} = (x_1, \ldots, x_n) \in C^n$, then

$$
f\left(\sum_{i=1}^{n} p_i x_i\right) \leq \sum_{i=1}^{n} p_i f(x_i), \qquad (17)
$$

is well known in the literature as Jensen's inequality.

In 1989, J. Pečarić and the author [30] obtained the following refinement of (17):

$$
f\left(\sum_{i=1}^{n} p_i x_i\right) \leq \sum_{i_1,\ldots,i_{k+1}=1}^{n} p_{i_1} \cdots p_{i_{k+1}} f\left(\frac{x_{i_1} + \cdots + x_{i_{k+1}}}{k+1}\right) \qquad (18)
$$

$$\le \sum_{i_1,\ldots,i_k=1}^{n} p_{i_1}\cdots p_{i_k} f\left(\frac{x_{i_1}+\cdots+x_{i_k}}{k}\right)$$
$$\le \cdots \le \sum_{i=1}^{n} p_i f(x_i),$$

for $k \ge 1$ and $\mathbf{p}$, $\mathbf{x}$ are as above.

If $q_1, \ldots, q_k \ge 0$ with $\sum_{j=1}^{k} q_j = 1$, then the following refinement obtained in 1994 by the author [2] also holds:

$$f\left(\sum_{i=1}^{n} p_i x_i\right) \le \sum_{i_1,\ldots,i_k=1}^{n} p_{i_1}\cdots p_{i_k} f\left(\frac{x_{i_1}+\cdots+x_{i_k}}{k}\right) \tag{19}$$
$$\le \sum_{i_1,\ldots,i_k=1}^{n} p_{i_1}\cdots p_{i_k} f\left(q_1 x_{i_1}+\cdots+q_k x_{i_k}\right) \le \sum_{i=1}^{n} p_i f(x_i),$$

where $1 \le k \le n$ and $\mathbf{p}$, $\mathbf{x}$ are as above.

For other refinements and applications related to Ky Fan's inequality, the arithmetic mean-geometric mean inequality, the generalised triangle inequality, etc., see [3–29].

General Results

The following result may be stated.

Theorem 6 (Dragomir [19]) *Let $f : C \to \mathbb{R}$ be a convex function on the convex subset C of the linear space X, $x_i \in C$, $p_i > 0$, $i \in \{1, \ldots, n\}$ with $\sum_{i=1}^{n} p_i = 1$. Then*

$$f\left(\sum_{j=1}^{n} p_j x_j\right) \le \min_{k\in\{1,\ldots,n\}} \left[(1-p_k) f\left(\frac{\sum_{j=1}^{n} p_j x_j - p_k x_k}{1-p_k}\right) + p_k f(x_k)\right] \tag{20}$$
$$\le \frac{1}{n}\left[\sum_{k=1}^{n}(1-p_k) f\left(\frac{\sum_{j=1}^{n} p_j x_j - p_k x_k}{1-p_k}\right) + \sum_{k=1}^{n} p_k f(x_k)\right]$$
$$\le \max_{k\in\{1,\ldots,n\}} \left[(1-p_k) f\left(\frac{\sum_{j=1}^{n} p_j x_j - p_k x_k}{1-p_k}\right) + p_k f(x_k)\right]$$
$$\le \sum_{j=1}^{n} p_j f(x_j).$$

In particular,

$$f\left(\frac{1}{n}\sum_{j=1}^{n}x_j\right) \le \frac{1}{n}\min_{k\in\{1,\dots,n\}}\left[(n-1)f\left(\frac{\sum_{j=1}^{n}x_j - x_k}{n-1}\right) + f(x_k)\right] \tag{21}$$
$$\le \frac{1}{n^2}\left[(n-1)\sum_{k=1}^{n} f\left(\frac{\sum_{j=1}^{n}x_j - x_k}{n-1}\right) + \sum_{k=1}^{n} f(x_k)\right]$$
$$\le \frac{1}{n}\max_{k\in\{1,\dots,n\}}\left[(n-1)f\left(\frac{\sum_{j=1}^{n}x_j - x_k}{n-1}\right) + f(x_k)\right]$$
$$\le \frac{1}{n}\sum_{j=1}^{n} f(x_j).$$

Proof For any $k \in \{1,\dots,n\}$, we have

$$\sum_{j=1}^{n}p_jx_j - p_kx_k = \sum_{\substack{j=1\\ j\neq k}}^{n}p_jx_j = \frac{\sum\limits_{\substack{j=1\\ j\neq k}}^{n}p_j}{\sum\limits_{\substack{j=1\\ j\neq k}}^{n}p_j}\sum_{\substack{j=1\\ j\neq k}}^{n}p_jx_j = (1-p_k)\cdot\frac{1}{\sum\limits_{\substack{j=1\\ j\neq k}}^{n}p_j}\sum_{\substack{j=1\\ j\neq k}}^{n}p_jx_j,$$

which implies that

$$\frac{\sum_{j=1}^{n}p_jx_j - p_kx_k}{1-p_k} = \frac{1}{\sum\limits_{\substack{j=1\\ j\neq k}}^{n}p_j}\sum_{\substack{j=1\\ j\neq k}}^{n}p_jx_j \in C \tag{22}$$

for each $k \in \{1,\dots,n\}$, since the right side of(22) is a convex combination of the elements $x_j \in C$, $j \in \{1,\dots,n\}\setminus\{k\}$.

Taking the function f on (22) and applying the Jensen inequality, we get successively

$$f\left(\frac{\sum_{j=1}^{n}p_jx_j - p_kx_k}{1-p_k}\right) = f\left(\frac{1}{\sum\limits_{\substack{j=1\\ j\neq k}}^{n}p_j}\sum_{\substack{j=1\\ j\neq k}}^{n}p_jx_j\right) \le \frac{1}{\sum\limits_{\substack{j=1\\ j\neq k}}^{n}p_j}\sum_{\substack{j=1\\ j\neq k}}^{n}p_jf(x_j)$$
$$= \frac{1}{1-p_k}\left[\sum_{j=1}^{n}p_jf(x_j) - p_kf(x_k)\right]$$

for any $k \in \{1, \ldots, n\}$, which implies

$$(1 - p_k) f\left(\frac{\sum_{j=1}^{n} p_j x_j - p_k x_k}{1 - p_k}\right) + p_k f(x_k) \le \sum_{j=1}^{n} p_j f(x_j) \tag{23}$$

for each $k \in \{1, \ldots, n\}$.

Utilising the convexity of f, we also have

$$\begin{aligned} &(1 - p_k) f\left(\frac{\sum_{j=1}^{n} p_j x_j - p_k x_k}{1 - p_k}\right) + p_k f(x_k) \\ &\quad \ge f\left[(1 - p_k) \cdot \frac{\sum_{j=1}^{n} p_j x_j - p_k x_k}{1 - p_k} + p_k x_k\right] = f\left(\sum_{j=1}^{n} p_j x_j\right) \end{aligned} \tag{24}$$

for each $k \in \{1, \ldots, n\}$.

Taking the minimum over k in (24), utilising the fact that

$$\min_{k \in \{1,\ldots,n\}} \alpha_k \le \frac{1}{n} \sum_{k=1}^{n} \alpha_k \le \max_{k \in \{1,\ldots,n\}} \alpha_k$$

and then taking the maximum in (23), we deduce the desired inequality (20). □

After setting $x_j = y_j - \sum_{l=1}^{n} q_l y_l$ and $p_j = q_j$, $j \in \{1, \ldots, n\}$, Theorem 6 becomes the following corollary:

Corollary 1 (Dragomir [19]) *Let $f : C \to \mathbb{R}$ be a convex function on the convex subset $C, 0 \in C, y_j \in X$ and $q_j > 0, j \in \{1, \ldots, n\}$ with $\sum_{j=1}^{n} q_j = 1$. If $y_j - \sum_{l=1}^{n} q_l y_l \in C$ for any $j \in \{1, \ldots, n\}$, then*

$$\begin{aligned} &f(0) \\ &\le \min_{k \in \{1,\ldots,n\}} \left\{(1 - q_k) f\left[\frac{q_k}{1 - q_k}\left(\sum_{l=1}^{n} q_l y_l - y_k\right)\right] + q_k f\left(y_k - \sum_{l=1}^{n} q_l y_l\right)\right\} \\ &\le \frac{1}{n}\left\{\sum_{l=1}^{n} (1 - q_k) f\left[\frac{q_k}{1 - q_k}\left(\sum_{l=1}^{n} q_l y_l - y_k\right)\right] + \sum_{l=1}^{n} q_k f\left(y_k - \sum_{l=1}^{n} q_l y_l\right)\right\} \\ &\le \max_{k \in \{1,\ldots,n\}} \left\{(1 - q_k) f\left[\frac{q_k}{1 - q_k}\left(\sum_{l=1}^{n} q_l y_l - y_k\right)\right] + q_k f\left(y_k - \sum_{l=1}^{n} q_l y_l\right)\right\} \\ &\le \sum_{j=1}^{n} q_j f\left(y_j - \sum_{l=1}^{n} q_l y_l\right). \end{aligned} \tag{25}$$

In particular, if $y_j - \frac{1}{n}\sum_{l=1}^{n} y_l \in C$ *for any* $j \in \{1, \ldots, n\}$, *then*

$$\begin{aligned} & f(0) \qquad (26) \\ & \leq \frac{1}{n} \min_{k\in\{1,\ldots,n\}} \left\{ (n-1) f\left[\frac{1}{n-1}\left(\frac{1}{n}\sum_{l=1}^{n} y_l - y_k\right)\right] + f\left(y_k - \frac{1}{n}\sum_{l=1}^{n} y_l\right)\right\} \\ & \leq \frac{1}{n^2}\left\{(n-1)\sum_{k=1}^{n} f\left[\frac{1}{n-1}\left(\frac{1}{n}\sum_{l=1}^{n} y_l - y_k\right)\right] + \sum_{k=1}^{n} f\left(y_k - \frac{1}{n}\sum_{l=1}^{n} y_l\right)\right\} \\ & \leq \frac{1}{n} \max_{k\in\{1,\ldots,n\}} \left\{ (n-1) f\left[\frac{1}{n-1}\left(\frac{1}{n}\sum_{l=1}^{n} y_l - y_k\right)\right] + f\left(y_k - \frac{1}{n}\sum_{l=1}^{n} y_l\right)\right\} \\ & \leq \frac{1}{n}\sum_{j=1}^{n} f\left(y_j - \frac{1}{n}\sum_{l=1}^{n} y_l\right). \end{aligned}$$

The above results can be applied for various convex functions related to celebrated inequalities as mentioned in the introduction.

Application 1. If $(X, \|\cdot\|)$ is a normed linear space and $p \geq 1$, then the function $f : X \to \mathbb{R}$, $f(x) = \|x\|^p$ is convex on X. Now, on applying Theorem 6 and Corollary 1 for $x_i \in X$, $p_i > 0$, $i \in \{1, \ldots, n\}$ with $\sum_{i=1}^{n} p_i = 1$, we get

$$\begin{aligned} \left\|\sum_{j=1}^{n} p_j x_j\right\|^p & \leq \min_{k\in\{1,\ldots,n\}} \left[(1-p_k)^{1-p}\left\|\sum_{j=1}^{n} p_j x_j - p_k x_k\right\|^p + p_k \|x_k\|^p\right] \qquad (27) \\ & \leq \frac{1}{n}\left[\sum_{k=1}^{n}(1-p_k)^{1-p}\left\|\sum_{j=1}^{n} p_j x_j - p_k x_k\right\|^p + \sum_{k=1}^{n} p_k \|x_k\|^p\right] \\ & \leq \max_{k\in\{1,\ldots,n\}} \left[(1-p_k)^{1-p}\left\|\sum_{j=1}^{n} p_j x_j - p_k x_k\right\|^p + p_k \|x_k\|^p\right] \\ & \leq \sum_{j=1}^{n} p_j \|x_j\|^p \end{aligned}$$

and

$$\begin{aligned} & \max_{k\in\{1,\ldots,n\}} \left\{\left[(1-p_k)^{1-p} p_k^p + p_k\right]\left\|x_k - \sum_{l=1}^{n} p_l x_l\right\|^p\right\} \\ & \leq \sum_{j=1}^{n} p_j \left\|x_k - \sum_{l=1}^{n} p_l x_l\right\|^p. \qquad (28) \end{aligned}$$

In particular, we have the inequality:

$$\left\|\frac{1}{n}\sum_{j=1}^{n}x_j\right\|^p \le \frac{1}{n}\min_{k\in\{1,\dots,n\}}\left[(n-1)^{1-p}\left\|\sum_{j=1}^{n}x_j - x_k\right\|^p + \|x_k\|^p\right] \tag{29}$$
$$\le \frac{1}{n^2}\left[(n-1)^{1-p}\sum_{k=1}^{n}\left\|\sum_{j=1}^{n}x_j - x_k\right\|^p + \sum_{k=1}^{n}\|x_k\|^p\right]$$
$$\le \frac{1}{n}\max_{k\in\{1,\dots,n\}}\left[(n-1)^{1-p}\left\|\sum_{j=1}^{n}x_j - x_k\right\|^p + \|x_k\|^p\right]$$
$$\le \frac{1}{n}\sum_{j=1}^{n}\|x_j\|^p$$

and

$$\left[(n-1)^{1-p}+1\right]\max_{k\in\{1,\dots,n\}}\left\|x_k - \frac{1}{n}\sum_{l=1}^{n}x_l\right\|^p \le \sum_{j=1}^{n}\left\|x_j - \frac{1}{n}\sum_{l=1}^{n}x_l\right\|^p. \tag{30}$$

If we consider the function $h_p(t) := (1-t)^{1-p}t^p + t$, $p \ge 1, t \in [0,1)$, then we observe that

$$h_p'(t) = 1 + pt^{p-1}(1-t)^{1-p} + (p-1)t^p(1-t)^{-p},$$

which shows that h_p is strictly increasing on $[0,1)$. Therefore,

$$\min_{k\in\{1,\dots,n\}}\left\{(1-p_k)^{1-p}p_k^p + p_k\right\} = p_m + (1-p_m)^{1-p}p_m^p,$$

where $p_m := \min_{k\in\{1,\dots,n\}} p_k$. By(28), we then obtain the following inequality:

$$\left[p_m + (1-p_m)^{1-p}\cdot p_m^p\right]\max_{k\in\{1,\dots,n\}}\left\|x_k - \sum_{l=1}^{n}p_l x_l\right\|^p$$
$$\le \sum_{j=1}^{n}p_j\left\|x_j - \sum_{l=1}^{n}p_l x_l\right\|^p. \tag{31}$$

Application 2. Let $x_i, p_i > 0, i \in \{1,\dots,n\}$ with $\sum_{i=1}^{n}p_i = 1$. The following inequality is well known in the literature as the *arithmetic mean-geometric mean* inequality:

$$\sum_{j=1}^{n}p_j x_j \ge \prod_{j=1}^{n}x_j^{p_j}. \tag{32}$$

The equality case holds in (32) iff $x_1 = \cdots = x_n$.

Applying the inequality (20) for the convex function $f : (0, \infty) \to \mathbb{R}$, $f(x) = -\ln x$ and performing the necessary computations, we derive the following refinement of (32):

$$\sum_{i=1}^{n} p_i x_i \geq \max_{k\in\{1,\ldots,n\}} \left\{ \left(\frac{\sum_{j=1}^{n} p_j x_j - p_k x_k}{1-p_k} \right)^{1-p_k} \cdot x_k^{p_k} \right\} \tag{33}$$

$$\geq \prod_{k=1}^{n} \left[\left(\frac{\sum_{j=1}^{n} p_j x_j - p_k x_k}{1-p_k} \right)^{1-p_k} \cdot x_k^{p_k} \right]^{\frac{1}{n}}$$

$$\geq \min_{k\in\{1,\ldots,n\}} \left\{ \left(\frac{\sum_{j=1}^{n} p_j x_j - p_k x_k}{1-p_k} \right)^{1-p_k} \cdot x_k^{p_k} \right\} \geq \prod_{i=1}^{n} x_i^{p_i}.$$

In particular, we have the inequality:

$$\frac{1}{n}\sum_{i=1}^{n} x_i \geq \max_{k\in\{1,\ldots,n\}} \left\{ \left(\frac{\sum_{j=1}^{n} x_j - x_k}{n-1} \right)^{\frac{n-1}{n}} \cdot x_k^{\frac{1}{n}} \right\}$$

$$\geq \prod_{k=1}^{n} \left[\left(\frac{\sum_{j=1}^{n} x_j - x_k}{n-1} \right)^{\frac{n-1}{n}} \cdot x_k^{\frac{1}{n}} \right]^{\frac{1}{n}}$$

$$\geq \min_{k\in\{1,\ldots,n\}} \left\{ \left(\frac{\sum_{j=1}^{n} x_j - x_k}{n-1} \right)^{\frac{n-1}{n}} \cdot x_k^{\frac{1}{n}} \right\} \geq \left(\prod_{i=1}^{n} x_i \right)^{\frac{1}{n}}.$$

Applications for f-Divergences

The following refinement of the positivity property of f-divergence may be stated.

Theorem 7 (Dragomir [19]) *For any* $\mathbf{p}, \mathbf{q} \in \mathbb{P}^n$*, namely,* $\mathbf{p}, \mathbf{q}$ *are probability distributions, we have the inequalities*

$$I_f(\mathbf{p}, \mathbf{q}) \geq \max_{k\in\{1,\ldots,n\}} \left[(1-q_k) f\left(\frac{1-p_k}{1-q_k} \right) + q_k f\left(\frac{p_k}{q_k} \right) \right] \tag{34}$$

$$\geq \frac{1}{n} \left[\sum_{k=1}^{n} (1-q_k) f\left(\frac{1-p_k}{1-q_k} \right) + \sum_{k=1}^{n} q_k f\left(\frac{p_k}{q_k} \right) \right]$$

$$\geq \min_{k\in\{1,\ldots,n\}} \left[(1-q_k) f\left(\frac{1-p_k}{1-q_k} \right) + q_k f\left(\frac{p_k}{q_k} \right) \right] \geq 0,$$

provided $f : [0, \infty) \to \mathbb{R}$ *is convex and normalised on* $[0, \infty)$.

The proof is obvious by Theorem 6 applied for the convex function $f : [0, \infty) \to \mathbb{R}$ and for the choice $x_i = \frac{p_i}{q_i}, i \in \{1, \ldots, n\}$ and the probabilities $q_i, i \in \{1, \ldots, n\}$.

If we consider a new divergence measure $R_f(\mathbf{p}, \mathbf{q})$ defined for $\mathbf{p}, \mathbf{q} \in \mathbb{P}^n$ by

$$R_f(\mathbf{p}, \mathbf{q}) := \frac{1}{n-1} \sum_{k=1}^{n} (1 - q_k) f\left(\frac{1 - p_k}{1 - q_k}\right) \tag{35}$$

and call it the *reverse f-divergence*, we observe that

$$R_f(\mathbf{p}, \mathbf{q}) = I_f(\mathbf{r}, \mathbf{t}) \tag{36}$$

with

$$\mathbf{r} = \left(\frac{1 - p_1}{n-1}, \ldots, \frac{1 - p_n}{n-1}\right), \quad \mathbf{t} = \left(\frac{1 - q_1}{n-1}, \ldots, \frac{1 - q_n}{n-1}\right) \qquad (n \geq 2).$$

With this notation, we can state the following corollary of the above proposition.

Corollary 2 *For any $\mathbf{p}, \mathbf{q} \in \mathbb{P}^n$, we have*

$$I_f(\mathbf{p}, \mathbf{q}) \geq R_f(\mathbf{p}, \mathbf{q}) \geq 0. \tag{37}$$

The proof is obvious by the second inequality in (34) and the details are omitted.

In what follows, we point out some particular inequalities for various instances of divergence measures such as the *total variation distance*, *χ^2-divergence*, *Kullback-Leibler divergence*, and *Jeffreys divergence*.

The *total variation distance* is defined by the convex function $f(t) = |t - 1|, t \in \mathbb{R}$ and given in

$$V(p, q) := \sum_{j=1}^{n} q_j \left|\frac{p_j}{q_j} - 1\right| = \sum_{j=1}^{n} |p_j - q_j|. \tag{38}$$

The following improvement of the positivity inequality for the total variation distance can be stated as follows.

Proposition 1 *For any $\mathbf{p}, \mathbf{q} \in \mathbb{P}^n$, we have the inequality:*

$$V(p, q) \geq 2 \max_{k \in \{1, \ldots, n\}} |p_k - q_k| \quad (\geq 0). \tag{39}$$

The proof follows by the first inequality in (34) for $f(t) = |t - 1|, t \in \mathbb{R}$.

The K. Pearson *χ^2-divergence* is obtained for the convex function $f(t) = (1 - t)^2, t \in \mathbb{R}$ and given by

$$\chi^2(p, q) := \sum_{j=1}^{n} q_j \left(\frac{p_j}{q_j} - 1\right)^2 = \sum_{j=1}^{n} \frac{(p_j - q_j)^2}{q_j}. \tag{40}$$

Proposition 2 *For any* $\mathbf{p}, \mathbf{q} \in \mathbb{P}^n$,

$$\chi^2(p,q) \geq \max_{k\in\{1,\dots,n\}} \left\{ \frac{(p_k - q_k)^2}{q_k(1-q_k)} \right\} \geq 4 \max_{k\in\{1,\dots,n\}} (p_k - q_k)^2 \quad (\geq 0). \tag{41}$$

Proof On applying the first inequality in (34) for the function $f(t) = (1-t)^2$, $t \in \mathbb{R}$, we get

$$\begin{aligned}\chi^2(p,q) &\geq \max_{k\in\{1,\dots,n\}} \left\{ (1-q_k)\left(\frac{1-p_k}{1-q_k} - 1\right)^2 + q_k\left(\frac{p_k}{q_k} - 1\right)^2 \right\} \\ &= \max_{k\in\{1,\dots,n\}} \left\{ \frac{(p_k - q_k)^2}{q_k(1-q_k)} \right\}.\end{aligned}$$

Since

$$q_k(1-q_k) \leq \frac{1}{4}[q_k + (1-q_k)]^2 = \frac{1}{4},$$

then

$$\frac{(p_k - q_k)^2}{q_k(1-q_k)} \geq 4(p_k - q_k)^2$$

for each $k \in \{1, \dots, n\}$, which proves the last part of (41). □

The *Kullback-Leibler divergence* can be obtained for the convex function $f : (0,\infty) \to \mathbb{R}$, $f(t) = t \ln t$ and is defined by

$$KL(p,q) := \sum_{j=1}^{n} q_j \cdot \frac{p_j}{q_j} \ln\left(\frac{p_j}{q_j}\right) = \sum_{j=1}^{n} p_j \ln\left(\frac{p_j}{q_j}\right). \tag{42}$$

Proposition 3 *For any* $\mathbf{p}, \mathbf{q} \in \mathbb{P}^n$, *we have*

$$KL(p,q) \geq \ln\left[\max_{k\in\{1,\dots,n\}} \left\{ \left(\frac{1-p_k}{1-q_k}\right)^{1-p_k} \cdot \left(\frac{p_k}{q_k}\right)^{p_k} \right\}\right] \geq 0. \tag{43}$$

Proof The first inequality is obvious by Theorem 7. Utilising the inequality between the *geometric mean and the harmonic mean*,

$$x^\alpha y^{1-\alpha} \geq \frac{1}{\frac{\alpha}{x} + \frac{1-\alpha}{y}}, \qquad x, y > 0,\ \alpha \in [0,1]$$

we have

$$\left(\frac{1-p_k}{1-q_k}\right)^{1-p_k}\cdot\left(\frac{p_k}{q_k}\right)^{p_k}\geq 1,$$

for any $k\in\{1,\ldots,n\}$, which implies the second part of (43). □

Another divergence measure that is of importance in information theory is the *Jeffreys divergence*:

$$J(p,q):=\sum_{j=1}^{n}q_j\cdot\left(\frac{p_j}{q_j}-1\right)\ln\left(\frac{p_j}{q_j}\right)=\sum_{j=1}^{n}(p_j-q_j)\ln\left(\frac{p_j}{q_j}\right),\tag{44}$$

which is an f-divergence for $f(t)=(t-1)\ln t$, $t>0$.

Proposition 4 *For any* $\mathbf{p},\mathbf{q}\in\mathbb{P}^n$, *we have*

$$\begin{aligned}J(p,q)&\geq\max_{k\in\{1,\ldots,n\}}\left\{(q_k-p_k)\ln\left[\frac{(1-p_k)q_k}{(1-q_k)p_k}\right]\right\}\\&\geq\max_{k\in\{1,\ldots,n\}}\left[\frac{(q_k-p_k)^2}{p_k+q_k-2p_kq_k}\right]\geq 0.\end{aligned}\tag{45}$$

Proof Writing the first inequality in Theorem 7 for $f(t)=(t-1)\ln t$, we have

$$\begin{aligned}J(p,q)&\geq\max_{k\in\{1,\ldots,n\}}\left\{(1-q_k)\left[\left(\frac{1-p_k}{1-q_k}-1\right)\ln\left(\frac{1-p_k}{1-q_k}\right)\right]+q_k\left(\frac{p_k}{q_k}-1\right)\ln\left(\frac{p_k}{q_k}\right)\right\}\\&=\max_{k\in\{1,\ldots,n\}}\left\{(q_k-p_k)\ln\left(\frac{1-p_k}{1-q_k}\right)-(q_k-p_k)\ln\left(\frac{p_k}{q_k}\right)\right\}\\&=\max_{k\in\{1,\ldots,n\}}\left\{(q_k-p_k)\ln\left[\frac{(1-p_k)q_k}{(1-q_k)p_k}\right]\right\},\end{aligned}$$

proving the first inequality in (45).

Utilising the elementary inequality for positive numbers,

$$\frac{\ln b-\ln a}{b-a}\geq\frac{2}{a+b},\qquad a,b>0$$

we have

$$\begin{aligned}&(q_k-p_k)\left[\ln\left(\frac{1-p_k}{1-q_k}\right)-\ln\left(\frac{p_k}{q_k}\right)\right]\\&=(q_k-p_k)\cdot\frac{\ln\left(\frac{1-p_k}{1-q_k}\right)-\ln\left(\frac{p_k}{q_k}\right)}{\frac{1-p_k}{1-q_k}-\frac{p_k}{q_k}}\cdot\left[\frac{1-p_k}{1-q_k}-\frac{p_k}{q_k}\right]\end{aligned}$$

$$= \frac{(q_k - p_k)^2}{q_k(1-q_k)} \cdot \frac{\ln\left(\frac{1-p_k}{1-q_k}\right) - \ln\left(\frac{p_k}{q_k}\right)}{\frac{1-p_k}{1-q_k} - \frac{p_k}{q_k}}$$

$$\geq \frac{(q_k - p_k)^2}{q_k(1-q_k)} \cdot \frac{2}{\frac{1-p_k}{1-q_k} + \frac{p_k}{q_k}} = \frac{2(q_k - p_k)^2}{p_k + q_k - 2p_k q_k} \geq 0,$$

for each $k \in \{1, \ldots, n\}$, giving the second inequality in (45). □

More General Results

Let C be a convex subset in the real linear space X and assume that $f : C \to \mathbb{R}$ is a convex function on C. If $x_i \in C$ and $p_i > 0, i \in \{1, \ldots, n\}$ with $\sum_{i=1}^n p_i = 1$, then for any nonempty subset J of $\{1, \ldots, n\}$, we put $\bar{J} := \{1, \ldots, n\} \setminus J \ (\neq \emptyset)$ and define $P_J := \sum_{i \in J} p_i$ and $\bar{P}_J := P_{\bar{J}} = \sum_{j \in \bar{J}} p_j = 1 - \sum_{i \in J} p_i$. For the convex function f and the n-tuples $\mathbf{x} := (x_1, \ldots, x_n)$ and $\mathbf{p} := (p_1, \ldots, p_n)$ as above, we can define the following functional:

$$D(f, \mathbf{p}, \mathbf{x}; J) := P_J f\left(\frac{1}{P_J}\sum_{i \in J} p_i x_i\right) + \bar{P}_J f\left(\frac{1}{\bar{P}_J}\sum_{j \in \bar{J}} p_j x_j\right) \tag{46}$$

where here and everywhere below $J \subset \{1, \ldots, n\}$ with $J \neq \emptyset$ and $J \neq \{1, \ldots, n\}$.

It is worth to observe that for $J = \{k\}, k \in \{1, \ldots, n\}$, we have the functional

$$D_k(f, \mathbf{p}, \mathbf{x}) := D(f, \mathbf{p}, \mathbf{x}; \{k\}) \tag{47}$$
$$= p_k f(x_k) + (1 - p_k) f\left(\frac{\sum_{i=1}^n p_i x_i - p_k x_k}{1 - p_k}\right)$$

that has been investigated in the paper [19].

Theorem 8 (Dragomir [20]) *Let C be a convex subset in the real linear space X and assume that $f : C \to \mathbb{R}$ is a convex function on C. If $x_i \in C$ and $p_i > 0, i \in \{1, \ldots, n\}$ with $\sum_{i=1}^n p_i = 1$, then for any nonempty subset J of $\{1, \ldots, n\}$, we have*

$$\sum_{k=1}^n p_k f(x_k) \geq D(f, \mathbf{p}, \mathbf{x}; J) \geq f\left(\sum_{k=1}^n p_k x_k\right). \tag{48}$$

Proof By the convexity of the function f, we have

$$D(f, \mathbf{p}, \mathbf{x}; J) = P_J f\left(\frac{1}{P_J}\sum_{i\in J} p_i x_i\right) + \bar{P}_J f\left(\frac{1}{\bar{P}_J}\sum_{j\in \bar{J}} p_j x_j\right)$$
$$\geq f\left[P_J\left(\frac{1}{P_J}\sum_{i\in J} p_i x_i\right) + \bar{P}_J\left(\frac{1}{\bar{P}_J}\sum_{j\in \bar{J}} p_j x_j\right)\right]$$
$$= f\left(\sum_{k=1}^{n} p_k x_k\right)$$

for any J, which proves the second inequality in (48).

By the Jensen inequality, we also have

$$\sum_{k=1}^{n} p_k f(x_k) = \sum_{i\in J} p_i f(x_i) + \sum_{j\in \bar{J}} p_j f(x_j)$$
$$\geq P_J f\left(\frac{1}{P_J}\sum_{i\in J} p_i x_i\right) + \bar{P}_J f\left(\frac{1}{\bar{P}_J}\sum_{j\in \bar{J}} p_j x_j\right)$$
$$= D(f, \mathbf{p}, \mathbf{x}; J)$$

for any J, which proves the first inequality in (48). □

Remark 4 We observe that the inequality (48) can be written in an equivalent form as

$$\sum_{k=1}^{n} p_k f(x_k) \geq \max_{\emptyset\neq J\subset\{1,\ldots,n\}} D(f, \mathbf{p}, \mathbf{x}; J) \tag{49}$$

and

$$\min_{\emptyset\neq J\subset\{1,\ldots,n\}} D(f, \mathbf{p}, \mathbf{x}; J) \geq f\left(\sum_{k=1}^{n} p_k x_k\right). \tag{50}$$

These inequalities imply the following results that have been obtained earlier by the author in [19] utilising a different method of proof slightly more complicated:

$$\sum_{k=1}^{n} p_k f(x_k) \geq \max_{k\in\{1,\ldots,n\}} D_k(f, \mathbf{p}, \mathbf{x}) \tag{51}$$

and

$$\min_{k\in\{1,\ldots,n\}} D_k(f, \mathbf{p}, \mathbf{x}) \geq f\left(\sum_{k=1}^{n} p_k x_k\right). \tag{52}$$

Moreover, since

$$\max_{\emptyset \neq J \subset \{1,\ldots,n\}} D(f, \mathbf{p}, \mathbf{x}; J) \geq \max_{k \in \{1,\ldots,n\}} D_k(f, \mathbf{p}, \mathbf{x})$$

and

$$\min_{k \in \{1,\ldots,n\}} D_k(f, \mathbf{p}, \mathbf{x}) \geq \min_{\emptyset \neq J \subset \{1,\ldots,n\}} D(f, \mathbf{p}, \mathbf{x}; J),$$

then the new inequalities (49) and (50) are better than the earlier results developed in [19].

The case of uniform distribution, namely, when $p_i = \frac{1}{n}$ for all $\{1, \ldots, n\}$, is of interest as well. If we consider a natural number m with $1 \leq m \leq n-1$ and if we define

$$D_m(f, \mathbf{x}) := \frac{m}{n} f\left(\frac{1}{m} \sum_{i=1}^{m} x_i\right) + \frac{n-m}{n} f\left(\frac{1}{n-m} \sum_{j=m+1}^{n} x_j\right) \tag{53}$$

then we can state the following result:

Corollary 3 (Dragomir [20]) *Let C be a convex subset in the real linear space X and assume that $f : C \to \mathbb{R}$ is a convex function on C. If $x_i \in C$, then for any $m \in \{1, \ldots, n-1\}$, we have*

$$\frac{1}{n} \sum_{k=1}^{n} f(x_k) \geq D_m(f, \mathbf{x}) \geq f\left(\frac{1}{n} \sum_{k=1}^{n} x_k\right). \tag{54}$$

In particular, we have the bounds

$$\frac{1}{n} \sum_{k=1}^{n} f(x_k) \geq \max_{m \in \{1,\ldots,n-1\}} \left[\frac{m}{n} f\left(\frac{1}{m} \sum_{i=1}^{m} x_i\right) + \frac{n-m}{n} f\left(\frac{1}{n-m} \sum_{j=m+1}^{n} x_j\right)\right] \tag{55}$$

and

$$\min_{m \in \{1,\ldots,n-1\}} \left[\frac{m}{n} f\left(\frac{1}{m} \sum_{i=1}^{m} x_i\right) + \frac{n-m}{n} f\left(\frac{1}{n-m} \sum_{j=m+1}^{n} x_j\right)\right] \geq f\left(\frac{1}{n} \sum_{k=1}^{n} x_k\right). \tag{56}$$

The following version of the inequality (48) may be useful for symmetric convex functions:

Corollary 4 (Dragomir [20]) *Let C be a convex function with the property that $0 \in C$. If $y_j \in X$ such that for $p_i > 0, i \in \{1, \ldots, n\}$ with $\sum_{i=1}^n p_i = 1$, we have $y_j - \sum_{i=1}^n p_i y_i \in C$ for any $j \in \{1, \ldots, n\}$, then for any nonempty subset J of $\{1, \ldots, n\}$, we have*

$$\sum_{k=1}^n p_k f\left(y_k - \sum_{i=1}^n p_i y_i\right) \geq P_J f\left[\bar{P}_J\left(\frac{1}{P_J}\sum_{i\in J} p_i y_i - \frac{1}{\bar{P}_J}\sum_{j\in \bar{J}} p_j y_j\right)\right]$$
$$+ \bar{P}_J f\left[P_J\left(\frac{1}{\bar{P}_J}\sum_{j\in \bar{J}} p_j y_j - \frac{1}{P_J}\sum_{i\in J} p_i y_i\right)\right] \geq f(0). \tag{57}$$

Remark 5 If C is as in Corollary 4 and $y_j \in X$ such that $y_j - \frac{1}{n}\sum_{i=1}^n y_i \in C$ for any $j \in \{1, \ldots, n\}$, then for any $m \in \{1, \ldots, n-1\}$, we have

$$\frac{1}{n}\sum_{k=1}^n f\left(y_k - \frac{1}{n}\sum_{i=1}^n y_i\right) \geq \frac{m}{n} f\left[\frac{n-m}{n}\left(\frac{1}{m}\sum_{i=1}^m y_i - \frac{1}{n-m}\sum_{j=m+1}^n y_j\right)\right]$$
$$+ \frac{n-m}{n} f\left[\frac{m}{n}\left(\frac{1}{n-m}\sum_{j=m+1}^n y_j - \frac{1}{m}\sum_{i=1}^m y_i\right)\right] \geq f(0). \tag{58}$$

Remark 6 It is also useful to remark that if $J = \{k\}$ where $k \in \{1, \ldots, n\}$, then the particular form we can derive from (57) can be written as

$$\sum_{\ell=1}^n p_\ell f\left(y_\ell - \sum_{i=1}^n p_i y_i\right)$$
$$\geq p_k f\left[(1-p_k)\left(y_k - \frac{1}{1-p_k}\left(\sum_{j=1}^n p_j y_j - p_k y_k\right)\right)\right]$$
$$+ (1-p_k) f\left[p_k\left(\frac{1}{1-p_k}\left(\sum_{j=1}^n p_j y_j - p_k y_k\right) - y_k\right)\right] \geq f(0), \tag{59}$$

which is equivalent with

$$\sum_{\ell=1}^n p_\ell f\left(y_\ell - \sum_{i=1}^n p_i y_i\right) \geq p_k f\left(y_k - \sum_{j=1}^n p_j y_j\right)$$

$$+ (1 - p_k) f \left[\frac{p_k}{1 - p_k} \left(\sum_{j=1}^{n} p_j y_j - y_k \right) \right] \geq f(0) \tag{60}$$

for any $k \in \{1, \ldots, n\}$.

A Lower Bound for Mean f-Deviation

Let X be a real linear space. For a convex function $f : X \to \mathbb{R}$ with the properties that $f(0) = 0$, define the *mean f-deviation* of an n-tuple of vectors $\mathbf{x} = (x_1, \ldots, x_n) \in X^n$ with the probability distribution $\mathbf{p} = (p_1, \ldots, p_n)$ by the nonnegative quantity

$$K_f(\mathbf{p}, \mathbf{x}) := \sum_{i=1}^{n} p_i f\left(x_i - \sum_{k=1}^{n} p_k x_k \right). \tag{61}$$

The fact that $K_f(\mathbf{p}, \mathbf{x})$ is nonnegative follows by Jensen's inequality, namely,

$$K_f(\mathbf{p}, \mathbf{x}) \geq f\left(\sum_{i=1}^{n} p_i \left(x_i - \sum_{k=1}^{n} p_k x_k \right) \right) = f(0) = 0.$$

A natural example of such deviations can be provided by the convex function $f(x) := \|x\|^r$ with $r \geq 1$ defined on a normed linear space $(X, \|\cdot\|)$. We denote this by

$$K_r(\mathbf{p}, \mathbf{x}) := \sum_{i=1}^{n} p_i \left\| x_i - \sum_{k=1}^{n} p_k x_k \right\|^r \tag{62}$$

and call it the *mean r-absolute deviation* of the n-tuple of vectors $\mathbf{x} = (x_1, \ldots, x_n) \in X^n$ with the probability distribution $\mathbf{p} = (p_1, \ldots, p_n)$.

The following result that provides a lower bound for the mean f-deviation holds:

Theorem 9 (Dragomir [20]) *Let $f : X \to [0, \infty)$ be a convex function with $f(0) = 0$. If $\mathbf{x} = (x_1, \ldots, x_n) \in X^n$ and $\mathbf{p} = (p_1, \ldots, p_n)$ is a probability distribution with all p_i nonzero, then*

$$K_f(\mathbf{p}, \mathbf{x}) \geq \max_{\emptyset \neq J \subset \{1,\ldots,n\}} \left\{ P_J f\left[\bar{P}_J \left(\frac{1}{P_J} \sum_{i \in J} p_i x_i - \frac{1}{\bar{P}_J} \sum_{j \in \bar{J}} p_j x_j \right) \right] \right.$$

$$\left. + P_J f\left(\frac{1}{\bar{P}_J} \sum_{j \in \bar{J}} p_j y_j - \frac{1}{P_J} \sum_{i \in J} p_i y_i \right) \right\} (\geq 0). \tag{63}$$

In particular, we have

$$K_f(\mathbf{p},\mathbf{x}) \geq \max_{k\in\{1,\dots,n\}} \left\{ (1-p_k) f\left[\frac{p_k}{1-p_k}\left(\sum_{l=1}^{n} p_l x_l - x_k\right)\right] + p_k f\left(x_k - \sum_{l=1}^{n} p_l x_l\right)\right\} (\geq 0). \tag{64}$$

The proof follows from Corollary 4 and Remark 6.

As a particular case of interest, we have the following:

Corollary 5 (Dragomir [20]) *Let* $(X, \|\cdot\|)$ *be a normed linear space. If* $\mathbf{x} = (x_1,\dots,x_n) \in X^n$ *and* $\mathbf{p} = (p_1,\dots,p_n)$ *is a probability distribution with all* p_i *nonzero, then for* $r \geq 1$ *we have*

$$K_r(\mathbf{p},\mathbf{x}) \geq \max_{\emptyset\neq J\subset\{1,\dots,n\}} \left\{ P_J \bar{P}_J \left(\bar{P}_J^{r-1} + P_J^{r-1}\right) \left\| \frac{1}{P_J}\sum_{i\in J} p_i x_i - \frac{1}{\bar{P}_J}\sum_{j\in \bar{J}} p_j x_j \right\|^r \right\} (\geq 0). \tag{65}$$

Remark 7 By the convexity of the power function $f(t) = t^r$, $r \geq 1$, we have

$$P_J \bar{P}_J \left(\bar{P}_J^{r-1} + P_J^{r-1}\right) = P_J \bar{P}_J^r + \bar{P}_J P_J^r \geq \left(P_J \bar{P}_J + \bar{P}_J P_J\right)^r = 2^r P_J^r \bar{P}_J^r$$

therefore

$$P_J \bar{P}_J \left(\bar{P}_J^{r-1} + P_J^{r-1}\right) \left\| \frac{1}{P_J}\sum_{i\in J} p_i x_i - \frac{1}{\bar{P}_J}\sum_{j\in \bar{J}} p_j x_j \right\|^r \geq 2^r P_J^r \bar{P}_J^r \left\| \frac{1}{P_J}\sum_{i\in J} p_i x_i - \frac{1}{\bar{P}_J}\sum_{j\in \bar{J}} p_j x_j \right\|^r = 2^r \left\| \bar{P}_J \sum_{i\in J} p_i x_i - P_J \sum_{j\in \bar{J}} p_j x_j \right\|^r. \tag{66}$$

Since

$$\bar{P}_J \sum_{i\in J} p_i x_i - P_J \sum_{j\in \bar{J}} p_j x_j = (1-P_J)\sum_{i\in J} p_i x_i - P_J\left(\sum_{k=1}^{n} p_k x_k - \sum_{i\in J} p_i x_i\right) \tag{67}$$

$$= \sum_{i\in J} p_i x_i - P_J \sum_{k=1}^{n} p_k x_k,$$

then by (65)–(67) we deduce the coarser but perhaps more useful lower bound

$$K_r(\mathbf{p}, \mathbf{x}) \geq 2^r \max_{\emptyset \neq J \subset \{1,\ldots,n\}} \left\{ \left\| \sum_{i \in J} p_i x_i - P_J \sum_{k=1}^{n} p_k x_k \right\|^r \right\} (\geq 0). \quad (68)$$

The case for mean r-absolute deviation is incorporated in:

Corollary 6 (Dragomir [20]) *Let $(X, \|\cdot\|)$ be a normed linear space. If $\mathbf{x} = (x_1, \ldots, x_n) \in X^n$ and $\mathbf{p} = (p_1, \ldots, p_n)$ is a probability distribution with all p_i nonzero, then for $r \geq 1$ we have*

$$K_r(\mathbf{p}, \mathbf{x}) \geq \max_{k \in \{1,\ldots,n\}} \left\{ \left[(1 - p_k)^{1-r} p_k^r + p_k \right] \left\| x_k - \sum_{l=1}^{n} p_l x_l \right\|^r \right\}. \quad (69)$$

Remark 8 Since the function $h_r(t) := (1 - t)^{1-r} t^r + t, r \geq 1, t \in [0, 1)$ is strictly increasing on $[0, 1)$, then

$$\min_{k \in \{1,\ldots,n\}} \left\{ (1 - p_k)^{1-r} p_k^r + p_k \right\} = p_m + (1 - p_m)^{1-r} p_m^r,$$

where $p_m := \min_{k \in \{1,\ldots,n\}} p_k$.

We then obtain the following simpler inequality:

$$K_r(\mathbf{p}, \mathbf{x}) \geq \left[p_m + (1 - p_m)^{1-r} \cdot p_m^r \right] \max_{k \in \{1,\ldots,n\}} \left\| x_k - \sum_{l=1}^{n} p_l x_l \right\|^p, \quad (70)$$

which is perhaps more useful for applications (see also [19]).

Applications for f-Divergence Measures

We endeavour to extend the concept of f-divergence for functions defined on a cone in a linear space as follows.

Firstly, we recall that the subset K in a linear space X is a *cone* if the following two conditions are satisfied:

1. for any $x, y \in K$ we have $x + y \in K$;
2. for any $x \in K$ and any $\alpha \geq 0$ we have $\alpha x \in K$.

For a given n-tuple of vectors $\mathbf{z} = (z_1, \ldots, z_n) \in K^n$ and a probability distribution $\mathbf{q} \in \mathbb{P}^n$ with all values nonzero, we can define, for the convex function $f : K \to \mathbb{R}$, the following *f-divergence of $\mathbf{z}$ with the distribution $\mathbf{q}$*:

$$I_f(\mathbf{z}, \mathbf{q}) := \sum_{i=1}^{n} q_i f\left(\frac{z_i}{q_i}\right). \quad (71)$$

It is obvious that if $X = \mathbb{R}$, $K = [0, \infty)$ and $\mathbf{x} = \mathbf{p} \in \mathbb{P}^n$, then we obtain the usual concept of the f-divergence associated with a function $f : [0, \infty) \to \mathbb{R}$.

Now, for a given n-tuple of vectors $\mathbf{x} = (x_1, \ldots, x_n) \in K^n$, a probability distribution $\mathbf{q} \in \mathbb{P}^n$ with all values nonzero and for any nonempty subset J of $\{1, \ldots, n\}$ we have

$$\mathbf{q}_J := \left(Q_J, \bar{Q}_J\right) \in \mathbb{P}^2$$

and

$$\mathbf{x}_J := \left(X_J, \bar{X}_J\right) \in K^2$$

where, as above

$$X_J := \sum_{i \in J} x_i, \quad \text{and} \quad \bar{X}_J := X_{\bar{J}}.$$

It is obvious that

$$I_f(\mathbf{x}_J, \mathbf{q}_J) = Q_J f\left(\frac{X_J}{Q_J}\right) + \bar{Q}_J f\left(\frac{\bar{X}_J}{\bar{Q}_J}\right).$$

The following inequality for the f-divergence of an n-tuple of vectors in a linear space holds:

Theorem 10 (Dragomir [20]) *Let $f : K \to \mathbb{R}$ be a convex function on the cone K. Then for any n-tuple of vectors $\mathbf{x} = (x_1, \ldots, x_n) \in K^n$, a probability distribution $\mathbf{q} \in \mathbb{P}^n$ with all values nonzero and for any nonempty subset J of $\{1, \ldots, n\}$ we have*

$$\begin{aligned} I_f(\mathbf{x}, \mathbf{q}) &\geq \max_{\emptyset \neq J \subset \{1,\ldots,n\}} I_f(\mathbf{x}_J, \mathbf{q}_J) \geq I_f(\mathbf{x}_J, \mathbf{q}_J) \\ &\geq \min_{\emptyset \neq J \subset \{1,\ldots,n\}} I_f(\mathbf{x}_J, \mathbf{q}_J) \geq f(X_n) \end{aligned} \tag{72}$$

where $X_n := \sum_{i=1}^n x_i$.

The proof follows by Theorem 8 and the details are omitted.

We observe that, for a given n-tuple of vectors $\mathbf{x} = (x_1, \ldots, x_n) \in K^n$, a sufficient condition for the positivity of $I_f(\mathbf{x}, \mathbf{q})$ for any probability distribution $\mathbf{q} \in \mathbb{P}^n$ with all values nonzero is that $f(X_n) \geq 0$. In the scalar case and if $\mathbf{x} = \mathbf{p} \in \mathbb{P}^n$, then a sufficient condition for the positivity of the f-divergence $I_f(\mathbf{p}, \mathbf{q})$ is that $f(1) \geq 0$.

The case of functions of a real variable that is of interest for applications is incorporated in:

Corollary 7 (Dragomir [20]) *Let $f : [0, \infty) \to \mathbb{R}$ be a normalised convex function. Then for any $\mathbf{p}, \mathbf{q} \in \mathbb{P}^n$ we have*

$$I_f(\mathbf{p}, \mathbf{q}) \geq \max_{\emptyset \neq J \subset \{1,\ldots,n\}} \left[Q_J f\left(\frac{P_J}{Q_J}\right) + (1 - Q_J) f\left(\frac{1 - P_J}{1 - Q_J}\right)\right] (\geq 0). \tag{73}$$

In what follows we provide some lower bounds for a number of f-divergences that are used in various fields of information theory, probability theory and statistics.

The *total variation distance* is defined by the convex function $f(t) = |t-1|, t \in \mathbb{R}$ and given in

$$V(p,q) := \sum_{j=1}^{n} q_j \left| \frac{p_j}{q_j} - 1 \right| = \sum_{j=1}^{n} |p_j - q_j| . \tag{74}$$

The following improvement of the positivity inequality for the total variation distance can be stated as follows.

Proposition 5 *For any* $\mathbf{p}, \mathbf{q} \in \mathbb{P}^n$, *we have the inequality:*

$$V(p,q) \geq 2 \max_{\emptyset \neq J \subset \{1,\ldots,n\}} |P_J - Q_J| \quad (\geq 0) . \tag{75}$$

The proof follows by the inequality (73) for $f(t) = |t-1|, t \in \mathbb{R}$.

The K. Pearson χ^2-*divergence* is obtained for the convex function $f(t) = (1-t)^2, t \in \mathbb{R}$ and given by

$$\chi^2(p,q) := \sum_{j=1}^{n} q_j \left(\frac{p_j}{q_j} - 1 \right)^2 = \sum_{j=1}^{n} \frac{(p_j - q_j)^2}{q_j} . \tag{76}$$

Proposition 6 *For any* $\mathbf{p}, \mathbf{q} \in \mathbb{P}^n$,

$$\begin{aligned} \chi^2(p,q) &\geq \max_{\emptyset \neq J \subset \{1,\ldots,n\}} \left\{ \frac{(P_J - Q_J)^2}{Q_J(1-Q_J)} \right\} \\ &\geq 4 \max_{\emptyset \neq J \subset \{1,\ldots,n\}} (P_J - Q_J)^2 \quad (\geq 0) . \end{aligned} \tag{77}$$

Proof On applying the inequality (73) for the function $f(t) = (1-t)^2, t \in \mathbb{R}$, we get

$$\begin{aligned} \chi^2(p,q) &\geq \max_{\emptyset \neq J \subset \{1,\ldots,n\}} \left\{ (1-Q_J) \left(\frac{1-P_J}{1-Q_J} - 1 \right)^2 + Q_J \left(\frac{P_J}{Q_J} - 1 \right)^2 \right\} \\ &= \max_{\emptyset \neq J \subset \{1,\ldots,n\}} \left\{ \frac{(P_J - Q_J)^2}{Q_J(1-Q_J)} \right\} . \end{aligned}$$

Since

$$Q_J(1-Q_J) \leq \frac{1}{4} [Q_J + (1-Q_J)]^2 = \frac{1}{4},$$

then

$$\frac{(P_J - Q_J)^2}{Q_J(1 - Q_J)} \geq 4(P_J - Q_J)^2$$

for each $J \subset \{1, \ldots, n\}$, which proves the last part of (77). □

The *Kullback-Leibler divergence* can be obtained for the convex function $f : (0, \infty) \to \mathbb{R}$, $f(t) = t \ln t$ and is defined by

$$KL(p, q) := \sum_{j=1}^{n} q_j \cdot \frac{p_j}{q_j} \ln\left(\frac{p_j}{q_j}\right) = \sum_{j=1}^{n} p_j \ln\left(\frac{p_j}{q_j}\right). \quad (78)$$

Proposition 7 *For any* $\mathbf{p}, \mathbf{q} \in \mathbb{P}^n$, *we have:*

$$KL(p, q) \geq \ln\left[\max_{\emptyset \neq J \subset \{1,\ldots,n\}} \left\{\left(\frac{1 - P_J}{1 - Q_J}\right)^{1-P_J} \cdot \left(\frac{P_J}{Q_J}\right)^{P_J}\right\}\right] \geq 0. \quad (79)$$

Proof The first inequality is obvious by Corollary 7. Utilising the inequality between the *geometric mean and the harmonic mean*,

$$x^{\alpha} y^{1-\alpha} \geq \frac{1}{\frac{\alpha}{x} + \frac{1-\alpha}{y}}, \qquad x, y > 0,\ \alpha \in [0, 1]$$

we have for $x = \frac{P_J}{Q_J}$, $y = \frac{1-P_J}{1-Q_J}$ and $\alpha = P_J$ that

$$\left(\frac{1 - P_J}{1 - Q_J}\right)^{1-P_J} \cdot \left(\frac{P_J}{Q_J}\right)^{P_J} \geq 1,$$

for any $J \subset \{1, \ldots, n\}$, which implies the second part of (79). □

Another divergence measure that is of importance in information theory is the *Jeffreys divergence*:

$$J(p, q) := \sum_{j=1}^{n} q_j \cdot \left(\frac{p_j}{q_j} - 1\right) \ln\left(\frac{p_j}{q_j}\right) = \sum_{j=1}^{n} (p_j - q_j) \ln\left(\frac{p_j}{q_j}\right), \quad (80)$$

which is an f-divergence for $f(t) = (t - 1) \ln t$, $t > 0$.

;

Proposition 8 *For any* $\mathbf{p}, \mathbf{q} \in \mathbb{P}^n$, *we have*

$$J(p, q) \geq \ln\left(\max_{\emptyset \neq J \subset \{1,\ldots,n\}} \left\{\left[\frac{(1 - P_J) Q_J}{(1 - Q_J) P_J}\right]^{(Q_J - P_J)}\right\}\right) \quad (81)$$

$$\geq \max_{\emptyset \neq J \subset \{1,\ldots,n\}} \left[\frac{(Q_J - P_J)^2}{P_J + Q_J - 2P_J Q_J}\right] \geq 0.$$

Proof On making use of the inequality (73) for $f(t) = (t-1)\ln t$, we have

$$
\begin{aligned}
&J(p,q) \\
&\geq \max_{k\in\{1,\dots,n\}} \left\{(1-Q_J)\left[\left(\frac{1-P_J}{1-Q_J}-1\right)\ln\left(\frac{1-P_J}{1-Q_J}\right)\right] + Q_J\left(\frac{P_J}{Q_J}-1\right)\ln\left(\frac{P_J}{Q_J}\right)\right\} \\
&= \max_{k\in\{1,\dots,n\}} \left\{(Q_J-P_J)\ln\left(\frac{1-P_J}{1-Q_J}\right) - (Q_J-P_J)\ln\left(\frac{P_J}{Q_J}\right)\right\} \\
&= \max_{k\in\{1,\dots,n\}} \left\{(Q_J-P_J)\ln\left[\frac{(1-P_J)\,Q_J}{(1-Q_J)\,P_J}\right]\right\},
\end{aligned}
$$

proving the first inequality in (81).

Utilising the elementary inequality for positive numbers,

$$
\frac{\ln b - \ln a}{b-a} \geq \frac{2}{a+b}, \qquad a, b > 0
$$

we have

$$
\begin{aligned}
&(Q_J-P_J)\left[\ln\left(\frac{1-P_J}{1-Q_J}\right) - \ln\left(\frac{P_J}{Q_J}\right)\right] \\
&\quad = (Q_J-P_J)\cdot\frac{\ln\left(\frac{1-P_J}{1-Q_J}\right)-\ln\left(\frac{P_J}{Q_J}\right)}{\frac{1-P_J}{1-Q_J}-\frac{P_J}{Q_J}}\cdot\left[\frac{1-P_J}{1-Q_J}-\frac{P_J}{Q_J}\right] \\
&\quad = \frac{(Q_J-P_J)^2}{Q_J(1-Q_J)}\cdot\frac{\ln\left(\frac{1-P_J}{1-Q_J}\right)-\ln\left(\frac{P_J}{Q_J}\right)}{\frac{1-P_J}{1-Q_J}-\frac{P_J}{Q_J}} \\
&\quad \geq \frac{(Q_J-P_J)^2}{Q_J(1-Q_J)}\cdot\frac{2}{\frac{1-P_J}{1-Q_J}+\frac{P_J}{Q_J}} = \frac{2(Q_J-P_J)^2}{P_J+Q_J-2P_JQ_J} \geq 0,
\end{aligned}
$$

for each $J \subset \{1,\dots,n\}$, giving the second inequality in (81). □

Inequalities in Terms of Gâteaux Derivatives

Gâteaux Derivatives

Assume that $f : X \to \mathbb{R}$ is a *convex function* on the real linear space X. Since for any vectors $x, y \in X$ the function $g_{x,y} : \mathbb{R} \to \mathbb{R}$, $g_{x,y}(t) := f(x+ty)$ is convex, it follows that the following limits exist:

$$
\nabla_{+(-)} f(x)(y) := \lim_{t\to 0+(-)} \frac{f(x+ty)-f(x)}{t}
$$

and they are called the *right (left) Gâteaux derivatives* of the function f in the point x over the direction y.

It is obvious that for any $t > 0 > s$ we have

$$\frac{f(x+ty) - f(x)}{t} \geq \nabla_{+} f(x)(y) = \inf_{t>0} \left[\frac{f(x+ty) - f(x)}{t} \right]$$
$$\geq \sup_{s<0} \left[\frac{f(x+sy) - f(x)}{s} \right] = \nabla_{-} f(x)(y) \geq \frac{f(x+sy) - f(x)}{s} \tag{82}$$

for any $x, y \in X$ and, in particular,

$$\nabla_{-} f(u)(u-v) \geq f(u) - f(v) \geq \nabla_{+} f(v)(u-v) \tag{83}$$

for any $u, v \in X$. We call this *the gradient inequality* for the convex function f. It will be used frequently in the sequel in order to obtain various results related to Jensen's inequality.

The following properties are also of importance:

$$\nabla_{+} f(x)(-y) = -\nabla_{-} f(x)(y), \tag{84}$$

and

$$\nabla_{+(-)} f(x)(\alpha y) = \alpha \nabla_{+(-)} f(x)(y) \tag{85}$$

for any $x, y \in X$ and $\alpha \geq 0$.

The right Gâteaux derivative is *subadditive*, while the left one is *superadditive*, i.e.

$$\nabla_{+} f(x)(y+z) \leq \nabla_{+} f(x)(y) + \nabla_{+} f(x)(z) \tag{86}$$

and

$$\nabla_{-} f(x)(y+z) \geq \nabla_{-} f(x)(y) + \nabla_{-} f(x)(z) \tag{87}$$

for any $x, y, z \in X$.

Some natural examples can be provided by the use of normed spaces.

Assume that $(X, \|\cdot\|)$ is a real normed linear space. The function $f : X \to \mathbb{R}$, $f(x) := \frac{1}{2}\|x\|^2$ is a convex function which generates *the superior* and *the inferior semi-inner products*:

$$\langle y, x \rangle_{s(i)} := \lim_{t \to 0+(-)} \frac{\|x+ty\|^2 - \|x\|^2}{t}.$$

For a comprehensive study of the properties of these mappings in the *Geometry of Banach Spaces*, see the monograph [16].

For the convex function $f_p : X \to \mathbb{R}$, $f_p(x) := \|x\|^p$ with $p > 1$, we have

$$\nabla_{+(-)} f_p(x)(y) = \begin{cases} p\|x\|^{p-2}\langle y, x\rangle_{s(i)} & \text{if } x \neq 0 \\ 0 & \text{if } x = 0 \end{cases}$$

for any $y \in X$.

If $p = 1$, then we have

$$\nabla_{+(-)} f_1(x)(y) = \begin{cases} \|x\|^{-1}\langle y, x\rangle_{s(i)} & \text{if } x \neq 0 \\ +(-)\|y\| & \text{if } x = 0 \end{cases}$$

for any $y \in X$.

This class of functions will be used to illustrate the inequalities obtained in the general case of convex functions defined on an entire linear space.

The following result holds:

Theorem 11 (Dragomir [21]) *Let $f : X \to \mathbb{R}$ be a convex function. Then for any $x, y \in X$ and $t \in [0, 1]$, we have*

$$\begin{aligned} & t(1-t)[\nabla_- f(y)(y-x) - \nabla_+ f(x)(y-x)] \\ & \quad \geq tf(x) + (1-t)f(y) - f(tx + (1-t)y) \\ & \quad \geq t(1-t)[\nabla_+ f(tx+(1-t)y)(y-x) - \nabla_- f(tx+(1-t)y)(y-x)] \geq 0. \end{aligned} \tag{88}$$

Proof Utilising the gradient inequality (83), we have

$$f(tx + (1-t)y) - f(x) \geq (1-t)\nabla_+ f(x)(y-x) \tag{89}$$

and

$$f(tx + (1-t)y) - f(y) \geq -t\nabla_- f(y)(y-x). \tag{90}$$

If we multiply (89) with t and (90) with $1-t$ and add the resultant inequalities, we obtain

$$\begin{aligned} & f(tx + (1-t)y) - tf(x) - (1-t)f(y) \\ & \quad \geq (1-t)t\nabla_+ f(x)(y-x) - t(1-t)\nabla_- f(y)(y-x) \end{aligned}$$

which is clearly equivalent with the first part of (88).

By the gradient inequality, we also have

$$(1-t)\nabla_- f(tx + (1-t)y)(y-x) \geq f(tx + (1-t)y) - f(x)$$

and

$$-t\nabla_{+}f\left(tx+\left(1-t\right)y\right)\left(y-x\right)\geq f\left(tx+\left(1-t\right)y\right)-f\left(y\right)$$

which by the same procedure as above yields the second part of (88). □

The following particular case for norms may be stated:

Corollary 8 (Dragomir [21]) *If x and y are two vectors in the normed linear space $(X,\|\cdot\|)$ such that $0\notin[x,y]:=\{(1-s)x+sy,s\in[0,1]\}$, then for any $p\geq 1$, we have the inequalities*

$$\begin{aligned}
&pt\left(1-t\right)\left[\|y\|^{p-2}\left\langle y-x,y\right\rangle_{i}-\|x\|^{p-2}\left\langle y-x,x\right\rangle_{s}\right]\\
&\geq t\|x\|^{p}+\left(1-t\right)\|y\|^{p}-\|tx+\left(1-t\right)y\|^{p}\\
&\geq pt\left(1-t\right)\|tx+\left(1-t\right)y\|^{p-2}\left[\left\langle y-x,tx+\left(1-t\right)y\right\rangle_{s}\right.\\
&\left.-\left\langle y-x,tx+\left(1-t\right)y\right\rangle_{i}\right]\geq 0
\end{aligned}\tag{91}$$

for any $t\in[0,1]$. If $p\geq 2$, the inequality holds for any x and y.

Remark 9 We observe that for $p=1$ in (91) we derive the result

$$\begin{aligned}
&t\left(1-t\right)\left[\left\langle y-x,\frac{y}{\|y\|}\right\rangle_{i}-\left\langle y-x,\frac{x}{\|x\|}\right\rangle_{s}\right]\\
&\geq t\|x\|+\left(1-t\right)\|y\|-\|tx+\left(1-t\right)y\|\\
&\geq t\left(1-t\right)\left[\left\langle y-x,\frac{tx+\left(1-t\right)y}{\|tx+\left(1-t\right)y\|}\right\rangle_{s}-\left\langle y-x,\frac{tx+\left(1-t\right)y}{\|tx+\left(1-t\right)y\|}\right\rangle_{i}\right]\geq 0
\end{aligned}\tag{92}$$

while for $p=2$ we have

$$\begin{aligned}
&2t\left(1-t\right)\left[\left\langle y-x,y\right\rangle_{i}-\left\langle y-x,x\right\rangle_{s}\right]\\
&\geq t\|x\|^{2}+\left(1-t\right)\|y\|^{2}-\|tx+\left(1-t\right)y\|^{2}\\
&\geq 2t\left(1-t\right)\left[\left\langle y-x,tx+\left(1-t\right)y\right\rangle_{s}-\left\langle y-x,tx+\left(1-t\right)y\right\rangle_{i}\right]\geq 0.
\end{aligned}\tag{93}$$

We notice that the inequality (93) holds for any $x,y\in X$, while in the inequality (92) we must assume that x, y and $tx+(1-t)y$ are not zero.

Remark 10 If the normed space is smooth, i.e. the norm is Gâteaux differentiable in any nonzero point, then the superior and inferior semi-inner products coincide with the Lumer-Giles semi-inner product $[\cdot,\cdot]$ that generates the norm and is linear in the first variable (see, for instance, [16]). In this situation the inequality (91) becomes

$$\begin{aligned}
&pt\left(1-t\right)\left(\|y\|^{p-2}\left[y-x,y\right]-\|x\|^{p-2}\left[y-x,x\right]\right)\\
&\geq t\|x\|^{p}+\left(1-t\right)\|y\|^{p}-\|tx+\left(1-t\right)y\|^{p}\geq 0
\end{aligned}\tag{94}$$

and holds for any nonzero x and y.

Moreover, if $(X, \langle\cdot,\cdot\rangle)$ is an inner product space, then (94) becomes

$$pt(1-t)\left\langle y-x, \|y\|^{p-2}y-\|x\|^{p-2}x\right\rangle \geq t\|x\|^{p}+(1-t)\|y\|^{p}-\|tx+(1-t)y\|^{p}\geq 0. \tag{95}$$

From (95) we deduce the particular inequalities of interest

$$t(1-t)\left\langle y-x, \frac{y}{\|y\|}-\frac{x}{\|x\|}\right\rangle \geq t\|x\|+(1-t)\|y\|-\|tx+(1-t)y\|\geq 0 \tag{96}$$

and

$$2t(1-t)\|y-x\|^{2}\geq t\|x\|^{2}+(1-t)\|y\|^{2}-\|tx+(1-t)y\|^{2}\geq 0. \tag{97}$$

Obviously, the inequality (97) can be proved directly on utilising the properties of the inner products.

A Refinement of Jensen's Inequality

The following refinement of Jensen's inequality holds:

Theorem 12 (Dragomir [21]) *Let $f: X \to \mathbb{R}$ be a convex function defined on a linear space X. Then for any n-tuple of vectors $\mathbf{x}=(x_1,\ldots,x_n)\in X^n$ and any probability distribution $\mathbf{p}=(p_1,\ldots,p_n)\in\mathbb{P}^n$, we have the inequality*

$$\sum_{i=1}^{n}p_i f(x_i)-f\left(\sum_{i=1}^{n}p_i x_i\right) \geq \sum_{k=1}^{n}p_k\nabla_{+}f\left(\sum_{i=1}^{n}p_i x_i\right)(x_k)-\nabla_{+}f\left(\sum_{i=1}^{n}p_i x_i\right)\left(\sum_{i=1}^{n}p_i x_i\right)\geq 0. \tag{98}$$

In particular, for the uniform distribution, we have

$$\frac{1}{n}\sum_{i=1}^{n}f(x_i)-f\left(\frac{1}{n}\sum_{i=1}^{n}x_i\right) \geq \frac{1}{n}\left[\sum_{k=1}^{n}\nabla_{+}f\left(\frac{1}{n}\sum_{i=1}^{n}x_i\right)(x_k)-\nabla_{+}f\left(\frac{1}{n}\sum_{i=1}^{n}x_i\right)\left(\sum_{i=1}^{n}x_i\right)\right]\geq 0. \tag{99}$$

Proof Utilising the gradient inequality (83), we have

$$f(x_k)-f\left(\sum_{i=1}^{n}p_i x_i\right)\geq \nabla_{+}f\left(\sum_{i=1}^{n}p_i x_i\right)\left(x_k-\sum_{i=1}^{n}p_i x_i\right) \tag{100}$$

for any $k\in\{1,\ldots,n\}$.

By the subadditivity of the functional $\nabla_+ f(\cdot)(\cdot)$ in the second variable, we also have

$$\nabla_+ f\left(\sum_{i=1}^n p_i x_i\right)\left(x_k - \sum_{i=1}^n p_i x_i\right)$$
$$\geq \nabla_+ f\left(\sum_{i=1}^n p_i x_i\right)(x_k) - \nabla_+ f\left(\sum_{i=1}^n p_i x_i\right)\left(\sum_{i=1}^n p_i x_i\right) \quad (101)$$

for any $k \in \{1, \ldots, n\}$.

Utilising the inequalities (100) and (101), we get

$$f(x_k) - f\left(\sum_{i=1}^n p_i x_i\right)$$
$$\geq \nabla_+ f\left(\sum_{i=1}^n p_i x_i\right)(x_k) - \nabla_+ f\left(\sum_{i=1}^n p_i x_i\right)\left(\sum_{i=1}^n p_i x_i\right) \quad (102)$$

for any $k \in \{1, \ldots, n\}$.

Now, if we multiply (102) with $p_k \geq 0$ and sum over k from 1 to n, then we deduce the first inequality in (98). The second inequality is obvious by the subadditivity property of the functional $\nabla_+ f(\cdot)(\cdot)$ in the second variable. □

The following particular case that provides a refinement for the generalised triangle inequality in normed linear spaces is of interest:

Corollary 9 (Dragomir [21]) *Let $(X, \|\cdot\|)$ be a normed linear space. Then for any $p \geq 1$, for any n-tuple of vectors $\mathbf{x} = (x_1, \ldots, x_n) \in X^n$ and for any probability distribution $\mathbf{p} = (p_1, \ldots, p_n) \in \mathbb{P}^n$ with $\sum_{i=1}^n p_i x_i \neq 0$, we have the inequality*

$$\sum_{i=1}^n p_i \|x_i\|^p - \left\|\sum_{i=1}^n p_i x_i\right\|^p$$
$$\geq p\left\|\sum_{i=1}^n p_i x_i\right\|^{p-2}\left[\sum_{k=1}^n p_k\left\langle x_k, \sum_{j=1}^n p_j x_j\right\rangle_s - \left\|\sum_{i=1}^n p_i x_i\right\|^2\right] \geq 0. \quad (103)$$

If $p \geq 2$, the inequality holds for any n-tuple of vectors and probability distribution.

In particular, we have the norm inequalities

$$\sum_{i=1}^n p_i \|x_i\| - \left\|\sum_{i=1}^n p_i x_i\right\|$$
$$\geq \left[\sum_{k=1}^n p_k\left\langle x_k, \frac{\sum_{i=1}^n p_i x_i}{\left\|\sum_{i=1}^n p_i x_i\right\|}\right\rangle_s - \left\|\sum_{i=1}^n p_i x_i\right\|\right] \geq 0. \quad (104)$$

and

$$\sum_{i=1}^{n} p_i \|x_i\|^2 - \left\|\sum_{i=1}^{n} p_i x_i\right\|^2$$
$$\geq 2\left[\sum_{k=1}^{n} p_k \left\langle x_k, \sum_{i=1}^{n} p_i x_i \right\rangle_s - \left\|\sum_{i=1}^{n} p_i x_i\right\|^2\right] \geq 0. \tag{105}$$

We notice that the first inequality in (105) is equivalent with

$$\sum_{i=1}^{n} p_i \|x_i\|^2 + \left\|\sum_{i=1}^{n} p_i x_i\right\|^2 \geq 2\sum_{k=1}^{n} p_k \left\langle x_k, \sum_{i=1}^{n} p_i x_i \right\rangle_s,$$

which provides the result

$$\frac{1}{2}\left[\sum_{i=1}^{n} p_i \|x_i\|^2 + \left\|\sum_{i=1}^{n} p_i x_i\right\|^2\right] \geq \sum_{k=1}^{n} p_k \left\langle x_k, \sum_{i=1}^{n} p_i x_i \right\rangle_s$$
$$\left(\geq \left\|\sum_{i=1}^{n} p_i x_i\right\|^2\right) \tag{106}$$

for any n-tuple of vectors and probability distribution.

Remark 11 If in the inequality (103) we consider the uniform distribution, then we get

$$\sum_{i=1}^{n} \|x_i\|^p - n^{1-p} \left\|\sum_{i=1}^{n} x_i\right\|^p$$
$$\geq pn^{1-p} \left\|\sum_{i=1}^{n} x_i\right\|^{p-2} \left[\sum_{k=1}^{n} \left\langle x_k, \sum_{i=1}^{n} x_i \right\rangle_s - \left\|\sum_{i=1}^{n} x_i\right\|^2\right] \geq 0. \tag{107}$$

A Reverse of Jensen's Inequality

The following result is of interest as well:

Theorem 13 (Dragomir [21]) *Let $f : X \to \mathbb{R}$ be a convex function defined on a linear space X. Then for any n-tuple of vectors $\mathbf{x} = (x_1, \ldots, x_n) \in X^n$ and any probability distribution $\mathbf{p} = (p_1, \ldots, p_n) \in \mathbb{P}^n$, we have the inequality*

$$\sum_{k=1}^{n} p_k \nabla_- f(x_k)(x_k) - \sum_{k=1}^{n} p_k \nabla_- f(x_k)\left(\sum_{i=1}^{n} p_i x_i\right)$$
$$\geq \sum_{i=1}^{n} p_i f(x_i) - f\left(\sum_{i=1}^{n} p_i x_i\right). \quad (108)$$

In particular, for the uniform distribution, we have

$$\frac{1}{n}\left[\sum_{k=1}^{n} \nabla_- f(x_k)(x_k) - \sum_{k=1}^{n} \nabla_- f(x_k)\left(\frac{1}{n}\sum_{i=1}^{n} x_i\right)\right]$$
$$\geq \frac{1}{n}\sum_{i=1}^{n} f(x_i) - f\left(\frac{1}{n}\sum_{i=1}^{n} x_i\right). \quad (109)$$

Proof Utilising the gradient inequality (83), we can state that

$$\nabla_- f(x_k)\left(x_k - \sum_{i=1}^{n} p_i x_i\right) \geq f(x_k) - f\left(\sum_{i=1}^{n} p_i x_i\right) \quad (110)$$

for any $k \in \{1, \ldots, n\}$.

By the superadditivity of the functional $\nabla_- f(\cdot)(\cdot)$ in the second variable, we also have

$$\nabla_- f(x_k)(x_k) - \nabla_- f(x_k)\left(\sum_{i=1}^{n} p_i x_i\right) \geq \nabla_- f(x_k)\left(x_k - \sum_{i=1}^{n} p_i x_i\right) \quad (111)$$

for any $k \in \{1, \ldots, n\}$.

Therefore, by (110) and (111), we get

$$\nabla_- f(x_k)(x_k) - \nabla_- f(x_k)\left(\sum_{i=1}^{n} p_i x_i\right) \geq f(x_k) - f\left(\sum_{i=1}^{n} p_i x_i\right) \quad (112)$$

for any $k \in \{1, \ldots, n\}$.

Finally, by multiplying (112) with $p_k \geq 0$ and summing over k from 1 to n, we deduce the desired inequality (108). □

Remark 12 If the function f is defined on the Euclidian space $\mathbb{R}^n$ and is differentiable and convex, then from (108) we get the inequality

$$\sum_{k=1}^{n} p_k \langle \nabla f(x_k), x_k\rangle - \left\langle \sum_{k=1}^{n} p_k \nabla f(x_k), \sum_{i=1}^{n} p_i x_i \right\rangle$$
$$\geq \sum_{i=1}^{n} p_i f(x_i) - f\left(\sum_{i=1}^{n} p_i x_i\right) \quad (113)$$

where, as usual, for $x_k = \left(x_k^1, \ldots, x_k^n\right), \nabla f(x_k) = \left(\frac{\partial f(x_k)}{\partial x^1}, \ldots, \frac{\partial f(x_k)}{\partial x^n}\right)$. This inequality was obtained firstly by Dragomir and Goh in 1996; see [26].

For one dimension we get the inequality

$$\sum_{k=1}^{n} p_k x_k f'(x_k) - \sum_{i=1}^{n} p_i x_i \sum_{k=1}^{n} p_k f'(x_k) \geq \sum_{i=1}^{n} p_i f(x_i) - f\left(\sum_{i=1}^{n} p_i x_i\right) \tag{114}$$

that was discovered in 1994 by Dragomir and Ionescu; see [28].

The following reverse of the generalised triangle inequality holds:

Corollary 10 (Dragomir [21]) *Let $(X, \|\cdot\|)$ be a normed linear space. Then for any $p \geq 1$, for any n-tuple of vectors $\mathbf{x} = (x_1, \ldots, x_n) \in X^n \setminus \{(0, \ldots, 0)\}$ and for any probability distribution $\mathbf{p} = (p_1, \ldots, p_n) \in \mathbb{P}^n$, we have the inequality*

$$p\left[\sum_{k=1}^{n} p_k \|x_k\|^p - \sum_{k=1}^{n} p_k \|x_k\|^{p-2} \left\langle \sum_{i=1}^{n} p_i x_i, x_k \right\rangle_i\right] \geq \sum_{i=1}^{n} p_i \|x_i\|^p - \left\|\sum_{i=1}^{n} p_i x_i\right\|^p. \tag{115}$$

In particular, we have the norm inequalities

$$\sum_{k=1}^{n} p_k \|x_k\| - \sum_{k=1}^{n} p_k \left\langle \sum_{i=1}^{n} p_i x_i, \frac{x_k}{\|x_k\|} \right\rangle_i \geq \sum_{i=1}^{n} p_i \|x_i\| - \left\|\sum_{i=1}^{n} p_i x_i\right\| \tag{116}$$

for $x_k \neq 0, k \in \{1, \ldots, n\}$ and

$$2\left[\sum_{k=1}^{n} p_k \|x_k\|^2 - \sum_{k=1}^{n} p_k \left\langle \sum_{j=1}^{n} p_j x_j, x_k \right\rangle_i\right] \geq \sum_{i=1}^{n} p_i \|x_i\|^2 - \left\|\sum_{i=1}^{n} p_i x_i\right\|^2, \tag{117}$$

for any x_k.

We observe that the inequality (117) is equivalent with

$$\sum_{i=1}^{n} p_i \|x_i\|^2 + \left\| \sum_{i=1}^{n} p_i x_i \right\|^2 \geq 2 \sum_{k=1}^{n} p_k \left\langle \sum_{j=1}^{n} p_j x_j, x_k \right\rangle_i,$$

which provides the interesting result

$$\frac{1}{2} \left[\sum_{i=1}^{n} p_i \|x_i\|^2 + \left\| \sum_{i=1}^{n} p_i x_i \right\|^2 \right] \geq \sum_{k=1}^{n} p_k \left\langle \sum_{j=1}^{n} p_j x_j, x_k \right\rangle_i$$
$$\left(\geq \sum_{k=1}^{n} \sum_{j=1}^{n} p_j p_k \langle x_j, x_k \rangle_i \right) \tag{118}$$

holding for any n-tuple of vectors and probability distribution.

Remark 13 If in the inequality (115) we consider the uniform distribution, then we get

$$p \left[\sum_{k=1}^{n} \|x_k\|^p - \frac{1}{n} \sum_{k=1}^{n} \|x_k\|^{p-2} \left\langle \sum_{j=1}^{n} x_j, x_k \right\rangle_i \right]$$
$$\geq \sum_{i=1}^{n} \|x_i\|^p - n^{1-p} \left\| \sum_{i=1}^{n} x_i \right\|^p. \tag{119}$$

For $p \in [1, 2)$ all vectors x_k should not be zero.

Bounds for the Mean f-Deviation

Utilising the result from [19], we can state then the following result providing a non-trivial lower bound for the mean f-deviation:

Theorem 14 *Let $f : X \to [0, \infty)$ be a convex function with $f(0) = 0$. If $y = (y_1, \ldots, y_n) \in X^n$ and $\mathbf{p} = (p_1, \ldots, p_n)$ is a probability distribution with all p_i nonzero, then*

$$K_f(\mathbf{p}, \mathbf{y})$$
$$\geq \max_{k \in \{1, \ldots, n\}} \left\{ (1 - p_k) f \left[\frac{p_k}{1 - p_k} \left(y_k - \sum_{l=1}^{n} p_l y_l \right) \right] + p_k f \left(y_k - \sum_{l=1}^{n} p_l y_l \right) \right\} (\geq 0). \tag{120}$$

The case for mean r-absolute deviation is incorporated in

Corollary 11 *Let $(X, \|\cdot\|)$ be a normed linear space. If $y = (y_1, \ldots, y_n) \in X^n$ and $\mathbf{p} = (p_1, \ldots, p_n)$ is a probability distribution with all p_i nonzero, then for $r \geq 1$ we have*

$$K_r(\mathbf{p}, \mathbf{y}) \geq \max_{k\in\{1,\ldots,n\}} \left\{ \left[(1-p_k)^{1-r} p_k^r + p_k \right] \left\| y_k - \sum_{l=1}^{n} p_l y_l \right\|^r \right\}. \tag{121}$$

Remark 14 Since the function $h_r(t) := (1-t)^{1-r} t^r + t, r \geq 1, t \in [0,1)$ is strictly increasing on $[0,1)$, then

$$\min_{k\in\{1,\ldots,n\}} \left\{ (1-p_k)^{1-r} p_k^r + p_k \right\} = p_m + (1-p_m)^{1-r} p_m^r,$$

where $p_m := \min_{k\in\{1,\ldots,n\}} p_k$. We then obtain the following simpler inequality:

$$K_r(\mathbf{p}, \mathbf{y}) \geq \left[p_m + (1-p_m)^{1-r} \cdot p_m^r \right] \max_{k\in\{1,\ldots,n\}} \left\| y_k - \sum_{l=1}^{n} p_l y_l \right\|^p, \tag{122}$$

which is perhaps more useful for applications.

We have the following double inequality for the mean f-mean deviation:

Theorem 15 (Dragomir [21]) *Let $f : X \to [0,\infty)$ be a convex function with $f(0) = 0$. If $y = (y_1, \ldots, y_n) \in X^n$ and $\mathbf{p} = (p_1, \ldots, p_n)$ is a probability distribution with all p_i nonzero, then*

$$K_{\nabla_- f(\cdot)(\cdot)}(\mathbf{p}, \mathbf{y}) \geq K_{f(\cdot)}(\mathbf{p}, \mathbf{y}) \geq K_{\nabla_+ f(0)(\cdot)}(\mathbf{p}, \mathbf{y}) \geq 0. \tag{123}$$

Proof If we use the inequality (98) for $x_i = y_i - \sum_{k=1}^{n} p_k y_k$, we get

$$\begin{aligned}
&\sum_{i=1}^{n} p_i f\left(y_i - \sum_{k=1}^{n} p_k y_k \right) - f\left(\sum_{i=1}^{n} p_i \left(y_i - \sum_{k=1}^{n} p_k y_k \right) \right) \\
&\quad \geq \sum_{j=1}^{n} p_j \nabla_+ f\left(\sum_{i=1}^{n} p_i \left(y_i - \sum_{k=1}^{n} p_k y_k \right) \right) \left(y_j - \sum_{k=1}^{n} p_k y_k \right) \\
&\quad - \nabla_+ f\left(\sum_{i=1}^{n} p_i \left(y_i - \sum_{k=1}^{n} p_k y_k \right) \right) \left(\sum_{i=1}^{n} p_i \left(y_i - \sum_{k=1}^{n} p_k y_k \right) \right) \geq 0,
\end{aligned}$$

which is equivalent with the second part of (123).

Now, by utilising the inequality (108) for the same choice of x_i, we get

$$\begin{aligned}
&\sum_{j=1}^{n} p_j \nabla_- f\left(y_j - \sum_{k=1}^{n} p_k y_k\right)\left(y_j - \sum_{k=1}^{n} p_k y_k\right) \\
&\quad - \sum_{k=1}^{n} p_j \nabla_- f\left(y_j - \sum_{k=1}^{n} p_k y_k\right)\left(\sum_{i=1}^{n} p_i\left(y_i - \sum_{k=1}^{n} p_k y_k\right)\right) \\
&\quad \geq \sum_{i=1}^{n} p_i f\left(y_i - \sum_{k=1}^{n} p_k y_k\right) - f\left(\sum_{i=1}^{n} p_i\left(y_i - \sum_{k=1}^{n} p_k y_k\right)\right),
\end{aligned}$$

which in its turn is equivalent with the first inequality in (123). □

We observe that as examples of convex functions defined on the entire normed linear space $(X, \|\cdot\|)$ that are convex and vanish in 0, we can consider the functions

$$f(x) := g(\|x\|), \quad x \in X$$

where $g : [0, \infty) \to [0, \infty)$ is a monotonic nondecreasing convex function with $g(0) = 0$.

For this kind of functions, we have by direct computation that

$$\nabla_+ f(0)(u) = g'_+(0)\|u\| \text{ for any } u \in X$$

and

$$\nabla_- f(u)(u) = g'_-(\|u\|)\|u\| \text{ for any } u \in X.$$

We then have the following norm inequalities that are of interest:

Corollary 12 (Dragomir [21]) *Let $(X, \|\cdot\|)$ be a normed linear space. If $g : [0, \infty) \to [0, \infty)$ is a monotonic nondecreasing convex function with $g(0) = 0$, then for any $y = (y_1, \ldots, y_n) \in X^n$ and $\mathbf{p} = (p_1, \ldots, p_n)$ a probability distribution, we have*

$$\begin{aligned}
&\sum_{i=1}^{n} p_i g'_-\left(\left\|y_i - \sum_{k=1}^{n} p_k y_k\right\|\right)\left\|y_i - \sum_{k=1}^{n} p_k y_k\right\| \\
&\quad \geq \sum_{i=1}^{n} p_i g\left(\left\|y_i - \sum_{k=1}^{n} p_k y_k\right\|\right) \geq g'_+(0) \sum_{i=1}^{n} p_i \left\|y_i - \sum_{k=1}^{n} p_k y_k\right\|. \qquad (124)
\end{aligned}$$

Bounds for f-Divergence Measures

The following inequality for the f-divergence of an n-tuple of vectors in a linear space holds [20]:

Theorem 16 *Let $f : K \to \mathbb{R}$ be a convex function on the cone K. Then for any n-tuple of vectors $\mathbf{x} = (x_1, \ldots, x_n) \in K^n$, a probability distribution $\mathbf{q} \in \mathbb{P}^n$ with all values nonzero and for any nonempty subset J of $\{1, \ldots, n\}$ we have*

$$I_f(\mathbf{x}, \mathbf{q}) \geq \max_{\emptyset \neq J \subset \{1,\ldots,n\}} I_f(\mathbf{x}_J, \mathbf{q}_J) \geq I_f(\mathbf{x}_J, \mathbf{q}_J) \geq \min_{\emptyset \neq J \subset \{1,\ldots,n\}} I_f(\mathbf{x}_J, \mathbf{q}_J) \geq f(X_n) \tag{125}$$

where $X_n := \sum_{i=1}^{n} x_i$.

We observe that, for a given n-tuple of vectors $\mathbf{x} = (x_1, \ldots, x_n) \in K^n$, a sufficient condition for the positivity of $I_f(\mathbf{x}, \mathbf{q})$ for any probability distribution $\mathbf{q} \in \mathbb{P}^n$ with all values nonzero is that $f(X_n) \geq 0$. In the scalar case and if $\mathbf{x} = \mathbf{p} \in \mathbb{P}^n$, then a sufficient condition for the positivity of the f-divergence $I_f(\mathbf{p}, \mathbf{q})$ is that $f(1) \geq 0$.

The case of functions of a real variable that is of interest for applications is incorporated in [20]:

Corollary 13 *Let $f : [0, \infty) \to \mathbb{R}$ be a normalised convex function. Then for any $\mathbf{p}, \mathbf{q} \in \mathbb{P}^n$, we have*

$$I_f(\mathbf{p}, \mathbf{q}) \geq \max_{\emptyset \neq J \subset \{1,\ldots,n\}} \left[Q_J f\left(\frac{P_J}{Q_J}\right) + (1 - Q_J) f\left(\frac{1 - P_J}{1 - Q_J}\right)\right] (\geq 0). \tag{126}$$

In what follows, by the use of the results in Theorem 12 and Theorem 13, we can provide an upper and a lower bound for the positive difference $I_f(\mathbf{x}, \mathbf{q}) - f(X_n)$.

Theorem 17 (Dragomir [21]) *Let $f : K \to \mathbb{R}$ be a convex function on the cone K. Then for any n-tuple of vectors $\mathbf{x} = (x_1, \ldots, x_n) \in K^n$ and a probability distribution $\mathbf{q} \in \mathbb{P}^n$ with all values nonzero, we have*

$$I_{\nabla_- f(\cdot)(\cdot)}(\mathbf{x}, \mathbf{q}) - I_{\nabla_- f(\cdot)(X_n)}(\mathbf{x}, \mathbf{q}) \geq I_f(\mathbf{x}, \mathbf{q}) - f(X_n) \geq I_{\nabla_+ f(X_n)(\cdot)}(\mathbf{x}, \mathbf{q}) - \nabla_+ f(X_n)(X_n) \geq 0. \tag{127}$$

The case of functions of a real variable that is useful for applications is as follows:

Corollary 14 *Let $f : [0, \infty) \to \mathbb{R}$ be a normalised convex function. Then for any $\mathbf{p}, \mathbf{q} \in \mathbb{P}^n$, we have*

$$I_{f'_-(\cdot)(\cdot)}(\mathbf{p}, \mathbf{q}) - I_{f'_-(\cdot)}(\mathbf{p}, \mathbf{q}) \geq I_f(\mathbf{p}, \mathbf{q}) \geq 0, \tag{128}$$

or, equivalently,

$$I_{f'_{-}(\cdot)[(\cdot)-1]}(\mathbf{p},\mathbf{q}) \geq I_f(\mathbf{p},\mathbf{q}) \geq 0. \tag{129}$$

The above corollary is useful to provide an upper bound in terms of the variational distance for the f-divergence $I_f(\mathbf{p},\mathbf{q})$ of normalised convex functions whose derivatives are bounded above and below.

Proposition 9 *Let $f:[0,\infty)\to\mathbb{R}$ be a normalised convex function and $\mathbf{p},\mathbf{q}\in\mathbb{P}^n$. If there exists the constants γ and Γ with*

$$-\infty < \gamma \leq f'_{-}\left(\frac{p_k}{q_k}\right) \leq \Gamma < \infty \text{ for all } k\in\{1,\dots,n\},$$

then we have the inequality

$$0 \leq I_f(\mathbf{p},\mathbf{q}) \leq \frac{1}{2}(\Gamma-\gamma)V(\mathbf{p},\mathbf{q}), \tag{130}$$

where

$$V(\mathbf{p},\mathbf{q}) = \sum_{i=1}^{n} q_i\left|\frac{p_i}{q_i}-1\right| = \sum_{i=1}^{n}|p_i-q_i|.$$

Proof By the inequality (129) we have successively that

$$\begin{aligned}
0 &\leq I_f(\mathbf{p},\mathbf{q}) \leq I_{f'_{-}(\cdot)[(\cdot)-1]}(\mathbf{p},\mathbf{q})\\
&= \sum_{i=1}^{n} q_i\left(\frac{p_i}{q_i}-1\right)\left[f'_{-}\left(\frac{p_i}{q_i}\right)-\frac{\Gamma+\gamma}{2}\right]\\
&\leq \sum_{i=1}^{n} q_i\left|\frac{p_i}{q_i}-1\right|\left|f'_{-}\left(\frac{p_i}{q_i}\right)-\frac{\Gamma+\gamma}{2}\right|\\
&\leq \frac{1}{2}(\Gamma-\gamma)\sum_{i=1}^{n} q_i\left|\frac{p_i}{q_i}-1\right|,
\end{aligned}$$

which proves the desired result (130). □

Corollary 15 *Let $f:[0,\infty)\to\mathbb{R}$ be a normalised convex function and $\mathbf{p},\mathbf{q}\in\mathbb{P}^n$. If there exist the constants r and R with*

$$0 < r \leq \frac{p_k}{q_k} \leq R < \infty \text{ for all } k\in\{1,\dots,n\},$$

then we have the inequality

$$0 \leq I_f(\mathbf{p},\mathbf{q}) \leq \frac{1}{2}\left[f'_{-}(R)-f'_{-}(r)\right]V(\mathbf{p},\mathbf{q}). \tag{131}$$

The K. Pearson χ^2-*divergence* is obtained for the convex function $f(t) = (1-t)^2, t \in \mathbb{R}$ and given by

$$\chi^2(p,q) := \sum_{j=1}^{n} q_j \left(\frac{p_j}{q_j} - 1\right)^2 = \sum_{j=1}^{n} \frac{(p_j - q_j)^2}{q_j}.$$

Finally, the following proposition giving another upper bound in terms of the χ^2-divergence can be stated:

Proposition 10 *Let $f : [0,\infty) \to \mathbb{R}$ be a normalised convex function and $\mathbf{p}, \mathbf{q} \in \mathbb{P}^n$. If there exists the constant $0 < \Delta < \infty$ with*

$$\left| \frac{f'_-\left(\frac{p_i}{q_i}\right) - f'_-(1)}{\frac{p_i}{q_i} - 1} \right| \le \Delta \text{ for all } k \in \{1,\ldots,n\}, \tag{132}$$

then we have the inequality

$$0 \le I_f(\mathbf{p},\mathbf{q}) \le \Delta \chi^2(p,q). \tag{133}$$

In particular, if $f'_-(\cdot)$ satisfies the local Lipschitz condition

$$\left| f'_-(x) - f'_-(1) \right| \le \Delta |x - 1| \text{ for any } x \in (0,\infty) \tag{134}$$

then (133) holds true for any $\mathbf{p}, \mathbf{q} \in \mathbb{P}^n$.

Proof We have from (129) that

$$\begin{aligned} 0 &\le I_f(\mathbf{p},\mathbf{q}) \le I_{f'_-(\cdot)[(\cdot)-1]}(\mathbf{p},\mathbf{q}) \\ &= \sum_{i=1}^{n} q_i \left(\frac{p_i}{q_i} - 1\right)\left[f'_-\left(\frac{p_i}{q_i}\right) - f'_-(1)\right] \\ &\le \sum_{i=1}^{n} q_i \left(\frac{p_i}{q_i} - 1\right)^2 \left| \frac{f'_-\left(\frac{p_i}{q_i}\right) - f'_-(1)}{\frac{p_i}{q_i} - 1} \right| \\ &\le \Delta \sum_{i=1}^{n} q_i \left(\frac{p_i}{q_i} - 1\right)^2 \end{aligned}$$

and the inequality (133) is obtained. □

Remark 15 It is obvious that if one chooses in the above inequalities particular normalised convex functions that generate the Kullback-Leibler, Jeffreys, Hellinger or other divergence measures or discrepancies, then one can obtain some results of interest. However the details are not provided here.

Inequalities of Slater's Type

Introduction

Suppose that I is an interval of real numbers with interior $\mathring{I}$ and $f : I \to \mathbb{R}$ is a convex function on I. Then f is continuous on $\mathring{I}$ and has finite left and right derivatives at each point of $\mathring{I}$. Moreover, if $x, y \in \mathring{I}$ and $x < y$, then $f'_{-}(x) \leq f'_{+}(x) \leq f'_{-}(y) \leq f'_{+}(y)$ which shows that both f'_{-} and f'_{+} are nondecreasing function on $\mathring{I}$. It is also known that a convex function must be differentiable except for at most countably many points.

For a convex function $f : I \to \mathbb{R}$, the subdifferential of f denoted by ∂f is the set of all functions $\varphi : I \to [-\infty, \infty]$ such that $\varphi\left(\mathring{I}\right) \subset \mathbb{R}$ and

$$f(x) \geq f(a) + (x-a)\varphi(a) \text{ for any } x, a \in I.$$

It is also well known that if f is convex on I, then ∂f is nonempty, $f'_{-}, f'_{+} \in \partial f$ and if $\varphi \in \partial f$, then

$$f'_{-}(x) \leq \varphi(x) \leq f'_{+}(x) \text{ for any } x \in \mathring{I}.$$

In particular, φ is a nondecreasing function.

If f is differentiable and convex on $\mathring{I}$, then $\partial f = \{f'\}$.

The following result is well known in the literature as *the Slater inequality:*

Theorem 18 (Slater [32]) *If $f : I \to \mathbb{R}$ is a nonincreasing (nondecreasing) convex function, $x_i \in I$, $p_i \geq 0$ with $P_n := \sum_{i=1}^{n} p_i > 0$ and $\sum_{i=1}^{n} p_i \varphi(x_i) \neq 0$, where $\varphi \in \partial f$, then*

$$\frac{1}{P_n}\sum_{i=1}^{n} p_i f(x_i) \leq f\left(\frac{\sum_{i=1}^{n} p_i x_i \varphi(x_i)}{\sum_{i=1}^{n} p_i \varphi(x_i)}\right). \tag{135}$$

As pointed out in [17, p. 208], the monotonicity assumption for the derivative φ can be replaced with the condition

$$\frac{\sum_{i=1}^{n} p_i x_i \varphi(x_i)}{\sum_{i=1}^{n} p_i \varphi(x_i)} \in I, \tag{136}$$

which is more general and can hold for suitable points in I and for not necessarily monotonic functions.

The main aim of the next section is to extend Slater's inequality for convex functions defined on general linear spaces. A reverse of the Slater's inequality is also obtained. Natural applications for norm inequalities and f-divergence measures are provided as well.

Slater's Inequality for Functions Defined on Linear Spaces

For a given convex function $f : X \to \mathbb{R}$ and a given n-tuple of vectors $\mathbf{x} = (x_1, \ldots, x_n) \in X^n$, we consider the sets

$$Sla_{+(-)}(f, \mathbf{x}) := \left\{ v \in X | \nabla_{+(-)} f(x_i)(v - x_i) \geq 0 \text{ for all } i \in \{1, \ldots, n\} \right\} \tag{137}$$

and

$$Sla_{+(-)}(f, \mathbf{x}, \mathbf{p}) := \left\{ v \in X | \sum_{i=1}^{n} p_i \nabla_{+(-)} f(x_i)(v - x_i) \geq 0 \right\} \tag{138}$$

where $\mathbf{p} = (p_1, \ldots, p_n) \in \mathbb{P}^n$ is a given probability distribution.

Since $\nabla_{+(-)} f(x)(0) = 0$ for any $x \in X$, then we observe that $\{x_1, \ldots, x_n\} \subset Sla_{+(-)}(f, \mathbf{x}, \mathbf{p})$, therefore the sets $Sla_{+(-)}(f, \mathbf{x}, \mathbf{p})$ are not empty for each $f, \mathbf{x}$ and $\mathbf{p}$ as above.

The following properties of these sets hold:

Lemma 1 (Dragomir [23]) *For a given convex function $f : X \to \mathbb{R}$, a given n-tuple of vectors $\mathbf{x} = (x_1, \ldots, x_n) \in X^n$ and a given probability distribution $\mathbf{p} = (p_1, \ldots, p_n) \in \mathbb{P}^n$, we have*

(i) $Sla_{-}(f, \mathbf{x}) \subset Sla_{+}(f, \mathbf{x})$ and $Sla_{-}(f, \mathbf{x}, \mathbf{p}) \subset Sla_{+}(f, \mathbf{x}, \mathbf{p})$;
(ii) $Sla_{-}(f, \mathbf{x}) \subset Sla_{-}(f, \mathbf{x}, \mathbf{p})$ and $Sla_{+}(f, \mathbf{x}) \subset Sla_{+}(f, \mathbf{x}, \mathbf{p})$ for all $\mathbf{p} = (p_1, \ldots, p_n) \in \mathbb{P}^n$;
(iii) The sets $Sla_{-}(f, \mathbf{x})$ and $Sla_{-}(f, \mathbf{x}, \mathbf{p})$ are convex.

Proof The properties (i) and (ii) follow from the definition and the fact that $\nabla_{+} f(x)(y) \geq \nabla_{-} f(x)(y)$ for any x, y.

(iii) Let us only prove that $Sla_{-}(f, \mathbf{x})$ is convex.

If we assume that $y_1, y_2 \in Sla_{-}(f, \mathbf{x})$ and $\alpha, \beta \in [0, 1]$ with $\alpha + \beta = 1$, then by the superadditivity and positive homogeneity of the Gâteaux derivative $\nabla_{-} f(\cdot)(\cdot)$ in the second variable, we have

$$\begin{aligned}
\nabla_{-} f(x_i)(\alpha y_1 + \beta y_2 - x_i) &= \nabla_{-} f(x_i)[\alpha(y_1 - x_i) + \beta(y_2 - x_i)] \\
&\geq \alpha \nabla_{-} f(x_i)(y_1 - x_i) + \beta \nabla_{-} f(x_i)(y_2 - x_i) \geq 0
\end{aligned}$$

for all $i \in \{1, \ldots, n\}$, which shows that $\alpha y_1 + \beta y_2 \in Sla_{-}(f, \mathbf{x})$.

The proof for the convexity of $Sla_{-}(f, \mathbf{x}, \mathbf{p})$ is similar and the details are omitted. □

For the convex function $f_p : X \to \mathbb{R}$, $f_p(x) := \|x\|^p$ with $p \geq 1$, defined on the normed linear space $(X, \|\cdot\|)$, and for the n-tuple of vectors $\mathbf{x} = (x_1, \ldots, x_n) \in X^n \setminus \{(0, \ldots, 0)\}$, we have, by the well-known property of the semi-inner products,

$$\langle y + \alpha x, x \rangle_{s(i)} = \langle y, x \rangle_{s(i)} + \alpha \|x\|^2 \text{ for any } x, y \in X \text{ and } \alpha \in \mathbb{R},$$

that

$$Sla_{+(-)}\left(\|\cdot\|^p, \mathbf{x}\right) = Sla_{+(-)}\left(\|\cdot\|, \mathbf{x}\right)$$
$$:= \left\{v \in X | \left\langle v, x_j \right\rangle_{s(i)} \geq \left\|x_j\right\|^2 \text{ for all } j \in \{1, \ldots, n\}\right\}$$

which, as can be seen, does not depend on p. We observe that, by the continuity of the semi-inner products in the first variable, $Sla_{+(-)}(\|\cdot\|, \mathbf{x})$ is closed in $(X, \|\cdot\|)$. Also, we should remark that if $v \in Sla_{+(-)}(\|\cdot\|, \mathbf{x})$, then for any $\gamma \geq 1$ we also have that $\gamma v \in Sla_{+(-)}(\|\cdot\|, \mathbf{x})$.

The larger classes, which are dependent on the probability distribution $\mathbf{p} \in \mathbb{P}^n$, are described by

$$Sla_{+(-)}\left(\|\cdot\|^p, \mathbf{x}, \mathbf{p}\right) := \left\{v \in X | \sum_{j=1}^{n} p_j \left\|x_j\right\|^{p-2} \left\langle v, x_j \right\rangle_{s(i)} \geq \sum_{j=1}^{n} p_j \left\|x_j\right\|^p \right\}.$$

If the normed space is smooth, i.e. the norm is Gâteaux differentiable in any nonzero point, then the superior and inferior semi-inner products coincide with the Lumer-Giles semi-inner product $[\cdot, \cdot]$ that generates the norm and is linear in the first variable (see, for instance, [16]). In this situation

$$Sla\left(\|\cdot\|, \mathbf{x}\right) = \left\{v \in X | \left[v, x_j\right] \geq \left\|x_j\right\|^2 \text{ for all } j \in \{1, \ldots, n\}\right\}$$

and

$$Sla\left(\|\cdot\|^p, \mathbf{x}, \mathbf{p}\right) = \left\{v \in X | \sum_{j=1}^{n} p_j \left\|x_j\right\|^{p-2} \left[v, x_j\right] \geq \sum_{j=1}^{n} p_j \left\|x_j\right\|^p \right\}.$$

If $(X, \langle\cdot, \cdot\rangle)$ is an inner product space, then $Sla\left(\|\cdot\|^p, \mathbf{x}, \mathbf{p}\right)$ can be described by

$$Sla\left(\|\cdot\|^p, \mathbf{x}, \mathbf{p}\right) = \left\{v \in X | \left\langle v, \sum_{j=1}^{n} p_j \left\|x_j\right\|^{p-2} x_j \right\rangle \geq \sum_{j=1}^{n} p_j \left\|x_j\right\|^p \right\}$$

and if the family $\left\{x_j\right\}_{j=1,\ldots,n}$ is orthogonal, then obviously, by the Pythagoras theorem, we have that the sum $\sum_{j=1}^{n} x_j$ belongs to $Sla\left(\|\cdot\|, \mathbf{x}\right)$ and therefore to $Sla\left(\|\cdot\|^p, \mathbf{x}, \mathbf{p}\right)$ for any $p \geq 1$ and any probability distribution $\mathbf{p} = (p_1, \ldots, p_n) \in \mathbb{P}^n$.

We can state now the following results that provide a generalisation of Slater's inequality as well as a counterpart for it.

Theorem 19 (Dragomir [23]) *Let $f : X \to \mathbb{R}$ be a convex function on the real linear space X, $\mathbf{x} = (x_1, \ldots, x_n) \in X^n$ an n-tuple of vectors and $\mathbf{p} = (p_1, \ldots, p_n) \in \mathbb{P}^n$ a probability distribution. Then for any $v \in Sla_+(f, \mathbf{x}, \mathbf{p})$, we have the inequalities*

$$\nabla_- f(v)(v) - \sum_{i=1}^{n} p_i \nabla_- f(v)(x_i) \geq f(v) - \sum_{i=1}^{n} p_i f(x_i) \geq 0. \tag{139}$$

Proof If we write the gradient inequality for $v \in Sla_+(f, \mathbf{x}, \mathbf{p})$ and x_i, then we have that

$$\nabla_- f(v)(v - x_i) \geq f(v) - f(x_i) \geq \nabla_+ f(x_i)(v - x_i) \tag{140}$$

for any $i \in \{1, \ldots, n\}$.

By multiplying (140) with $p_i \geq 0$ and summing over i from 1 to n, we get

$$\sum_{i=1}^{n} p_i \nabla_- f(v)(v - x_i) \geq f(v) - \sum_{i=1}^{n} p_i f(x_i) \geq \sum_{i=1}^{n} p_i \nabla_+ f(x_i)(v - x_i). \tag{141}$$

Now, since $v \in Sla_+(f, \mathbf{x}, \mathbf{p})$, then the right hand side of (141) is nonnegative, which proves the second inequality in (139).

By the superadditivity of the Gâteaux derivative $\nabla_- f(\cdot)(\cdot)$ in the second variable, we have

$$\nabla_- f(v)(v) - \nabla_- f(v)(x_i) \geq \nabla_- f(v)(v - x_i),$$

which, by multiplying with $p_i \geq 0$ and summing over i from 1 to n, produces the inequality

$$\nabla_- f(v)(v) - \sum_{i=1}^{n} p_i \nabla_- f(v)(x_i) \geq \sum_{i=1}^{n} p_i \nabla_- f(v)(v - x_i). \tag{142}$$

Utilising (141) and (142), we deduce the desired result (139). □

Remark 16 The above result has the following form for normed linear spaces. Let $(X, \|\cdot\|)$ be a normed linear space, $\mathbf{x} = (x_1, \ldots, x_n) \in X^n$ an n-tuple of vectors from X and $\mathbf{p} = (p_1, \ldots, p_n) \in \mathbb{P}^n$ a probability distribution. Then for any vector $v \in X$ with the property

$$\sum_{j=1}^{n} p_j \|x_j\|^{p-2} \langle v, x_j \rangle_s \geq \sum_{j=1}^{n} p_j \|x_j\|^p, \quad p \geq 1, \tag{143}$$

we have the inequalities

$$p\left[\|v\|^{p}-\sum_{j=1}^{n}p_{j}\left\|x_{j}\right\|^{p-2}\left\langle v,x_{j}\right\rangle_{i}\right]\geq\|v\|^{p}-\sum_{j=1}^{n}p_{j}\left\|x_{j}\right\|^{p}\geq 0. \tag{144}$$

Rearranging the first inequality in (144), we also have that

$$(p-1)\|v\|^{p}+\sum_{j=1}^{n}p_{j}\left\|x_{j}\right\|^{p}\geq p\sum_{j=1}^{n}p_{j}\left\|x_{j}\right\|^{p-2}\left\langle v,x_{j}\right\rangle_{i}. \tag{145}$$

If the space is smooth, then the condition (143) becomes

$$\sum_{j=1}^{n}p_{j}\left\|x_{j}\right\|^{p-2}\left[v,x_{j}\right]\geq\sum_{j=1}^{n}p_{j}\left\|x_{j}\right\|^{p},\ p\geq 1, \tag{146}$$

implying the inequality

$$p\left[\|v\|^{p}-\sum_{j=1}^{n}p_{j}\left\|x_{j}\right\|^{p-2}\left[v,x_{j}\right]\right]\geq\|v\|^{p}-\sum_{j=1}^{n}p_{j}\left\|x_{j}\right\|^{p}\geq 0. \tag{147}$$

Notice also that the first inequality in (147) is equivalent with

$$\begin{aligned}(p-1)\|v\|^{p}+\sum_{j=1}^{n}p_{j}\left\|x_{j}\right\|^{p}&\geq p\sum_{j=1}^{n}p_{j}\left\|x_{j}\right\|^{p-2}\left[v,x_{j}\right]\\ &\left(\geq p\sum_{j=1}^{n}p_{j}\left\|x_{j}\right\|^{p}\geq 0\right).\end{aligned} \tag{148}$$

The following corollary is of interest:

Corollary 16 (Dragomir [23]) *Let $f: X\to\mathbb{R}$ be a convex function on the real linear space X, $\mathbf{x}=(x_1,\ldots,x_n)\in X^n$ an n-tuple of vectors and $\mathbf{p}=(p_1,\ldots,p_n)\in\mathbb{P}^n$ a probability distribution. If*

$$\sum_{i=1}^{n}p_{i}\nabla_{+}f\left(x_{i}\right)\left(x_{i}\right)\geq(<)\,0 \tag{149}$$

and there exists a vector $s\in X$ with

$$\sum_{i=1}^{n}p_{i}\nabla_{+(-)}f\left(x_{i}\right)(s)\geq(\leq)\,1 \tag{150}$$

then

$$\begin{aligned}
&\nabla_- f\left(\sum_{j=1}^n p_j \nabla_+ f\left(x_j\right)\left(x_j\right) s\right)\left(\sum_{j=1}^n p_j \nabla_+ f\left(x_j\right)\left(x_j\right) s\right) \\
&\quad -\sum_{i=1}^n p_i \nabla_- f\left(\sum_{j=1}^n p_j \nabla_+ f\left(x_j\right)\left(x_j\right) s\right)\left(x_i\right) \\
&\quad \geq f\left(\sum_{j=1}^n p_j \nabla_+ f\left(x_j\right)\left(x_j\right) s\right)-\sum_{i=1}^n p_i f\left(x_i\right) \geq 0. \qquad (151)
\end{aligned}$$

Proof Assume that $\sum_{i=1}^n p_i \nabla_+ f\left(x_i\right)\left(x_i\right) \geq 0$ and $\sum_{i=1}^n p_i \nabla_+ f\left(x_i\right)(s) \geq 1$, and define $v:=\sum_{j=1}^n p_j \nabla_+ f\left(x_j\right)\left(x_j\right) s$. We claim that $v \in Sla_+(f, \mathbf{x}, \mathbf{p})$.

By the subadditivity and positive homogeneity of the mapping $\nabla_+ f(\cdot)(\cdot)$ in the second variable, we have

$$\begin{aligned}
&\sum_{i=1}^n p_i \nabla_+ f\left(x_i\right)\left(v-x_i\right) \\
&\quad \geq \sum_{i=1}^n p_i \nabla_+ f\left(x_i\right)(v)-\sum_{i=1}^n p_i \nabla_+ f\left(x_i\right)\left(x_i\right) \\
&\quad =\sum_{i=1}^n p_i \nabla_+ f\left(x_i\right)\left(\sum_{j=1}^n p_j \nabla_+ f\left(x_j\right)\left(x_j\right) s\right)-\sum_{i=1}^n p_i \nabla_+ f\left(x_i\right)\left(x_i\right) \\
&\quad =\sum_{j=1}^n p_j \nabla_+ f\left(x_j\right)\left(x_j\right) \sum_{i=1}^n p_i \nabla_+ f\left(x_i\right)(s)-\sum_{i=1}^n p_i \nabla_+ f\left(x_i\right)\left(x_i\right) \\
&\quad =\sum_{j=1}^n p_j \nabla_+ f\left(x_j\right)\left(x_j\right)\left[\sum_{i=1}^n p_i \nabla_+ f\left(x_i\right)(s)-1\right] \geq 0,
\end{aligned}$$

as claimed. Applying Theorem 19 for this v, we get the desired result.

If $\sum_{i=1}^n p_i \nabla_+ f\left(x_i\right)\left(x_i\right)<0$ and $\sum_{i=1}^n p_i \nabla_- f\left(x_i\right)(s) \leq 1$, then for

$$w:=\sum_{j=1}^n p_j \nabla_+ f\left(x_j\right)\left(x_j\right) s$$

we also have that

$$\sum_{i=1}^{n} p_i \nabla_+ f(x_i)(w - x_i)$$

$$\geq \sum_{i=1}^{n} p_i \nabla_+ f(x_i)\left(\sum_{j=1}^{n} p_j \nabla_+ f(x_j)(x_j)\, s\right) - \sum_{i=1}^{n} p_i \nabla_+ f(x_i)(x_i)$$

$$= \sum_{i=1}^{n} p_i \nabla_+ f(x_i)\left(\left(-\sum_{j=1}^{n} p_j \nabla_+ f(x_j)(x_j)\right)(-s)\right) - \sum_{i=1}^{n} p_i \nabla_+ f(x_i)(x_i)$$

$$= \left(-\sum_{j=1}^{n} p_j \nabla_+ f(x_j)(x_j)\right)\sum_{i=1}^{n} p_i \nabla_+ f(x_i)(-s) - \sum_{i=1}^{n} p_i \nabla_+ f(x_i)(x_i)$$

$$= \left(-\sum_{j=1}^{n} p_j \nabla_+ f(x_j)(x_j)\right)\left(1 + \sum_{i=1}^{n} p_i \nabla_+ f(x_i)(-s)\right)$$

$$= \left(-\sum_{j=1}^{n} p_j \nabla_+ f(x_j)(x_j)\right)\left(1 - \sum_{i=1}^{n} p_i \nabla_- f(x_i)(s)\right) \geq 0.$$

Therefore $w \in Sla_+(f, \mathbf{x}, \mathbf{p})$ and by Theorem 19 we get the desired result. □

It is natural to consider the case of normed spaces.

Remark 17 Let $(X, \|\cdot\|)$ be a normed linear space, $\mathbf{x} = (x_1, \ldots, x_n) \in X^n$ an n -tuple of vectors from X and $\mathbf{p} = (p_1, \ldots, p_n) \in \mathbb{P}^n$ a probability distribution. Then for any vector $s \in X$ with the property that

$$p \sum_{i=1}^{n} p_i \|x_i\|^{p-2} \langle s, x_i \rangle_s \geq 1, \tag{152}$$

we have the inequalities

$$p^p \|s\|^{p-1}\left(\sum_{j=1}^{n} p_j \|x_j\|^p\right)^{p-1}\left(p\|s\| \sum_{j=1}^{n} p_j \|x_j\|^p - \sum_{j=1}^{n} p_j \langle x_j, s \rangle_i\right)$$

$$\geq p^p \|s\|^p \left(\sum_{j=1}^{n} p_j \|x_j\|^p\right)^p - \sum_{j=1}^{n} p_j \|x_j\|^p \geq 0.$$

The case of smooth spaces can be easily derived from the above; however the details are left to the interested reader.

The Case of Finite Dimensional Linear Spaces

Consider now the finite dimensional linear space $X = \mathbb{R}^m$, and assume that C is an open convex subset of $\mathbb{R}^m$. Assume also that the function $f : C \to \mathbb{R}$ is differentiable and convex on C. Obviously, if $x = \left(x^1, \ldots, x^m\right) \in C$, then for any $y = \left(y^1, \ldots, y^m\right) \in \mathbb{R}^m$, we have

$$\nabla f(x)(y) = \sum_{k=1}^{m} \frac{\partial f\left(x^1, \ldots, x^m\right)}{\partial x^k} \cdot y^k$$

For the convex function $f : C \to \mathbb{R}$ and a given n-tuple of vectors $\mathbf{x} = (x_1, \ldots, x_n) \in C^n$ with $x_i = \left(x_i^1, \ldots, x_i^m\right)$ with $i \in \{1, \ldots, n\}$, we consider the sets

$$Sla(f, \mathbf{x}, C) := \left\{ v \in C \mid \sum_{k=1}^{m} \frac{\partial f\left(x_i^1, \ldots, x_i^m\right)}{\partial x^k} \cdot v^k \right.$$
$$\left. \geq \sum_{k=1}^{m} \frac{\partial f\left(x_i^1, \ldots, x_i^m\right)}{\partial x^k} \cdot x_i^k \text{ for all } i \in \{1, \ldots, n\} \right\} \quad (153)$$

and

$$Sla(f, \mathbf{x}, \mathbf{p}, C) := \left\{ v \in C \mid \sum_{i=1}^{n} \sum_{k=1}^{m} p_i \frac{\partial f\left(x_i^1, \ldots, x_i^m\right)}{\partial x^k} \cdot v^k \right.$$
$$\left. \geq \sum_{i=1}^{n} \sum_{k=1}^{m} p_i \frac{\partial f\left(x_i^1, \ldots, x_i^m\right)}{\partial x^k} \cdot x_i^k \right\} \quad (154)$$

where $\mathbf{p} = (p_1, \ldots, p_n) \in \mathbb{P}^n$ is a given probability distribution.

As in the previous section, the sets $Sla(f, \mathbf{x}, C)$ and $Sla(f, \mathbf{x}, \mathbf{p}, C)$ are convex and closed subsets of $\operatorname{clo}(C)$, the closure of C. Also $\{x_1, \ldots, x_n\} \subset Sla(f, \mathbf{x}, C) \subset Sla(f, \mathbf{x}, \mathbf{p}, C)$ for any $\mathbf{p} = (p_1, \ldots, p_n) \in \mathbb{P}^n$ a probability distribution.

Proposition 11 *Let $f : C \to \mathbb{R}$ be a convex function on the open convex set C in the finite dimensional linear space $\mathbb{R}^m$, $(x_1, \ldots, x_n) \in C^n$ an n-tuple of vectors and $(p_1, \ldots, p_n) \in \mathbb{P}^n$ a probability distribution. Then for any $v = \left(v^1, \ldots, v^n\right) \in Sla(f, \mathbf{x}, \mathbf{p}, C)$, we have the inequalities*

$$\sum_{k=1}^{m} \frac{\partial f\left(v^1, \ldots, v^m\right)}{\partial x^k} \cdot v^k - \sum_{i=1}^{n} \sum_{k=1}^{m} p_i \frac{\partial f\left(x_i^1, \ldots, x_i^m\right)}{\partial x^k} \cdot v^k$$
$$\geq f\left(v^1, \ldots, v^n\right) - \sum_{i=1}^{n} p_i f\left(x_i^1, \ldots, x_i^m\right) \geq 0. \quad (155)$$

The unidimensional case, i.e. $m = 1$, is of interest for applications. We will state this case with the general assumption that $f : I \to \mathbb{R}$ is a convex function on an *open* interval I. For a given n-tuple of vectors $\mathbf{x} = (x_1, \ldots, x_n) \in I^n$, we have

$$Sla_{+(-)}(f, \mathbf{x}, I) := \left\{ v \in I | f'_{+(-)}(x_i) \cdot (v - x_i) \geq 0 \text{ for all } i \in \{1, \ldots, n\} \right\}$$

and

$$Sla_{+(-)}(f, \mathbf{x}, \mathbf{p}, \mathbf{I}) := \left\{ v \in I | \sum_{i=1}^{n} p_i f'_{+(-)}(x_i) \cdot (v - x_i) \geq 0 \right\},$$

where $(p_1, \ldots, p_n) \in \mathbb{P}^n$ is a probability distribution. These sets inherit the general properties pointed out in Lemma 1. Moreover, if we make the assumption that $\sum_{i=1}^{n} p_i f'_+(x_i) \neq 0$, then for $\sum_{i=1}^{n} p_i f'_+(x_i) > 0$, we have

$$Sla_+(f, \mathbf{x}, \mathbf{p}, \mathbf{I}) = \left\{ v \in I | v \geq \frac{\sum_{i=1}^{n} p_i f'_+(x_i) x_i}{\sum_{i=1}^{n} p_i f'_+(x_i)} \right\}$$

while for $\sum_{i=1}^{n} p_i f'_+(x_i) < 0$, we have

$$v = \left\{ v \in I | v \leq \frac{\sum_{i=1}^{n} p_i f'_+(x_i) x_i}{\sum_{i=1}^{n} p_i f'_+(x_i)} \right\}.$$

Also, if we assume that $f'_+(x_i) \geq 0$ for all $i \in \{1, \ldots, n\}$ and $\sum_{i=1}^{n} p_i f'_+(x_i) > 0$, then

$$v_s := \frac{\sum_{i=1}^{n} p_i f'_+(x_i) x_i}{\sum_{i=1}^{n} p_i f'_+(x_i)} \in I$$

due to the fact that $x_i \in I$ and I is a convex set.

Proposition 12 *Let $f : I \to \mathbb{R}$ be a convex function on an open interval I. For a given n-tuple of vectors $\mathbf{x} = (x_1, \ldots, x_n) \in I^n$ and $(p_1, \ldots, p_n) \in \mathbb{P}^n$ a probability distribution, we have*

$$f'_-(v) \left(v - \sum_{i=1}^{n} p_i x_i \right) \geq f(v) - \sum_{i=1}^{n} p_i f(x_i) \geq 0 \tag{156}$$

for any $v \in Sla_+(f, \mathbf{x}, \mathbf{p}, \mathbf{I})$.

In particular, if we assume that $\sum_{i=1}^{n} p_i f'_+(x_i) \neq 0$ and

$$\frac{\sum_{i=1}^{n} p_i f'_+(x_i) x_i}{\sum_{i=1}^{n} p_i f'_+(x_i)} \in I$$

then

$$f'_{-}\left(\frac{\sum_{i=1}^{n} p_i f'_{+}(x_i) x_i}{\sum_{i=1}^{n} p_i f'_{+}(x_i)}\right)\left[\frac{\sum_{i=1}^{n} p_i f'_{+}(x_i) x_i}{\sum_{i=1}^{n} p_i f'_{+}(x_i)} - \sum_{i=1}^{n} p_i x_i\right]$$
$$\geq f\left(\frac{\sum_{i=1}^{n} p_i f'_{+}(x_i) x_i}{\sum_{i=1}^{n} p_i f'_{+}(x_i)}\right) - \sum_{i=1}^{n} p_i f(x_i) \geq 0 \tag{157}$$

Moreover, if $f'_{+}(x_i) \geq 0$ *for all* $i \in \{1, \ldots, n\}$ *and* $\sum_{i=1}^{n} p_i f'_{+}(x_i) > 0$, *then (157) holds true as well.*

Remark 18 We remark that the first inequality in (157) provides a reverse inequality for the classical result due to Slater.

Some Applications for f-divergences

It is obvious that the above definition of $I_f(\mathbf{p}, \mathbf{q})$ can be extended to any function $f : [0, \infty) \to \mathbb{R}$; however the positivity condition will not generally hold for normalised functions and $\mathbf{p}, \mathbf{q} \in R_{+}^{n}$ *with* $\sum_{i=1}^{n} p_i = \sum_{i=1}^{n} q_i$.

For a normalised convex function $f : [0, \infty) \to \mathbb{R}$ and two probability distributions $\mathbf{p}, \mathbf{q} \in \mathbb{P}^n$, we define the set

$$Sla_{+}(f, \mathbf{p}, \mathbf{q}) := \left\{ v \in [0, \infty) | \sum_{i=1}^{n} q_i f'_{+}\left(\frac{p_i}{q_i}\right) \cdot \left(v - \frac{p_i}{q_i}\right) \geq 0 \right\}. \tag{158}$$

Now, observe that

$$\sum_{i=1}^{n} q_i f'_{+}\left(\frac{p_i}{q_i}\right) \cdot \left(v - \frac{p_i}{q_i}\right) \geq 0$$

is equivalent with

$$v \sum_{i=1}^{n} q_i f'_{+}\left(\frac{p_i}{q_i}\right) \geq \sum_{i=1}^{n} p_i f'_{+}\left(\frac{p_i}{q_i}\right). \tag{159}$$

If $\sum_{i=1}^{n} q_i f'_{+}\left(\frac{p_i}{q_i}\right) > 0$, then (159) is equivalent with

$$v \geq \frac{\sum_{i=1}^{n} p_i f'_{+}\left(\frac{p_i}{q_i}\right)}{\sum_{i=1}^{n} q_i f'_{+}\left(\frac{p_i}{q_i}\right)}$$

therefore in this case

$$Sla_{+}(f,\mathbf{p},\mathbf{q})=\begin{cases}[0,\infty) & \text{if } \sum_{i=1}^{n}p_if_{+}'\left(\frac{p_i}{q_i}\right)<0,\\ \left[\frac{\sum_{i=1}^{n}p_if_{+}'\left(\frac{p_i}{q_i}\right)}{\sum_{i=1}^{n}q_if_{+}'\left(\frac{p_i}{q_i}\right)},\infty\right) & \text{if } \sum_{i=1}^{n}p_if_{+}'\left(\frac{p_i}{q_i}\right)\geq 0.\end{cases} \tag{160}$$

If $\sum_{i=1}^{n}q_if_{+}'\left(\frac{p_i}{q_i}\right)<0$, then (159) is equivalent with

$$v\leq\frac{\sum_{i=1}^{n}p_if_{+}'\left(\frac{p_i}{q_i}\right)}{\sum_{i=1}^{n}q_if_{+}'\left(\frac{p_i}{q_i}\right)}$$

therefore

$$Sla_{+}(f,\mathbf{p},\mathbf{q})=\begin{cases}\left[0,\frac{\sum_{i=1}^{n}p_if_{+}'\left(\frac{p_i}{q_i}\right)}{\sum_{i=1}^{n}q_if_{+}'\left(\frac{p_i}{q_i}\right)}\right] & \text{if } \sum_{i=1}^{n}p_if_{+}'\left(\frac{p_i}{q_i}\right)\leq 0,\\ \emptyset & \text{if } \sum_{i=1}^{n}p_if_{+}'\left(\frac{p_i}{q_i}\right)>0.\end{cases} \tag{161}$$

Utilising the extended f-divergences notation, we can state the following result:

Theorem 20 (Dragomir [23]) *Let $f:[0,\infty)\to\mathbb{R}$ be a normalised convex function and $\mathbf{p},\mathbf{q}\in\mathbb{P}^n$ two probability distributions. If $v\in Sla_{+}(f,\mathbf{p},\mathbf{q})$, then we have*

$$f_{-}'(v)(v-1)\geq f(v)-I_f(\mathbf{p},\mathbf{q})\geq 0. \tag{162}$$

In particular, if we assume that $I_{f_{+}'}(\mathbf{p},\mathbf{q})\neq 0$ and

$$\frac{I_{f_{+}'(\cdot)(\cdot)}(\mathbf{p},\mathbf{q})}{I_{f_{+}'}(\mathbf{p},\mathbf{q})}\in[0,\infty)$$

then

$$\begin{aligned}f_{-}'&\left(\frac{I_{f_{+}'(\cdot)(\cdot)}(\mathbf{p},\mathbf{q})}{I_{f_{+}'}(\mathbf{p},\mathbf{q})}\right)\left[\frac{I_{f_{+}'(\cdot)(\cdot)}(\mathbf{p},\mathbf{q})}{I_{f_{+}'}(\mathbf{p},\mathbf{q})}-1\right]\\ &\geq f\left(\frac{I_{f_{+}'(\cdot)(\cdot)}(\mathbf{p},\mathbf{q})}{I_{f_{+}'}(\mathbf{p},\mathbf{q})}\right)-I_f(\mathbf{p},\mathbf{q})\geq 0. \end{aligned}\tag{163}$$

Moreover, if $f_{+}'\left(\frac{p_i}{q_i}\right)\geq 0$ for all $i\in\{1,\ldots,n\}$ and $I_{f_{+}'}(\mathbf{p},\mathbf{q})>0$, then (163) holds true as well.

The proof follows immediately from Proposition 12 and the details are omitted.

The K. Pearson χ^2*-divergence* is obtained for the convex function $f(t) = (1-t)^2$, $t \in \mathbb{R}$ and given by

$$\chi^2(\mathbf{p},\mathbf{q}) := \sum_{j=1}^{n} q_j \left(\frac{p_j}{q_j} - 1\right)^2 = \sum_{j=1}^{n} \frac{(p_j - q_j)^2}{q_j} = \sum_{j=1}^{n} \frac{p_i^2}{q_i} - 1. \tag{164}$$

The *Kullback-Leibler divergence* can be obtained for the convex function $f: (0,\infty) \to \mathbb{R}$, $f(t) = t \ln t$ and is defined by

$$KL(\mathbf{p},\mathbf{q}) := \sum_{j=1}^{n} q_j \cdot \frac{p_j}{q_j} \ln\left(\frac{p_j}{q_j}\right) = \sum_{j=1}^{n} p_j \ln\left(\frac{p_j}{q_j}\right). \tag{165}$$

If we consider the convex function $f: (0,\infty) \to \mathbb{R}$, $f(t) = -\ln t$, then we observe that

$$\begin{aligned} I_f(\mathbf{p},\mathbf{q}) &:= \sum_{i=1}^{n} q_i f\left(\frac{p_i}{q_i}\right) = -\sum_{i=1}^{n} q_i \ln\left(\frac{p_i}{q_i}\right) \\ &= \sum_{i=1}^{n} q_i \ln\left(\frac{q_i}{p_i}\right) = KL(\mathbf{q},\mathbf{p}). \end{aligned} \tag{166}$$

For the function $f(t) = -\ln t$, we obviously have that

$$\begin{aligned} Sla(-\ln,\mathbf{p},\mathbf{q}) &:= \left\{ v \in [0,\infty) \middle| -\sum_{i=1}^{n} q_i \left(\frac{p_i}{q_i}\right)^{-1} \cdot \left(v - \frac{p_i}{q_i}\right) \geq 0 \right\} \\ &= \left\{ v \in [0,\infty) \middle| v \sum_{i=1}^{n} \frac{q_i^2}{p_i} - 1 \leq 0 \right\} \\ &= \left[0, \frac{1}{\chi^2(\mathbf{q},\mathbf{p}) + 1}\right]. \end{aligned} \tag{167}$$

Utilising the first part of Theorem 20, we can state the following:

Proposition 13 *Let* $\mathbf{p},\mathbf{q} \in \mathbb{P}^n$ *two probability distributions. If* $v \in \left[0, \frac{1}{\chi^2(\mathbf{q},\mathbf{p})+1}\right]$, *then we have*

$$\frac{1-v}{v} \geq -\ln(v) - KL(\mathbf{q},\mathbf{p}) \geq 0. \tag{168}$$

In particular, for $v = \frac{1}{\chi^2(\mathbf{q},\mathbf{p})+1}$, *we get*

$$\chi^2(\mathbf{q},\mathbf{p}) \geq \ln\left[\chi^2(\mathbf{q},\mathbf{p}) + 1\right] - KL(\mathbf{q},\mathbf{p}) \geq 0. \tag{169}$$

If we consider now the function $f : (0, \infty) \to \mathbb{R}$, $f(t) = t \ln t$, then $f'(t) = \ln t + 1$ and

$$\begin{aligned} & Sla\left((\cdot)\ln(\cdot), \mathbf{p}, \mathbf{q}\right) \qquad (170) \\ & := \left\{ v \in [0,\infty) | \sum_{i=1}^{n} q_i \left(\ln\left(\frac{p_i}{q_i}\right) + 1 \right) \cdot \left(v - \frac{p_i}{q_i} \right) \geq 0 \right\} \\ & = \left\{ v \in [0,\infty) | v \sum_{i=1}^{n} q_i \left(\ln\left(\frac{p_i}{q_i}\right) + 1 \right) - \sum_{i=1}^{n} p_i \cdot \left(\ln\left(\frac{p_i}{q_i}\right) + 1 \right) \geq 0 \right\} \\ & = \left\{ v \in [0,\infty) | v \left(1 - KL(\mathbf{q}, \mathbf{p})\right) \geq 1 + KL(\mathbf{p}, \mathbf{q}) \right\}. \end{aligned}$$

We observe that if $\mathbf{p}, \mathbf{q} \in \mathbb{P}^n$ two probability distributions such that $0 < KL(\mathbf{q}, \mathbf{p}) < 1$, then

$$Sla\left((\cdot)\ln(\cdot), \mathbf{p}, \mathbf{q}\right) = \left[\frac{1 + KL(\mathbf{p}, \mathbf{q})}{1 - KL(\mathbf{q}, \mathbf{p})}, \infty \right).$$

If $KL(\mathbf{q}, \mathbf{p}) \geq 1$, then $Sla\left((\cdot)\ln(\cdot), \mathbf{p}, \mathbf{q}\right) = \emptyset$.

By the use of Theorem 20, we can state now the following:

Proposition 14 *Let* $\mathbf{p}, \mathbf{q} \in \mathbb{P}^n$ *two probability distributions such that* $0 < KL(\mathbf{q}, \mathbf{p}) < 1$. *If* $v \in \left[\frac{1+KL(\mathbf{p},\mathbf{q})}{1-KL(\mathbf{q},\mathbf{p})}, \infty \right)$, *then we have*

$$(\ln v + 1)(v - 1) \geq v \ln(v) - KL(\mathbf{p}, \mathbf{q}) \geq 0. \qquad (171)$$

In particular, for $v = \frac{1+KL(\mathbf{p},\mathbf{q})}{1-KL(\mathbf{q},\mathbf{p})}$, *we get*

$$\begin{aligned} & \left(\ln\left[\frac{1 + KL(\mathbf{p}, \mathbf{q})}{1 - KL(\mathbf{q}, \mathbf{p})} \right] + 1 \right) \left(\frac{1 + KL(\mathbf{p}, \mathbf{q})}{1 - KL(\mathbf{q}, \mathbf{p})} - 1 \right) \\ & \geq \frac{1 + KL(\mathbf{p}, \mathbf{q})}{1 - KL(\mathbf{q}, \mathbf{p})} \ln\left[\frac{1 + KL(\mathbf{p}, \mathbf{q})}{1 - KL(\mathbf{q}, \mathbf{p})} \right] - KL(\mathbf{p}, \mathbf{q}) \geq 0. \qquad (172) \end{aligned}$$

Similar results can be obtained for other divergence measures of interest such as the *Jeffreys divergence, Hellinger discrimination,* etc .... However the details are left to the interested reader.

References

1. Cerone, P., Dragomir, S.S.: A refinement of the Grüss inequality and applications. Tamkang J. Math. **38**(1), 37–49 (2007). Preprint RGMIA Res. Rep. Coll. **5**(2) (2002). Article 14. http://rgmia.org/papers/v5n2/RGIApp.pdf
2. Dragomir, S.S.: An improvement of Jensen's inequality. Bull. Math. Soc. Sci. Math. Roum. **34**(82)(4), 291–296 (1990)

3. Dragomir, S.S.: Some refinements of Ky Fan's inequality. J. Math. Anal. Appl. **163**(2), 317–321 (1992)
4. Dragomir, S.S.: Some refinements of Jensen's inequality. J. Math. Anal. Appl. **168**(2), 518–522 (1992)
5. Dragomir, S.S.: A further improvement of Jensen's inequality. Tamkang J. Math. **25**(1), 29–36 (1994)
6. Dragomir, S.S.: A new improvement of Jensen's inequality. Indian J. Pure Appl. Math. **26**(10), 959–968 (1995)
7. Dragomir, S.S.: New estimation of the remainder in Taylor's formula using Grüss' type inequalities and applications. Math. Inequal. Appl. **2**(2), 183–193 (1999)
8. Dragomir, S.S.: On the Ostrowski's integral inequality for mappings of bounded variation and applications. RGMIA Res. Rep. Coll. **2**(1), 63–70 (1999). http://rgmia.org/papers/v2n1/v2n1-7.pdf
9. Dragomir, S.S.: An improvement of the remainder estimate in the generalised Taylor's formula. RGMIA Res. Rep. Coll. **3**(1) (2000). Article 1. http://rgmia.org/papers/v3n1/IREGTF.pdf
10. Dragomir, S.S.: Other inequalities for Csiszár divergence and applications. Preprint, RGMIA Monographs, Victoria University (2000). http://rgmia.org/papers/Csiszar/ICDApp.pdf
11. Dragomir, S.S.: Some inequalities for (m, M)-convex mappings and applications for the Csiszár f-divergence in Information Theory. Math. J. Ibaraki Univ. **33**, 35–50 (2001). http://rgmia.org/papers/Csiszar/ImMCMACFDIT.pdf
12. Dragomir, S.S.: Some inequalities for two Csiszár divergences and applications. Mat. Bilten (Macedonia) **25**, 73–90 (2001). http://rgmia.org/papers/Csiszar/In2CsisDivApp.pdf
13. Dragomir, S.S.: On a reverse of Jessen's inequality for isotonic linear functionals. J. Inequal. Pure Appl. Math. **2**(3) (2001). Article 36
14. Dragomir, S.S.: A Grüss type inequality for isotonic linear functionals and applications. Demonstratio Math. **36**(3), 551–562 (2003). Preprint RGMIA Res. Rep. Coll. **5**(2002). Supplement, Article 12. http://rgmia.org/papers/v5e/GTIILFApp.pdf
15. Dragomir, S.S.: A converse inequality for the Csiszár Φ-divergence. Tamsui Oxf. J. Math. Sci. **20**(1), 35–53 (2004). Preprint in Dragomir, S.S. (ed.) Inequalities for Csiszár f-Divergence in Information Theory, RGMIA Monographs. Victoria University (2000). http://rgmia.org/papers/Csiszar/Csiszar.pdf
16. Dragomir, S.S.: Semi-inner Products and Applications. Nova Science Publishers Inc., New York (2004)
17. Dragomir, S.S.: Discrete Inequalities of the Cauchy-Bunyakovsky-Schwarz Type. Nova Science Publishers, New York (2004)
18. Dragomir, S.S.: Bounds for the normalized Jensen functional. Bull. Aust. Math. Soc. **74**(3), 471–476 (2006)
19. Dragomir, S.S.: A refinement of Jensen's inequality with applications for f-divergence measures. Taiwanese J. Math. **14**(1), 153–164 (2010)
20. Dragomir, S.S.: A new refinement of Jensen's inequality in linear spaces with applications. Math. Comput. Model. **52**(9–10), 1497–1505 (2010)
21. Dragomir, S.S.: Inequalities in terms of the Gâteaux derivatives for convex functions on linear spaces with applications. Bull. Aust. Math. Soc. **83**(3), 500–517 (2011)
22. Dragomir, S.S.: A refinement and a divided difference reverse of Jensen's inequality with applications. Rev. Colomb. Mat. **50**(1), 17–39 (2016). Preprint RGMIA Res. Rep. Coll. **14** (2011). Article 74. http://rgmia.org/papers/v14/v14a74.pdf
23. Dragomir, S.S.: Some Slater's type inequalities for convex functions defined on linear spaces and applications. Abstr. Appl. Anal. **2012**, 16 pp. (2012). Article ID 168405
24. Dragomir, S.S.: Some reverses of the Jensen inequality with applications. Bull. Aust. Math. Soc. **87**(2), 177–194 (2013)
25. Dragomir, S.S.: Reverses of the Jensen inequality in terms of first derivative and applications. Acta Math. Vietnam. **38**(3), 429–446 (2013)

26. Dragomir, S.S., Goh, C.J.: A counterpart of Jensen's discrete inequality for differentiable convex mappings and applications in information theory. Math. Comput. Model. **24**(2), 1–11 (1996)
27. Dragomir, S.S., Goh, C.J.: Some bounds on entropy measures in information theory. Appl. Math. Lett., **10**(3), 23–28 (1997)
28. Dragomir, S.S., Ionescu, N.M.: Some converse of Jensen's inequality and applications. Rev. Anal. Numér. Théor. Approx. **23**(1), 71–78 (1994)
29. Dragomir, S.S., Pečarić, J., Persson, L.E.: Properties of some functionals related to Jensen's inequality. Acta Math. Hung. **70**(1–2), 129–143 (1996)
30. Pečarić, J., Dragomir, S.S.: A refinements of Jensen inequality and applications. Studia Univ. Babeş-Bolyai, Math. **24**(1), 15–19 (1989)
31. Simić, S.: On a global upper bound for Jensen's inequality. J. Math. Anal. Appl. **343**, 414–419 (2008)
32. Slater, M.S.: A companion inequality to Jensen's inequality. J. Approx. Theory **32**, 160–166 (1981)

Extrapolation Methods for Estimating the Trace of the Matrix Inverse

Paraskevi Fika

Abstract The evaluation of the trace of the matrix inverse, $Tr(A^{-1})$, arises in many applications and an efficient approximation of it without evaluating explicitly the matrix A^{-1} is very important, especially for large matrices that appear in real applications. In this work, we compare and analyze the performance of two numerical methods for the estimation of the trace of the matrix A^{-1}, where $A \in \mathbb{R}^{p\times p}$ is a symmetric matrix. The applied numerical methods are based on extrapolation techniques and can be adjusted for the trace estimation either through a stochastic approach or via the diagonal approximation. Numerical examples illustrating the performance of these methods are presented and a useful application of them in problems stemming from real-world networks is discussed. Through the presented numerical results, the methods are compared in terms of accuracy and efficiency.

Keywords Trace · Matrix inverse · Extrapolation · Prediction · Aitken's process · Moments

Introduction

The necessity of evaluating the trace of the matrix A^{-1}, where $A \in \mathbb{R}^{p\times p}$ is a large symmetric matrix, appears in several applications arising in statistics, physics, fractals, lattice quantum chromodynamics, crystals, machine learning, network analysis and graph theory [10, 11, 20]. In particular, in network analysis the "resolvent Estrada index" is evaluated by estimating the trace $\mathrm{Tr}((I_p - aA)^{-1})$, where A is the adjacency matrix of the graph and a an appropriate parameter [7, 13, 14]. In most of the aforementioned applications, the matrix A is symmetric and in most cases is also positive definite.

P. Fika (✉)
National and Kapodistrian University of Athens, Department of Mathematics, Athens, Greece
e-mail: pfika@math.uoa.gr

N. J. Daras, Th. M. Rassias (eds.), *Modern Discrete Mathematics and Analysis*,
Springer Optimization and Its Applications 131,
https://doi.org/10.1007/978-3-319-74325-7_7

The computation of the trace of a matrix inverse can have a high computational cost, especially for large matrices that arise in real-world problems, such as social networks, military networks, and other complex networks that provide models for physical, biological, engineered, or social systems (e.g., molecular structure, gene and protein interaction, food webs, telecommunication and transportation networks, power grids, energy, economic networks). Therefore, it is necessary to study estimates for the trace $Tr(A^{-1})$, aiming to the avoidance of the explicit computation of the matrix inverse. Various methods are proposed in the literature for this estimation. In particular, in [22, 24] and [25], stochastic methods and probing techniques are proposed for the estimation of the diagonal or the trace of the matrix inverse. Also, randomized algorithms for the trace estimation are proposed in [3], and an approach based on the interpolation from the diagonal of an approximate inverse is developed in [27]. Moreover, methods based on Gauss quadrature rules [2, 4–6, 20, 23, 26] and an extrapolation approach [10, 11] can be also applied for this estimation.

In this paper, we compare two extrapolation procedures for the approximation of the trace of matrix inverses. These methods actually give estimates for the quadratic form $x^T A^{-1} x, x \in \mathbb{R}^p$. In the first approach, the aforementioned quadratic form is approximated by extrapolating the moments of the matrix A, whereas in the second approach, the Aitken's extrapolation algorithm is adopted. Then, for specific selection of the vector x, the estimations of $x^T A^{-1} x$ lead to the approximation of the trace of the matrix A^{-1}.

A brief presentation of the employed numerical methods and their adjustment to the estimation of the trace $Tr(A^{-1})$ are introduced in sections "Overview of the Methods" and "Estimation of $Tr(A^{-1})$," respectively. In section "Numerical Examples" several numerical experiments and comparisons are reported. Useful conclusions and remarks concerning the implementation of the employed methods are derived in section "Concluding Remarks."

Throughout the paper $(\cdot, \cdot)$ is the inner product, $\lceil a \rceil$ is the least integer greater than or equal to a, $|a|$ denotes the absolute value of a, the superscript T denotes the transpose, and the δ_i stands for the ith column of the identity matrix of dimension p, denoted as I_p. The maximum and the minimum eigenvalues of a matrix are denoted by λ_1 and λ_p, respectively, and the symbol $\simeq$ means "approximately equal to." All the denominators are assumed to be different from zero.

Overview of the Methods

Let $A \in \mathbb{R}^{p \times p}$ be a symmetric positive definite matrix. In this section, we describe the extrapolation methods that are adjusted for the estimation of the trace of the matrix A^{-1}. Actually, these two methods give approximations of the quadratic form $x^T A^{-1} x, x \in \mathbb{R}^p$ method. In section "Estimation of $Tr(A^{-1})$" we see how these estimates can lead to the trace estimation. A fundamental tool in these extrapolation methods is the notion of moments $c_r, r \in \mathbb{R}$, of the matrix A with respect to a vector x, which are defined as $c_r = (x, A^r x)$.

Extrapolation of Moments

The first approach concerns an extrapolation procedure which was introduced in [9] for estimating the norm of the error when solving a linear system. This approach was extended in [11] and [10] to the estimation of the trace of powers of linear operators on a Hilbert space. Generalizations of this approach were presented in [18] and [16] for deriving families of estimates for the bilinear forms $x^T A^{-1} y$, for any real matrix A, and $y^* f(A)x$, for a Hermitian matrix $A \in \mathbb{C}^{p \times p}$ and given vectors $x, y \in \mathbb{C}^p$. In the sequel, we specify how this extrapolation procedure can lead to a direct estimation of the trace of the matrix A^{-1}.

In the aforementioned extrapolation method, a mathematical tool is the spectral decomposition of the symmetric matrix $A \in \mathbb{R}^{p \times p}$, which is given by $A = V \Lambda V^T = \sum_{k=1}^{p} \lambda_k v_k v_k^T$, where $V = [v_1, \ldots, v_p]$ is an orthonormal matrix, whose columns v_k are the normalized eigenvectors of the matrix A, and Λ is a diagonal matrix, whose diagonal elements are the eigenvalues $\lambda_k \in \mathbb{R}$ of the matrix A [12].

Then, for $x \in \mathbb{R}^p$, it holds $Ax = \sum_{k=1}^{p} \lambda_k (v_k, x) v_k$ and $A^r x = \sum_{k=1}^{p} \lambda_k^r (v_k, x) v_k$, $r \in \mathbb{R}$. Therefore, the moments of the matrix A can be expressed as sums:

$$c_r = x^T A^r x = \sum_k \lambda_k^r (x, v_k)^2. \tag{1}$$

In the sequel, keeping one term in these sums, the moment c_{-1} is approximated by considering some moments with nonnegative integer index r as interpolation conditions. In [16] the following family of estimates for the quadratic form $(x, A^{-1}x), x \in \mathbb{R}^p$, was obtained.

$$(x, A^{-1}x) \simeq e_\nu = \frac{c_0^2}{\rho^\nu c_1}, \quad \text{where } \rho = c_0 c_2 / c_1^2 \geq 1, \quad \nu \in \mathbb{C}. \tag{2}$$

There exists an optimal value of ν, denoted as ν_o, such that $e_{\nu_o} = (x, A^{-1}x)$. This value is given by $\nu_o = log(c_0^2/c_1 c_{-1})/log(\rho), \rho \neq 1$. In practice, this optimal value ν_o is not computed, since it requires the a priori knowledge of the exact value of c_{-1}. However, bounds and an approximation of ν_o for symmetric positive definite matrices are given in [16]. Moreover, if $\rho = 1$, then the estimate becomes exact, i.e., $e_\nu = (x, A^{-1}x), \forall\, \nu \in \mathbb{C}$.

Formula (2) has a low computational cost, since it requires only the computation of the moments c_0, c_1, c_2, i.e., only one matrix vector product.

Prediction by the Aitken's Process

The second approach concerns the prediction of the moment $(x, A^{-1}x)$ via an extrapolation method based on the Aitken's process [1]. The prediction properties of some extrapolation methods for approximating the unknown terms of a sequence were initially discussed in [8]. In [17], families of estimates for the quadratic form $(x, A^{-1}x)$ were derived through the Aitken's extrapolation process, for any symmetric positive definite matrix $A \in \mathbb{R}^{p\times p}$ and $x \in \mathbb{R}^p$.

Let us consider the sequence of moments $\{c_i,\ i \in \mathbb{Z}\}$. Supposing that the terms $c_0, \ldots, c_n, n \in \mathbb{N}$ are known, we can predict possible unknown terms of this sequence by employing the Aitken's algorithm [17]. In other words, we use the Aitken's process, which is actually a convergence acceleration method, in order to obtain approximate values of some unknowns terms, i.e., to predict them, from a restricted number of terms of it, as interpolation conditions.

Let us denote by $g_{-1}^{(n+1)}$, $n \in \mathbb{N}$, the predicted value of the moment c_{-1}, which requires the computation of the moments c_n, c_{n-1}, c_{n+1}. In [17] the following prediction scheme is derived.

$$g_i^{(n+1)} = c_i,\ \ i \le n+1$$

$$g_{-i}^{(n+1)} = c_{n-1} - \frac{(c_n - g_{-i+1}^{(n+1)})\Delta c_{n-1}}{\Delta c_n},\ \ i, n \in \mathbb{N}. \tag{3}$$

The recursive formula (3) for $i = 1$ leads to the following prediction formula for the quadratic form $(x, A^{-1}x), x \in \mathbb{R}^p$, which requires only the moments $c_{n-1},\ c_n,\ c_{n+1}, n \in \mathbb{N}$:

$$c_{-1} \simeq g_{-1}^{(n+1)} = c_{n-1} - \frac{(c_n - c_0)}{a},\ \ a = \frac{\Delta c_n}{\Delta c_{n-1}},\ \ n \in \mathbb{N}. \tag{4}$$

where Δ is the forward difference operator defined by $\Delta c_n = c_{n+1} - c_n,\ n \ge 0$ and $\Delta^k c_n = \Delta^{k-1}(\Delta c_n) = \Delta^k c_{n+1} - \Delta^k c_n,\ n \ge 2$.

The computational complexity of the method depends on the parameter n of formula (4). Actually, this formula requires $\lceil \frac{n+1}{2} \rceil$ matrix vector products. Formula (4) attains the lowest complexity for $n = 1$, since it requires only the computation of the moments $c_0,\ c_1,\ c_2$, i.e., only one matrix vector product. For $n = 1$, formula (4) gives

$$g_{-1}^{(2)} = c_0 - \frac{(c_1 - c_0)^2}{c_2 - c_1}. \tag{5}$$

Estimation of $Tr(A^{-1})$

In this section we discuss two concrete approaches for approximating the trace $Tr(A^{-1})$ by employing the estimates for the moment c_{-1} of formulae (2) and (4). Firstly, a stochastic approach is presented, whereas the second approach concerns the trace estimation through the efficient approximation of the whole diagonal of the matrix inverse.

Hutchinson's Trace Estimation

For a symmetric matrix A, the $\text{Tr}(A^{-1})$ is connected with the moment c_{-1} due to a stochastic result proved by Hutchinson [21]. According to this, it holds $E((x, A^{-1}x)) = \text{Tr}(A^{-1})$, for $x \in X^p$ (the vector x has entries 1 and -1 with equal probability), where $E(\cdot)$ denotes the expected value. Also, it is proved that $Var((x, A^{-1}x)) = 2\sum_{i\neq j}(A^{-1})_{ij}^2$ where $Var(\cdot)$ denotes the variance.

In practice, the aforementioned expected value is evaluated as the mean value of some moments $c_{-1}(x_i)$ for different sample vectors $x_i \in X^p$. In this sense, estimates for the $\text{Tr}(A^{-1})$ are obtained by considering the mean value of the computed estimates for $c_{-1}(x_i)$, for N sample vectors $x_i \in X^p$, where N is the sample size. Therefore, we have the next estimates for the trace

$$\widetilde{t}_\nu = \frac{1}{N}\sum_{i=1}^{N} e_\nu(x_i),\ \nu \in \mathbb{C},\quad \hat{t}_n = \frac{1}{N}\sum_{i=1}^{N} g_{-1}^{(n+1)}(x_i),\ n \in \mathbb{Z},\ \ x_i \in X^p,$$

using the estimations of formulae (2) and (4), respectively. Similarly, estimates for the variance are given by

$$\widetilde{v}_\nu = \frac{\sum_{i=1}^{N}(e_\nu(x_i) - \widetilde{t}_\nu)^2}{N-1},\ \text{ and }\ \hat{v}_n = \frac{\sum_{i=1}^{N}(g_{-1}^{(n+1)}(x_i) - \hat{t}_n)^2}{N-1}. \tag{6}$$

The following result specifies the confidence interval for the trace estimates $\widetilde{t}_\nu$ and $\hat{t}_n$. Let us remind that the amount of evidence required to accept that an event is unlikely to arise by chance is known as the *significance level*. The lower the significance level, the stronger the evidence. The choice of the level of significance is arbitrary, but for many applications, a value of 5% is chosen, for no better reason than that it is conventional.

Let $Z_{a/2}$ be the upper $a/2$ percentage point of the normal distribution $\mathcal{N}(0, 1)$. Then, the following result is a classical one about the probability of having a good estimate.

$$\Pr\left(\left|\frac{t - Tr(A^{-1})}{\sqrt{Var((x, A^{-1}x))/N}}\right| < Z_{a/2}\right) = 1 - a, \tag{7}$$

where N is the number of trials, a is the significance level, t denotes the trace estimates $\widetilde{t}_\nu$ or $\hat{t}_n$, and $Z_{a/2}$ is the *critical value* of the standard normal distribution defined above.

For a significance level $a = 0.01$, we have $Z_{a/2} = 2.58$, and we obtain a confidence interval for $Tr(A^{-1})$ with probability $100(1 - a)\% = 99\%$. Thus, we expect, for any sample's size, the trace estimates t to be in this interval with a probability of 99%.

If, in the probabilistic bound (7), $Var((x, A^{-1}x))$ be replaced by $\widetilde{v}_\nu$ or $\hat{v}_n$ given by formulae (6), an approximation of the confidence interval is obtained.

Trace Estimation Through the Diagonal Approximation

Estimates for the trace $Tr(A^{-1})$ can be also obtained by approximating all the diagonal elements of the matrix A^{-1} and then by summing these approximations. The value of the moment c_{-1} with respect to the vector $x = \delta_i$, i.e., the ith column of the identity matrix, equals to the entry $(A^{-1})_{ii}$. Therefore, the estimations of the quadratic form $(x, A^{-1}x)$ for this specific selection of the vector x give approximations of the ith diagonal element of the matrix A^{-1}.

Let $A = [a_{ij}] \in \mathbb{R}^{p \times p}$, $i, j = 1, \ldots, p$. Using the estimations of formulae (2) and (5) for $x = \delta_i$, we obtain the following estimates $\widetilde{d}_i$ and $\hat{d}_i$ for the ith diagonal entry of the matrix inverse.

$$\widetilde{d}_i = 1/(\rho^\nu a_{ii}),\ \ \hat{d}_i = 1 - \frac{(a_{ii} - 1)^2}{s_i - a_{ii}},\quad \rho = s_i/a_{ii}^2,\ \ s_i = \sum_{k=1}^{p} a_{ki}^2, \tag{8}$$

$$i = 1, 2, \ldots, p,\quad \nu \in \mathbb{C}.$$

The whole diagonal of the matrix A^{-1} can be efficiently approximated by a vectorized implementation of the estimates for the diagonal elements of the matrix inverse given by formulae (8) [19]. In particular, vector estimations of the whole diagonal of the matrix A^{-1}, which are denoted as $\widetilde{d} = [\widetilde{d}_1, \ldots, \widetilde{d}_p]$ and $\hat{d} = [\hat{d}_1, \ldots, \hat{d}_p] \in \mathbb{R}^p$, are given by the following one command expression written in MATLAB notation. The .^ is the element-wise power operation, the .* is the element-wise product, and the ./ is the element-wise division .

$$\widetilde{d} = 1./(\rho.\hat{\ }\nu.*d),\ \ \hat{d} = 1 - ((d-1).\hat{\ }2)./(S - d), \tag{9}$$

where $\rho = S./(d.\hat{\ }2)$, $d = \texttt{diag}(A)$, and $S = \texttt{sum}(A.\hat{\ }2)'$.

Then, the summation of the entries of the vector estimates $\widetilde{d}$ and $\hat{d}$ yields estimations of the trace of the matrix A^{-1}.

Numerical Examples

In this section, numerical experiments are presented, comparing in terms of accuracy and efficiency the two extrapolation methods described in section "Overview of the Methods," which are adjusted for estimating the trace of the matrix inverse either by using the Hutchinson's method or through the diagonal approximation, as presented in section "Estimation of $Tr(A^{-1})$." Throughout the examples, we present the relative error and the execution time in seconds for the trace estimation. The method of the extrapolation of the moments is referred to as *Extrapolation*, the method of the prediction of the moments through the Aitken's process is referred to as *Prediction*, the Hutchinson's trace estimation is denoted as *Hutch*, and the trace estimation through the diagonal approximation is denoted as *Diag*.

Concerning the *Extrapolation* method, the value of ν is selected using the result of [16, Corollary 4]. In the *Prediction* method, the value of $n = 1$ is considered. For this value of n, the two methods have the same computational complexity, and therefore, they are compared in terms of their accuracy and execution time having the same computational complexity. For the Hutchinson's trace estimators, we used $N = 30$ sample vectors $x_i \in X^p$.

All computations are performed in MATLAB R2015b 64-bit (win64), on an Intel Core i7-5820k with 16 GB DDR4 RAM at 2133 MHz. The so-called *exact* values reported in this section are those given by the internal functions *inv* and *trace* of MATLAB.

Example 1: A Dense Matrix

Let us consider a symmetric positive definite matrix $Q = [q_{ij}]$ with entries $q_{ij} = exp(-2 \cdot |i-j|)$. Table 1 reports the relative error and the execution time in seconds for estimating the trace of the matrix Q^{-1}, for different values of its dimension p.

In Table 1 we notice that the two methods have a comparable behavior in general. In particular, the *Extrapolation* method has better accuracy, attaining relative errors of order $\mathcal{O}(10^{-3})$, whereas the *Prediction* method attains relative errors of order $\mathcal{O}(10^{-2})$. This behavior is stable as the matrix dimension increases. Also, we notice that the methods are efficient, and although the execution time increases as the matrix dimension increases, it remains satisfactory. Furthermore, we see that the trace estimation through the diagonal approximation by a vectorized implementation is more efficient than the Hutchinson's method.

Table 1 Estimating the $Tr(Q^{-1})$

	Extrapolation		Prediction	
p	Hutch	Diag	Hutch	Diag
500	2.1025e−3	2.0225e−3	4.6119e−2	3.5903e−2
	4.3877e−3 s	8.2168e−4 s	3.0345e−3 s	6.7274e−4 s
1000	3.2388e−3	2.0251e−3	3.7442e−2	3.5938e−2
	1.2484e−2 s	2.6313e−3 s	6.5402e−3 s	2.3660e−3 s
2000	2.5275e−3	2.0265e−3	3.6846e−2	3.5955e−2
	8.1356e−2 s	1.1202e−2 s	4.0656e−2 s	1.0986e−2 s
3000	2.1045e−3	2.0269e−3	3.6849e−2	3.5961e−2
	1.5516e−1 s	2.4224e−2 s	7.4029e−2 s	2.2231e−2 s
4000	1.9780e−3	2.0271e−3	3.6614e−2	3.5964e−2
	2.9527e−1 s	3.9707e−2 s	1.3178e−1 s	3.8113e−2 s
5000	2.6438e−3	2.0273e−3	3.6344e−2	3.5965e−2
	4.2377e−1 s	5.9784e−2 s	2.1385e−1 s	5.9362e−2 s
6000	2.2005e−3	2.0273e−3	3.6323e−2	3.5967e−2
	5.9859e−1 s	8.6477e−2 s	2.9758e−1 s	8.6176e−2 s
7000	3.2718e−3	2.0274e−3	3.6184e−2	3.5967e−2
	8.1186e−1 s	1.1838e−1 s	4.0076e−1 s	1.1681e−1 s
8000	1.9664e−3	2.0275e−3	3.6202e−2	3.5968e−2
	1.0870e0 s	1.5253e−1 s	5.2068e−1 s	1.5532e−1 s

Example 2: A Block Tridiagonal Matrix

Let us consider the *heat flow* matrix H [5]. This matrix is symmetric, block tridiagonal (sparse) and comes from the discretization of the linear heat flow problem using the simplest implicit finite difference method. The coefficient matrix H of the resulted linear system of equations is a $m^2 \times m^2$ block tridiagonal matrix $H = \texttt{tridiag}(C, D, C)$, where D is a $m \times m$ tridiagonal matrix $D = \texttt{tridiag}(-u, 1+4u, -u)$, $C = \texttt{diag}\left(-u, -u, \cdots, -u\right)$, $u = \frac{\Delta t}{h^2}$, Δt is the timestep, and h is the spacing interval. The matrix H is symmetric positive definite for $u > 0$. We tested this matrix for $u = 0.2$.

In Table 2 we report the relative error and the execution time in seconds for estimating the trace of the matrix H^{-1}, for different values of its dimension p.

In Table 2 we see that the *Extrapolation* method gives relative error of order $\mathcal{O}(10^{-3})$, whereas the *Prediction* method attains relative error of order $\mathcal{O}(10^{-2})$. Concerning the efficiency, we see that the execution time of the trace estimation through the diagonal approximation is better in comparison with the Hutchinson's method. However, both methods are very efficient, even when the dimension of the test matrix increases. It is worth to mention that the execution time for the $Tr(H^{-1})$ is generally better than the execution time for the $Tr(Q^{-1})$ presented in Table 1. This is due to sparsity of the matrix H.

Table 2 Estimating the $Tr(H^{-1})$

p	Extrapolation		Prediction	
	Hutch	Diag	Hutch	Diag
400	2.1414e−3	1.3400e−3	2.3765e−2	2.2453e−2
	7.6644e−4 s	2.3179e−4 s	4.7041e−4 s	7.5403e−5 s
900	3.7939e−3	1.3600e−3	2.3242e−2	2.2628e−2
	1.3122e−3 s	4.1301e−4 s	7.1990e−4 s	1.0488e−4 s
1600	3.4930e−4	1.3698e−3	1.9063e−2	2.2713e−2
	1.7072e−3 s	6.0974e−4 s	1.1580e−3 s	1.8432e−4 s
2500	2.1968e−3	1.3756e−3	2.2136e−2	2.2764e−2
	2.1594e−3 s	8.0834e−4 s	1.5574e−3 s	2.5041e−4 s
3600	1.2968e−4	1.3794e−3	2.1670e−2	2.2797e−2
	3.3556e−3 s	1.1257e−3 s	2.3620e−3 s	3.5622e−4 s
4900	1.9772e−3	1.3822e−3	2.2267e−2	2.2821e−2
	4.2663e−3 s	1.4298e−3 s	3.0633e−3 s	4.6111e−4 s
6400	1.3016e−3	1.3842e−3	2.2790e−2	2.2839e−2
	5.3940e−3 s	1.9766e−3 s	3.9045e−3 s	5.9516e−4 s
8100	1.6492e−3	1.3858e−3	2.2913e−2	2.2853e−2
	6.4059e−3 s	2.3654e−3 s	4.7612e−3 s	7.5186e−4 s

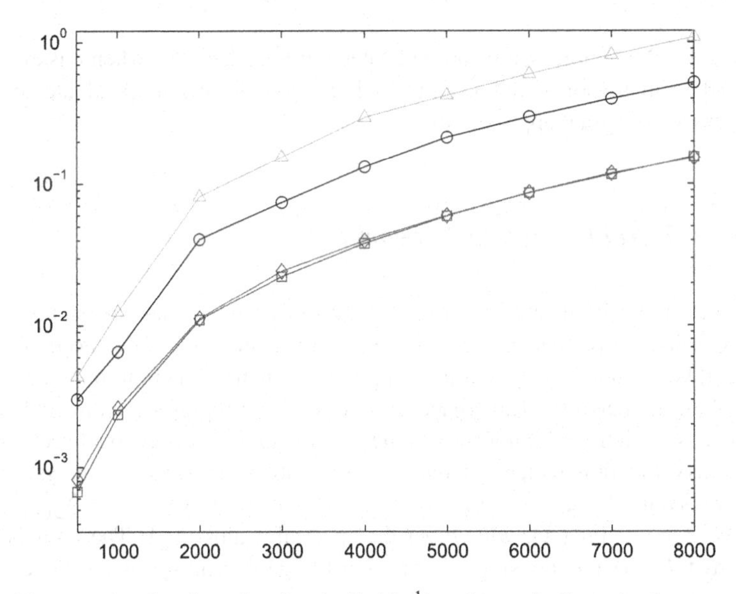

Fig. 1 The execution time for estimating the $Tr(Q^{-1})$ as the matrix dimension increases

A plot of the execution times of Tables 1 and 2 can be seen in Figs. 1 and 2. The green triangles represent the results of the *Extrapolation* method evaluated through the Hutchinson's method for the trace estimation, whereas the red diamonds represent the same method evaluated through the vectorized diagonal approximation. The

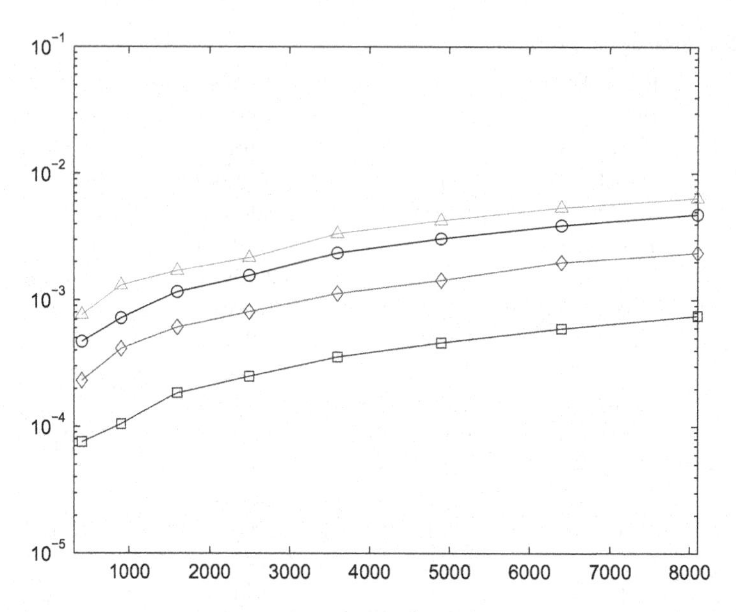

Fig. 2 The execution time for estimating the $Tr(H^{-1})$ as the matrix dimension increases

results of the *Prediction* method are indicated with black circles when it is evaluated through the Hutchinson's method and with blue squares when it is evaluated through the vectorized diagonal approximation.

Example 3: Application to Networks

In network analysis, it is necessary to evaluate certain numerical quantities that describe some properties of the graph of a given network. The properties of a graph can be obtained by applying an appropriate matrix function to the adjacency matrix A of the network. The adjacency matrices that represent undirected graphs are sparse, symmetric with nonzero elements corresponding only to adjacent nodes.

A matrix function which evaluates useful notions in network analysis is the matrix resolvent $(I - \alpha A)^{-1} = I + \alpha A + \alpha^2 A^2 + \cdots + \alpha^k A^k + \cdots = \sum_{k=0}^{\infty} \alpha^k A^k$, where $0 < \alpha < 1/\rho(A)$, with $\rho(A)$ the spectral radius of A. The bounds on α ensure that $I - \alpha A$ is nonsingular and that the geometric series converges to its inverse [14].

Let us consider the test networks *Email*, *Power*, *Internet* and *Facebook* that represent large real-world networks [15]. Tables 3, 4, 5, and 6 display the relative error and the execution time for the estimation of the resolvent Estrada index,

Table 3 Estimating the resolvent Estrada index of the *Email* network, $p = 1,133$

	Extrapolation		Prediction	
	Hutch	Diag	Hutch	Diag
Relative error	4.6633e-3	2.9695e-3	2.4236e-2	2.3148e-2
Execution time	1.6790e-3	1.2935e-4	9.6101e-4	2.2373e-4

Table 4 Estimating the resolvent Estrada index of the *Power* network, $p = 4,941$

	Extrapolation		Prediction	
	Hutch	Diag	Hutch	Diag
Relative error	8.5863e-5	1.6619e-3	3.9360e-2	3.9039e-2
Execution time	5.4883e-3	4.8889e-4	3.7115e-3	4.4342e-4

Table 5 Estimating the resolvent Estrada index of the *Internet* network, $p = 22{,}963$

	Extrapolation		Prediction	
	Hutch	Diag	Hutch	Diag
Relative error	1.8208e-4	9.2380e-5	9.3101e-4	8.1986e-4
Execution time	3.9814e-2	2.2897e-3	2.5061e-2	3.9452e-3

Table 6 Estimating the resolvent Estrada index of the *Facebook* network, $p = 63{,}731$

	Extrapolation		Prediction	
	Hutch	Diag	Hutch	Diag
Relative error	3.0980e-5	8.2147e-5	1.2735e-3	1.2497e-3
Execution time	3.2572e-1	3.8869e-2	1.5081e-1	3.2805e-2

which is defined as $\mathrm{Tr}((I_p - aA)^{-1})$, using the matrix resolvent with parameter $a = 0.85/\lambda_1$ [7, 13]. A brief description of these networks is reported below. Also, the total number of nodes is reported, i.e., the dimension p of the corresponding adjacency matrix.

- *Email* (1,133 nodes, 10,902 edges): It is a representation of email interchanges between members of the University Rovira i Virgili (Tarragona).
- *Power* (4,941 nodes, 13,188 edges): It is an undirected unweighted representation of the topology of the western states power grid of the United States.
- *Internet* (22,963 nodes, 96,872 edges): It is a symmetrized snapshot of the structure of the Internet at the level of autonomous systems, reconstructed from BGP (Border Gateway Protocol) tables posted by the University of Oregon Route Views Project.
- *Facebook* (63,731 nodes, 1,545,686 edges): It describes all the user-to-user links (friendship) from the Facebook New Orleans networks. Each edge between two users means that the second user appeared in the first user's friend list.

In this example, we notice that the two extrapolation methods attain accuracy of order $\mathcal{O}(10^{-5})$–$\mathcal{O}(10^{-2})$ for the approximation of the resolvent Estrada index for the test networks. Also, we notice that the execution time does not increase rapidly as the network size grows and that it is satisfactory even for large networks. Therefore, the tested extrapolation methods can be used even for large networks and generally for high-dimensional data.

Concluding Remarks

In this paper, two extrapolation methods, i.e., the extrapolation of the moments and the prediction through the Aitken's process, were compared for the approximation of the trace of the matrix inverse, either by using the Hutchinson's method or via approximating the whole diagonal of the matrix inverse. The extrapolation methods give simple elegant formulae of low computational complexity having total operations count of the same order with the matrix vector multiplication, which can be implemented without requiring any additional software. Also, due to their nature, the attained formulae can be executed in vectorized form, which speeds up the whole computation for the approximation of all the diagonal entries of the matrix inverse, leading to an efficient implementation.

These approaches were compared in terms of their execution time and accuracy for the trace estimation. Based on the numerical examples that were performed, the following remarks were derived. Firstly, the attained accuracy is mostly of order $\mathcal{O}(10^{-3})$ which is satisfactory for most of the applications where a high accuracy is not required, such as in statistics and in network analysis. Also, for dense matrices, although the execution time increases as the matrix dimension increases, it still remains satisfactory. On the other hand, for sparse matrices the execution time has a more stable behavior, and it slightly increases as the matrix dimension increases. Due to their efficiency, the extrapolation methods can be applied to large sparse matrices, such as large real-world networks.

To conclude, the methods described in this paper for the approximation of the trace of the matrix inverse attain a satisfactory execution time and a fair accuracy, and they can be applied in large-scale problems in applications where a high accuracy is not required.

Acknowledgements The author acknowledges financial support from the Foundation for Education and European Culture (IPEP).

References

1. Atkinson, K.: An Introduction to Numerical Analysis, 2nd edn. Wiley, Hoboken, NJ (1989)
2. August, M., Banuls, M.C., Huckle, T.: On the approximation of functionals of very large Hermitian matrices represented as matrix product operators (2016, preprint). arXiv:1610.06086
3. Avron, H., Toledo, S.: Randomized algorithms for estimating the trace of an implicit symmetric positive semi-definite matrix. J. ACM **58**, article 8 (2011)

4. Bai, Z., Golub, G.H.: Bounds for the trace of the inverse and the determinant of symmetric positive definite matrices. Ann. Numer. Math. **4**, 29–38 (1997)
5. Bai, Z., Fahey, M., Golub, G.: Some large scale computation problems. J. Comput. Appl. Math. **74**, 71–89 (1996)
6. Bellalij, M., Reichel, L., Rodriguez, G., Sadok, H.: Bounding matrix functionals via partial global block Lanczos decomposition. Appl. Numer. Math. **94** 127–139 (2015)
7. Benzi, M., Klymko, C.: Total communicability as a centrality measure. J. Complex Netw. **1**(2), 124–149 (2013)
8. Brezinski, C.: Prediction properties of some extrapolation methods. Appl. Numer. Math. **61**, 457–462 (1985)
9. Brezinski, C.: Error estimates for the solution of linear systems. SIAM J. Sci. Comput. **21**, 764–781 (1999)
10. Brezinski, C., Fika, P., Mitrouli, M.: Estimations of the trace of powers of positive self-adjoint operators by extrapolation of the moments. Electron. Trans. Numer. Anal. **39**, 144–155 (2012)
11. Brezinski, C., Fika, P., Mitrouli, M.: Moments of a linear operator, with applications to the trace of the inverse of matrices and the solution of equations. Numer. Linear Algebra Appl. **19**, 937–953 (2012)
12. Datta, B.N.: Numerical Linear Algebra and Applications, 2nd edn. SIAM, Philadelphia (2010)
13. Estrada, E., Higham, D.J.: Network properties revealed through matrix functions. SIAM Rev. **52**(4), 696–714 (2010)
14. Estrada, E., Rodríguez-Velázquez, J.A.: Subgraph centrality in complex networks. Phys. Rev. E **71**, 056103 (2005)
15. Fenu, C., Martin, D., Reichel, L., Rodriguez, G.: Network analysis via partial spectral factorization and Gauss quadrature. SIAM J. Sci. Comput. **35**(4), 2046–2068 (2013)
16. Fika, P., Mitrouli, M.: Estimation of the bilinear form $y^*f(A)x$ for Hermitian matrices. Linear Algebra Appl. **502**, 140–158 (2016)
17. Fika, P., Mitrouli, M.: Aitken's method for estimating bilinear forms arising in applications. Calcolo **54**, 455–470 (2017)
18. Fika, P., Mitrouli, M., Roupa, P.: Estimates for the bilinear form $x^T A^{-1} y$ with applications to linear algebra problems. Electron. Trans. Numer. Anal. **43**, 70–89 (2014)
19. Fika, P., Mitrouli, M., Roupa, P.: Estimating the diagonal of matrix functions. Math. Methods Appl. Sci. **41**, 1083–1088 (2018). https://doi.org/10.1002/mma.4228
20. Golub, G.H., Meurant, G.: Matrices, Moments and Quadrature with Applications. Princeton University Press, Princeton (2010)
21. Hutchinson, M.: A stochastic estimator of the trace of the influence matrix for Laplacian smoothing splines. Commun. Stat. Simul. Comput. **18**, 1059–1076 (1989)
22. Kalantzis, V., Bekas, C., Curioni, A., Gallopoulos, E.: Accelerating data uncertainty quantification by solving linear systems with multiple right-hand sides. Numer. Algorithms **62**, 637–653 (2013)
23. Meurant, G.: Estimates of the trace of the inverse of a symmetric matrix using the modified Chebyshev algorithms. Numer. Algorithms **51**, 309–318 (2009)
24. Stathopoulos, A., Laeuchli, J., Orginos, K.: Hierarchical probing for estimating the trace of the matrix inverse on toroidal lattices. SIAM J. Sci. Comput. **35**, S299–S322 (2013)
25. Tang, J., Saad, Y.: A probing method for computing the diagonal of a matrix inverse. Numer. Linear Algebra Appl. **19**, 485–501 (2012)
26. Ubaru, S., Chen, J., Saad,Y.: Fast estimation of $tr(f(A))$ via stochastic Lanczos quadrature. Preprint ys-2016–04, Department of Computer Science and Engineering, University of Minnesota, Minneapolis, MN (2016)
27. Wu, L., Laeuchli, J., Kalantzis, V., Stathopoulos, A., Gallopoulos, E.: Estimating the trace of the matrix inverse by interpolating from the diagonal of an approximate inverse. J. Comput. Phys. **326**, 828–844 (2016)

Moment Generating Functions and Moments of Linear Positive Operators

Vijay Gupta, Neha Malik, and Themistocles M. Rassias

Abstract In the theory of approximation, moments play an important role in order to study the convergence of sequence of linear positive operators. Several new operators have been discussed in the past decade and their moments have been obtained by direct computation or by attaining the recurrence relation to get the higher moments. Using the concept of moment generating function, we provide an alternate approach to estimate the higher order moments. The present article deals with the m.g.f. of some of the important operators. We estimate the moments up to order six for some of the discrete operators and their Kantorovich variants.

Preliminaries

In the last century, Weierstrass approximation theorem came into existence and thereafter, several sequences of positive linear operators have been introduced and their convergence behaviour has been studied by several mathematicians (cf. [8, 9, 11, 14, 22, 24, 25]). Also, some well-known operators have been extensively discussed in generalized form (cf. [1, 3, 4, 7, 10, 12, 13, 15–17, 20, 27]).

The other important theorem in the theory of approximation concerning convergence is the result due to Korovkin. For any sequence of linear positive operators, the first three moments are required in order to guarantee the convergence. However, for quantitative estimates, higher-order moments are also important as they are required

The original version of this chapter was revised. A correction to this chapter is available at https://doi.org/10.1007/978-3-319-74325-7_25

V. Gupta · N. Malik
Department of Mathematics, Netaji Subhas Institute of Technology, New Delhi, India

Th. M. Rassias (✉)
National Technical University of Athens, Department of Mathematics, Athens, Greece
e-mail: trassias@math.ntua.gr

N. J. Daras, Th. M. Rassias (eds.), *Modern Discrete Mathematics and Analysis*,
Springer Optimization and Its Applications 131,
https://doi.org/10.1007/978-3-319-74325-7_8

for analysis. In literature, there are several methods available to find the moments of well-known operators. Here, we apply a different approach, viz. moment-generating function (abbrev. m.g.f.) to find the moments of well-known operators. The moment-generating function, as its name suggests, can be used to generate moments. The moments are obtained from the derivatives of the generating function. This approach is not new in statistics and special functions, but in approximation theory, the method was not popular earlier due to complicated form and it was not possible to find expansion and derivatives of m.g.f. Here,we use the software Mathematica to overcome this difficulty.

It is known that the mathematical expectation of $g(x)$ is $E(g(x))$. If we consider $g(x) = e^{tx}$, then we have $E(e^{tx}) = \sum\limits_k e^{tx_k} f_k(x)$ and $\int_{-\infty}^{\infty} e^{\theta x} f(x)\,dx$ in case of discrete and continuous distributions, respectively. Usually, the m.g.f. is denoted by $M_x(t) = E(e^{tx})$.

In the present article, we provide the m.g.f. of some well-known discretely defined operators and their Kantorovich variants. We write the m.g.f. in expanded form and are thus able to find moments of any order.

M.G.F. and Moments of Exponential Type Operators

The present section deals with the moment-generating functions and moments up to sixth order of some discretely defined operators. We write the m.g.f. in expanded form and obtain moments, which are important in theory of approximation from the convergence point of view.

Bernstein Operators

In the year 1912, S.N. Bernstein in his famous paper [6] established a probabilistic proof of the well-known Weierstrass approximation theorem and introduced what we today call now the Bernstein polynomials. For $f \in C[0,1]$, the n-th degree Bernstein polynomials are defined as

$$B_n(f,x) := \sum_{k=0}^{n} \binom{n}{k} x^k (1-x)^{n-k} f\left(\frac{k}{n}\right). \tag{1}$$

The operators defined by (1) are linear and positive and preserve linear and constant functions. Let us consider $f(t) = e^{\theta t}, \theta \in \mathbb{R}$, then we have

$$B_n(e^{\theta t}, x) = \sum_{k=0}^{n} \binom{n}{k} x^k (1-x)^{n-k} e^{\frac{k\theta}{n}}$$

$$= (1-x)^n \sum_{k=0}^{n} \frac{(-1)^k (-n)_k}{k!} \left(\frac{x\, e^{\frac{\theta}{n}}}{1-x} \right)^k$$

$$= (1-x)^n \sum_{k=0}^{n} \frac{(-n)_k}{k!} \left(\frac{-x\, e^{\frac{\theta}{n}}}{1-x} \right)^k .$$

Now, using the binomial series $\sum_{k=0}^{\infty} \frac{(c)_k}{k!} z^k = (1-z)^{-c}$, $|z| < 1$, we get

$$B_n(e^{\theta t}, x) = (1-x)^n \left(1 - \frac{(-x\, e^{\frac{\theta}{n}})}{1-x} \right)^n$$

$$= \left(1 - x + x e^{\frac{\theta}{n}} \right)^n . \tag{2}$$

It may be observed that $B_n(e^{\theta t}, x)$ may be treated as m.g.f. of the operators B_n, which may be utilized to obtain the moments of (1). Let $\mu_r^{B_n}(x) = B_n(e_r, x)$, where $e_r(t) = t^r, r \in \mathbb{N} \cup \{0\}$. The moments are given by

$$\mu_r^{B_n}(x) = \left[\frac{\partial^r}{\partial \theta^r} B_n(e^{\theta t}, x) \right]_{\theta=0}$$

$$= \left[\frac{\partial^r}{\partial \theta^r} \left\{ \left(1 - x + x e^{\frac{\theta}{n}} \right)^n \right\} \right]_{\theta=0} .$$

Using the software Mathematica, we get the expansion of (2) in powers of θ as follows:

$$B_n(e^{\theta t}, x)$$

$$= 1 + x\,\theta + \left(\frac{nx^2 - x^2 + x}{n} \right) \frac{\theta^2}{2!} + \left(\frac{x^3(n-1)(n-2) + 3x^2(n-1) + x}{n^2} \right) \frac{\theta^3}{3!}$$

$$+ \left(\frac{x^4(n-1)(n-2)(n-3) + 6x^3(n-1)(n-2) + 7x^2(n-1) + x}{n^3} \right) \frac{\theta^4}{4!}$$

$$+ \frac{\left[\begin{array}{c} x^5(n-1)(n-2)(n-3)(n-4) + 10x^4(n-1)(n-2)(n-3) \\ +25x^3(n-1)(n-2) + 15x^2(n-1) + x \end{array} \right]}{n^4} \frac{\theta^5}{5!}$$

$$+ \frac{\left[\begin{array}{c} x^6(n-1)(n-2)(n-3)(n-4)(n-5) + 15x^5(n-1)(n-2)(n-3)(n-4) \\ +65x^4(n-1)(n-2)(n-3) + 90x^3(n-1)(n-2) + 31x^2(n-1) + x \end{array} \right]}{n^5} \frac{\theta^6}{6!}$$

$$+ \mathcal{O}(\theta^7).$$

In particular, the first few moments may be obtained as

$$\mu_0^{B_n}(x) = 1,$$
$$\mu_1^{B_n}(x) = x,$$
$$\mu_2^{B_n}(x) = \frac{nx^2 - x^2 + x}{n},$$
$$\mu_3^{B_n}(x) = \frac{x^3(n-1)(n-2) + 3x^2(n-1) + x}{n^2},$$
$$\mu_4^{B_n}(x) = \frac{x^4(n-1)(n-2)(n-3) + 6x^3(n-1)(n-2) + 7x^2(n-1) + x}{n^3},$$
$$\mu_5^{B_n}(x) = \frac{\left[\begin{array}{c} x^5(n-1)(n-2)(n-3)(n-4) + 10x^4(n-1)(n-2)(n-3) \\ +25x^3(n-1)(n-2) + 15x^2(n-1) + x \end{array}\right]}{n^4},$$
$$\mu_6^{B_n}(x) = \frac{\left[\begin{array}{c} x^6(n-1)(n-2)(n-3)(n-4)(n-5) + 15x^5(n-1)(n-2)(n-3)(n-4) \\ +65x^4(n-1)(n-2)(n-3) + 90x^3(n-1)(n-2) + 31x^2(n-1) + x \end{array}\right]}{n^5}.$$

Post-Widder Operators

In the year 1941, Widder [28] defined the following operators for $f \in C[0,\infty)$:

$$P_n(f,x) := \frac{1}{n!}\left(\frac{n}{x}\right)^{n+1} \int_0^\infty t^n\, e^{-\frac{nt}{x}}\, f(t)\, dt. \tag{3}$$

Let us consider $f(t) = e^{\theta t}$, $\theta \in \mathbb{R}$, then we have

$$\begin{aligned} P_n(e^{\theta t},x) &= \frac{1}{n!}\left(\frac{n}{x}\right)^{n+1} \int_0^\infty t^n\, e^{-\frac{nt}{x}}\, e^{\theta t}\, dt \\ &= \frac{1}{n!}\left(\frac{n}{x}\right)^{n+1} \int_0^\infty t^n\, e^{-(\frac{n}{x}-\theta)t}\, dt. \end{aligned}$$

Substituting $\left(\frac{n}{x} - \theta\right) t = u$, we can write

$$\begin{aligned} P_n(e^{\theta t},x) &= \frac{1}{n!}\left(\frac{n}{x}\right)^{n+1} \frac{1}{\left(\frac{n}{x}-\theta\right)^{n+1}} \int_0^\infty u^n\, e^{-u}\, du \\ &= \frac{1}{n!}\left(\frac{n}{x}\right)^{n+1} \frac{1}{\left(\frac{n}{x}-\theta\right)^{n+1}}\, \Gamma(n+1) \\ &= \left(\frac{n}{x}\right)^{n+1} \left(\frac{n}{x}-\theta\right)^{-(n+1)}. \end{aligned} \tag{4}$$

It may be observed that $P_n(e^{\theta t}, x)$ may be treated as m.g.f. of the operators P_n, which may be utilized to obtain the moments of (3). Let $\mu_r^{P_n}(x) = P_n(e_r, x)$, where $e_r(t) = t^r, r \in \mathbb{N} \cup \{0\}$. The moments are given by

$$\begin{aligned}\mu_r^{P_n}(x) &= \left[\frac{\partial^r}{\partial \theta^r} P_n(e^{\theta t}, x)\right]_{\theta=0} \\ &= \left[\frac{\partial^r}{\partial \theta^r}\left\{\left(\frac{n}{x}\right)^{n+1}\left(\frac{n}{x}-\theta\right)^{-(n+1)}\right\}\right]_{\theta=0}.\end{aligned}$$

Using the software Mathematica, we get the expansion of (4) in powers of θ as follows:

$$\begin{aligned}&P_n(e^{\theta t}, x) \\ &= 1 + \left(\frac{x(n+1)}{n}\right)\theta + \left(\frac{x^2(n+1)(n+2)}{n^2}\right)\frac{\theta^2}{2!} \\ &+ \left(\frac{x^3(n+1)(n+2)(n+3)}{n^3}\right)\frac{\theta^3}{3!} + \left(\frac{x^4(n+1)(n+2)(n+3)(n+4)}{n^4}\right)\frac{\theta^4}{4!} \\ &+ \left(\frac{x^5(n+1)(n+2)(n+3)(n+4)(n+5)}{n^5}\right)\frac{\theta^5}{5!} \\ &+ \left(\frac{x^6(n+1)(n+2)(n+3)(n+4)(n+5)(n+6)}{n^6}\right)\frac{\theta^6}{6!} + \mathcal{O}(\theta^7).\end{aligned}$$

In particular, first few moments may be obtained as

$$\begin{aligned}\mu_0^{P_n}(x) &= 1, \\ \mu_1^{P_n}(x) &= \frac{x(n+1)}{n}, \\ \mu_2^{P_n}(x) &= \frac{x^2(n+1)(n+2)}{n^2}, \\ \mu_3^{P_n}(x) &= \frac{x^3(n+1)(n+2)(n+3)}{n^3}, \\ \mu_4^{P_n}(x) &= \frac{x^4(n+1)(n+2)(n+3)(n+4)}{n^4}, \\ \mu_5^{P_n}(x) &= \frac{x^5(n+1)(n+2)(n+3)(n+4)(n+5)}{n^5}, \\ \mu_6^{P_n}(x) &= \frac{x^6(n+1)(n+2)(n+3)(n+4)(n+5)(n+6)}{n^6}.\end{aligned}$$

Szász-Mirakyan Operators

The Szász-Mirakyan operators [23, 26] are generalizations of the Bernstein polynomials to infinite intervals. For $f \in C[0, \infty)$, the Szász operators are defined as

$$S_n(f, x) := \sum_{k=0}^{\infty} e^{-nx} \frac{(nx)^k}{k!} f\left(\frac{k}{n}\right). \tag{5}$$

These operators are related to Poisson distribution in probability and statistics. Also (5) preserves linear functions. Let $f(t) = e^{\theta t}, \theta \in \mathbb{R}$, then

$$\begin{aligned} S_n(e^{\theta t}, x) &= \sum_{k=0}^{\infty} e^{-nx} \frac{(nx)^k}{k!} e^{\frac{k\theta}{n}} \\ &= e^{-nx} \sum_{k=0}^{\infty} \frac{\left(nxe^{\frac{\theta}{n}}\right)^k}{k!} \\ &= e^{-nx} e^{nxe^{\frac{\theta}{n}}} \\ &= e^{nx\left(e^{\frac{\theta}{n}}-1\right)}. \end{aligned} \tag{6}$$

It may be observed that $S_n(e^{\theta t}, x)$ may be treated as m.g.f. of the operators S_n, which may be utilized to obtain the moments of (5). Let $\mu_r^{S_n}(x) = S_n(e_r, x)$, where $e_r(t) = t^r, r \in \mathbb{N} \cup \{0\}$. The moments are given by

$$\begin{aligned} \mu_r^{S_n}(x) &= \left[\frac{\partial^r}{\partial\theta^r} S_n(e^{\theta t}, x)\right]_{\theta=0} \\ &= \left[\frac{\partial^r}{\partial\theta^r} \left\{e^{nx\left(e^{\frac{\theta}{n}}-1\right)}\right\}\right]_{\theta=0}. \end{aligned}$$

Using the software Mathematica, we get the expansion of (6) in powers of θ as follows:

$$\begin{aligned} S_n(e^{\theta t}, x) &= 1 + x\,\theta + \left(x^2 + \frac{x}{n}\right)\frac{\theta^2}{2!} + \left(x^3 + \frac{3x^2}{n} + \frac{x}{n^2}\right)\frac{\theta^3}{3!} \\ &+ \left(x^4 + \frac{6x^3}{n} + \frac{7x^2}{n^2} + \frac{x}{n^3}\right)\frac{\theta^4}{4!} \\ &+ \left(x^5 + \frac{10x^4}{n} + \frac{25x^3}{n^2} + \frac{15x^2}{n^3} + \frac{x}{n^4}\right)\frac{\theta^5}{5!} \end{aligned}$$

$$+\left(x^6+\frac{15x^5}{n}+\frac{65x^4}{n^2}+\frac{90x^3}{n^3}+\frac{31x^2}{n^4}+\frac{x}{n^5}\right)\frac{\theta^6}{6!}$$
$$+\mathcal{O}(\theta^7).$$

In particular, the first few moments may be obtained as

$$\mu_0^{S_n}(x)=1,$$
$$\mu_1^{S_n}(x)=x,$$
$$\mu_2^{S_n}(x)=x^2+\frac{x}{n},$$
$$\mu_3^{S_n}(x)=x^3+\frac{3x^2}{n}+\frac{x}{n^2},$$
$$\mu_4^{S_n}(x)=x^4+\frac{6x^3}{n}+\frac{7x^2}{n^2}+\frac{x}{n^3},$$
$$\mu_5^{S_n}(x)=x^5+\frac{10x^4}{n}+\frac{25x^3}{n^2}+\frac{15x^2}{n^3}+\frac{x}{n^4},$$
$$\mu_6^{S_n}(x)=x^6+\frac{15x^5}{n}+\frac{65x^4}{n^2}+\frac{90x^3}{n^3}+\frac{31x^2}{n^4}+\frac{x}{n^5}.$$

Baskakov Operators

In the year 1957, V.A. Baskakov [5] proposed a generalization of the Bernstein polynomials based on negative binomial distribution. For $f \in C[0,\infty)$, the Baskakov operators are defined as

$$V_n(f,x):=\sum_{k=0}^{\infty}\binom{n+k-1}{k}\frac{x^k}{(1+x)^{n+k}}\,f\left(\frac{k}{n}\right). \tag{7}$$

These operators (7) also reproduce constant as well as linear functions. Let $f(t)=e^{\theta t}, \theta\in\mathbb{R}$, then we have

$$V_n(e^{\theta t},x)=\sum_{k=0}^{\infty}\binom{n+k-1}{k}\frac{x^k}{(1+x)^{n+k}}\,e^{\frac{k\theta}{n}}$$
$$=\frac{1}{(1+x)^n}\sum_{k=0}^{\infty}\frac{(n)_k}{k!}\left(\frac{x\,e^{\frac{\theta}{n}}}{1+x}\right)^k.$$

Now, using the binomial series $\sum_{k=0}^{\infty} \frac{(c)_k}{k!} z^k = (1-z)^{-c}$, $|z|<1$, we get

$$V_n(e^{\theta t}, x) = \frac{1}{(1+x)^n} \left(1 - \frac{x\, e^{\frac{\theta}{n}}}{1+x}\right)^{-n}$$
$$= \left(1 + x - x e^{\frac{\theta}{n}}\right)^{-n}. \tag{8}$$

It may be observed that $V_n(e^{\theta t}, x)$ may be treated as m.g.f. of the operators V_n, which may be utilized to obtain the moments of (7). Let $\mu_r^{V_n}(x) = V_n(e_r, x)$, where $e_r(t) = t^r$, $r \in \mathbb{N} \cup \{0\}$. The moments are given by

$$\mu_r^{V_n}(x) = \left[\frac{\partial^r}{\partial \theta^r} V_n(e^{\theta t}, x)\right]_{\theta=0}$$
$$= \left[\frac{\partial^r}{\partial \theta^r} \left\{\left(1 + x - x e^{\frac{\theta}{n}}\right)^{-n}\right\}\right]_{\theta=0}.$$

Using the software Mathematica, we get the expansion of (8) in powers of θ as follows:

$$V_n(e^{\theta t}, x) = 1 + x\,\theta + \left(\frac{nx^2 + x^2 + x}{n}\right)\frac{\theta^2}{2!}$$
$$+ \left(\frac{n^2x^3 + 3nx^3 + 2x^3 + 3nx^2 + 3x^2 + x}{n^2}\right)\frac{\theta^3}{3!}$$
$$+ \left(\frac{n^3x^4 + 6n^2x^4 + 11nx^4 + 6x^4 + 6n^2x^3 + 18nx^3 + 12x^3 + 7nx^2 + 7x^2 + x}{n^3}\right)\frac{\theta^4}{4!}$$
$$+ \frac{\left[\begin{array}{c} x^5(n+1)(n+2)(n+3)(n+4) + 10x^4(n+1)(n+2)(n+3) \\ +25x^3(n+1)(n+2) + 15x^2(n+1) + x \end{array}\right]}{n^4}\frac{\theta^5}{5!}$$
$$+ \frac{\left[\begin{array}{c} x^6(n+1)(n+2)(n+3)(n+4)(n+5) + 15x^5(n+1)(n+2)(n+3)(n+4) \\ +65x^4(n+1)(n+2)(n+3) + 90x^3(n+1)(n+2) + 31x^2(n+1) + x \end{array}\right]}{n^5}\frac{\theta^6}{6!}$$
$$+ \mathcal{O}(\theta^7).$$

In particular, the first few moments may be obtained as

$$\mu_0^{V_n}(x) = 1,$$

$$\mu_1^{V_n}(x) = x,$$

$$\mu_2^{V_n}(x) = \frac{nx^2 + x^2 + x}{n},$$

$$\mu_3^{V_n}(x) = \frac{n^2x^3 + 3nx^3 + 2x^3 + 3nx^2 + 3x^2 + x}{n^2},$$

$$\mu_4^{V_n}(x) = \frac{n^3x^4 + 6n^2x^4 + 11nx^4 + 6x^4 + 6n^2x^3 + 18nx^3 + 12x^3 + 7nx^2 + 7x^2 + x}{n^3},$$

$$\mu_5^{V_n}(x) = \frac{\begin{bmatrix} x^5(n+1)(n+2)(n+3)(n+4) + 10x^4(n+1)(n+2)(n+3) \\ +25x^3(n+1)(n+2) + 15x^2(n+1) + x \end{bmatrix}}{n^4},$$

$$\mu_6^{V_n}(x) = \frac{\begin{bmatrix} x^6(n+1)(n+2)(n+3)(n+4)(n+5) + 15x^5(n+1)(n+2)(n+3)(n+4) \\ +65x^4(n+1)(n+2)(n+3) + 90x^3(n+1)(n+2) + 31x^2(n+1) + x \end{bmatrix}}{n^5}.$$

Lupaş Operators

In the year 1995, A. Lupaş [19] proposed a discrete operator and established some direct results. For $f \in C[0, \infty)$, the Lupaş operators are defined as

$$L_n(f, x) := \sum_{k=0}^{\infty} 2^{-nx} \frac{(nx)_k}{k!\, 2^k} f\left(\frac{k}{n}\right). \tag{9}$$

It was observed that these operators are linear and positive and preserve linear functions. Suppose $f(t) = e^{\theta t}$, $\theta \in \mathbb{R}$, then

$$\begin{aligned} L_n(e^{\theta t}, x) &= \sum_{k=0}^{\infty} 2^{-nx} \frac{(nx)_k}{k!\, 2^k} e^{\frac{k\theta}{n}} \\ &= 2^{-nx} \sum_{k=0}^{\infty} \frac{(nx)_k}{k!} \left(\frac{e^{\frac{\theta}{n}}}{2}\right)^k. \end{aligned}$$

Now, using the binomial series $\sum\limits_{k=0}^{\infty} \frac{(c)_k}{k!} z^k = (1-z)^{-c}$, $|z| < 1$, we get

$$\begin{aligned} L_n(e^{\theta t}, x) &= 2^{-nx} \left(1 - \frac{e^{\frac{\theta}{n}}}{2}\right)^{-nx} \\ &= \left(2 - e^{\frac{\theta}{n}}\right)^{-nx}. \end{aligned} \tag{10}$$

It may be observed that $L_n(e^{\theta t}, x)$ may be treated as m.g.f. of the operators L_n, which may be utilized to obtain the moments of (9). Let $\mu_r^{L_n}(x) = L_n(e_r, x)$, where $e_r(t) = t^r, r \in \mathbb{N} \cup \{0\}$. The moments are given by

$$\begin{aligned}
\mu_r^{L_n}(x) &= \left[\frac{\partial^r}{\partial \theta^r} L_n(e^{\theta t}, x)\right]_{\theta=0} \\
&= \left[\frac{\partial^r}{\partial \theta^r} \left\{\left(2 - e^{\frac{\theta}{n}}\right)^{-nx}\right\}\right]_{\theta=0}.
\end{aligned}$$

Using the software Mathematica, we get the expansion of (10) in powers of θ as follows:

$$\begin{aligned}
L_n(e^{\theta t}, x) &= 1 + x\,\theta + \left(x^2 + \frac{2x}{n}\right)\frac{\theta^2}{2!} + \left(x^3 + \frac{6x^2}{n} + \frac{6x}{n^2}\right)\frac{\theta^3}{3!} \\
&+ \left(x^4 + \frac{12x^3}{n} + \frac{36x^2}{n^2} + \frac{26x}{n^3}\right)\frac{\theta^4}{4!} \\
&+ \left(x^5 + \frac{20x^4}{n} + \frac{120x^3}{n^2} + \frac{250x^2}{n^3} + \frac{150x}{n^4}\right)\frac{\theta^5}{5!} \\
&+ \left(x^6 + \frac{30x^5}{n} + \frac{300x^4}{n^2} + \frac{1230x^3}{n^3} + \frac{2040x^2}{n^4} + \frac{1082x}{n^5}\right)\frac{\theta^6}{6!} \\
&+ \mathcal{O}(\theta^7).
\end{aligned}$$

In particular, the first few moments may be obtained as

$$\begin{aligned}
\mu_0^{L_n}(x) &= 1, \\
\mu_1^{L_n}(x) &= x, \\
\mu_2^{L_n}(x) &= x^2 + \frac{2x}{n}, \\
\mu_3^{L_n}(x) &= x^3 + \frac{6x^2}{n} + \frac{6x}{n^2}, \\
\mu_4^{L_n}(x) &= x^4 + \frac{12x^3}{n} + \frac{36x^2}{n^2} + \frac{26x}{n^3}, \\
\mu_5^{L_n}(x) &= x^5 + \frac{20x^4}{n} + \frac{120x^3}{n^2} + \frac{250x^2}{n^3} + \frac{150x}{n^4}, \\
\mu_6^{L_n}(x) &= x^6 + \frac{30x^5}{n} + \frac{300x^4}{n^2} + \frac{1230x^3}{n^3} + \frac{2040x^2}{n^4} + \frac{1082x}{n^5}.
\end{aligned}$$

Jain and Pethe Operators

In the year 1977, Jain and Pethe [18] proposed a generalization of the well-known Szász-Mirakyan operators for $f \in C[0, \infty)$ as

$$P_n^{\alpha}(f, x) = (1 + n\alpha)^{-\frac{x}{\alpha}} \sum_{k=0}^{\infty} \left(\alpha + \frac{1}{n}\right)^{-k} \frac{x^{[k,-\alpha]}}{k!} f\left(\frac{k}{n}\right), \tag{11}$$

where $x^{[k,-\alpha]} = x(x + \alpha)(x + 2\alpha) \dots [x + (k - 1)\alpha]$, $x^{[0,-\alpha]} = 1$.

In the year 2007, Abel and Ivan in [2] considered slightly modified form of the operators of Jain and Pethe (11) by taking $\alpha = (nc)^{-1}$ and established complete asymptotic expansion. The operators discussed in [2] are defined as

$$U_n^c(f, x) = \sum_{k=0}^{\infty} \left(\frac{c}{1+c}\right)^{ncx} \frac{(ncx)_k}{k!(1+c)^k} f\left(\frac{k}{n}\right). \tag{12}$$

Also, in the year 1998, V. Miheşan [21] proposed generalized operators, which contain some of the well-known operators as special cases. For $f \in C[0, \infty)$ and $a \in \mathbb{R}$, the operators studied by Miheşan are defined as

$$M_n(f, x) := \sum_{k=0}^{\infty} \frac{(a)_k}{k!} \frac{\left(\frac{nx}{a}\right)^k}{\left(1 + \frac{nx}{a}\right)^{a+k}} f\left(\frac{k}{n}\right). \tag{13}$$

As special cases, these operators reduce to the following:

- For $a = -n$, $x \in [0, 1]$, (13) reduces to the operators (1).
- For $a = n$, (13) reduces to the operators (7).
- For $a = nx$, $x > 0$, (13) reduces to the operators (9).
- For $a \to \infty$, (13) reduces to the operators (5).

It may be observed that the operators (13) are very much similar to the operators considered by Abel and Ivan (12) by substituting $a = ncx$ in (13). Intuitively, the above two operators are slightly modified form of the operators introduced by Jain and Pethe [18].

Let $f(t) = e^{\theta t}$, $\theta \in \mathbb{R}$, then

$$\begin{aligned} M_n(e^{\theta t}, x) &= \sum_{k=0}^{\infty} \frac{(a)_k}{k!} \frac{\left(\frac{nx}{a}\right)^k}{\left(1 + \frac{nx}{a}\right)^{a+k}} e^{\frac{k\theta}{n}} \\ &= \left(1 + \frac{nx}{a}\right)^{-a} \sum_{k=0}^{\infty} \frac{(a)_k}{k!} \left(\frac{\frac{nxe^{\frac{\theta}{n}}}{a}}{1 + \frac{nx}{a}}\right)^k. \end{aligned}$$

Now, using the binomial series $\sum_{k=0}^{\infty} \frac{(c)_k}{k!} z^k = (1-z)^{-c}$, $|z|<1$, we get

$$M_n(e^{\theta t}, x) = \left(1+\frac{nx}{a}\right)^{-a} \left(1 - \frac{\frac{nxe^{\frac{\theta}{n}}}{a}}{1+\frac{nx}{a}}\right)^{-a}$$

$$= \left(1 + \frac{nx\left(1-e^{\frac{\theta}{n}}\right)}{a}\right)^{-a}. \tag{14}$$

It may be observed that $M_n(e^{\theta t}, x)$ may be treated as m.g.f. of the operators M_n, which may be utilized to obtain the moments of (13). Let $\mu_r^{M_n}(x) = M_n(e_r, x)$, where $e_r(t) = t^r, r \in \mathbb{N} \cup \{0\}$. The moments are given by

$$\mu_r^{M_n}(x) = \left[\frac{\partial^r}{\partial \theta^r} M_n(e^{\theta t}, x)\right]_{\theta=0}$$

$$= \left[\frac{\partial^r}{\partial \theta^r}\left\{\left(1 + \frac{nx\left(1-e^{\frac{\theta}{n}}\right)}{a}\right)^{-a}\right\}\right]_{\theta=0}.$$

Using the software Mathematica, we get the expansion of (14) in powers of θ as follows:

$$M_n(e^{\theta t}, x) = 1 + x\,\theta + \left(\frac{x^2(a+1)}{a} + \frac{x}{n}\right)\frac{\theta^2}{2!}$$

$$+\left(\frac{x^3(a+1)(a+2)}{a^2} + \frac{3x^2(a+1)}{an} + \frac{x}{n^2}\right)\frac{\theta^3}{3!}$$

$$+\left(\frac{x^4(a+1)(a+2)(a+3)}{a^3} + \frac{6x^3(a+1)(a+2)}{a^2 n} + \frac{7x^2(a+1)}{an^2} + \frac{x}{n^3}\right)\frac{\theta^4}{4!}$$

$$+\left(\frac{x^5(a+1)(a+2)(a+3)(a+4)}{a^4} + \frac{10x^4(a+1)(a+2)(a+3)}{a^3 n}\right.$$

$$\left.+\frac{25x^3(a+1)(a+2)}{a^2n^2} + \frac{15x^2(a+1)}{an^3} + \frac{x}{n^4}\right)\frac{\theta^5}{5!}$$

$$+\left(\frac{x^6(a+1)(a+2)(a+3)(a+4)(a+5)}{a^5}\right.$$

$$+\frac{15x^5(a+1)(a+2)(a+3)(a+4)}{a^4 n} + \frac{65x^4(a+1)(a+2)(a+3)}{a^3n^2}$$

$$+\frac{90x^3(a+1)(a+2)}{a^2n^3}+\frac{31x^2(a+1)}{an^4}+\frac{x}{n^5}\Big)\frac{\theta^6}{6!}$$
$$+\mathcal{O}(\theta^7).$$

In particular, the first few moments may be obtained as

$$\mu_0^{M_n}(x)=1,$$
$$\mu_1^{M_n}(x)=x,$$
$$\mu_2^{M_n}(x)=\frac{x^2(a+1)}{a}+\frac{x}{n},$$
$$\mu_3^{M_n}(x)=\frac{x^3(a+1)(a+2)}{a^2}+\frac{3x^2(a+1)}{an}+\frac{x}{n^2},$$
$$\mu_4^{M_n}(x)=\frac{x^4(a+1)(a+2)(a+3)}{a^3}+\frac{6x^3(a+1)(a+2)}{a^2n}+\frac{7x^2(a+1)}{an^2}+\frac{x}{n^3},$$
$$\mu_5^{M_n}(x)=\frac{x^5(a+1)(a+2)(a+3)(a+4)}{a^4}+\frac{10x^4(a+1)(a+2)(a+3)}{a^3n}$$
$$+\frac{25x^3(a+1)(a+2)}{a^2n^2}+\frac{15x^2(a+1)}{an^3}+\frac{x}{n^4},$$
$$\mu_6^{M_n}(x)=\frac{x^6(a+1)(a+2)(a+3)(a+4)(a+5)}{a^5}$$
$$+\frac{15x^5(a+1)(a+2)(a+3)(a+4)}{a^4n}+\frac{65x^4(a+1)(a+2)(a+3)}{a^3n^2}$$
$$+\frac{90x^3(a+1)(a+2)}{a^2n^3}+\frac{31x^2(a+1)}{an^4}+\frac{x}{n^5}.$$

Modified Baskakov Operators

Mihesan [21] considered, for a non-negative constant a (independent of n) and $x\in[0,\infty)$, the following modification of the Baskakov operators as

$$W_n^a(f,x)=\sum_{k=0}^{\infty}e^{-\frac{ax}{1+x}}\frac{\sum\limits_{i=0}^{k}\binom{k}{i}(n)_i\,a^{k-i}}{k!}\frac{x^k}{(1+x)^{n+k}}\,f\left(\frac{k}{n}\right),\tag{15}$$

where $(n)_i = n(n+1)\dots(n+i-1), (n)_0 = 1$ denotes the rising factorial. Obviously, if $a = 0$, then these operators (15) reduce to the classical Baskakov operators (7). For non-negative constant $a \geqslant 0$ and for some finite number θ, we have

$$W_n^a(e^{\theta t}, x) = e^{\frac{ax}{1+x}\left(e^{\frac{\theta}{n}}-1\right)}\left(1+x-xe^{\frac{\theta}{n}}\right)^{-n}. \tag{16}$$

The r-th order moment can be obtained directly as

$$\begin{aligned}\mu_r(x) &= \left[\frac{\partial^r}{\partial\theta^r}W_n^a(e^{\theta t}, x)\right]_{\theta=0}\\ &= \left[\frac{\partial^r}{\partial\theta^r}\left\{e^{\frac{ax}{1+x}\left(e^{\frac{\theta}{n}}-1\right)}\left(1+x-xe^{\frac{\theta}{n}}\right)^{-n}\right\}\right]_{\theta=0},\end{aligned}$$

where $\mu_r^{W_n^a}(x) = W_n^a(e_r, x), e_r(t) = t^r, r \in \mathbb{N} \cup \{0\}$.
Using the software Mathematica, we get the expansion of (16) in powers of θ as follows:

$$\begin{aligned}W_n^a(e^{\theta t}, x) &= 1 + \left(x + \frac{ax}{n(1+x)}\right)\theta\\
&+\left(x^2 + \frac{2ax^2}{n(1+x)} + \frac{ax+ax^2+a^2x^2}{n^2(1+x)^2} + \frac{x+x^2}{n}\right)\frac{\theta^2}{2!}\\
&+\left(\frac{n^2x^3+3nx^3+2x^3+3nx^2+3x^2+x}{n^2}\right.\\
&+\frac{a^3x^3+3a^2x^3+ax^3+3a^2x^2+2ax^2+ax}{n^3(1+x)^3}\\
&\left.+\frac{3ax(nx^2+x^2+x)}{n^2(1+x)} + \frac{3x(a^2x^2+ax^2+ax)}{n^2(1+x)^2}\right)\frac{\theta^3}{3!}\\
&+\left(\frac{n^3x^4+6n^2x^4+11nx^4+6x^4+6n^2x^3+18nx^3+12x^3+7nx^2+7x^2+x}{n^3}\right.\\
&+\frac{a^4x^4+6a^3x^4+7a^2x^4+ax^4+6a^3x^3+14a^2x^3+3ax^3+7a^2x^2+3ax^2+ax}{n^4(1+x)^4}\\
&+\frac{4ax(n^2x^3+3nx^3+2x^3+3nx^2+3x^2+x)}{n^3(1+x)}\\
&+\frac{4x(a^3x^3+3a^2x^3+ax^3+3a^2x^2+2ax^2+ax)}{n^3(1+x)^3}\\
&\left.+\frac{6(a^2x^2+ax^2+ax)(nx^2+x^2+x)}{n^3(1+x)^2}\right)\frac{\theta^4}{4!}\end{aligned}$$

$$+\left(\begin{array}{c}\dfrac{\left(\begin{array}{c}a^5x^5+10a^4x^5+25a^3x^5+15a^2x^5+ax^5+10a^4x^4+50a^3x^4\\+45a^2x^4+4ax^4+25a^3x^3+45a^2x^3+6ax^3+15a^2x^2+4ax^2+ax\end{array}\right)}{n^5(1+x)^5}\\+\dfrac{5ax(n^3x^4+6n^2x^4+11nx^4+6x^4+6n^2x^3+18nx^3+12x^3+7nx^2+7x^2+x)}{n^4(1+x)}\\+\dfrac{5x(a^4x^4+6a^3x^4+7a^2x^4+ax^4+6a^3x^3+14a^2x^3+3ax^3+7a^2x^2+3ax^2+ax)}{n^4(1+x)^4}\\+\dfrac{10(a^2x^2+ax^2+ax)(n^2x^3+3nx^3+2x^3+3nx^2+3x^2+x)}{n^4(1+x)^2}\\+\dfrac{10(nx^2+x^2+x)(a^3x^3+3a^2x^3+ax^3+3a^2x^2+2ax^2+ax)}{n^4(1+x)^3}\\+\dfrac{x^5(n+1)(n+2)(n+3)(n+4)}{n^4}+\dfrac{10x^4(n+1)(n+2)(n+3)}{n^4}\\+\dfrac{25x^3(n+1)(n+2)}{n^4}+\dfrac{15x^2(n+1)}{n^4}+\dfrac{x}{n^4}\end{array}\right)\frac{\theta^5}{5!}$$

$$+\left(\begin{array}{c}\dfrac{\left(\begin{array}{c}a^6x^6+15a^5x^6+65a^4x^6+90a^3x^6+31a^2x^6+ax^6\\+15a^5x^5+130a^4x^5+270a^3x^5+124a^2x^5+5ax^5+65a^4x^4+270a^3x^4\\+186a^2x^4+10ax^4+90a^3x^3+124a^2x^3+10ax^3+31a^2x^2+5ax^2+ax\end{array}\right)}{n^6(1+x)^6}\\+\dfrac{ax\left(\begin{array}{c}6x^5(n+1)(n+2)(n+3)(n+4)+60x^4(n+1)(n+2)(n+3)+150x^3(n+1)(n+2)\\+90x^2(n+1)+6x\end{array}\right)}{n^5(1+x)}\\+\dfrac{\begin{array}{c}6x(a^5x^5+10a^4x^5+25a^3x^5+15a^2x^5+ax^5+10a^4x^4+50a^3x^4+45a^2x^4+4ax^4\\+25a^3x^3+45a^2x^3+6ax^3+15a^2x^2+4ax^2+ax)\end{array}}{n^5(1+x)^5}\\+\dfrac{15(a^2x^2+ax^2+ax)(n^3x^4+6n^2x^4+11nx^4+6x^4+6n^2x^3+18nx^3+12x^3+7nx^2+7x^2+x)}{n^5(1+x)^2}\\+\dfrac{15(nx^2+x^2+x)(a^4x^4+6a^3x^4+7a^2x^4+ax^4+6a^3x^3+14a^2x^3+3ax^3+7a^2x^2+3ax^2+ax)}{n^5(1+x)^4}\\+\dfrac{20(a^3x^3+3a^2x^3+ax^3+3a^2x^2+2ax^2+ax)(n^2x^3+3nx^3+2x^3+3nx^2+3x^2+x)}{n^5(1+x)^3}\\+\dfrac{x^6(n+1)(n+2)(n+3)(n+4)(n+5)}{n^5}+\dfrac{15x^5(n+1)(n+2)(n+3)(n+4)}{n^5}\\+\dfrac{65x^4(n+1)(n+2)(n+3)}{n^5}+\dfrac{90x^3(n+1)(n+2)}{n^5}+\dfrac{31x^2(n+1)}{n^5}+\dfrac{x}{n^5}\end{array}\right)\frac{\theta^6}{6!}$$

$$+\mathcal{O}(\theta^7).$$

In particular, the first few moments may be obtained as

$$\mu_0^{W_n^a}(x) = 1,$$

$$\mu_1^{W_n^a}(x) = x + \frac{ax}{n(1+x)},$$

$$\mu_2^{W_n^a}(x) = x^2 + \frac{2ax^2}{n(1+x)} + \frac{ax + ax^2 + a^2x^2}{n^2(1+x)^2} + \frac{x+x^2}{n},$$

$$\mu_3^{W_n^a}(x) = \frac{n^2x^3 + 3nx^3 + 2x^3 + 3nx^2 + 3x^2 + x}{n^2}$$
$$+ \frac{a^3x^3 + 3a^2x^3 + ax^3 + 3a^2x^2 + 2ax^2 + ax}{n^3(1+x)^3}$$
$$+ \frac{3ax(nx^2 + x^2 + x)}{n^2(1+x)} + \frac{3x(a^2x^2 + ax^2 + ax)}{n^2(1+x)^2},$$

$$\mu_4^{W_n^a}(x) = \frac{n^3x^4 + 6n^2x^4 + 11nx^4 + 6x^4 + 6n^2x^3 + 18nx^3 + 12x^3 + 7nx^2 + 7x^2 + x}{n^3}$$
$$+ \frac{a^4x^4 + 6a^3x^4 + 7a^2x^4 + ax^4 + 6a^3x^3 + 14a^2x^3 + 3ax^3 + 7a^2x^2 + 3ax^2 + ax}{n^4(1+x)^4}$$
$$+ \frac{4ax(n^2x^3 + 3nx^3 + 2x^3 + 3nx^2 + 3x^2 + x)}{n^3(1+x)}$$
$$+ \frac{4x(a^3x^3 + 3a^2x^3 + ax^3 + 3a^2x^2 + 2ax^2 + ax)}{n^3(1+x)^3}$$
$$+ \frac{6(a^2x^2 + ax^2 + ax)(nx^2 + x^2 + x)}{n^3(1+x)^2},$$

$$\mu_5^{W_n^a}(x) = \frac{\begin{pmatrix} a^5x^5 + 10a^4x^5 + 25a^3x^5 + 15a^2x^5 + ax^5 + 10a^4x^4 + 50a^3x^4 \\ +45a^2x^4 + 4ax^4 + 25a^3x^3 + 45a^2x^3 + 6ax^3 + 15a^2x^2 + 4ax^2 + ax \end{pmatrix}}{n^5(1+x)^5}$$
$$+ \frac{5ax(n^3x^4 + 6n^2x^4 + 11nx^4 + 6x^4 + 6n^2x^3 + 18nx^3 + 12x^3 + 7nx^2 + 7x^2 + x)}{n^4(1+x)}$$
$$+ \frac{5x(a^4x^4 + 6a^3x^4 + 7a^2x^4 + ax^4 + 6a^3x^3 + 14a^2x^3 + 3ax^3 + 7a^2x^2 + 3ax^2 + ax)}{n^4(1+x)^4}$$
$$+ \frac{10(a^2x^2 + ax^2 + ax)(n^2x^3 + 3nx^3 + 2x^3 + 3nx^2 + 3x^2 + x)}{n^4(1+x)^2}$$
$$+ \frac{10(nx^2 + x^2 + x)(a^3x^3 + 3a^2x^3 + ax^3 + 3a^2x^2 + 2ax^2 + ax)}{n^4(1+x)^3}$$
$$+ \frac{x^5(n+1)(n+2)(n+3)(n+4)}{n^4} + \frac{10x^4(n+1)(n+2)(n+3)}{n^4}$$
$$+ \frac{25x^3(n+1)(n+2)}{n^4} + \frac{15x^2(n+1)}{n^4} + \frac{x}{n^4},$$

$$\mu_6^{W_n^a}(x) = \frac{\begin{pmatrix} a^6x^6 + 15a^5x^6 + 65a^4x^6 + 90a^3x^6 + 31a^2x^6 + ax^6 \\ +15a^5x^5 + 130a^4x^5 + 270a^3x^5 + 124a^2x^5 + 5ax^5 + 65a^4x^4 + 270a^3x^4 \\ +186a^2x^4 + 10ax^4 + 90a^3x^3 + 124a^2x^3 + 10ax^3 + 31a^2x^2 + 5ax^2 + ax \end{pmatrix}}{n^6(1+x)^6}$$
$$+ \frac{ax\begin{pmatrix} 6x^5(n+1)(n+2)(n+3)(n+4) + 60x^4(n+1)(n+2)(n+3) \\ +150x^3(n+1)(n+2) + 90x^2(n+1) + 6x \end{pmatrix}}{n^5(1+x)}$$
$$+ \frac{6x\begin{pmatrix} a^5x^5 + 10a^4x^5 + 25a^3x^5 + 15a^2x^5 + ax^5 + 10a^4x^4 + 50a^3x^4 \\ +45a^2x^4 + 4ax^4 + 25a^3x^3 + 45a^2x^3 + 6ax^3 + 15a^2x^2 + 4ax^2 + ax \end{pmatrix}}{n^5(1+x)^5}$$
$$+ \frac{15(a^2x^2 + ax^2 + ax)(n^3x^4 + 6n^2x^4 + 11nx^4 + 6x^4 + 6n^2x^3 + 18nx^3 + 12x^3 + 7nx^2 + 7x^2 + x)}{n^5(1+x)^2}$$
$$+ \frac{15(nx^2 + x^2 + x)(a^4x^4 + 6a^3x^4 + 7a^2x^4 + ax^4 + 6a^3x^3 + 14a^2x^3 + 3ax^3 + 7a^2x^2 + 3ax^2 + ax)}{n^5(1+x)^4}$$
$$+ \frac{20(a^3x^3 + 3a^2x^3 + ax^3 + 3a^2x^2 + 2ax^2 + ax)(n^2x^3 + 3nx^3 + 2x^3 + 3nx^2 + 3x^2 + x)}{n^5(1+x)^3}$$
$$+ \frac{x^6(n+1)(n+2)(n+3)(n+4)(n+5)}{n^5}$$
$$+ \frac{15x^5(n+1)(n+2)(n+3)(n+4)}{n^5} + \frac{65x^4(n+1)(n+2)(n+3)}{n^5}$$
$$+ \frac{90x^3(n+1)(n+2)}{n^5} + \frac{31x^2(n+1)}{n^5} + \frac{x}{n^5}.$$

Kantorovich-Type Operators

In this section, we deal with the Kantorovich-type modifications of some well-known operators and obtain m.g.f. and moments up to sixth order.

Bernstein-Kantorovich Operators

For $f \in C[0,1]$, the Bernstein-Kantorovich operators are defined as

$$\overline{B}_n(f,x) := (n+1)\sum_{k=0}^{n}\binom{n}{k}x^k(1-x)^{n-k}\int_{\frac{k}{n+1}}^{\frac{k+1}{n+1}} f(t)\,dt. \tag{17}$$

Let $f(t)=e^{\theta t}, \theta \in \mathbb{R}$, then $\int_{\frac{k}{n+1}}^{\frac{k+1}{n+1}} e^{\theta t}\,dt = \frac{1}{\theta}\, e^{\frac{k\theta}{n+1}}\left(e^{\frac{\theta}{n+1}}-1\right)$.

Therefore,

$$\begin{aligned}
\overline{B}_n(e^{\theta t},x) &= \frac{1}{\theta}(n+1)\left(e^{\frac{\theta}{n+1}}-1\right)\sum_{k=0}^{n}\binom{n}{k}\left(xe^{\frac{\theta}{n+1}}\right)^k(1-x)^{n-k}\\
&= \frac{1}{\theta}(1-x)^n(n+1)\left(e^{\frac{\theta}{n+1}}-1\right)\sum_{k=0}^{n}\frac{(-1)^k(-n)_k}{k!}\left(\frac{x\,e^{\frac{\theta}{n+1}}}{1-x}\right)^k\\
&= \frac{1}{\theta}(1-x)^n(n+1)\left(e^{\frac{\theta}{n+1}}-1\right)\sum_{k=0}^{n}\frac{(-n)_k}{k!}\left(\frac{-x\,e^{\frac{\theta}{n+1}}}{1-x}\right)^k.
\end{aligned}$$

Now, using the binomial series $\sum_{k=0}^{\infty}\frac{(c)_k}{k!}z^k=(1-z)^{-c}$, $|z|<1$, we get

$$\begin{aligned}
\overline{B}_n(e^{\theta t},x) &= \frac{1}{\theta}(1-x)^n(n+1)\left(e^{\frac{\theta}{n+1}}-1\right)\left(1-\frac{(-x\,e^{\frac{\theta}{n+1}})}{1-x}\right)^n\\
&= \frac{1}{\theta}(n+1)\left(e^{\frac{\theta}{n+1}}-1\right)\left(1-x+xe^{\frac{\theta}{n+1}}\right)^n.
\end{aligned} \tag{18}$$

It may be observed that $\overline{B}_n(e^{\theta t}, x)$ may be treated as m.g.f. of the operators $\overline{B}_n$, which may be utilized to obtain the moments of (17). Let $\mu_r^{\overline{B}_n}(x) = \overline{B}_n(e_r, x)$, where $e_r(t) = t^r, r \in \mathbb{N} \cup \{0\}$. The moments are given by

$$\begin{aligned}\mu_r^{\overline{B}_n}(x) &= \left[\frac{\partial^r}{\partial\theta^r}\overline{B}_n(e^{\theta t}, x)\right]_{\theta=0}\\ &= \left[\frac{\partial^r}{\partial\theta^r}\left\{\frac{1}{\theta}(n+1)\left(e^{\frac{\theta}{n+1}}-1\right)\left(1-x+xe^{\frac{\theta}{n+1}}\right)^n\right\}\right]_{\theta=0}.\end{aligned}$$

Using the software Mathematica, we get the expansion of (18) in powers of θ as follows:

$$\begin{aligned}\overline{B}_n(e^{\theta t}, x) &= 1 + \left(\frac{2nx+1}{2(n+1)}\right)\theta + \left(\frac{3n^2x^2 - 3nx^2 + 6nx + 1}{3(n+1)^2}\right)\frac{\theta^2}{2!}\\ &+ \left(\frac{4n^3x^3 - 12n^2x^3 + 8nx^3 + 18n^2x^2 - 18nx^2 + 14nx + 1}{4(n+1)^3}\right)\frac{\theta^3}{3!}\\ &+ \frac{\begin{bmatrix} 5n^4x^4 - 30n^3x^4 + 55n^2x^4 - 30nx^4 + 40n^3x^3\\ -120n^2x^3 + 80nx^3 + 75n^2x^2 - 75nx^2 + 30nx + 1\end{bmatrix}}{5(n+1)^4}\frac{\theta^4}{4!}\\ &+ \frac{\begin{bmatrix} 6n^5x^5 - 60n^4x^5 + 210n^3x^5 - 300n^2x^5 + 144nx^5 + 75n^4x^4\\ -450n^3x^4 + 825n^2x^4 - 450nx^4 + 260n^3x^3 - 780n^2x^3\\ +520nx^3 + 270n^2x^2 - 270nx^2 + 62nx + 1\end{bmatrix}}{6(n+1)^5}\frac{\theta^5}{5!}\\ &+ \frac{\begin{bmatrix} 7n^6x^6 - 105n^5x^6 + 595n^4x^6 - 1575n^3x^6 + 1918n^2x^6 - 840nx^6\\ +126n^5x^5 - 1260n^4x^5 + 4410n^3x^5 - 6300n^2x^5 + 3024nx^5\\ +700n^4x^4 - 4200n^3x^4 + 7700n^2x^4 - 4200nx^4 + 1400n^3x^3\\ -4200n^2x^3 + 2800nx^3 + 903n^2x^2 - 903nx^2 + 126nx + 1\end{bmatrix}}{7(n+1)^6}\frac{\theta^6}{6!}\\ &+ \mathcal{O}(\theta^7).\end{aligned}$$

In particular, the first few moments may be obtained as

$$\mu_0^{\overline{B}_n}(x) = 1,$$

$$\mu_1^{\overline{B}_n}(x) = \frac{2nx+1}{2(n+1)},$$

$$\mu_2^{\overline{B}_n}(x) = \frac{3n^2x^2 - 3nx^2 + 6nx + 1}{3(n+1)^2},$$

$$\mu_3^{\overline{B}_n}(x) = \frac{4n^3x^3 - 12n^2x^3 + 8nx^3 + 18n^2x^2 - 18nx^2 + 14nx + 1}{4(n+1)^3},$$

$$\mu_4^{\overline{B}_n}(x) = \frac{\begin{bmatrix} 5n^4x^4 - 30n^3x^4 + 55n^2x^4 - 30nx^4 + 40n^3x^3 \\ -120n^2x^3 + 80nx^3 + 75n^2x^2 - 75nx^2 + 30nx + 1 \end{bmatrix}}{5(n+1)^4},$$

$$\mu_5^{\overline{B}_n}(x) = \frac{\begin{bmatrix} 6n^5x^5 - 60n^4x^5 + 210n^3x^5 - 300n^2x^5 + 144nx^5 + 75n^4x^4 \\ -450n^3x^4 + 825n^2x^4 - 450nx^4 + 260n^3x^3 - 780n^2x^3 \\ +520nx^3 + 270n^2x^2 - 270nx^2 + 62nx + 1 \end{bmatrix}}{6(n+1)^5},$$

$$\mu_6^{\overline{B}_n}(x) = \frac{\begin{bmatrix} 7n^6x^6 - 105n^5x^6 + 595n^4x^6 - 1575n^3x^6 + 1918n^2x^6 - 840nx^6 \\ +126n^5x^5 - 1260n^4x^5 + 4410n^3x^5 - 6300n^2x^5 + 3024nx^5 \\ +700n^4x^4 - 4200n^3x^4 + 7700n^2x^4 - 4200nx^4 + 1400n^3x^3 \\ -4200n^2x^3 + 2800nx^3 + 903n^2x^2 - 903nx^2 + 126nx + 1 \end{bmatrix}}{7(n+1)^6}.$$

Szász-Mirakyan-Kantorovich Operators

For $f \in C[0,\infty)$, the Szász-Mirakyan-Kantorovich operators are defined as

$$\overline{S}_n(f,x) := n\sum_{k=0}^{\infty} e^{-nx}\frac{(nx)^k}{k!}\int_{\frac{k}{n}}^{\frac{k+1}{n}} f(t)\,dt. \tag{19}$$

Let $f(t) = e^{\theta t}, \theta \in \mathbb{R}$, then $\displaystyle\int_{\frac{k}{n}}^{\frac{k+1}{n}} e^{\theta t}\,dt = \frac{1}{\theta}e^{\frac{k\theta}{n}}\left(e^{\frac{\theta}{n}} - 1\right)$.

Therefore,

$$\begin{aligned}\overline{S}_n(e^{\theta t},x) &= \frac{1}{\theta}\,ne^{-nx}\left(e^{\frac{\theta}{n}}-1\right)\sum_{k=0}^{\infty}\frac{\left(nx\,e^{\frac{\theta}{n}}\right)^k}{k!}\\ &= \frac{1}{\theta}\,ne^{-nx}\left(e^{\frac{\theta}{n}}-1\right)\,e^{nx\,e^{\frac{\theta}{n}}}\\ &= \frac{1}{\theta}\,n\left(e^{\frac{\theta}{n}}-1\right)e^{nx\left(e^{\frac{\theta}{n}}-1\right)}.\end{aligned} \tag{20}$$

It may be observed that $\overline{S}_n(e^{\theta t},x)$ may be treated as m.g.f. of the operators $\overline{S}_n$, which may be utilized to obtain the moments of (19). Let $\mu_r^{\overline{S}_n}(x) = \overline{S}_n(e_r,x)$, where $e_r(t)=t^r, r\in\mathbb{N}\cup\{0\}$. The moments are given by

$$\begin{aligned}\mu_r^{\overline{S}_n}(x) &= \left[\frac{\partial^r}{\partial\theta^r}\overline{S}_n(e^{\theta t},x)\right]_{\theta=0}\\ &= \left[\frac{\partial^r}{\partial\theta^r}\left\{\frac{1}{\theta}\,n\left(e^{\frac{\theta}{n}}-1\right)e^{nx\left(e^{\frac{\theta}{n}}-1\right)}\right\}\right]_{\theta=0}.\end{aligned}$$

Using the software Mathematica, we get the expansion of (20) in powers of θ as follows:

$$\begin{aligned}\overline{S}_n(e^{\theta t},x) &= 1+\left(\frac{2nx+1}{2n}\right)\theta+\left(\frac{3n^2x^2+6nx+1}{3n^2}\right)\frac{\theta^2}{2!}\\ &+\left(\frac{4n^3x^3+18n^2x^2+14nx+1}{4n^3}\right)\frac{\theta^3}{3!}\\ &+\left(\frac{5n^4x^4+40n^3x^3+75n^2x^2+30nx+1}{5n^4}\right)\frac{\theta^4}{4!}\\ &+\left(\frac{6n^5x^5+75n^4x^4+260n^3x^3+270n^2x^2+62nx+1}{6n^5}\right)\frac{\theta^5}{5!}\\ &+\left(\frac{7n^6x^6+126n^5x^5+700n^4x^4+1400n^3x^3+903n^2x^2+126nx+1}{7n^6}\right)\frac{\theta^6}{6!}\\ &+\mathcal{O}(\theta^7).\end{aligned}$$

In particular, the first few moments may be obtained as

$$\mu_0^{\overline{S}_n}(x) = 1,$$

$$\mu_1^{\overline{S}_n}(x) = \frac{2nx+1}{2n},$$

$$\mu_2^{\overline{S}_n}(x) = \frac{3n^2x^2+6nx+1}{3n^2},$$

$$\mu_3^{\overline{S}_n}(x) = \frac{4n^3x^3+18n^2x^2+14nx+1}{4n^3},$$

$$\mu_4^{\overline{S}_n}(x) = \frac{5n^4x^4+40n^3x^3+75n^2x^2+30nx+1}{5n^4},$$

$$\mu_5^{\overline{S}_n}(x) = \frac{6n^5x^5+75n^4x^4+260n^3x^3+270n^2x^2+62nx+1}{6n^5},$$

$$\mu_6^{\overline{S}_n}(x) = \frac{7n^6x^6+126n^5x^5+700n^4x^4+1400n^3x^3+903n^2x^2+126nx+1}{7n^6}.$$

Baskakov-Kantorovich Operators

For $f \in C[0,\infty)$, the Baskakov-Kantorovich operators are defined as

$$\overline{V}_n(f,x) := (n-1)\sum_{k=0}^{\infty}\binom{n+k-1}{k}\frac{x^k}{(1+x)^{n+k}}\int_{\frac{k}{n-1}}^{\frac{k+1}{n-1}} f(t)\,dt. \tag{21}$$

Let $f(t)=e^{\theta t}, \theta \in \mathbb{R}$, then $\displaystyle\int_{\frac{k}{n-1}}^{\frac{k+1}{n-1}} e^{\theta t}\,dt = \frac{1}{\theta}\,e^{\frac{k\theta}{n-1}}\left(e^{\frac{\theta}{n-1}}-1\right).$

Therefore,

$$\begin{aligned}
\overline{V}_n(e^{\theta t},x) &= \frac{1}{\theta}\,(n-1)\left(e^{\frac{\theta}{n-1}}-1\right)\sum_{k=0}^{\infty}\binom{n+k-1}{k}\frac{x^k}{(1+x)^{n+k}}\,e^{\frac{k\theta}{n-1}} \\
&= \frac{(n-1)\left(e^{\frac{\theta}{n-1}}-1\right)}{\theta\,(1+x)^n}\sum_{k=0}^{\infty}\frac{(n)_k}{k!}\left(\frac{x\,e^{\frac{\theta}{n-1}}}{1+x}\right)^k.
\end{aligned}$$

Now, using the binomial series $\sum_{k=0}^{\infty} \frac{(c)_k}{k!} z^k = (1-z)^{-c}$, $|z|<1$, we get

$$\overline{V}_n(e^{\theta t}, x) = \frac{(n-1)\left(e^{\frac{\theta}{n-1}}-1\right)}{\theta\,(1+x)^n}\left(1-\frac{x\,e^{\frac{\theta}{n-1}}}{1+x}\right)^{-n}$$

$$= \frac{1}{\theta}\,(n-1)\left(e^{\frac{\theta}{n-1}}-1\right)\left(1+x\left(1-e^{\frac{\theta}{n-1}}\right)\right)^{-n}. \tag{22}$$

It may be observed that $\overline{V}_n(e^{\theta t}, x)$ may be treated as m.g.f. of the operators $\overline{V}_n$, which may be utilized to obtain the moments of (21). Let $\mu_r^{\overline{V}_n}(x) = \overline{V}_n(e_r, x)$, where $e_r(t) = t^r, r \in \mathbb{N} \cup \{0\}$. The moments are given by

$$\mu_r^{\overline{V}_n}(x) = \left[\frac{\partial^r}{\partial\theta^r}\overline{V}_n(e^{\theta t}, x)\right]_{\theta=0}$$

$$= \left[\frac{\partial^r}{\partial\theta^r}\left\{\frac{1}{\theta}\,(n-1)\left(e^{\frac{\theta}{n-1}}-1\right)\left(1+x\left(1-e^{\frac{\theta}{n-1}}\right)\right)^{-n}\right\}\right]_{\theta=0}.$$

Using the software Mathematica, we get the expansion of (22) in powers of θ as follows:

$$\overline{V}_n(e^{\theta t}, x) = 1 + \left(\frac{2nx+1}{2(n-1)}\right)\theta + \left(\frac{3n^2x^2+3nx^2+6nx+1}{3(n-1)^2}\right)\frac{\theta^2}{2!}$$

$$+\left(\frac{4n^3x^3+12n^2x^3+8nx^3+18n^2x^2+18nx^2+14nx+1}{4(n-1)^3}\right)\frac{\theta^3}{3!}$$

$$+\frac{\left[\begin{array}{c}5n^4x^4+30n^3x^4+55n^2x^4+30nx^4+40n^3x^3+120n^2x^3\\ +80nx^3+75n^2x^2+75nx^2+30nx+1\end{array}\right]}{5(n-1)^4}\frac{\theta^4}{4!}$$

$$+\frac{\left[\begin{array}{c}6n^5x^5+60n^4x^5+210n^3x^5+300n^2x^5+144nx^5+75n^4x^4\\ +450n^3x^4+825n^2x^4+450nx^4+260n^3x^3+780n^2x^3\\ +520nx^3+270n^2x^2+270nx^2+62nx+1\end{array}\right]}{6(n-1)^5}\frac{\theta^5}{5!}$$

$$+\frac{\left[\begin{array}{c}7n^6x^6+105n^5x^6+595n^4x^6+1575n^3x^6+1918n^2x^6+840nx^6\\ +126n^5x^5+1260n^4x^5+4410n^3x^5+6300n^2x^5+3024nx^5+700n^4x^4\\ +4200n^3x^4+7700n^2x^4+4200nx^4+1400n^3x^3+4200n^2x^3\\ +2800nx^3+903n^2x^2+903nx^2+126nx+1\end{array}\right]}{7(n-1)^6}\frac{\theta^6}{6!}$$

$$+\mathcal{O}(\theta^7).$$

In particular, the first few moments may be obtained as

$$\mu_0^{\overline{V}_n}(x) = 1,$$

$$\mu_1^{\overline{V}_n}(x) = \frac{2nx+1}{2(n-1)},$$

$$\mu_2^{\overline{V}_n}(x) = \frac{3n^2x^2+3nx^2+6nx+1}{3(n-1)^2},$$

$$\mu_3^{\overline{V}_n}(x) = \frac{4n^3x^3+12n^2x^3+8nx^3+18n^2x^2+18nx^2+14nx+1}{4(n-1)^3},$$

$$\mu_4^{\overline{V}_n}(x) = \frac{\begin{bmatrix} 5n^4x^4+30n^3x^4+55n^2x^4+30nx^4+40n^3x^3 \\ +120n^2x^3+80nx^3+75n^2x^2+75nx^2+30nx+1 \end{bmatrix}}{5(n-1)^4},$$

$$\mu_5^{\overline{V}_n}(x) = \frac{\begin{bmatrix} 6n^5x^5+60n^4x^5+210n^3x^5+300n^2x^5+144nx^5+75n^4x^4 \\ +450n^3x^4+825n^2x^4+450nx^4+260n^3x^3+780n^2x^3 \\ +520nx^3+270n^2x^2+270nx^2+62nx+1 \end{bmatrix}}{6(n-1)^5},$$

$$\mu_6^{\overline{V}_n}(x) = \frac{\begin{bmatrix} 7n^6x^6+105n^5x^6+595n^4x^6+1575n^3x^6+1918n^2x^6+840nx^6 \\ +126n^5x^5+1260n^4x^5+4410n^3x^5+6300n^2x^5+3024nx^5+700n^4x^4 \\ +4200n^3x^4+7700n^2x^4+4200nx^4+1400n^3x^3+4200n^2x^3 \\ +2800nx^3+903n^2x^2+903nx^2+126nx+1 \end{bmatrix}}{7(n-1)^6}.$$

Lupaş-Kantorovich Operators

For $f \in C[0,\infty)$, the Lupaş-Kantorovich operators are defined as

$$\overline{L}_n(f,x) := (n+1)\sum_{k=0}^{\infty} 2^{-nx}\frac{(nx)_k}{k!\,2^k}\int_{\frac{k}{n+1}}^{\frac{k+1}{n+1}} f(t)\,dt. \tag{23}$$

Let $f(t)=e^{\theta t}, \theta \in \mathbb{R}$, then $\displaystyle\int_{\frac{k}{n+1}}^{\frac{k+1}{n+1}} e^{\theta t}\,dt = \frac{1}{\theta}\,e^{\frac{k\theta}{n+1}}\left(e^{\frac{\theta}{n+1}}-1\right).$

Therefore,

$$\overline{L}_n(e^{\theta t},x) = \frac{1}{\theta}(n+1)\,2^{-nx}\left(e^{\frac{\theta}{n+1}}-1\right)\sum_{k=0}^{\infty}\frac{(nx)_k}{k!}\left(\frac{e^{\frac{\theta}{n+1}}}{2}\right)^k.$$

Now, using the binomial series $\sum_{k=0}^{\infty} \frac{(c)_k}{k!} z^k = (1-z)^{-c}$, $|z|<1$, we get

$$\overline{L}_n(e^{\theta t}, x) = \frac{1}{\theta}(n+1)\, 2^{-nx} \left(e^{\frac{\theta}{n+1}} - 1\right) \left(1 - \frac{e^{\frac{\theta}{n+1}}}{2}\right)^{-nx}$$

$$= \frac{1}{\theta}(n+1) \left(e^{\frac{\theta}{n+1}} - 1\right) \left(2 - e^{\frac{\theta}{n+1}}\right)^{-nx}. \tag{24}$$

It may be observed that $\overline{L}_n(e^{\theta t}, x)$ may be treated as m.g.f. of the operators $\overline{L}_n$, which may be utilized to obtain the moments of (23). Let $\mu_r^{\overline{L}_n}(x) = \overline{L}_n(e_r, x)$, where $e_r(t) = t^r, r \in \mathbb{N} \cup \{0\}$. The moments are given by

$$\mu_r^{\overline{L}_n}(x) = \left[\frac{\partial^r}{\partial \theta^r} \overline{L}_n(e^{\theta t}, x)\right]_{\theta=0}$$

$$= \left[\frac{\partial^r}{\partial \theta^r} \left\{\frac{1}{\theta}(n+1) \left(e^{\frac{\theta}{n+1}} - 1\right) \left(2 - e^{\frac{\theta}{n+1}}\right)^{-nx}\right\}\right]_{\theta=0}.$$

Using the software Mathematica, we get the expansion of (24) in powers of θ as follows:

$$\overline{L}_n(e^{\theta t}, x) = 1 + \left(\frac{2nx+1}{2(n+1)}\right)\theta + \left(\frac{3nx^2+9nx+1}{3(n+1)^2}\right)\frac{\theta^2}{2!}$$
$$+ \left(\frac{4nx^3+30nx^2+40nx+1}{4(n+1)^3}\right)\frac{\theta^3}{3!}$$
$$+ \left(\frac{5nx^4+70nx^3+250nx^2+215nx+1}{5(n+1)^4}\right)\frac{\theta^4}{4!}$$
$$+ \left(\frac{6nx^5+135nx^4+920nx^3+2175nx^2+1446nx+1}{6(n+1)^5}\right)\frac{\theta^5}{5!}$$
$$+ \left(\frac{7nx^6+231nx^5+2555nx^4+11585nx^3+21021nx^2+11893nx+1}{7(n+1)^6}\right)\frac{\theta^6}{6!}$$
$$+\mathcal{O}(\theta^7).$$

In particular, the first few moments may be obtained as

$$\mu_0^{\overline{L}_n}(x) = 1,$$
$$\mu_1^{\overline{L}_n}(x) = \frac{2nx+1}{2(n+1)},$$
$$\mu_2^{\overline{L}_n}(x) = \frac{3nx^2+9nx+1}{3(n+1)^2},$$
$$\mu_3^{\overline{L}_n}(x) = \frac{4nx^3+30nx^2+40nx+1}{4(n+1)^3},$$
$$\mu_4^{\overline{L}_n}(x) = \frac{5nx^4+70nx^3+250nx^2+215nx+1}{5(n+1)^4},$$
$$\mu_5^{\overline{L}_n}(x) = \frac{6nx^5+135nx^4+920nx^3+2175nx^2+1446nx+1}{6(n+1)^5},$$
$$\mu_6^{\overline{L}_n}(x) = \frac{7nx^6+231nx^5+2555nx^4+11585nx^3+21021nx^2+11893nx+1}{7(n+1)^6}.$$

Miheşan-Kantorovich Operators

For $f \in C[0,\infty)$ and $a \in \mathbb{R}$, the Miheşan-Kantorovich operators are defined as

$$\overline{M}_n(f,x) := n\left(1-\frac{1}{a}\right)\sum_{k=0}^{\infty}\frac{(a)_k}{k!}\frac{\left(\frac{nx}{a}\right)^k}{\left(1+\frac{nx}{a}\right)^{a+k}}\int_{\frac{k}{n\left(1-\frac{1}{a}\right)}}^{\frac{k+1}{n\left(1-\frac{1}{a}\right)}} f(t)\,dt. \tag{25}$$

Let $f(t)=e^{\theta t}, \theta \in \mathbb{R}$, then $\int_{\frac{k}{n\left(1-\frac{1}{a}\right)}}^{\frac{k+1}{n\left(1-\frac{1}{a}\right)}} e^{\theta t}\,dt = \frac{1}{\theta}\,e^{\frac{k\theta}{n\left(1-\frac{1}{a}\right)}}\left(e^{\frac{\theta}{n\left(1-\frac{1}{a}\right)}}-1\right).$

Therefore,

$$\overline{M}_n(e^{\theta t},x) = \frac{1}{\theta}\,n\left(1-\frac{1}{a}\right)\left(e^{\frac{\theta}{n\left(1-\frac{1}{a}\right)}}-1\right)\sum_{k=0}^{\infty}\frac{(a)_k}{k!}\frac{\left(\frac{nx}{a}\right)^k}{\left(1+\frac{nx}{a}\right)^{a+k}}\,e^{\frac{k\theta}{n\left(1-\frac{1}{a}\right)}}$$
$$= \frac{1}{\theta}\,n\left(1-\frac{1}{a}\right)\left(e^{\frac{\theta}{n\left(1-\frac{1}{a}\right)}}-1\right)\left(1+\frac{nx}{a}\right)^{-a}\sum_{k=0}^{\infty}\frac{(a)_k}{k!}\left(\frac{\frac{nxe^{\frac{\theta}{n\left(1-\frac{1}{a}\right)}}}{a}}{1+\frac{nx}{a}}\right)^k.$$

Now, using the binomial series $\sum_{k=0}^{\infty}\frac{(c)_k}{k!}z^k = (1-z)^{-c}, |z|<1$, we get

$$\overline{M}_n(e^{\theta t},x)=\frac{1}{\theta}n\left(1-\frac{1}{a}\right)\left(e^{\frac{\theta}{n\left(1-\frac{1}{a}\right)}}-1\right)\left(1+\frac{nx}{a}\right)^{-a}\left(1-\frac{\frac{nxe^{\frac{\theta}{n\left(1-\frac{1}{a}\right)}}}{a}}{1+\frac{nx}{a}}\right)^{-a}$$

$$=\frac{1}{\theta}n\left(1-\frac{1}{a}\right)\left(e^{\frac{\theta}{n\left(1-\frac{1}{a}\right)}}-1\right)\left[1+\frac{nx}{a}\left(1-e^{\frac{\theta}{n\left(1-\frac{1}{a}\right)}}\right)\right]^{-a}. \quad (26)$$

It may be observed that $\overline{M}_n(e^{\theta t},x)$ may be treated as m.g.f. of the operators $\overline{M}_n$, which may be utilized to obtain the moments of (25). Let $\mu_r^{\overline{M}_n}(x)=\overline{M}_n(e_r,x)$, where $e_r(t)=t^r, r\in\mathbb{N}\cup\{0\}$. The moments are given by

$$\mu_r^{\overline{M}_n}(x)=\left[\frac{\partial^r}{\partial\theta^r}\overline{M}_n(e^{\theta t},x)\right]_{\theta=0}$$

$$=\left[\frac{\partial^r}{\partial\theta^r}\left\{\frac{1}{\theta}n\left(1-\frac{1}{a}\right)\left(e^{\frac{\theta}{n\left(1-\frac{1}{a}\right)}}-1\right)\left[1+\frac{nx}{a}\left(1-e^{\frac{\theta}{n\left(1-\frac{1}{a}\right)}}\right)\right]^{-a}\right\}\right]_{\theta=0}.$$

Using the software Mathematica, we get the expansion of (26) in powers of θ as follows:

$$\overline{M}_n(e^{\theta t},x)=1+\left(\frac{(2nx+1)a}{2n(a-1)}\right)\theta+\left(\frac{(3anx^2+3nx^2+6anx+a)a}{3n^2(a-1)^2}\right)\frac{\theta^2}{2!}$$

$$+\left(\frac{(4a^2nx^3+12anx^3+8nx^3+18a^2nx^2+18anx^2+14a^2nx+a^2)a}{4n^3(a-1)^3}\right)\frac{\theta^3}{3!}$$

$$+\frac{\begin{bmatrix}5a^3nx^4+30a^2nx^4+55anx^4+30nx^4+40a^3nx^3\\+120a^2nx^3+80anx^3+75a^3nx^2+75a^2nx^2+30a^3nx+a^3\end{bmatrix}a}{5n^4(a-1)^4}\frac{\theta^4}{4!}$$

$$+\frac{\begin{bmatrix}6a^4nx^5+60a^3nx^5+210a^2nx^5+300anx^5+144nx^5+75a^4nx^4\\+450a^3nx^4+825a^2nx^4+450anx^4+260a^4nx^3+780a^3nx^3\\+520a^2nx^3+270a^4nx^2+270a^3nx^2+62a^4nx+a^4\end{bmatrix}a}{6n^5(a-1)^5}\frac{\theta^5}{5!}$$

$$+\frac{\begin{bmatrix}7a^5nx^6+105a^4nx^6+595a^3nx^6+1575a^2nx^6+1918anx^6+840nx^6\\+126a^5nx^5+1260a^4nx^5+4410a^3nx^5+6300a^2nx^5+3024anx^5\\+700a^5nx^4+4200a^4nx^4+7700a^3nx^4+4200a^2nx^4+1400a^5nx^3\\+4200a^4nx^3+2800a^3nx^3+903a^5nx^2+903a^4nx^2+126a^5nx+a^5\end{bmatrix}a}{7n^6(a-1)^6}\frac{\theta^6}{6!}$$

$$+\mathcal{O}(\theta^7).$$

In particular, the first few moments may be obtained as

$$\mu_0^{\overline{M}_n}(x) = 1,$$

$$\mu_1^{\overline{M}_n}(x) = \frac{(2nx+1)a}{2n(a-1)},$$

$$\mu_2^{\overline{M}_n}(x) = \frac{(3anx^2+3nx^2+6anx+a)a}{3n^2(a-1)^2},$$

$$\mu_3^{\overline{M}_n}(x) = \frac{(4a^2nx^3+12anx^3+8nx^3+18a^2nx^2+18anx^2+14a^2nx+a^2)a}{4n^3(a-1)^3},$$

$$\mu_4^{\overline{M}_n}(x) = \frac{\begin{bmatrix} 5a^3nx^4+30a^2nx^4+55anx^4+30nx^4+40a^3nx^3 \\ +120a^2nx^3+80anx^3+75a^3nx^2+75a^2nx^2+30a^3nx+a^3 \end{bmatrix} a}{5n^4(a-1)^4},$$

$$\mu_5^{\overline{M}_n}(x) = \frac{\begin{bmatrix} 6a^4nx^5+60a^3nx^5+210a^2nx^5+300anx^5+144nx^5+75a^4nx^4 \\ +450a^3nx^4+825a^2nx^4+450anx^4+260a^4nx^3+780a^3nx^3 \\ +520a^2nx^3+270a^4nx^2+270a^3nx^2+62a^4nx+a^4 \end{bmatrix} a}{6n^5(a-1)^5},$$

$$\mu_6^{\overline{M}_n}(x) = \frac{\begin{bmatrix} 7a^5nx^6+105a^4nx^6+595a^3nx^6+1575a^2nx^6+1918anx^6+840nx^6 \\ +126a^5nx^5+1260a^4nx^5+4410a^3nx^5+6300a^2nx^5+3024anx^5 \\ +700a^5nx^4+4200a^4nx^4+7700a^3nx^4+4200a^2nx^4+1400a^5nx^3 \\ +4200a^4nx^3+2800a^3nx^3+903a^5nx^2+903a^4nx^2+126a^5nx+a^5 \end{bmatrix} a}{7n^6(a-1)^6}.$$

References

1. Abel, U.: Asymptotic approximation with Kantorovich polynomials. Approx. Theory Appl. **14**(3), 106–116 (1998)
2. Abel, U., Ivan, M.: On a generalization of an approximation operator defined by A. Lupaş. Gen. Math. **15**(1), 21–34 (2007)
3. Aral, A., Gupta, V.: On q-Baskakov type operators. Demons. Math. **42**(1), 109–122 (2009)
4. Aral, A., Gupta, V., Agarwal, R.P.: Applications of q-Calculus in Operator Theory. Springer, New York (2013)
5. Baskakov, V.A.: An instance of a sequence of linear positive operators in the space of continuous functions. Dokl. Akad. Nauk SSSR **113**(2), 249–251 (1957) (in Russian)
6. Bernstein, S.N.: Demonstration du Théoréme de Weierstrass fondée sur le calcul des Probabilités. Comm. Soc. Math. Kharkov 2. **13**(1), 1–2 (1912)
7. Boyanov, B.D., Veselinov, V.M.: A note on the approximation of functions in an infinite interval by linear positive operators. Bull. Math. Soc. Sci. Math. Roum. **14**(62), 9–13 (1970)
8. DeVore, R.A., Lorentz, G.G.: Constructive Approximation. Grundlehren der Mathematischen Wissenschaften, Band 303. Springer, Berlin (1993)
9. Ditzian, Z., Totik, V.: Moduli of Smoothness. Springer, New York (1987)

10. Finta, Z., Govil, N.K., Gupta, V.: Some results on modified Szász-Mirakjan operators. J. Math. Anal. Appl. **327**(2), 1284–1296 (2007)
11. Gal, S.G.: Overconvergence in Complex Approximation. Springer, New York (2013)
12. Gonska, H., Pitul, P., Rasa, I.: General king-type operators. Result. Math. **53**(3–4), 279–296 (2009)
13. Gupta, V.: A note on modified Bernstein polynomials. Pure Appl. Math. Sci. **44**, 1–4 (1996)
14. Gupta, V., Agarwal, R.P.: Convergence Estimates in Approximation Theory. Springer, Cham (2014)
15. Gupta, V., Malik, N.: Approximation for genuine summation-integral type link operators. Appl. Math. Comput. **260**, 321–330 (2015)
16. Gupta, V., Rassias, T.M.: Direct estimates for certain Szász type operators. Appl. Math. Comput. **251**, 469–474 (2015)
17. Holhoş, A.: The rate of approximation of functions in an infinite interval by positive linear operators. Stud. Univ. Babeş-Bolyai, Math. **2**, 133–142 (2010)
18. Jain, G.C., Pethe, S.: On the generalizations of Bernstein and Szász-Mirakjan operators. Nanta Math. **10**, 185–193 (1977)
19. Lupaş, A.: The approximation by means of some linear positive operators. In: Muller, M.W., Felten, M., Mache, D.H. (eds.) Approximation Theory (Proceedings of the International Dortmund Meeting IDoMAT 95, Held in Witten, March 13–17, 1995). Mathematical Research, vol. 86, pp.201–229. Akademie Verlag, Berlin (1995)
20. Malik, N.: Some approximation properties for generalized Srivastava-Gupta operators. Appl. Math. Comput. **269**, 747–758 (2015)
21. Miheşan, V.: Uniform approximation with positive linear operators generated by generalized Baskakov method. Autom. Comput. Appl. Math. **7**(1), 34–37 (1998)
22. Milovanović, G.V., Rassias, M.T.: Analytic Number Theory, Approximation Theory, and Special Functions. Springer, New York (2014)
23. Mirakjan, G.M.: Approximation des fonctions continues au moyen de polynmes de la forme $e^{-nx}\sum_{k=0}^{m_n} C_{k,n}x^k$. Comptes rendus de l'Acadmie des sciences de l'URSS (in French). **31**, 201–205 (1941)
24. Păltănea, R.: Approximation Theory Using Positive Linear Operators. Birkhäuser Basel, Boston (2004)
25. Rassias, T.M., Gupta, V. (eds.) Mathematical Analysis, Approximation Theory and Their Applications Series. Springer Optimization and Its Applications, vol. 111. Springer, Berlin (2016). ISBN:978-3-319-31279-8
26. Szász, O.: Generalizations of S. Bernstein's polynomials to the infinite interval. J. Res. Natl. Bur. Stand. **45**(3), 239–245 (1950)
27. Tachev, G.T.: The complete asymptotic expansion for Bernstein operators. J. Math. Anal. Appl. **385**(2), 1179–1183 (2012)
28. Widder, D.V.: The Laplace Transform. Princeton Mathematical Series. Princeton University Press, Princeton (1941)

Approximation by Lupaș–Kantorovich Operators

Vijay Gupta, Themistocles M. Rassias, and Deepika Agrawal

Abstract The present article deals with the approximation properties of certain Lupaș-Kantorovich operators preserving e^{-x}. We obtain uniform convergence estimates which also include an asymptotic formula in quantitative sense. In the end, we provide the estimates for another modification of such operators, which preserve the function e^{-2x}.

Introduction

In the year 1995, Lupaș [9] proposed the Lupaș operators:

$$L_n(f,x) = \sum_{k=0}^{\infty} \frac{2^{-nx}(nx)_k}{k!2^k} f\left(\frac{k}{n}\right),$$

where $(nx)_k$ is the rising factorial given by

$$(nx)_k = nx(nx+1)(nx+2)\cdots(nx+k-1), \quad (nx)_0 = 1.$$

Four years later, Agratini [2] introduced the Kantorovich-type generalization of the operators L_n. After a decade Erençin and Taşdelen [4] considered a generalization of the operators discussed in [2] based on some parameters and established some approximation properties. We start here with the Kantorovich variant of Lupaș operators defined by

V. Gupta · D. Agrawal
Department of Mathematics, Netaji Subhas Institute of Technology, New Delhi, India

Th. M. Rassias (✉)
Department of Mathematics, National Technical University of Athens, Athens, Greece
e-mail: trassias@math.ntua.gr

N. J. Daras, Th. M. Rassias (eds.), *Modern Discrete Mathematics and Analysis*, Springer Optimization and Its Applications 131,
https://doi.org/10.1007/978-3-319-74325-7_9

$$K_n(f,x) = n\sum_{k=0}^{\infty} \frac{2^{-na_n(x)}(na_n(x))_k}{k!2^k}\int_{k/n}^{(k+1)/n} f(t)dt \tag{1}$$

with the hypothesis that these operators preserve the function e^{-x}. Then using

$$\sum_{k=0}^{\infty}\frac{(a)_k}{k!}z^k = (1-z)^{-a},\ |z|<1,$$

we write

$$\begin{aligned}
e^{-x} &= n\sum_{k=0}^{\infty} \frac{2^{-na_n(x)}(na_n(x))_k}{k!2^k}\int_{k/n}^{(k+1)/n} e^{-t}dt \\
&= n\sum_{k=0}^{\infty} \frac{2^{-na_n(x)}(na_n(x))_k}{k!2^k} e^{-k/n}(1-e^{-1/n}) \\
&= n(1-e^{-1/n})(2-e^{-1/n})^{-na_n(x)},
\end{aligned}$$

which concludes

$$a_n(x) = \frac{x+\ln\left(n(1-e^{-1/n})\right)}{n\ln\left(2-e^{-1/n}\right)}. \tag{2}$$

Therefore the operators defined by (1) take the following alternate form

$$\begin{aligned}
K_n(f,x) = n\sum_{k=0}^{\infty}\frac{1}{k!2^k} 2^{-\frac{x+\ln\left(n(1-e^{-1/n})\right)}{\ln\left(2-e^{-1/n}\right)}}\left(\frac{x+\ln\left(n(1-e^{-1/n})\right)}{\ln\left(2-e^{-1/n}\right)}\right)_k \\
\int_{k/n}^{(k+1)/n} f(t)dt.
\end{aligned}$$

These operators preserve constant and the function e^{-x}. The quantitative direct estimate for a sequence of linear positive operators was discussed and proved in [8] as the following result:

Theorem A ([8]) *If a sequence of linear positive operators $L_n : C^*[0,\infty) \to C^*[0,\infty)$, (where $C^*[0,\infty)$ be the subspace of all real-valued continuous functions, which has finite limit at infinity) satisfy the equalities*

$$\begin{aligned}
||L_n(e_0) - 1||_{[0,\infty)} &= \alpha_n \\
||L_n(e^{-t}) - e^{-x}||_{[0,\infty)} &= \beta_n \\
||L_n(e^{-2t}) - e^{-2x}||_{[0,\infty)} &= \gamma_n
\end{aligned}$$

then

$$||L_n f - f||_{[0,\infty)} \leq 2\omega^*\left(f, \sqrt{\alpha_n + 2\beta_n + \gamma_n}\right), f \in C^*[0,\infty),$$

where the norm is the uniform norm and the modulus of continuity is defined by

$$\omega^*(f,\delta) = \sup_{|e^{-x}-e^{-t}|\leq\delta, x,t>0} |f(t) - f(x)|.$$

Very recently Acar et al. [1] used the above theorem and established quantitative estimates for the modification of well-known Szász–Mirakyan operators, which preserve the function $e^{2ax}, a > 0$. Actually such a modification may be important to discuss approximation properties, but if the operators preserve e^{-x} or e^{-2x}, then such results may provide better approximation in the sense of reducing the error. In the present paper, we study Kantorovich variant of Lupaș operators defined by (1) with $a_n(x)$ as given by (2) preserving e^{-x}. We calculate a uniform estimate and establish a quantitative asymptotic result for the modified operators.

Auxiliary Results

In order to prove the main results, the following lemmas are required.

Lemma 1 *The following representation holds*

$$K_n(e^{At}, x) = \frac{n(e^{A/n} - 1)}{A}(2 - e^{A/n})^{-na_n(x)}.$$

Proof We have

$$\begin{aligned}
K_n(e^{At}, x) &= n\sum_{k=0}^{\infty} \frac{2^{-na_n(x)}(na_n(x))_k}{k!2^k} \int_{k/n}^{(k+1)/n} e^{At}dt \\
&= n\sum_{k=0}^{\infty} \frac{2^{-na_n(x)}(na_n(x))_k}{k!2^k} \left[e^{At}(e^{A/n} - 1)\right] \\
&= \frac{n(e^{A/n} - 1)}{A}\left(2 - e^{A/n}\right)^{-na_n(x)}.
\end{aligned}$$

Lemma 2 *If $e_r(t) = t^r, r \in N^0$, then the moments of the operators (1) are given as follows:*

$$K_n(e_0, x) = 1,$$

$$K_n(e_1, x) = a_n(x) + \frac{1}{2n},$$

$$K_n(e_2,x) = (a_n(x))^2 + \frac{3a_n(x)}{n} + \frac{1}{3n^2},$$

$$K_n(e_3,x) = (a_n(x))^3 + \frac{15(a_n(x))^2}{2n} + \frac{10a_n(x)}{n^2} + \frac{1}{4n^3},$$

$$K_n(e_4,x) = (a_n(x))^4 + \frac{14(a_n(x))^3}{n} + \frac{50(a_n(x))^2}{n^2} + \frac{53a_n(x)}{n^3} + \frac{1}{5n^4}.$$

Lemma 3 *If* $\mu_{n,m}(x) = K_n\left((t-x)^m, x\right)$, *then by using Lemma 2, we have*

$$\mu_{n,0}(x) = 1,$$

$$\mu_{n,1}(x) = a_n(x) + \frac{1}{2n} - x,$$

$$\mu_{n,2}(x) = (a_n(x)-x)^2 + \frac{3a_n(x)}{n} - \frac{x}{n} + \frac{1}{3n^2},$$

$$\mu_{n,4}(x) = (a_n(x)-x)^4 + \frac{14(a_n(x))^3 - 30x(a_n(x))^2 + 18x^2a_n(x) - 2x^3}{n}$$
$$+\frac{50(a_n(x))^2 - 40xa_n(x) + 2x^2}{n^2} + \frac{53a_n(x)-x}{n^3} + \frac{1}{5n^4}.$$

Furthermore,

$$\lim_{n\to\infty} n\left[\frac{x+\ln\left(n(1-e^{-1/n})\right)}{n\ln\left(2-e^{-1/n}\right)} + \frac{1}{2n} - x\right] = x$$

and

$$\lim_{n\to\infty} n\left[\left(\frac{x+\ln\left(n(1-e^{-1/n})\right)}{n\ln\left(2-e^{-1/n}\right)} - x\right)^2 + \frac{3\left[x+\ln\left(n(1-e^{-1/n})\right)\right]}{n^2\ln\left(2-e^{-1/n}\right)}\right.$$
$$\left. -\frac{x}{n} + \frac{1}{3n^2}\right] = 2x.$$

Main Results

In this section, we present the quantitative estimates.

Theorem 1 *For* $f \in C^*[0,\infty)$, *we have*

$$||K_nf - f||_{[0,\infty)} \leq 2\omega^*\left(f, \sqrt{\gamma_n}\right),$$

where

$$\gamma_n = ||K_n(e^{-2t}) - e^{-2x}||_{[0,\infty)}$$
$$= \left|\left|\frac{2xe^{-2x}}{n} + \frac{(24x^2 - 48x - 11)e^{-2x}}{12n^2} + O\left(\frac{1}{n^3}\right)\right|\right|_{[0,\infty)}.$$

Proof The operators K_n preserve the constant and e^{-x}. Thus $\alpha_n = \beta_n = 0$. We only have to evaluate γ_n. In view of Lemma 1, we have

$$K_n(e^{-2t}, x) = n\sum_{k=0}^{\infty}\frac{2^{-na_n(x)}(na_n(x))_k}{k!2^k}\int_{k/n}^{(k+1)/n} e^{-2t}dt$$
$$= \frac{n(1-e^{-2/n})}{2}\left(2-e^{-2/n}\right)^{-na_n(x)},$$

where $a_n(x)$ is given as

$$a_n(x) = \frac{x + \ln\left(n(1-e^{-1/n})\right)}{n\ln\left(2-e^{-1/n}\right)}.$$

Thus using the software Mathematica, we get at once

$$K_n(e^{-2t}, x) = \frac{n(1-e^{-2/n})}{2}\left(2-e^{-2/n}\right)^{\left[-\frac{x+\ln\left(n(1-e^{-1/n})\right)}{\ln\left(2-e^{-1/n}\right)}\right]}$$
$$= e^{-2x} + \frac{2xe^{-2x}}{n} + \frac{(24x^2-48x-11)e^{-2x}}{12n^2} + O\left(\frac{1}{n^3}\right).$$

This completes the proof of the theorem.

Theorem 2 *Let $f, f'' \in C^*[0,\infty)$. Then the inequality*

$$\left|n\left[K_n(f,x) - f(x)\right] - x[f'(x) + f''(x)]\right|$$
$$\leq |p_n(x)||f'| + |q_n(x)||f''| + 2\left(2q_n(x) + 2x + r_n(x)\right)\omega^*\left(f'', n^{-1/2}\right)$$

holds for any $x \in [0,\infty)$, where

$$p_n(x) = n\mu_{n,1}(x) - x,$$
$$q_n(x) = \frac{1}{2}\left(n\mu_{n,2}(x) - 2x\right),$$
$$r_n(x) = n^2\sqrt{K_n\left((e^{-x}-e^{-t})^4, x\right)}\sqrt{\mu_{n,4}(x)},$$

and $\mu_{n,1}(x)$, $\mu_{n,2}(x)$, and $\mu_{n,4}(x)$ are given in Lemma 3.

Proof By Taylor's expansion, we have

$$f(t) = f(x) + (t-x)f'(x) + \frac{1}{2}(t-x)^2 f''(x) + \varepsilon(t,x)(t-x)^2, \tag{3}$$

where

$$\varepsilon(t,x) = \frac{f''(\eta) - f''(x)}{2}$$

and η is a number lying between x and t. If we apply the operator K_n to both sides of (3), we have

$$\left| K_n(f,x) - f(x) - \mu_{n,1}(x)f'(x) - \frac{1}{2}\mu_{n,2}(x)f''(x) \right|$$
$$\leq |K_n(\varepsilon(t,x)(t-x)^2, x)|,$$

Applying Lemma 2, we get

$$\left| n[K_n(f,x) - f(x)] - x[f'(x) + f''(x)] \right|$$
$$\leq \left| n\mu_{n,1}(x) - x \right| |f'(x)| + \frac{1}{2} \left| n\mu_{n,2}(x) - 2x \right| |f''(x)|$$
$$+ |nK_n(\varepsilon(t,x)(t-x)^2, x)|.$$

Put $p_n(x) := n\mu_{n,1}(x) - x$ and $q_n(x) := \frac{1}{2}[n\mu_{n,2}(x) - 2x]$. Thus

$$\left| n[K_n(f,x) - f(x)] - x[f'(x) + f''(x)] \right|$$
$$\leq |p_n(x)|.|f'(x)| + |q_n(x)|.|f''(x)| + |nK_n(\varepsilon(t,x)(t-x)^2, x)|.$$

In order to complete the proof of the theorem, we must estimate the term $|nK_n(\varepsilon(t,x)(t-x)^2, x)|$. Using the property

$$|f(t) - f(x)| \leq \left(1 + \frac{(e^{-t} - e^{-x})^2}{\delta^2}\right) \omega^*(f, \delta), \delta > 0,$$

we get

$$|\varepsilon(t,x)| \leq \left(1 + \frac{(e^{-t} - e^{-x})^2}{\delta^2}\right) \omega^*(f'', \delta).$$

For $|e^{-x} - e^{-t}| \leq \delta$, one has $|\varepsilon(t,x)| \leq 2\omega^*(f'', \delta)$. In case $|e^{-x} - e^{-t}| > \delta$, then $|\varepsilon(t,x)| < 2\frac{(e^{-x}-e^{-t})^2}{\delta^2}\omega^*(f'', \delta)$. Thus

$$|\varepsilon(t,x)| \leq 2\left(1 + \frac{(e^{-x} - e^{-t})^2}{\delta^2}\omega^*(f'', \delta)\right).$$

Obviously using this and Cauchy–Schwarz inequality after choosing $\delta = n^{-1/2}$, we get

$$\begin{aligned} nK_n(|\varepsilon(t,x)|(t-x)^2, x) &\leq 2\omega^*(f''(x), n^{-1/2})\left[n\mu_{n,2}(x) + r_n(x)\right] \\ &= 2\omega^*(f''(x), n^{-1/2})\left[2q_n(x) + 2x + r_n(x)\right], \end{aligned}$$

where $r_n(x) = n^2[K_n((e^{-x} - e^{-t})^4, x).\mu_{n,4}(x)]^{1/2}$ and

$$\begin{aligned} K_n((e^{-x} - e^{-t})^4, x) = &-\frac{n}{4}(e^{-4/n} - 1)(2 - e^{-4/n})^{-na_n(x)} \\ &+\frac{4n}{3}e^{-x}(e^{-3/n} - 1)(2 - e^{-3/n})^{-na_n(x)} \\ &-3ne^{-2x}(e^{-2/n} - 1)(2 - e^{-2/n})^{-na_n(x)} \\ &+4ne^{-3x}(e^{-1/n} - 1)(2 - e^{-1/n})^{-na_n(x)} + e^{-4x}. \end{aligned}$$

This completes the proof of the result.

Remark 1 From the Lemma 3, $p_n(x) \to 0, q_n(x) \to 0$ as $n \to \infty$ and using Mathematica, we get

$$\lim_{n\to\infty} n^2\mu_{n,4}(x) = 12x^2.$$

Furthermore

$$\lim_{n\to\infty} n^2K_n\left((e^{-t} - e^{-x})^4, x\right) = 12e^{-4x}x^2.$$

Thus in the above Theorem 2, convergence occurs for sufficiently large n.

Corollary 1 *Let $f, f'' \in C^*[0,\infty)$. Then, the inequality*

$$\lim_{n\to\infty} n\left[K_n(f,x) - f(x)\right] = x[f'(x) + f''(x)]$$

holds for any $x \in [0,\infty)$.

Remark 2 In case the operators (1) preserve the function e^{-2x}, then in that case using Lemma 1, we have

$$e^{-2x} = \frac{n(1-e^{-2/n})}{2}\left(2-e^{-2/n}\right)^{-na_n(x)},$$

which implies

$$a_n(x) = \frac{2x + \ln\left(\frac{n(1-e^{-2/n})}{2}\right)}{n\ln(2-e^{-2/n})} \tag{4}$$

Also, for this preservation corresponding limits of Lemma 3 takes the following forms:

$$\lim_{n\to\infty} n\left[\frac{2x + \ln\left(\frac{n(1-e^{-2/n})}{2}\right)}{n\ln(2-e^{-2/n})} + \frac{1}{2n} - x\right] = 2x$$

and

$$\lim_{n\to\infty} n\left[\left(\frac{2x + \ln\left(\frac{n(1-e^{-2/n})}{2}\right)}{n\ln(2-e^{-2/n})} - x\right)^2 + \frac{3(2x + \ln\left(\frac{n(1-e^{-2/n}))}{2}\right)}{n^2\ln(2-e^{-2/n})} - \frac{x}{n} + \frac{1}{3n^2}\right] = 2x$$

and we have the following Theorems 1and 2 and Corollary 1 taking the following forms:

Theorem 3 *For $f \in C^*[0,\infty)$, we have*

$$||K_n f - f||_{[0,\infty)} \le 2\omega^*\left(f, \sqrt{2\beta_n}\right),$$

where

$$\begin{aligned}\beta_n &= ||K_n(e^{-t}) - e^{-x}||_{[0,\infty)}\\ &= \left|\left|\frac{-xe^{-x}}{n} + \frac{(12x^2+24x+11)e^{-x}}{24n^2} + O\left(\frac{1}{n^3}\right)\right|\right|_{[0,\infty)}.\end{aligned}$$

Theorem 4 *Let $f, f'' \in C^*[0,\infty)$. Then the inequality*

$$\begin{aligned}&\left|n\left[K_n(f,x) - f(x)\right] - x[2f'(x) + f''(x)]\right|\\ &\le |\hat{p}_n(x)||f'| + |\hat{q}_n(x)||f''| + 2\left(2\hat{q}_n(x) + 2x + \hat{r}_n(x)\right)\omega^*\left(f'', n^{-1/2}\right)\end{aligned}$$

holds for any $x \in [0, \infty)$, *where*

$$\hat{p}_n(x) = n\mu_{n,1}(x) - x,$$

$$\hat{q}_n(x) = \frac{1}{2}\left(n\mu_{n,2}(x) - 4x\right),$$

$$\hat{r}_n(x) = n^2\sqrt{K_n\left((e^{-x} - e^{-t})^4, x\right)}\sqrt{\mu_{n,4}(x)}.$$

and $\mu_{n,1}(x)$, $\mu_{n,2}(x)$ *and* $\mu_{n,4}(x)$ *are given in Lemma 3, with values of* $a_n(x)$, *given by (4).*

Corollary 2 *Let* $f, f'' \in C^*[0, \infty)$. *Then, the inequality*

$$\lim_{n\to\infty} n\left[K_n(f, x) - f(x)\right] = x[2f'(x) + f''(x)]$$

holds for any $x \in [0, \infty)$.

Remark 3 Several other operators, which are linear and positive, can be applied to establish analogous results. Also, some other approximation properties for the operators studied in [3, 5–7, 10] and references therein may be considered for these operators.

References

1. Acar, T., Aral, A., Gonska, H.: On Szász-Mirakyan operators preserving $e^{2ax}, a > 0$, H. Mediterr. J. Math. **14**(6) (2017). https://doi.org/10.1007/s00009-016-0804-7
2. Agratini, O.: On a sequence of linear positive operators. Facta Universitatis (Nis), Ser. Math. Inform. **14**, 41–48 (1999)
3. Aral, A., Gupta, V., Agarwal, R.P.: Applications of q-Calculus in Operator Theory. Springer, Berlin (2013)
4. Erençin, A., Taşdelen, F.: On certain Kantorovich type operators. Fasciculi Math. **41**, 65–71 (2009)
5. Gupta, V., Agarwal, R.P.: Convergence Estimates in Approximation Theory. Springer, Berlin (2014)
6. Gupta, V., Lupaș, A.: On the rate of approximation for the Bézier variant of Kantorovich-Balasz operators. Gen. Math. **12**(3), 3–18 (2004)
7. Gupta, V., Rassias, Th.M., Sinha, J.: A survey on Durrmeyer operators. In: Pardalos, P.M., Rassias, T.M. (eds.) Contributions in Mathematics and Engineering, pp. 299–312. Springer International Publishing, Cham (2016). https://doi.org/10.1007/978-3-319-31317-7-14
8. Holhoş, A.: The rate of approximation of functions in an infinite interval by positive linear operators. Stud. Univ. Babe-Bolyai. Math. (2), 133–142 (2010)
9. Lupaș, A.: The approximation by means of some linear positive operators. In: Muller, M.W., Felten, M., Mache, D.H. (eds.) Approximation Theory. Proceedings of the International Dortmund Meeting IDoMAT 95, held in Witten, March 13–17, 1995. Mathematical Research, vol. 86, pp. 201–229 Akademie, Berlin (1995)
10. Rassias, Th.M., Gupta, V. (eds.): Mathematical Analysis, Approximation Theory and Their Applications Series. Springer Optimization and Its Applications, vol. 111. Springer, Berlin (2016). ISBN:978-3-319-31279-8

Enumeration by e

Mehdi Hassani

Abstract We obtain some formulas, explicit bounds, and an asymptotic approximation for the number of all distinct paths between a specific pair of vertices in a complete graph on n vertices. Also, we give further enumerative formulas related to e, concerning the number derangements.

Introduction and Summary of the Results

Let w_n denote the number of all distinct paths between a specific pair of vertices in a complete graph on n vertices. Each path of length k between two distinct vertices u and v consists of $k+1$ distinct vertices initiated by u and terminated in v. Thus, the number of paths of length k with $1 \leq k \leq n-1$ is $\frac{(n-2)!}{(n-1-k)!}$, and

$$w_n = \sum_{k=1}^{n-1} \frac{(n-2)!}{(n-1-k)!} = \sum_{k=0}^{n-2} \frac{(n-2)!}{k!} = (n-2)!\mathrm{e}_{n-2}, \tag{1}$$

where

$$\mathrm{e}_n = \sum_{k=0}^{n} \frac{1}{k!}.$$

Since $\mathrm{e}_n \to \mathrm{e}$ as $n \to \infty$, naturally we seek a connection to the number e, for which we observe that

$$0 < \mathrm{e} - \mathrm{e}_n = \sum_{k=1}^{\infty} \frac{1}{(n+k)!} = \frac{1}{n!} \sum_{k=1}^{\infty} \prod_{j=1}^{k} \frac{1}{n+j} < \frac{1}{n!} \sum_{k=1}^{\infty} \frac{1}{(n+1)^k} = \frac{1}{n.n!}.$$

M. Hassani (✉)
Department of Mathematics, University of Zanjan, Zanjan, Iran
e-mail: mehdi.hassani@znu.ac.ir

N. J. Daras, Th. M. Rassias (eds.), *Modern Discrete Mathematics and Analysis*, Springer Optimization and Its Applications 131,
https://doi.org/10.1007/978-3-319-74325-7_10

Thus, for each $n \geq 1$, we obtain

$$0 < \mathrm{e} - \mathrm{e}_n < \frac{1}{n.n!}, \tag{2}$$

and as a consequence, for each $n \geq 1$, we get

$$n!\mathrm{e}_n = \lfloor \mathrm{e}n! \rfloor, \tag{3}$$

where $\lfloor x \rfloor$ denotes the largest integer not exceeding x. Hence, for each $n \geq 3$, we get

$$w_n = \lfloor \mathrm{e}(n-2)! \rfloor.$$

This is one of several enumerative formulas concerning the complete graphs, considered in [2]. In the present paper, we continue our study on w_n to obtain more similar formulas, explicit bounds, and an asymptotic approximation for w_n. More precisely, we prove the following results.

Theorem 1 *Assume that $m \geq 3$ is a fixed integer, and let $\eta \in [-m + \frac{m}{m-2}, 1]$ be an arbitrary real. Then, for each $n \geq m$, we have*

$$w_n = \left\lfloor \mathrm{e}(n-2)! - \frac{\eta}{n} \right\rfloor. \tag{4}$$

More precisely (4) is valid for each arbitrary real $\eta \in [0, 1]$ and for each $n \geq 3$.

Theorem 2 *For each integer $n \geq 5$, we have*

$$\left(\frac{n}{\mathrm{e}}\right)^{n-1} \sqrt{\frac{2\pi}{n}} \left(1 + \frac{1.08}{n}\right) < w_n < \left(\frac{n}{\mathrm{e}}\right)^{n-1} \sqrt{\frac{2\pi}{n}} \left(1 + \frac{1.28}{n}\right). \tag{5}$$

Moreover, as $n \to \infty$, we have

$$w_n = \left(\frac{n}{\mathrm{e}}\right)^{n-1} \sqrt{\frac{2\pi}{n}} \left(1 + \frac{13}{12n} + o\left(\frac{1}{n}\right)\right). \tag{6}$$

Note that we use the small oh and the big oh notations with their usual meaning from analysis. The key point to imply the above results is studying the difference

$$\mathrm{d}_n = \mathrm{e}(n-2)! - w_n.$$

While the inequality (2) implies validity of $0 < (n-2)\mathrm{d}_n < 1$ for each $n \geq 3$, to prove the above results, we require some sharp approximations for d_n, in particular guaranteeing that $(n-2)\mathrm{d}_n \to 1$ as $n \to \infty$. More precisely, we require the following auxiliary result, which is a corollary of Theorem 1 of [3].

Lemma 1 *For each $n \geq 3$, we have*

$$\frac{1}{n-2+\alpha} \leq \mathrm{d}_n < \frac{1}{n-2}, \tag{7}$$

with

$$\alpha = \frac{1}{\mathrm{e}-2} - 1.$$

Meanwhile, as an immediate consequence of the above lemma, we obtain the following average result.

Corollary 1 *For each real $x \geq 3$ we have*

$$\sum_{3 \leq n \leq x} \mathrm{d}_n = \log x + \left(\gamma + \delta - \frac{3}{2}\right) + O\left(\frac{1}{x}\right), \tag{8}$$

where γ is Euler's constant, and $\delta = \sum_{n=3}^{\infty}(\mathrm{d}_n - \frac{1}{n})$ is an absolute constant satisfying $1 < \delta < 1.5$.

Proofs

Proof of Theorem 1 We consider the inequality (7), and we take the real constant η such that the inequalities

$$0 \leq \frac{1}{n-2+\alpha} - \frac{\eta}{n} \leq \mathrm{d}_n - \frac{\eta}{n} < \frac{1}{n-2} - \frac{\eta}{n} \leq 1, \tag{9}$$

all are valid for each $n \geq 3$. The leftmost inequality in (9) is equivalent by the inequality $\eta \leq \frac{n}{n-2+\alpha} := f_1(n)$, say. The function $f_1(n)$ is strictly decreasing for $n \geq 3$; hence, we need to have $\eta \leq \lim_{n\to\infty} f_1(n) = 1$. The rightmost inequality in (9) is equivalent by $\eta \geq \frac{n}{n-2} - n := f_2(n)$, say. The function $f_2(n)$ is strictly decreasing for $n \geq 3$. Thus, we should have $\eta \geq f_2(3) = 0$ for each $n \geq 3$, and $\eta \geq f_2(m)$ for each $n \geq m \geq 3$. This completes the proof.

Proof of Theorem 2 The approximation (7) implies that

$$w_n = \frac{\mathrm{e}n!}{n(n-1)} + O\left(\frac{1}{n}\right).$$

Now, we use Stirling's approximation asserting that

$$n! = \left(\frac{n}{\mathrm{e}}\right)^n \sqrt{2\pi n}\left(1 + O\left(\frac{1}{n}\right)\right),$$

and we consider the relation $\frac{\sqrt{n}}{n-1} = \frac{1}{\sqrt{n}}(1 + O(\frac{1}{n}))$ to obtain

$$w_n = \left(\frac{n}{\mathrm{e}}\right)^{n-1}\sqrt{\frac{2\pi}{n}}\left(1 + O\left(\frac{1}{n}\right)\right).$$

Motivated by determining the constant in the last O-term, we rewrite (7) in the form

$$\frac{\mathrm{e}n!}{n(n-1)} - \frac{1}{n-2} < w_n \le \frac{\mathrm{e}n!}{n(n-1)} - \frac{1}{n-2+\alpha}. \tag{10}$$

Now, we recall that as a corollary of Theorem 8 of [1], one may obtain validity of the double-sided inequality

$$\left(\frac{n}{\mathrm{e}}\right)^n \sqrt{2\pi n}\, \mathrm{e}^{\frac{1}{12n} - \frac{1}{180n^3}} \le n! \le \left(\frac{n}{\mathrm{e}}\right)^n \sqrt{2\pi n}\, \mathrm{e}^{\frac{1}{12n}}, \tag{11}$$

for each $n \ge 1$. We let

$$M_n = \left(\frac{n}{\mathrm{e}}\right)^{n-1}\sqrt{\frac{2\pi}{n}}.$$

By combining the bounds (10) and (11), we get validity of $1 + \frac{L_n}{n} < \frac{w_n}{M_n} < 1 + \frac{U_n}{n}$ for each $n \ge 3$ with

$$L_n = n\left(\frac{n}{n-1}\mathrm{e}^{\frac{1}{12n} - \frac{1}{180n^3}} - \frac{1}{(n-2)M_n} - 1\right),$$

and

$$U_n = n\left(\frac{n}{n-1}\mathrm{e}^{\frac{1}{12n}} - \frac{1}{(n-2+\alpha)M_n} - 1\right).$$

The limit relations

$$\lim_{n\to\infty} L_n = \lim_{n\to\infty} U_n = \frac{13}{12}$$

give the expansion (6). Moreover, the sequence L_n is strictly decreasing for $n \ge 6$. Thus, for each $n \ge 6$, we deduce $L_n > \frac{13}{12} > 1.08$, from which we get validity of the left-hand side of (5). By computation we confirm it for $n = 5$, too. On the other hand, we have

$$\max_{n\ge 3} U_n = U_6 < 1.28.$$

This gives the right-hand side of (5) for each $n \ge 3$ and completes the proof.

Proof of Corollary 1 To deduce (8) we let $S(x) = \sum_{3\le n\le x} \mathrm{d}_n$, and also $H(x) = \sum_{1\le n\le x} \frac{1}{n}$. We rewrite (7) in the form

$$\frac{2-\alpha}{n(n-2+\alpha)} \le \mathrm{d}_n - \frac{1}{n} < \frac{2}{n(n-2)}.$$

By summing over the integers $n \ge 3$, we obtain

$$c \le \sum_{n=3}^{\infty}\left(\mathrm{d}_n - \frac{1}{n}\right) < \sum_{n=3}^{\infty}\frac{2}{n(n-2)} = \frac{3}{2},$$

with

$$c = \sum_{n=3}^{\infty}\frac{2-\alpha}{n(n-2+\alpha)} = \frac{2-\alpha}{2(\alpha+1)(3\mathrm{e}-7)}(3-2\gamma-2\psi(\alpha+1)) = 0.9922\ldots,$$

where $\psi(x) = \frac{\Gamma'(x)}{\Gamma(x)}$. Hence, the series $\sum_{n=3}^{\infty}(\mathrm{d}_n - \frac{1}{n})$ converges, and we denote its value by δ. Now, we use the approximation $\mathrm{d}_n - \frac{1}{n} = O(\frac{1}{n^2})$ to write

$$\delta = \sum_{3\le n\le x}\left(\mathrm{d}_n - \frac{1}{n}\right) + \sum_{n>x}\left(\mathrm{d}_n - \frac{1}{n}\right) = S(x) - H(x) + \frac{3}{2} + O\left(\frac{1}{x}\right).$$

The Euler–Maclaurin summation formula (see [5]) gives

$$H(x) = \log x + \gamma + O\left(\frac{1}{x}\right),$$

where $\log x$ denotes the natural logarithm of positive real x. By combining the above approximations, we get (8). Meanwhile, the above argument implies that $c \le \delta < \frac{3}{2}$. We note that for each $n \ge 3$, one has $\mathrm{d}_n - \frac{1}{n} > 0$. Thus, the partial sum $E(N) = \sum_{n=3}^{N}(\mathrm{d}_n - \frac{1}{n})$ is strictly increasing. By computation we observe that $E(20) < 1 < E(21)$, and hence $1 < \delta < 1.5$. This completes the proof.

Further Enumerative Formulas Related to e

In [4], analogue to (3), we prove for each $n \ge 1$ that

$$n!\sum_{k=0}^{n}\frac{(-1)^k}{k!} = \begin{cases} \lfloor \frac{n!}{\mathrm{e}} + \lambda_1 \rfloor, & n \text{ odd}, \lambda_1 \in [0, \frac{1}{2}], \\ \lfloor \frac{n!}{\mathrm{e}} + \lambda_2 \rfloor, & n \text{ even}, \lambda_2 \in [\frac{1}{3}, 1]. \end{cases}$$

This implies that for each positive integer $n \geq 1$ and for each real $\lambda \in [\frac{1}{3}, \frac{1}{2}]$, we have

$$D_n = \left\lfloor \frac{n!}{\mathrm{e}} + \lambda \right\rfloor, \tag{12}$$

where D_n is the number of derangements (permutations with no fixed point) of n distinct objects. Indeed

$$D_n = \frac{n!}{\mathrm{e}} + (-1)^n \left(\frac{1}{n+1} - \frac{1}{(n+1)(n+2)} + \cdots \right). \tag{13}$$

Hence, for each $n \geq 1$, we have

$$\left| D_n - \frac{n!}{\mathrm{e}} \right| < \frac{1}{n+1}.$$

If n is even, $D_n > \frac{n!}{\mathrm{e}}$ and $D_n = \lfloor \frac{n!}{\mathrm{e}} + \lambda \rfloor$ provided $\frac{1}{n+1} \leq \lambda \leq 1$. If n is odd, $D_n < \frac{n!}{\mathrm{e}}$ and $D_n = \lfloor \frac{n!}{\mathrm{e}} + \lambda \rfloor$ provided $0 < \frac{1}{n+1} + \lambda \leq 1$. So we require $\frac{1}{3} \leq \lambda \leq 1$ and $0 \leq \lambda \leq \frac{1}{2}$, giving (12). More precisely, for each $n \geq 1$, we obtain

$$D_n = \left\lfloor \frac{n! + 1}{\mathrm{e}} \right\rfloor.$$

On the other hand, the expansion (13) gives

$$\left| \frac{n!}{\mathrm{e}} - D_n \right| \leq \frac{1}{(n+1)} + \frac{1}{(n+1)(n+2)} + \frac{1}{(n+1)(n+2)(n+3)} + \cdots.$$

If we denote the right side of the above inequality by $M(n)$, then we get

$$M(n) < \frac{1}{(n+1)} + \frac{1}{(n+1)^2} + \cdots = \frac{1}{n},$$

from which for each $n \geq 2$, we obtain

$$D_n = \left\lfloor \frac{n!}{\mathrm{e}} + \frac{1}{n} \right\rfloor.$$

A better bound for $M(n)$ is as follows

$$M(n) < \frac{1}{n+1} \left(1 + \frac{1}{(n+2)} + \frac{1}{(n+2)^2} + \cdots \right) = \frac{n+2}{(n+1)^2},$$

from which for each $n \geq 2$, we get

$$D_n = \left\lfloor \frac{n!}{e} + \frac{n+2}{(n+1)^2} \right\rfloor.$$

More generally, assuming that m is an integer with $m \geq 3$, then for each $n \geq 2$, we get

$$D_n = \left\lfloor \left(\frac{\lfloor e(n+m-2)! \rfloor}{(n+m-2)!} + \frac{n+m}{(n+m-1)(n+m-1)!} + e^{-1} \right) n! \right\rfloor - \lfloor en! \rfloor,$$

and as an immediate corollary, for each $n \geq 2$, we obtain

$$D_n = \lfloor (e + e^{-1}) n! \rfloor - \lfloor en! \rfloor.$$

References

1. Alzer, H.: On some inequalities for the gamma and psi functions. Math. Comput. **66**, 373–389 (1997)
2. Hassani, M.: Cycles in graphs and derangements. Math. Gaz. **88**, 123–126 (2004)
3. Hassani, M., Sofo, A.: Sharp bounds for the constant e. Vietnam J. Math. **43**, 629–633 (2015)
4. Hassani, M.: Derangements and applications. J. Integer Seq. **6**(1) (2003). Article 03.1.2
5. Odlyzko, A. M.: Asymptotic enumeration methods. In: Handbook of Combinatorics, vols. 1, 2, pp. 1063–1229. Elsevier, Amsterdam (1995)

Fixed Point and Nearly m-Dimensional Euler–Lagrange-Type Additive Mappings

Hassan Azadi Kenary

Abstract In this paper, using the fixed point alternative approach, we prove the generalized Hyers–Ulam–Rassias stability of the following Euler–Lagrange-type additive functional equation

$$\sum_{j=1}^{m} f\left(\sum_{1\leq i\leq m, i\neq j} r_i x_i - r_j x_j\right) = mf\left(\sum_{i=1}^{m} r_i x_i\right) - 2\sum_{i=1}^{m} r_i f(x_i) \tag{1}$$

where $r_1, \ldots, r_m \in \mathbb{R}$, $\sum_{i=1}^{m} r_i \neq 0$, and $r_i, r_j \neq 0$ for some $1 \leq i < j \leq m$ in random normed spaces.

Introduction and Preliminaries

The stability problem of functional equations originated from a question of Ulam [25] concerning the stability of group homomorphisms. Hyers [9] gave a first affirmative partial answer to the question of Ulam for Banach spaces. Hyers' theorem was generalized by Aoki [1] for additive mappings and by Rassias [18] for linear mappings by considering an unbounded Cauchy difference.

The paper of Rassias has provided a lot of influence in the development of what we call the *generalized Hyers–Ulam stability* of functional equations. In 1994, a generalization of Rassias' theorem was obtained by Găvruta [4] by replacing the bound $\varepsilon(\|x\|^p + \|y\|^p)$ by a general control function $\varphi(x, y)$.

The functional equation

$$f(x + y) + f(x - y) = 2f(x) + 2f(y) \tag{2}$$

H. A. Kenary (✉)
Department of Mathematics, College of Sciences, Yasouj University, Yasouj, Iran
e-mail: azadi@yu.ac.ir

N. J. Daras, Th. M. Rassias (eds.), *Modern Discrete Mathematics and Analysis*, Springer Optimization and Its Applications 131,
https://doi.org/10.1007/978-3-319-74325-7_11

is called a *quadratic functional equation*. In particular, every solution of the quadratic functional equation is said to be a *quadratic mapping*. The generalized Hyers–Ulam stability problem for the quadratic functional equation was proved by Skof [24] for mappings $f : X \to Y$, where X is a normed space and Y is a Banach space. Cholewa [2] noticed that the theorem of Skof is still true if the relevant domain X is replaced by an Abelian group. Czerwik [3] proved the generalized Hyers–Ulam stability of the quadratic functional equation.

The stability problems of several functional equations have been extensively investigated by a number of authors, and there are many interesting results concerning this problem (see [4–22]).

In the sequel, we shall adopt the usual terminology, notions, and conventions of the theory of random normed spaces as in [23]. Throughout this paper, the spaces of all probability distribution functions are denoted by Δ^+. Elements of Δ^+ are functions $F : \mathbb{R} \cup \{-\infty, +\infty\} \to [0, 1]$, such that F is left continuous and nondecreasing on $\mathbb{R}$, $F(0) = 0$ and $F(+\infty) = 1$. It's clear that the subset $D^+ = \{F \in \Delta^+ : l^- F(+\infty) = 1\}$, where $l^- f(x) = \lim_{t \to x^-} f(t)$, is a subset of Δ^+. The space Δ^+ is partially ordered by the usual point-wise ordering of functions, that is, for all $t \in R$, $F \leq G$ if and only if $F(t) \leq G(t)$. For every $a \geq 0$, $H_a(t)$ is the element of D^+ defined by

$$H_a(t) = \begin{cases} 0 \text{ if } t \leq a \\ 1 \text{ if } t > a \end{cases}.$$

One can easily show that the maximal element for Δ^+ in this order is the distribution function $H_0(t)$.

Definition 1 A function $T : [0, 1]^2 \to [0, 1]$ is a continuous triangular norm (briefly a t-norm) if T satisfies the following conditions:

(i) T is commutative and associative;
(ii) T is continuous;
(iii) $T(x, 1) = x$ for all $x \in [0, 1]$;
(iv) $T(x, y) \leq T(z, w)$ whenever $x \leq z$ and $y \leq w$ for all $x, y, z, w \in [0, 1]$.

Three typical examples of continuous $t-$norms are $T(x, y) = xy$, $T(x, y) = max\{a + b - 1, 0\}$, and $T(x, y) = min(a, b)$. Recall that, if T is a $t-$norm and $\{x_n\}$ is a given of numbers in $[0, 1]$, $T_{i=1}^n x_i$ is defined recursively by $T_{i=1}^1 x_1$ and $T_{i=1}^n x_i = T(T_{i=1}^{n-1} x_i, x_n)$ for $n \geq 2$.

Definition 2 A random normed space (briefly RN-space) is a triple (X, μ', T), where X is a vector space, T is a continuous t-norm, and $\mu' : X \to D^+$ is a mapping such that the following conditions hold:

(i) $\mu'_x(t) = H_0(t)$ for all $t > 0$ if and only if $x = 0$.
(ii) $\mu'_{\alpha x}(t) = \mu'_x\left(\frac{t}{|\alpha|}\right)$ for all $\alpha \in \mathbb{R}$, $\alpha \neq 0$, $x \in X$ and $t \geq 0$.
(iii) $\mu'_{x+y}(t + s) \geq T(\mu'_x(t), \mu'_y(s))$, for all $x, y \in X$ and $t, s \geq 0$.

Definition 3 Let (X, μ', T) be an RN-space.

(i) A sequence $\{x_n\}$ in X is said to be convergent to $x \in X$ in X if for all $t > 0$, $lim_{n\to\infty}\mu'_{x_n-x}(t) = 1$.
(ii) A sequence $\{x_n\}$ in X is said to be Cauchy sequence in X if for all $t > 0$, $lim_{n\to\infty}\mu'_{x_n-x_m}(t) = 1$.
(iii) The RN-space (X, μ', T) is said to be complete if every Cauchy sequence in X is convergent.

Theorem 1 *If (X, μ', T) is RN-space and $\{x_n\}$ is a sequence such that $x_n \to x$, then* $\lim_{n\to\infty} \mu'_{x_n}(t) = \mu'_x(t)$.

Definition 4 Let X be a set. A function $d : X \times X \to [0, \infty]$ is called a generalized metric on X if d satisfies the following conditions:

(a) $d(x, y) = 0$ if and only if $x = y$ for all $x, y \in X$;
(b) $d(x, y) = d(y, x)$ for all $x, y \in X$;
(c) $d(x, z) \le d(x, y) + d(y, z)$ for all $x, y, z \in X$.

Theorem 2 *Let (X,d) be a complete generalized metric space and $J : X \to X$ be a strictly contractive mapping with Lipschitz constant $L < 1$. Then, for all $x \in X$, either*

$$d(J^n x, J^{n+1}x) = \infty \tag{3}$$

for all nonnegative integers n or there exists a positive integer n_0 such that:

(a) $d(J^n x, J^{n+1}x) < \infty$ *for all* $n_0 \ge n_0$*;*
(b) *the sequence* $\{J^n x\}$ *converges to a fixed point* y^* *of* J*;*
(c) y^* *is the unique fixed point of* J *in the set* $Y = \{y \in X : d(J^{n_0}x, y) < \infty\}$*;*
(d) $d(y, y^*) \le \frac{1}{1-L} d(y, Jy)$ *for all* $y \in Y$.

In this paper, we investigate the generalized Hyers–Ulam stability of the following additive functional equation of Euler–Lagrange type:

$$\sum_{j=1}^{m} f\left(-r_j x_j + \sum_{1\le i\le m, i\ne j} r_i x_i\right) + 2\sum_{i=1}^{m} r_i f(x_i) = mf\left(\sum_{i=1}^{m} r_i x_i\right), \tag{4}$$

where $r_1, \ldots, r_n \in \mathbb{R}$, $\sum_{i=1}^{m} r_i \ne 0$, and $r_i, r_j \ne 0$ for some $1 \le i < j \le m$, in random normed spaces.

Every solution of the functional equation (4) is said to be a *generalized Euler–Lagrange-type additive mapping.*

RN-Approximation of Functional Equation (4)

Remark 1 Throughout this paper, $r_1, \ldots, r_m$ will be real numbers such that r_i, $r_j \ne 0$ for fixed $1 \le i < j \le m$.

In this section, using the fixed point alternative approach, we prove the generalized Hyers–Ulam stability of functional equation (4) in random normed spaces.

Theorem 3 *Let X be a linear space, (Y, μ, T_M) be a complete RN-space, and Φ be a mapping from X^m to D^+ such that there exists $0 < \alpha < \frac{1}{2}$ such that*

$$\Phi(x_1, \ldots, x_m)(t) \leq \Phi\left(\frac{x_1}{2}, \ldots, \frac{x_m}{2}\right)(\alpha t) \tag{5}$$

for all $x_1, \ldots, x_m \in X$ and $t > 0$. Let $f : X \to Y$ be a mapping with $f(0) = 0$ and satisfying

$$\begin{aligned}
&\mu_{\sum_{j=1}^m f\left(-r_j x_j + \sum_{1\leq i\leq m, i\neq j} r_i x_i\right) + 2\sum_{i=1}^m r_i f(x_i) - mf\left(\sum_{i=1}^m r_i x_i\right)}(t) \\
&\geq \Phi(x_1, \ldots, x_m)(t)
\end{aligned} \tag{6}$$

for all $x_1, \ldots, x_m \in X$ and $t > 0$. Then, for all $x \in X$

$$\mathrm{EL}(x) := \lim_{n\to\infty} 2^n f\left(\frac{x}{2^n}\right)$$

exists and $\mathrm{EL} : X \to Y$ is a unique Euler–Lagrange additive mapping such that

$$\begin{aligned}
&\mu_{f(x)-\mathrm{EL}(x)}(t) \\
&\geq T_M\Bigg(T_M\Bigg(\Phi_{i,j}\left(\frac{x}{2r_i}, -\frac{x}{2r_j}\right)\left(\frac{(1-2\alpha)t}{6\alpha}\right), \Phi_{i,j}\left(\frac{x}{2r_i}, 0\right)\left(\frac{(1-2\alpha)t}{6\alpha}\right), \\
&\quad \Phi_{i,j}\left(0, -\frac{x}{2r_j}\right)\left(\frac{(1-2\alpha)t}{6\alpha}\right)\Bigg), T_M\Bigg(\Phi_{i,j}\left(\frac{x}{r_i}, \frac{x}{r_j}\right)\left(\frac{(1-2\alpha)t}{3\alpha}\right), \\
&\quad \Phi_{i,j}\left(\frac{x}{r_i}, 0\right)\left(\frac{(1-2\alpha)t}{3\alpha}\right), \Phi_{i,j}\left(0, \frac{x}{r_j}\right)\left(\frac{(1-2\alpha)t}{3\alpha}\right)\Bigg)\Bigg)
\end{aligned} \tag{7}$$

for all $x \in X$ and $t > 0$.

Proof For each $1 \leq k \leq m$ with $k \neq i, j$, let $x_k = 0$ in (6). Then we get the following inequality

$$\mu_{f(-r_i x_i + r_j x_j) + f(r_i x_i - r_j x_j) - 2f(r_i x_i + r_j x_j) + 2r_i f(x_i) + 2r_j f(x_j)}(t) \geq \Phi_{i,j}(x_i, x_j)(t) \tag{8}$$

for all $x_i, x_j \in X$, where

$$\Phi_{i,j}(x, y)(t) := \Phi(0, \ldots, 0, \underbrace{x}_{ith}, 0, \ldots, 0, \underbrace{y}_{jth}, 0, \ldots, 0)(t)$$

for all $x, y \in X$ and all $1 \leq i < j \leq m$. Letting $x_i = 0$ in (8), we get

$$\mu_{f(-r_j x_j)-f(r_j x_j)+2r_j f(x_j)}(t) \geq \Phi_{i,j}(0, x_j)(t) \tag{9}$$

for all $x_j \in X$. Similarly, letting $x_j = 0$ in (8), we get

$$\mu_{f(-r_i x_i)-f(r_i x_i)+2r_i f(x_i)}(t) \geq \Phi_{i,j}(x_i, 0)(t) \tag{10}$$

for all $x_i \in X$. It follows from (8) to (10) that for all $x_i, x_j \in X$

$$\mu_{f(-r_i x_i+r_j x_j)+f(r_i x_i-r_j x_j)-2f(r_i x_i+r_j x_j)+f(r_i x_i)+f(r_j x_j)-f(-r_i x_i)-f(-r_j x_j)}(t)$$
$$T_M\left(\Phi_{i,j}(x_i, x_j)\left(\frac{t}{3}\right), \Phi_{i,j}(x_i, 0)\left(\frac{t}{3}\right), \Phi_{i,j}(0, x_j)\left(\frac{t}{3}\right)\right). \tag{11}$$

Replacing x_i and x_j by $\frac{x}{r_i}$ and $\frac{y}{r_j}$ in (11), we get that

$$\begin{aligned}&\mu_{f(-x+y)+f(x-y)-2f(x+y)+f(x)+f(y)-f(-x)-f(-y)}(t)\\&\geq T_M\left(\Phi_{i,j}\left(\frac{x}{r_i}, \frac{y}{r_j}\right)\left(\frac{t}{3}\right), \Phi_{i,j}\left(\frac{x}{r_i}, 0\right)\left(\frac{t}{3}\right), \Phi_{i,j}\left(0, \frac{y}{r_j}\right)\left(\frac{t}{3}\right)\right),\end{aligned} \tag{12}$$

for all $x, y \in X$. Putting $y = x$ in (12), we get

$$\begin{aligned}\mu_{2f(x)-2f(-x)-2f(2x)}(t) \geq T_M\Bigg(&\Phi_{i,j}\left(\frac{x}{r_i}, \frac{x}{r_j}\right)\left(\frac{t}{3}\right),\\&\Phi_{i,j}\left(\frac{x}{r_i}, 0\right)\left(\frac{t}{3}\right), \Phi_{i,j}\left(0, \frac{x}{r_j}\right)\left(\frac{t}{3}\right)\Bigg)\end{aligned} \tag{13}$$

for all $x \in X$. Replacing x and y by $\frac{x}{2}$ and $-\frac{x}{2}$ in (12), respectively, we get

$$\begin{aligned}\mu_{f(x)+f(-x)}(t) \geq T_M\Bigg(&\Phi_{i,j}\left(\frac{x}{2r_i}, -\frac{x}{2r_j}\right)\left(\frac{t}{3}\right),\\&\Phi_{i,j}\left(\frac{x}{2r_i}, 0\right)\left(\frac{t}{3}\right), \Phi_{i,j}\left(0, -\frac{x}{2r_j}\right)\left(\frac{t}{3}\right)\Bigg)\end{aligned} \tag{14}$$

for all $x \in X$. It follows from (13) and (14) that

$$\begin{aligned}&\mu_{f(2x)-2f(x)}(t)\\&\quad= \mu_{f(x)+f(-x)+\frac{2f(x)-2f(-x)-2f(2x)}{2}}(t)\\&\quad\geq T_M\left(\mu_{f(x)+f(-x)}\left(\frac{t}{2}\right), \mu_{2f(x)-2f(-x)-2f(2x)}(t)\right)\end{aligned} \tag{15}$$

$$\geq T_M\Bigg(T_M\Bigg(\Phi_{i,j}\left(\frac{x}{2r_i},-\frac{x}{2r_j}\right)\left(\frac{t}{6}\right),\Phi_{i,j}\left(\frac{x}{2r_i},0\right)\left(\frac{t}{6}\right),$$
$$\Phi_{i,j}\left(0,-\frac{x}{2r_j}\right)\left(\frac{t}{6}\right)\Bigg),T_M\Bigg(\Phi_{i,j}\left(\frac{x}{r_i},\frac{x}{r_j}\right)\left(\frac{t}{3}\right),\Phi_{i,j}\left(\frac{x}{r_i},0\right)\left(\frac{t}{3}\right),$$
$$\Phi_{i,j}\left(0,\frac{x}{r_j}\right)\left(\frac{t}{3}\right)\Bigg)\Bigg)$$

for all $x \in X$. Replacing x by $\frac{x}{2}$ in (15), we have

$$\mu_{f(x)-2f(\frac{x}{2})}(t) \geq T_M\Bigg(T_M\Bigg(\Phi_{i,j}\left(\frac{x}{4r_i},-\frac{x}{4r_j}\right)\left(\frac{t}{6}\right),\Phi_{i,j}\left(\frac{x}{4r_i},0\right)\left(\frac{t}{6}\right),$$
$$\Phi_{i,j}\left(0,-\frac{x}{4r_j}\right)\left(\frac{t}{6}\right)\Bigg),T_M\Bigg(\Phi_{i,j}\left(\frac{x}{2r_i},\frac{x}{2r_j}\right)\left(\frac{t}{3}\right),$$
$$\Phi_{i,j}\left(\frac{x}{2r_i},0\right)\left(\frac{t}{3}\right),\Phi_{i,j}\left(0,\frac{x}{2r_j}\right)\left(\frac{t}{3}\right)\Bigg)\Bigg). \tag{16}$$

Consider the set

$$S := \{g : X \to Y;\, g(0) = 0\}$$

and the generalized metric d in S defined by

$$d(f,g) = \inf\Bigg\{u \in \mathbb{R}^+ : \mu_{g(x)-h(x)}(ut) \geq T_M\Bigg(T_M\Bigg(\Phi_{i,j}\left(\frac{x}{2r_i},-\frac{x}{2r_j}\right)\left(\frac{t}{6}\right),$$
$$\Phi_{i,j}\left(\frac{x}{2r_i},0\right)\left(\frac{t}{6}\right),\Phi_{i,j}\left(0,-\frac{x}{2r_j}\right)\left(\frac{t}{6}\right)\Bigg),T_M\Bigg(\Phi_{i,j}\left(\frac{x}{r_i},\frac{x}{r_j}\right)\left(\frac{t}{3}\right),$$
$$\Phi_{i,j}\left(\frac{x}{r_i},0\right)\left(\frac{t}{3}\right),\Phi_{i,j}\left(0,\frac{x}{r_j}\right)\left(\frac{t}{3}\right)\Bigg)\Bigg),\ \forall x \in X,\ t > 0\Bigg\},$$

where $\inf \emptyset = +\infty$. It is easy to show that (S, d) is complete (see [15], Lemma 2.1). Now, we consider a linear mapping $J : S \to S$ such that

$$Jh(x) := 2h\left(\frac{x}{2}\right)$$

for all $x \in X$. First, we prove that J is a strictly contractive mapping with the Lipschitz constant 2α.

In fact, let $g, h \in S$ be such that $d(g, h) < \varepsilon$. Then we have

$$\begin{aligned}
&\mu_{g(x)-h(x)}(\varepsilon t)\\
&\geq T_M\Bigg(T_M\bigg(\Phi_{i,j}\left(\frac{x}{2r_i}, -\frac{x}{2r_j}\right)\left(\frac{t}{6}\right), \Phi_{i,j}\left(\frac{x}{2r_i}, 0\right)\left(\frac{t}{6}\right),\\
&\quad \Phi_{i,j}\left(0, -\frac{x}{2r_j}\right)\left(\frac{t}{6}\right)\bigg), T_M\bigg(\Phi_{i,j}\left(\frac{x}{r_i}, \frac{x}{r_j}\right)\left(\frac{t}{3}\right), \Phi_{i,j}\left(\frac{x}{r_i}, 0\right)\left(\frac{t}{3}\right),\\
&\quad \Phi_{i,j}\left(0, \frac{x}{r_j}\right)\left(\frac{t}{3}\right)\bigg)\Bigg)
\end{aligned}$$

for all $x \in X$ and $t > 0$ and so

$$\begin{aligned}
&\mu_{Jg(x)-Jh(x)}(2\alpha\varepsilon t)\\
&\geq T_M\Bigg(T_M\bigg(\Phi_{i,j}\left(\frac{x}{4r_i}, -\frac{x}{4r_j}\right)\left(\frac{\alpha t}{6}\right), \Phi_{i,j}\left(\frac{x}{4r_i}, 0\right)\left(\frac{\alpha t}{6}\right),\\
&\quad \Phi_{i,j}\left(0, -\frac{x}{4r_j}\right)\left(\frac{\alpha t}{6}\right)\bigg), T_M\bigg(\Phi_{i,j}\left(\frac{x}{2r_i}, \frac{x}{2r_j}\right)\left(\frac{\alpha t}{3}\right),\\
&\quad \Phi_{i,j}\left(\frac{x}{2r_i}, 0\right)\left(\frac{\alpha t}{3}\right), \Phi_{i,j}\left(0, \frac{x}{2r_j}\right)\left(\frac{\alpha t}{3}\right)\bigg)\Bigg)\\
&\geq T_M\Bigg(T_M\bigg(\Phi_{i,j}\left(\frac{x}{2r_i}, -\frac{x}{2r_j}\right)\left(\frac{t}{6}\right), \Phi_{i,j}\left(\frac{x}{2r_i}, 0\right)\left(\frac{t}{6}\right),\\
&\quad \Phi_{i,j}\left(0, -\frac{x}{2r_j}\right)\left(\frac{t}{6}\right)\bigg), T_M\bigg(\Phi_{i,j}\left(\frac{x}{r_i}, \frac{x}{r_j}\right)\left(\frac{t}{3}\right), \Phi_{i,j}\left(\frac{x}{r_i}, 0\right)\left(\frac{t}{3}\right),\\
&\quad \Phi_{i,j}\left(0, \frac{x}{r_j}\right)\left(\frac{t}{3}\right)\bigg)\Bigg)
\end{aligned}$$

for all $x \in X$ and $t > 0$. Thus $d(g, h) < \varepsilon$ implies that

$$d(Jg, Jh) = d\left(2g\left(\frac{x}{2}\right), 2h\left(\frac{x}{2}\right)\right) < 2\alpha\varepsilon.$$

This means that

$$d(Jg, Jh) = d\left(2g\left(\frac{x}{2}\right), 2h\left(\frac{x}{2}\right)\right) \leq 2\alpha d(g, h)$$

for all $g, h \in S$. It follows from (16) that

$$d(f, Jf) = d\left(f, 2f\left(\frac{x}{2}\right)\right) \leq \alpha.$$

By Theorem 2, there exists a mapping $\mathrm{EL}: X \to Y$ satisfying the following:

(1) EL is a fixed point of J, that is,

$$\mathrm{EL}\left(\frac{x}{2}\right) = \frac{1}{2}\mathrm{EL}(x) \tag{17}$$

for all $x \in X$.

The mapping EL is a unique fixed point of J in the set

$$\Omega = \{h \in S : d(g, h) < \infty\}.$$

This implies that EL is a unique mapping satisfying (17) such that there exists $u \in (0, \infty)$ satisfying

$$\begin{aligned}
&\mu_{f(x)-\mathrm{EL}(x)}(ut) \\
&\geq T_M\Bigg(T_M\left(\Phi_{i,j}\left(\frac{x}{2r_i}, -\frac{x}{2r_j}\right)\left(\frac{t}{6}\right), \Phi_{i,j}\left(\frac{x}{2r_i}, 0\right)\left(\frac{t}{6}\right),\right. \\
&\quad \left.\Phi_{i,j}\left(0, -\frac{x}{2r_j}\right)\left(\frac{t}{6}\right)\right), T_M\left(\Phi_{i,j}\left(\frac{x}{r_i}, \frac{x}{r_j}\right)\left(\frac{2t}{3}\right), \Phi_{i,j}\left(\frac{x}{r_i}, 0\right)\left(\frac{t}{3}\right),\right. \\
&\quad \left.\Phi_{i,j}\left(0, \frac{x}{r_j}\right)\left(\frac{t}{3}\right)\right)\Bigg)
\end{aligned}$$

for all $x \in X$ and $t > 0$.

(2) $d(J^n f, \mathrm{EL}) \to 0$ as $n \to \infty$. This implies the equality

$$\lim_{n\to\infty} 2^n f\left(\frac{x}{2^n}\right) = \mathrm{EL}(x)$$

for all $x \in X$.

(3) $d(f, \mathrm{EL}) \leq \frac{d(f, Jf)}{1-2\alpha}$ with $f \in \Omega$, which implies the inequality $d(f, \mathrm{EL}) \leq \frac{\alpha}{1-2\alpha}$ and so

$$\begin{aligned}
&\mu_{f(x)-\mathrm{EL}(x)}\left(\frac{\alpha t}{1-2\alpha}\right) \\
&\geq T_M\Bigg(T_M\left(\Phi_{i,j}\left(\frac{x}{2r_i}, -\frac{x}{2r_j}\right)\left(\frac{t}{6}\right), \Phi_{i,j}\left(\frac{x}{2r_i}, 0\right)\left(\frac{t}{6}\right),\right. \\
&\quad \left.\Phi_{i,j}\left(0, -\frac{x}{2r_j}\right)\left(\frac{t}{6}\right)\right), T_M\left(\Phi_{i,j}\left(\frac{x}{r_i}, \frac{x}{r_j}\right)\left(\frac{t}{3}\right), \Phi_{i,j}\left(\frac{x}{r_i}, 0\right)\left(\frac{t}{3}\right),\right. \\
&\quad \left.\Phi_{i,j}\left(0, \frac{x}{r_j}\right)\left(\frac{t}{3}\right)\right)\Bigg)
\end{aligned}$$

for all $x \in X$ and $t > 0$. This implies that the inequality (7) holds.

Replacing x_i by $\frac{x_i}{2^n}$, for all $1 \leq i \leq n$, in (6), we have

$$\begin{aligned}
&\mu_{2^n \sum_{j=1}^m f\left(\frac{-r_j x_j}{2^n}+\sum_{1\leq i\leq m, i\neq j}\frac{r_i x_i}{2^n}\right)+2^{n+1}\sum_{i=1}^m r_i f\left(\frac{x_i}{2^n}\right)-m2^n f\left(\sum_{i=1}^m \frac{r_i x_i}{2^n}\right)}(t) \\
&\geq \Phi\left(\frac{x_1}{2^n}, \dots, \frac{x_m}{2^n}\right)\left(\frac{t}{2^n}\right) \\
&\geq \Phi(x_1, \dots, x_m)\left(\frac{t}{2^n \alpha^n}\right)
\end{aligned}$$

for all $x_1, \dots, x_m \in X$, $t > 0$. Since

$$\lim_{n\to\infty} \Phi(x_1, \dots, x_m)\left(\frac{t}{2^n \alpha^n}\right) = 1$$

for all $x_1, \dots, x_m \in X$ and $t > 0$, we have

$$\mu_{\sum_{j=1}^m \mathrm{EL}\left(-r_j x_j+\sum_{1\leq i\leq m, i\neq j} r_i x_i\right)+2\sum_{i=1}^m r_i \mathrm{EL}(x_i)-m\mathrm{EL}\left(\sum_{i=1}^m r_i x_i\right)}(t) = 1$$

for all $x_1, \dots, x_m \in X$ and $t > 0$.

On the other hand,

$$2\mathrm{EL}\left(\frac{x}{2}\right) - \mathrm{EL}(x) = \lim_{n\to\infty} 2^{n+1} f\left(\frac{x}{2^{n+1}}\right) - \lim_{n\to\infty} 2^n f\left(\frac{x}{2^n}\right) = 0.$$

Thus the mapping $\mathrm{EL} : X \to Y$ is Euler–Lagrange-type additive mapping. This completes the proof.

Corollary 1 *Let X be a real normed space, $\theta \geq 0$, and p be a real number with $p > 1$. Let $f : X \to Y$ be a mapping with $f(0) = 0$ and satisfying*

$$\begin{aligned}
&\mu_{\sum_{j=1}^m f\left(-r_j x_j+\sum_{1\leq i\leq m, i\neq j} r_i x_i\right)+2\sum_{i=1}^m r_i f(x_i)-mf\left(\sum_{i=1}^m r_i x_i\right)}(t) \\
&\geq \frac{t}{t+\theta\left(\sum_{i=1}^m \|x_i\|^p\right)}
\end{aligned} \tag{18}$$

for all $x_1, \dots, x_m \in X$ and $t > 0$. Then, for all $x \in X$, the limit $\mathrm{EL}(x) = \lim_{n\to\infty} 2^n f\left(\frac{x}{2^n}\right)$ exists, and $\mathrm{EL} : X \to Y$ is a unique Euler–Lagrange-type additive mapping such that

$$\begin{aligned}
\mu_{f(x)-\mathrm{EL}(x)}(t) \geq T_M\Bigg(T_M\Bigg(&\frac{|r_i r_j|^p(2^{2p}-2^{p+1})t}{|r_i r_j|^p(2^{2p}-2^{p+1})t+6\theta\|x\|^p(|r_i|^p+|r_j|^p)}, \\
&\frac{|r_i|^p(2^{2p}-2^{p+1})t}{|r_i|^p(2^{2p}-2^{p+1})t+6\theta\|x\|^p}, \frac{|r_j|^p(2^{2p}-2^{p+1})t}{|r_j|^p(2^{2p}-2^{p+1})t+6\theta\|x\|^p}\Bigg),
\end{aligned}$$

$$T_M\left(\frac{|r_ir_j|^p(2^p-2)t}{|r_ir_j|^p(2^p-2)t+3\theta\|x\|^p(|r_i|^p+|r_j|^p)},\right.$$
$$\left.\frac{|r_i|^p(2^p-2)t}{|r_i|^p(2^p-2)t+3\theta\|x\|^p},\frac{|r_j|^p(2^p-2)t}{|r_j|^p(2^p-2)t+3\theta\|x\|^p}\right)\Bigg)$$

for all $x \in X$ and $t > 0$.

Proof The proof follows from Theorem 3 if we take

$$\Phi(x_1,\ldots,x_m)(t)=\frac{t}{t+\theta\left(\sum_{k=1}^m\|x_k\|^p\right)}$$

for all $x_1,\ldots,x_m \in X$ and $t > 0$. In fact, if we choose $\alpha = 2^{-p}$, then we get the desired result.

Corollary 2 *Let X be a real normed space, $\theta \geq 0$, and p be a real number with $p > 1$. Let $f : X \to Y$ be a mapping with $f(0) = 0$ and satisfying (18). Then, for all $x \in X$, the limit $\mathrm{EL}(x) = \lim_{n\to\infty} 2^n f\left(\frac{x}{2^n}\right)$ exists, and $\mathrm{EL} : X \to Y$ is a unique Euler–Lagrange-type additive mapping such that*

$$\mu_{f(x)-\mathrm{EL}(x)}(t)$$
$$\geq T_M\left(T_M\left(\frac{2^p|r_ir_j|^pt}{2^p|r_ir_j|^pt+3\theta\|x\|^p(|r_i|^p+|r_j|^p)},\frac{2^p|r_i|^pt}{2^p|r_i|^pt+3\theta\|x\|^p},\right.\right.$$
$$\left.\frac{2^p|r_j|^pt}{2^p|r_j|^pt+3\theta\|x\|^p}\right),T_M\left(\frac{|r_ir_j|^pt}{|r_ir_j|^pt+3\theta\|x\|^p(|r_i|^p+|r_j|^p)},\right.$$
$$\left.\left.\frac{|r_i|^pt}{|r_i|^pt+3\theta\|x\|^p},\frac{|r_j|^pt}{|r_j|^pt+3\theta\|x\|^p}\right)\right)$$

for all $x \in X$ and $t > 0$.

Proof The proof follows from Theorem 3 if we take

$$\Phi(x_1,\ldots,x_m)(t)=\frac{t}{t+\theta\left(\sum_{k=1}^m\|x_k\|^p\right)}$$

for all $x_1,\ldots,x_m \in X$ and $t > 0$. In fact, if we choose $\alpha = \frac{1}{4}$, then we get the desired result.

Theorem 4 *Let X be a linear space, (Y, μ, T_M) be a complete RN-space, and Φ be a mapping from X^m to D^+ such that for some $0 < \alpha < 2$*

$$\Phi\left(\frac{x_1}{2},\ldots,\frac{x_m}{2}\right)(t)\leq\Phi(x_1,\ldots,x_m)(\alpha t)$$

for all $x_1, \ldots, x_m \in X$ and $t > 0$. Let $f : X \to Y$ be a mapping with $f(0) = 0$ and satisfying (6). Then, for all $x \in X$, the limit $\mathrm{EL}(x) := \lim_{n\to\infty} \frac{f(2^n x)}{2^n}$ exists, and $\mathrm{EL} : X \to Y$ is a unique Euler–Lagrange-type additive mapping such that

$$\begin{aligned}
&\mu_{f(x)-\mathrm{EL}(x)}(t)\\
&\geq T_M\Bigg(T_M\Bigg(\Phi_{i,j}\left(\frac{x}{2r_i}, -\frac{x}{2r_j}\right)\left(\frac{(2-\alpha)t}{6}\right), \Phi_{i,j}\left(\frac{x}{2r_i}, 0\right)\left(\frac{(2-\alpha)t}{6}\right),\\
&\quad \Phi_{i,j}\left(0, -\frac{x}{2r_j}\right)\left(\frac{(2-\alpha)t}{6}\right)\Bigg), T_M\Bigg(\Phi_{i,j}\left(\frac{x}{r_i}, \frac{x}{r_j}\right)\left(\frac{(2-\alpha)t}{3}\right),\\
&\quad \Phi_{i,j}\left(\frac{x}{r_i}, 0\right)\left(\frac{(2-\alpha)t}{3}\right), \Phi_{i,j}\left(0, \frac{x}{r_j}\right)\left(\frac{(2-\alpha)t}{3}\right)\Bigg)\Bigg)
\end{aligned}$$

for all $x \in X$ and $t > 0$.

Proof Let (S, d) be the generalized metric space defined in the proof of Theorem 3. Now, we consider a linear mapping $J : S \to S$ such that

$$Jh(x) := \frac{1}{2}h(2x)$$

for all $x \in X$.

Let $g, h \in S$ be such that $d(g, h) < \varepsilon$. Then we have

$$\begin{aligned}
&\mu_{g(x)-h(x)}(\varepsilon t)\\
&\geq T_M\Bigg(T_M\Bigg(\Phi_{i,j}\left(\frac{x}{2r_i}, -\frac{x}{2r_j}\right)\left(\frac{t}{3}\right), \Phi_{i,j}\left(\frac{x}{2r_i}, 0\right)\left(\frac{t}{3}\right),\\
&\quad \Phi_{i,j}\left(0, -\frac{x}{2r_j}\right)\left(\frac{t}{3}\right)\Bigg), T_M\Bigg(\Phi_{i,j}\left(\frac{x}{r_i}, \frac{x}{r_j}\right)\left(\frac{2t}{3}\right), \Phi_{i,j}\left(\frac{x}{r_i}, 0\right)\left(\frac{2t}{3}\right),\\
&\quad \Phi_{i,j}\left(0, \frac{x}{r_j}\right)\left(\frac{2t}{3}\right)\Bigg)\Bigg)
\end{aligned}$$

for all $x \in X$ and $t > 0$ and so

$$\begin{aligned}
&\mu_{Jg(x)-Jh(x)}\left(\frac{\alpha\varepsilon t}{2}\right)\\
&= \mu_{\frac{1}{2}g(2x)-\frac{1}{2}h(2x)}\left(\frac{\alpha\varepsilon t}{2}\right)\\
&= \mu_{g(2x)-h(2x)}(\alpha\varepsilon t)\\
&\geq T_M\Bigg(T_M\Bigg(\Phi_{i,j}\left(\frac{x}{r_i}, -\frac{x}{r_j}\right)\left(\frac{\alpha t}{3}\right), \Phi_{i,j}\left(\frac{x}{r_i}, 0\right)\left(\frac{\alpha t}{3}\right),
\end{aligned}$$

$$
\begin{aligned}
&\Phi_{i,j}\left(0,-\frac{x}{r_j}\right)\left(\frac{\alpha t}{3}\right)\Bigg), T_M\Bigg(\Phi_{i,j}\left(\frac{2x}{r_i},\frac{2x}{r_j}\right)\left(\frac{2\alpha t}{3}\right),\\
&\Phi_{i,j}\left(\frac{2x}{r_i},0\right)\left(\frac{2\alpha t}{3}\right),\Phi_{i,j}\left(0,\frac{2x}{r_j}\right)\left(\frac{2\alpha t}{3}\right)\Bigg)\Bigg)\\
\geq\ & T_M\Bigg(T_M\Bigg(\Phi_{i,j}\left(\frac{x}{2r_i},-\frac{x}{2r_j}\right)\left(\frac{t}{3}\right),\Phi_{i,j}\left(\frac{x}{2r_i},0\right)\left(\frac{t}{3}\right),\\
&\Phi_{i,j}\left(0,-\frac{x}{2r_j}\right)\left(\frac{t}{3}\right)\Bigg), T_M\Bigg(\Phi_{i,j}\left(\frac{x}{r_i},\frac{x}{r_j}\right)\left(\frac{2t}{3}\right),\Phi_{i,j}\left(\frac{x}{r_i},0\right)\left(\frac{2t}{3}\right),\\
&\Phi_{i,j}\left(0,\frac{x}{r_j}\right)\left(\frac{2t}{3}\right)\Bigg)\Bigg)
\end{aligned}
$$

for all $x \in X$ and $t > 0$. Thus $d(g, h) < \varepsilon$ implies that

$$d(Jg, Jh) = d\left(\frac{g(2x)}{2}, \frac{h(2x)}{2}\right) < \frac{\alpha\varepsilon}{2}.$$

This means that

$$d(Jg, Jh) = d\left(\frac{g(2x)}{2}, \frac{h(2x)}{2}\right) \leq \frac{\alpha}{2} d(g, h)$$

for all $g, h \in S$.

It follows from (15) that for all $x \in X$,

$$
\begin{aligned}
\mu_{\frac{f(2x)}{2}-f(x)}(t) \geq\ & T_M\Bigg(T_M\Bigg(\Phi_{i,j}\left(\frac{x}{2r_i},-\frac{x}{2r_j}\right)\left(\frac{t}{3}\right),\Phi_{i,j}\left(\frac{x}{2r_i},0\right)\left(\frac{t}{3}\right),\\
&\Phi_{i,j}\left(0,-\frac{x}{2r_j}\right)\left(\frac{t}{3}\right)\Bigg), T_M\Bigg(\Phi_{i,j}\left(\frac{x}{r_i},\frac{x}{r_j}\right)\left(\frac{2t}{3}\right),\\
&\Phi_{i,j}\left(\frac{x}{r_i},0\right)\left(\frac{2t}{3}\right),\Phi_{i,j}\left(0,\frac{x}{r_j}\right)\left(\frac{2t}{3}\right)\Bigg)\Bigg)
\end{aligned}
$$

for all $x \in X$. So

$$d(f, Jf) = d\left(f, \frac{f(2x)}{2}\right) \leq 1.$$

By Theorem 2, there exists a mapping $\mathrm{EL} : X \to Y$ satisfying the following:

(1) EL is a fixed point of J, that is,

$$\mathrm{EL}(2x) = 2\mathrm{EL}(x) \tag{19}$$

for all $x \in X$. The mapping EL is a unique fixed point of J in the set

$$\Omega = \{h \in S : d(g, h) < \infty\}.$$

This implies that EL is a unique mapping satisfying (19) such that there exists $u \in (0, \infty)$ satisfying

$$\begin{aligned}
&\mu_{f(x)-\mathrm{EL}(x)}(ut) \\
&\geq T_M\Bigg(T_M\Bigg(\Phi_{i,j}\left(\frac{x}{2r_i}, -\frac{x}{2r_j}\right)\left(\frac{t}{3}\right), \Phi_{i,j}\left(\frac{x}{2r_i}, 0\right)\left(\frac{t}{3}\right), \\
&\quad \Phi_{i,j}\left(0, -\frac{x}{2r_j}\right)\left(\frac{t}{3}\right)\Bigg), T_M\Bigg(\Phi_{i,j}\left(\frac{x}{r_i}, \frac{x}{r_j}\right)\left(\frac{2t}{3}\right), \Phi_{i,j}\left(\frac{x}{r_i}, 0\right)\left(\frac{2t}{3}\right), \\
&\quad \Phi_{i,j}\left(0, \frac{x}{r_j}\right)\left(\frac{2t}{3}\right)\Bigg)\Bigg)
\end{aligned}$$

for all $x \in X$ and $t > 0$.

(2) $d(J^n f, \mathrm{EL}) \to 0$ as $n \to \infty$. This implies the equality

$$\lim_{n\to\infty} \frac{f(2^n x)}{2^n} = \mathrm{EL}(x)$$

for all $x \in X$.

(3) $d(f, \mathrm{EL}) \leq \frac{d(f,Jf)}{1-\frac{\alpha}{2}}$ with $f \in \Omega$, which implies the inequality $d(f, \mathrm{EL}) \leq \frac{2}{2-\alpha}$ and so

$$\begin{aligned}
&\mu_{f(x)-\mathrm{EL}(x)}\left(\frac{2t}{2-\alpha}\right) \\
&\geq T_M\Bigg(T_M\Bigg(\Phi_{i,j}\left(\frac{x}{2r_i}, -\frac{x}{2r_j}\right)\left(\frac{t}{3}\right), \Phi_{i,j}\left(\frac{x}{2r_i}, 0\right)\left(\frac{t}{3}\right), \\
&\quad \Phi_{i,j}\left(0, -\frac{x}{2r_j}\right)\left(\frac{t}{3}\right)\Bigg), T_M\Bigg(\Phi_{i,j}\left(\frac{x}{r_i}, \frac{x}{r_j}\right)\left(\frac{2t}{3}\right), \Phi_{i,j}\left(\frac{x}{r_i}, 0\right)\left(\frac{2t}{3}\right), \\
&\quad \Phi_{i,j}\left(0, \frac{x}{r_j}\right)\left(\frac{2t}{3}\right)\Bigg)\Bigg)
\end{aligned}$$

for all $x \in X$ and $t > 0$. The rest of the proof is similar to the proof of Theorem 3.

Corollary 3 *Let X be a real normed space, $\theta \geq 0$, and p be a real number with $p \in (0, 1)$. Let $f : X \to Y$ be a mapping with $f(0) = 0$ and satisfying (18). Then, for all $x \in X$, the limit $\mathrm{EL}(x) = \lim_{n\to\infty} \frac{f(2^n x)}{2^n}$ exists, and $\mathrm{EL} : X \to Y$ is a unique Euler–Lagrange-type additive mapping such that*

$$
\begin{aligned}
&\mu_{f(x)-\mathrm{EL}(x)}(t)\\
&\geq T_M\Bigg(T_M\bigg(\frac{2^{p-1}|r_ir_j|^p(2-2^p)t}{2^{p-1}|r_ir_j|^p(2-2^p)t+3\theta\|x\|^p(|r_i|^p+|r_j|^p)},\\
&\frac{2^{p-1}|r_i|^p(2-2^p)t}{2^{p-1}|r_i|^p(2-2^p)t+3\theta\|x\|^p},\frac{2^{p-1}|r_j|^p(2-2^p)t}{2^{p-1}|r_j|^p(2-2^p)t+3\theta\|x\|^p}\bigg),\\
&T_M\bigg(\frac{|r_ir_j|^p(2-2^p)t}{|r_ir_j|^p(2-2^p)t+3\theta\|x\|^p(|r_i|^p+|r_j|^p)},\frac{|r_i|^p(2-2^p)t}{|r_i|^p(2-2^p)t+3\theta\|x\|^p},\\
&\frac{|r_j|^p(2-2^p)t}{|r_j|^p(2-2^p)t+3\theta\|x\|^p}\bigg)\Bigg)
\end{aligned}
$$

for all $x \in X$ and $t > 0$.

Proof The proof follows from Theorem 4 if we take

$$\Phi(x_1,\ldots,x_m)(t)=\frac{t}{t+\theta\left(\sum_{k=1}^{m}\|x_k\|^p\right)}$$

for all $x_1,\ldots,x_m \in X$ and $t > 0$. In fact, if we choose $\alpha = 2^p$, then we get the desired result.

Corollary 4 *Let X be a real normed space, $\theta \geq 0$, and p be a real number with $p \in (0,1)$. Let $f : X \to Y$ be a mapping with $f(0) = 0$ and satisfying (18). Then, for all $x \in X$, the limit $\mathrm{EL}(x) = \lim_{n\to\infty}\frac{f(2^n x)}{2^n}$ exists, and $\mathrm{EL} : X \to Y$ is a unique Euler–Lagrange-type additive mapping such that*

$$
\begin{aligned}
&\mu_{f(x)-\mathrm{EL}(x)}(t)\\
&\geq T_M\Bigg(T_M\bigg(\frac{4^p|r_ir_j|^pt}{4^p|r_ir_j|^pt+3.2^{p+1}\theta\|x\|^p(|r_i|^p+|r_j|^p)},\frac{2^{p-1}|r_i|^pt}{2^{p-1}|r_i|^pt+3\theta\|x\|^p},\\
&\frac{2^{p-1}|r_j|^pt}{2^{p-1}|r_j|^pt+3\theta\|x\|^p}\bigg),T_M\bigg(\frac{|r_ir_j|^pt}{|r_ir_j|^pt+3\theta\|x\|^p(|r_i|^p+|r_j|^p)},\\
&\frac{|r_i|^pt}{|r_i|^pt+3\theta\|x\|^p},\frac{|r_j|^pt}{|r_j|^pt+3\theta\|x\|^p}\bigg)\Bigg)
\end{aligned}
$$

for all $x \in X$ and $t > 0$.

Proof The proof follows from Theorem 4 if we take

$$\Phi(x_1,\ldots,x_m)(t)=\frac{t}{t+\theta\left(\sum_{k=1}^{m}\|x_k\|^p\right)}$$

for all $x_1,\ldots,x_m \in X$ and $t > 0$. In fact, if we choose $\alpha = 1$, then we get the desired result.

References

1. Aoki, T.: On the stability of the linear transformation in Banach spaces. J. Math. Soc. Jpn. **2**, 64–66 (1950)
2. Cholewa, P.W.: Remarks on the stability of functional equations. Aequationes Math. **27**, 76–86 (1984)
3. Czerwik, S.: Functional Equations and Inequalities in Several Variables. World Scientific, River Edge (2002)
4. Găvruta, P.: A generalization of the Hyers-Ulam-Rassias stability of approximately additive mappings. J. Math. Anal. Appl. **184**, 431–436 (1994)
5. Gordji, M.E., Khodaei, H.: Stability of Functional Equations. Lap Lambert Academic Publishing, Tehran (2010)
6. Gordji, M.E., Savadkouhi, M.B.: Stability of mixed type cubic and quartic functional equations in random normed spaces. J. Inequal. Appl. **2009**, 9 pp. (2009). Article ID 527462
7. Gordji, M.E., Zolfaghari, S., Rassias, J.M., Savadkouhi, M.B.: Solution and stability of a mixed type cubic and quartic functional equation in quasi-Banach spaces. Abstr. Appl. Anal. **2009**, 14 pp. (2009). Article ID 417473
8. Gordji, M.E., Savadkouhi, M.B., Park, C.: Quadratic-quartic functional equations in RN-spaces. J. Inequal. Appl. **2009**, 14 pp. (2009). Article ID 868423
9. Hyers, D.H.: On the stability of the linear functional equation. Proc. Natl. Acad. Sci. U. S. A. **27**, 222–224 (1941)
10. Jung, S.M.: Hyers-Ulam-Rassias stability of Jensen's equation and its application. Proc. Am. Math. Soc. **126**, 3137–3143 (1998)
11. Jung, S.-M., Rassias, M.Th.: A linear functional equation of third order associated to the Fibonacci numbers. Abstr. Appl. Anal. **2014**, 7 pp. (2014). Article ID 137468
12. Jung, S.-M., Popa, D., Rassias, M.Th.: On the stability of the linear functional equation in a single variable on complete metric groups. J. Glob. Optim. **59**, 165–171 (2014)
13. Jung, S.-M., Rassias, M.Th., Mortici, C.: On a functional equation of trigonometric type. Appl. Math. Comput. **252**, 294–303 (2015)
14. Kenary, H.A., Cho, Y.J.: Stability of mixed additive-quadratic Jensen type functional equation in various spaces. Comput. Math. Appl. **61**, 2704–2724 (2011)
15. Mihet, D., Radu, V.: On the stability of the additive Cauchy functional equation in random normed spaces. J. Math. Anal. Appl. **343**, 567–572 (2008)
16. Park, C.: Generalized Hyers-Ulam-Rassias stability of n-sesquilinear-quadratic mappings on Banach modules over C^*-algebras. J. Comput. Appl. Math. **180**, 279–291 (2005)
17. Park, C., Hou, J., Oh, S.: Homomorphisms between JC -algebras and Lie C^*-algebras. Acta Math. Sin. (Engl. Ser.) **21**(6), 1391–1398 (2005)
18. Rassias, Th.M.: On the stability of the linear mapping in Banach spaces. Proc. Am. Math. Soc. **72**, 297–300 (1978)
19. Rassias, Th.M.: On the stability of functional equations and a problem of Ulam. Acta Appl. Math. **62**(1), 23–130 (2000)
20. Saadati, R., Park, C.: Non-Archimedean $\mathscr{L}$-fuzzy normed spaces and stability of functional equations. Comput. Math. Appl. **60**, 2488–2496 (2010)
21. Saadati, R., Vaezpour, M., Cho, Y.J.: A note to paper "On the stability of cubic mappings and quartic mappings in random normed spaces". J. Inequal. Appl. **2009** (2009). Article ID 214530. https://doi.org/10.1155/2009/214530
22. Saadati, R., Zohdi, M.M., Vaezpour, S.M.: Nonlinear L-random stability of an ACQ functional equation. J. Inequal. Appl. **2011**, 23 pp. (2011). Article ID 194394. https://doi.org/10.1155/2011/194394
23. Schewizer, B., Sklar, A.: Probabilistic Metric Spaces. North-Holland Series in Probability and Applied Mathematics. North-Holland, New York (1983)
24. Skof, F.: Local properties and approximation of operators. Rend. Sem. Mat. Fis. Milano **53**, 113–129 (1983)
25. Ulam, S.M.: Problems in Modern Mathematics, Science Editions. Wiley, New York (1964)

Discrete Mathematics for Statistical and Probability Problems

Christos P. Kitsos and Thomas L. Toulias

Abstract This paper offers a compact presentation of the solid involvement of Discrete Mathematics in various fields of Statistics and Probability Theory. As far as the discrete methodologies in Statistics are concerned, our interest is focused on the foundations and applications of the Experimental Design Theory. The set-theoretic approach of the foundations of Probability Theory is also presented, while the notions of concepts and fuzzy logic are formulated and discussed.

Introduction

The aim of this paper is to provide a Discrete Mathematics point of view of some statistical applications. Two are our main lines of thought: Design Theory and statistical distances. The Design Theory attracts interest from the group theory and projective geometry. Design Theory is discussed, while some emphasis is given to the Latin squares. We also recall the theory of ideals and provide some aspects from the probability theory that, we believe, deserves more attention.

The notion of distance is fundamental in Statistics. In mathematical analysis, especially in metric spaces, the distance serves as a criterion to check the convergence of a sequence, while a sequence in Statistics (with typical example being the Robbins-Monro iterative scheme) is asked to converge *in distribution*, which is usually the normal distribution; see [13] for details.

The reason is that such sequences can provide maximum likelihood estimators (MLE), being within the classical statistical framework, while other methods might not.

The notion of "concept" associated with the "objects" and "attributes" is introduced, from which the idea of a set-theoretic distance, in a Discrete Mathematics sense, is emerged.

C. P. Kitsos · T. L. Toulias (✉)
Technological Educational Institute of Athens, Athens, Greece
e-mail: xkitsos@teiath.gr

N. J. Daras, Th. M. Rassias (eds.), *Modern Discrete Mathematics and Analysis*, Springer Optimization and Its Applications 131,
https://doi.org/10.1007/978-3-319-74325-7_12

Geometric methods, and therefore distance metrics methods, are adopted in various problems in Statistics. In optimal Experimental Design Theory for the continuous case, geometric methodologies are considered on the induced design space and the relative geometrical aspects have been discussed by Kitsos et al. [20]. For the discrete case, the geometrical approach is tackled in a compact form in this paper.

In principle, the usual (geometrical) distance metric in Statistics is considered to be the Euclidean distance, based on the ℓ_2 norm, but this is not the case for Discrete Mathematics; see section "Finite Geometry and Design Theory." In section "Discrete Mathematics and Statistics," the relation between Discrete Mathematics and Statistics is developed. The Experimental Design Theory is also discussed, especially the Latin squares. Moreover, a finite geometry approach is also developed in a compact form. In section "Discrete Mathematics and Probability," the applications of Discrete Mathematics to Probability is presented, while in section "Discrete Distance Measures," certain distance measures are discussed.

Discrete Mathematics and Statistics

Introduction

Discrete Mathematics offers a strong background to statistical problems, especially to the Design Theory. We shall trace some of these applications, bringing practice with theory. Consider the practical problem where a manufacturer is developing a new product. Let us assume that he/she wishes to evaluate v varieties (of the product) and asks a number of consumers to test them. However, it seems impossible in practice to ask each consumer to test all the varieties. Therefore, two lines of thought might be adopted:

1. Each consumer tests the same number of varieties, say $k < v$.
2. Each variety should be tested by r consumers.

The above problem gives rise to the following generalization: Let X be any set of size v, i.e., $v := |X|$. We say that a set $\mathscr{B}$ of k-subsets of X is a *design*, denoted by $D(v, k, r)$ with parameters $v, k, r \in \mathbb{N}^* := \mathbb{N} \setminus \{0\}$, when each member $x \in X$ belongs to exactly $r \leq v$ sets of $\mathscr{B}$. In Design Theory, a subset $B \in \mathscr{B}$ is called a *block* of the design under investigation.

Suppose now that C denotes any set of k-subsets of X with $v := |X|$. In general, we say that the marks (i.e., the readings of an experiment) are members of the set

$$\mathfrak{C} := \{(x, C)\}_{x \in C}. \tag{1}$$

What in Statistics is known as a *replication* of a value x is the row total $r(x) := \#\big(\{C : \ x \text{ occurs in } C\}\big)$. The column total is k in all the cases by the definition

Table 1 The table of Example 1 with $k = 3$, $|C| = 4$

x	C_1	C_2	C_3	C_4	$r(x)$
1	✓	✓		✓	3
2	✓		✓		3
3		✓			1
4	✓		✓	✓	3
5			✓		1
6		✓		✓	2
k	3	3	3	3	12

of C. Therefore, it is easy to see that

$$\sum_{x \in X} r(x) = k|C|. \tag{2}$$

Example 1 The above discussion can be visualized with Table 1 where $X := \{1, 2, \ldots, 6\}$. Notice that $\sum r(x) = 12$.

In principle, when we are working on a statistical design $D(v, k, r)$ then $r(x) := r$ and as we are working with blocks B, $b := |B|$, relation (2) is then reduced to $vr = bk$, and hence

$$b = \frac{vr}{k} \leq \binom{v}{k}. \tag{3}$$

In general, it can be proved that there is a design $D(v, k, r)$ if and only if (iff) $k \mid vr$,

$$\frac{vr}{k} \leq \binom{v}{k}.$$

The condition that each object (i.e., element of X) belongs to the same number of blocks can be strengthened. In particular, it can be required that a pair of objects or, even more, that t objects are taken at a time: this is known as *t-design*, $t \in \mathbb{Z}^+$.

Let X be a set with $v := |X|$. Then a set B of k-subsets of X is said to be a *t-design*, denoted with $D(v, k, r_t)$, iff for each t-subset T of X, the number of blocks which contain T is constant r_t. It is interesting to notice that

i. If B is a t-design, it is also an s-design, $1 \leq s \leq t - 1$, $s \in \mathbb{Z}^+$.
ii. If B is a $D(v, k, r_t)$ t-design, it is then also a $D(v, k, r_s)$ s-design, with

$$r_s = r_t \frac{(v - s)(v - s - 1) \cdots (v - t + 1)}{(k - s)(k - s - 1) \cdots (k - t + 1)}. \tag{4}$$

iii. For $0 \leq s \leq t - 1$, it is required that

$$(k - s)(k - s - 1) \cdots (k - t + 1) \mid r_t(v - s)(v - s - 1) \cdots (v - t + 1).$$

iv. A recursive formula holds:

Table 2 A 4 × 4 Latin square L_1

A	B	C	D
B	A	D	C
C	D	A	B
D	C	B	A

$$r_{t-1} = r_t \frac{v - t + 1}{k - t + 1}, \quad t \in \mathbb{N}.$$

v. If $b = |B|$, then

$$b = r_0 = r_1 \frac{v}{k}.$$

Usually a 2-design with $k = 3$ and $r_2 = 1$ is called as *Steiner Triple System* (STS). Such a system can be seen by considering two words, say u_1 and u_2, both of length n in the alphabet $\{0, 1\}$. Let u_1+u_2 denote the word obtained adding the corresponding digits of u_1 and u_2. Then, if we consider X to be the set of all such words with the exception of $00\ldots0$, the set of all three subsets of X, formed by $\{u_1, u_2, u_1 + u_2\}$, is a 2-design such as $D(2^{n-1}, 3, 1)$ which is an STS design with $2^n - 1$ varieties.

Latin Squares

In Statistics, and especially in Experimental Design Theory, the Latin squares (LS), as proposed by Fisher in [5], play an important role; see [2, 23, 24] and [3] among others. The traditional example is the following: Suppose we have to plan an agriculture experiment in which four new kinds of fertilizers, say A, B, C, and D, are to be tested in a square field. The "scheme" has to be as in Table 2 in order to describe an LS.

In principle, an LS of order n is an $n \times n$ array in which each of the n elements occurs once in each row and once in each column. The statistical term LS comes from the fact that R.A. Fisher used "Latin letters" to cover the "square" field in his agricultural experiments.

We shall denote with $L(i, j)$ the (i, j) entry of an LS L, while the labels for the rows and columns of L as well as for all its (i, j) elements shall be considered as $\mathbb{Z}_m$-valued, with $\mathbb{Z}_m$ being the set of nonnegative integers modulo m.

Given two Latin squares of the same order n, say L_1 and L_2, the relation for the orthogonality between L_1 and L_2 can be defined. Indeed, $L_1 \perp L_2$ iff for every $k, l \in \mathbb{Z}_m$ there are just two $i, j \in \mathbb{Z}_m$ for which

$$L_1(i, j) = k, \text{ and } L_2(i, j) = l.$$

Following this terminology, Euler in 1786 was the first who suggests constructing an orthogonal pair of arrays of a six-order LS. He was unable to solve this problem. It is now known that no such pair exists. Actually, the problem that Euler suggested was

> Given 36 officers of 6 different ranks from 6 different regiments, can they be arranged in a square in such a way that each row and each column contains one officer of each rank and one officer from each regiment?

What to admire? The physical problem itself or the underlined mathematical insight which tries to tackle the problem? We shall come to this problem later.

Despite Euler's problem that has no solution, there is a way of constructing orthogonal LS. Theorem 2 below offers a method of constructing orthogonal LS (OLS) based on properties of $\mathbb{Z}_p$ with p being a prime.

Theorem 1 *For each $v \geq 2$, the $v \times v$ array defined by $L(i, j) = i + j$, $i, j \in \mathbb{Z}_v$ is an LS.*

See Appendix for the proof.

Theorem 2 *Let p be a prime and $0 \neq a \in \mathbb{Z}_p$ given. Then, the rule*

$$L_a(i, j) = ai + j, \quad i, j \in \mathbb{Z}_p, \tag{5}$$

defines an LS. Furthermore, for given $b \neq a$, $b \in \mathbb{Z}_p$, it holds that

$$L_b \perp L_a. \tag{6}$$

See Appendix for the proof.

Example 2 The LS, say L_2, of Table 3 below is orthogonal to LS L_1 of Table 2, as the 16 pairs $(A, A), (A, B) \ldots, (D, D)$ occur in one of the 16 positions, i.e., $L_2 \perp L_1$ by the LS orthogonality definition.

Based on Theorem 2, we say that we can obtain a set of $p-1$ *mutually orthogonal LS* (MOLS) of order p (MOLS$_p$) for each prime p.

Example 3 For $p = 3$ we get two MOLS$_3$, i.e.,

$$L_1 : \begin{matrix} 0 & 1 & 2 \\ 1 & 2 & 0 \\ 2 & 0 & 1 \end{matrix} \quad \text{and} \quad L_2 : \begin{matrix} 0 & 1 & 2 \\ 2 & 0 & 1 \\ 1 & 2 & 0 \end{matrix}$$

Table 3 A 4×4 Latin square L_2

A	B	C	D
C	D	A	B
D	C	B	A
B	A	D	C

So far, we discussed that for a prime p, using the properties of the field $\mathbb{Z}_p$, it is possible to construct a set of $p-1$ MOLS_p. The same result holds if we replace p with a prime power $q := p^r$, $r \in \mathbb{Z}^+$. Indeed:

Theorem 3 *For q being a prime r-power of p, it is possible to construct $q-1$ mutually orthogonal Latin squares of order q (MOLS_q).*

Proof Apply Theorem 2 where $a \in \mathbb{Z}_p$ is now replaced by $a \in \mathbb{F}_q$, with $q-1$ nonzero and $\mathbb{F}_q$ being a finite field in place of $\mathbb{Z}_p$.

In practice, given any field with n elements, we would like to construct $n-1$ MOLS. Due to Theorem 3 the arisen question is:

Question Is it possible to construct $n-1$ mutually orthogonal Latin squares of order n (MOLS_n) when n is not a prime power?

Answer In Discrete Mathematics and Statistics, this is one of the most well-known unsolved problems. For the case of $n = 6$, it is known already that there is not a set of 5 MOLS_6 (recall Euler's problem mentioned earlier which is unsolved).

Another approach to Design Theory, under the context of Discrete Mathematics, is through geometry and particular through finite geometry on projective planes. Recall that there are two main approaches of finite plane geometry: affine and projective. In an affine plane, the normal sense of parallel lines is valid. In a projective plane, by contrast, any two lines intersect at a unique point, so parallel lines do not exist. The finite affine plane geometry and finite projective plane geometry can be described by simple axioms. An affine plane geometry is a nonempty set X (whose elements are called "points"), along with a nonempty collection L of subsets of X (whose elements are called "lines"), such that:

(a) For every two distinct points, there is exactly one line that contains both points.
(b) Given a line ℓ and a point p outside ℓ, there exists exactly one line ℓ' containing p such that $\ell \cap \ell' = \varnothing$ (Playfair's axiom).
(c) There exists a set of four points, no three of which belong to the same line.

The last axiom ensures that the geometry is not trivial (either empty or too simple to be of interest, such as a single line with an arbitrary number of points on it), while the first two specify the nature of the geometry. Note that the simplest affine plane contains only four points; it is called the *affine plane of order* 2, where the order of an affine plane is the number of points on any line. Since no three are collinear, any pair of points determines a unique line, and so this plane (of four points) contains six lines. It corresponds to a tetrahedron where nonintersecting edges are considered "parallel" or a square where not only opposite sides but also diagonals are considered "parallel"; see Fig. 1.

In optimal Experimental Design Theory, a geometry is constructed via the induced design space; see [12, 27] and [18] among others. In this paper, the geometric approach is realized in the following through the finite field $\mathbb{F}_q$.

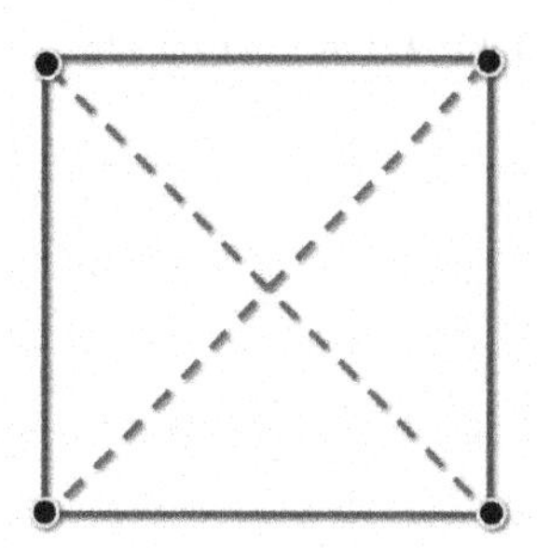

Fig. 1 A finite affine plane of order 2. The two diagonal lines do not intersect

Finite Geometry and Design Theory

Let $x, y \in \mathbb{F}_q$, where $\mathbb{F}_q$ being a finite field. Then, the "coordinate" or analytic plane geometry for $(x, y) \in \mathbb{R}^2$ is still valid for the elements of $\mathbb{F}_q$. As all the algebraic "manipulations" hold in $\mathbb{R}^2$, the sense of "line" and "plane" for a finite number of "points" $x, y \in \mathbb{F}_q$ is also parent in $\mathbb{F}_q$. In particular, the lines in a 2-design are the blocks on the set points; see Theorem 4 below. Thus, a line satisfies the analytic expression $ax + by + c = 0$ where $x, y, a, b, c \in \mathbb{F}_q$ with $a^2 + b^2 \neq 0$.

Theorem 4 *Consider a finite field $\mathbb{F}_q$ equipped with lines and planes as above. Then, the lines of $\mathbb{F}_q$ are the blocks of a 2-design $D(v, k, r_2)$ of the set of points of $\mathbb{F}_q$. In particular, the design is $D\big(q^2, q, 1\big)$.*

See Appendix for the proof.

The $D\big(q^2, q, 1\big)$ design, described in Theorem 4, is usually known as the *affine plane over* $\mathbb{F}_q$ (see also the proof in Appendix). Recall the point (iv) in sub-Section 2.1. For the 2-design above, i.e., for $t := 2$, it is

$$r_1 = r_2 \frac{v - 2 + 1}{k - 2 + 1} = r_2 \frac{v - 1}{k - 1} = 1 \times \frac{q^2 - 1}{q - 1} = q + 1. \tag{7}$$

According to property (v), as in Section 3.2, it is

$$b = r_0 = r_1 \frac{v}{k} = (q + 1) \frac{q^2}{q} = q(q + 1). \tag{8}$$

Example 4 Let $q = 3$, i.e., $\mathbb{F}_3 \equiv \mathbb{Z}_3$ is under consideration. There are $v = q^2 = 9$ points and $k = q = 3$. Those nine points, say P_i, $i = 1, 2, \ldots, 9$, are

$$\begin{array}{lll} P_1 = (0, 0), & P_2 = (0, 1), & P_3 = (0, 2), \\ P_4 = (1, 0), & P_5 = (1, 1), & P_6 = (1, 2), \\ P_7 = (2, 0), & P_8 = (2, 1), & P_9 = (2, 2). \end{array}$$

The $b = r_0 = q(q + 1) = 12$ lines, say ℓ_i, $i = 1, 2, \ldots, 12$, are presented in Table 4:

Table 4 Lines of Example 4

Line	Equation	Points
ℓ_1	$x = 0$	P_1, P_2, P_3
ℓ_2	$x = 1$	P_4, P_5, P_6
ℓ_3	$x = 2$	P_7, P_8, P_9
ℓ_4	$y = 0$	P_1, P_4, P_7
ℓ_5	$y = 1$	P_2, P_5, P_8
ℓ_6	$y = 2$	P_3, P_6, P_9
ℓ_7	$x + y = 0$	P_1, P_6, P_8
ℓ_8	$x + y = 1$	P_2, P_4, P_9
ℓ_9	$x + y = 2$	P_3, P_5, P_7
ℓ_{10}	$2x + y = 0$	P_1, P_5, P_9
ℓ_{11}	$2x + y = 1$	P_2, P_6, P_7
ℓ_{12}	$2x + y = 2$	P_3, P_4, P_8

Notice that there are four classes with three lines as

$$\big\{(\ell_1, \ell_2, \ell_3), (\ell_4, \ell_5, \ell_6), (\ell_7, \ell_8, \ell_9), (\ell_{10}, \ell_{11}, \ell_{12})\big\}.$$

Each class of parallel lines has no intersection (common point) between them, while when there is an intersection, one common point exists (as in the Euclidean case of $\mathbb{R}^2$). However, if we adopt the projective geometry's approach, i.e., assume that every two lines have always one common point, we are in a finite version of projective geometry [4], and its relation with the Design Theory. Considering a prime power, Theorem 5 holds where the projective plane property over $\mathbb{F}_q$ is demonstrated in comparison with the affine plane over $\mathbb{F}_q$, according to Theorem 4.

Theorem 5 *For any prime power q, there is a 2-design $D(v, k, r_2) = D\big(q^2 + q + 1, q + 1, 1\big)$. This particular design has the additional property that any two blocks have just one member in common.*

See Appendix for the proof.

Calculating r_1 and $r_0 = b$, due to the relations (iv) and (v) in Section 3.2, it holds

$$r_1 = \frac{v-1}{k-1} r_2 = \frac{\big(q^2 + q + 1\big) - 1}{(q+1) - 1} - 1 = q + 1, \tag{9a}$$

$$b = r_0 = \left(\frac{v}{k}\right) r_1 = q^2 + q + 1. \tag{9b}$$

See the similarity between (7)–(9b) and (9a)–(9b). Moreover, we can notice that:

- There are $q^2 + q + 1$ points and $q^2 + q + 1$ lines.
- Each line contains $q + 1$ points and each point belongs to $q + 1$ lines.
- any two points belong to one common line and any two lines have one common point.

Example 5 Consider the affine plane over $\mathbb{F}_3$, as in Example 4. We need the parallel lines of $\mathbb{F}_3$ to meet, so we add to each line a new (arbitrary) point, which corresponds to the projective geometry's point at infinity or *infinity point*. In particular,

$$\begin{aligned}
\ell_i' &:= \ell_i \cup \{X_1\}, \quad i = 1, 2, 3,\\
\ell_i' &:= \ell_i \cup \{X_2\}, \quad i = 4, 5, 6,\\
\ell_i' &:= \ell_i \cup \{X_3\}, \quad i = 7, 8, 9,\\
\ell_i' &:= \ell_i \cup \{X_4\}, \quad i = 10, 11, 12.
\end{aligned}$$

A new line $\ell_\infty \supseteq \{X_1, X_2, X_3, X_4\}$, containing the newly introduced points X_i, $i = 1, 2, 3, 4$, is then introduced and called as the *line at infinity* or *infinity-line*. Therefore, the projective plane over $\mathbb{F}_3$ has in total 13 lines, i.e., $\ell_i', i = 1, 2, \ldots, 12$, and ℓ_∞, and 13 points, i.e., the given P_i, $i = 1, 2, \ldots, 9$, and X_j, $j = 1, 2, 3, 4$. Each line contains four points, i.e., each block contains four elements, and each pair of points belongs exactly to one line. Hence, the 2-design $D(13, 4, 1)$ is obtained.

Applications of Experimental Design

In practice, a *complete randomized block design* (CRBD) is analyzed as a two-way ANalysis Of VAriance (ANOVA); see the pioneering work of [26]. The *incomplete balanced* needs a special ANOVA to analyze the collected data while an *incomplete general block design* is analyzed through regression analysis; see [8] among others. A (complete) Latin square is analyzed through a three-way ANOVA in industry. That is, ANOVA and regression analysis are adopted to analyze real data with the assistance of an appropriate software; see [15] among others.

The theoretical insight of experimental design provides food for thought for different branches of mathematics. We tried to present some of them in a compact form. The experimenter faces often the need of a strong mathematical background when analyzing a 2^n *factorial experiment*, defined as in [30], and especially for a portion of it. When we are talking about a *confounded experiment*, one may consider the number-theoretic Kempthorne method [11]; see [9] among others, for a number-theoretic development. A "traditional" example is the following:

Example 6 Construct two blocks in a 2^3 factorial experiment, so that the term ABC is confounded. We then have the linear combination $L = X_1 + X_2 + X_3$ with the value of L is evaluated as follows:

$$\begin{aligned}
(1) &= \mathrm{A}^0\mathrm{B}^0\mathrm{C}^0,\ L = 0 + 0 + 0 = 0,\\
a &= \mathrm{A}^1\mathrm{B}^0\mathrm{C}^0,\ L = 1 + 0 + 0 = 1,\\
b &= \mathrm{A}^1\mathrm{B}^1\mathrm{C}^0,\ L = 1,\\
ab &= \mathrm{A}^1\mathrm{B}^1\mathrm{C}^0,\ L = 2 = 0\ (\mathrm{mod} 2),
\end{aligned}$$

$$\begin{aligned} c &= \mathrm{A}^0\mathrm{B}^0\mathrm{C}^1,\ L = 1, \\ ac &= \mathrm{A}^1\mathrm{B}^0\mathrm{C}^1,\ L = 2 = 0\ (\text{mod}2), \\ bc &= \mathrm{A}^0\mathrm{B}^1\mathrm{C}^1,\ L = 2 = 0\ (\text{mod}2), \\ abc &= \mathrm{A}^1\mathrm{B}^1\mathrm{C}^1,\ L = 3 = 1\ (\text{mod}2). \end{aligned}$$

So, for

$$L = 0 \ \text{ the block is } \ (1), ab, ac, bc,$$

$$L = 1 \ \text{ the block is } \ (1), a, b, c, abc.$$

Therefore, if we decide to apply a $\frac{1}{2}2^3 = 2^{3-1}$ experiment, i.e., a half 2^3 factorial experiment, we have to choose one of the two blocks as above.

Different rules have been proposed to overpass confounding. For Fisher's multiple confounding rule, see [23]. The violation of the structure of a 2^n factorial design, by adding center points, dominates EVolutionary OPeration (EVOP), [1]. Then we moved to a new "model" by adding more points, i.e., $2^n +$ center $+$ "star", to study response surface exploration; see [25] among others.

The nonlinear (optimal) experimental design, as it was studied by Kitsos [12], has no relation with the elegant mathematical approach of Fisher. The nonlinear Design Theory suffers from parameter dependence [12], and the practical solution is to adopt sequential design; see [6]. The induced design space offers the possibility of a geometrical consideration, [18, 20]. The compromise of a quasi-sequential approach [14] was also proposed, while some technics based on polynomial root-finding, for nonlinear problems, were studied in [28].

This section offers a quick attempt to complete the development of the experimental design topic, the backbone of Statistics.

Discrete Mathematics and Probability

As we already discussed in section "Discrete Mathematics and Statistics," discrete mathematical ideas appear to have an aesthetic appeal in Statistics. Especially during the Fisher's and Kolmogorov's era, there was an attempt to develop a theoretical discrete framework to shed a new light in statistical problems. In this section, we show influence of Discrete Mathematics in other research areas related to Statistics. The measure-theoretic approach of Kolmogorov had a competitor from algebra, i.e., probability algebra, which was never successfully formulated. The measure theory context of probability was accepted completely. But, still, the algebraic definition of probability space deserves some further attention, especially when the foundation needs some discrete treatment.

Probability algebra (PA), as introduced in [10], is defined to be the pair (α, P), with $\alpha \neq \varnothing$ where its set elements $A, B, C, \ldots \in \alpha$ are called *events*. P is a real function on α and is called *probability*. The binary operations $A \vee B$ and $A \wedge B$ and the unitary operation A^{c} (not in A) equip α with the algebraic structure of a Boolean algebra. For the probability P we assume that:

i. P is positive definite, i.e.,

$$\mathrm{P}(A) \geq 0 \text{ for every } A \in \alpha, \text{ and } \mathrm{P}(A) = 0 \Longleftrightarrow A = \varnothing \in \alpha.$$

ii. P is normed i.e.,

$$\mathrm{P}(E) = 1 \text{ where } E \in \alpha \text{ is the unitary element.}$$

iii. P is additive, i.e.,

$$\mathrm{P}(A \vee B) = \mathrm{P}(A) + \mathrm{P}(B) \text{ when } A \wedge B = \varnothing.$$

Let β be a Boolean sub-algebra of α. Then the restriction of the function P to β is probability on β. If $\alpha := \{\varnothing, E\}$ with $\mathrm{P}(E) = 1$, $\mathrm{P}(\varnothing) = 0$, the probability algebra (α, P) is called *improper*.

The terms probability space and Borel probability field, introduced by Kolmogorov in his pioneering work [22], are also constructed through the algebraic approach, while the distribution function was also well defined; see [10, Theorem 5.4].

For given probability algebras (α_1, P_1) and (α_2, P_2) with α_i, $i = 1, 2$, Boolean algebras, consider the isomorphism

$$\varphi : \alpha_1 \longrightarrow \alpha_2, \text{ where } A \longmapsto \varphi(A).$$

Then, we say that the two probability algebras are *isometric* iff

$$P_1(A) = P_2\big(\varphi(A)\big), \quad A \in \alpha_1.$$

Example 7 Let $A := \{\alpha_1, \alpha_2, \ldots, \alpha_n\}$, $n \geq 2$. We define α_n to be the class of all subsets of A forming a Boolean algebra. Let P_i, $i = 1, 2, \ldots, n$, $0 < \mathrm{P}_i < 1$ with $\sum_i \mathrm{P}_i = 1$ be associated with α_i, $i = 1, 2, \ldots, n$. For every subset of A of the form $\{\alpha_{\ell_1}, \alpha_{\ell_2}, \ldots, \alpha_{\ell_k}\}$, we define the probability P as follows:

$$\mathrm{P}\Big(\big\{\alpha_{\ell_1}, \alpha_{\ell_2}, \ldots, \alpha_{\ell_n}\big\}\Big) := \mathrm{P}_{\ell_1} + \mathrm{P}_{\ell_2} + \cdots + \mathrm{P}_{\ell_k}.$$

Then (α_n, P) is a probability algebra with 2^n elements, provided that $\mathrm{P}(\varnothing) = 0$.

With the above, we tried to provide some elements of the algebraic foundations of probability. Problems such us convergence in stochastic spaces, expectations of random variables, moments, etc. can be defined appropriately this algebraic approach to probability, [10]. The multivariate problem, the sequential line of thought, and other statistical fields have not been tackled yet.

Algebraic Approach to Concept

We introduce now the term *concept* through lattice theory. Recall that a *lattice* is an abstract order structure. It consists of a partially ordered set in which every two elements have a unique supremum (also called a *least upper bound* or *join*) and a unique infimum (also called a *greatest lower bound* or *meet*). An example is given by the natural numbers, partially ordered by divisibility, for which the unique supremum is the least common multiple and the unique infimum is the greatest common divisor.

The main question, from a statistical point of view, and not only, might be: why lattice? When the study of hierarchies is one of the target of the research, the hierarchy of concept can be proved dominant, using *subconcepts* and *superconcepts*. A typical example of a subconcept is "human with a disease" where the superconcept is "healthy human being" which is a subconcept of the superconcept "being." The concept has to be determined from all the objects belonging to the concept under consideration as well as from all attributes necessarily valid for the defined objects. Usually it is not expected the experimenter to consider all objects and attributes describing the given concept, i.e., a certain amount of sets of objects and attributes is initially fixed.

A *concept* is every set function ϕ of a set O, called the *objects*, to another set A, called the *attributes*. We shall use the notation (O, A).

The above definition of concept provides the mathematical insight of the expression: the "object" O has "attributes" A. Let now all the objects under consideration form the set $\mathfrak{O}$, which has a finite number of elements. Similarly, all the attributes form the set $\mathfrak{A}$ which also has a finite number of elements. We emphasize that $\phi(O)$ does not define a unique set A while, equivalently, $\phi^{-1}(A)$ does not define a unique set O. Based on the collected data, A, O and φ can be defined appropriately.

The data we examine (e.g., any qualitative attributes, "yes" or "no" to a given disease, to human or animals, or the level of quality of an industrial product) act as a generator ϕ of concepts:

$$\phi : \mathfrak{O} \longmapsto \mathfrak{A}, \quad \varphi(O) = A. \tag{10}$$

From the above discussion, the set $\mathfrak{O}$ represents a set of "humans" or "animals," and O can represent "humans with a disease" and $\varphi(O) = A = \{0, 1\}$ with $0 :=$ "no" and $1 :=$ "yes".

When we are interested to create a new concept, we must consider the simple laws of set theory.

The *concept union* $\dot{\cup}$ of two concepts is a new concept of the form $(O_1, A_1) \ \dot{\cup} \ (O_2, A_2) := (O_1 \cup O_2, A_1 \cap A_2)$, $(O_1, A_1), (O_2, A_2) \in \mathfrak{C}$. The *concept intersection* $\dot{\cap}$ is defined, respectively, as $(O_1, A_1) \ \dot{\cap} \ (O_2, A_2) := (O_1 \cap O_2, A_1 \cup A_2)$, $(O_1, A_1), (O_2, A_2) \in \mathfrak{C}$; see [17].

It is easy to verify that the concept $(\varnothing, A)$ is the neutral (zero) element for the union in the sense that $(O, A) \ \dot{\cup} \ (\varnothing, \mathfrak{A}) = (O, A)$.

The definition of the union between two concepts is not only mathematically valid but also practical, as you can assign to the empty set any attribute. The neutral element for the intersection $\dot{\cap}$ is the element $(\mathfrak{O}, \varnothing)$. It can be proved that the set of all concepts with operation either $\dot{\cup}$ or $\dot{\cap}$ is a commutative Abelian semigroup with the appropriate neutral element. The set of all concepts $\mathfrak{C}$ it can be proved, as far as the union and intersection of concepts are concerned, to be commutative and associative and, therefore, a lattice.

Two concepts (O_1, A_1) and (O_2, A_2) belonging to $\mathfrak{C}$ are *equivalent* if and only if $A_1 = A_2$. We shall write $(O_1, A_1) \cong (O_2, A_2) \stackrel{\text{def.}}{\Longleftrightarrow} A_1 = A_2$.

Proposition 1 *The equivalence between two concepts is a genuine equivalence relation among concepts. Therefore, we can create a partition of the concepts (equivalent with each other) coming from the collected data* $(\mathfrak{C}/\cong)$.

See Appendix for the proof.

Therefore, all the concepts of the form (O_i, A) are equivalent to $(O_i \cup O_j, A)$, $O_i, O_j \in \mathfrak{C}$ under the relation "$\cong$," and the whole class of equivalence is formed by taking the "concept union" $\dot{\cup}$. Consequently, from the objects $O_i \in \mathfrak{O}$, we create the new object elements $O_i \,\dot{\cup}\, O_j$ of the power set $\mathscr{P}(\mathfrak{O})$. This is a way to classify concepts depending on attributes.

To classify concepts depending on concepts themselves, we define another equivalence relation of the form $(O_1, A_1) \equiv (O_2, A_2) \Longleftrightarrow O_1 = O_2$.

Relation "$\equiv$" is an equivalence relation as it can be proved similarly to Proposition 1.

We call the set O as the *extension* of a concept, while set A shall be called as the *intension* of it.

Now, from the definition of the concept union $\dot{\cup}$, we realize that by taking the union of two concepts, we find common attributes (similarities) of another "greater" object. Correspondingly, thinking about the concept intersection, we find that "less extension implies greater intension." Lattice means order, as we mentioned already; so for every two elements of it, there exists another "upper" or "preceding" element and another "lower" or "following" element. It is not a hierarchy (a tree), but it is a network as in Figs. 2 and 3.

We now define the order relations "$\preceq$" of the lattice for the already existing operations $\dot{\cup}$ and $\dot{\cap}$. Indeed:

The concept (O_1, A_1) *follows* concept (O_2, A_2) or, equivalently, the concept (O_2, A_2) *precedes* concept (O_1, A_1), if and only if $O_1 \subseteq O_2$ and $A_1 \supseteq A_2$, i.e., $(O_1, A_1) \preceq (O_2, A_2) \stackrel{\text{def.}}{\Longleftrightarrow} O_1 \subseteq O_2$ and $A_1 \supseteq A_2$.

Moreover a ring structure can be proved for the set $\mathfrak{C}$ of concepts. The complement $(O, A)^{\mathrm{c}}$ of the concept $(O, A) \in \mathfrak{C}$ is the concept consisted by the usual set-theoretic complements of $O \in \mathfrak{O}$ and $A \in \mathfrak{A}$, i.e., $(O, A)^{\mathrm{c}} := \left(O^{\mathrm{c}}, A^{\mathrm{c}}\right)$; see [17].

The complement of a concept as above is well defined because:

(a) $O^{\mathrm{c}} \subseteq \mathfrak{O}$ and $A^{\mathrm{c}} \subseteq \mathfrak{A}$ imply that $\left(O^{\mathrm{c}}, A^{\mathrm{c}}\right) \in \mathfrak{C}$.
(b) There is only one complement O^{c} of O and only A^{c} of A; hence, there is only one complement of the concept $(O, A) \in \mathfrak{C}$.

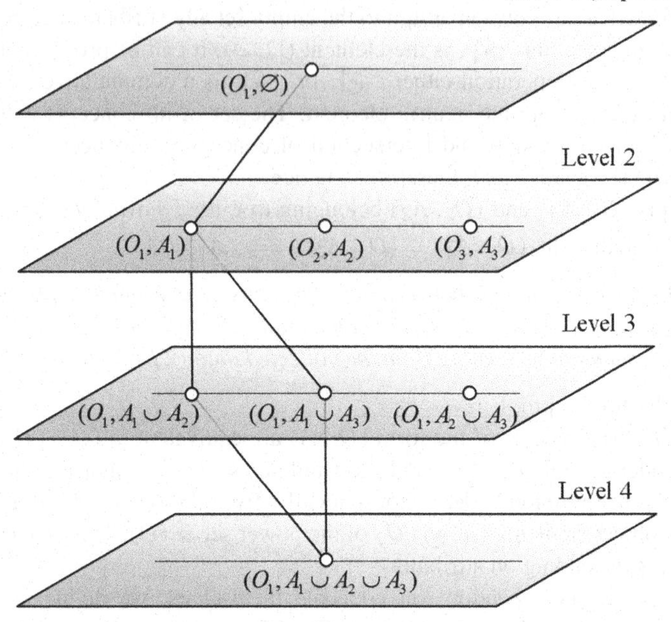

Fig. 2 From fourth level to super-ordinated

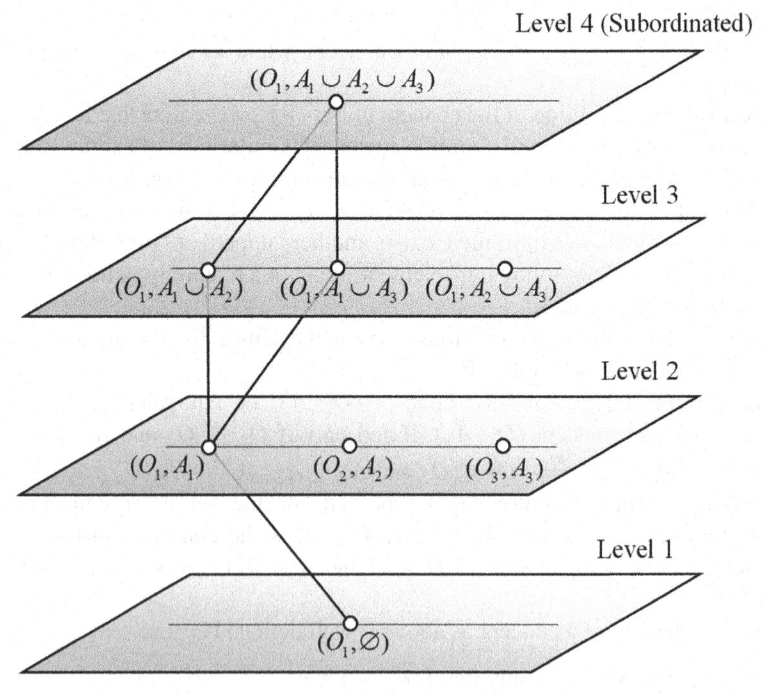

Fig. 3 The construction of concept $\left(O_1, \bigcup_{i=1}^{3} A_i\right)$ from $(O_1, \varnothing)$

The *symmetric difference* or *disjunctive union* $(O_1, A_1) \odot (O_2, A_2)$ between two concepts (O_1, A_1) and (O_2, A_2) belonging to $\mathfrak{C}$ is the concept $\big(O_1 \ominus O_2, (A_1 \ominus A_2)^c\big) \in \mathfrak{C}$ where $O_1 \ominus O_2 := (O_1 \cup O_2) \setminus (O_1 \cap O_2)$ and $A_1 \ominus A_2 := (A_1 \cup A_2) \setminus (A_1 \cap A_2)$ are the usual set-theoretic symmetric differences (or disjunctive unions) of $O_1, O_2 \in \mathfrak{O}$ and $A_1, A_2 \in \mathfrak{A}$, respectively. The following can be proved; see [17].

Theorem 6 *The set $\mathfrak{C}$ enriched with the operation $\odot$ is a group.*

Moreover, the set C enriched with the operations $\odot$ and ⋒ is a ring, where $\odot$ plays the role of "addition" and ⋒ of "multiplication." It is commutative (due to the commutative property of the operation ⋒) with unit (because of the neutral element $(\mathfrak{O}, \varnothing)$ of the operation ⋒) and distributive from both sides.

Since "$\cong$" is an equivalence relation, we can define equivalence classes of concepts, through this similarity relation in which classes are disjoint. Therefore, one can define the "orbit" in the geometrical sense, and not only, as we are moving from one class to another class; see [7] for a general affine geometric point of view of Statistics and [16] for an affine geometric approach for the logit problem. As we know, the classes are disjoint sets and their union makes the set of reference, $\mathfrak{C}$ in this paper. So, in this case, we have a partitioning of $\mathfrak{C}$ according to the defined equivalence relation.

Discrete Distance Measures

It is known that the number of coefficients in which vectors X and Y differ is a distance $d(X, Y)$, known as Hamming distance. If we let $X := (0, 0, 1, 1, 1)$ and $Y := (1, 1, 0, 0, 0)$ with $X, Y \in \mathbb{F}_2^{(5)}$, then $d(X, Y) = 5$. Such a definition is used in binary codes where the minimum distance is always desired. In particular, if we define the weight $w := w(a)$ of a word a, to be the number of ones in a, then $w(a) = d(a, 0)$ and $d(a, y) := w(a - y)$ with y being another word.

The above discussion provides us food for thought to work on the introduction of a discrete distance measure between two given concepts.

The *object distance* $d(O_1, O_2)$, i.e., the distance between two finite objects $O_1, O_2 \in \mathfrak{O}$, is defined to be the nonnegative integer expressing the number of elements of their symmetric difference $O_1 \ominus O_2$, i.e.,

$$d(O_1, O_2) := |O_1 \ominus O_2|. \tag{11}$$

Proposition 2 *The defined object distance $d(O_1, O_2)$ as above is a genuine distance metric, i.e., it satisfies the three properties of a metric: positive definiteness, symmetricity, and triangularity.*

Proof Trivial.

Recall the symmetric difference between two concepts, i.e., $C_1 \odot C_2 = \big(O_1 \ominus O_2, (A_1 \ominus A_2)^{\mathrm{c}}\big)$. The distance d between objects, as in (11), is not that informative, since it is a quantitative but not a qualitative one: two sets of objects may have many different elements, coming from the same (we assume homogenous) population, but we are not measuring the data differences qualitatively but quantitatively. Besides, we are not working with objects or attributes, but with both of them, i.e., with concepts.

The symmetric difference $O_1 \ominus O_2$ between two objects acts between attributes to create a new one of the form $(A_1 \ominus A_2)$. Thus, if we want a qualitative distance between $O_1, O_2 \in \mathfrak{O}$, we must check $(A_1 \ominus A_2)$. In such a case, we can then define

$$d(A_1, A_2) := |A_1 \ominus A_2| = |\mathfrak{A}| - \big|(A_1 \ominus A_2)^{\mathrm{c}}\big|, \quad A_1, A_2 \in \mathfrak{A}. \tag{12}$$

Note that if the distance between two attributes is increasing, i.e., there are many noncommon attributes, then (12) yields that the number of elements of $(A_1 \ominus A_2)^{\mathrm{c}}$ is decreasing.

Suppose now we have two objects, O_1 and O_2. As a measure of "comparison," we introduce the normalized distance

$$d_{\mathrm{n}}(O_1, O_2) := \frac{|O_1 \ominus O_2|}{|O_1| + |O_2|}, \quad O_1, O_2 \in \mathfrak{O}. \tag{13}$$

For a discussion and a number of calculations for different cases of objects, see the Proof of Claim in Appendix.

Claim The normalized distance between objects is a function depending on the number of different elements between them, ranging from value 0 (no differences) to 1 (everything is different).

See Appendix for the proof.

Fuzzy Logic Approach

The fuzzy logic extends the classical binary response: a sentence is either true or false (not true), and hence it belongs to the {0, 1} binary set. This binary set can be extended to the [0, 1] interval. In strictly mathematical terms, the *characteristic function* of a given set $Q \in \Omega$, i.e.,

$$I_Q : Q \longrightarrow \{0, 1\}, \quad \text{with } Q \ni x \longmapsto I_Q(x) \in \{0, 1\},$$

is now considered as the *membership function* of a fuzzy set $Q \subseteq \Omega$, i.e.,

$$M_Q : Q \longrightarrow [0, 1], \quad \text{with } Q \ni x \longmapsto M_Q(x) \in [0, 1].$$

The value of $M_Q(x)$ declares the degree of participation of the element $x \in Q$ which belongs/participates to the fuzzy set $Q \in \Omega$. In particular,

$$M_Q(x) := \begin{cases} 1, & \text{declares that } x \text{ belongs to } Q, \\ 0, & \text{declares that } x \text{ does not belongs to } Q, \\ q \in (0,\ 1), & \text{declares that } x \text{ belongs "partially" (in some degree) to } Q. \end{cases}$$

The above introductory elements are useful to realize the extensions succeeded by fuzzy logic, i.e., adopting an interval of values rather than a single binary value to describe a phenomenon.

The interval mathematics, as described in [29], offers another approach to develop intervals, different than the fuzzy logic one; see [21] and [31] for the corresponding applications.

Recall that, under the fuzzy logic approach, the subset-hood between two sets A and B, subsets of the "universe" set Ω, is

$$S(A, B) = \frac{k(A \cap B)}{k(A)},$$

where $k(A) := \text{card}(A) = |A|$ is the generalized cardinal number and represents the degree to which B is a subset of A. If $A \subseteq B$ then $S(A, B) = 1$. Based on this, the *fuzzy entropy* of a set A, denoted with $E_{\text{F}}(A)$, can be defined as

$$E_{\text{F}}(A) := \frac{k(A \cap A^{\text{c}})}{k(A \cup A^{\text{c}})}.$$

The fuzzy entropy of A measures "how much" underlap $A \cup A^{\text{c}}$ and overlap $A \cap A^{\text{c}}$ violate the existent laws $A \cap A^{\text{c}} = \varnothing$, $A \cup A^{\text{c}} = \Omega$. That is, the fuzzy entropy measures eventually "how much" of $A \cup A^{\text{c}}$ is included to $A \cap A^{\text{c}}$.

The following theorem rules the fuzzy entropy theory:

Theorem 7 (of Fuzzy Entropy) *It holds that*

$$E_{\text{F}}(A) = S\big(A \cup A^{\text{c}}, A \cap A^{\text{c}}\big).$$

Proof From the definition (14), it holds that

$$S\big(A \cup A^{\text{c}}, A \cap A^{\text{c}}\big) = \frac{k\Big(\big(A \cup A^{\text{c}}\big) \cap \big(A \cap A^{\text{c}}\big)\Big)}{k(A \cup A^{\text{c}})} = \frac{k(A \cap A^{\text{c}})}{k(A \cup A^{\text{c}})} = E_{\text{F}}(A).$$

Example 8 It holds

$$E_{\text{F}}(\Omega) = S\big(\Omega \cup \Omega^{\text{c}}, \Omega \cap \Omega^{\text{c}}\big) = S\big(\Omega \cup \varnothing, \Omega \cap \varnothing\big) = S(\Omega, \varnothing) = 0.$$

Therefore, the universe set Ω has fuzzy entropy 0, while for the middle point M, it holds

$$E_{\mathrm{F}}(M) = S\big(M \cup M^{\mathrm{c}}, M \cap M^{\mathrm{c}}\big) = 1.$$

Based on the above discussion, we can define the *fuzzy entropy deviance*, i.e.,

$$\delta(O_1, O_2) := E_{\mathrm{F}}(O_1) - E_{\mathrm{F}}(O_2) = \frac{k\big(O_1 \cap O_1^{\mathrm{c}}\big)}{k\big(O_1 \cup O_1^{\mathrm{c}}\big)} - \frac{k\big(O_2 \cap O_2^{\mathrm{c}}\big)}{k\big(O_2 \cup O_2^{\mathrm{c}}\big)}.$$

It is always a problem to define a distance measure when information divergences are under consideration; see [19] for the continuous case of the Kullback–Leibler divergence.

We would like to emphasize that fuzziness and randomness are different ideas. They seem similar, but they are not identical. The randomness concerns problems where the event is well defined but it is uncertain if it will take place or not. The fuzziness concerns situations which are not well defined and can only be described in a sufficient way when it is known how we shall move between different classes. That is, we are moving under a fuzzy event to the "probability of a fuzzy event," which is still (even to such a probability oriented procedure) closer to a measure-theoretic approach than to an algebraic approach (via probability algebras). Moreover, in fuzzy logic, the additivity due to a "measure" is not existing, so it is not a defined probability measure, while the term "possibility" replaces the term "probability."

Discussion

This paper studied the general influence of the Discrete Mathematics line of thought to Statistics and probability. In particular, the Experimental Design Theory employs many discrete statistical concepts, which offer a solid framework, although the real data analysis is mainly performed by the ANOVA approach. Furthermore, geometry plays an important role in all the mathematical scenarios—so does in the Experimental Design Theory in its finite formulation.

The foundations of probability theory are mainly based on measure-theoretic concepts. But still, there is also a set-theoretic approach. Lattice theory is applied in concepts and Boolean algebra supports the fuzzy logic extension. Distance measures offer criteria to decide "how close" two estimates or two sets or two concepts are. A brief discussion to the subject was also offered in this paper.

Appendix

Proof (of Proposition 1) We shall prove that the introduced relation $\cong$ is reflective, symmetric, and transitive. Indeed:

i. Reflexivity of relation "$\cong$". Indeed, $(O, A) \cong (O, A) \Leftrightarrow A = A$, which is true due to the reflexivity of the equality relation "=" for sets.
ii. Symmetricity of relation "$\cong$". Indeed, $(O_1, A1) \cong (O_2, A_2) \Leftrightarrow A_1 = A_2 \Leftrightarrow A_2 = A_1$ (symmetricity of the equality relation "=" for sets) $\Leftrightarrow (O_2, A_2) = (O_1, A_1)$.
iii. Transitivity of relation "$\cong$". Indeed, it holds that

$$(O_1, A_1) \cong (O_2, A_2) \Leftrightarrow A_1 = A_2 \text{ and} \tag{14a}$$

$$(O_2, A_2) \cong (O_3, A_3) \Leftrightarrow A_2 = A_3, \tag{14b}$$

which are both equivalent to $A_1 = A_3$, due to the transitivity of the equality relation "=" for sets, and hence $(O_1, A_1) \cong (O_3, A_3)$.

Proof (of Theorem 1) Let us consider (i, j) and $(i.j')$ be the same symbols for the position. Then,

$$L(i, j) = L(i, j') \Longrightarrow i + j = i + j'.$$

As $i, j, j' \in \mathbb{Z}_V$, then $-i$ exists and thus

$$(-i) + i + j = (-i) + i + j' \Longrightarrow j = j'.$$

This means each symbol occurs once in row i. Since there are v symbols and v positions, each symbol occurs exactly once. The same line of thought is followed for the columns. Therefore, $L(i, j)$ is an LS.

Proof (of Theorem 2) Following the same line of thought of Theorem 1, we prove that $L_a = L_a(i, j)$ is an LS. Indeed, for $L_a(i, j) = L_a(i, j')$, it holds $ai + j = ai + j'$. Since $a, i, j \in \mathbb{Z}_p$, then $a^{-1}, -j \in \mathbb{Z}_p$, and hence $j = j'$. Similarly, $L_a(i, j) = L_a(i, j')$ yields $j' = j$. Consider now the position, say (i_1, j_1) of L_a, and a different position, say (i_2, j_2) of L_b. Moreover, let k_1, k_2 be the symbols for both positions. Then, for

$$L_a(i_1, j_1) = k_1, \quad L_b(i_1, j_1) = k_2, \quad \text{it is}$$

$$ai_1 + j_1 = k_1, \quad bi_1 + j_1 = k_2,$$

and for

$$L_a(i_2, j_2) = k_1, \quad L_b(i_2, j_2) = k_2, \text{ it is}$$
$$ai_2 + j_2 = k_1, \quad bi_2 + j_2 = k_2.$$

Thus,

$$a(i_1 - i_2) = j_2 - j_1 \text{ and } b(i_1 - i_2) = j_2 - j_1.$$

Assuming that $i_1 \neq i_2$ then $(i_1 - i_2)^{-1} \in \mathbb{Z}_p$ and

$$a = b = (i_1 - i_2)^{-1}(j_2 - j_1).$$

However, we assumed that $a \neq b$; thus, k_1 and k_2 are equal in only one position. Therefore, $L_a \perp L_b$.

Proof (of Theorem 4) Since x, y are elements of $\mathbb{F}_q$, they can take q different values. So, there are $v = q^2$ points. As far as the block is concerned, we must prove that every line has exactly q points and that any two points of $\mathbb{F}_q$ belong to exactly one line. Indeed:

Consider the line $ax + by + c = 0$, $b \neq 0$. Then,

$$y = -b^{-1}(c + ax),$$

such that (x, y) is on the line, and hence the line has q points. If $b = 0$, $a \neq 0$, it holds

$$x = -a^{-1}c.$$

In such a case, there are q possible values of y in $\mathbb{F}_q$ and q points of the form $\big(-a^{-1}c, y\big)$ lie on the line.

Now, suppose that (x_1, y_1) and (x_2, y_2) are two given distinct points, and hence $x_2 - x_1$ and $y_1 - y_2$ are not both zero. The equation of the line "passing" (actually "containing") the two points is

$$\ell: \; (y_1 - y_2)x + (x_2 - x_1) = x_2 y_1 - x_1 y_2,$$

is the equation of a line. Moreover, it contains the two given points. If another line is containing the two given points and described by the analytic form

$$ax + by + c = 0,$$

then it holds

$$ax_1 + by_1 + c = 0 \text{ and } ax_2 + by_2 + c = 0, \text{ i.e.,}$$
$$a(x_2 - x_1) = b(y_1 - y_2).$$

The value $(x_2 - x_1)^{-1}$ exists in $\mathbb{F}_q$, provided $x_1 \neq x_2$, and hence

$$a = b(x_2 - x_1)^{-1}(y_1 - y_2) = \lambda(y_1 - y_2).$$

So we have

$$\begin{aligned} b &= \lambda(x_2 - x_1) \text{ and} \\ c &= -ax_1 - by_1 = -ax_1 - \lambda(x_2 - x_1)y_1 = -\lambda(y_1 - y_2)x_1 - \lambda(x_2 - x_1)y_1 \\ &= \lambda(x_1y_2 - x_2y_1). \end{aligned}$$

Thus, the line is the same with the above defined since lines ℓ and $\lambda\ell$ coincide, in finite geometry. Therefore, only one line exists and "contains" the two points.

Proof (of Theorem 5) In the affine plane over $\mathbb{F}_q$, the lines

$$ax + by + c = 0 \text{ and } a'x + b'y + c' = 0,$$

are said to be *parallel* if $ab' = a'b$ in $\mathbb{F}_q$. There are $q + 1$ equivalence classes of parallel lines of the form

$$x + \lambda y = 0, \quad \lambda \in \mathbb{F}_q,$$

and the $y = 0$ line. Any point of the affine plane belongs to just one line in each class.

We introduce now $q + 1$ points X_λ, $\lambda \in \mathbb{F}_q$, and X_∞, all belonging to a new line ℓ_∞; see also Example 5 for the discussion on ℓ_∞ line. The X_λ points lie to each line parallel to line $x + \lambda y = 0$, while X_∞ to each line parallel to $y = 0$. We have to prove that the lines are blocks of a design with parameters stated above. There are $q^2 + q + 1$ points and each line contains $q + 1$ points. Thus, we proved that any two distinct points, say H and I, belong to just one line. Then, the following cases can be true:

1. The points H and I are both parts of the initial affine plane; let us called them "old" points. So, H and I belong to a unique line of this plane, which corresponds uniquely to a line on the extended plane (with infinity point).
2. If H is an "old point" and I is a "newly added" point, i.e., $I := X_\lambda$, then H belongs already to precisely one "old" line in the parallel class represented by X_∞, and, therefore, the corresponding new line in the unique line ℓ' contains points H and I. The same is certainly true if $X := X_\infty$.
3. If H and I are both "new" points, then they belong to ℓ_∞ by definition. Therefore, any two points belong to just one line (i.e., to one block). Now, from the above construction, any two lines have just one common point.

Proof (of Claim) We distinguish the following cases:

- *Case* $O_1 \cap O_2 = \varnothing$ (disjointed objects). In general, it holds

$$O_1 \ominus O_2 = (O_1 \setminus O_2) \cup (O_2 \setminus O_1) = O_1 \cap O_2, \quad O_1, O_2 \in \mathfrak{D}, \tag{15}$$

and, hence, in this case. we obtain

$$|O_1 \ominus O_2| = |O_1 \cup O_2| = |O_1| + |O_2| - |O_1 \cap O_2| = |O_1| + |O_2|, \tag{16}$$

which is—in principle—the maximum possible number of elements of the symmetric difference between two objects. Thus, the normalized distance d_n between $O_1, O_2 \in \mathfrak{D}$ equals 1. Indeed, via (15),

$$\begin{aligned} |O_1 \ominus O_2| &= \big|(O_1 \setminus O_2) \cup (O_2 \setminus O_1)\big| \\ &= \|(O_1 \setminus O_2| + |O_2 \setminus O_1| - \big|(O_1 \setminus O_2) \cap (O_2 \setminus O_1)\big|) \\ &= |O_1 \setminus O_2| + |O_2 \setminus O_1| - |O_1 \cap O_2| \\ &\leq |O_1 \setminus O_2| + |O_2 \setminus O_1| \leq |O_1| + |O_2|, \end{aligned} \tag{17}$$

where the equality holds iff $O_1 \setminus O_2 = O_1$ and $O_2 \setminus O_1 = O_2$, which is equivalent to $O_1 \cap O_2 = \varnothing$. Furthermore, the normalized distance between O_1 and O_2 is confirmed to be 1 since

$$d_\text{n}(O_1, O_2) := \frac{|O_1 \ominus O_2}{|O_1| + |O_2|} = \frac{|O_1| + |O_2|}{|O_1| + |O_2|} = 1. \tag{18}$$

- *Case* $O_1 \subseteq O_2$ (included objects). If O_1 is a subset of O_2, then

$$\begin{aligned} |O_1 \ominus O_2| &= |O_1 \setminus O_2| + |O_2 \setminus O_1| - \big|(O_1 \setminus O_2) \cap (O_2 \setminus O_1)\big| \\ &= |\varnothing| + |O_2 \setminus O_1| - \big|\varnothing \cap (O_2 \setminus O_1)\big| \\ &= |O_2 \setminus O_1| - |\varnothing| \leq |O_2|, \end{aligned} \tag{19}$$

where the equality holds iff $O_1 = \varnothing$. In principle, the normalized distance d_n between objects is less than or equal to 1. Indeed, for $O_1, O_2 \in \mathfrak{D}$,

$$\begin{aligned} d_\text{n}(O_1, O_2) &= \frac{|O_1 \setminus O_2| + |O_2 \setminus O_1| - \big|(O_1 \setminus O_2) \cap (O_2 \setminus O_1)\big|}{|O_1| + |O_2|} \\ &\leq \frac{|O_1 \setminus O_2| + |O_2 \setminus O_1|}{|O_1| + |O_2|} \leq \frac{|O_1| + |O_2|}{|O_1| + |O_2|} = 1. \end{aligned} \tag{20}$$

The above is also confirmed, via (19), for the specific case of $O_1 \subseteq O_2$, as

$$d_\text{n}(O_1, O_2) := \frac{|O_1 \ominus O_2}{|O_1| + |O_2|} = \frac{|O_2 \setminus O_1|}{|O_1| + |O_2|} \leq \frac{|O_2|}{|O_1| + |O_2|} \leq 1, \tag{21}$$

where, again, the equality holds iff $O_1 = \varnothing$.

- *Case* $O_2 \subseteq O_1$ (included objects). If O_2 is a subset of O_1, then we obtain dually that

$$d_{\mathrm{n}}(O_1, O_2) := \frac{|O_1 \ominus O_2|}{|O_1| + |O_2|} = \frac{|O_1 \setminus O_2|}{|O_1| + |O_2|} \leq \frac{|O_1|}{|O_1| + |O_2|} \leq 1, \tag{22}$$

where the equality holds iff $O_2 = \varnothing$.

- *Case* $O_1 = O_2$ (equated objects). If object O_1 coincides with object O_2, then

$$\begin{aligned} d_{\mathrm{n}}(O_1, O_2) &= \frac{|O_1 \setminus O_2 + |O_2 \setminus O_1| - \big|(O_1 \setminus O_2) \cap (O_2 \setminus O_1)\big|}{|O_1| + |O_2|} \\ &= \frac{|\varnothing| + |\varnothing| - |\varnothing \cap \varnothing|}{|O_1| + |O_2|} = 0. \end{aligned} \tag{23}$$

References

1. Box, G.E.P.: Evolutionary operation: a method of increasing industrial productivity. J. R. Soc. Ser. B **26**, 211–252 (1957)
2. Box, I.F.: R.A. Fisher: The Life of a Scientist. Wiley, New York (1978)
3. Cochran, W.C., Box, G.M.: Experimental Design. Wiley, New York (1957)
4. Coxeter, H.S.M.: Projective Geometry. Springer, New York (1987)
5. Fisher, R.A.: The Design of Experiments. Oliver and Boy, London (1947)
6. Ford, I., Kitsos, C.P., Titterington, D.M.: Recent advances in nonlinear experimental design. Technometrics **31**, 49–60 (1989)
7. Fraser, D.A.S.: Structural Inference. Wiley, New York (1968)
8. Hicks, C.R.: Fundumental Concepts in the Design of Experiments. Holt, Rinehart and Winston, New York (1973)
9. Hunter, J.: Numder Theory. Oliver and Boyd, Clasgow (1964)
10. Kappos, D.: Probability Algebra and Stochastic Spaces. Academic, New York, London (1969)
11. Kempthorne, O.: The Design and Analysis of Experiments. Wiley, New York (1952)
12. Kitsos, C.P.: Design and Inference for Nonlinear Problems. PhD Thesis, University of Glasgow (1986)
13. Kitsos, C.P.: Fully sequential procedures in nonlinear design problems. Comput. Stat. Data Anal. **8**, 13–19 (1989)
14. Kitsos, C.P.: Quasi-sequential procedures for the calibration problem. In: Dodge, Y., Whittaker, J. (eds.) COMPSTAT 92 Proceedings, vol. 2, pp. 227–231 (1992)
15. Kitsos, C.P.: Statistical Analysis of Experimental Design (in Greek). New Technologies, Athens (1994)
16. Kitsos, C.P.: The SPRT for the Poisson process. In: e-Proceedings of the XII-th Applied Stochastic Models and Data Analysis (ASMDA 2007), Chania, Crete, 29 May–1 June 2007
17. Kitsos, C.P., Sotiropoulos, M.: Distance methods for bioassays. Biometrie und Medizinische Informatik Greifswalder Seminarberichte **15**, 55–74 (2009)
18. Kitsos, C.P., Toulias, T.L.: On the information matrix of the truncated cosinor model. Br. J. Math. Comput. Sci. **4**(6), 759–773 (2014)
19. Kitsos, C.P., Toulias, T.L.: Hellinger distance between generalized normal distributions. Br. J. Math. Comput. Sci. **21**(2), 1–16 (2017)
20. Kitsos, C.P., Titterington, D.M., Torsney, B.: An optimal design problem in rhythmometry. Biometrics **44**, 657–671 (1988)

21. Klir, G., Yan, B.: Fuzzy Logic. Prentice Hall, New Jersey (1995)
22. Kolmogorov, A.N.: Grundbegriffe der Wahrscheinlichkeitsrechung. Springer, Berlin (1933)
23. Mead, R.: The Design of Experiments. Cambridge University Press, Cambridge (1988)
24. Montgomery, D.G.: Design and Analysis of Experiments. Wiley, New York (1991)
25. Myers, R.H.: Response Surface Methodology. Allyn and Bacon Inc., Boston (1971)
26. Scheffe, H.: The Analysis of Variance. Wiley, New York (1959)
27. Silvey, S.D.: Optimal Design - An Introduction to the Theory for Parameter Estimation. Springer Netherlands, Dordrecht (1980)
28. Toulias, T.L., Kitsos, C.P.: Fitting the Michaelis-Menten model. J. Comput. Appl. Math. **296**, 303–319 (2016)
29. Wolfe, M.A.: Interval mathematics, algebraic equations and optimization. J. Comput. Appl. Math. **124**, 263–280 (2000)
30. Yates, F.: Design and Analysis of Factorial Experiment. Imperial Bureau of Soil Sciences, London (1937)
31. Zandeh, L.A.: Toward a theory of fuzzy information granulation and its centrality in human reasoning and fuzzy logic. Fuzzy Sets Syst. **90**, 111–127 (1997)

On the Use of the Fractal Box-Counting Dimension in Urban Planning

Konstantina Lantitsou, Apostolos Syropoulos, and Basil K. Papadopoulos

Abstract Fractal geometry has found many applications in science and technology. Some time ago, it was used to study urban development. However, something that has not been addressed so far, to the best of our knowledge, is whether a drastic extension of some urban area also changes drastically the box-counting dimension of the area. In addition, it is not known if it is possible to predict any change of the urban or neighborhood character of a specific area by just comparing the box-counting dimensions of the city or the neighborhood before and after the suggested extension. This is a first attempt to answer these questions.

Introduction

Typically, most educated people associate the term *fractal* (see [4] for a good introduction to fractal geometry and chaos theory) with images, shapes, or structures that are usually static. Although there are many beautiful static images and shapes, still there are many fractals that are part of a dynamic system that evolves in the course of time. For example, such a system can be a plant or a town.

Benoit Mandelbrot is the father of fractal geometry since he coined the term "fractal" and made many contributions to fractal geometry; nevertheless, he was not the first one to recognize that there are geometrical shapes and mathematical structures that are *self-similar*. Such geometric shapes were introduced in the beginning of the twentieth century and include the Sierpiński triangle, which is shown in Fig. 1, the Sierpiński carpet, the Koch snowflake, etc. In all fractal shapes, parts of the whole are similar, if not identical, to the whole. For example, note that parts of Sierpiński triangle are similar to the whole. Shapes like this were considered

K. Lantitsou · B. K. Papadopoulos
Democritus University of Thrace, Xanthi, Greece
e-mail: klantits@civil.duth.gr; papadob@civil.duth.gr

A. Syropoulos (✉)
Greek Molecular Computing Group, Xanthi, Greece

N. J. Daras, Th. M. Rassias (eds.), *Modern Discrete Mathematics and Analysis*,
Springer Optimization and Its Applications 131,
https://doi.org/10.1007/978-3-319-74325-7_13

Fig. 1 The Sierpiński triangle

Fig. 2 Romanesco broccoli: a "natural" fractal

as mathematical "monsters," but soon they got forgotten by the mathematical community. Later on, Mandelbrot realized that many shapes and solids we encounter in nature have many things in common with these mathematical "monsters." For example, Fig. 2 is an image of Romanesco broccoli which exhibits self-similarity.

It is an undeniable fact that fractals are extremely complicated shapes or solids. What is even more remarkable is the fact that it is not easy to measure their length, area, or volume. In particular, Mandelbrot [2], while trying to measure the coastline of Britain, concluded that the smaller the yardstick is used to measure the coastline, the larger the measured length becomes. Instead of measuring the length, the area, or the volume of an object, we can measure the object's *dimension*. However, the term dimension does not have the usual meaning as in "the three dimensions of Euclidean space". Here we are interested in what is called the *box-counting* dimension of objects. When dealing with objects that "live" in two-dimensional Euclidean space, one can calculate the box-counting dimension by covering the object with a grid and then count the number of boxes that cover the object, hence the name box-counting dimension. Next, we use a new grid with smaller boxes, cover the object, and count the boxes, etc. If for a grid that contains boxes whose length is s and an object is covered by such $N(s)$ boxes, then we use the data to make a log-log plot. In particular, the horizontal axis uses the $\log \frac{1}{s}$ values, and the vertical axis uses the $\log N(s)$ values. Next, we compute the slope of the line of best fit. This number is the box-counting dimension of this particular object.

Although the box-counting dimension might have found many applications, still we are interested only in applications related to the urban planning. In particular, the box-counting dimension has been used for predicting the growth of large urbanized area [5]. The same method has been applied for Xanthi, Greece [3]. This small urban area hosts the Democritus University of Thrace. Unfortunately the data available for this study were not quite adequate; nevertheless, the prediction made is not far from reality.

If it makes sense to use the box-counting dimension to make predictions about the growth of urban areas, then we should be able to use it to guide urban planning. Provided that the box-counting dimension represents to a certain degree the character of an urban area, can we use it to see whether a new building will alter the character of the urban area? For example, when there are plans to build a shopping mall in particular area, then a natural question is: can we predict if the shopping mall will alter the character of urban area?

In what follows we will describe how to use the box-counting dimension in urban planning.

Box-Counting Dimension and Urban Planning

In many cases, consortia and/or governments decide to build football stadia or shopping malls that alter the character of the surrounding urban area. For example, the certainty that the urban development plan for Istanbul's Taksim Gezi Park would alter the character of the area led to a protest that eventually led to the Gezi Park protests.[1] We propose that one should use the box-counting dimension to predict

[1]Of course there were deeper political reasons behind this uprising; nevertheless, a political analysis of this event is far beyond the scope of this paper.

whether any change in an urban area will eventually alter the character of the urban area. In particular, we compute the box-counting dimension of an urban area before the proposed change and then do the same using an architectural model. If the two dimensions are not that different, then this is an indication that the proposed urban development will not alter the architectural character of the area. We say *indication* and not *proof* because we need to make this check on a big amount of data, something that is extremely time-consuming.

In many cases people cannot prevent the construction of a new building, and all they can do is to check afterward whether the new building has altered the character of the urban area. For example, it has been argued that the New Acropolis Museum in Athens has altered the architectural character of the area [1]; nevertheless, this argument is based mainly on aesthetic principles and not on some "scientific" research. In order to provide such a "scientific" argument, we will compute the box-counting dimension of the New Acropolis Museum and its surroundings and then compare them.

Figures 3 and 4 have been used to evaluate the box-counting dimension of the New Acropolis Museum and its surrounding area. In order to compute the box-counting dimension of a geometrical object, we use the following formula:

$$-\frac{\log N_{a/2} - \log N_a}{\log \frac{a}{2} - \log a},$$

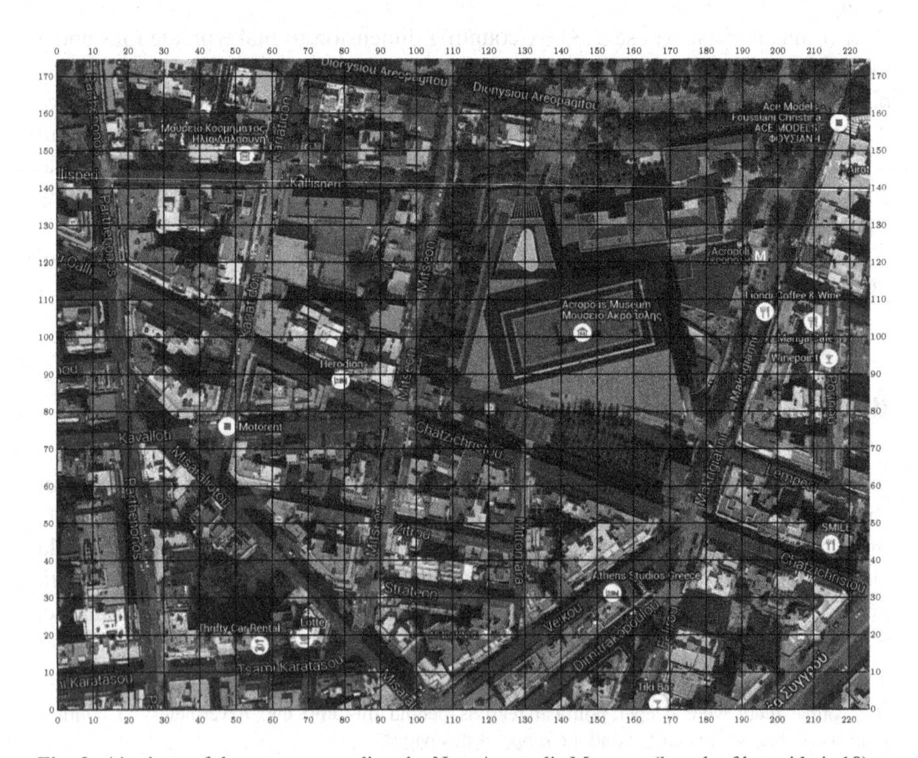

Fig. 3 Air photo of the area surrounding the New Acropolis Museum (length of box side is 10)

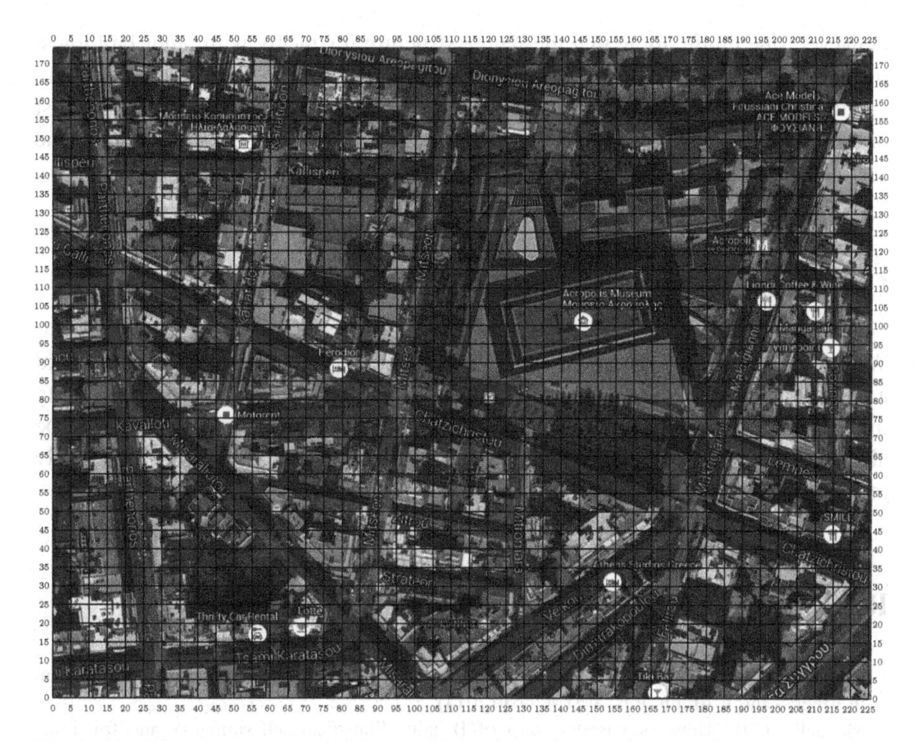

Fig. 4 Air photo of the area surrounding the New Acropolis Museum (length of box side is 5)

where a is the length of the side of a box and N_a is the number of boxes that cover the particular object. In order to compute the box-counting dimension of the museum, we have used the following:

$$-\frac{\log 116 - \log 33}{\log 5 - \log 10} \approx 1.813.$$

When we say surrounding area, we mean the area that is delimited by Mitseon street, Chatzichristou street, Makrigianni street, and Dionysiou Areopagitou street. The box-counting dimension of this block is calculated as follows:

$$-\frac{\log 286 - \log 97}{\log 5 - \log 10} \approx 1.559.$$

Clearly, the two dimensions are rather different. Naturally, one could use a wider area to compute the box-counting dimension in order to get "better" results; nevertheless, the aim of this note is to demonstrate that fractal dimension can be used in urban planning.

The fact that the two dimensions are rather different is a "scientific" proof that the building of the Acropolis Museum has altered the character of the whole area.

However, one might argue that the whole structure fits nicely with other buildings. In this case, one needs to compute the "cube-counting" dimension of the building and of the surrounding buildings in order to see if the new building fits with the rest of the buildings. Since we do not have volumetric data for either the museum or the surrounding buildings, we cannot say anything.

Conclusions

We have discussed the relationship between the box-counting fractal dimension and a possible urban development project. Also, we briefly touched the idea of using the "cube-counting" dimension for exactly the same purpose. Provided we will be able to get volumetric data for the museum and its surroundings, we plan to fully investigate the use of the two dimensions in urban development.

References

1. Horáček, M.: Museum of art vs. the city as a work of art: a case of the new acropolis museum in Athens. Int. J. Archit. Res. **8**(2), 47–61 (2014)
2. Mandelbrot, B.: How long is the coast of Britain? Statistical self-similarity and fractional dimension. Science **156**(3775), 636–638 (1967)
3. Pappa, E., Syropoulos, A., Papadopoulos, B.K.: An application of fractal geometry in urban growth. Technika Chronika **25**(2–3), 73–80 (2005). Paper in Greek
4. Peitgen, H.O., Jürgens, H., Saupe, D.: Chaos and Fractals: New Frontiers of Science. Springer, New York (2004)
5. Shen, G.: Fractal dimension and fractal growth of urbanized areas. Int. J. Geogr. Inf. Sci. **16**(5), 419–437 (2002)

Additive-Quadratic ρ-Functional Equations in Banach Spaces

Jung Rye Lee, Choonkil Park, and Themistocles M. Rassias

Abstract Let

$$M_1 f(x,y) := \frac{3}{4} f(x+y) - \frac{1}{4} f(-x-y) + \frac{1}{4} f(x-y) + \frac{1}{4} f(y-x) - f(x) - f(y)$$

and

$$M_2 f(x,y) := 2f\left(\frac{x+y}{2}\right) + f\left(\frac{x-y}{2}\right) + f\left(\frac{y-x}{2}\right) - f(x) - f(y).$$

We solve the additive-quadratic ρ-functional equations

$$M_1 f(x,y) = \rho M_2 f(x,y) \tag{1}$$

and

$$M_2 f(x,y) = \rho M_1 f(x,y), \tag{2}$$

where ρ is a fixed nonzero number with $\rho \neq 1$.

Using the direct method, we prove the Hyers–Ulam stability of the additive-quadratic ρ-functional equations (1) and (2) in Banach spaces.

Keywords Hyers-Ulam stability · Additive-quadratic ρ-functional equation · Banach space

---

J. R. Lee
Department of Mathematics, Daejin University, Pocheon, Republic of Korea
e-mail: jrlee@daejin.ac.kr

C. Park
Research Institute for Natural Sciences, Hanyang University, Seoul, Republic of Korea
e-mail: baak@hanyang.ac.kr

Th. M. Rassias (✉)
Department of Mathematics, National Technical University of Athens, Athens, Greece
e-mail: trassias@math.ntua.gr

N. J. Daras, Th. M. Rassias (eds.), *Modern Discrete Mathematics and Analysis*,
Springer Optimization and Its Applications 131,
https://doi.org/10.1007/978-3-319-74325-7_14

Introduction and Preliminaries

The stability problem of functional equations originated from a question of Ulam [20] concerning the stability of group homomorphisms.

The functional equation $f(x+y) = f(x) + f(y)$ is called the *Cauchy equation*. In particular, every solution of the Cauchy equation is said to be an *additive mapping*. Hyers [8] gave a first affirmative partial answer to the question of Ulam for Banach spaces. Hyers' theorem was generalized by Aoki [2] for additive mappings and by Rassias [12] for linear mappings by considering an unbounded Cauchy difference. A generalization of the Rassias theorem was obtained by Găvruta [7] by replacing the unbounded Cauchy difference by a general control function in the spirit of Rassias' approach.

The functional equation $f(x + y) + f(x - y) = 2f(x) + 2f(y)$ is called the quadratic functional equation. In particular, every solution of the quadratic functional equation is said to be a *quadratic mapping*. The stability of quadratic functional equation was proved by Skof [19] for mappings $f : E_1 \to E_2$, where E_1 is a normed space and E_2 is a Banach space. Cholewa [5] noticed that the theorem of Skof is still true if the relevant domain E_1 is replaced by an Abelian group. The stability problems of various functional equations have been extensively investigated by a number of authors (see [1, 3, 4, 6, 9–11, 13–18, 21, 22]).

In section "Additive-Quadratic ρ-Functional Equation (1) in Banach Spaces," we solve the additive-quadratic ρ-functional equation (1) and prove the Hyers–Ulam stability of the additive-quadratic ρ-functional equation (1) in Banach spaces.

In section "Additive-Quadratic ρ-Functional Equation (2) in Banach Spaces," we solve the additive-quadratic ρ-functional equation (2) and prove the Hyers–Ulam stability of the additive-quadratic ρ-functional equation (2) in Banach spaces.

Throughout this paper, assume that X is a normed space and that Y is a Banach space. Let ρ be a nonzero number with $\rho \neq 1$.

Additive-Quadratic ρ-Functional Equation (1) in Banach Spaces

We solve and investigate the additive-quadratic ρ-functional equation (1) in normed spaces.

Lemma 1

(i) *If a mapping* $f : X \to Y$ *satisfies* $M_1 f(x, y) = 0$, *then* $f = f_o + f_e$, *where* $f_o(x) := \frac{f(x)-f(-x)}{2}$ *is the Cauchy additive mapping and* $f_e(x) := \frac{f(x)+f(-x)}{2}$ *is the quadratic mapping.*

(ii) *If a mapping* $f : X \to Y$ *satisfies* $M_2 f(x, y) = 0$, *then* $f = f_o + f_e$, *where* $f_o(x) := \frac{f(x)-f(-x)}{2}$ *is the Cauchy additive mapping and* $f_e(x) := \frac{f(x)+f(-x)}{2}$ *is the quadratic mapping.*

Proof

(i)

$$M_1 f_o(x, y) = f_o(x + y) - f_o(x) - f_o(y) = 0$$

for all $x, y \in X$. So f_o is the Cauchy additive mapping.

$$M_1 f_e(x, y) = \frac{1}{2} f_e(x + y) + \frac{1}{2} f_e(x - y) - f_e(x) - f_e(y) = 0$$

for all $x, y \in X$. So f_o is the quadratic mapping.

(ii)

$$M_2 f_o(x, y) = 2 f_o\left(\frac{x + y}{2}\right) - f_o(x) - f_o(y) = 0$$

for all $x, y \in X$. Since $M_2 f(0, 0) = 0$, $f(0) = 0$ and f_o is the Cauchy additive mapping.

$$M_2 f_e(x, y) = 2 f_e\left(\frac{x + y}{2}\right) + 2 f_e\left(\frac{x - y}{2}\right) - f_e(x) - f_e(y) = 0$$

for all $x, y \in X$. Since $M_2 f(0, 0) = 0$, $f(0) = 0$ and f_e is the quadratic mapping.

Therefore, the mapping $f : X \to Y$ is the sum of the Cauchy additive mapping and the quadratic mapping.

From now on, for a given mapping $f : X \to Y$, define $f_o(x) := \frac{f(x) - f(-x)}{2}$ and $f_e(x) := \frac{f(x) + f(-x)}{2}$ for all $x \in X$. Then f_o is an odd mapping and f_e is an even mapping.

Lemma 2 *If a mapping $f : X \to Y$ satisfies $f(0) = 0$ and*

$$M_1 f(x, y) = \rho M_2 f(x, y) \tag{3}$$

for all $x, y \in X$, then $f : X \to Y$ is the sum of the Cauchy additive mapping f_o and the quadratic mapping f_e.

Proof Letting $y = x$ in (3) for f_o, we get $f_o(2x) - 2 f_o(x) = 0$ and so $f_o(2x) = 2 f_o(x)$ for all $x \in X$. Thus

$$f_o\left(\frac{x}{2}\right) = \frac{1}{2} f_o(x) \tag{4}$$

for all $x \in X$.

It follows from (3) and (4) that

$$f_o(x+y) - f_o(x) - f_o(y) = \rho\left(2f_o\left(\frac{x+y}{2}\right) - f_o(x) - f_o(y)\right)$$
$$= \rho(f_o(x+y) - f_o(x) - f_o(y))$$

and so

$$f_o(x+y) = f_o(x) + f_o(y)$$

for all $x, y \in X$.

Letting $y = x$ in (3) for f_e, we get $\frac{1}{2}f_e(2x) - 2f_e(x) = 0$ and so $f_e(2x) = 4f_e(x)$ for all $x \in X$. Thus

$$f_e\left(\frac{x}{2}\right) = \frac{1}{4}f_e(x) \tag{5}$$

for all $x \in X$.

It follows from (3) and (5) that

$$\frac{1}{2}f_e(x+y) + \frac{1}{2}f_e(x-y) - f_e(x) - f_e(y)$$
$$= \rho\left(2f_e\left(\frac{x+y}{2}\right) + 2f_e\left(\frac{x-y}{2}\right) - f_e(x) - f_e(y)\right)$$
$$= \rho\left(\frac{1}{2}f_e(x+y) + \frac{1}{2}f_e(x-y) - f_e(x) - f_e(y)\right)$$

and so

$$f_e(x+y) + f_e(x-y) = 2f_e(x) + 2f_e(y)$$

for all $x, y \in X$.

Therefore, the mapping $f : X \to Y$ is the sum of the Cauchy additive mapping f_o and the quadratic mapping f_e.

We prove the Hyers–Ulam stability of the additive-quadratic ρ-functional equation (3) in Banach spaces.

Theorem 1 *Let $\varphi : X^2 \to [0, \infty)$ be a function and let $f : X \to Y$ be a mapping satisfying $f(0) = 0$ and*

$$\sum_{j=1}^{\infty} 4^j \varphi\left(\frac{x}{2^j}, \frac{y}{2^j}\right) < \infty \tag{6}$$

$$\|M_1 f(x,y) - \rho M_2 f(x,y)\| \le \varphi(x,y) \tag{7}$$

for all $x, y \in X$. Then there exist a unique additive mapping $A : X \to Y$ and a unique quadratic mapping $Q : X \to Y$ such that

$$\|f_o(x) - A(x)\| \leq \frac{1}{4}\Psi(x, x) + \frac{1}{4}\Psi(-x, -x), \tag{8}$$

$$\|f_e(x) - Q(x)\| \leq \frac{1}{4}\Phi(x, x) + \frac{1}{4}\Phi(-x, -x) \tag{9}$$

for all $x \in X$, where $\Psi(x, y) := \sum_{j=1}^{\infty} 2^j \varphi\left(\frac{x}{2^j}, \frac{y}{2^j}\right)$ and $\Phi(x, y) := \sum_{j=1}^{\infty} 4^j \varphi\left(\frac{x}{2^j}, \frac{y}{2^j}\right)$ for all $x, y \in X$.

Proof Letting $y = x$ in (7) for f_o, we get

$$\|f_o(2x) - 2f_o(x)\| \leq \frac{1}{2}\varphi(x, x) + \frac{1}{2}\varphi(-x, -x) \tag{10}$$

for all $x \in X$. So

$$\left\|f_o(x) - 2f_o\left(\frac{x}{2}\right)\right\| \leq \frac{1}{2}\varphi\left(\frac{x}{2}, \frac{x}{2}\right) + \frac{1}{2}\varphi\left(-\frac{x}{2}, -\frac{x}{2}\right)$$

for all $x \in X$. Hence

$$\begin{aligned}\left\|2^l f_o\left(\frac{x}{2^l}\right) - 2^m f_o\left(\frac{x}{2^m}\right)\right\| &\leq \sum_{j=l}^{m-1}\left\|2^j f_o\left(\frac{x}{2^j}\right) - 2^{j+1} f_o\left(\frac{x}{2^{j+1}}\right)\right\| \\ &\leq \frac{1}{2}\sum_{j=l}^{m-1}\left(2^j\varphi\left(\frac{x}{2^{j+1}}, \frac{x}{2^{j+1}}\right) + 2^j\varphi\left(-\frac{x}{2^{j+1}}, -\frac{x}{2^{j+1}}\right)\right)\end{aligned} \tag{11}$$

for all nonnegative integers m and l with $m > l$ and all $x \in X$. It follows from (11) that the sequence $\{2^k f_o(\frac{x}{2^k})\}$ is Cauchy for all $x \in X$. Since Y is a Banach space, the sequence $\{2^k f_o(\frac{x}{2^k})\}$ converges. So one can define the mapping $A : X \to Y$ by

$$A(x) := \lim_{k\to\infty} 2^k f_o\left(\frac{x}{2^k}\right)$$

for all $x \in X$. Since f_o is an odd mapping, A is an odd mapping. Moreover, letting $l = 0$ and passing the limit $m \to \infty$ in (11), we get (8).

It follows from (6) and (7) that

$$\left\|A(x + y) - A(x) - A(y) - \rho\left(2A\left(\frac{x + y}{2}\right) - A(x) - A(y)\right)\right\|$$

$$= \lim_{n\to\infty} \left\| 2^n \left(f_o\left(\frac{x+y}{2^n}\right) - f_o\left(\frac{x}{2^n}\right) - f_o\left(\frac{y}{2^n}\right)\right) \right.$$
$$\left. -2^n \rho \left(2 f_o\left(\frac{x+y}{2^{n+1}}\right) - f_o\left(\frac{x}{2^n}\right) - f_o\left(\frac{y}{2^n}\right)\right) \right\|$$
$$\leq \frac{1}{2} \lim_{n\to\infty} \left(2^n \varphi\left(\frac{x}{2^n}, \frac{y}{2^n}\right) + 2^n \varphi\left(-\frac{x}{2^n}, -\frac{y}{2^n}\right)\right) = 0$$

for all $x, y \in X$. So

$$A(x+y) - A(x) - A(y) = \rho\left(2A\left(\frac{x+y}{2}\right) - A(x) - A(y)\right)$$

for all $x, y \in X$. By Lemma 2, the mapping $A : X \to Y$ is additive.

Now, let $T : X \to Y$ be another additive mapping satisfying (8). Then we have

$$\begin{aligned} \|A(x) - T(x)\| &= \left\| 2^q A\left(\frac{x}{2^q}\right) - 2^q T\left(\frac{x}{2^q}\right)\right\| \\ &\leq \left\| 2^q A\left(\frac{x}{2^q}\right) - 2^q f_o\left(\frac{x}{2^q}\right)\right\| \\ &\quad + \left\| 2^q T\left(\frac{x}{2^q}\right) - 2^q f_o\left(\frac{x}{2^q}\right)\right\| \\ &\leq \frac{1}{4}\left(2^q \Psi\left(\frac{x}{2^q}, \frac{x}{2^q}\right) + 2^q \Psi\left(-\frac{x}{2^q}, -\frac{x}{2^q}\right)\right), \end{aligned}$$

which tends to zero as $q \to \infty$ for all $x \in X$. So we can conclude that $A(x) = T(x)$ for all $x \in X$. This proves the uniqueness of A.

Letting $y = x$ in (7) for f_e, we get

$$\left\| \frac{1}{2} f_e(2x) - 2 f_e(x)\right\| \leq \frac{1}{2}\varphi(x, x) + \frac{1}{2}\varphi(-x, -x) \tag{12}$$

for all $x \in X$. So

$$\left\| f_e(x) - 4 f_e\left(\frac{x}{2}\right)\right\| \leq \varphi\left(\frac{x}{2}, \frac{x}{2}\right) + \varphi\left(-\frac{x}{2}, -\frac{x}{2}\right)$$

for all $x \in X$. Hence

$$\left\| 4^l f_e\left(\frac{x}{2^l}\right) - 4^m f_e\left(\frac{x}{2^m}\right)\right\| \leq \sum_{j=l}^{m-1} \left\| 4^j f_e\left(\frac{x}{2^j}\right) - 4^{j+1} f_e\left(\frac{x}{2^{j+1}}\right)\right\|$$
$$\leq \sum_{j=l}^{m-1} \left(4^j \varphi\left(\frac{x}{2^{j+1}}, \frac{x}{2^{j+1}}\right) + 4^j \varphi\left(-\frac{x}{2^{j+1}}, -\frac{x}{2^{j+1}}\right)\right) \tag{13}$$

for all nonnegative integers m and l with $m > l$ and all $x \in X$. It follows from (13) that the sequence $\{4^k f_e(\frac{x}{2^k})\}$ is Cauchy for all $x \in X$. Since Y is a Banach space, the sequence $\{4^k f_e(\frac{x}{2^k})\}$ converges. So one can define the mapping $Q : X \to Y$ by

$$Q(x) := \lim_{k\to\infty} 4^k f_e\left(\frac{x}{2^k}\right)$$

for all $x \in X$. Since f_e is an even mapping, Q is an even mapping. Moreover, letting $l = 0$ and passing the limit $m \to \infty$ in (13), we get (9).

It follows from (6) and (7) that

$$\begin{aligned}
&\left\| \frac{1}{2}Q\left(\frac{x+y}{2}\right) + \frac{1}{2}Q\left(\frac{x-y}{2}\right) - Q(x) - Q(y) \right.\\
&\qquad \left. - \rho\left(2Q\left(\frac{x+y}{2}\right) + 2Q\left(\frac{x-y}{2}\right) - Q(x) - Q(y)\right)\right\| \\
&= \lim_{n\to\infty} \left\| 4^n\left(\frac{1}{2}f_e\left(\frac{x+y}{2^n}\right) + \frac{1}{2}f_e\left(\frac{x-y}{2^n}\right) - f_e\left(\frac{x}{2^n}\right) - f_e\left(\frac{y}{2^n}\right)\right)\right.\\
&\qquad \left. - 4^n\rho\left(2f_e\left(\frac{x+y}{2^{n+1}}\right) + 2f_e\left(\frac{x-y}{2^{n+1}}\right) - f_e\left(\frac{x}{2^n}\right) - f_e\left(\frac{y}{2^n}\right)\right)\right\| \\
&\le \frac{1}{2}\lim_{n\to\infty}\left(4^n\varphi\left(\frac{x}{2^n}, \frac{y}{2^n}\right) + 4^n\varphi\left(-\frac{x}{2^n}, -\frac{y}{2^n}\right)\right) = 0
\end{aligned}$$

for all $x, y \in X$. So

$$\begin{aligned}
&\frac{1}{2}Q\left(\frac{x+y}{2}\right) + \frac{1}{2}Q\left(\frac{x-y}{2}\right) - Q(x) - Q(y) \\
&\quad = \rho\left(2Q\left(\frac{x+y}{2}\right) + 2Q\left(\frac{x-y}{2}\right) - Q(x) - Q(y)\right)
\end{aligned}$$

for all $x, y \in X$. By Lemma 2, the mapping $Q : X \to Y$ is quadratic.

Now, let $T : X \to Y$ be another quadratic mapping satisfying (9). Then we have

$$\begin{aligned}
\|Q(x) - T(x)\| &= \left\|4^q Q\left(\frac{x}{2^q}\right) - 4^q T\left(\frac{x}{2^q}\right)\right\| \\
&\le \left\|4^q Q\left(\frac{x}{2^q}\right) - 4^q f_e\left(\frac{x}{2^q}\right)\right\| \\
&\quad + \left\|4^q T\left(\frac{x}{2^q}\right) - 4^q f_e\left(\frac{x}{2^q}\right)\right\| \\
&\le 4^{q-1}\Phi\left(\frac{x}{2^q}, \frac{x}{2^q}\right) + 4^{q-1}\Phi\left(-\frac{x}{2^q}, -\frac{x}{2^q}\right),
\end{aligned}$$

which tends to zero as $q \to \infty$ for all $x \in X$. So we can conclude that $Q(x) = T(x)$ for all $x \in X$. This proves the uniqueness of Q, as desired.

Corollary 1 *Let $r > 2$ and θ be nonnegative real numbers, and let $f : X \to Y$ be a mapping satisfying $f(0) = 0$ and*

$$\|M_1 f(x, y) - \rho M_2 f(x, y)\| \le \theta(\|x\|^r + \|y\|^r) \tag{14}$$

for all $x, y \in X$. Then there exist a unique additive mapping $A : X \to Y$ and a unique quadratic mapping $Q : X \to Y$ such that

$$\|f_o(x) - A(x)\| \le \frac{2\theta}{2^r - 2}\|x\|^r,$$
$$\|f_e(x) - Q(x)\| \le \frac{4\theta}{2^r - 4}\|x\|^r$$

for all $x \in X$.

Theorem 2 *Let $\varphi : X^2 \to [0, \infty)$ be a function and let $f : X \to Y$ be a mapping satisfying $f(0) = 0$, (7) and*

$$\sum_{j=0}^{\infty} \frac{1}{2^j}\varphi(2^j x, 2^j y) < \infty$$

for all $x, y \in X$. Then there exist a unique additive mapping $A : X \to Y$ and a unique quadratic mapping $Q : X \to Y$ such that

$$\|f_o(x) - A(x)\| \le \frac{1}{4}\Psi(x, x) + \frac{1}{4}\Psi(-x, -x), \tag{15}$$
$$\|f_e(x) - Q(x)\| \le \frac{1}{4}\Phi(x, x) + \frac{1}{4}\Phi(-x, -x) \tag{16}$$

for all $x \in X$, where $\Psi(x, y) := \sum_{j=0}^{\infty} \frac{1}{2^j}\varphi(2^j x, 2^j y)$ and $\Phi(x, y) := \sum_{j=0}^{\infty} \frac{1}{4^j}\varphi(2^j x, 2^j y)$ for all $x, y \in X$.

Proof It follows from (10) that

$$\left\| f_o(x) - \frac{1}{2} f_o(2x) \right\| \le \frac{1}{4}\varphi(x, x) + \frac{1}{4}\varphi(-x, -x)$$

for all $x \in X$. Hence

$$\left\| \frac{1}{2^l} f_o(2^l x) - \frac{1}{2^m} f_o(2^m x) \right\| \le \sum_{j=l}^{m-1} \left\| \frac{1}{2^j} f_o\left(2^j x\right) - \frac{1}{2^{j+1}} f_o\left(2^{j+1} x\right) \right\|$$
$$\le \sum_{j=l}^{m-1} \left(\frac{1}{2^{j+2}}\varphi(2^j x, 2^j x) + \frac{1}{2^{j+2}}\varphi(-2^j x, -2^j x) \right) \tag{17}$$

for all nonnegative integers m and l with $m > l$ and all $x \in X$. It follows from (17) that the sequence $\{\frac{1}{2^n} f_o(2^n x)\}$ is a Cauchy sequence for all $x \in X$. Since Y is complete, the sequence $\{\frac{1}{2^n} f_o(2^n x)\}$ converges. So one can define the mapping $A : X \to Y$ by

$$A(x) := \lim_{n\to\infty} \frac{1}{2^n} f_o(2^n x)$$

for all $x \in X$. Moreover, letting $l = 0$ and passing the limit $m \to \infty$ in (17), we get (15).

It follows from (12) that

$$\left\| f_e(x) - \frac{1}{4} f_e(2x) \right\| \le \frac{1}{4}\varphi(x, x) + \frac{1}{4}\varphi(-x, -x)$$

for all $x \in X$. Hence

$$\begin{aligned} \left\| \frac{1}{4^l} f_e(2^l x) - \frac{1}{4^m} f_e(2^m x) \right\| &\le \sum_{j=l}^{m-1} \left\| \frac{1}{4^j} f_e\left(2^j x\right) - \frac{1}{4^{j+1}} f_e\left(2^{j+1} x\right) \right\| \\ &\le \sum_{j=l}^{m-1} \left(\frac{1}{4^{j+1}} \varphi(2^j x, 2^j x) + \frac{1}{4^{j+1}} \varphi(-2^j x, -2^j x) \right) \end{aligned} \tag{18}$$

for all nonnegative integers m and l with $m > l$ and all $x \in X$. It follows from (18) that the sequence $\{\frac{1}{4^n} f_e(2^n x)\}$ is a Cauchy sequence for all $x \in X$. Since Y is complete, the sequence $\{\frac{1}{4^n} f_e(2^n x)\}$ converges. So one can define the mapping $Q : X \to Y$ by

$$Q(x) := \lim_{n\to\infty} \frac{1}{4^n} f_e(2^n x)$$

for all $x \in X$. Moreover, letting $l = 0$ and passing the limit $m \to \infty$ in (18), we get (16).

The rest of the proof is similar to the proof of Theorem 1.

Corollary 2 *Let $r < 1$ and θ be nonnegative real numbers, and let $f : X \to Y$ be a mapping satisfying $f(0) = 0$ and (14). Then there exist a unique additive mapping $A : X \to Y$ and a unique quadratic mapping $Q : X \to Y$ such that*

$$\begin{aligned} \|f_o(x) - A(x)\| &\le \frac{2\theta}{2 - 2^r} \|x\|^r, \\ \|f_e(x) - Q(x)\| &\le \frac{4\theta}{4 - 2^r} \|x\|^r \end{aligned}$$

for all $x \in X$.

Additive-Quadratic ρ-Functional Equation (2) in Banach Spaces

We solve and investigate the additive-quadratic ρ-functional equation (2) in normed spaces.

Lemma 3 *If a mapping $f : X \to Y$ satisfies $f(0) = 0$ and*

$$M_2 f(x, y) = \rho M_1 f(x, y) \tag{19}$$

for all $x, y \in X$, then $f : X \to Y$ is the sum of the Cauchy additive mapping f_o and the quadratic mapping f_e.

Proof Letting $y = 0$ in (19) for f_o, we get

$$f_o\left(\frac{x}{2}\right) = \frac{1}{2} f_o(x) \tag{20}$$

for all $x \in X$.

It follows from (19) and (20) that

$$\begin{aligned} f_o(x+y) - f_o(x) - f_o(y) &= 2 f_o\left(\frac{x+y}{2}\right) - f_o(x) - f_o(y) \\ &= \rho(f_o(x+y) - f_o(x) - f_o(y)) \end{aligned}$$

and so

$$f_o(x+y) = f_o(x) + f_o(y)$$

for all $x, y \in X$.

Letting $y = 0$ in (19) for f_e, we get

$$f_e\left(\frac{x}{2}\right) = \frac{1}{4} f_e(x) \tag{21}$$

for all $x \in X$.

It follows from (19) and (21) that

$$\begin{aligned} &\frac{1}{2} f_e(x+y) + \frac{1}{2} f_e(x-y) - f_e(x) - f_e(y) \\ &= 2 f_e\left(\frac{x+y}{2}\right) + 2 f_e\left(\frac{x-y}{2}\right) - f_e(x) - f_e(y) \\ &= \rho\left(\frac{1}{2} f_e(x+y) + \frac{1}{2} f_e(x-y) - f_e(x) - f_e(y)\right) \end{aligned}$$

and so

$$f_e(x+y) + f_e(x-y) = 2f_e(x) + 2f_e(y)$$

for all $x, y \in X$.

We prove the Hyers–Ulam stability of the additive-quadratic ρ-functional equation (19) in Banach spaces.

Theorem 3 *Let $\varphi : X^2 \to [0, \infty)$ be a function and let $f : X \to Y$ be a mapping satisfying $f(0) = 0$ and*

$$\sum_{j=0}^{\infty} 4^j \varphi\left(\frac{x}{2^j}, \frac{y}{2^j}\right) < \infty,$$

$$\|M_2 f(x,y) - \rho M_1 f(x,y)\| \le \varphi(x,y) \tag{22}$$

for all $x, y \in X$. Then there exist a unique additive mapping $A : X \to Y$ and a unique quadratic mapping $Q : X \to Y$ such that

$$\|f_o(x) - A(x)\| \le \frac{1}{2}\Psi(x,0) + \frac{1}{2}\Psi(-x,0), \tag{23}$$

$$\|f_e(x) - Q(x)\| \le \frac{1}{2}\Phi(x,0) + \frac{1}{2}\Phi(-x,0) \tag{24}$$

for all $x \in X$, where $\Psi(x,y) := \sum_{j=0}^{\infty} 2^j \varphi\left(\frac{x}{2^j}, \frac{y}{2^j}\right)$ and $\Phi(x,y) := \sum_{j=0}^{\infty} 4^j \varphi\left(\frac{x}{2^j}, \frac{y}{2^j}\right)$ for all $x, y \in X$.

Proof Letting $y = 0$ in (22) for f_o, we get

$$\left\| f_o(x) - 2f_o\left(\frac{x}{2}\right)\right\| = \left\|2f_o\left(\frac{x}{2}\right) - f_o(x)\right\| \le \frac{1}{2}\varphi(x,0) + \frac{1}{2}\varphi(x,0) \tag{25}$$

for all $x \in X$. So

$$\left\|2^l f_o\left(\frac{x}{2^l}\right) - 2^m f_o\left(\frac{x}{2^m}\right)\right\| \le \sum_{j=l}^{m-1} \left\|2^j f_o\left(\frac{x}{2^j}\right) - 2^{j+1} f_o\left(\frac{x}{2^{j+1}}\right)\right\|$$

$$\le \frac{1}{2}\sum_{j=l}^{m-1}\left(2^j\varphi\left(\frac{x}{2^j}, 0\right) + 2^j\varphi\left(-\frac{x}{2^j}, 0\right)\right) \tag{26}$$

for all nonnegative integers m and l with $m > l$ and all $x \in X$. It follows from (26) that the sequence $\{2^k f_o(\frac{x}{2^k})\}$ is Cauchy for all $x \in X$. Since Y is a Banach space, the sequence $\{2^k f_o(\frac{x}{2^k})\}$ converges. So one can define the mapping $A : X \to Y$ by

$$A(x) := \lim_{k\to\infty} 2^k f_o\left(\frac{x}{2^k}\right)$$

for all $x \in X$. Since f_o is an odd mapping, A is an odd mapping. Moreover, letting $l = 0$ and passing the limit $m \to \infty$ in (26), we get (23).

Letting $y = 0$ in (22) for f_e, we get

$$\left\| f_e(x) - 4f_e\left(\frac{x}{2}\right)\right\| = \left\|4f_e\left(\frac{x}{2}\right) - f_e(x)\right\| \le \frac{1}{2}\varphi(x,0) + \frac{1}{2}\varphi(-x,0) \quad (27)$$

for all $x \in X$. So

$$\begin{aligned}\left\|4^l f_e\left(\frac{x}{2^l}\right) - 4^m f_e\left(\frac{x}{2^m}\right)\right\| &\le \sum_{j=l}^{m-1}\left\|4^j f_e\left(\frac{x}{2^j}\right) - 4^{j+1} f_e\left(\frac{x}{2^{j+1}}\right)\right\| \\ &\le \frac{1}{2}\sum_{j=l}^{m-1}\left(4^j\varphi\left(\frac{x}{2^j},0\right) + 4^j\varphi\left(-\frac{x}{2^j},0\right)\right) \quad (28)\end{aligned}$$

for all nonnegative integers m and l with $m > l$ and all $x \in X$. It follows from (28) that the sequence $\{4^k f_e(\frac{x}{2^k})\}$ is Cauchy for all $x \in X$. Since Y is a Banach space, the sequence $\{4^k f_e(\frac{x}{2^k})\}$ converges. So one can define the mapping $Q : X \to Y$ by

$$Q(x) := \lim_{k\to\infty} 4^k f_e\left(\frac{x}{2^k}\right)$$

for all $x \in X$. Since f_e is an even mapping, Q is an even mapping. Moreover, letting $l = 0$ and passing the limit $m \to \infty$ in (28), we get (24).

The rest of the proof is similar to the proof of Theorem 1.

Corollary 3 *Let $r > 2$ and θ be nonnegative real numbers, and let $f : X \to Y$ be a mapping satisfying $f(0) = 0$ and*

$$\|M_2 f(x,y) - \rho M_1 f(x,y)\| \le \theta(\|x\|^r + \|y\|^r) \quad (29)$$

for all $x, y \in X$. Then there exist a unique additive mapping $A : X \to Y$ and a unique quadratic mapping $Q : X \to Y$ such that

$$\|f_o(x) - A(x)\| \le \frac{2^r\theta}{2^r - 2}\|x\|^r,$$

$$\|f_e(x) - Q(x)\| \le \frac{2^r\theta}{2^r - 4}\|x\|^r$$

for all $x \in X$.

Theorem 4 *Let $\varphi : X^2 \to [0, \infty)$ be a function and let $f : X \to Y$ be a mapping satisfying $f(0) = 0$, (22) and*

$$\sum_{j=1}^{\infty} \frac{1}{2^j} \varphi(2^j x, 2^j y) < \infty$$

for all $x, y \in X$. Then there exist a unique additive mapping $A : X \to Y$ and a unique quadratic mapping $Q : X \to Y$ such that

$$\|f_o(x) - A(x)\| \leq \frac{1}{2}\Psi(x, 0) + \frac{1}{2}\Psi(-x, 0), \tag{30}$$

$$\|f_e(x) - Q(x)\| \leq \frac{1}{2}\Phi(x, 0) + \frac{1}{2}\Phi(-x, 0) \tag{31}$$

for all $x \in X$, where $\Psi(x, y) := \sum_{j=1}^{\infty} \frac{1}{2^j} \varphi(2^j x, 2^j y)$ and $\Phi(x, y) := \sum_{j=1}^{\infty} \frac{1}{4^j} \varphi(2^j x, 2^j y)$ for all $x \in X$.

Proof It follows from (25) that

$$\left\| f_o(x) - \frac{1}{2} f_o(2x) \right\| \leq \frac{1}{4}\varphi(2x, 0) + \frac{1}{4}\varphi(-2x, 0)$$

for all $x \in X$. Hence

$$\begin{aligned} \left\| \frac{1}{2^l} f_o(2^l x) - \frac{1}{2^m} f_o(2^m x) \right\| &\leq \sum_{j=l}^{m-1} \left\| \frac{1}{2^j} f_o\left(2^j x\right) - \frac{1}{2^{j+1}} f_o\left(2^{j+1} x\right) \right\| \\ &\leq \frac{1}{2} \sum_{j=l+1}^{m} \left(\frac{1}{2^j}\varphi(2^j x, 0) + \frac{1}{2^j}\varphi(-2^j x, 0) \right) \end{aligned} \tag{32}$$

for all nonnegative integers m and l with $m > l$ and all $x \in X$. It follows from (32) that the sequence $\{\frac{1}{2^n} f_o(2^n x)\}$ is a Cauchy sequence for all $x \in X$. Since Y is complete, the sequence $\{\frac{1}{2^n} f_o(2^n x)\}$ converges. So one can define the mapping $A : X \to Y$ by

$$A(x) := \lim_{n\to\infty} \frac{1}{2^n} f_o(2^n x)$$

for all $x \in X$. Moreover, letting $l = 0$ and passing the limit $m \to \infty$ in (32), we get (30).

It follows from (27) that

$$\left\| f_e(x) - \frac{1}{4} f_e(2x) \right\| \leq \frac{1}{8}\varphi(2x, 0) + \frac{1}{8}\varphi(-2x, 0)$$

for all $x \in X$. Hence

$$\left\| \frac{1}{4^l} f_e(2^l x) - \frac{1}{4^m} f_e(2^m x) \right\| \le \sum_{j=l}^{m-1} \left\| \frac{1}{4^j} f_e\left(2^j x\right) - \frac{1}{4^{j+1}} f_e\left(2^{j+1} x\right) \right\|$$

$$\le \frac{1}{2} \sum_{j=l+1}^{m} \left(\frac{1}{4^j} \varphi(2^j x, 0) + \frac{1}{4^j} \varphi(-2^j x, 0) \right) \tag{33}$$

for all nonnegative integers m and l with $m > l$ and all $x \in X$. It follows from (33) that the sequence $\{\frac{1}{4^n} f_e(2^n x)\}$ is a Cauchy sequence for all $x \in X$. Since Y is complete, the sequence $\{\frac{1}{4^n} f_e(2^n x)\}$ converges. So one can define the mapping $Q : X \to Y$ by

$$Q(x) := \lim_{n \to \infty} \frac{1}{4^n} f_e(2^n x)$$

for all $x \in X$. Moreover, letting $l = 0$ and passing the limit $m \to \infty$ in (33), we get (31).

The rest of the proof is similar to the proof of Theorem 1.

Corollary 4 *Let $r < 1$ and θ be positive real numbers, and let $f : X \to Y$ be a mapping satisfying (29). Then there exist a unique additive mapping $A : X \to Y$ and a unique quadratic mapping $Q : X \to Y$ such that*

$$\|f_o(x) - A(x)\| \le \frac{2^r \theta}{2 - 2^r} \|x\|^r,$$

$$\|f_e(x) - Q(x)\| \le \frac{2^r \theta}{4 - 2^r} \|x\|^r$$

for all $x \in X$.

References

1. Adam, M.: On the stability of some quadratic functional equation. J. Nonlinear Sci. Appl. **4**, 50–59 (2011)
2. Aoki, T.: On the stability of the linear transformation in Banach spaces. J. Math. Soc. Jpn. **2**, 64–66 (1950)
3. Cădariu, L., Găvruta L., Găvruta, P.: On the stability of an affine functional equation. J. Nonlinear Sci. Appl. **6**, 60–67 (2013)
4. Chahbi, A., Bounader, N.: On the generalized stability of d'Alembert functional equation. J. Nonlinear Sci. Appl. **6**, 198–204 (2013)
5. Cholewa, P.W.: Remarks on the stability of functional equations. Aequationes Math. **27**, 76–86 (1984)

6. Eskandani, G.Z., Găvruta, P.: Hyers-Ulam-Rassias stability of pexiderized Cauchy functional equation in 2-Banach spaces. J. Nonlinear Sci. Appl. **5**, 459–465 (2012)
7. Găvruta, P.: A generalization of the Hyers-Ulam-Rassias stability of approximately additive mappings. J. Math. Anal. Appl. **184**, 431–436 (1994)
8. Hyers, D.H.: On the stability of the linear functional equation. Proc. Natl. Acad. Sci. U. S. A. **27**, 222–224 (1941)
9. Park, C.: Orthogonal stability of a cubic-quartic functional equation. J. Nonlinear Sci. Appl. **5**, 28–36 (2012)
10. Park, C., Ghasemi, K., Ghaleh, S.G., Jang, S.: Approximate n-Jordan $*$-homomorphisms in C^*-algebras. J. Comput. Anal. Appl. **15**, 365–368 (2013)
11. Park, C., Najati, A., Jang, S.: Fixed points and fuzzy stability of an additive-quadratic functional equation. J. Comput. Anal. Appl. **15**, 452–462 (2013)
12. Rassias, T.M.: On the stability of the linear mapping in Banach spaces. Proc. Am. Math. Soc. **72**, 297–300 (1978)
13. Ravi, K., Thandapani, E., Senthil Kumar, B.V.: Solution and stability of a reciprocal type functional equation in several variables. J. Nonlinear Sci. Appl. **7**, 18–27 (2014)
14. Schin, S., Ki, D., Chang, J., Kim, M.: Random stability of quadratic functional equations: a fixed point approach. J. Nonlinear Sci. Appl. **4**, 37–49 (2011)
15. Shagholi, S., Bavand Savadkouhi, M., Eshaghi Gordji, M.: Nearly ternary cubic homomorphism in ternary Fréchet algebras. J. Comput. Anal. Appl. **13**, 1106–1114 (2011)
16. Shagholi, S., Eshaghi Gordji, M., Bavand Savadkouhi, M.: Stability of ternary quadratic derivation on ternary Banach algebras. J. Comput. Anal. Appl. **13**, 1097–1105 (2011)
17. Shin, D., Park, C., Farhadabadi, S.: On the superstability of ternary Jordan C^*-homomorphisms. J. Comput. Anal. Appl. **16**, 964–973 (2014)
18. Shin, D., Park, C., Farhadabadi, S.: Stability and superstability of J^*-homomorphisms and J^*-derivations for a generalized Cauchy-Jensen equation. J. Comput. Anal. Appl. **17**, 125–134 (2014)
19. Skof, F.: Proprietà locali e approssimazione di operatori. Rend. Sem. Mat. Fis. Milano **53**, 113–129 (1983)
20. Ulam, S.M.: Problems in Modern Mathematics. Chapter VI, Science edn. Wiley, New York (1940)
21. Zaharia, C.: On the probabilistic stability of the monomial functional equation. J. Nonlinear Sci. Appl. **6**, 51–59 (2013)
22. Zolfaghari, S.: Approximation of mixed type functional equations in p-Banach spaces. J. Nonlinear Sci. Appl. **3**, 110–122 (2010)

De Bruijn Sequences and Suffix Arrays: Analysis and Constructions

Konstantinos Limniotis, Nicholas Kolokotronis, and Dimitrios Kotanidis

Abstract Binary De Bruijn sequences are studied in this work, by means of analysing the properties of their suffix arrays. More precisely, specific properties that are present in the suffix array of any binary De Bruijn sequence are proved; by this way, it is shown that building an appropriate array based on these properties is equivalent to constructing a De Bruijn sequence. Therefore, simultaneous construction of a De Bruijn sequence and its suffix array is possible, whereas any possible De Bruijn sequence can be theoretically obtained. It is also shown that known constructions of De Bruijn sequences, such as the method of cross-join pairs, can be facilitated by the use of suffix arrays.

Introduction

De Bruijn sequences have a prominent role in many areas, including robotics, bioinformatics (e.g. DNA sequence assembly) and cryptography. Especially in cryptography, De Bruijn sequences received great attention during the past years, since they are strongly associated with maximum-period nonlinear feedback shift registers (NLFSRs)—i.e. specific-type finite state machines generating sequences with the maximum possible period [14]. More precisely, NLFSRs form the basic building blocks of contemporary stream ciphers (e.g. Grain, Trivium) [28].

K. Limniotis (✉)
School of Pure and Applied Sciences, Open University of Cyprus, Latsia, Cyprus

Hellenic Data Protection Authority, Athens, Greece
e-mail: konstantinos.limniotis@ouc.ac.cy; klimniotis@dpa.gr

N. Kolokotronis
Department of Informatics and Telecommunications, University of Peloponnese, Tripoli, Greece
e-mail: nkolok@uop.gr

D. Kotanidis
School of Pure and Applied Sciences, Open University of Cyprus, Latsia, Cyprus
e-mail: dkotanid@sch.gr

N. J. Daras, Th. M. Rassias (eds.), *Modern Discrete Mathematics and Analysis*, Springer Optimization and Its Applications 131,
https://doi.org/10.1007/978-3-319-74325-7_15

There are several known constructions of De Bruijn sequences [8]; many of them are based on forming De Bruijn sequences of some order n starting from De Bruijn sequences of lower order (see, e.g. [9, 13, 19, 27]), whereas several other primary constructions are also known (see, e.g. [6, 12]). The known constructions though do not cover the entire space of binary De Bruijn sequences, which has cardinality $2^{2^{n-1}-n}$ [10], i.e. the number of Hamiltonian paths in a De Bruijn graph. Therefore, it remains an open problem to describe a method that is capable to generate any possible De Bruijn sequence. Similarly, from a cryptographic point of view, it is still unknown how to construct a NLFSR that generates a De Bruijn sequence.

In this work, a new approach to study binary De Bruijn sequences is proposed, based on the properties of their suffix arrays; such arrays are generally known as a convenient tool to handle sequences (e.g. to search for specific patterns within the sequence) [22]. More precisely, since any sequence is uniquely associated with its suffix array, we prove specific properties that the suffix array of an arbitrary De Bruijn sequence possesses. It is shown that constructing a suffix array with these properties is equivalent to constructing a De Bruijn sequence. Moreover, we show that any possible binary De Bruijn sequence can be obtained, by constructing an appropriate suffix array with the desired properties. Apart from its theoretical interest, this approach directly leads to an algorithmic procedure for constructing binary De Bruijn sequences, performing operations only on the suffix array that is being built; by this way, the simultaneous construction of a De Bruijn sequence and its suffix array can be achieved. In addition, we illustrate a method to appropriately modify a given suffix array of a De Bruijn sequence, so as to produce the suffix array of another different De Bruijn sequence; it is shown that such an approach is strongly associated with the well-known method for obtaining new De Bruijn sequences that is based on the so-called cross-join pairs [7]. Since querying suffix arrays is known to be fast in practice (see, e.g. [30]), our analysis indicates that suffix arrays of De Bruijn sequences can be used to facilitate the construction of new De Bruijn sequences by employing cross-join pairs. The clear advantage is that new De Bruijn sequences that can be subsequently used as a seed to other known recursive or joining constructions can be provided.

The chapter is organized as follows: first, the basic definitions and notation are introduced in section "Preliminaries". In section "Properties of Suffix Arrays of De Bruijn Sequences", the properties that are present in the suffix array of a binary De Bruijn sequence are proved, whilst properties of the corresponding longest common prefix array are also shown. Such properties of the suffix array form the basis to develop a new algorithm for deriving De Bruijn sequences that is described in section "Construction of De Bruijn Sequences Through Suffix Arrays"; it is shown that any possible binary De Bruijn sequence can be obtained by this approach. Section "Suffix Arrays and Cross-Join Pairs" focuses on illustrating how suffix arrays can be utilized to efficiently find out cross-join pairs within a De Bruijn sequence, thus facilitating the construction of new De Bruijn sequences. Finally, concluding remarks are given in section "Conclusions".

Preliminaries

Boolean Functions

A Boolean function with n variables is a mapping $f : \mathbb{F}_2^n \rightarrow \mathbb{F}_2$, where $\mathbb{F}_2 = \{0, 1\}$ is the binary Galois field and $\mathbb{F}_2^n$ is the nth dimensional vector space over $\mathbb{F}_2$. Let $[n] \triangleq \{1, \ldots, n\}$ and $\boldsymbol{x} = (x_1 \cdots x_n)$; any $f \in \mathbb{B}_n$ can be uniquely expressed in its *algebraic normal form* (ANF) as [21]

$$f(\boldsymbol{x}) = \sum_{I \subseteq [n]} v_I \boldsymbol{x}^I, \qquad v_I \in \mathbb{F}_2 \tag{1}$$

where $\boldsymbol{x}^I = \prod_{i \in I} x_i$ (by convention $\boldsymbol{x}^{\emptyset} = 1$) and the sum is taken modulo 2. The degree of f, denoted by $\deg(f)$, is the highest number of variables that appear in a monomial of its ANF; in the special case that $\deg(f) = 1$, we say that f is an *affine* function, and if it additionally satisfies that the constant term in its ANF is zero, then f is *linear*; otherwise, f is *nonlinear* (i.e. if $\deg(f) \geq 2$). The binary vector of length 2^n

$$f = \big(f(0 \cdots 0) \cdots f(1 \cdots 1)\big)$$

comprised by values of f on vectors of $\mathbb{F}_{2^n}$ lexicographically ordered is the truth table of $f \in \mathbb{B}_n$, and the set $\mathsf{supp}(f) = \{\boldsymbol{x} \in \mathbb{F}_2^n : f(\boldsymbol{x}) \neq 0\}$ is called its support. The function $f \in \mathbb{B}_n$ is said to be balanced if its Hamming weight $\mathsf{wt}(f) = |\mathsf{supp}(f)|$ is equal to 2^{n-1}.

Boolean functions constitute important building blocks for several cryptographic primitives; such a use of Boolean functions requires that several cryptographic criteria need to be met [1]. For example, balancedness and high algebraic degree are often prerequisites to withstand specific cryptanalytic attacks.

Keystream Generators

Generating sequences with good cryptographic (i.e. pseudorandomness) properties is of high importance in cryptography; a notable example is the class of stream ciphers, which aims to provide confidentiality in environments characterized by a limited computing power or memory capacity, e.g. when highly constrained devices are interconnected. In the classical model of a synchronous stream cipher (Fig. 1), the encryption (*resp.* decryption) operation is based on xor-ing the plaintext (*resp.* ciphertext) with a so-called keystream, which needs to possess specific cryptographic properties so as to ensure its unpredictability.

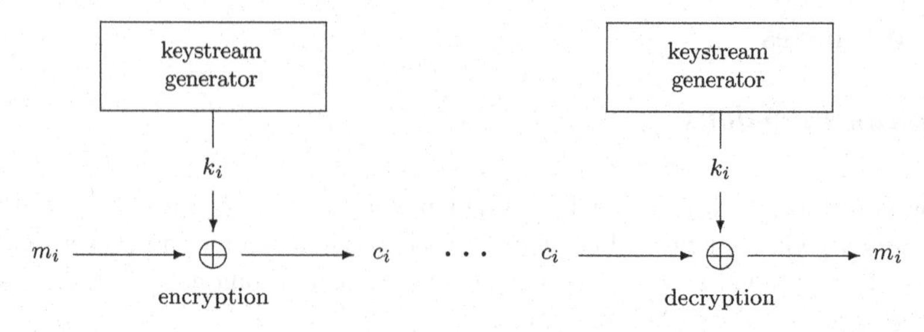

Fig. 1 A typical stream cipher operation

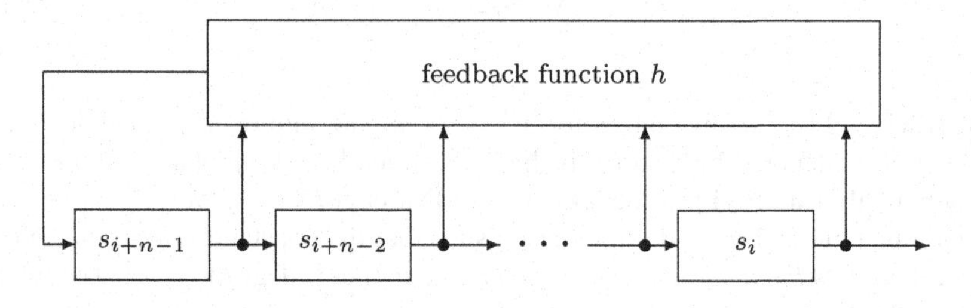

Fig. 2 A diagram of a feedback shift register

Therefore, a keystream generator should suffice to produce cryptographically strong sequences. A basic building block in many keystream generators is a (nonlinear) feedback shift register (NLFSR). A typical diagram of a NLFSR is depicted in Fig. 2. Each such circuit consists of n consecutive 2-state storage units; at each clock pulse, the content of each stage (0 or 1) is shifted to the next stage, whilst the rightmost stage provides the output; the content of the first (leftmost) stage is computed according to a feedback Boolean function $h \in \mathbb{B}_n$

$$s_{i+n} = h(s_{i+n-1}, \ldots, s_i), \qquad i \geq 0 . \tag{2}$$

If h is linear, then we refer to such register as linear feedback shift register (LFSR). LFSRs have been extensively studied throughout the literature and seem to have many appealing properties; however, the underlying linearity of LFSRs has given rise to several successful cryptanalytic attacks (e.g. algebraic attacks that have been mounted against keystream generators consisting of nonlinear filters applied to LFSR), and, thus, nonlinear feedback shift registers are preferable in contemporary ciphers. NLFSRs though have not been studied to the same extent as LFSRs.

At each time instant, the contents of the n stages of the (N)LFSR constitute its current *state*; for any $i \geq 0$, the state at the i-th time instant is denoted by $S_i = (s_{i+n-1} \ \ldots \ s_{i+1} \ s_i)$—i.e. it is an element (vector) of $\mathbb{F}_2^n$. Any such state is subsequently updated by

$$\big(h(s_{i+n-1}, \ \ldots, \ s_{i+1}, \ s_i) \ s_{i+n-1} \ \ldots \ s_{i+1}\big)$$

and the bit s_i is the output. There are 2^n possible initial states, whereas any output sequence $s_0s_1s_2\ldots$ satisfies (2). The set of all possible sequences that can be obtained by any FSR (linear or nonlinear) with feedback function h is denoted by $\mathbb{FSR}(h)$. If h has the form

$$h(s_{i+n-1}, \ldots, s_{i+1}, s_i) = s_i + g(s_{i+n-1}, \ldots, s_{i+1})$$

for given $g \in \mathbb{B}_{n-1}$, then the FSR is called *non-singular*. It is well known [10] that if the function h is non-singular, then starting from any initial state $\boldsymbol{S}_0 = (s_{n-1} \ldots s_1\ s_0)$, consecutive clock pulses will generate a cycle $C = [\boldsymbol{S}_0, \boldsymbol{S}_1, \ldots \boldsymbol{S}_{L-1}]$ for some integer $1 \leq L \leq 2^n$, such that $\boldsymbol{S}_0$ is the next state of $\boldsymbol{S}_{L-1}$—i.e. the sequence of states is periodic. Hence, for any non-singular feedback function h, the space $\mathbb{F}_2^n$ is divided into separate cycles induced by h, and each state—which can be seen as an element of $\mathbb{F}_2^n$—belongs to exactly one cycle. Clearly, the maximum possible size of a cycle is 2^n (containing all the elements of $\mathbb{F}_2^n$), which means that the FSR runs through all the possible 2^n states regardless the choice of the initial state; such an FSR is called a *maximum-period* FSR. Such FSRs are of high cryptographic importance; however, it is generally not known how to choose a proper nonlinear feedback function corresponding to a maximum-period FSR.

For any state $\boldsymbol{S}_i = (s_{i+n-1}, \ldots, s_{i+1}, s_i)$, we define its conjugate state $\hat{\boldsymbol{S}}_i = (s_{i+n-1}, \ldots, s_{i+1}, s_i')$, where s_i' is the binary complement of s_i; clearly, the conjugate of $\hat{\boldsymbol{S}}_i$ is $\boldsymbol{S}_i$. Two cycles C_1 and C_2 are said to be *adjacent* if there exists $\boldsymbol{S}_i \in C_1$ such that $\hat{\boldsymbol{S}}_i \in C_2$. Given those C_1 and C_2, it is easy to see that interchanging the successors of $\boldsymbol{S}_i$ and $\hat{\boldsymbol{S}}_i$ results in obtaining a cycle that contains all the states of C_1 and C_2; such a procedure is known as the cycle joining method [13].

Since any FSR uniquely determines a set of cycles, the following definition has been introduced (see, e.g. [11]).

Definition 1 For given FSR, its adjacency graph is an undirected graph whose vertices correspond to the cycles in it, whereas there exists an edge labeled with an integer $m > 0$ between two vertexes if and only if the two vertices share m conjugate pairs. □

Clearly, the adjacency graph provides information on the distribution of conjugate pairs.

Example 1 Let us consider the NLFSR with length 4 and feedback function

$$h(s_{i+3}, s_{i+2}, s_{i+1}, s_i) = s_i + s_{i+1}s_{i+2} + s_{i+3}, \quad i \geq 0\,.$$

The cycles of this NLFSR are

$$C_1 = \{0000\},$$

$$C_2 = \{0001, 1000, 1100, 1110, 0111, 0011, 1001, 0100, 0010\},$$

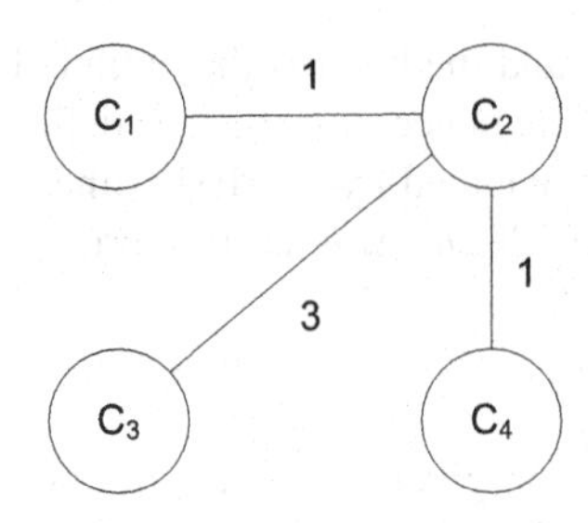

Fig. 3 The adjacency graph of NLFR with feedback function $h(s_{i+3}, s_{i+2}, s_{i+1}, s_i) = s_i + s_{i+1}s_{i+2} + s_{i+3}$ for $i \geq 0$

$$C_3 = \{1010, 1101, 0110, 1011, 0101\},$$

$$C_4 = \{1111\}.$$

Therefore, its adjacency graph is illustrated in Fig. 3. □

Sequences and Suffix Arrays

Let $y = y_0y_1y_2\cdots$ be a sequence over the binary field $\mathbb{F}_2$. We denote by y_i^j, with $i \leq j$, the subsequence (tuple) $y_i \ldots y_j$. Thus, if y has finite-length N, then $y^N \triangleq y_0^{N-1}$ denotes the whole sequence. For any $0 \leq j < N$, the tuple y_0^j is a *prefix* of y^N; for the special case that $j < N-1$, such a prefix is called *proper prefix*. A *suffix* of y^N is any tuple y_j^{N-1}, $0 \leq j < N$, where we can similarly define the proper suffix.

Any n-tuple y_i^{i+n-1} is uniquely associated with an element of the vector space $\mathbb{F}_2^n$—namely, with $(y_i \ y_{i+1} \ \ldots \ y_{i+n-1})$. The corresponding *inverse vector* is defined as $(y_{i+n-1} \ \ldots \ y_{i+1} \ y_i)$.

A periodic sequence is any sequence of infinite length such that there exists $N > 0$ satisfying $y_{i+N} = y_i$ for all $i \geq 0$; the least such integer is called the *period* of y. The aforementioned definitions of prefixes and suffixes are also trivially applied to the part y_0^{N-1} of the periodic sequence y. It is well known [10] that any periodic sequence can be generated by a feedback shift register (FSR) satisfying

$$y_{i+n} = h(y_{i+n-1}, \ldots, y_i), \qquad i \geq 0$$

where $n > 0$ equals the *length* of the FSR and h is an appropriate feedback function. An important indexing data structure for sequences is the so-called suffix array, defined as follows.

Definition 2 ([22]) The *suffix array* S of a binary sequence y^N holds the starting positions (ranging from 0 to $N-1$) of its N lexicographically ordered suffixes; i.e. for $0 \leq i < j < N$, it holds $y_{S[i]}^{N-1} < y_{S[j]}^{N-1}$.

Table 1 The suffix array S of $y = 00010111$

Index k of S	Value of $S[k]$	$y_{S[k]}^{S[k]+n-1}$	Value of $P[k]$
0	0	000	0
1	1	001	2
2	2	010	1
3	4	011	2
4	7	100	0
5	3	101	2
6	6	110	1
7	5	111	2

The suffix array stores the same information as the suffix tree, which is also a well-known indexing data structure [31]; however, suffix arrays are more economical, since suffix trees need too much space [22]. Apart from the suffix array, the *inverse suffix array* S^{-1} is also frequently used; namely, $S^{-1}[i]$ indicates the number of suffixes that are lexicographically smaller than y_i^{N-1}; by definition $S[S^{-1}[i]] = i$, for all $i = 0, 1, \ldots, N-1$. Both data structures have size $\mathcal{O}(N \log N)$ and can be constructed in linear time [15, 17, 18].

Another important structure, which also facilitates pattern searching, is the *longest common prefix* array P which gives the length of the longest common prefix between two subsequences lying in consecutive places of S; more precisely, $P[i]$ equals the length of the longest common prefix between $y_{S[i]}^{N-1}$ and $y_{S[i-1]}^{N-1}$. Linear-time algorithms for computing the longest common prefix array are given in [16], whereas improvements in the space complexity are achieved in [24].

In the sequel, we shall slightly modify the classical definition of the suffix and longest common prefix array, to accommodate our analysis. More precisely, since we shall consider only periodic sequences of period $N = 2^n$, the corresponding suffix array S is being constructed from y^{N+n-1}, and, in particular, we only allow the suffixes y_i^{N+n-2} for $0 \leq i < N$ of length at least n. The longest common prefix array P is similarly defined; note that, with regard to the value of $P[0]$, we consider the length of the longest common prefix between $y_{S[0]}^{n+N-2}$ and $y_{S[N-1]}^{n+N-2}$.

Example 2 Let us consider a binary sequence y with period $N = 8$ for $n = 3$ and $y_0^N = 00010111$. The corresponding suffix array S is given by Table 1, whereas the same table also illustrates, for each suffix $y_{S[k]}^{N+n-2}$, the first n elements—i.e. $y_{S[k]}^{S[k]+2}$. Note that Table 1 also illustrates the longest common prefix array P of y. □

De Bruijn Sequences

A binary De Bruijn sequence of order n is a 2^n-periodic sequence that contains each n-tuple exactly once within one period. For instance, the sequence in Example 2 is a De Bruijn sequence of order 3. Any De Bruijn sequence is balanced, and, moreover,

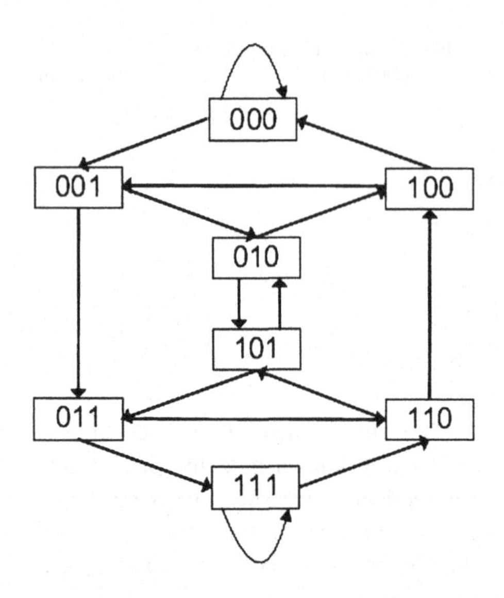

Fig. 4 The De Bruijn graph of order 3

it also satisfies several other pseudorandom properties. Clearly, any maximum-period FSR generates a De Bruijn sequence. It is easy to see that if h is the feedback function of an FSR that generates a De Bruijn sequence, it holds $h(0, 0, \ldots, 0) = 1$ (to avoid the all-zeros cycle) and $h(1, 1, \ldots, 1) = 0$ (to avoid the all-ones cycle). Choosing though a NLFSR that generates De Bruijn sequence is still an open problem as also stated in section "Keystream Generators", since only a few such feedback function families are known [4].

Binary De Bruijn sequences are strongly associated with the so-called binary de Bruijn graphs; as an example, the De Bruijn graph of order $n = 3$ is illustrated in Fig. 4. More precisely, a binary De Bruijn graph G_n of order n is a directed graph with 2^n nodes, each corresponding to an element of $\mathbb{F}_2^n$ and having an edge from node $S = (y_{i_0}\ y_{i_1}\ \ldots\ y_{i_{n-1}})$ to node $T = (y_{j_0}\ y_{j_1}\ \ldots\ y_{j_{n-1}})$ if and only if

$$(y_{i_1}\ y_{i_2}\ \ldots\ y_{i_{n-1}}) = (y_{j_0}\ y_{j_1}\ \ldots\ y_{j_{n-2}})\,.$$

Finding out a Hamiltonian cycle in a De Bruijn graph is equivalent to form a De Bruijn sequence, since such a cycle corresponds to a sequence of successive states of a maximum-period FSR of length n (note that there is a one-to-one correspondence between the nodes of the De Bruijn graph of order n and the possible states of an FSR of length n).

There are many algorithms for constructing de Bruijn sequences. A classical method is proposed by Lempel in [19], based on the so-called D-morphism defined as the mapping $D : \mathbb{F}_2^n \to \mathbb{F}_2^n$ that satisfies

$$D(e_0\ e_1\ \ldots\ e_{n-1}) = (e_0 + e_1\ \ldots\ e_{n-2} + e_{n-1}),$$

where the addition is taken over $\mathbb{F}_2$. More precisely, having a De Bruijn sequence of order n, we can obtain a De Bruijn sequence of order $n+1$ by first computing the two D-morphic preimages of the sequence and subsequently concatenating them at a conjugate pair of states of the corresponding cycles. In [27], the construction of [19] is presented in the form of composing feedback functions, so as to produce a long stage maximum-period NLFSR from a short stage maximum-period NLFSR; more recently, a refinement of this method is given in [23]. The idea of D-morphism is also applied in [2], whereas in [9], the D-morphism is appropriately used to derive a De Bruijn sequence of order $n+1$ from two different De Bruijn sequences of order n.

Another approach to construct De Bruijn sequences is based on the cycle joining method that is described in section "Keystream Generators". This is the case in [13], where the feedback functions of maximum-period NLFSRs are being developed; an improvement of the algorithm in [13] has been recently proposed in [20]. In the same framework, [11] describes a technique that is based on an irreducible polynomial and its adjacency graph, whilst more recently De Bruijn sequences of high order have been constructed [3].

A third approach to derive De Bruijn sequences is based on splitting a cycle C into two cycles C_1, C_2 and subsequently appropriately joining them into another cycle. Two distinct such pairs are employed for these two cases (splitting and joining) that are called *cross-join pairs* [5, 7, 26]. To define the notion of cross-join pairs, we follow the notation of [26]; let $y = y_0 y_1 \ldots y_{2^n-1}$ be a de Bruijn sequence. For any n-tuple $\boldsymbol{a} = (a_0 a_1 \ldots a_{n-1}) \in \mathbb{F}_2^n$, we define its conjugate $\hat{\boldsymbol{a}} = (a_0' a_1 \ldots a_{n-1}) \in \mathbb{F}_2^n$ where $a_0' = a_0 + 1$. Clearly, any $\boldsymbol{a} \in \mathbb{F}^n$ is uniquely associated with a state of maximum-period NLFSR of length n—i.e. the corresponding state of the NLFSR, being considered as an element of $\mathbb{F}_2^n$, is the inverse of $\boldsymbol{a}$.

Definition 3 Two conjugate pairs $(\boldsymbol{a}, \hat{\boldsymbol{a}})$ and $(\boldsymbol{b}, \hat{\boldsymbol{b}})$ constitute a cross-join pair for the sequence y if these four n-tuples occur, under a cyclic shift of y, in the order

$$\ldots \boldsymbol{a}, \ldots, \boldsymbol{b}, \ldots, \hat{\boldsymbol{a}}, \ldots, \hat{\boldsymbol{b}} \ldots$$

within y; that is, $\boldsymbol{b}$ appears after $\boldsymbol{a}$, $\hat{\boldsymbol{a}}$ occurs after $\boldsymbol{b}$ and $\hat{\boldsymbol{b}}$ occurs after $\hat{\boldsymbol{a}}$.

Hence, interchanging the successors of $\boldsymbol{a}, \hat{\boldsymbol{a}}$ results in splitting y into two cycles. Next, we join these two cycles into a new one by interchanging the successors of $\boldsymbol{b}, \hat{\boldsymbol{b}}$. Note that this joining operation, when it is expressed in terms of feedback functions of NLFSRs, is equivalent to the aforementioned cycle joining method in NLFSRs. The significance of using cross-join pairs rests with the following result.

Theorem 1 ([26]) *Let y, w be two de Bruijn sequences of order n. Then w can be obtained from y by repeated application of the cross-join pair operation.*

Finally, there are several other algorithmic approaches to construct De Bruijn sequences. The following is one of the easiest [12]: start with n zeros and append an 1 whenever the n-tuple thus formed has not appeared in the sequence so far, otherwise append a 0. Constructions of De Bruijn sequences with minimal linear complexity are given in [6], whereas methods for obtaining decodable De Bruijn sequences (i.e. sequences with the property that any particular subsequence can be easily found), via interleaving smaller De Bruijn sequences, are given in [25]. More recently, a new construction of De Bruijn sequences based on a simple shift rule is given in [29].

Properties of Suffix Arrays of De Bruijn Sequences

In this section, we establish some general properties that are inherent in the suffix arrays of De Bruijn sequences. The key point in our analysis stems from the observation that for any index k, the tuple $y_{S[k]}^{S[k]+n-1}$ coincides with the binary representation of k (*see*, e.g. Table 1). Hence, we state the following.

Lemma 1 *For any binary De Bruijn sequence y of order n, the tuple y_i^{i+n-1} for any $i \geq 0$ is equal to the binary representation of $S^{-1}[i]$.*

Proof Straightforward from the fact that if y_i^{i+n-1} coincides with the binary representation of the integer k, then it necessarily holds $S[k] = i$, since each possible n-tuple appears exactly once within y and the tuples are lexicographically ordered within S. □

Lemma 1 results to the fact that, in a De Bruijn sequence y, knowledge of y_m^{m+n-1} for any $0 \leq m \leq 2^n - 1$ is adequate for directly computing the integer k such that $S[k] = m$ (i.e. $S^{-1}[m]$); conversely, if it is known that $S[k] = m$ for any k, m, then the subsequence y_m^{m+n-1} is fully determined. In the sequel, for any $0 \leq k < 2^n$, we denote by

$$\boldsymbol{k}^{(n)} \triangleq (k_0\ k_1\ \ldots\ k_{n-1})$$

its binary representation of length n.

Proposition 1 *If y is a binary De Bruijn sequence of order n and S is its corresponding suffix array, then the following hold:*

1. $S[1] = S[0] + 1$, $S[2^{n-1}] = S[0] - 1$,
2. $S[2^n - 2] = S[2^n - 1] + 1$, $S[2^{n-1} - 1] = S[2^n - 1] - 1$,
3. $S[2^n - 1] - S[0] \geq n$,
4. $y_i = 0$ *if and only if* $S^{-1}[i] < 2^{n-1}$,

where all operations are performed mod 2^n. □

Proof First note that $S[0] = m$ implies that y_m^{m+n-1} is the all-zero tuple; therefore, since y is De Bruijn, y_{m+1}^{m+n} equals the n-tuple $00\ldots01$ and y_{m-1}^{m+n-2} equals $10\ldots0$, and, thus, Property 1 follows. Property 2 can be similarly proved, via noticing that $S[2^n - 1] = \ell$ implies that $y_\ell^{\ell+n-1}$ is the all-ones tuple, which in turn leads to the result that $y_{\ell+1}^{\ell+n}$ is equal to the n-tuple $11\ldots10$ and $y_{\ell-1}^{\ell+n-2}$ equals to $01\ldots1$.

Moreover, the fact that y_{m+1}^{m+n} equals the n-tuple $00\ldots01$ leads to the result that, for any $m+1 < i < m+n$, it holds $y_i = 0$ and, thus, $i \neq \ell$, and property 3 follows.

Finally, property 4 is a direct result from the observation that, for any $k < 2^{n-1}$, if $S[k] = i$, then it necessarily holds $y_i = 0$ (since the first half of S corresponds to those suffixes of y that have 0 as a starting element), whereas for any $k \geq 2^{n-1}$, if $S[k] = i$, then it necessarily holds $y_i = 1$ (since the second half of S corresponds to those suffixes of y starting with 1). □

Next we prove a more powerful property on the suffix array of the De Bruijn sequence y.

Proposition 2 *With the notation of Proposition 1, for all* $0 \leq k < 2^n$, *the integer* k' *such that* $S[k'] = S[k] + n - 1 \pmod{2^n}$ *satisfies:*

- $k' < 2^{n-1}$ *if* $k \equiv 0 \pmod 2$,
- $k' \geq 2^{n-1}$ *if* $k \equiv 1 \pmod 2$.

Proof Due to Proposition 1 (Property 4), it suffices to show that if k is even, then $y_{m+n-1} = 0$, as well as if k is odd, then $y_{m+n-1} = 1$. The claim follows immediately from Lemma 1.

A direct generalization of Proposition 2 is the following:

Proposition 3 *With the notation of Proposition 1, for all* $0 \leq k < 2^n$, *the integer* k' *such that* $S[k'] = S[k] + n - i \pmod{2^n}$ *satisfies:*

- $k' < 2^{n-1}$ *if* $k \pmod{2^i} < 2^{i-1}$,
- $k' \geq 2^{n-1}$ *if* $k \pmod{2^i} \geq 2^{i-1}$.

Proof The proof is similar with the one of Proposition 2; more precisely, we simply note that, due to Lemma 1, if $k \mod 2^i < 2^{i-1}$, then $y_{m+n-i} = 0$, whilst if $k \mod 2^i \geq 2^{i-1}$, then $y_{m+n-i} = 1$. □

Proposition 3 illustrates the fact that knowledge of $S^{-1}[m]$ implies knowledge of each element y_i, $m \leq i \leq m+n-1$ and, thus, knowledge of whether $S^{-1}[i]$ lies in the upper or the down half of the S (by property 4 of Proposition 1). However, knowledge of each element y_i, $m \leq i \leq m+n-1$, actually implies a more powerful property, as seen below:

Theorem 2 *With the notation of Proposition 1, for all* $0 \leq k, m < 2^n$ *with* $S[k] = m$, *we have either*

$$S^{-1}[m+1] = 2k' \quad or \quad S^{-1}[m+1] = 2k' + 1$$

where $k' = k - y_m 2^{n-1}$.

Proof Since $S[k] = m$, we get—due to Lemma 1—that the n-tuple $y_m y_{m+1} \ldots y_{m+n-1}$ coincides with the binary representation of k, i.e. with $\boldsymbol{k}^{(n)}$. Consequently, the value of $S^{-1}[m+1]$ is the number $\tilde{k}$ whose binary representation $\tilde{\boldsymbol{k}}^{(n)}$ is given by $y_{m+1} \ldots y_{m+n-1} e$, where e is either 0 or 1. The claim follows from the observation that if $y_m = 0$, then $\tilde{k} = 2k + e$, whereas if $y_m = 1$, then $\tilde{k} = 2(k - 2^{n-1}) + e$. □

Remark 1 The above Theorem states that given $S[k] = m$, then there are exactly two possible values for $S^{-1}[m+1]$. Conversely, it is easy to see that there are exactly two possible values for $S^{-1}[m-1]$—i.e. either $\lfloor \frac{k}{2} \rfloor$ or $\lfloor \frac{k}{2} \rfloor + 2^{n-1}$.

The property of Theorem 2 is stronger than the properties of Propositions 2 and 3, since it ensures that a uniquely defined De Bruijn sequence is associated with the suffix array S having this property. This is illustrated next.

Lemma 2 *Let S be an integer-valued array of length 2^n such that each pair (k, m) with $S[k] = m$, $0 \le k, m < 2^n$, satisfies Theorem 2, for $y_m = k_0$ (where $(k_0 k_1 \ldots k_{n-1}) = \boldsymbol{k}^{(n)}$). Then, S constitutes the suffix array of a uniquely determined binary De Bruijn sequence y.*

Proof Having such a S, we construct a sequence y via the following rule: the value of $S[k]$, for any $0 \le k \le 2^n - 1$, indicates the position of the n-tuple within y which coincides with $\boldsymbol{k}^{(n)}$. Starting from an arbitrary pair (k, m), such a construction is always possible. Indeed, for any pair k, m such that $S[k] = m$, which means that y_m^{m+n-1} coincides with $\boldsymbol{k}^{(n)}$, the above property implies that y_{m+n} is either 0 or 1 (depending on whether $S[2k']$ or $S[2k'+1]$ is $m+1$, where $k' = k - y_m 2^{n-1}$). Since all possible n-tuples are present in the y (as Theorem 2 implies that any $0 \le m < 2^n$ is present in S), we get that y is a De Bruijn sequence. □

The Longest Common Prefix Array of a De Bruijn Sequence

Apart from the suffix array, the longest common prefix array of a De Bruijn sequence has also a well-determined structure, as stated next.

Lemma 3 *Let y be a binary De Bruijn sequence of order n, with P being its corresponding longest common prefix array. Then, for any $0 \le k < 2^n$, the value $P[k]$ is given by m, where $0 \le m < n$ is the largest integer satisfying $k_m = 1$ (where $(k_0 k_1 \ldots k_{n-1}) = \boldsymbol{k}^{(n)}$); if none such m exists (i.e. $k = 0$), then $P[k] = 0$. Conversely, any array P satisfying the above property corresponds to a longest common prefix array of a De Bruijn sequence.*

Proof For any De Bruijn sequence y, the value $P[k]$ is directly proved via recalling Lemma 1 and the observation that $P[k]$ is equal to the maximum m such that

$$k_0 = (k-1)_0, \; k_1 = (k-1)_1, \; \ldots, \; k_{m-1} = (k-1)_{m-1}.$$

Conversely, let us assume that there exists a binary sequence y, either with finite-length 2^n or periodic with period 2^n, not being De Bruijn, and assume that P is its longest common prefix array. Since y is not De Bruijn, there necessarily exists $0 \le i < j < 2^n$ such that $y_i^{i+n-1} = y_j^{j+n-1}$ (i.e. a specific n-tuple is present at least twice within y). This implies that $P[S^{-1}[j]] \ge n$—a contradiction (since all values of P are smaller than n). □

From the above analysis, we get that all binary De Bruijn sequences of the same order n have the same longest common prefix array P (note that this is clearly not true for the suffix array), whereas any such P does not correspond to any other sequence apart from these De Bruijn sequences.

Construction of De Bruijn Sequences Through Suffix Arrays

We next provide a new approach for constructing binary De Bruijn sequences, based on the properties of their suffix arrays. In contrast to other algebraic approaches which build the sequence on a per-bit basis, the new method builds its suffix array S step by step; this is clearly equivalent to putting all possible n-tuples within the sequence (since setting $S[k] = m$ for any $0 \leq k, m < 2^n$ results in forcing y_m^{m+n-1} to coincide with $\boldsymbol{k}^{(n)}$). Therefore, under a proper choice on the values of S, any possible De Bruijn sequence can be generated. Of course, an assignment $S[k] = m$ should be made only if it is compliant with the restrictions posed by Theorem 2; therefore, these properties guide the whole process. However, for any $1 < m < 2^n$, it might be impossible to assign m to some $S[k]$, due to conflict(s) with previous assignment(s). As a consequence, an assignment $S[k'] = m'$ may change later on, if for an $m > m'$ there is no possible k that can be set as a value of $S^{-1}[m]$ without leading to a conflict. Hence, a backtracking approach is needed to ensure that the suffix array S will be properly constructed so that it corresponds to a De Bruijn sequence.

Putting all the above together, we get Algorithm 1. Note that we assume, without loss of generality, that the first n bits of the n-th order De Bruijn sequence are all

Algorithm 1 Construct De Bruijn sequence

input: n
initialization: $S \leftarrow \varnothing$, $S[0] \leftarrow 0$, $S[1] \leftarrow 1$, $S[2^{n-1}] \leftarrow 2^n - 1$

```
 1: m ← 2
 2: while m < 2^n − 1 do                                  \\ S^{-1}[2^n − 1] = 2^{n−1}
 3:     f ← assign(n, m, S)                               \\ see Theorem 2
 4:     while f = 0 do                                    \\ conflict
 5:         m ← m − 1
 6:         if only one option of S^{-1}[m] has been checked then
 7:             f ← reassign(n, m, S)
 8:         else
 9:             S[S^{-1}[m]] ← ∅
10:         end
11:     end
12:     m ← m + 1
13: end
```

output: suffix array S corresponding to a De Bruijn sequence

zeros—i.e. the n-tuple y_0^{n-1} equals to the all-zeros tuple $00 \cdots 0$. Hence we always have, according to Proposition 2, $S[0] = 0$, $S[1] = 1$ and $S[2^{n-1}] = 2^n - 1$.

The main building block of Algorithm 1 is the function assign, which assigns any given m to some $S[k]$ according to Theorem 2. Only two are the possible options for k, provided that $S[k]$ has not been assigned any value yet. In such a case, choosing such a k, we get $S[k] = m$, $f = 1$; otherwise, if none such k can be chosen, we have $f = 0$, and the algorithm reexamines the previous assignments, starting from $m - 1$. However, reassigning a value to $S^{-1}[m - 1]$ leaves only one option available, as the previous one led to a conflict. Thus, the function reassign is almost the same with assign, with the exception that it checks, for input m, only one possible value for $S^{-1}[m]$ (the one that has not been previously chosen); hence, the only difference between assign and reassign can be found in Line 1 of Algorithm 3, where k gets the value that has not been chosen earlier (out of the two possible ones). Note that reassign is executed if we have checked only one out of the two possible options for $S^{-1}[m]$ (see Line 6 of Algorithm 1); otherwise, we get that both options of $S^{-1}[m]$ resulted in a conflict, and, thus, we proceed similarly with reexamining previous assignments.

The algorithm is not deterministic, since in general the pairs k, m such that $S[k] = m$ are randomly chosen (see Line 8 of Algorithm 2). The algorithm can be slightly modified in case that we seek for De Bruijn sequences with specific consecutive n-tuples or patterns (i.e. by forcing specific assignments for some $S^{-1}[m]$).

The computational complexity of each step of the main loop in Algorithm 1, which generates the next bit in the De Bruijn sequence, simply rests with quering S—that is, to check the value of S at one specific position. Since the algorithm constructs a De Bruijn sequence of period 2^n, it has linear computational complexity with respect to the length of the sequence, as any other algorithm that constructs a sequence step by step. The algorithm passes through 2^n steps, whereas some steps

Algorithm 2 Function assign

input: n, m, S
initialization: $K \leftarrow \varnothing$, $f \leftarrow 0$ $\quad$ \\ f is a flag

1: $k' \leftarrow S^{-1}[m-1] - y_{m-1} 2^{n-1}$ $\quad$ \\ y_{m-1} is known from S
2: **for** $e \leftarrow 0 : 1$ **do**
3: $\quad$ **if** $S[2k' + e] = \varnothing$ **then**
4: $\quad\quad$ $K \leftarrow K \cup \{2k' + e\}$
5: $\quad$ **end**
6: **end**
7: **if** $K \neq \varnothing$ **then**
8: $\quad$ $k \leftarrow_R K$ $\quad$ \\ random choice
9: $\quad$ $S[k] \leftarrow m$
10: $\quad$ $f \leftarrow 1$
11: **end**

output: flag f

Algorithm 3 Function reassign

input: n, m, S
initialization: $f \leftarrow 0$ \\ f is a flag

```
1: k ← S^{-1}[m] + (-1)^{S^{-1}[m]}
2: S[S^{-1}[m]] ← ∅
3: if S[k] = ∅ then
4:     S[k] ← m
5:     f ← 1
6: end
```

output: flag f

may be visited more than once due to the backtracking procedure. However, it can be used for the construction of De Bruijn sequences for relatively small order n (i.e. $n \approx 15$) that cannot be obtained by other known constructions. The following example illustrates all the above issues.

Example 3 In this example, we present the steps of Algorithm 1 for $n = 4$. Due to the initialization process, we have (see the first three rows of Table 2)

$$S[0] = 0, \;\; S[1] = 1, \;\; S[8] = 15.$$

For each m, there are two possible options for $k = S^{-1}[m]$ (second column of Table 2). In some cases, although there are two options for $k = S^{-1}[m]$, only one can be chosen since the other has been already assigned a value previously; these are the cases where the chosen k (third column) is written in bold. It should be also pointed out that the cases where the possible value of k is only one correspond to the function reassign—i.e. a conflict occurred and the algorithm backtracked to a previous step, in order to try another option.

The third column simply shows, for each step, the current values of k such that $S[k]$ has been already assigned a value—hence, the last entry of this column coincides with the values of S^{-1}. The constructed S is equal to

$$S = (0, 1, 2, 7, 5, 3, 8, 11, 15, 6, 4, 10, 14, 9, 13, 12),$$

and the corresponding De Bruijn sequence (from Lemma 2) is given by 0000101001101111. □

Suffix Arrays and Cross-Join Pairs

In this section we further focus on the information that the suffix arrays store, in terms of providing the means to identify cross-join pairs within a De Bruijn sequence. To this end, we first show the following.

Table 2 An execution of Algorithm 1 for $n = 4$

m	Valid k	Chosen k	Suffix array S																
			$k =$	0	1	2	3	4	5	6	7	8	9	10	11	12	13	14	15
Initialization																			
0	0	0		0	*	*	*	*	*	*	*	*	*	*	*	*	*	*	*
1	1	1		0	1	*	*	*	*	*	*	*	*	*	*	*	*	*	*
15	8	8		0	1	*	*	*	*	*	*	15	*	*	*	*	*	*	*
Main loop																			
2	2, 3	2		0	1	2	*	*	*	*	*	15	*	*	*	*	*	*	*
3	4, 5	5		0	1	2	*	*	3	*	*	15	*	*	*	*	*	*	*
4	10, 11	10		0	1	2	*	*	3	*	*	15	*	4	*	*	*	*	*
5	4, 5	**4**		0	1	2	*	5	3	*	*	15	*	4	*	*	*	*	*
6	8, 9	**9**		0	1	2	*	5	3	*	*	15	6	4	*	*	*	*	*
7	2, 3	**3**		0	1	2	7	5	3	*	*	15	6	4	*	*	*	*	*
8	6, 7	6		0	1	2	7	5	3	8	*	15	6	4	*	*	*	*	*
9	12, 13	12		0	1	2	7	5	3	8	*	15	6	4	*	9	*	*	*
10	8, 9	None		0	1	2	7	5	3	8	*	15	6	4	*	*	*	*	*
9	13	13		0	1	2	7	5	3	8	*	15	6	4	*	*	9	*	*
10	10, 11	**11**		0	1	2	7	5	3	8	*	15	6	4	10	*	9	*	*
11	6, 7	**7**		0	1	2	7	5	3	8	11	15	6	4	10	*	9	*	*
12	14, 15	14		0	1	2	7	5	3	8	11	15	6	4	10	*	9	12	*
13	12, 13	**12**		0	1	2	7	5	3	8	11	15	6	4	10	13	9	12	*
14	8, 9	None		0	1	2	7	5	3	8	11	15	6	4	10	*	9	12	*
13	13	None		0	1	2	7	5	3	8	11	15	6	4	10	*	9	*	*
12	15	15		0	1	2	7	5	3	8	11	15	6	4	10	*	9	*	12
13	14, 15	**14**		0	1	2	7	5	3	8	11	15	6	4	10	*	9	13	12
14	12, 13	**12**		0	1	2	7	5	3	8	11	15	6	4	10	14	9	13	12

Proposition 4 *Let y be a De Bruijn sequence of order n. Then, the tuple y_m^{m+n-1} is the conjugate of $y_{m'}^{m'+n-1}$, for $m \neq m'$, if and only if it holds*

$$|S^{-1}[m] - S^{-1}[m+1]| = 2^{n-1}.$$

Proof The claim follows directly from Lemma 1. □

Proposition 5 *Let y be a binary De Bruijn sequence of order n with suffix array S, and let $0 \leq m_1 < m_2 \leq 2^n - 1$ with*

$$k_i = S^{-1}[m_i], \quad i = 1, 2.$$

Let also $\{2k_1', 2k_1'+1\}$ and $\{2k_2', 2k_2'+1\}$ be defined as in Theorem 2 for the integers k_1 and k_2, respectively. Then, $\{2k_1', 2k_1'+1\}$ and $\{2k_2', 2k_2'+1\}$ coincide if and only if the n-tuple $y_{m_1}^{m_1+n-1}$ is the conjugate of $y_{m_2}^{m_2+n-1}$

Proof Note that the sets $\{2k'_1, 2k'_1 + 1\}$ and $\{2k'_2, 2k'_2 + 1\}$ coincide if and only if $y_{m_1+1}^{m_1+n-1} = y_{m_2+1}^{m_2+n-1}$, and, recalling the definition of the conjugate tuples, the claim follows. □

The importance of the above results rests with the following observation. First note that, due to Theorem 2, for a given De Bruijn sequence y of order n, if $S[k] = m$ for some $0 \leq k, m < 2^n - 1$, then $S^{-1}[m+1]$ may have one out of the two possible values $2k', 2k' + 1$, where k' is defined in Theorem 2 (see also the function assign in section "Construction of De Bruijn Sequences Through Suffix Arrays"). Hence, Proposition 5 implies that if $S^{-1}[m+1] = 2k'$ (resp. $S^{-1}[m+1] = 2k' + 1$), then it necessarily holds $S^{-1}[m'+1] = 2k' + 1$ (resp. $S^{-1}[m'+1] = 2k'$) for some $0 \leq m' < 2^n - 1$ if and only if $y_{m'}^{m'+n-1}$ is the conjugate of y_m^{m+n-1}. Therefore, transforming a suffix array S of a de Bruijn sequence y into another suffix array S' corresponding to another De Bruijn sequence y' necessarily involves a swapping of the values of $S^{-1}[m+1]$ and $S^{-1}[m'+1]$ for a conjugate pair y_m^{m+n-1} and $y_{m'}^{m'+n-1}$ since any other transformation of S would not satisfy the properties implied by Theorem 2. Recalling the discussion in section "De Bruijn Sequences", we get that a transformation of a suffix array into another array which still satisfies the properties of Theorem 2—and, thus, it corresponds to another De Bruijn sequence according to Lemma 2—is equivalent to the cross-join pairs method. Hence, identifying cross-join pairs enables appropriate modifications of S^{-1} and, consequently, of S; to this end, we proceed with the following observation.

Lemma 4 *Let y be a binary De Bruijn sequence of order n, and let*

$$(y_{m_1}^{m_1+n-1}, y_{m_2}^{m_2+n-1}) \quad and \quad (y_{\ell_1}^{\ell_1+n-1}, y_{\ell_2}^{\ell_2+n-1})$$

be a cross-join pair, where $y_{m_2}^{m_2+n-1}$ and $y_{\ell_2}^{\ell_2+n-1}$ are the conjugates of $y_{m_1}^{m_1+n-1}$ and $y_{\ell_1}^{\ell_1+n-1}$, respectively. Then, m_1, m_2, ℓ_1, ℓ_2 occur in the following order

$$\ldots m_1 \ldots \ell_1 \ldots m_2 \ldots \ell_2 \ldots$$

within S^{-1}, where S^{-1} is the inverse suffix array of y.

Proof Straightforward from the Definition 3, as well as from the definition of the inverse suffix array. □

Hence, recalling that all the conjugate pairs are uniquely determined by Proposition 4, Lemma 4 implies that the inverse suffix array S^{-1} allows for finding out cross-join pairs. This is illustrated in the following example.

Example 4 Let us revisit the De Bruijn sequence y of order $n = 4$ constructed in Example 3. It is easy to see that its inverse suffix array S^{-1} is equal to

$$(0\ 1\ 2\ 5\ 10\ 4\ 9\ 3\ 6\ 13\ 11\ 7\ 15\ 14\ 12\ 8)\,.$$

According to Proposition 4, the conjugate pairs are indexed as follows:

$$(0, 8), (1, 9), (2, 10), (3, 11), (4, 12), (5, 13), (6, 14), (7, 15).$$

Examining for cross-join pairs within S^{-1} according to Lemma 4, we see that such a cross-join pair is, e.g. (1, 9), (5, 13). Hence, interchanging the successors of these pairs (i.e. the successors of 1, 9 and the successors of 5, 13), we get a new array

$$(0\ 1\ 3\ 6\ 13\ 10\ 4\ 9\ 2\ 5\ 11\ 7\ 15\ 14\ 12\ 8)\,.$$

It is easy to see that the latter is the inverse suffix array of the De Bruijn sequence $y' = 0000110100101111$. We may similarly proceed with other cross-join pairs to derive new De Bruijn sequences.

Moreover, if $f \in \mathbb{B}_n$ is the feedback function of the maximum-period NLFSR that generates y, then the corresponding function $h \in \mathbb{B}_n$ of the maximum-period NLFSR that generates y' is clearly determined as follows:

$$h(\boldsymbol{x}) = \begin{cases} f(\boldsymbol{x}) + 1, & \text{if } x \in \{(1\,0\,0\,0), (1\,0\,0\,1), (1\,0\,1\,0), (1\,0\,1\,1)\} \\ f(\boldsymbol{x}), & \text{otherwise} \end{cases}$$

where $\{(1\ \ 0\ \ 0\ \ 0), (1\ \ 0\ \ 0\ \ 1), (1\ \ 0\ \ 1\ \ 0), (1\ \ 0\ \ 1\ \ 1)\}$ are the inverses of $\{(0\ \ 0\ \ 0\ \ 1), (1\ \ 0\ \ 0\ \ 1), (0\ \ 1\ \ 0\ \ 1), (1\ \ 1\ \ 0\ \ 1)\}$ which correspond to the binary representations of 1, 9, 5, 13, respectively. □

Hence, from the above analysis, we get that the (inverse) suffix array allows for performing the cross-join pairs method for obtaining new De Bruijn sequences; this observation is of high importance, since suffix arrays are known to be very convenient for performing such queries on the corresponding sequences (see, e.g.[30]).

Conclusions

In this work, a new approach is proposed for studying binary De Bruijn sequences, based on the properties of their suffix arrays. It is shown that such suffix arrays have specific, well-determined, properties which allow for primary constructions of De Bruijn sequences, as well as for secondary constructions having as starting point another De Bruijn sequence; most importantly, with regard to the secondary constructions, our analysis indicates that suffix arrays facilitate the deployment of currently known techniques for deriving De Bruijn sequences, such as the well-known cross-join pair method.

The results obtained open many directions for further research. First, it is expected that the proposed algorithm can be more effective via proper handling of suffix arrays to improve search complexity or space requirements—and thus, sequences of higher orders may be obtained. Moreover, it is of interest to further examine other open problems on De Bruijn sequences—such as the computation of the number of the cross-join pairs in a De Bruijn sequence derived by a maximum-period NLFSR—based on this new approach. Finally, a very challenging task is to associate the suffix array of a De Bruijn sequence with cryptographic properties of the corresponding feedback function of the maximum-period NLFSR.

References

1. Carlet, C.: Boolean functions for cryptography and error correction codes. In: Boolean Models and Methods in Mathematics, Computer Science and Engineering, pp. 257–397. Cambridge University Press, Cambridge (2010)
2. Chang, T., Park, B., Kim, Y.H., Song, I.: An Efficient Implementation of the D-Homomorphism for Generation of de Bruijn Sequences. IEEE Trans. Inf. Theory **45**, 1280–1283 (1999)
3. Dong, J. and Pei, D.: Construction for de Bruijn sequences with large stage. Des. Codes Cryptogr. **85**, 343–385, Springer (2017)
4. Dubrova, E.: A list of maximum period NLFSRs. Cryptology ePrint Archive, Report 2012/166 (2012). Available: http://eprint.iacr.org
5. Dubrova, E.: A scalable method for constructing Galois NLFSRs with period $2^n - 1$ using cross-join pairs. IEEE Trans. Inf. Theory **59**, 703–709 (2013)
6. Etzion, T., Lempel, A.: Construction of de Bruijn sequences of minimal complexity. IEEE Trans. Inf. Theory **30**, 705–708 (1984)
7. Fredricksen, H.: A Class of nonlinear de Bruijn cycles. J. Comb. Theory Ser. A **19**, 192–199 (1975)
8. Fredricksen, H.: A survey of full length nonlinear shift register cycle algorithms. SIAM Rev. **24**, 195–221 (1982)
9. Games, R. A.: A generalized recursive construction for de Bruijn sequences. IEEE Trans. Inf. Theory **29**, 843–849 (1983)
10. Golomb, S.W.: Shift Register Sequences. Aegean Park Press, Laguna Hills (1981)
11. Hauge, E.R., Helleseth, T.: De Bruijn sequences, irreducible codes and cyclotomy. Discret. Math. **159**, 143–154 (1996)
12. Helleseth, T.: De Bruijn sequence. In: Encyclopedia of Cryptography and Security, pp. 138–140. Springer, Heidelberg (2005)
13. Jansen, C.J.A., Franx, W.G., Boekee, D.E.: An efficient algorithm for the generation of de Bruijn cycles. IEEE Trans. Inf. Theory **37**, 1475–1478 (1991)
14. Kalouptsidis, N., Limniotis, K.: Nonlinear span, minimal realizations of sequences over finite fields and De Bruijn generators. In: Proceedings of International Symposium on Information Theory and Application, pp. 794–799 (2004)
15. Kärkkäinen, J., Sanders, P.: Simple linear work suffix array construction. In: ICALP 2003. Lecture Notes in Computer Science, vol. 2719, pp. 943–955. Springer, Heidelberg (2003)
16. Kasai, T., Lee, G., Arimura, H., Arikawa, S., Park, K.: Linear-time longest-common prefix computation in suffix arrays and its applications. In: CPM 2001, Lecture Notes in Computer Science, vol. 2089, pp. 181–192. Springer, Heidelberg (2001)
17. Kim, D.K., Sim, J.S., Park, H., Park, K.: Constructing suffix arrays in linear time. J. Discret. Algorithms Elsevier **3**, 126–142 (2005)
18. Ko, P., Aluru, S.: Space efficient linear time construction of suffix arrays. J. Discret. Algorithms Elsevier **3**, 143–156 (2005)
19. Lempel, A.: On a homomorphism of the de Bruijn graph and its applications to the design of feedback shift registers. IEEE Trans. Comput. **C-19**, 1204–1209 (1970)

20. Li, M., Lin, D.: De Bruijn sequences from nonlinear feedback shift registers. Cryptology ePrint Archive, Report 2015/666 (2015) Available: http://eprint.iacr.org
21. MacWilliams, F., Sloane, N.: The Theory of Error Correcting Codes. North-Holland, Amsterdam (1977)
22. Manber, U., Myers, G.: Suffix arrays: a new method for on-line string searches. SIAM J. Comput. **22**, 935–948 (1993)
23. Mandal, K., Gong, G.: Cryptographically strong de Bruijn sequences with large periods. In: Selected Areas in Cryptography (SAC), Revised Selected Papers, pp. 104–118 (2012)
24. Manzini, G.: Two space-saving tricks for linear-time LCP computation. In: SWAT 2004. Lecture Notes in Computer Science, vol. 3111, pp. 372–383. Springer, Heidelberg (2004)
25. Mitchell, C.J., Etzion, T., Paterson, K.G.: A method for constructing decodable de Bruijn sequences. IEEE Trans. Inf. Theory **42**, 1472–1478 (1996)
26. Mykkeltveit, J., Szmidt J.: On cross joining de Bruijn sequences. Contemp. Math. **632**, 333–344 (2015)
27. Mykkeltveit, J., Siu, M.K., Tong, T.: On the cycle structure of some nonlinear shift register sequences. Inform. Control **43**, 202–215 (1979)
28. Robshaw, M.J.B., Billet, O.: New Stream Cipher Designs – The eSTREAM Finalists. Lecture Notes in Computer Science, vol. 4986. Springer, Heidelberg (2008)
29. Sawada, J., Williams, A., Wong, D.: A surprisingly simple de Bruijn sequence construction. J. Discret. Math. Elsevier **339**, 127–131 (2016)
30. Sinha, R., Puglisi, S.J., Moffat, A., Turpin, A.: Improving suffix array locality for fast pattern matching on disk. In: ACM SIGMOD International Conference Management of Data, pp. 661–678 (2008)
31. Weiner. P.: Linear pattern matching algorithms. In: 14th Annual Symposium on Switching and Automata Theory, pp. 1–14 (1973)

Fuzzy Empiristic Implication, A New Approach

Konstantinos Mattas and Basil K. Papadopoulos

Abstract The present paper is a brief introduction to logical fuzzy implication operators, the basic properties of a fuzzy implication function, and ways to construct new fuzzy implication functions. It is also argued that logical implication functions are defined in a rather rationalistic manner. Thus a new, empiristic approach is proposed, defining implication relations that are derived from data observation and with no regard to any preexisting constrains. A number of axioms are introduced to define a fuzzy empiristic implication relation, and a method of computing such a relation is proposed. It is argued that the proposed method is easy and with small time requirement even for very large data sets. Finally an application of the empiristic fuzzy implication relation is presented, the choice of a suitable logical fuzzy implication function to describe an "If... then..." fuzzy rule, when observed data exists. An empiristic fuzzy implication relation is computed according to the data, and through schemas of approximate reasoning, the difference of it to any logical fuzzy implication function is measured. The fuzzy implication function that is closer to the empiristic best resembles the observed "If... then..." fuzzy rule.

Keywords Fuzzy implication · Approximate reasoning

Introduction

The theory of fuzzy logic that has been presented by Zadeh [12] has developed rapidly both in theoretical and application basis. The main characteristic of this development is the abandonment of the binary classical logic of zero and one. In that sense a logical proposition can be true with any degree of truth from one, meaning it is absolutely true, to zero, meaning it is absolutely false. It also defines fuzzy sets, to which any element can belong with any value from zero to one. Special interest

K. Mattas (✉) · B. K. Papadopoulos
Democritus University of Thrace, Department of Civil Engineering, Xanthi, Greece
e-mail: kmattas@civil.duth.gr; papadob@civil.duth.gr

N. J. Daras, Th. M. Rassias (eds.), *Modern Discrete Mathematics and Analysis*, Springer Optimization and Its Applications 131,
https://doi.org/10.1007/978-3-319-74325-7_16

is the generalization of operations between fuzzy propositions and fuzzy sets with the generalization of the classical implication to fuzzy logic which is significant.

The present paper describes the basic properties of fuzzy implication functions. Also a new type of fuzzy implication relationship constructed entirely from observation, which is called empiricist fuzzy implication relationship, is proposed. It is theoretically founded with proposed axioms, and a method of evaluating the relationship has been developed. Also a method of choosing a logical implication function is proposed in order to be suitable to observed data, using the empiristic fuzzy implication relationship.

Preliminaries

Classical Implication

In classical logic there are propositions, e.g., "The number three is odd" or "the number four is odd," which may be true or false, respectively. These propositions can be combined using logical operations as "and," "or," etc. A very important operation between the two proposals is the implication. For two logical propositions p and q, it is denoted that p implies q ($p \Rightarrow q$) if any time p is true, q must be also true. In the above case, proposition p is called cause or antecedent, and proposition p is called consequent. It represents an "If. . . Then. . . " rule [3].

Abiding classical logic, any proposition can be valid or not valid, so it is evaluated with one or zero, respectively. Thus the implication between two propositions, being a proposition itself, can be valid or not valid. Furthermore any implication ($p \Rightarrow q$) is equivalent to the proposition "negation of q or p." In the following Table 1, the Boolean table of the classical implication operation is presented.

Classical implication operation is directly applied on logical reasoning schemas as modus ponens, modus tollens, and hypothetical reasoning [1]. Modus ponens from the Latin modus ponendo ponens is proof inductive reasoning and has the following format: IF $p \Rightarrow q$ is true AND p is true, THEN q is true. A modus tollens schema (from Latin modus tollendo tollens) or rebuttal process follows these steps: IF $p \Rightarrow q$ is true and $n(q)$ is true, THEN $n(p)$ is true, where $n(p)$ is the negation of p, i.e., the proposal p is false. Finally, an example of hypothetical reasoning is this: IF $p \Rightarrow q$ is true and $q \Rightarrow r$ is true, THEN $p \Rightarrow r$ is true.

Table 1 Boolean table of classical implication

p	q	$p \Rightarrow q$
0	0	1
0	1	1
1	0	0
1	1	1

Fuzzy Implication

Fuzzy logic is based on the idea that a proposition may be true to some degree of truth. It is immediately apparent that a generalization of the classical implication is required, such that firstly it is possible to evaluate the implication between fuzzy propositions, and secondly it is possible for an implication to be valid to a degree of truth. Thus, fuzzy implication can be defined as a function I (Eq. (1)):

$$I : [0, 1] \times [0, 1] \rightarrow [0, 1] \tag{1}$$

For the definition of a fuzzy logic implication operation, a set of axioms have been proposed in literature that any function has to fulfill in order to be considered as a fuzzy implication function. A fuzzy implication function is a function $I : [0, 1] \times [0, 1] \rightarrow [0, 1]$ which satisfies the maximum of the following axioms [3]:

1. if $a \leq b$, then $I(a, x) \geq I(b, x)$ (left antitonicity)
2. if $a \leq b$ then $I(x, a) \leq I(x, b)$ (right isotonicity)
3. $I(0, a) = 1$
4. $I(1, a) = a$
5. $I(a, a) = 1$
6. $I(a, I(b, x)) = I(b, I(a, x))$
7. $I(a, b) \Leftrightarrow a \leq b$
8. $I(a, b) = I(n(b), n(a))$
9. A fuzzy implication function must be continuous on its domain

Construction of Fuzzy Implication Functions

In the literature there are many functions that have been proposed as fuzzy implication functions, satisfying some or all of the aforementioned axioms. In addition there are several ways to create implication functions based on logical rules of classical logic and using *t-norms*, *t-conorms*, or even other implications. Furthermore it is possible to create functions that satisfy the axiomatic restrictions on an algebraic basis. Some ways to construct a fuzzy implication function are as follows:

S-N Implications

As mentioned, the proposition p implies q in classical logic is identical with the proposition negation of p or q as in (Eq. (2)):

$$p \Rightarrow q \equiv n(p) \vee q \tag{2}$$

Thus, a fuzzy implication may be produced by a fuzzy negation N and a t-$conormS$ by the aforementioned tautology as follows [1] (Eq. (3)):

$$I(x, y) = S(N(x), y) \tag{3}$$

Reciprocal Implications

When for a given pair of fuzzy implication and negation operations, the rule of contraposition is not satisfied, it is possible for a new fuzzy implication function to be constructed with the following formula. The new fuzzy implication function paired with the negation utilized satisfies the rule of contraposition [1] (Eq. (4)).

$$I_N(x, y) = I(N(x), N(y)) \tag{4}$$

R Implications

The following formula of set theory, where A and B are subsets of a universal set X, obtained from the isomorphism of the classical set theory with the binary Boolean logic, can be used to construct new fuzzy implication functions (Eq. (5)).

$$A' \cup B = (A \setminus B)' = \cup\{C \subseteq X|A \cap C \subseteq B\} \tag{5}$$

Fuzzy implications created as a generalization of this rule in fuzzy logic are widely used in the intuitionistic fuzzy logic and are often found in the literature as R implications. For this purpose a t-$normT$ is needed, and the implications are calculated as follows (Eq. (6)):

$$I(x, y) = \sup\{t \in [0, 1]|T(x, t) < y\} \tag{6}$$

QL Implications

In quantum theory, a new tautology has prevailed to describe the implication operation. It is shown in the following equation (Eq. (7)), and it yields the same results with the rest known ways to describe the operation in classical logic.

$$p \Rightarrow q \equiv n(p) \vee (p \wedge q) \tag{7}$$

From the generalization (Eq. (7)) to fuzzy set theory arises a new class of implications that are found in literature as QL implications (quantum logic implications). These implication functions are created as in (Eq. (8)), when a fuzzy negation N, a fuzzy t-$conormS$, and a fuzzy t-$normT$ are applied.

$$I(x, y) = S(N(x)T(x, y)) \tag{8}$$

f and g Implications

For a function to be appropriate to be a fuzzy implication, it has to satisfy a set of axioms that have been mentioned above in the definition of fuzzy implication. So it is possible to create algebraic implication production techniques designed based on the restrictions, without any further association with logical rules as the other ways that have been reported [1]. Suppose a function $f : [0, 1] \rightarrow [0, \infty]$ that is strictly decreasing and continuous and that satisfies $f(1) = 0$. A fuzzy implication $I_f : [0, 1]^2 \rightarrow [0, 1]$ is defined as follows (Eq. (9)):

$$I_f(x, y) = f^{-1}(xf(y)) \tag{9}$$

The function f is called f-*generator* and the implication f-*generated.*

Respectively, g-*generator* generator functions and g-*generated* generated implications can be defined where function $g : [0, 1] \rightarrow [0, \infty]$ is strictly increasing and continuous with $g(0) = 0$. The implication $I_g : [0, 1]^2 \rightarrow [0, 1]$ is computed as follows (Eq. (10)):

$$I_g(x, y) = g^{(-1)}\left(\frac{1}{x}g(y)\right) \tag{10}$$

where the function $g^{(-1)}$ pseudo-inverse of g, given by (Eq. (11)):

$$g^{(-1)}(x) = \begin{cases} g^{-1}(x) & \text{if } x \in [0, g(1)] \\ 1 & \text{if } x \in [g(1), \infty] \end{cases} \tag{11}$$

Convex Combinations

Another very usual way to produce a new fuzzy implication is from a convex combination of two old ones. Two fuzzy implication functions I and J and any real number λ from the closed interval [0, 1] can be combined in order to produce a new implication I^{λ} as follows (Eq. (12)):

$$I^{\lambda}(x, y) = \lambda I(x, y) + (1 - \lambda)J(x, y) \tag{12}$$

It turns out that since I and J implications satisfy the axioms of fuzzy implication definition, the same is true for the produced I^{λ} [7].

Symmetric Implication Functions

It is noteworthy that apart from the classical implication functions, symmetric implication functions are widely used and have been proposed in the literature [3]. Such examples are Mamdani and Larson implications that use *t-norms* to evaluate the degree of truth of the implication [8].

These functions are very useful for fuzzy control systems, particularly in some implemented to engineering problems. It is common for these problems to involve correlated variables, where it is not always clear to distinct the antecedent from the consequent. Furthermore in such a system, a decrease in the degree of truth of the antecedent should be followed by a decrease in the degree of truth of the consequent. Hence this kind of fuzzy implications, although cannot satisfy the established axioms, has significant advantages when modeling engineering problems. Hence they can be found in the literature as engineering implications (Eqs. (13)–(14)):

$$I_M(x, y) = \min(x, y) \tag{13}$$

$$I_L(x, y) = xy \tag{14}$$

Approximate Reasoning

In nature, as in engineering applications, a need emerges to make decisions when accurate data are not available. This can be due to failure of accurate forecasting (e.g., seismic actions on structures, road traffic volume), difficulty or cost of measurements (e.g., accurate traffic volume data), and even in some cases when there are no models to explain a phenomenon sufficiently (e.g., the seismic behavior of reinforced concrete frame structures with infill masonry). These inconsistencies make it necessary to develop approximate reasoning to judge the data and make decisions in a scientific and responsible manner.

Approximate reasoning is used to relate fuzzy propositions. The fuzzy inference schemas that have been used are an extension of the classical inference schemas. So generalized modus ponens (GMP), generalized modus tollens (GMT), and hypothetical reasoning can be used. For calculating these schemas, fuzzy implications are as significant as the classical implication is to classical inference.

The schemas are generalized as follows. Assume two variables x and y set in X and Y sets, respectively. Furthermore assume a relation R between x and y. Assume now that it is known that $x \in A$. Then we can reasonably assume that $y \in B$ where $B = \{y \in Y | \langle x, y \rangle \in R, x \in A\}$ This also applies when the sets A, B are fuzzy and R is a fuzzy relation with membership functions x_A, x_B, x_R, respectively, and the formula transformed as follows [6] (Eq. (15)):

$$x_B(y) = \sup_{x \in X} \min[x_A(x), x_R(x, y)] \tag{15}$$

This formula is known as compositional rule of inference [13] when $R(x, y)$ is a fuzzy implication relation $I(A(x), B(y))$.

Generalized modus ponens or any other approximate reason scheme can be calculated with the use of the compositional rule of inference. Assume for x and y that A implies B and that it is known that in a given instance $x \in A'$. Then $y \in B'$ and B' can be computed using the rule, as follows. It is apparent that if all involved sets are crisp, the formula is equivalent to classical modus ponens (Eq. (16)).

$$B'(y) = \sup_{x \in X} \min[A'(x), R(x, y)] \tag{16}$$

Choosing Fuzzy Implication Function

In the process of approximate reasoning, the selection of fuzzy operations can cause large variations in the results. Especially when constructing a fuzzy inference system for which an "IF…Then…" fuzzy rule base is the cornerstone, the rules are essentially fuzzy implication operators. The choice has to be from a large set of alternative functions. There are some well-known functions that can be found in most handbooks [6], and there are many methods to construct new functions. Some of the most significant have been noted.

Although there is no unique and exact way of selecting the implication, several attempts have been made. It should be made clear that the fuzzy implication functions can be divided into groups. Thus, the literature is possible to give guidance regarding the selection of the initial implication group that matches according to theory in an optimal way. Thus the selection of the group may be based on the type of propositional reasoning schema that is utilized [6]. Additionally the implication function can be chosen from the *t-norm*, *t-conorm*, or negation utilized, or it could be chosen to fit optimal the observed data [9].

Special mention needs to be made to the fuzzy implication selection method based on data that was applied to describe the relationship of a country's economic status to the number of air transport movements in this country, after the creation of appropriate linguistic variables [4]. Recall that an implication function can be described as a relationship between two sets, the domain set of the antecedent and the consequent. Thus, each pair of observations can be considered as an element that certainly belongs to the relationship, as it is verified by the data.

Empiristic Fuzzy Implications

Empiricism and Fuzzy Implication

Fuzzy implication is perceived as a generalization of classical implication operator. Therefore any function utilized as a fuzzy implication relation between two linguistic variables should follow the axioms that have been determined as an extension of the rules of classical Boolean logic to fuzzy logic a priori. Those rules do not arise from observation, data collection, and professional experience but from philosophical reasoning. Thus it can be argued that those implication operators have been founded in a rationalistic manner. Rationalism in philosophy is the view that regards reason as the chief source and test of knowledge [2].

In the history of philosophy, starting from the controversy of Plato and Aristotle, there was a controversy between the two movements of rationalism and empiricism. Empiricism is the view that all concepts originate in experience [5]. Knowledge originates in what one can sense and there is no a priori knowledge. In this view human is described by John Locke as a blank table (*tabula rasa*), a phrase inspired by the works of Aristotle. From this scope, since a fuzzy implication function resembles knowledge about a correlation between two variables, it is possible the implication to be defined empiristicly.

Establishing a fuzzy implication relation between two fuzzy sets with the empiristic method requires data collection. The properties of this relationship must be constructed from the data, without a priori knowledge and constraints. Thus any implication function that abides to the aforementioned axioms cannot be considered as empiristic, but only rationalistic as the axioms effect the function significantly. Observed data can only come second, calibrating the function or distinguishing the most efficient from the available choices. Therefore any implication that is defined with the empiristic method will not satisfy the axioms, and so it will not be a logical implication function in the classical way of thinking.

Defining Empiristic Fuzzy Implication

Assume two variables $x \in X$ and $y \in Y$ and two linguistic variables $A : X \to [0, 1]$ and $B : Y \to [0, 1]$, with their corresponding membership functions. Suppose even that the variables x and y have some degree of correlation and there are n observed pairs of values, (x_i, y_i). It is possible to study the implication $x = A \Rightarrow y = B$ with the empiristic method, based on the data. If, for example, it is observed that for $a, b, c, d \in [0, 1]$ any time that the membership value of $A(x)$ resides in the closed interval $[a, b]$, $A(x) \in [a, b]$, then $B(y) \in [c, d]$, it is reasonable to assume that the fuzzy implication $A \Rightarrow B$ for $[a, b] \Rightarrow [c, d]$ is true with a degree of truth one, as it happens every time. Accordingly if there is no observation where $A(x) \in [a, b]$ and $B(y) \in [c, d]$, then the implication $[a, b] \Rightarrow [c, d]$ does not hold, i.e., it is true with degree zero. Furthermore if for $a, b, c, d, e, f \in [0, 1]$, when $A(x) \in [a, b]$, there are more observations for $B(y) \in [c, d]$ than for $B(y) \in [e, f]$, then the degree of truth for $[a, b] \Rightarrow [c, d]$ is greater than the degree of truth for $[a, b] \Rightarrow [e, f]$.

Definition 1 Let two variables $x \in X$ and $y \in Y$, two linguistic variables $A : X \to [0, 1]$ and $B : Y \to [0, 1]$, and there is n number of data pairs (x_i, y_i). Define E_{AB} the empiricist implication of linguistic variables based on the sample, a fuzzy relation between two sets of intervals that are inside the interval [0, 1] such that:

α) If for A' and B' intervals inside [0, 1]

$$E_{AB}(A', B') = 1 \Leftrightarrow \forall x_i : A(x_i) \in A' \Rightarrow B(y_i) \in B'$$

β) If for A' and B' intervals inside [0, 1]

$$E_{AB}(A', B') = 0 \Leftrightarrow \forall x_i : A(x_i) \in A' \Rightarrow B(y_i) \notin B'$$

γ) If for A', B', and C' intervals inside [0, 1]

$$E_{AB}(A', B') < E_{AB}(A', C')$$
$$\Leftrightarrow |\{x_i : A(x_i) \in A' \text{ and } B(y_i) \in B'\}| < |\{x_i : A(x_i) \in A' \text{ and } B(y_i) \in C'\}|$$

The first two of the axioms are based on classical logic, which must be verified by the fuzzy empiricist implication relation. Specifically, the first axiom means that two propositions $p(A(x_i) \in A')$ and $q(B(y_i) \in B')$ if it is true that whenever p is true, q is also true, then based on the data, the implication $p \Rightarrow q$ is true. In contrast if whenever p is true, q is not true, the implication $p \Rightarrow q$ is not true, and it stands for the second of the axioms. The third axiom includes three logical propositions $p(A(x_i) \in A')$, $q(B(y_i) \in B')$, and $r(B(y_i) \in C')$. If based on observations, when p is true, it is more likely r to be true than q to be true, then the implication $p \Rightarrow r$ is valid with a greater degree of truth than the implication $p \Rightarrow q$.

Computation of Fuzzy Empiristic Implication Relation

A method to calculate the membership function of a fuzzy empiristic implication relation derived from data observation is proposed. The computation is easy to comprehend and use and can be evaluated in short time when using a computer, even for large amounts of data.

Assume two variables $x \in X$ and $y \in Y$ and two linguistic variables $A : X \rightarrow [0, 1]$ and $B : Y \rightarrow [0, 1]$. Also there are pairs of observations (x_i, y_i) for $i = 1, 2 \ldots n$. In order to evaluate the implication relation $A \Rightarrow B$, the observation pairs should be transformed to pairs of membership values $(A(x_i), B(y_i))$. The characteristic functions $A(x)$, $B(y)$ are considered known.

The process can be done even if the original data are not in the form (x_i, y_i), but they are already in the form $(A(x_i), B(y_i))$. The reason is that the implication does not examine directly the correlation between the variables x and y, but the correlation between the linguistic variables that arises. Furthermore, an empiristic implication relation can be calculated for linguistic variables that are based on qualitative rather than quantitative data, e.g., bad weather implies a desire for hot beverage. In such cases it is possible to find a connection directly to the fuzzy data that can be collected, for example, from questionnaires, without the study of the relationship of the linguistic variable "bad weather" to quantitative data such as temperature, sunshine, humidity, etc. So from qualitative data, it is possible to obtain useful conclusions.

Initially the pairs of membership values $(A(x_i), B(y_i))$ are calculated if they are not available, from the observation pairs (x_i, y_i). Afterward the pairs of membership values $(A(x_i), B(y_i))$ are divided into k and l in incompatible intervals shaping rectangular areas of the $[0, 1] \times [0, 1]$ The numbers of intervals k and l can be equal or not, and the determination of k and l can be made with binning techniques. The division is shown in Fig. 1, where the dots resemble pairs of membership values of observations and the lines form a possible division. So for $A(x)$ there has been a fragmentation in $A_1, A_2 \ldots A_k$ intervals inside $[0, 1]$, and for $B(y)$ there has been a fragmentation in $B_1, B_2 \ldots B_l$ intervals inside $[0, 1]$. The intervals are proposed to be of the same diameter, but it is not necessary, so if there is a

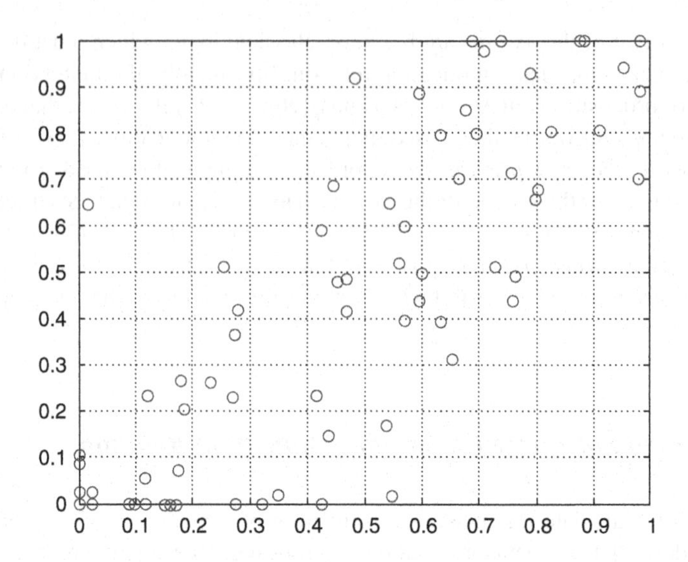

Fig. 1 Membership values of observation pairs and division

Table 2 Fuzzy empiristic implication relation matrix

	B_1	B_2	...	B_l
A_1	$I(A_1, B_1)$	$I(A_1, B_2)$		$I(A_1, B_l)$
A_2	$I(A_2, B_1)$	$I(A_2, B_2)$		$I(A_2, B_l)$
...				
A_k	$I(A_k, B_1)$	$I(A_k, B_2)$		$I(A_k, B_l)$

practical purpose, it can be uneven. Afterward for every possible pair of $A_i, B_j, i = 1, 2 \ldots k, j = 1, 2 \ldots l$, the degree of truth of the implication $E_{AB}(A_i, B_j)$ is computed as shown in (Eq. (17)). The formula chosen is quite similar to the law of conditional probability (Eq. (18)). The implications that are created this way satisfy the proposed axioms of the definition of empiristic implication relation.

$N((x, y) : A(x) \in A_i) \cap B(y) \in B_j$ is the number of pairs of observations for which $A(x) \in A_i$ and $B(y) \in B_j$, while $N(x : A(x) \in A_i)$ is the number of pairs of observations for which $A(x) \in A_i$. Thus the empiristic implication can be evaluated, and the matrix of the fuzzy relation is shown in Table 2.

$$E_{AB}(A_i, B_j) = \frac{N((x, y) : (A(x) \in A_i) \cap (B(y) \in B_j))}{N(x : A(x) \in A_i)} \tag{17}$$

$$P(A/B) = \frac{P(A \cap B)}{P(A)} \tag{18}$$

To carry out the required calculations, first it is necessary to create another matrix, counting the number of the instances for any possible pair of A_i, B_j. The divided data matrix is shown in Table 3 where $N(A_i, B_j)$ is the number of instances

Table 3 Divided data matrix

	B_1	B_2	...	B_l
A_1	$N(A_1, B_1)$	$N(A_1, B_2)$		$N(A_1, B_l)$
A_2	$N(A_2, B_1)$	$N(A_2, B_2)$		$N(A_2, B_l)$
...				
A_k	$N(A_k, B_1)$	$N(A_k, B_2)$		$N(A_k, B_l)$

where $A(x) \in A_i$ and $B(y) \in B_j$. Subsequently every element of the matrix is divided by the sum of the column, and thus the empiristic implication operator is calculated.

The separation in intervals as mentioned is another important step for calculating an implication that adequately describes the relationship between cause and effect. It is natural that in every case, the ideal number of intervals, and perhaps their diameters, when we have unequal intervals, cannot be determined a priori. This is because the ideal division depends on the number of the observations and on the relation between antecedent and consequent. If the division is done to a small number of intervals, the implication constructed becomes too biased and cannot provide much information. On the other hand, if the division is done to a larger number of intervals, the constructed implication may have significant variance, making any resulting model over-fitted, and thus unable to generalize and interpolate.

An efficient way to make the choice is with the sample split into two groups randomly and calculating the relation for both groups separately. Starting from small k and l numbers, the two matrices that describe the same phenomenon should have few differences. In large numbers of k and l, the two matrices should be very different from one another indicating that the created relation is over-fitted. By successively increasing the numbers k and l, calculating the relations and their differences, it is possible to find the ideal division for the data.

If the sample is limited, the use of rule of Sturges is proposed to calculate the appropriate number of intervals based on (Eqs. (19)–(20)), where n is the sample size [10, 11]. In such problems it is also suggested for k and l to be equal and for the intervals to be of equal diameter. In this case, when separating the [0, 1] interval into same length intervals, the diameter d of each interval is given by (Eq. (21)):

$$k = 1 + \log_2 n \tag{19}$$

or

$$k = 1 + 3.32 \log n \tag{20}$$

$$d = \frac{1}{k} \tag{21}$$

Finally, it should be clear that the results of such a process cannot be considered absolutely safe in every case because the sample used has an important role in the outcome. So the quality of the sample firstly and the quality of the fuzzification secondly impact the quality of the resulting implication relation. Of course, most methods that have to do with data are vulnerable to biased or incorrect data.

Application

Selection of Logical Fuzzy Implication Through the Empiricist

As it has been noted, the empiristic fuzzy implication relation that has been defined does not satisfy the established logical fuzzy implication axioms, and therefore it cannot resemble a logical fuzzy implication operator. Furthermore the empiristic implication is a relation between intervals and not a relation between numbers as the logical fuzzy implication functions are. So there can be no direct comparison between the two. However the two can be compared in the procedures of approximate reasoning.

Assume that there are two variables $x \in X$, $y \in Y$ and two linguistic variables $A(x) : X \rightarrow [0,1]$, $B(y) : Y \rightarrow [0,1]$. Also there is a relation $R(x, y) = I(A(x), B(y))$, where $I(A(x), B(y))$ is a logical fuzzy implication function. Then using the generalized modus ponens scheme, approximate reasoning has provided the compositional rule of inference, so if $x \in A'$, where A' is a fuzzy set, y belongs to a fuzzy set B' computed by (Eq. (22)).

$$B'(y) = \sup_{x \in X} \min[A'(x), R(x, y)] \tag{22}$$

Assume that for the aforementioned variables x, y and for the linguistic variables $A(x)$, $B(y)$ there is a data set of pairs of observations. In this case it is possible to calculate an empiristic fuzzy implication relation. Utilizing the empiristic implication, it is possible to argue that when $\forall x : A(x) \in A'$, y is such that $B(y) \in B_1$ with a degree of truth $E_{AB}(A', B_1)$, $B(y) \in B_2$ with a degree of truth $E_{AB}(A', B_2)$, etc. While the values calculated by the compositional rule of inference are results of a rationalistic manner of reasoning, the corresponding values of the empiristic implication, when the data set is proper, are results of observation. Thus the empiristic implication better resembles reality. So it is possible to measure the difference between the logical and the empiristic implication, for any A_i and B_j pair of the empiristic relation.

It is plausible that this process can be a criterion for choice of logical fuzzy implication functions. So to make such a choice between some logical fuzzy implication functions in a particular problem, for which observation data is available, the difference of each of the logical implication functions to the empiristic implication can be measured, and the one that is closer to the empiristic implication should be the appropriate logical implication function. Such a calculation can be made with results of the compositional rule of inference and any approximate reasoning scheme as modus ponens, modus tollens, hypothetical reasoning, etc. Assuming that the implication to be chosen will be used on a real-life problem, it is preferable to use the reasoning scheme that is useful for the specific problem.

This implication selection process is of a great theoretical and practical interest. The logical implication decided by this process results is something of a compromise between rationalistic and empiricist logic, since it is a relationship structured with all the constraints dictated by the logical reasoning, but selected for good behavior with respect to the data observed in the physical world.

Algorithm

First step of the process is to calculate the empiristic fuzzy implication relationship based on the data as discussed in the previous subchapter. Result of the process is the relationship matrix of the implication.

Second step is to calculate a corresponding table for each of the logical implication functions in order to be compared. The calculation will be done using the compositional rule of inference, for a reasoning scheme as generalized modus ponens. The table has to have the same dimensions as the empiricist implication relation.

Third step is to count the difference between the matrices using some metric and chose the optimal implication function.

In the relation matrix of the empiristic fuzzy implication, the element that resides in the ith row and the jth column is the degree of truth in implying that $E(A_i, B_j)$. So when x is such that $A(x) \in A_i$, it implies that y is such that $B(y) \in B_j$, with a degree of truth $E(A_i, B_j)$. It is noted that any A_i, B_j set for $i = 1, 2 \ldots k, j = 1, 2 \ldots l$ is a classical, crisp set. When $A(x)]inA_i$, from the CRI (Eq. (23)) is:

$$B'(y) = \sup_{x \in X} \min[A_i', I(A(x) \in A_i, B(y))] \tag{23}$$

This formula, for crisp sets A_i, becomes (Eq. (24)):

$$B'(y) = \sup_{x \in X}[I(A(x), B(y))] \tag{24}$$

Note that for the compositional rule of inference formula, any *t-norm* is applicable. Since sets A_i are crisp and for any *t-norm* $T(0, a) = a$, the selection of *t-norm* is of no consequence.

Most logical implication functions meet the first two axioms (1 and 2), so they are decreasing for the first variable and increasing for the second. The algorithm is developed for implication functions for who this is valid. Otherwise the complexity of operations is increased, although not very much. Thus, because any A_i interval is bounded, with an infimum (assume a_i) on the left boarder, (Eq. (24)) becomes (Eq. (25)):

$$B'(y) = I(a_i, B(y)) \tag{25}$$

So when $x \in A_i'$, $B'(y)$ is calculated according to the formula above. But to describe the implication relation matrix corresponding to each element of the empiricist implication relation matrix the question: "When $x \in A'_i$, with what degree of truth is it implied that $y \in B'_j$?". So the degree of truth $I(A_i, B_j)$ is calculated as follows (Eq. (26)):

$$I(A_i, B_j) = \cup_{B(y) \in B_j} I(a_i, B(y)) \tag{26}$$

As *t-conorm* can be used the maximum. This maximum is very easy to find for logical implication functions for which the second axiom is satisfied, i.e., increasing the second variable. Then, since $B(y)$ belongs in a bounded interval with supremum being the right boarder (assume b_j), (Eq. (26)) becomes:

$$I(A_i, B_j) = I(a_i, b_j) \tag{27}$$

Thus, the appropriate matrix is calculated for each logical implication function from (Eq. (27)), so that the element located at the ith row and jth column has the value $I(a_i, b_j)$.

Finally the deviation of any logical implication function from the empiristic has to be calculated. It is proposed that the norm of the difference of the two matrices is calculated. The implication function with the smallest deviation from the empiristic is selected to be the one that represents more accurately the relationship between linguistic variables according to the given data set.

Conclusions and Further Research

The fuzzy implication functions that can be used are numerous, and their properties have important influence on the quality of the models and the understanding of the phenomena described. A new class of fuzzy implication relations is founded, named empiristic fuzzy implication relations. These relations are inspired by the philosophical movement of empiricism, constructed only from observed data, with no a priori restrictions and conditions. An easy and clear way of calculating an empiristic fuzzy implication relation is described that can be functional even for huge volumes of data.

Finally fuzzy implication function selection problem was analyzed. The problem has been studied by many researchers in the world, and there is no unanimous solution to address it. A fuzzy implication function selection method is recommended, based on empiristic implication relations. For comparison, observational data is needed based on which the empiricist implication relations are calculated, and the logical implication relationship with the less deviation from the empiristic is chosen to be the one that accurately represents the data.

In the future the construction of fuzzy inference systems will be studied, using only the fuzzy empiristic implication relation. These systems will consist of a base of fuzzy rules "If ... then ...," each one of which will be an empiristic implication, computed from data observations.

References

1. Baczynski, M., Jayaram, B.: Fuzzy Implications. Studies in Fuzziness and Soft Computing, vol. 231. Springer, Berlin (2008)
2. Blanshard, B.: Rationalism. https://www.britannica.com/topic/rationalism
3. Botzoris, G., Papadopoulos, B.: Fuzzy Sets: Applications for the Design and Operation of Civil Engineering Projects (in Greek). Sofia Editions, Thessaloniki (2015)
4. Botzoris, G., Papadopoulos, K., Papadopoulos, V.: A method for the evaluation and selection of an appropriate fuzzy implication by using statistical data. Fuzzy Econ. Rev. **XX**(2), 19–29 (2015). 7th International Mathematical Week, Thessaloniki
5. Fumerton, R., Quinton, A.M., Quinton, B.: Empiricism. https://www.britannica.com/topic/empiricism
6. Klir, G., Yuan, B.: Fuzzy Sets and Fuzzy Logic: Theory and Applications. Prentice Hall, Upper Saddle River (1995)
7. Massanet, S., Torrens, J.: An overview of construction methods of fuzzy implications. In: Advances in Fuzzy Implication Functions. Studies in Fuzziness and Soft Computing. Springer, Berlin (2013)
8. Mendel, J.M.: Fuzzy logic systems for engineering: a tutorial. Proc. IEEE **83**(3), 345–377 (1995)
9. Revault d'Allonnes, A., Akdag, H., Bouchon-Meunier, B.: Selecting implications in fuzzy abductive problems. In: Proceedings of the 2007 IEEE Symposium on Foundations of Computational Intelligence (FOCI) (2007)
10. Scott, D.W.: Multivariate Density Estimation: Theory, Practice, and Visualization. Wiley, New York (1992)
11. Sturges, H.A.: The choice of a class interval. J. Am. Stat. Assoc. **21**, 65–66 (1926)
12. Zadeh, L.A.: Fuzzy sets. Inf. Control. **8**, 338–353 (1965)
13. Zadeh, L.A.: Outline of a new approach to the analysis complex systems and decision processes. IEEE Trans. Syst. Man Cybern. **3**, 28–44 (1973)

Adaptive Traffic Modelling for Network Anomaly Detection

Vassilios C. Moussas

Abstract With the rapid expansion of computer networks, security has become a crucial issue, either for small home networks or large corporate intranets. A standard way to detect illegitimate use of a network is through traffic monitoring. Consistent modelling of typical network activity can help separate the normal use of the network from an intruder activity or an unusual user activity. In this work an adaptive traffic modelling and estimation method for detecting network unusual activity, network anomaly or intrusion is presented. The proposed method uses simple and widely collected sets of traffic data, such as bandwidth utilization. The advantage of the method is that it builds the traffic patterns using data found easily by polling a network node MIB. The method was tested using real traffic data from various network segments in our university campus. The method performed equally well either offline or in real time, running at a fraction of the smallest sampling interval set by the network monitoring programs. The implemented adaptive multi-model partitioning algorithm was able to identify successfully all typical or unusual activities contained in the test datasets.

Keywords Traffic modelling · Fault detection · Anomaly detection · Network simulation · Adaptive estimation · Forecasting · SARIMA models · Nonlinear time series · State-space models · Kalman filter · Multi-model

---

V. C. Moussas (✉)
Laboratory of Applied Informatics, Department of Civil Engineering, University of West Attica, Egaleo-Athens, Greece
e-mail: vmouss@teiath.gr

N. J. Daras, Th. M. Rassias (eds.), *Modern Discrete Mathematics and Analysis*, Springer Optimization and Its Applications 131,
https://doi.org/10.1007/978-3-319-74325-7_17

Introduction

In order to separate the normal use of a network from an intruder activity or an unusual user activity, consistent models of typical network activity or abuse are required. Traffic monitoring and modelling is also essential in order to determine the network's current state (normal or faulty) and also to predict its future trends [9].

Intrusion detection systems (IDS) are being designed to protect such critical networked systems. There are two major approaches in intrusion detection: anomaly detection and misuse detection. Misuse detection is first recording and modelling specific patterns of intrusions and then monitoring and reporting if any matches are found. Anomaly detection, on the other hand, first records and models the normal behaviour of the network and then detects any variations from the normal model in the observed data. The main advantage with anomaly intrusion is that it can detect new forms of attacks or network misuse, as they will probably deviate from any other normal behaviour [5].

Anomaly detection systems apply various methods to model the normal behaviour of the network. Some systems utilize artificial neural networks (ANN) [4]and self-organizing maps (SOM) [26]. The NN is fed initially by normal traffic to learn the normal conditions and then by the observed traffic to detect anomalies. Other systems collect statistics from certain system parameters into a profile, and they construct a distance vector for the observed traffic and the specified profile [25].

Most methods of intrusion detection are based on hand-coded rule sets or predicting commands online, they are laborious to build, and they require a very large amount of special traffic data (detailed static logs, protocols, ports, connections, etc.) provided by hubs, routers, firewalls, hosts and network sniffers. In addition, most of these algorithms require that the data used for training is purely normal and does not contain any attacks. The process of manually cleaning the data is quite time-consuming, and a large set of clean data can be very expensive, although some algorithms may tolerate mixed data [6].

Network Monitoring

Traffic monitoring, traffic prediction and anomaly detection are crucial for today's networks, and they all play a significant role when designing a network or network upgrades [7, 11, 30].

When planning or designing a network, good forecasts of the network traffic workload are required. Early detection of a traffic anomaly is also crucial when controlling or managing LAN, MAN and WAN networks. Both forecasts and detections can be calculated using various models of the network behaviour in combination with a corresponding simulation or identification technique [15, 21, 27].

Traditionally, network fault management emphasizes the detection and the processing of network and element failures in the form of alarms. Regarding network fault detection, the past years have witnessed much progress in anomaly detection in Ethernet segments [14], anomaly and performance change detection in small networks and proactive anomaly detection of network and service faults [8]. In the last case, proactive anomaly detection can infer the presence of non-monitored failures (i.e. no MIB nor trap information access) from the monitored performance data of the networks. In addition to the online anomaly detection, the same models can also be applied either for network simulation or prediction.

Traffic Modelling Detail Levels

Network model selection depends mainly on the applied technique and the available network data. There are several types of data available to collect, when studying a network. Almost any traffic characteristic may be measured and logged, i.e. bit or packet rate, packet size, protocols, ports, connections, addresses, server load, applications, etc. Routers, firewalls, servers or managers (servers with agents) can all be used for this task.

Each modelling method represents the behaviour of the network at a different level of detail. More abstract models, based only on the overall line utilization, are less precise, but they are also very fast and less demanding. On the other hand, more detailed models represent the exact packet exchange procedure in a network, but they are very slow and resource demanding. Measuring and archiving all traffic data at full detail for potential future use is not a regular procedure. Based on the level of traffic detail observed, the traffic models may be divided in two main groups:

More Abstract Traffic Models Usually most networks log only the load of their lines and the utilization of some critical resources, while a more detailed monitoring is used only when a resource requires specific attention. On almost any network, traffic rate and utilization are the only data collections that are always available and with long history records. These data are easily taken from the router MIB or from server logs, and they can be used to create global or aggregate traffic models [17, 18, 31].

More Detailed Traffic Models When more detailed models of the network behaviour are needed, special traffic data provided by agents, switches, routers, firewalls, hosts or network sniffers must be used. Moreover, when modelling user behaviour, other types of data such as transaction duration, user habits, skills or position may be required [13, 22].

Traffic Modelling Categories and Uses

In [16] an effort to classify the traffic models by their detailed or abstract view of the inherent network mechanism resulted to three (3) major model categories:

1. *Overall utilization (OU) modelling category*: The most abstract models that observe only the overall utilization of network lines or components.

   OU models describe the network load of each segment or component in packets per time unit (pps) or bytes per time unit. They may also use any other characteristic (e.g. processor load) found in the MIB of a network component and captured by a monitoring application.

   The required data are easily collected from the router interface and stored in a dedicated machine. This is done online without delays and it does not require any special HW or SW capabilities. In addition, OU models have a large database of past records waiting to be used for model training. This happens just because most networks use these sets of data for every day monitoring and they keep them for reference or for their network load history.

   OU models take also into account the periodic nature of the network utilization, its stochastic properties and any known anomalies observed in the past. These abstract models require much less processing time and can be applied online on any machine, either for simulation and prediction or for anomaly detection. Due to their abstract nature, these models offer an online first warning for any network anomaly, even if they cannot be more precise about the type or cause of the problem.
2. *Packet pattern (PP) modelling category*: The most detailed models that describe the network traffic at packet level in full detail.

   PP models attempt to describe the network traffic at packet or signalling level. Each action is analysed in full detail, and the exact packet type, size and exchange procedure are defined. PP models may detect suspicious packets or other port pattern anomalies from, e.g. their TCP flag combinations, timing or matching to a certain pattern library. Typical examples in this category are the packet-spoofing attacks, such as SYN flood, Smurf, TCP spoofing, Bounce Scan, Zombie control, etc. Most IDS or anomaly detection systems in this category belong in the 'signature analysis' class where detailed descriptions of known attacks or anomalies are encoded, e.g. in the form of rules, and stored for comparison and reference.

   Data at this level are usually collected by a packet capturing tool. Packet capturing is a very intensive and hardware-demanding task. The network adapter is usually working in promiscuous mode capturing all network traffic and storing it in long files for further analysis. Packet analysis and statistics are then done offline by other programs. Statistics and/or patterns for typical packets are often stored in a pattern library and subsequently used to detect anomalies in the same type of traffic.

   Network traffic analysis or simulation using very detailed models can be slow and resource demanding. The processed or simulated time is often a small fraction of the physical time, and therefore it is difficult to apply the models online for long periods.

   PP models can identify an anomaly with high accuracy. They are able to distinguish between different types of network misuse or attacks and trigger the correct reaction. Due to their complexity, they are usually activated for detailed detection after a global anomaly detection by another less detailed model (TP or OU).

3. *Task pattern (TP) modelling category*: Less detailed models that distinguish the various categories of network traffic, e.g. by application, protocol or user behaviour.

   TP models attempt to describe the network traffic per service, protocol or user task. Each type of traffic is characterized mainly by the protocol used, the originating service, the network path between client and server and the task objective or duration. Most IDS or anomaly detection systems in this category belong in the 'statistical systems' class that intend to 'learn' normal behaviour from the data and then issue alerts for suspected anomalies.

   Data for this category of models are provided by the various application logs, the network component MIBs and the firewall logs or by specialized applications running on a host or server, possibly with agents on other machines. The network traffic data collected report, usually, the total number of packets or bytes per time unit, the average size, or statistics on the size, the frequency, headers, origin and other characteristics of the observed messages.

   In order to detect network anomalies, a library of normal or expected behaviour is created, and all newer arrivals are compared to the stored patterns and classified accordingly. This is repeated for any application under consideration and for any type of service or protocol.

   Although it is possible to observe and analyse online at this level of network traffic, it is still impossible to store long records in such detail due to space and time limitations. Therefore, it is difficult to find sufficient past records in order to create adequate models for any type of traffic, unless there has been a specific preparation for such methods.

4. *Combined PP and TP models*: It is not a model category itself, but it is mentioned here as many applications used for network simulation combine both PP and TP models.

   Applications used for network simulation may combine PP and TP models. The network load is modelled as a set of tasks (TC) producing packets that travel across lines and nodes according to the network protocols used (PP). Such applications require a model for each node of the network and a model for each application served by the network. They superimpose all generated traffic (user tasks, applications, etc.) on the underlying network model (lines, nodes, servers, etc.) and take into account the network type properties and limitations (congestion, retransmissions, etc.), thus producing a simulated network response. The accuracy of these simulations depends on the accuracy of the network component models and application models introduced by the user.

   Combined PP-TP models have higher computational requirements, both in processing power and storage, and they are running much slower than the real-world events. The simulated time is only a fraction of the real time passed, and often a computer cluster is required just to keep real time below 1000× the simulation physical time. Consequently, these combined models, although they are a good technique for simulation and prediction, are not always suitable for online detection of network anomalies.

Each of the above model categories (OU, PP and TP) represents the network behaviour from a different point of view and requires different types of network data. The selection of a model category should be based on the available resources, the available datasets and the desired results of the modelling procedure, i.e. the use case where it is applied.

In this work we distinguish two major uses where the above models can be applied: *simulation and prediction* or *fault and anomaly detection*. In both cases, the models need some training from past data and known cases. The final model selection depends on the desired outcome, the available infrastructure, the knowledge base of past records and the desired level of detail. Each of the three categories (OU, PP and TP) may offer different pros and cons per use case:

- *Using the traffic models for simulation and prediction*: For simulation and prediction purposes, all three model categories can be applied, provided that enough past data are available for training. More precisely:
  - OU models require only the default data stored in component MIBs. These data are available on almost any network. If there is no other monitoring tool applied on a network, the OU model is the only option. The almost certain existence of past utilization data guaranties that the training of the model will be mostly accurate. An OU model should also be used when general utilization forecasts are required, using periodic models that incorporate current trends and future uncertainties.
  - PP models require detailed records from packet capturing applications and precise knowledge of the packet exchange procedures of the network. In addition, there must be plenty of real time available in order to run the simulations for a sufficient simulated time. If real time is of essence, a faster and less detailed model should be considered.
  - TP models require data from different sources such as server application logs, manager-agent monitoring tools and component MIBs. Statistical analysis of the past data provides the TP models with probabilistic information and distributions that will be used during data regeneration and prediction. The availability of these data will eventually define the exact form and the applicability of the model.
- *Using the traffic models for fault or anomaly detection*: For fault or anomaly detection purposes, all three model categories can also be applied, provided that the required past and present data are available. More precisely:
  - OU models can be applied easily on any network, and they are reporting the presence of a fault or anomaly rather than its nature or source. They are abstract but also much faster and less demanding. Their main advantage is that there is always a past utilization record available to train them, and

in combination with a good identification technique, they perform quite satisfactorily. OU models combined with seasonal ARMA or multi-model techniques can detect successfully known (trained) situations and also isolate any other detected faults or anomalies.
- PP models are not quite suitable for 24/7 all-purpose anomaly detection. They should be used at a second stage for finer more detailed identification of an attack or a fault cause. These models are usually applied offline to post-process and analyse the collected data. PP models can be used online only when they focus to a narrow subset of the overall traffic and usually for a limited time period.
- TP models can be more suitable than PP models for anomaly or fault detection. When past data are available, TP models with the necessary statistical information for the various applications or transactions may be calculated and stored. Libraries or knowledge bases with such models are then used by the detection application as a reference for the classification of the incoming traffic. This category includes a wide range of models including rule models, statistical models, lookup tables, neural networks, stochastic and multiple models, etc. TP detection models may vary from more detailed (closer to PP) to less detailed (closer to OU). When detail is increased, speed and ease of implementation are decreased and vice versa.

Network Traffic Model Identification

In this section, the components of the adaptive traffic modelling and estimation method, for network unusual activity and intrusion detection, are presented. The first aim of the proposed method is to use very common, simple and widely found traffic datasets, such as overall bandwidth utilization data (OU). Bandwidth use is the most common set of network traffic data; almost all network administrators monitor periodically and store the bandwidth utilization for their servers, routers, LAN users or network connections. The second aim of the proposed method is to take advantage of the time-series techniques [3] that have been applied already successfully in almost any research field, such as economy, signal processing, computer networks [18, 32], wireless communication [27], BW management [24] or even structural safety [29], and today, it is a well-established tool. It is also clear that the time-series models perform satisfactorily under conditions or circumstances similar to those of the training dataset. In our case this leads to the creation of many different models, each one fitting best a different pattern of traffic flow [28]. Normal traffic, congestion, weekdays, weekends, works or accidents, they all require different modelling schemes. As a result, there are many models available to describe the various status of traffic flow, and we could have better forecasting results if we are able to combine them optimally under one global method.

The proposed method is using past traffic data to learn and model (by ARMA, state-space, or other models) the normal or typical periodic behaviour of a network connection. In addition, any known faulty, abuse or anomaly state can be modelled and stored in this continuously updated model base.

An adaptive identification mechanism based on a powerful multi-model partitioning algorithm, (MMPA), proposed by Lainiotis [12] known for its stability and well established in identification and modelling, is then applying the collected OU data to the candidate traffic models. If the traffic pattern does not match an expected behaviour of the network connection, an anomaly is detected, and furthermore, if the traffic pattern matches a known anomaly case, the type of anomaly can be identified.

In the following sections, we first present some ARMA and state-space models of network traffic that can be used by MMPA, then we present the multi-model partitioning algorithm (MMPA), and finally, we present detection results from MMPA, using real datasets collected at the campus network of the Technical Educational Institute (TEI) of Athens.

SARIMA Traffic Modelling

As shown in Fig. 1, the recorded OU network traffic and bandwidth utilization demonstrate a remarkable periodicity, daily, weekly and also yearly. One method to model such ‘seasonal’ behaviour is to use a set of seasonal ARIMA (SARIMA) time-series models. In an earlier work [17], the network bandwidth utilization of the

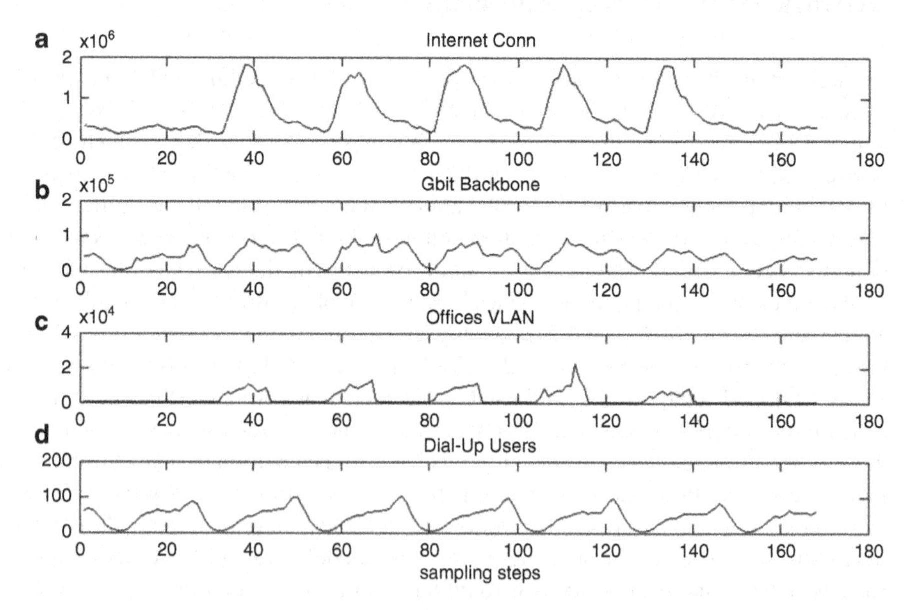

Fig. 1 Average utilization data from the TEI of Athens campus network (weekly data): (**a**) the campus Internet connection, (**b**) an educational premise backbone, (**c**) an administration office VLAN, and (**d**) the remote users’ connections

TEI of Athens campus network was successfully modelled using such SARIMA models. In this contribution the same method is applied in order to model the periodic behaviour observed in the daily and weekly repeated OU patterns.

The principle underlying this methodology is that traffic data occur in a form of a time series where observations are dependent. This dependency is not necessarily limited to one step (Markov assumption), but it can extend to many steps in the past of the series. Thus in general the current value X_t (= network traffic at time t) of the process X can be expressed as a finite linear aggregate of previous values of the process and the present and previous values of a random input u, i.e. [3]:

$$X_t = \phi_1 X_{t-1} + \phi_2 X_{t-2} + \cdots + \phi_p X_{t-p} + u_t - \theta_1 u_{t-1} - \theta_2 u_{t-2} - \cdots - \theta_q u_{t-q} \quad (1)$$

In Eq. (1), X_t and u_t represent, respectively, the traffic volume and the random input at equally spaced time intervals $(t,\ t-1,\ t-2, \ldots)$. The random input u constitutes a white noise stochastic process, whose distribution is assumed to be Gaussian with zero mean and standard deviation σ_u.

Equation (1) can be economically rewritten as (4) by defining the autoregressive (AR) operator of order p and the moving-average (MA) operator of order q by Eqs. (2) and (3), respectively:

$$\varphi_p(B) = 1 - \phi_1 B - \phi_2 B^2 - \cdots - \phi_p B^p \quad (2)$$

$$\vartheta_q(B) = 1 - \theta_1 B - \theta_2 B^2 - \cdots - \theta_q B^q \quad (3)$$

$$\varphi_p(B) X_t = \vartheta_q(B) u_t \quad (4)$$

where B stands for the backward shift operator defined as $B^s X_t = X_{t-s}$. Another closely related operator is the backward difference operator ∇ defined as $\nabla X_t = X_t - X_{t-1}$ and thus $\nabla = 1 - B$, $\nabla^d = (1-B)^d$ and $\nabla_s^D = (1-B^s)^D$.

The autoregressive moving-average model (ARMA) as formulated above is limited to modelling phenomena exhibiting stationarity. Clearly this is not the case for the network traffic data of Fig. 1. It is possible though that the processes still possess homogeneity of some kind. It is usually the case that the dth difference of the original time series exhibits stationary characteristics. The previous ARMA model could then be applied to the new stationary process ∇X, and Eq. (4) will correspondingly read

$$\varphi_p(B) \nabla^d X_t = \vartheta_q(B) u_t \quad (5)$$

This equation represents the general model used in this work. Clearly, it can describe stationary ($d = 0$) or nonstationary ($d \neq 0$), purely autoregressive ($q = 0$) or purely moving-average ($p = 0$) processes. It is called autoregressive integrated moving-average (ARIMA) process of order (p, d, q). It employs $p+q+1$ unknown parameters $\phi_1, \ldots, \phi_p; \theta_1, \ldots, \theta_p; u$, which will have to be estimated from the data.

Starting from ARIMA model of Eq. (5), it can be deducted [3] that a seasonal series can be mathematically represented by the general multiplicative model often called seasonal ARIMA or SARIMA of order $(p, d, q)x(P, D, Q)_s$:

$$\varphi_p(B)\phi_P(B^s)\nabla^d\nabla_s^D X_t = \vartheta_q(B)\theta_Q(B^s)u_t \tag{6}$$

The general scheme for determining these traffic models includes three phases, which are:

1. Model identification, where the values of the model order parameters p, d, q, P, D, Q and s are defined.
2. Parameter estimation, where all ϕ and θ coefficients in $\varphi_p, \phi_P, \vartheta_q, \theta_Q$ are determined in some optimal way, and,
3. Diagnostic checking for verifying the model's performance over the collected data.

As is stated however in [3], there is no uniqueness in the ARIMA models for a particular physical problem. In the selection procedure, among potentially good candidates, one is aided by certain criteria. Although more advanced methods for model selection have been proposed [10, 23], the most common and classic criteria remain the Akaike information criterion (AIC) and Schwartz's Bayesian information criterion (SBC or BIC) [1, 3]. Proper choice of the parameters at phase 1 calls for a minimization of the AIC and SBC.

By analysing our campus OU traffic data from different subnets and VLANs, recorded hourly for several months, we verified the periodicity of the data. As shown in Fig. 2 by taking the autocorrelation function (ACF), two major seasonal components were identified, a daily and a weekly one, every 24 h and 168 h, respectively. In this weekly repeated pattern of OU, the observed daily network behaviour is then classified in two categories: (a) the OU traffic during normal working days and (b) the OU traffic during weekends and holidays.

After several tests on datasets collected at various time intervals (5–30 min), a common SARIMA model was identified that can satisfy both categories. Provided that its past period data belong to the same category with the forecasting period, the seasonal ARIMA $(1,1,1) \times (0,1,1)_{48}$ model predicts satisfactorily the future network traffic, as shown in Fig. 3.

Equation (7) represents the above SARIMA model mathematically. The autoregressive (AR) and moving-average (MA) parameters of the model are: $\phi_1 = 0.413027$, $\theta_1 = 0.942437$, $\theta_1 = 0.959323$.

$$\phi(B)\nabla^1\nabla_{48}^1 X_k = \theta(B)\theta\left(B^{48}\right)u_k \tag{7}$$

where $\phi(B) = 1 - \phi_1 B$, $\theta(B) = 1 - \theta_1 B$, $\theta\left(B^{48}\right) = 1 - \theta_1 B^{48}$, and the analytic expression for model equation (7) will be:

$$(1 - \phi_1 B)(1 - B)\left(1 - B^{48}\right)X_k = (1 - \theta_1 B)\left(1 - \theta_1 B^{48}\right)u_k$$

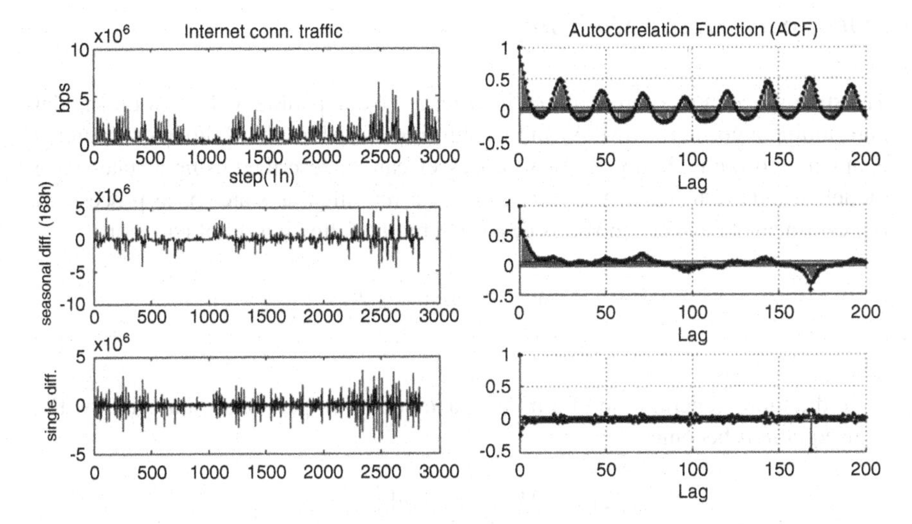

Fig. 2 TEI campus internet connection OU: 4 months of traffic data, the single and seasonal differences and their ACFs, demonstrating the 24 and 168 h periodicity

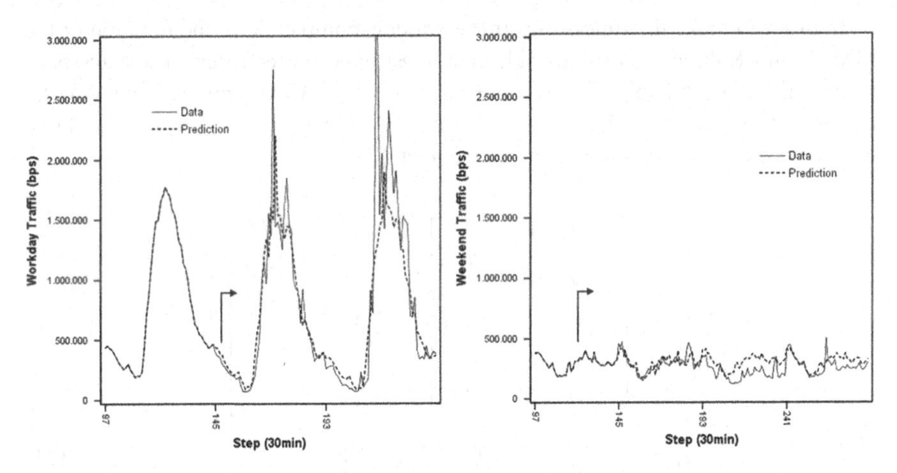

Fig. 3 Daily traffic prediction using the seasonal ARIMA (1,1,1) × (0,1,1)$_{48}$ model: (left) working day traffic and (right) weekend or holiday traffic. Prediction starts at step 145. The previous period (steps 97–144) is replaced by the average of all past periods (days) of the same type (weekends or working days)

$$\Rightarrow X_k - (1+\phi_1)X_{k-1} + \phi_1 X_{k-2} - X_{k-48} + (1+\phi_1)X_{k-49} - \phi_1 X_{k-50}$$
$$= u_k - \theta_1 u_{k-1} - \theta_1 u_{k-48} + \theta_1\theta_1 u_{k-49} \quad (8)$$

State-Space Traffic Modelling

The state-space models are required in order to be compatible with the multi-model partitioning algorithm (MMPA) and its sub-filters, such as the Kalman or extended Kalman algorithms. Many physical processes can be described using a state-space model. In addition, ARMA processes can be rewritten as state-space process. A typical linear state-space model is described by the following set of equations:

$$x_{k+1} = F \cdot x_k + G \cdot w_k,$$
$$z_k = H \cdot x_k + v_k \tag{9}$$

In the more general case of a nonlinear model with parametric uncertainty, the state equations become:

$$x_{k+1} = f[k, x_k; n] + g[k, x_k] \cdot w_k,$$
$$z_k = h[k, x_k; n] + v_k \tag{10}$$

In order to make the time-series traffic models compatible to the notation of the MMPA and Kalman algorithms [2], model (8) must be rewritten in a state-space format. Based on the innovations representation of an ARMA process, any ARMA model of the type, $z_k + a_1 z_{k-1} + \cdots + a_n z_{k-n} = b_0 u_k + \cdots + b_m u_{k-m}$, can be written in the following linear state-space form [2]:

$$x_{k+1} = \begin{bmatrix} -a_1 & I \cdots 0\ 0 \\ -a_2 & \vdots \ddots \vdots\ \vdots \\ \vdots & \vdots \ldots I\ 0 \\ -a_{n-1} & 0 \ldots 0\ I \\ -a_n & 0 \ldots 0\ 0 \end{bmatrix} x_k + \begin{bmatrix} b_1 - a_1 b_0 \\ b_2 - a_2 b_0 \\ \vdots \\ \vdots \\ \vdots \end{bmatrix} u_k,$$
$$z_k = \begin{bmatrix} I & 0 & \ldots & 0 & 0 \end{bmatrix} x_k + b_0 u_k \tag{11}$$

By substituting the model coefficients of Eq. (8) to the state-space representation (11), the SARIMA model can be directly implemented by a typical state-space algorithm such as the Kalman filter. The corresponding (non-zero) coefficients are:

$$\begin{aligned} a_0 = 1, \quad a_1 = -(1+\phi_1), \quad a_2 = \phi_1, \\ a_{48} = -1, \quad a_{49} = (1+\phi_1), \quad a_{50} = -\phi_1 \\ b_0 = 1, \quad b_1 = -\theta_1, \quad b_{48} = -\theta_1, \quad b_{49} = \theta_1\theta_1 \end{aligned} \tag{12}$$

In addition, state-space models can also be used to describe any non-periodic OU traffic patterns. There are numerous traffic conditions, such as line failures or network misuse, that cannot be modelled by an ARMA process. These events are not periodic, they occur at random instances, and therefore, the above seasonal models are not very helpful.

Typical cases can be a sudden rise (peak) in traffic (due of an attack or misuse), a zero traffic rate (due to a failure), a very constant (usually high) traffic (due to congestion or a DoS attack), etc. Below we present the state-space equations corresponding to each of these unusual cases:

$$\begin{array}{lll} \text{Traffic rise} \times 10: & z_k = x_k + v_k, & x_{k+1} = 10 \cdot x_k \\ \text{Line failure}: & z_k = x_k + v_k, & x_{k+1} = x_k, \quad (x_0 = 0) \\ \text{Line saturation}: & z_k = x_k + v_k, & x_{k+1} = x_k, \quad (x_0 = maxBW) \end{array} \tag{13}$$

The traffic models described above (ARIMA or state-space) are the adaptive method's candidates that will be matched, each one to a Kalman filter, and subsequently, introduced to the MMPA algorithm in order to detect adaptively the correct model of network utilization. A sample containing four various modelled traffic sequences is shown in Fig. 4.

The Multi-Model Partitioning Algorithm (MMPA)

The adaptive method applied here is based on the multi-model partitioning algorithm originally presented by Lainiotis [2, 12]. MMPA consists of a parallel bank of N sub-filters (i.e. Kalman, extended Kalman, etc.), operating concurrently on the measurements (Fig. 5).

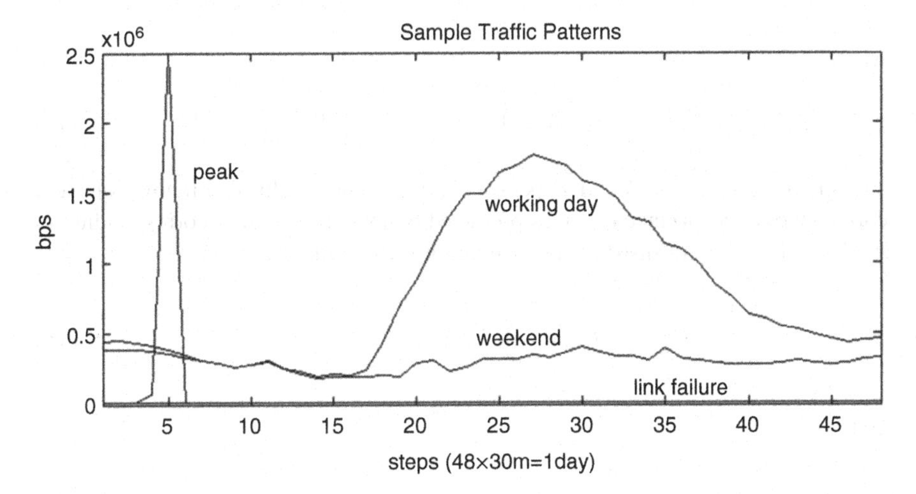

Fig. 4 Four samples of traffic sequences representing different traffic conditions and modelled using Eqs. (11) and (13)

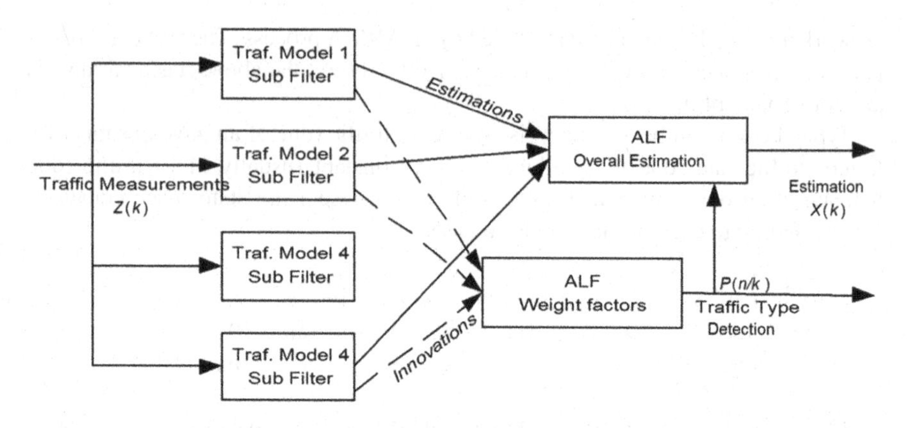

Fig. 5 Structure of the multi-model partitioning algorithm for network anomaly detection

Each sub-filter is tuned to a state-space modelling a different traffic behaviour and described by Eqs. (11) and (13). At time step k, first, each filter processes the measurement z_k and produces a state estimate $x(k/k; n)$ of the state x_k, conditioned on the hypothesis that the corresponding model is the correct one, and then the MMPA uses the output of all elemental filters to select the most likely model as the one that maximizes the a posteriori probability density $p(n/k)$. This density can be calculated recursively by Eq. (14) [12]:

$$p(n/k) = \frac{L(k/k; n)}{\sum_{i=1}^{N} L(k/k; i) p(i/k-1)} p(n/k-1) \tag{14}$$

where

$$L(k/k; n) = |P_{\tilde{z}}(k/k-1; n)|^{-\frac{1}{2}} e^{-\frac{1}{2}\|\tilde{z}(k/k-1;n)\|^2 P_{\tilde{z}}^{-1}(k/k-1;n)} \tag{15}$$

and where $\tilde{z}(k/k-1; n)$ and $P_{\tilde{z}}(k/k-1; n)$ are the conditional innovations and corresponding covariance matrices produced by the Kalman filter corresponding to model n. The overall MMPA state estimation is then calculated by:

$$\hat{x}(k/k) = \sum_{i=1}^{N} \hat{x}(k/k; n) p(n/k) \tag{16}$$

and

$$P(k/k) = \sum_{i=1}^{N} \left[P(k/k; n) + \left\| \hat{x}(k/k) - \hat{x}(k/k; n) \right\|^2 \right] p(n/k) \tag{17}$$

At each iteration, the MMPA algorithm identifies the model that corresponds to the maximum a posteriori probability as the correct one. This probability tends (asymptotically) to one, while the remaining probabilities tend to zero. If the model changes, the algorithm senses the variation and increases the corresponding a posteriori probability while decreasing the remaining ones. Thus the algorithm is adaptive in the sense of being able to track model changes in real time. This procedure incorporates the algorithm's intelligence.

The above presented multi-model partitioning algorithm (MMPA) possesses several interesting properties:

- Its structure is a natural parallel distributed processing architecture, and hence it is more suitable to current computer clusters.
- By breaking a large and/or nonlinear model to smaller sub-cases, the algorithm has a much smaller dimensionality and hence much less architectural complexity.
- Although computationally intensive, it works faster due to parallelism, and hence it is much more appropriate for real-time applications.
- It is more robust than any single filter as it is capable to isolate any diverging sub-filter. Numerous applications and simulations in the literature also show this.
- The algorithm is well structured and modular, and it is easy to implement and modify on any standard programming environment (e.g. MATLAB).

Detection Results Using Real Traffic Data

In order to test the efficiency of the MMPA method, we use real data from the TEI of Athens campus network. The test dataset was created from real cases, and as shown in Fig. 6, the dataset represents a week of traffic, i.e. five working days and a weekend.

Our test traffic data were collected from our router's standard MIB and/or the server's typical logs. In order to avoid any device- or system-specific problem, the data were taken via a monitoring tool. The earlier Multi Router Traffic Grapher (MRTG) tool [19] and its current version Round Robin Database Tool (RRDtool) [20] have been applied for over a decade in our campus network for continuous monitoring and utilization data collection. These tools are computationally efficient, widely applied and easy to implement software packages for collecting and/or monitoring utilization data from any router or server MIB. They produce standard log files with current and past data that can be downloaded and saved by any browser or simple GET commands. Our adaptive method first reads these standard log files and then performs the model identification steps. This is done repeatedly every 5 min which is the default MIB sampling period or even faster, provided that the network responses arrive in time.

In order to test the MMPA performance, we introduced in our dataset link failures and sudden high traffic peaks. The MMPA was equipped with four (4) Kalman sub-filters that were tuned to the four types of traffic we are investigating as already shown in Fig. 4.

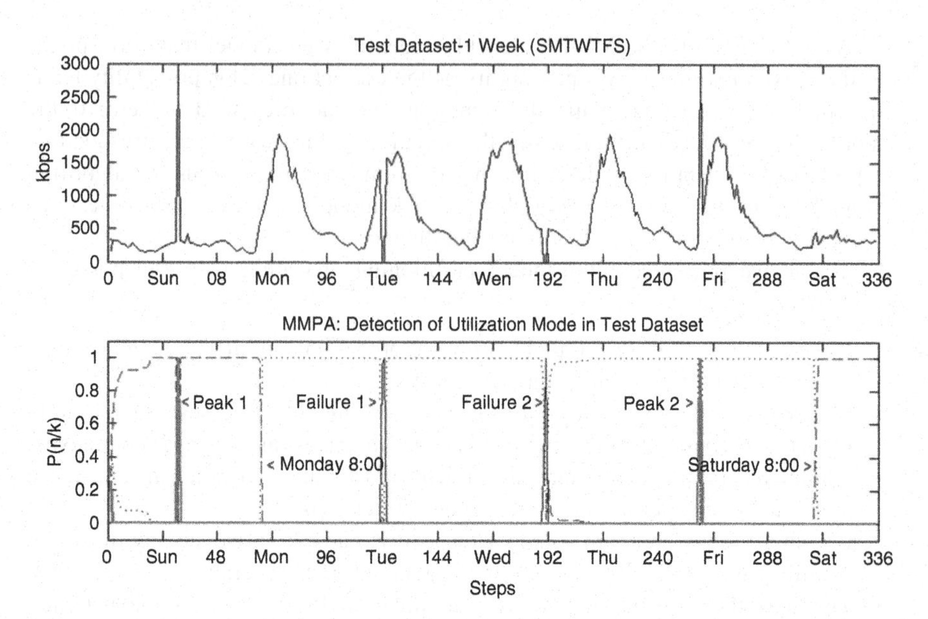

Fig. 6 Test dataset (upper) for 1 week (S-M-T-W-T-F-S) of data containing peaks and failures and (lower) the MMPA's successful detection of the pattern changes and traffic anomalies in the test dataset

The aim of MMPA is to select the correct model n among the N various 'candidates'. By identifying the correct model, MMPA detects the current type of traffic and, consequently, if this matches the normal behaviour or a potential traffic anomaly. In our example the elemental filters are based on the family of models described by Eqs. (11)–(13).

The a posteriori probability density $p(n/k)$ of each model is used to identify the type of the observed traffic. The model that maximizes this quantity is selected. If the selected model is also the correct day pattern of the current day, then we have normal traffic conditions; otherwise, an anomaly is detected.

As shown in Fig. 6, the proposed method detects successfully both the changes from weekend to working days and vice versa. On Saturday at 8:00, offices remain closed, and the traffic pattern changes and matches the weekend pattern. The MMPA detects the difference, and the probability of the weekend model (dashed line) is increased versus 1, while the probability of the working day model (dotted line) falls versus 0. After the weekend, on Monday at 8:00, employees start using the network, and the usage pattern changes back and matches the working day model. The method detects equally well traffic peaks (misuse) and traffic zeros (i.e. link failures).

In addition to the successful detection, the adaptive algorithm also completes the detection in a matter of seconds, thus permitting us to increase the sampling rate of the data collection and the online response of the system. The default sampling rate to collect measurements from the routers' MIB is usually set to 5 min. The proposed method is so fast that does not pose any restrictions on the sampling rate. On the contrary it is the network nodes that may not be able to reply in time if an increased sampling rate is used.

Further work that is currently in progress, based on the above results, investigates monitoring and modelling of other MIB quantities related to network faults or misuse, further increase of the sampling rate to obtain faster reaction times, modelling of end-user behaviour and enriching the MMPA model bank with more patterns of unusual activities or network problems.

Note that, although in the presented work the elemental Kalman filters were tuned to the state-space models describing the various traffic patterns, as required by the Kalman filter structure, this is not obligatory for the MMPA. MMPA can use any type of sub-filter and its corresponding model (e.g. artificial neural network), provided that it is accompanied by the corresponding estimator/predictor component that will interface the algorithm and handle the sub-filter inputs and results.

Conclusions

In this paper an adaptive multi-model method is presented for modelling network traffic flow and detecting any network unusual activity or misuse. The method is based on standard bandwidth utilization data found in the MIB and does not require specialized data collection tools. The proposed method uses the past traffic data to model all normal periodic behaviours of a network connection. ARMA and state-space models are mainly used for traffic pattern modelling without excluding other models such as neural nets. An adaptive multi-model partitioning algorithm processes the collected traffic data through a set of filters, each matching a traffic pattern. The method was tested using real datasets from the campus network, and it detected correctly all pattern changes, failures or unusual activities contained in the test datasets. The method is also very fast, and it can perform equally well in real time even in a fraction of the default 5 min sampling interval that was used to poll the devices and the campus network segments.

References

1. Akaike, H.: Fitting autoregressive models for prediction. Ann. Inst. Stat. Math. **21**, 243–247 (1969)
2. Anderson, B.D.O., Moore, J.B.: Optimal Filtering. Prentice Hall, Englewood Cliffs, NJ (1979)
3. Box, G., Jenkins, G.M., Reinsel, G.: Time Series Analysis: Forecasting and Control, 3rd edn. Prentice Hall, Englewood Cliffs, NJ (1994)
4. Debar, H., Becker, M., Siboni, D.: A neural network component for an intrusion detection system. In: IEEE Computer Society Symposium on Research in Security and Privacy, Oakland, CA (1992)
5. Denning, D.E.: An intrusion-detection model. IEEE Trans. Softw. Eng. **13**, 222–232 (1987)

6. Eskin, E.: Anomaly detection over noisy data using learned probability distributions. In: ICML 2000, Menlo Park, CA. AAAI Press (2000)
7. Halsall, F.: Data Communications, Computer Networks and Open Systems. Addison-Wesley, Harlow (1996)
8. Hood, C., Ji, C.: Proactive network fault detection. IEEE Trans. Reliab. **46**, 333 (1997)
9. Katris, C., Daskalaki, S.: Comparing forecasting approaches for internet traffic. Expert Syst. Appl. **42**(21), 8172–8183 (2015)
10. Katsikas, S.K., Likothanassis, S.D., Lainiotis, D.G.: AR model identification with unknown process order. IEEE Trans. Acoust. Speech Signal Process. **38**(5), 872–876 (1990)
11. Keshav, S.: An Engineering Approach to Computer Networking: ATM, Internet and Telephone Network. Addison-Wesley, Reading, MA (1997)
12. Lainiotis, D.G.: Partitioning: a unifying framework for adaptive systems, I: estimation. Proc. IEEE **64**, 1126–1142 (1976)
13. Lawrence, L.H., Cavuto, D.J., Papavassiliou, S., Zawadzki, A.G.: Adaptive and automated detection of service anomalies in transaction-oriented WAN's: network analysis, algorithms, implementation, and deployment. IEEE J. Sel. Areas Commun. **18**(5), 744–757 (2000)
14. Maxion, R., Feather, F.E.: A case study of ethernet anomalies in a distributed computing environment. IEEE Trans. Reliab. **39**, 433–443 (1990)
15. Moussas, V.C.: Network traffic flow prediction using multi-model partitioning algorithms. In: Tsahalis, D.T. (ed.) 2nd International Conference SCCE, Athens. Patras University Press, Patras (2006)
16. Moussas, V.C.: Traffic and user behaviour model classification for network simulation and anomaly detection. In: Tsahalis, D.T. (ed.) 2nd International Conference EPSMSO, Athens. Patras University Press, Patras (2007)
17. Moussas, V.C., Pappas, Sp.St.: Adaptive network anomaly detection using bandwidth utilization data. In: Tsahalis, D.T. (ed.) 1st International Conference EPSMSO, Athens. Patras University Press (2005)
18. Moussas, V.C., Daglis, M., Kolega, E.: Network traffic modeling and prediction using multiplicative seasonal ARIMA models. In: Tsahalis, D.T. (ed.) 1st International Conference EPSMSO, Athens, 6–9 July 2005
19. Oetiker, T.: Multi Router Traffic Grapher (MRTG) tool - Software Package and Manuals. At: oss.oetiker.ch/mrtg (2005)
20. Oetikerr, T.: Round Robin Database tool (RRDtool) - Software Package and Manuals. At: oss.oetiker.ch/rrdtool (2016)
21. Papagiannaki, K., Taft, N., Zhang, Z., Diot, C.: Long-term forecasting of internet backbone traffic: observations and initial models. In: IEEE Infocom (2003)
22. Papazoglou, P.M., Karras, D.A., Papademetriou, R.C.: High performance novel hybrid DCA algorithms for efficient channel allocation in cellular communications modelled and evaluated through a Java simulation system. WSEAS Trans. Comput. **5**(11), 2078–2085 (2006)
23. Pappas, S.Sp., Katsikas, S.K., Moussas, V.C.: MV-ARMA order estimation via multi-model partition theory. In: Tsahalis, D.T. (ed.) 2nd International Conference EPSMSO, Athens, vol. II, pp. 688–698. Patras University Press, Patras (2007)
24. Permanasari, A.E., Hidayah, I., Bustoni, I.A.: Forecasting model for hotspot bandwidth management at Department of Electrical Engineering and Information Technology UGM. Int. J. Appl. Math. Stat. **53**(4), 227 (2015)
25. Porras, P., Neumann, P.: Emerald: event monitoring enabling responses to anomalous live disturbances. In: Proceedings of the 20th National Information Systems Security Conference, Baltimore, MD (1997)
26. Rhodes, B., Mahafey, J., Cannady, J.: Multiple self-organizing maps for intrusion detection. In: Proceedings of NISSC 2000 Conference (2000)
27. Shu, Y., Yu, M., Liu, J., Yang, O.W.W.: Wireless traffic modeling and prediction using seasonal ARIMA models. In: IEEE International Conference Communication May 2003, ICC03, vol. 3 (2003)

28. Smith, L.B.: Comparison of parametric and nonparametric models for traffic flow forecasting. Transp. Res. C **10**, 303–321 (2002)
29. Solomos, G.P., Moussas, V.C.: A time series approach to fatigue crack propagation. Struct. Saf. **9**, 211–226 (1991)
30. Tanenbaum, A.S.: Computer Networks. Prentice-Hall, Englewood Cliffs, NJ (1996)
31. Thottan, M., Ji, C.: Anomaly detection in IP networks. IEEE Trans. Signal Process. **51**(8), 2191–2204 (2003)
32. You, C., Chandra, K.: Time series models for internet data traffic. In: 24th Conference on Local Computer Networks. LCN-99 (1999)

Bounds Involving Operator s-Godunova-Levin-Dragomir Functions

Muhammad Aslam Noor, Muhammad Uzair Awan, and Khalida Inayat Noor

Abstract The objective of this chapter is to introduce a new class of operator s-Godunova-Levin-Dragomir convex functions. We also derive some new Hermite-Hadamard-like inequalities for operator s-Godunova-Levin-Dragomir convex functions of positive operators in Hilbert spaces.

Introduction

The theory of convexity has received considerable attention by many researchers in the last few years. This is because it has numerous applications in various branches of pure and applied sciences. Due to this fact, the classical concepts of convex sets and convex functions have been generalized in different directions (see [2, 3, 5, 9, 10, 13]). Recently Dragomir [5] introduced the notion of s-Godunova-Levin-Dragomir type of convex functions. It has been shown that this class of convex functions unifies the class of P-functions and Godunova-Levin functions. For some recent studies on s-Godunova-Levin type of convex functions, see [5, 12–14].

An important fact which makes the theory of convexity more fascinating is its close relationship with theory of inequalities. Many inequalities known in the literature are established via convex functions. An important result in this regard is Hermite-Hadamard's inequality, which reads as:

M. A. Noor (✉)
Mathematics Department, King Saud University, Riyadh, Saudi Arabia

COMSATS Institute of Information Technology, Islamabad, Pakistan

M. U. Awan
GC University, Faisalabad, Pakistan

K. I. Noor
COMSATS Institute of Information Technology, Islamabad, Pakistan

N. J. Daras, Th. M. Rassias (eds.), *Modern Discrete Mathematics and Analysis*, Springer Optimization and Its Applications 131,
https://doi.org/10.1007/978-3-319-74325-7_18

Let $f : I = [a, b] \subset \mathbb{R} \to \mathbb{R}$ be a convex function, then

$$f\left(\frac{a+b}{2}\right) \leq \frac{1}{b-a}\int_a^b f(x)\mathrm{d}x \leq \frac{f(a)+f(b)}{2}.$$

For some useful details on Hermite-Hadamard inequalities, see [4–7, 9, 10, 13, 15].

In this paper, we introduce the class of operator s-Godunova-Levin-Dragomir type of convex functions. We establish some new Hermite-Hadamard-like inequalities for this new class of operator convex functions. This is the main motivation of this paper.

Preliminary Results

In this section, we recall some basic results.

Let X be a vector space, $x, y \in X, x \neq y$. Define the segment

$$[x, y] := (1-t)x + ty; \quad t \in [0, 1].$$

We consider the function $f : [x, y] \to \mathbb{R}$ and the associated function

$$g(x, y) : [0, 1] \to \mathbb{R};$$
$$g(x, y)(t) := f((1-t)x + ty), \quad t \in [0, 1].$$

Note that f is convex on $[x, y]$ if and only if $g(x, y)$ is convex on $[0, 1]$. For any convex function defined on a segment $[x, y] \in X$, we have the Hermite-Hadamard inequality

$$f\left(\frac{x+y}{2}\right) \leq \int_0^1 f((1-t)x + ty)\mathrm{d}t \leq \frac{f(x)+f(y)}{2},$$

which can be derived from the classical Hermite-Hadamard inequality for the convex function $g(x, y) : [0, 1] \to \mathbb{R}$.

Let $\mathbf{A}$ be a bounded self-adjoint linear operator on a complex Hilbert space $(H; \langle ., . \rangle)$. The Gelfand map establishes a $*$-isometrically isomorphism Φ between set $C(Sp(\mathbf{A}))$ of all continuous functions defined on the spectrum of $\mathbf{A}$, denoted $Sp(\mathbf{A})$, and the C^*-algebra $C^*(\mathbf{A})$ generated by $\mathbf{A}$ and the identity operator 1_H on H as follows:

For any $f, g \in C(Sp(\mathbf{A}))$ and any $\alpha, \beta \in \mathbb{C}$, we have

1. $\Phi(\alpha f + \beta g) = \alpha\phi(f) + \beta\phi(g)$;
2. $\Phi(fg) = \Phi(f)\Phi(g)$ and $\Phi(f^*) = \Phi(f)^*$;

3. $\|\Phi(f)\| = \|f\| := \sup_{t \in Sp(\mathbf{A})} |f(t)|$;
4. $\Phi(f_0) = 1$ and $\Phi(f_1) = \mathbf{A}$, where $f_0(t) = 1$ and $f_1(t) = t$, for $t \in Sp(\mathbf{A})$,

With this

$$f(\mathbf{A}) := \Phi(f) \quad \forall f \in C(Sp(\mathbf{A})),$$

we call this continuous functional calculus for a bounded self-adjoint operator $\mathbf{A}$. If $\mathbf{A}$ is bounded self-adjoint operator and f is real-valued continuous function on $Sp(\mathbf{A})$, then $f(t) \geq 0$ for any $t \in Sp(\mathbf{A})$ implies that $f(\mathbf{A}) \geq 0$, that is, $f(\mathbf{A})$ is a positive operator on H. If both f and g are real-valued functions on $Sp(\mathbf{A})$, then

$$f(t) \geq g(t) \quad \text{for any } t \in Sp(\mathbf{A}) \Longrightarrow f(\mathbf{A}) \geq g(\mathbf{A}),$$

in the operator order in $B(\mathbb{H})$.

Definition 1 ([8]) A real-valued continuous function f on an interval I is said to be operator convex, if

$$f((1-t)\mathbf{A} + t\mathbf{B}) \leq (1-t)f(\mathbf{A}) + tf(\mathbf{B}), \quad \forall t \in [0, 1], \tag{1}$$

in the operator order in $B(\mathbb{H})$, and for every self-adjoint operators $\mathbf{A}$ and $\mathbf{B}$ in $B(\mathbb{H})$ whose spectra is contained in I.

Example 1 ([8]) The convex function $f(t) = \alpha t^2 + \beta t + \gamma (\alpha \geq 0, \beta, \gamma \in \mathbb{R})$ is operator convex on every interval.

$$\begin{aligned}
&\frac{f(\mathbf{A}) + f(\mathbf{B})}{2} - f\left(\frac{\mathbf{A}+\mathbf{B}}{2}\right) \\
&= \alpha\left[\frac{\mathbf{A}^2 + \mathbf{B}^2}{2} - \left(\frac{\mathbf{A}+\mathbf{B}}{2}\right)^2\right] \\
&+\beta\left[\frac{\mathbf{A}+\mathbf{B}}{2} - \frac{\mathbf{A}+\mathbf{B}}{2}\right] + (\gamma - \gamma) \\
&= \frac{\alpha}{4}[\mathbf{A}^2 + \mathbf{B}^2 - \mathbf{AB} - \mathbf{BA}] = \frac{\alpha}{4}(\mathbf{A}-\mathbf{B})^2 \geq 0.
\end{aligned}$$

Dragomir [4] has proved Hermite-Hadamard-like inequality for operator convex functions, which reads as:

$$\begin{aligned}
&f\left(\frac{\mathbf{A}+\mathbf{B}}{2}\right) \\
&\leq \frac{1}{2}\left[f\left(\frac{3\mathbf{A}+\mathbf{B}}{4}\right) + f\left(\frac{\mathbf{A}+3\mathbf{B}}{4}\right)\right]
\end{aligned}$$

$$
\leq \int_0^1 f((1-t)\mathbf{A}+t\mathbf{B})dt
$$

$$
\leq \frac{1}{2}\left[f\left(\frac{\mathbf{A}+\mathbf{B}}{2}\right)+\frac{f(\mathbf{A})+f(\mathbf{B})}{2}\right] \leq \frac{f(\mathbf{A})+f(\mathbf{B})}{2}.
$$

For more details interested readers are referred to [1, 8, 10, 11].

Let us denote by $B(\mathbb{H})^+$ the set of all positive operators in $B(\mathbb{H})$.

Now we are in a position to define the class of operator s-Godunova-Levin-Dragomir convex function.

Definition 2 Let $I \subset [0, \infty)$ and K be a convex subset of $B(\mathbb{H})^+$. A continuous function $f : I \to \mathbb{R}$ is said to be operator s-Godunova-Levin-Dragomir convex function on I for operators in K, if

$$
f((1-t)\mathbf{A}+t\mathbf{B}) \leq (1-t)^{-s} f(\mathbf{A}) + t^{-s} f(\mathbf{B}), \quad \forall t \in [0,1], s \in [0,1], \tag{2}
$$

in the operator order in $B(\mathbb{H})$, and for every positive operators $\mathbf{A}$ and $\mathbf{B}$ in K whose spectra is contained in I. For $K = B(\mathbb{H})^+$ we say f is operator s-Godunova-Levin-Dragomir convex function on I.

For the reader's convenience, we recall here the definitions of the gamma function

$$
\Gamma(x) = \int_0^\infty e^{-x} t^{x-1} dt
$$

and the Beta function

$$
\mathrm{B}(x, y) = \int_0^1 t^{x-1}(1-t)^{y-1}\, dt = \frac{\Gamma(x)\Gamma(y)}{\Gamma(x+y)}.
$$

Main Results

Proposition 1 *Let $f : I \subset [0, \infty) \to \mathbb{R}$ be a continuous function on I. Then for every two positive operators $\mathbf{A}, \mathbf{B} \in K \subseteq B(\mathbb{H})^+$ with spectra in I, the function f is operator s-Godunova-Levin-Dragomir function for operators in $[\mathbf{A}, \mathbf{B}] := \{(1-t)\mathbf{A}+t\mathbf{B} : t \in [0,1]\}$ if and only if the function $\varphi_{x,\mathbf{A},\mathbf{B}} : [0,1] \to \mathbb{R}$ defined by*

$$
\varphi_{x,\mathbf{A},\mathbf{B}} = \langle f((1-t)\mathbf{A}+t\mathbf{B})x, x\rangle,
$$

is s-Godunova-Levin convex function on $[0,1]$, for every $x \in H$ with $\|x\| = 1$.

Proof Let f be operator s-Godunova-Levin-Dragomir convex function for operators in $[\mathbf{A}, \mathbf{B}]$, and then, for any $t_1, t_2 \in [0, 1]$ and $\alpha, \beta \geq 0$ with $\alpha + \beta = 1$, we have

$$\begin{aligned}
&\varphi_{x,\mathbf{A},\mathbf{B}}(\alpha t_1 + \beta t_2) \\
&\quad = \langle f((1 - (\alpha t_1 + \beta t_2))\mathbf{A} + (\alpha t_1 + \beta t_2)\mathbf{B})x, x\rangle \\
&\quad = \langle f(\alpha[(1 - t_1)\mathbf{A} + t_1\mathbf{B}] + \beta[(1 - t_2)\mathbf{A} + t_2\mathbf{B}])x, x\rangle \\
&\quad \leq \alpha^{-s}\langle f((1 - t_1)\mathbf{A} + t_1\mathbf{B})x, x\rangle + \beta^{-s}\langle f((1 - t_2)\mathbf{A} + t_2\mathbf{B})x, x\rangle \\
&\quad = \alpha^{-s}\varphi_{x,\mathbf{A},\mathbf{B}}(t_1) + \beta^{-s}\varphi_{x,\mathbf{A},\mathbf{B}}(t_2),
\end{aligned}$$

which shows that $\varphi_{x,\mathbf{A},\mathbf{B}}$ is s-Godunova-Levin-Dragomir convex function.

Now to prove converse, let $\varphi_{x,\mathbf{A},\mathbf{B}}$ be s-Godunova-Levin-Dragomir convex function; we show that f is operator s-Godunova-Levin-Dragomir convex functions in $[\mathbf{A}, \mathbf{B}]$. For every $\mathbf{C} = (1 - t_1)\mathbf{A} + t_1\mathbf{B}$ and $\mathbf{D} = (1 - t_2)\mathbf{A} + t_2\mathbf{B}$ in $[\mathbf{A}, \mathbf{B}]$, we have

$$\begin{aligned}
&\langle f(t\mathbf{C} + (1 - t)\mathbf{D})x, x\rangle \\
&\quad = \langle f[t((1 - t_1)\mathbf{A} + t_1\mathbf{B}) + (1 - t)((1 - t_2)\mathbf{A} + t_2\mathbf{B})]x, x\rangle \\
&\quad = f[(1 - (tt_1 + (1 - t)t_2))\mathbf{A} + (tt_1 + (1 - t)t_2)\mathbf{B}]x, x\rangle \\
&\quad = \varphi_{x,\mathbf{A},\mathbf{B}}(tt_1 + (1 - t)t_2) \\
&\quad \leq t^{-s}\varphi_{x,\mathbf{A},\mathbf{B}}(t_1) + (1 - t)^{-s}\varphi_{x,\mathbf{A},\mathbf{B}}(t_2) \\
&\quad = t^{-s}\langle f((1 - t_1)\mathbf{A} + t_1\mathbf{B})x, x\rangle + (1 - t)^{-s}\langle f((1 - t_2)\mathbf{A} + t_2\mathbf{B})x, x\rangle \\
&\quad \leq t^{-s}\langle f(\mathbf{C})x, x\rangle + (1 - t)^{-s}\langle f(\mathbf{D})x, x\rangle.
\end{aligned}$$

This completes the proof. □

Our next result is generalization of Hermite-Hadamard-like inequality for operator s-Godunova-Levin-Dragomir convex functions.

Theorem 1 *Let $f : I \to \mathbb{R}$ be an operator s-Godunova-Levin-Dragomir convex function on $I \subseteq [0, \infty)$ for operators in $K \subseteq B(\mathbb{H})^+$. Then, for all positive operators $\mathbf{A}$ and $\mathbf{B}$ in K with spectra in I, we have*

$$\frac{1}{2^{1+s}} f\left(\frac{\mathbf{A} + \mathbf{B}}{2}\right) \leq \int_0^1 f((1 - t)\mathbf{A} + t\mathbf{B})\mathrm{d}t \leq \frac{f(\mathbf{A}) + f(\mathbf{B})}{1 - s}.$$

Proof For $x \in H$ with $\|x\| = 1$ and $t \in [0, 1]$, we have

$$\langle((1 - t)\mathbf{A} + t\mathbf{B})x, x\rangle = (1 - t)\langle Ax, x\rangle + t\langle Bx, x\rangle \in I. \tag{3}$$

Since $\langle Ax, x\rangle \in Sp(\mathbf{A}) \subseteq I$ and $\langle Bx, x\rangle \in Sp(\mathbf{B}) \subseteq I$.

Now continuity of f and (3) implies that the operator valued integral

$$\int_0^1 f((1-t)\mathbf{A}+t\mathbf{B})\mathrm{d}t$$

exists.

Since f is operator s-Godunova-Levin-Dragomir convex function, thus for $t \in [0, 1]$ and $\mathbf{A}, \mathbf{B} \in K$, we have

$$f((1-t)\mathbf{A}+t\mathbf{B}) \leq (1-t)^{-s} f(\mathbf{A}) + t^{-s} f(\mathbf{B}).$$

Integrating above inequality with respect to t on $[0, 1]$, we have

$$\int_0^1 f((1-t)\mathbf{A}+t\mathbf{B})\mathrm{d}t \leq \frac{f(\mathbf{A})+f(\mathbf{B})}{1-s}.$$

Now, we prove the left-hand side. For this

$$\begin{aligned} & f\left(\frac{\mathbf{A}+\mathbf{B}}{2}\right) \\ & = f\left(\frac{t\mathbf{A}+(1-t)\mathbf{B}+(1-t)\mathbf{A}+t\mathbf{B}}{2}\right) \\ & \leq \frac{1}{2^{-s}}[f(t\mathbf{A}+(1-t)\mathbf{B})+f((1-t)\mathbf{A}+t\mathbf{B})]. \end{aligned}$$

Integrating above inequality with respect to t on $[0, 1]$ completes the proof. □

For product of two operator s-Godunova-Levin-Dragomir convex functions, we have the following result.

Theorem 2 *Let $f : I \to \mathbb{R}$ be operator s_1-Godunova-Levin-Dragomir convex and $g : I \to \mathbb{R}$ be operator s_2-Godunova-Levin-Dragomir convex function on the interval I for operators in $K \subseteq B(\mathbb{H})^+$. Then for all positive operators $\mathbf{A}$ and $\mathbf{B}$ in K with spectra in I, we have*

$$\begin{aligned} & \frac{1}{2^{1+s_1+s_2}}\left\langle f\left(\frac{\mathbf{A}+\mathbf{B}}{2}\right)x, x\right\rangle\left\langle g\left(\frac{\mathbf{A}+\mathbf{B}}{2}\right)x, x\right\rangle \\ & \leq \int_0^1 \langle f(t\mathbf{A}+(1-t)\mathbf{B})x, x\rangle\langle g(t\mathbf{A}+(1-t)\mathbf{B})x, x\rangle \mathrm{d}t \\ & \leq \frac{1}{1-s_1-s_2}M(\mathbf{A},\mathbf{B})(x) + \frac{\Gamma(1-s_1)\Gamma(1-s_2)}{\Gamma(2-s_1-s_2)}N(\mathbf{A},\mathbf{B})(x), \end{aligned}$$

holds for any $x \in H$ *with* $\|x\| = 1$*, where*

$$M(\mathbf{A},\mathbf{B})(x) = \langle f(\mathbf{A})x, x\rangle\langle g(\mathbf{A})x, x\rangle + \langle f(\mathbf{B})x, x\rangle\langle g(\mathbf{B})x, x\rangle, \tag{4}$$

$$N(\mathbf{A},\mathbf{B})(x) = \langle f(\mathbf{A})x, x\rangle\langle g(\mathbf{B})x, x\rangle + \langle f(\mathbf{B})x, x\rangle\langle g(\mathbf{A})x, x\rangle. \tag{5}$$

Proof Since f and g are operator s_1-Godunova-Levin-Dragomir convex and operator s_2-Godunova-Levin-Dragomir convex, respectively, then, for $x \in H$ with $\|x\| = 1$ and $t \in [0, 1]$, we have

$$\begin{aligned}
&\left\langle f\left(\frac{\mathbf{A}+\mathbf{B}}{2}\right)x, x\right\rangle\left\langle g\left(\frac{\mathbf{A}+\mathbf{B}}{2}\right)x, x\right\rangle\\
&= \left\langle f\left(\frac{t\mathbf{A}+(1-t)\mathbf{B}}{2} + \frac{(1-t)\mathbf{A}+t\mathbf{B}}{2}\right)x, x\right\rangle\\
&\quad\times\left\langle g\left(\frac{t\mathbf{A}+(1-t)\mathbf{B}}{2} + \frac{(1-t)\mathbf{A}+t\mathbf{B}}{2}\right)x, x\right\rangle\\
&\le \frac{1}{2^{-s_1-s_2}}\Big\{[\langle f(t\mathbf{A}+(1-t)\mathbf{B})x, x\rangle + \langle f((1-t)\mathbf{A}+t\mathbf{B})x, x\rangle]\\
&\quad\times[\langle g(t\mathbf{A}+(1-t)\mathbf{B})x, x\rangle + \langle g((1-t)\mathbf{A}+t\mathbf{B})x, x\rangle]\Big\}\\
&\le \frac{1}{2^{-s_1-s_2}}\Big\{[\langle f(t\mathbf{A}+(1-t)\mathbf{B})x, x\rangle\langle g(t\mathbf{A}+(1-t)\mathbf{B})x, x\rangle\\
&\quad+\langle f((1-t)\mathbf{A}+t\mathbf{B})x, x\rangle\langle g((1-t)\mathbf{A}+t\mathbf{B})x, x\rangle]\\
&\quad+[(t^{-s_1}(1-t)^{-s_2} + t^{-s_2}(1-t)^{-s_1})[\langle f(\mathbf{A})x, x\rangle\langle g(\mathbf{A})x, x\rangle\\
&\quad+\langle f(\mathbf{B})x, x\rangle\langle g(\mathbf{B})x, x\rangle]\\
&\quad+[(t^{-s_1-s_2}(1-t)^{-s_1-s_2} + t^{-s_2}(1-t)^{-s_1})[\langle f(\mathbf{A})x, x\rangle\langle g(\mathbf{B})x, x\rangle\\
&\quad+\langle f(\mathbf{A})x, x\rangle\langle g(\mathbf{A})x, x\rangle]\Big\}.
\end{aligned}$$

Integrating the above inequality with respect t on $[0, 1]$ completes the proof. □

Theorem 3 *Let* $f : I \to \mathbb{R}$ *be operator* s_1*-Godunova-Levin-Dragomir convex and* $g : I \to \mathbb{R}$ *be operator* s_2*-Godunova-Levin-Dragomir convex function on the interval* I *for operators in* $K \subseteq \mathbf{B}(\mathbb{H})^+$*. Then for all positive operators* $\mathbf{A}$ *and* $\mathbf{B}$ *in* K *with spectra in* I*, we have*

$$\begin{aligned}
&\int_0^1 \langle f(t\mathbf{A}+(1-t)\mathbf{B})x, x\rangle\langle g(t\mathbf{A}+(1-t)\mathbf{B})x, x\rangle \mathrm{d}t\\
&\le \frac{1}{1-s_1-s_2}M(\mathbf{A},\mathbf{B})(x) + \frac{\Gamma(1-s_1)\Gamma(1-s_2)}{\Gamma(2-s_1-s_2)}N(\mathbf{A},\mathbf{B})(x),
\end{aligned}$$

holds for any $x \in H$ *with* $\|x\| = 1$*, where* $M(\mathbf{A}, \mathbf{B})(x)$ *and* $N(\mathbf{A}, \mathbf{B})(x)$ *are given by (4) and (5), respectively.*

Proof Since f and g are operator s_1-Godunova-Levin-Dragomir convex and operator s_2-Godunova-Levin-Dragomir convex, respectively, then, for $x \in H$ with $\|x\| = 1$ and $t \in [0, 1]$, we have

$$\langle (t\mathbf{A} + (1-t)\mathbf{B})x, x\rangle = t\langle Ax, x\rangle + (1-t)\langle Bx, x\rangle \in I, \tag{6}$$

since $\langle Ax, x\rangle \in Sp(\mathbf{A}) \subseteq I$ and $\langle Bx, x\rangle \in Sp(\mathbf{B}) \subseteq I$. Now continuity of f, g and (6) implies that operator-valued integrals $\int_0^1 f(t\mathbf{A}+(1-t)\mathbf{B})\mathrm{d}t$, $\int_0^1 g(t\mathbf{A}+(1-t)\mathbf{B})\mathrm{d}t$ and $\int_0^1 (fg)(t\mathbf{A}+(1-t)\mathbf{B})\mathrm{d}t$ exists.

Since f is operator s_1-Godunova-Levin-Dragomir convex and g is operator s_2-Godunova-Levin-Dragomir convex, then, for $t \in [0, 1]$ and $x \in H$, we have

$$\langle f(t\mathbf{A} + (1-t)\mathbf{B})x, x\rangle \le \langle (t^{-s_1} f(\mathbf{A}) + (1-t)^{-s_2} f(\mathbf{B}))x, x\rangle;$$
$$\langle g(t\mathbf{A} + (1-t)\mathbf{B})x, x\rangle \le \langle (t^{-s_1} g(\mathbf{A}) + (1-t)^{-s_2} g(\mathbf{B}))x, x\rangle.$$

This implies

$$\begin{aligned}&\langle f(t\mathbf{A} + (1-t)\mathbf{B})x, x\rangle\langle g(t\mathbf{A} + (1-t)\mathbf{B})x, x\rangle \\ &\quad\le t^{-s_1-s_2}\langle f(\mathbf{A})x, x\rangle\langle g(\mathbf{A})x, x\rangle + (1-t)^{-s_1-s_2}\langle f(\mathbf{B})x, x\rangle\langle g(\mathbf{B})x, x\rangle \\ &\qquad+t^{-s_1}(1-t)^{-s_2}\langle f(\mathbf{A})x, x\rangle\langle g(\mathbf{B})x, x\rangle+t^{-s_2}(1-t)^{-s_1}\langle f(\mathbf{B})x, x\rangle\langle g(\mathbf{A})x, x\rangle.\end{aligned}$$

Integrating above inequalities with respect to t on $[0, 1]$ completes the proof. □

Acknowledgements Authors would like to express their gratitude to Prof. Dr. Themistocles M. Rassias for his kind invitation. Authors are pleased to acknowledge the "support of Distinguished Scientist Fellowship Program(DSFP), King Saud University," Riyadh, Saudi Arabia.

References

1. Bacak, V., Turkmen, R.: New inequalities for operator convex functions. J. Inequal. Appl. **2013**, 190 (2013)
2. Breckner, W.W.: Stetigkeitsaussagen ir eine Klasse verallgemeinerter convexer funktionen in topologischen linearen Raumen. Pupl. Inst. Math. **23**, 13–20 (1978)
3. Cristescu, G., Lupsa, L.: Non-connected Convexities and Applications. Kluwer Academic Publishers, Dordrecht (2002)
4. Dragomir, S.S.: The Hermite-Hadamard type inequalities for operator convex functions. Appl. Math. Comput. **218**(3), 766–772 (2011). https://doi.org/10.1016/j.amc.2011.01.056

5. Dragomir, S.S.: Inequalities of Hermite-Hadamard type for h-convex functions on linear spaces (2014, preprint)
6. Dragomir, S.S., Fitzpatrick, S.: The Hadamard's inequality for s-convex functions in the second sense. Demonstratio Math. **32**(4), 687–696 (1999)
7. Dragomir, S.S., Pearce, C.E.M.: Selected Topics on Hermite-Hadamard Inequalities and Applications. (RGMIA Monographs http://rgmia.vu.edu.au/monographs/hermite_hadamard.html), Victoria University (2000)
8. Furuta, T., Micic Hot, J., Pecaric, J., Seo, Y.: Mond-Pecaric Method in Operator Inequalities, Inequalities for Bounded Selfadjoint Operators on a Hilbert Space. Element, Zagreb (2005)
9. Ghazanfari, A.G.: The Hermite-Hadamard type inequalities for operator s-convex functions. J. Adv. Res. Pure Math. **6**(3), 52–61 (2014)
10. Ghazanfari, A.G., Shakoori, M., Barani, A., Dragomir, S.S.: Hermite-Hadamard type inequality for operator preinvex functions, arXiv:1306.0730v1 [math.FA] 4 Jun 2013
11. Mond, B., Pecaric, J.: On some operator inequalities. Indian J. Math. **35**, 221–232 (1993)
12. Noor, M.A., Noor, K.I., Awan, M.U.: Fractional Ostrowski inequalities for s-Godunova-Levin functions. Int. J. Anal. Appl. **5**, 167–173 (2014)
13. Noor, M.A., Noor, K.I., Awan, M.U., Khan, S.: Fractional Hermite-Hadamard inequalities for some new classes of Godunova-Levin functions. Appl. Math. Inf. Sci. **8**(6), 2865–2872 (2014)
14. Pachpatte, B.G.: On some inequalities for convex functions. RGMIA Research Report Collection 6(E) (2003)
15. Pecaric, J.E., Proschan, F., Tong, Y.L.: Convex Functions, Partial Orderings, and Statistical Applications. Academic, San Diego (1992)

Closed-Form Solutions for Some Classes of Loaded Difference Equations with Initial and Nonlocal Multipoint Conditions

I. N. Parasidis and E. Providas

Abstract We state solvability criteria and derive closed-form solutions to nth-order difference equations and loaded difference equations involving initial and nonlocal discrete multipoint, homogeneous and nonhomogeneous, conditions by using the extension operator method.

Keywords Difference equations · Loaded differential equations · Nonlocal problems · Discrete multipoint conditions · Exact solutions

Introduction

Loaded differential and integrodifferential equations describe processes where states occurring at given individual domain or/and time points influence the process over the whole domain or/and the entire course of time. The terminology *loaded equations* seems to have been first introduced by Nakhushev [11, 12] who defined and classified various types of loaded equations. They emerge in mathematical models in biology [13], ground fluid mechanics [14], fluid dynamics [20], heat problems [6, 18], inverse problems [9, 19], and elsewhere. They are mainly solved approximately by finite-difference methods [1, 2, 4, 5].

Difference equations usually appear in modeling physical processes, or in discretizing analogous continuous problems, depending on one or more variables that can only take a discrete set of possible values. Difference equations involving, in addition, values of the unknown function at some specific given points of a domain or/and time are called *loaded difference equations*.

I. N. Parasidis
Department of Electrical Engineering, TEI of Thessaly, Larissa, Greece
e-mail: paras@teilar.gr

E. Providas (✉)
Department of Mechanical Engineering, TEI of Thessaly, Larissa, Greece
e-mail: providas@teilar.gr

N. J. Daras, Th. M. Rassias (eds.), *Modern Discrete Mathematics and Analysis*, Springer Optimization and Its Applications 131,
https://doi.org/10.1007/978-3-319-74325-7_19

In this article we are concerned with the solvability and finding closed-form solutions to nth-order loaded difference equations subject to initial and nonlocal discrete multipoint conditions, specifically,

$$\sum_{i=0}^{n} a_i u_{k+n-i} - \sum_{s=1}^{m} g_k^s \psi_s(u_k) = f_k,$$
$$\sum_{j=0}^{l-1} \mu_{ij} u_j = \beta_i, \quad i = 0, \ldots, n-1, \quad n \leq l, \tag{1}$$

where $u_k = u(k)$ is the unknown function, $k \in \mathbf{Z}^+$, $a_i,\ \mu_{ij},\ \beta_i \in \mathbf{R}$, with $a_0 a_n \neq 0$, ψ_s are linear real-valued functionals, and $g_k^s = g_s(k),\ f_k = f(k)$ are given real-valued functions. We benefit of the vast literature in studying difference equations, see, for example, [7, 10], and in particular the papers [3, 8, 16, 17] on multipoint boundary value problems which are of relevance here, and the extension operator method presented by the authors [15].

The rest of the article is organized in three sections. In section "Linear Difference Equations", we present some preliminary results and provide solution formulas for difference equations with initial and nonlocal multipoint, homogeneous and nonhomogeneous, conditions. In section "Loaded Difference Equations" we derive closed-form solutions for loaded difference equations with initial and nonlocal multipoint, homogeneous and nonhomogeneous, conditions. Lastly, several example problems are solved in section "Example Problems" to explain the theory and demonstrate its effectiveness.

Linear Difference Equations

An nth-order difference equation is a relation between the present value of a function and a discrete set of its n previous values, and it can be written in the general form

$$F(k,\ u_{k+n},\ u_{k+n-1},\ \ldots,\ u_k) = 0, \tag{2}$$

where F is a well-defined function of its arguments and $k \in \mathbf{Z}^+$. A solution to difference equation is a function $u : \mathbf{Z}^+ \to \mathbf{R}$ such that $u(k)$ or u_k satisfies Eq. (2) for all $k \in \mathbf{Z}^+$.

A linear difference equation of nth-order with constant coefficients has the structure

$$u_{k+n} + a_1 u_{k+n-1} + \cdots + a_n u_k = f_k, \tag{3}$$

where $a_i \in \mathbf{R},\ i = 1, \ldots, n$ are known with $a_n \neq 0$, and $f : \mathbf{Z}^+ \to \mathbf{R}$ is a given function such that $f(k)$ or f_k realizes Eq. (3) for all $k \in \mathbf{Z}^+$. For $f_k \neq 0$, Eq. (3)

is said to be nonhomogeneous. If $f_k = 0$ for all $k \in \mathbf{Z}^+$, then Eq. (3) is termed homogeneous equation, viz.,

$$u_{k+n} + a_1 u_{k+n-1} + \cdots + a_n u_k = 0. \tag{4}$$

Let $u_k^{(i)}$, $i = 1, \ldots, n$ be n solutions to the homogeneous difference equation (4). If these solutions are linearly independent, then we say that they form a fundamental set of solutions to Eq. (4). For a fundamental set of solutions, the determinant of the Casorati matrix, known as Casoratian, is nonzero for all $k \in \mathbf{Z}^+$, specifically

$$\det C_k = \det \begin{pmatrix} u_k^{(1)} & u_k^{(2)} & \ldots & u_k^{(n)} \\ u_{k+1}^{(1)} & u_{k+1}^{(2)} & \ldots & u_{k+1}^{(n)} \\ \vdots & \vdots & & \vdots \\ u_{k+n-1}^{(1)} & u_{k+n-1}^{(2)} & \ldots & u_{k+n-1}^{(n)} \end{pmatrix} \neq 0. \tag{5}$$

The general solution to the homogeneous difference equation (4) is given by

$$u_k^{(H)} = c_1 u_k^{(1)} + c_2 u_k^{(2)} + \cdots + c_n u_k^{(n)}, \tag{6}$$

where $c_1, c_2, \ldots, c_n$ are arbitrary constants. If $u_k^{(f_k)}$ is a particular solution to the nonhomogeneous difference equation (3), then the general solution to Eq. (3) is obtained by superposition, namely,

$$u_k = u_k^{(H)} + u_k^{(f_k)}. \tag{7}$$

The constants $c_1, c_2, \ldots, c_n$ in (6) can be determined by requiring the solution to satisfy some initial or nonlocal boundary conditions. If the first n discrete values of the solution are specified, then we have the initial value problem

$$\begin{aligned} &u_{k+n} + a_1 u_{k+n-1} + \cdots + a_n u_k = f_k, \\ &u_i = \beta_i, \quad i = 0, \ldots, n-1, \end{aligned} \tag{8}$$

where $\beta_i \in \mathbf{R}$. The initial value problem (8) has a unique solution, see, e.g., [7] and [10]. In the case when nonlocal discrete multipoint conditions are prescribed, then we have the nonlocal problem

$$\begin{aligned} &u_{k+n} + a_1 u_{k+n-1} + \cdots + a_n u_k = f_k, \\ &\sum_{j=0}^{l-1} \mu_{ij} u_j = \beta_i, \quad i = 0, \ldots, n-1, \quad n \le l, \end{aligned} \tag{9}$$

where in addition $\mu_{ij} \in \mathbf{R}$, $i = 0, \ldots, n-1$, $j = 0, \ldots, l-1$.

In the sequel we make use also of the matrix

$$U_k = \begin{pmatrix} u_k^{(1)} & u_k^{(2)} & \dots & u_k^{(n)} \\ u_{k+1}^{(1)} & u_{k+1}^{(2)} & \dots & u_{k+1}^{(n)} \\ \vdots & \vdots & & \vdots \\ u_{k+l-1}^{(1)} & u_{k+l-1}^{(2)} & \dots & u_{k+l-1}^{(n)} \end{pmatrix}, \tag{10}$$

which we call the $l \times n$, $n < l$, generalized Casorati matrix. Furthermore, we use for convenience the term nonlocal conditions instead of the nonlocal multipoint conditions. Finally, the procedure followed here for solving the above problems requires their reformulation in an operator form. Thus, we define S to be the set of all functions $u : \mathbf{Z}^+ \to \mathbf{R}$ or, equivalently, all real sequences $u(k)$ or u_k indexed by $\mathbf{Z}^+$. By defining the addition and scalar multiplication as usual, it turns out that S forms a real vector space. Hat quantities like $\widehat{A} : S \to S$ are used to define a linear operator while $D(\widehat{A})$ denotes its domain and $R(\widehat{A})$ its range being here $R(\widehat{A}) = S$.

Initial Value Problems

We define the operator $A : S \to S$ by

$$\begin{aligned} Au_k &= u_{k+n} + a_1 u_{k+n-1} + \dots + a_n u_k, \\ D(A) &= \{u_k \in S : u_i = \beta_i, \ i = 0, \dots, n-1\}, \end{aligned} \tag{11}$$

and express the initial value problem (8) in the operator form

$$Au_k = f_k, \quad f_k \in S. \tag{12}$$

Notice that if homogeneous conditions $u_i = \beta_i \equiv 0$, $i = 0, \dots, n-1$ are specified, then the operator A is linear. For the case of nonhomogeneous conditions $u_i = \beta_i \neq 0$, the operator A is nonlinear.

First, we consider the initial value problem with homogeneous conditions.

Theorem 1 *Let $\widehat{A} : S \to S$ be the linear difference operator defined as*

$$\begin{aligned} \widehat{A}u_k &= u_{k+n} + a_1 u_{k+n-1} + \dots + a_n u_k = f_k, \\ D(\widehat{A}) &= \{u_k \in S : u_i = 0, \ i = 0, \dots, n-1\}, \end{aligned} \tag{13}$$

$u_k^{(i)}$, $i = 1, \dots, n$ be a fundamental set of solutions to the homogeneous equation (4) and $u_k^{(f_k)}$ be a particular solution to the nonhomogeneous equation (3). Then, the operator $\widehat{A}$ is injective, and the unique solution to the problem (13) for any $f_k \in S$ is given by

$$u_k = \widehat{A}^{-1} f_k = u_k^{(f_k)} - \mathbf{u}_k^{(H)} C_0^{-1} \mathbf{u}^{(f_k)}, \tag{14}$$

where the vector of functions $\mathbf{u}_k^{(H)} = (u_k^{(1)}, u_k^{(2)}, \ldots, u_k^{(n)})$*, the algebraic vector* $\mathbf{u}^{(f_k)} = \mathrm{col}(u_0^{(f_k)}, u_1^{(f_k)}, \ldots, u_{n-1}^{(f_k)})$*, and* C_0 *designates the Casorati matrix for* $k = 0$*.*

Proof Let any $u_k,\ v_k \in D(\widehat{A})$ and suppose $\widehat{A}u_k = \widehat{A}v_k$. Since $\widehat{A}$ is linear, we have

$$\widehat{A}u_k - \widehat{A}v_k = \widehat{A}(u_k - v_k) = \widehat{A}w_k = 0, \tag{15}$$

where $w_k = u_k - v_k$. Any solution to the homogeneous equation (15) can be written according to Eq. (6) as

$$w_k = c_1 u_k^{(1)} + c_2 u_k^{(2)} + \cdots + c_n u_k^{(n)}, \tag{16}$$

where $u_k^{(i)}$, $i = 1, \ldots, n$ are a fundamental set of solutions and $c_1, c_2, \ldots, c_n$ are constants. This solution has to fulfill the n initial conditions, viz.,

$$w_i = c_1 u_i^{(1)} + c_2 u_i^{(2)} + \cdots + c_n u_i^{(n)} = 0, \quad i = 0, \ldots, n-1, \tag{17}$$

or in matrix form

$$C_0 \mathbf{c} = 0, \tag{18}$$

where

$$C_0 = \begin{pmatrix} u_0^{(1)} & u_0^{(2)} & \ldots & u_0^{(n)} \\ u_1^{(1)} & u_1^{(2)} & \ldots & u_1^{(n)} \\ \vdots & \vdots & & \vdots \\ u_{n-1}^{(1)} & u_{n-1}^{(2)} & \ldots & u_{n-1}^{(n)} \end{pmatrix}, \quad \mathbf{c} = \begin{pmatrix} c_1 \\ c_2 \\ \vdots \\ c_n \end{pmatrix}. \tag{19}$$

The $n \times n$ matrix C_0 is the Casorati matrix C_k for $k = 0$. Since $\det C_0 \neq 0$, it is implied that $c_1 = c_2 \ldots = c_n = 0$ and hence $w_k = u_k - v_k = 0$ or $u_k = v_k$. This proves that the operator $\widehat{A}$ is injective. In constructing the inverse $\widehat{A}^{-1}$, we employ the general solution to the nonhomogeneous equation (3) in Eqs. (7), (6) which must satisfy the initial conditions

$$u_i = u_i^{(H)} + u_i^{(f_k)} = 0, \quad i = 0, \ldots, n-1. \tag{20}$$

or in matrix form

$$C_0 \mathbf{c} = -\mathbf{u}^{(f_k)}, \tag{21}$$

where $\mathbf{u}^{(f_k)} = \operatorname{col}(u_0^{(f_k)}, u_1^{(f_k)}, \ldots, u_{n-1}^{(f_k)})$. Since $\det C_0 \neq 0$, we get $\mathbf{c} = -C_0^{-1}\mathbf{u}^{(f_k)}$. Substitution into (6) and (7) produces Eq. (14). □

Next, we look upon the initial value problem implicating nonhomogeneous conditions.

Theorem 2 *Let $A : S \to S$ be the nonlinear difference operator defined by*

$$\begin{aligned} Au_k &= u_{k+n} + a_1 u_{k+n-1} + \cdots + a_n u_k = f_k, \\ D(A) &= \{u_k \in S : u_i = \beta_i,\ i = 0, \ldots, n-1\}, \end{aligned} \tag{22}$$

$u_k^{(i)}$, $i = 1, \ldots, n$ be a fundamental set of solutions to the homogeneous equation (4) and $u_k^{(f_k)}$ be a particular solution to the nonhomogeneous equation (3). Then, the operator A is injective, and the unique solution to the problem (22) for any $f_k \in S$ is given by

$$u_k = A^{-1} f_k = u_k^{(f_k)} + \mathbf{u}_k^{(H)} C_0^{-1}\left(\mathbf{b} - \mathbf{u}^{(f_k)}\right), \tag{23}$$

where the vector of functions $\mathbf{u}_k^{(H)} = (u_k^{(1)}, u_k^{(2)}, \ldots, u_k^{(n)})$, the algebraic vector $\mathbf{u}^{(f_k)} = \operatorname{col}(u_0^{(f_k)}, u_1^{(f_k)}, \ldots, u_{n-1}^{(f_k)})$, $\mathbf{b} = \operatorname{col}(\beta_0, \beta_1, \ldots, \beta_{n-1})$, and C_0 is the Casorati matrix for $k = 0$.

Proof The general solution to the nonhomogeneous equation (3) given by (7) must comply with the initial conditions, namely,

$$\begin{aligned} u_i &= u_i^{(H)} + u_i^{(f_k)} \\ &= c_1 u_i^{(1)} + c_2 u_i^{(2)} + \cdots + c_n u_i^{(n)} + u_i^{(f_k)} = \beta_i, \quad i = 0, \ldots, n-1, \end{aligned} \tag{24}$$

where $u_k^{(i)}$, $i = 1, \ldots, n$ are a fundamental set of solutions to the homogeneous equation (4) and $u_k^{(f_k)}$ is a particular solution to the nonhomogeneous equation (3). Writing (24) in matrix form, we have

$$C_0 \mathbf{c} = \mathbf{b} - \mathbf{u}^{(f_k)}, \tag{25}$$

where

$$C_0 = \begin{pmatrix} u_0^{(1)} & u_0^{(2)} & \ldots & u_0^{(n)} \\ u_1^{(1)} & u_1^{(2)} & \ldots & u_1^{(n)} \\ \vdots & \vdots & & \vdots \\ u_{n-1}^{(1)} & u_{n-1}^{(2)} & \ldots & u_{n-1}^{(n)} \end{pmatrix}, \quad \mathbf{c} = \begin{pmatrix} c_1 \\ c_2 \\ \vdots \\ c_n \end{pmatrix}, \quad \mathbf{b} = \begin{pmatrix} \beta_0 \\ \beta_1 \\ \vdots \\ \beta_{n-1} \end{pmatrix}, \quad \mathbf{u}^{(f_k)} = \begin{pmatrix} u_0^{(f_k)} \\ u_1^{(f_k)} \\ \vdots \\ u_{n-1}^{(f_k)} \end{pmatrix}. \tag{26}$$

The Casorati matrix C_0 is nonsingular and therefore $\mathbf{c} = C_0^{-1}(\mathbf{b} - \mathbf{u}^{(f_k)})$, while substitution into (6) and (7) yields Eq. (23). In order to prove the uniqueness of the solution, we take $u_k,\ v_k$ to be two solutions of Eq. (22) and we will show that $u_k = v_k$. By assumption we have $Au_k - Av_k = 0$ and $u_i - v_i = 0,\ i = 0, \ldots, n-1$ or, explicitly,

$$\begin{aligned} &(u_{k+n} - v_{k+n}) + a_1(u_{k+n-1} - v_{k+n-1}) + \cdots + a_n(u_k - v_k) = 0, \\ &u_i - v_i = 0, \quad i = 0, \ldots, n-1. \end{aligned} \tag{27}$$

By putting $w_k = u_k - v_k$, we get

$$\begin{aligned} &w_{k+n} + a_1 w_{k+n-1} + \cdots + a_n w_k = 0, \\ &w_i = 0, \quad i = 0, \ldots, n-1. \end{aligned} \tag{28}$$

This means that w_k is a solution to the homogeneous equation (4) and hence it can be expressed as

$$w_k = c_1 u_k^{(1)} + c_2 u_k^{(2)} + \cdots + c_n u_k^{(n)}, \tag{29}$$

by (6). Substituting (29) into the second equation in (28), we have

$$C_0 \mathbf{c} = 0. \tag{30}$$

Since $\det C_0 \neq 0$, it is implied that $c_1 = c_2 \ldots = c_n = 0$ and hence $w_k = u_k - v_k = 0$ or $u_k = v_k$. This completes the proof. □

Problems with Nonlocal Conditions

Let now the operator $A : S \to S$ be defined by

$$\begin{aligned} &Au_k = u_{k+n} + a_1 u_{k+n-1} + \cdots + a_n u_k, \\ &D(A) = \{u_k \in S : \mu_{i0} u_0 + \cdots + \mu_{il-1} u_{l-1} = \beta_i,\ i = 0, \ldots, n-1,\ n \leq l\}, \end{aligned} \tag{31}$$

and reformulate the nonlocal problem (9) in the operator form

$$Au_k = f_k, \quad f_k \in S. \tag{32}$$

The nonlocal conditions may be homogeneous, $\sum_{j=0}^{l-1} \mu_{ij} u_j = \beta_i \equiv 0,\ i = 0, \ldots, n-1$, or they can be nonhomogeneous, $\sum_{j=0}^{l-1} \mu_{ij} u_j = \beta_i \neq 0$. In the first case, the operator A is linear whereas in the second is nonlinear.

The next theorem specifies the prerequisites under which the nonlocal problem (32) with homogeneous conditions has a unique solution and yields it in closed-form.

Theorem 3 *Let $\widehat{A} : S \to S$ be the linear nonlocal difference operator defined by*

$$\begin{aligned} \widehat{A}u_k &= u_{k+n} + a_1 u_{k+n-1} + \cdots + a_n u_k = f_k, \\ D(\widehat{A}) &= \{u_k \in S : \mu_{i0}u_0 + \cdots + \mu_{il-1}u_{l-1} = 0,\ i = 0, \ldots, n-1,\ n \leq l\}, \end{aligned} \tag{33}$$

$u_k^{(i)}$, $i = 1, \ldots, n$ be a fundamental set of solutions of the homogeneous equation (4) and $u_k^{(f_k)}$ be a particular solution of the nonhomogeneous equation (3). Moreover, let the vectors $\mathbf{u}_k^{(H)} = (u_k^{(1)}, u_k^{(2)}, \ldots, u_k^{(n)})$ and $\mathbf{u}_l^{(f_k)} = col(u_0^{(f_k)}, u_1^{(f_k)}, \ldots, u_{l-1}^{(f_k)})$ as well as the nonlocal conditions matrix $M = (\mu_{ij}) \in \mathbf{R}^{n\times l}$, $i = 0, \ldots, n-1$, $j = 0, \ldots, l-1$ and the generalized Casorati matrix $U = (u_j^{(i)}) \in \mathbf{R}^{l\times n}$, $j = 0, \ldots, l-1$, $i = 1, \ldots, n$ for $k = 0$. The following hold:

(i) If $\det(MU) \neq 0$, then the operator $\widehat{A}$ is injective and the unique solution to the problem (33) for any $f_k \in S$ is given by

$$u_k = \widehat{A}^{-1} f_k = u_k^{(f_k)} - \mathbf{u}_k^{(H)} (MU)^{-1} M\mathbf{u}_l^{(f_k)}. \tag{34}$$

(ii) If $\det(MU) = 0$ and $\operatorname{rank}(MU) = \operatorname{rank}\left(MU \quad - M\mathbf{u}_l^{(f_k)}\right)$, then the problem (33) has infinitely many solutions.

(iii) If $\operatorname{rank}(MU) < \operatorname{rank}\left(MU \quad - M\mathbf{u}_l^{(f_k)}\right)$, then the problem (33) possesses no solution.

Proof

(i) The general solution of the related nonhomogeneous equation (3) is provided by

$$u_k = \mathbf{u}^{(H)}\mathbf{c} + u_k^{(f_k)}, \tag{35}$$

where $\mathbf{u}^{(H)} = (u_k^{(1)}, u_k^{(2)}, \ldots, u_k^{(n)})$ is a fundamental set of solutions of the homogeneous equation (4) and $\mathbf{c} = \operatorname{col}(c_1, c_2, \ldots, c_n)$ is a vector of constants. Taking (35) for $k = 0, 1, \ldots, l-1$, we have

$$\mathbf{u}_l = U\mathbf{c} + \mathbf{u}_l^{(f_k)}, \tag{36}$$

where

$$\mathbf{u}_l = \begin{pmatrix} u_0 \\ u_1 \\ \vdots \\ u_{l-1} \end{pmatrix}, \quad U = \begin{pmatrix} u_0^{(1)} & u_0^{(2)} & \dots & u_0^{(n)} \\ u_1^{(1)} & u_1^{(2)} & \dots & u_1^{(n)} \\ \vdots & \vdots & & \vdots \\ u_{l-1}^{(1)} & u_{l-1}^{(2)} & \dots & u_{l-1}^{(n)} \end{pmatrix} = (u_j^{(i)}), \quad \mathbf{u}_l^{(f_k)} = \begin{pmatrix} u_0^{(f_k)} \\ u_1^{(f_k)} \\ \vdots \\ u_{l-1}^{(f_k)} \end{pmatrix}. \tag{37}$$

The matrix U is the $l \times n$ generalized Casorati matrix for $k = 0$. Writing the homogeneous nonlocal conditions in matrix form, we obtain

$$M\mathbf{u}_l = 0, \tag{38}$$

where the nonlocal conditions matrix

$$M = \begin{pmatrix} \mu_{00} & \mu_{01} & \dots & \mu_{0l-1} \\ \mu_{10} & \mu_{11} & \dots & \mu_{1l-1} \\ \vdots & \vdots & & \vdots \\ \mu_{n-10} & \mu_{n-11} & \dots & \mu_{n-1l-1} \end{pmatrix} = (\mu_{ij}). \tag{39}$$

From (36) and (38), we acquire

$$MU\mathbf{c} = -M\mathbf{u}_l^{(f_k)}. \tag{40}$$

If $\det(MU) \neq 0$, then

$$\mathbf{c} = -(MU)^{-1}M\mathbf{u}_l^{(f_k)}. \tag{41}$$

Substituting $\mathbf{c}$ into (35), we get the solution in (34). Let now u_k and v_k be two solutions of Eq. (33). Then $\widehat{A}u_k - \widehat{A}v_k = \widehat{A}(u_k - v_k) = 0$ since $\widehat{A}$ is linear. By placing $w_k = u_k - v_k$, we have

$$\begin{gathered} w_{k+n} + a_1 w_{k+n-1} + \cdots + a_n w_k = 0, \\ \mu_{i0}w_0 + \mu_{i1}w_1 + \cdots + \mu_{il-1}w_{l-1} = 0, \quad i = 0, \dots, n-1. \end{gathered} \tag{42}$$

It follows that w_k is a solution of the corresponding homogeneous equation (4), which by means of (6) it can be written as

$$w_k = c_1 u_k^{(1)} + c_2 u_k^{(2)} + \cdots + c_n u_k^{(n)}, \tag{43}$$

and hence by inserting (43) into the second equation in (42), we get

$$MU\mathbf{c} = 0. \tag{44}$$

Since $\det(MU) \neq 0$, then it is implied that $\mathbf{c} = 0$ and subsequently from (43) follows that $w_k = u_k - v_k = 0$ or $u_k = v_k$ which proves that the operator $\widehat{A}$ is injective and that the solution in (34) is unique.

(ii) It is well known from linear algebra that if $\det(MU) = 0$, i.e., the $\text{rank}(MU)$ is strictly less than n, and $\text{rank}(MU) = \text{rank}\left(MU \;\; -M\mathbf{u}_l^{(f_k)}\right)$, then the system of equations (40) has infinitely many solutions. Hence, from (35) the problem (33) has an infinite number of solutions.

(iii) As in (ii), if $\text{rank}(MU) < \text{rank}\left(MU \;\; -M\mathbf{u}_l^{(f_k)}\right)$, then the system (40) and consequently the problem (33) has no solution. □

Remark 1 In Theorem 3 when $l = n$ the matrix $U = (u_j^{(i)}) \in \mathbf{R}^{n \times n}$, $j = 0, \ldots, n-1$, $i = 1, \ldots, n$ becomes the Casorati matrix C_0, and if $\det M \neq 0$, then the unique solution to the problem (33) for any $f_k \in S$ is given by

$$u_k = \widehat{A}^{-1} f_k = u_k^{(f_k)} - \mathbf{u}_k^{(H)} C_0^{-1} \mathbf{u}_l^{(f_k)}. \tag{45}$$

Finally, for the case of the nonlocal problem (32) with nonhomogeneous conditions, we prove the following theorem which identifies additional requirements for the existence of the unique solution and presents an exact formula for its computation.

Theorem 4 *Let $A : S \to S$ be the nonlinear nonlocal difference operator defined as*

$$\begin{aligned} &Au_k = u_{k+n} + a_1 u_{k+n-1} + \cdots + a_n u_k = f_k, \\ &D(A) = \{u_k \in S : \mu_{i0} u_0 + \cdots + \mu_{il-1} u_{l-1} = \beta_i,\ i = 0, \ldots, n-1,\ n \leq l\}, \end{aligned} \tag{46}$$

$u_k^{(i)}$, $i = 1, \ldots, n$ be a fundamental set of solutions of the homogeneous equation (4) and $u_k^{(f_k)}$ be a particular solution of the nonhomogeneous equation (3). Furthermore, let the vectors $\mathbf{u}_k^{(H)} = (u_k^{(1)}, u_k^{(2)}, \ldots, u_k^{(n)})$, $\mathbf{u}_l^{(f_k)} = \text{col}(u_0^{(f_k)}, u_1^{(f_k)}, \ldots, u_{l-1}^{(f_k)})$, and $\mathbf{b} = \text{col}(\beta_0, \beta_1, \ldots, \beta_{n-1})$ as well as the nonlocal conditions matrix $M = (\mu_{ij}) \in \mathbf{R}^{n \times l}$, $i = 0, \ldots, n-1$, $j = 0, \ldots, l-1$ and the generalized Casorati matrix $U = (u_j^{(i)}) \in \mathbf{R}^{l \times n}$, $j = 0, \ldots, l-1$, $i = 1, \ldots, n$ for $k = 0$. The following hold:

(i) If $\det(MU) \neq 0$, then the operator A is injective and the unique solution to the problem (46) for any $f_k \in S$ is given by

$$u_k = A^{-1} f_k = u_k^{(f_k)} + \mathbf{u}_k^{(H)} (MU)^{-1} \left(\mathbf{b} - M\mathbf{u}_l^{(f_k)}\right). \tag{47}$$

(ii) If $\det(MU) = 0$ and $\text{rank}\,(MU) = \text{rank}\left(MU \;\; \mathbf{b} - M\mathbf{u}_l^{(f_k)}\right)$, then the problem (46) has infinitely many solutions.

(iii) *If* $\operatorname{rank}(MU) < \operatorname{rank}\left(MU \;\; \mathbf{b} - M\mathbf{u}_l^{(f_k)}\right)$, *then the problem (46) possesses no solution.*

Proof

(i) By hypothesis the general solution to the corresponding nonhomogeneous equation (3) is given by (7), namely,

$$u_k = c_1 u_k^{(1)} + c_2 u_k^{(2)} + \cdots + c_n u_k^{(n)} + u_k^{(f_k)} = \mathbf{u}_k^{(H)}\mathbf{c} + u_k^{(f_k)} \tag{48}$$

where $\mathbf{u}_k^{(H)} = (u_k^{(1)}, u_k^{(2)}, \ldots, u_k^{(n)})$ and $\mathbf{c} = \operatorname{col}(c_1, c_2, \ldots, c_n)$. Taking (48) for $k = 0, 1, \ldots, l-1$, we have

$$u_j = \mathbf{u}_j^{(H)}\mathbf{c} + u_j^{(f_k)}, \quad j = 0, \ldots, l-1 \tag{49}$$

or

$$\mathbf{u}_l = U\mathbf{c} + \mathbf{u}_l^{(f_k)}, \tag{50}$$

where

$$\mathbf{u}_l = \begin{pmatrix} u_0 \\ u_1 \\ \vdots \\ u_{l-1} \end{pmatrix}, \quad U = \begin{pmatrix} u_0^{(1)} & u_0^{(2)} & \ldots & u_0^{(n)} \\ u_1^{(1)} & u_1^{(2)} & \ldots & u_1^{(n)} \\ \vdots & \vdots & & \vdots \\ u_{l-1}^{(1)} & u_{l-1}^{(2)} & \ldots & u_{l-1}^{(n)} \end{pmatrix} = (u_j^{(i)}), \quad \mathbf{u}_l^{(f_k)} = \begin{pmatrix} u_0^{(f_k)} \\ u_1^{(f_k)} \\ \vdots \\ u_{l-1}^{(f_k)} \end{pmatrix}. \tag{51}$$

The matrix U is the $l \times n$ generalized Casorati matrix for $k = 0$. The nonlocal conditions in (46) can be put in the matrix form

$$M\mathbf{u}_l = \mathbf{b}, \tag{52}$$

where the nonlocal conditions matrix

$$M = \begin{pmatrix} \mu_{00} & \mu_{01} & \ldots & \mu_{0l-1} \\ \mu_{10} & \mu_{11} & \ldots & \mu_{1l-1} \\ \vdots & \vdots & & \vdots \\ \mu_{n-10} & \mu_{n-11} & \ldots & \mu_{n-1l-1} \end{pmatrix} = \left(\mu_{ij}\right), \quad \mathbf{b} = \begin{pmatrix} \beta_0 \\ \beta_1 \\ \vdots \\ \beta_{n-1} \end{pmatrix}. \tag{53}$$

From (50) and (52), we obtain

$$MU\mathbf{c} = \mathbf{b} - M\mathbf{u}_l^{(f_k)}. \tag{54}$$

If $\det(MU) \neq 0$, then

$$\mathbf{c} = (MU)^{-1}\left(\mathbf{b} - M\mathbf{u}_l^{(f_k)}\right). \quad (55)$$

Substituting $\mathbf{c}$ into (48), we get the solution in (47). Let now u_k and v_k be two solutions of Eq. (46). Then $Au_k - Av_k = 0$ or

$$\begin{aligned} &(u_{k+n} - v_{k+n}) + a_1(u_{k+n-1} - v_{k+n-1}) + \cdots + a_n(u_k - v_k) = 0, \\ &\mu_{i0}(u_0 - v_0) + \mu_{i1}(u_1 - v_1) + \cdots + \mu_{il-1}(u_{l-1} - v_{l-1}) = 0, \\ &i = 0, \ldots, n-1. \end{aligned} \quad (56)$$

By substituting $w_k = u_k - v_k$ we have

$$\begin{aligned} w_{k+n} + a_1 w_{k+n-1} + \cdots + a_n w_k = 0, \\ \mu_{i0}w_0 + \mu_{i1}w_1 + \cdots + \mu_{il-1}w_{l-1} = 0, \quad i = 0, \ldots, n-1. \end{aligned} \quad (57)$$

It follows that w_k is a solution of the corresponding homogeneous equation (4), which by means of (6) it can be written as

$$w_k = c_1 u_k^{(1)} + c_2 u_k^{(2)} + \cdots + c_n u_k^{(n)}, \quad (58)$$

and hence

$$MU\mathbf{c} = 0. \quad (59)$$

Since $\det(MU) \neq 0$, then it is implied that $\mathbf{c} = 0$ and subsequently from (58) follows that $w_k = u_k - v_k = 0$ or $u_k = v_k$ which proves that the operator A is injective and that the solution in (47) is unique.

(ii) It is well known from linear algebra that if $\det(MU) = 0$, i.e., the $\text{rank}(MU)$ is strictly less than n, and $\text{rank}(MU) = \text{rank}\left(MU \;\; \mathbf{b} - M\mathbf{u}_l^{(f_k)}\right)$, then the system of equations (54) has infinitely many solutions. Hence, from (48) the problem (46) has an infinite number of solutions.

(iii) As in (ii), if $\text{rank}(MU) < \text{rank}\left(MU \;\; \mathbf{b} - M\mathbf{u}_l^{(f_k)}\right)$, then the system (54) and consequently the problem (46) has no solution. □

Remark 2 In Theorem 4 when $l = n$ the matrix $U = (u_j^{(i)}) \in \mathbf{R}^{n\times n}$, $j = 0, \ldots, n-1$, $i = 1, \ldots, n$ is the Casorati matrix C_0. Thus, if $\det M \neq 0$, then the operator A is injective, and the unique solution to the problem (46) for any $f_k \in S$ is given by the simplified formula

$$u_k = A_1^{-1} f_k = u_k^{(f_k)} + \mathbf{u}_k^{(H)} C_0^{-1}\left(M^{-1}\mathbf{b} - \mathbf{u}_l^{(f_k)}\right). \quad (60)$$

Loaded Difference Equations

A linear loaded difference equation of nth-order is a difference equation incorporating values of some functionals of the solution named loads, viz.,

$$u_{k+n} + a_1 u_{k+n-1} + \cdots + a_n u_k - \sum_{s=1}^{m} g_k^s \psi_s(u_k) = f_k, \tag{61}$$

where the coefficients $a_i \in \mathbf{R}$, $i = 1, \ldots, n$ are known with $a_n \neq 0$, $k \in \mathbf{Z}^+$, $\psi_s : S \to \mathbf{R}$, $s = 1, \ldots, m$ are linear functionals, $g_k^s = g_s(k)$, $s = 1, \ldots, m$ and $f_k = f(k) \in S$ are given real-valued functions and $u_k = u(k) \in S$ is unknown. Equation (61) is usually accompanied with initial conditions

$$u_i = \beta_i, \quad i = 0, \ldots, n-1, \tag{62}$$

or nonlocal conditions, in fact discrete multipoint conditions,

$$\sum_{j=0}^{l-1} \mu_{ij} u_j = \beta_i, \quad i = 0, \ldots, n-1, \quad n \leq l. \tag{63}$$

For the solution of the above loaded initial value or nonlocal problems, we use the extension operator method presented by the authors in [15] for problems involving continuous functions. This method adjusted here to solving loaded difference equations requires the solution to the corresponding linear difference equations.

In this section, the separation to two categories of problems seems to be appropriate, those having homogeneous conditions and those with nonhomogeneous conditions.

Loaded Problems with Homogeneous Conditions

By employing the linear operator $\widehat{A}$, expounded in (13) or (33), we define the linear operator $\widehat{B} : S \to S$ as a perturbation of $\widehat{A}$ as in [15], succinctly

$$\begin{aligned} \widehat{B}u_k &= \widehat{A}u_k - \sum_{s=1}^{m} g_k^s \psi_s(u_k) = \widehat{A}u_k - \mathbf{g}_k \Psi(u_k), \\ D(\widehat{B}) &= D(\widehat{A}), \end{aligned} \tag{64}$$

where the vectors $\Psi = \text{col}\,(\psi_1, \psi_2, \ldots, \psi_m)$ and $\mathbf{g}_k = (g_k^1, g_k^2, \ldots, g_k^m)$. Without any loss of generality, we may assume that each of the sets $\{\psi_s\}$ and $\{g_k^s\}$ is linearly independent; otherwise, we could decrease the number of their elements. Thus, the loaded difference equation (61) along with homogeneous initial or nonlocal conditions can be put in the operator form

$$\widehat{B}u_k = \widehat{A}u_k - \mathbf{g}_k\Psi(u_k) = f_k,$$
$$D(\widehat{B}) = D(\widehat{A}). \tag{65}$$

The exact solution to the loaded problem (65) can be now constructed by means of Theorems 1 and 3 and specially the formulas (14) and (34) as it is asserted by the following general theorem.

Theorem 5 *Let $\widehat{A} : S \to S$ be an injective linear difference operator, the vector $\Psi = \mathrm{col}(\psi_1, \psi_2 \ldots, \psi_m)$ be a linearly independent set of linear functionals, the vector $\mathbf{g}_k = (g_k^1, g_k^2, \ldots, g_k^m)$ be a linearly independent set of real functions $g_k^s \in S$, $s = 1, \ldots, m$ and $f_k \in S$. Then:*

(i) The linear operator $\widehat{B} : S \to S$ defined by

$$\widehat{B}u_k = \widehat{A}u_k - \mathbf{g}_k\Psi(u_k) = f_k,$$
$$D(\widehat{B}) = D(\widehat{A}), \tag{66}$$

is injective if and only if

$$\det W = \det\left(I_m - \Psi(\widehat{A}^{-1}\mathbf{g}_k)\right) \neq 0, \tag{67}$$

where I_m denotes the $m \times m$ identity matrix.

(ii) If (i) is true, then the unique solution to the problem (66) for any $f_k \in S$ is given by

$$u_k = \widehat{B}^{-1}f_k = \widehat{A}^{-1}f_k + (\widehat{A}^{-1}\mathbf{g}_k)W^{-1}\Psi(\widehat{A}^{-1}f_k). \tag{68}$$

Proof

(i) Let any $u_k, v_k \in D(\widehat{B})$ and assume $\widehat{B}u_k = \widehat{B}v_k$. Since the operator $\widehat{B}$ is linear, we have

$$\widehat{B}u_k - \widehat{B}v_k = \widehat{B}(u_k - v_k) = \widehat{B}w_k = \widehat{A}w_k - \sum_{s=1}^{m} g_k^s\psi_s(w_k) = \widehat{A}w_k - \mathbf{g}_k\Psi(w_k) = 0, \tag{69}$$

where $w_k = u_k - v_k$. By hypothesis $\widehat{A}$ is linear and injective, and therefore there exists the inverse $\widehat{A}^{-1}$. By applying $\widehat{A}^{-1}$ on both sides of (69), we get

$$w_k - \sum_{s=1}^{m} \widehat{A}^{-1}g_k^s\psi_s(w_k) = \widehat{A}^{-1}0 = 0. \tag{70}$$

Employing the functionals $\psi_\ell,\ \ell = 1, \dots m$ on (70), we obtain

$$\psi_\ell(w_k) - \sum_{s=1}^{m} \psi_\ell(\widehat{A}^{-1} g_k^s)\psi_s(w_k) = \psi_\ell(0) = 0, \quad \ell = 1, \dots, m, \tag{71}$$

or in matrix form

$$\begin{pmatrix} 1-\psi_1(\widehat{A}^{-1}g_k^1) & -\psi_1(\widehat{A}^{-1}g_k^2) & \dots & -\psi_1(\widehat{A}^{-1}g_k^m) \\ -\psi_2(\widehat{A}^{-1}g_k^1) & 1-\psi_2(\widehat{A}^{-1}g_k^2) & \dots & -\psi_2(\widehat{A}^{-1}g_k^m) \\ \vdots & \vdots & & \vdots \\ -\psi_m(\widehat{A}^{-1}g_k^1) & -\psi_m(\widehat{A}^{-1}g_k^2) & \dots & 1-\psi_m(\widehat{A}^{-1}g_k^m) \end{pmatrix} \begin{pmatrix} \psi_1(w_k) \\ \psi_2(w_k) \\ \vdots \\ \psi_m(w_k) \end{pmatrix} = \begin{pmatrix} 0 \\ 0 \\ \vdots \\ 0 \end{pmatrix}, \tag{72}$$

or

$$\left(I_m - \Psi(\widehat{A}^{-1}\mathbf{g}_k)\right)\Psi(w_k) = W\Psi(w_k) = 0. \tag{73}$$

If Eq. (67) holds true, then from (73) follows that $\Psi(w_k) = 0$. Substitution into (69) yields $\widehat{A}w_k = 0$, and since $\widehat{A}$ is injective, we get $w_k = u_k - v_k = 0$ or $u_k = v_k$. This proves that the operator $\widehat{B}$ is injective.

Conversely, suppose that $\widehat{B}$ is injective and $\det W = 0$. Let any $u_k,\ v_k \in D(\widehat{B})$ for which holds $\widehat{B}u_k = \widehat{B}v_k$. Repeating the same steps as above, we acquire (73). If $\det W = 0$, then the vector $\Psi(w_k)$ may be nonzero, i.e., $\Psi(w_k) = \Psi(u_k - v_k) = \Psi(u_k) - \Psi(v_k) \neq 0$ or $\Psi(u_k) \neq \Psi(v_k)$, which implies that $u_k \neq v_k$ since the sets $\{g_k^s\}$ and $\{\psi_s\}$ are linearly independent. This contradicts the assumption that $\widehat{B}$ is injective, and therefore $\det W \neq 0$.

(ii) By applying the inverse operator $\widehat{A}^{-1}$ on (66), we get

$$u_k - \widehat{A}^{-1}\mathbf{g}_k\Psi(u_k) = \widehat{A}^{-1}f_k. \tag{74}$$

Acting by the vector Ψ on both sides of (74), we obtain

$$\Psi(u_k) - \Psi(\widehat{A}^{-1}\mathbf{g}_k)\Psi(u_k) = \Psi(\widehat{A}^{-1}f_k),$$
$$\left(I_m - \Psi(\widehat{A}^{-1}\mathbf{g}_k)\right)\Psi(u_k) = \Psi(\widehat{A}^{-1}f_k).$$

Since $\det W = \det\left(I_m - \Psi(\widehat{A}^{-1}\mathbf{g}_k)\right) \neq 0$, we have

$$\Psi(u_k) = W^{-1}\Psi(\widehat{A}^{-1}f_k). \tag{75}$$

Substitution of (75) into (74) yields Eq. (68). □

Loaded Problems with Nonhomogeneous Conditions

Loaded problems involving nonhomogeneous initial or nonlocal conditions are more difficult to deal with. Here, we adopt an indirect approach where initial value or nonlocal problems with nonhomogeneous conditions are first reduced to corresponding ones having homogeneous conditions, and then the solution can be obtained by Theorem 5 in the previous subsection.

We commence by studying the loaded initial value problems implicating nonhomogeneous conditions.

Theorem 6 *Let $A : S \to S$ be the nonlinear operator,*

$$\begin{aligned} Au_k &= u_{k+n} + a_1 u_{k+n-1} + \cdots + a_n u_k, \\ D(A) &= \{u_k \in S : u_i = \beta_i,\ i = 0, \ldots, n-1\}, \end{aligned} \tag{76}$$

and $B : S \to S$ be the operator defined by

$$\begin{aligned} Bu_k &= Au_k - \sum_{s=1}^{m} g_k^s \psi_s(u_k) = Au_k - \mathbf{g}_k \Psi(u_k) = f_k, \\ D(B) &= D(A), \end{aligned} \tag{77}$$

where $\Psi = \operatorname{col}(\psi_1, \psi_2, \ldots, \psi_m)$ is a vector of linear functionals $\psi_s : S \to R$, $\mathbf{g}_k = (g_k^1, g_k^2, \ldots, g_k^m)$ is a vector of functions $g_k^s \in S$, each of the sets $\{\psi_s\}$ and $\{g_k^s\}$ is taken to be linearly independent, and $f_k \in S$. Consider

$$u_k = v_k + p^{n-1}(k) = v_k + \mathbf{p}_k^{n-1} D_0^{-1} \mathbf{b}, \tag{78}$$

where $v : \mathbf{Z}^+ \to \mathbf{R}$, $p^{n-1}(k)$ is a real polynomial of degree less or equal to $n-1$, $\mathbf{p}_k^{n-1} = (1, k, \ldots, k^{n-1})$ is a vector of n polynomial functions, D_0 is the Casorati matrix for the n functions k^{i-1}, $i = 1, \ldots, n$ for $k = 0$ and $\mathbf{b} = \operatorname{col}(\beta_0, \beta_1, \ldots, \beta_{n-1})$. Then, the problem (77) can be reduced to the linear problem with homogeneous initial conditions:

$$\begin{aligned} \widehat{B} v_k &= \widehat{A} v_k - \mathbf{g}_k \Psi(v_k) = \varphi_k, \\ D(\widehat{B}) &= D(\widehat{A}), \end{aligned} \tag{79}$$

where

$$\begin{aligned} \varphi_k = f_k &- \left(p^{n-1}(k+n) + a_1 p^{n-1}(k+n-1) + \cdots + a_n p^{n-1}(k)\right) \\ &+ \sum_{s=1}^{m} g_k^s \psi_s\left(p^{n-1}(k)\right), \end{aligned} \tag{80}$$

$\varphi_k \in S$, $\widehat{A} : S \to S$ is the linear operator

$$\widehat{A}v_k = v_{k+n} + a_1 v_{k+n-1} + \cdots + a_n v_k,$$
$$D(\widehat{A}) = \{v_k \in S : v_i = 0,\ i = 0, \ldots, n-1\}, \tag{81}$$

$\widehat{B} : S \to S$ is a linear operator and the unknown is now v_k. The solution to the problem (79) follows from Theorem 5.

Proof We take

$$u_k = v_k + \mathbf{p}_k^{n-1}\mathbf{d}, \tag{82}$$

where $v : \mathbf{Z}^+ \to \mathbf{R}$, $p^{n-1}(k)$ is a real polynomial of degree less or equal to n−1, $\mathbf{p}_k^{n-1} = (1, k, \ldots, k^{n-1})$ is a vector of n polynomial functions, and the vector $\mathbf{d} = \mathrm{col}(d_0, d_1, \ldots, d_{n-1})$ contains the n coefficients of the polynomial $p^{n-1}(k)$ to be determined. We demand u_k to fulfill the given nonhomogeneous initial conditions whereas v_k to satisfy the new homogeneous initial conditions. This is to say that

$$\mathbf{p}_i^{n-1}\mathbf{d} = \beta_i, \quad i = 0, \ldots, n-1, \tag{83}$$

or in matrix form

$$D_0\mathbf{d} = \mathbf{b}, \tag{84}$$

where

$$D_0 = \begin{pmatrix} 1 & 0 & 0 & \ldots & 0 \\ 1 & 1 & 1 & \ldots & 1 \\ 1 & 2 & 2^2 & \ldots & 2^{n-1} \\ \vdots & \vdots & \vdots & \vdots & \vdots \\ 1 & n-1 & (n-1)^2 & \ldots & (n-1)^{n-1} \end{pmatrix}, \quad \mathbf{b} = \begin{pmatrix} \beta_0 \\ \beta_1 \\ \beta_2 \\ \vdots \\ \beta_{n-1} \end{pmatrix}. \tag{85}$$

The matrix D_0 is the Casorati matrix for the n functions k^{i-1}, $i = 1, \ldots, n$ for $k = 0$. Because $\det D_0 \neq 0$, we get $\mathbf{d} = D_0^{-1}\mathbf{b}$. By substituting into (82), we obtain Eq. (78) and then from (77), after some algebra is performed, follows the reduced linear problem (79) involving homogeneous initial conditions. This can be solved by using Theorem 5. □

Next, we contemplate the loaded nonlocal problems incorporating nonhomogeneous conditions.

Theorem 7 *Let $A : S \to S$ be the nonlinear operator,*

$$Au_k = u_{k+n} + a_1 u_{k+n-1} + \cdots + a_n u_k,$$
$$D(A) = \{u_k \in S : \mu_{i0}u_0 + \cdots + \mu_{il-1}u_{l-1} = \beta_i,\ i=0, \ldots, n-1,\ n \leq l\}, \tag{86}$$

and $B : S \to S$ be the operator defined by

$$\begin{aligned} Bu_k &= Au_k - \sum_{s=1}^{m} g_k^s \psi_s(u_k) = Au_k - \mathbf{g}_k \Psi(u_k) = f_k, \\ D(B) &= D(A), \end{aligned} \tag{87}$$

where $\Psi = \text{col}\,(\psi_1, \psi_2, \ldots, \psi_m)$ is a vector of linear functionals, $\psi_s : S \to R$, $\mathbf{g}_k = (g_k^1, g_k^2, \ldots, g_k^m)$ is a vector of functions $g_k^s \in S$, each of the sets $\{\psi_s\}$ and $\{g_k^s\}$ is taken to be linearly independent, and $f_k \in S$. Further, let $\mathbf{p}_k^{n-1} = (1, k, \ldots, k^{n-1})$, $P = \left(\mathbf{p}_j^{n-1}\right)$, $j = 0, \ldots, l-1$ be the $l \times n$ generalized Casorati matrix for the n functions k^{i-1}, $i = 1, \ldots, n$ for $k = 0$, $M = \left(\mu_{ij}\right)$ be the $n \times l$ matrix associated with the nonlocal conditions, and $\mathbf{b} = \text{col}(\beta_0, \beta_1, \ldots, \beta_{n-1})$. If

$$\det\,(MP) \neq 0, \tag{88}$$

then by means of

$$u_k = v_k + p^{n-1}(k) = v_k + \mathbf{p}_k^{n-1}\,(MP)^{-1}\,\mathbf{b}, \tag{89}$$

where $v : \mathbf{Z}^+ \to \mathbf{R}$, the problem (87) can be reduced to the linear problem with homogeneous nonlocal conditions:

$$\begin{aligned} \widehat{B}v_k &= \widehat{A}v_k - \mathbf{g}_k \Psi(v_k) = \varphi_k, \\ D(\widehat{B}) &= D(\widehat{A}), \end{aligned} \tag{90}$$

where

$$\begin{aligned} \varphi_k = f_k &- \left(p^{n-1}(k+n) + a_1 p^{n-1}(k+n-1) + \cdots + a_n p^{n-1}(k)\right) \\ &+ \sum_{s=1}^{m} g_k^s \psi_s\left(p^{n-1}(k)\right), \end{aligned} \tag{91}$$

$\varphi_k \in S$, $\widehat{A} : S \to S$ is the linear operator

$$\begin{aligned} \widehat{A}v_k &= v_{k+n} + a_1 v_{k+n-1} + \cdots + a_n v_k, \\ D(\widehat{A}) &= \{v_k \in S : \mu_{i0}v_0 + \cdots + \mu_{il-1}v_{l-1} = 0,\ i = 0, \ldots, n-1,\ n \le l\}, \end{aligned} \tag{92}$$

$\widehat{B} : S \to S$ is a linear operator and the unknown is now v_k. The solution to the problem (90) is obtained by Theorem 5.

Proof We put

$$u_k = v_k + \mathbf{p}_k^{n-1}\mathbf{d}, \tag{93}$$

where $v : \mathbf{Z}^+ \to \mathbf{R}$, $p^{n-1}(k)$ is a real polynomial of degree less or equal to n−1, $\mathbf{p}_k^{n-1} = (1, k, \ldots, k^{n-1})$ is a vector with n polynomial functions, and the vector $\mathbf{d} = \text{col}(d_0, d_1, \ldots, d_{n-1})$ contains the n coefficients of the polynomial $p^{n-1}(k)$ to be determined. We want u_k to satisfy the given nonhomogeneous nonlocal conditions and v_k to fulfill the new homogeneous nonlocal conditions. Subject to this requirement, from the nonhomogeneous nonlocal conditions, we have

$$\mu_{i0}p^{n-1}(0)+\mu_{i1}p^{n-1}(1)+\cdots+\mu_{il-1}p^{n-1}(l-1) = \beta_i,\ \ i=0,\ldots,n-1,\ n \le l \tag{94}$$

or in matrix form

$$MP\mathbf{d} = \mathbf{b}, \tag{95}$$

where

$$M = \begin{pmatrix} \mu_{00} & \mu_{01} & \mu_{02} & \cdots & \mu_{0l-1} \\ \mu_{10} & \mu_{11} & \mu_{12} & \cdots & \mu_{1l-1} \\ \mu_{20} & \mu_{21} & \mu_{22} & \cdots & \mu_{2l-1} \\ \vdots & \vdots & \vdots & \vdots & \vdots \\ \mu_{n-10} & \mu_{n-11} & \mu_{n-12} & \cdots & \mu_{n-1l-1} \end{pmatrix} = (\mu_{ij}),$$

$$P = \begin{pmatrix} 1 & 0 & 0 & \ldots & 0 \\ 1 & 1 & 1 & \ldots & 1 \\ 1 & 2 & 2^2 & & 2^{n-1} \\ \vdots & \vdots & \vdots & & \vdots \\ 1 & l-1 & (l-1)^2 & \ldots & (l-1)^{n-1} \end{pmatrix} = (p_{ji}), \quad \mathbf{b} = \begin{pmatrix} \beta_0 \\ \beta_1 \\ \beta_2 \\ \vdots \\ \beta_{n-1} \end{pmatrix}. \tag{96}$$

The matrix P is the $l \times n$ generalized Casorati matrix for the n functions k^{i-1}, $i = 1, \ldots, n$ for $k = 0$. If $\det(MP) \neq 0$, then $\mathbf{d} = (MP)^{-1}\,\mathbf{b}$. By substituting into (93) and then into (87) and after some algebra, we acquire the reduced linear problem (90) with homogeneous nonlocal conditions. This can be solved by Theorem 5. □

Example Problems

In this section, we solve several problems involving difference equations and loaded difference equations with initial and nonlocal conditions.

Problem 1 Solve the following second-order difference equation with nonlocal homogeneous boundary conditions:

$$u_{k+2} - 3u_{k+1} + 2u_k = k, \quad u_0 = 0, \quad u_1 = 2u_2. \tag{97}$$

Theorem 3 is applicable, and therefore we formulate the problem as in Eq. (33) where we take

$$\begin{aligned} \widehat{A}u_k &= u_{k+2} - 3u_{k+1} + 2u_k, \\ D(\widehat{A}) &= \{u_k \in S : u_0 = 0,\ u_1 = 2u_2\}, \end{aligned} \tag{98}$$

and $f_k = k$. It is easy to verify that $u_k^{(1)} = 1$, $u_k^{(2)} = 2^k$ are a fundamental set of solutions to homogeneous equation $u_{k+2}-3u_{k+1}+2u_k = 0$, and $u_k^{(f_k)} = -\frac{1}{2}(k^2+k)$ is a particular solution to nonhomogeneous equation $u_{k+2} - 3u_{k+1} + 2u_k = k$. Furthermore, $n = 2$, $l = 3$ and

$$\mathbf{u}_k^{(H)} = \begin{pmatrix} 1 & 2^k \end{pmatrix}, \quad U = \begin{pmatrix} 1 & 1 \\ 1 & 2 \\ 1 & 4 \end{pmatrix}, \quad \mathbf{u}_l^{(f_k)} = \begin{pmatrix} 0 \\ -1 \\ -3 \end{pmatrix}, \quad M = \begin{pmatrix} 1 & 0 & 0 \\ 0 & 1 & -2 \end{pmatrix}. \tag{99}$$

Since $\det(MU) = -5 \neq 0$, the operator $\widehat{A}$ is injective and the exact solution to the problem (97) follows by substituting (99) into (34), viz.,

$$u_k = \widehat{A}^{-1}k = 2^k - \frac{1}{2}\left(k^2 + k\right) - 1. \tag{100}$$

Problem 2 Find the solution to the succeeding third-order difference equation subject to nonlocal nonhomogeneous conditions

$$\begin{aligned} &u_{k+3} - 7u_{k+2} + 16u_{k+1} - 12u_k = k2^k, \\ &u_0 - u_1 = 3, \quad u_0 + u_1 - 3u_2 = \frac{1}{3}, \quad 2u_1 - u_2 = \frac{4}{3}. \end{aligned} \tag{101}$$

Notice that $l = n = 3$ and hence Theorem 4 and in particular Remark 2 are pertinent. We formulate the problem as in Eq. (46) where we define the operator

$$\begin{aligned} Au_k &= u_{k+3} - 7u_{k+2} + 16u_{k+1} - 12u_k, \\ D(A) &= \left\{u_k \in S : u_0 - u_1 = 3,\ u_0 + u_1 - 3u_2 = \frac{1}{3},\ 2u_1 - u_2 = \frac{4}{3}\right\}, \end{aligned} \tag{102}$$

and $f_k = k2^k$. It can be easily shown that $u_k^{(1)} = 2^k$, $u_k^{(2)} = k2^k$, $u_k^{(3)} = 3^k$ constitute a fundamental set of solutions to the homogeneous equation $u_{k+3} -$

$7u_{k+2} + 16u_{k+1} - 12u_k = 0$, whereas $u_k^{(f_k)} = -\frac{1}{24}(3+k)k^2 2^k$ is a particular solution to nonhomogeneous equation $u_{k+3} - 7u_{k+2} + 16u_{k+1} - 12u_k = k2^k$. Thus, we have

$$\mathbf{u}_k^{(H)} = \begin{pmatrix} 2^k & k2^k & 3^k \end{pmatrix}, \quad C_0 = \begin{pmatrix} 1 & 0 & 1 \\ 2 & 2 & 3 \\ 4 & 8 & 9 \end{pmatrix}, \quad \mathbf{u}_l^{(f_k)} = -\begin{pmatrix} 0 \\ \frac{1}{3} \\ \frac{10}{3} \end{pmatrix}, \tag{103}$$

and

$$M = \begin{pmatrix} 1 & -1 & 0 \\ 1 & 1 & -3 \\ 0 & 2 & -1 \end{pmatrix} \quad \mathbf{b} = \begin{pmatrix} 3 \\ \frac{1}{3} \\ \frac{4}{3} \end{pmatrix}. \tag{104}$$

Since $\det M = 4 \neq 0$, it follows from Remark 2 that the operator A in (102) is injective and the unique solution to the problem (101) is obtained by substituting (103), (104) into (60), specifically,

$$u_k = 16 \cdot 3^k - \frac{1}{24}\left(k^3 + 3k^2 + 280k + 272\right)2^k. \tag{105}$$

Problem 3 Solve the following second-order loaded difference equation along with initial homogeneous conditions

$$u_{k+2} - 2u_{k+1} + u_k - ku_7 = 2^k, \quad u_0 = u_1 = 0. \tag{106}$$

We put the problem (106) into the operator form (66) in Theorem 5 where $n = 2$ and the operator $\widehat{B}$ is defined as a perturbation of the operator

$$\begin{aligned} \widehat{A}u_k &= u_{k+2} - 2u_{k+1} + u_k, \\ D(\widehat{A}) &= \{u_k \in S : u_0 = 0,\ u_1 = 0\}, \end{aligned} \tag{107}$$

and

$$\psi(u_k) = u_7, \quad g_k = k, \quad f_k = 2^k. \tag{108}$$

By Theorem 1 it is implied that the operator $\widehat{A}$ is injective and that the solution to the problems $\widehat{A}u_k = f_k$ and $\widehat{A}u_k = g_k$ are

$$\begin{aligned} \widehat{A}^{-1}f_k &= 2^k - \begin{pmatrix} 1 & k \end{pmatrix}\begin{pmatrix} 1 & 0 \\ -1 & 1 \end{pmatrix}\begin{pmatrix} 1 \\ 2 \end{pmatrix} = 2^k - k - 1, \\ \widehat{A}^{-1}g_k &= \frac{1}{6}k^3 - \frac{1}{2}k^2 - \begin{pmatrix} 1 & k \end{pmatrix}\begin{pmatrix} 1 & 0 \\ -1 & 1 \end{pmatrix}\begin{pmatrix} 0 \\ -\frac{1}{3} \end{pmatrix} = \frac{1}{6}k(k-1)(k-2), \end{aligned} \tag{109}$$

respectively. From (108) and (109), we get

$$\psi(\widehat{A}^{-1} f_k) = \widehat{A}^{-1} f_7 = \widehat{A}^{-1} 2^7 = 2^7 - 7 - 1 = 120,$$
$$\psi(\widehat{A}^{-1} g_k) = \widehat{A}^{-1} g_7 = \widehat{A}^{-1} 7 = \frac{1}{6} 7(7-1)(7-2) = 35, \tag{110}$$

Hence, $\det W = \left[I_m - \psi(\widehat{A}^{-1} g_k)\right] = -34 \neq 0$. Then, from Theorem 5 follows that the operator $\widehat{B}$ is injective, while substitution of (109) and (110) into (68) yields the solution to the problem (106), viz.,

$$u_k = 2^k - \frac{1}{17}(10k^3 - 30k^2 + 37k + 17). \tag{111}$$

Problem 4 Find the solution to the loaded nonlocal boundary problem incorporating nonhomogeneous conditions

$$u_{k+2} - 4u_{k+1} + 3u_k - k u_3 - 2^k u_5 = 2^{k+1} + \frac{1}{2} k + 6,$$
$$u_0 + 2u_1 - u_2 = 1, \quad 2u_0 - u_1 + u_2 = -1. \tag{112}$$

We begin by converting the problem (112) to a corresponding one with homogeneous conditions by using Theorem 7. Observe that $n = 2$ and $l = 3$. Hence, we take $\mathbf{p}_k^{n-1} = (1 \;\, k)$ and create the 3×2 generalized Casorati matrix P and the nonlocal conditions matrix M as follows:

$$P = \begin{pmatrix} 1 & 0 \\ 1 & 1 \\ 1 & 2 \end{pmatrix}, \quad M = \begin{pmatrix} 1 & 2 & -1 \\ 2 & -1 & 1 \end{pmatrix}. \tag{113}$$

Since $\det(MP) = 2 \neq 0$, from (89), we get

$$u_k = v_k - 2k + \frac{1}{2}. \tag{114}$$

Hence, the problem (112) can be reduced to a loaded nonlocal boundary problem having homogeneous conditions which can be put in an operator form as in (90), namely,

$$\widehat{B} v_k = \widehat{A} v_k - \sum_{s=1}^{m} g_k^s \psi_s(v_k) = \widehat{A} v_k - \mathbf{g}_k \Psi(v_k) = \varphi_k,$$
$$D(\widehat{B}) = D(\widehat{A}) \tag{115}$$

where the operator $\widehat{A}$ is defined as

$$\widehat{A}v_k = v_{k+2} - 4v_{k+1} + 3v_k,$$
$$D(\widehat{A}) = \{v_k \in S : v_0 + 2v_1 - v_2 = 0,\ 2v_0 - v_1 + v_2 = 0\}, \tag{116}$$

and

$$\Psi(v_k) = \begin{pmatrix} \psi_1(v_k) \\ \psi_2(v_k) \end{pmatrix} = \begin{pmatrix} v_3 \\ v_5 \end{pmatrix},$$
$$\mathbf{g}_k = \begin{pmatrix} g_k^1 & g_k^2 \end{pmatrix} = \begin{pmatrix} k & 2^k \end{pmatrix},$$
$$\varphi_k = -15 \cdot 2^{k-1} - 5k + 2, \tag{117}$$

in which φ_k is obtained by (91). Notice that the operators $\widehat{A}$, $\widehat{B}$ are linear, ψ_1, ψ_2 are linear functionals, and both sets $\{\psi_1, \psi_2\}$ and $\{g_k^1, g_k^2\}$ are linearly independent.

Next, we solve the problem $\widehat{A}v_k = \varphi_k$ according to Theorem 3. It is easy to show that $v_k^1 = 1$, $v_k^2 = 3^k$ constitute a set of fundamental solutions to the homogeneous difference equation $v_{k+2} - 4v_{k+1} + 3v_k = 0$ and that $v_k^{(\varphi_k)} = 15 \cdot 2^{k-1} + \frac{k}{4}(5k-4)$ is a particular solution to the nonhomogeneous difference equations $v_{k+2} - 4v_{k+1} + 3v_k = \varphi_k$. By noticing that $n = 2$ and $l = 3$, form the matrices

$$\mathbf{v}_k^{(H)} = \begin{pmatrix} 1 & 3^k \end{pmatrix}, \quad U = \begin{pmatrix} 1 & 1 \\ 1 & 3 \\ 1 & 9 \end{pmatrix}, \quad M = \begin{pmatrix} 1 & 2 & -1 \\ 2 & -1 & 1 \end{pmatrix} \tag{118}$$

and $\mathbf{v}_l^{(\varphi_k)} = \text{col}\left(\frac{15}{2}\ \frac{61}{4}\ 33\right)$. The $\det(MU) = 20 \neq 0$ and hence $\widehat{A}$ is injective, and the unique solution to the problem $\widehat{A}v_k = \varphi_k$ is given by (34), viz.,

$$\widehat{A}^{-1}\varphi_k = -\frac{1}{40}\left(111 \cdot 3^k - 300 \cdot 2^k - 50k^2 + 40k + 211\right). \tag{119}$$

We can now apply Theorem 5. By recognizing that $v_k^{(g_k^1)} = -\frac{1}{4}k^2$ is a particular solution to nonhomogeneous difference equation $v_{k+2} - 4v_{k+1} + 3v_k = g_k^1$ and hence $\mathbf{v}_l^{(g_k^1)} = \text{col}\left(0\ -\frac{1}{4}\ -1\right)$, we obtain the unique solution to the problem $\widehat{A}v_k = g_k^1$ as

$$\widehat{A}^{-1}g_k^1 = \frac{1}{8}\left(3^k - 2k^2 - 1\right). \tag{120}$$

Similarly, by seeing that $v_k^{(g_k^2)} = -2^k$ is a particular solution to nonhomogeneous difference equation $v_{k+2} - 4v_{k+1} + 3v_k = g_k^2$ and thus $\mathbf{v}_l^{(g_k^2)} = \text{col}\,(-1\ -2\ -4)$, the unique solution to the problem $\widehat{A}v_k = g_k^2$ is given by

$$\widehat{A}^{-1} g_k^2 = \frac{1}{10}\left(3^{k+1} - 10 \cdot 2^k + 8\right). \tag{121}$$

Further,

$$\begin{aligned}\widehat{A}^{-1}\mathbf{g}_k &= \left(\widehat{A}^{-1} g_k^1 \;\; \widehat{A}^{-1} g_k^2\right) \\ &= \left(\tfrac{1}{8}\left(3^k - 2k^2 - 1\right) \;\; \tfrac{1}{10}\left(3^{k+1} - 10 \cdot 2^k + 8\right)\right),\end{aligned} \tag{122}$$

and

$$\begin{aligned}\Psi(\widehat{A}^{-1}\varphi_k) &= \begin{pmatrix} \psi_1(\widehat{A}^{-1}\varphi_k) \\ \psi_2(\widehat{A}^{-1}\varphi_k) \end{pmatrix} = \begin{pmatrix} -\frac{239}{20} \\ -\frac{8267}{20} \end{pmatrix}, \\ \Psi(\widehat{A}^{-1}\mathbf{g}_k) &= \begin{pmatrix} \psi_1(\widehat{A}^{-1}g_k^1) & \psi_1(\widehat{A}^{-1}g_k^2) \\ \psi_2(\widehat{A}^{-1}g_k^1) & \psi_2(\widehat{A}^{-1}g_k^2) \end{pmatrix} = \begin{pmatrix} 1 & \frac{9}{10} \\ 24 & \frac{417}{10} \end{pmatrix}.\end{aligned} \tag{123}$$

The det $W = \det\left[I_m - \Psi(\widehat{A}^{-1}\mathbf{g}_k)\right] = -\frac{108}{5} \neq 0$, and therefore $\widehat{B}$ is injective and the unique solution to the problem (115) is obtained by substituting (119), (122) and (123) into (68), specifically

$$v_k = \frac{1}{3456}\left(1889 \cdot 3^k - 19968 \cdot 2^k + 8894k^2 - 3456k + 20767\right). \tag{124}$$

By means of (114), we get

$$u_k = \frac{1}{3456}\left(1889 \cdot 3^k - 19968 \cdot 2^k + 8894k^2 - 10368k + 22495\right), \tag{125}$$

which is the unique solution to the given problem (112).

References

1. Abdullaev, V.M., Aida-Zade, K.R.: Numerical method of solution to loaded nonlocal boundary value problems for ordinary differential equations. Comput. Math. Math. Phys. **54**(7), 1096–1109 (2014)
2. Abdullaev, V.M., Aida-Zade, K.R.: Finite-difference methods for solving loaded parabolic equations. Comput. Math. Math. Phys. **56**(1), 93–105 (2016)
3. Agarwal, R.P., ORegan, D.: Boundary value problems for discrete equations. Appl. Math. Lett. **10**, 83–89 (1997)
4. Alikhanov, A.A., Berezgov, A.M., Shkhanukov-Lafishev, M.X.: Boundary value problems for certain classes of loaded differential equations and solving them by finite difference methods. Comput. Math. Math. Phys. **48**, 1581–1590 (2008)

5. Bondarev, E.A., Voevodin, A.F.: A finite-difference method for solving initial-boundary value problems for loaded differential and integro-differential equations. Differ. Equ. **36**(11), 1711–1714 (2000). https://doi.org/10.1007/BF02757374
6. Dzhenaliev, M.T., Ramazanov, M.I.: On a boundary value problem for a spectrally loaded heat operator: I. Differ. Equ. **43**(4), 513–524 (2007). https://doi.org/10.1134/S0012266107040106
7. Elaydi, S.: An Introduction to Difference Equations. Springer, New York (2005)
8. Eloe, P.W.: Difference equations and multipoint boundary value problems. Proc. Am. Math. Soc. **86**, 253–259 (1982)
9. Kozhanov, A.I.: A nonlinear loaded parabolic equation and a related inverse problem. Math. Notes **76**(5), 784–795 (2004). https://doi.org/10.1023/B:MATN.0000049678.16540.a5
10. Mickens, R.E.: Difference Equations: Theory, Applications and Advanced Topics. CRC Press, Boca Raton (2015)
11. Nakhushev, A.M.: The Darboux problem for a certain degenerate second order loaded integrodifferential equation. Differ. Uravn. **12**(1), 103–108 (1976). (in Russian)
12. Nakhushev, A.M.: Loaded equations and their applications. Differ. Uravn. **19**(1), 86–94 (1983). (in Russian)
13. Nakhushev, A.M.: Equations of Mathematical Biology. Vysshaya Shkola, Moscow (1995). (in Russian)
14. Nakhushev, A.M., Borisov, V.N.: Boundary value problems for loaded parabolic equations and their applications to the prediction of ground water level. Differ. Uravn. **13**(1), 105–110 (1977). (in Russian)
15. Parasidis, I.N., Providas, E.: Extension operator method for the exact solution of integro-differential equations. In: Pardalos, P., Rassias, T. (eds.) Contributions in Mathematics and Engineering: In Honor of Constantin Carathéodory, pp. 473–496. Springer, Cham (2016)
16. Paukštaitė, G., Štikonas, A.: Nullity of the second order discrete problem with nonlocal multipoint boundary conditions. Proc. Lithuanian Math. Soc. Ser. A **56**, 72–77 (2015). https://doi.org/10.15388/LMR.A.2015.13
17. Rodríguez, J., Abernathy, K.K.: Nonlocal boundary value problems for discrete systems. J. Math. Anal. Appl. **385**, 49–59 (2012). https://doi.org/10.1016/j.jmaa.2011.06.028
18. Shkhanukov-Lafishev, M.K.: Locally one-dimensional scheme for a loaded heat equation with Robin boundary conditions. Comput. Math. Math. Phys. **49**(7), 1167–1174 (2009). https://doi.org/10.1134/S0965542509070094
19. Wiener, J., Debnath, L.: Partial differential equations with piecewise constant delay. Int. J. Math. Math. Sci. **14**, 485–496 (1991)
20. Zhuravleva, E.N., Karabut, E.A.: Loaded complex equations in the jet collision problem. Comput. Math. Math. Phys. **51**(5), 876–894 (2011). https://doi.org/10.1134/S0965542511050186

Cauchy's Functional Equation, Schur's Lemma, One-Dimensional Special Relativity, and Möbius's Functional Equation

Teerapong Suksumran

Abstract This article explores a remarkable connection between Cauchy's functional equation, Schur's lemma in representation theory, the one-dimensional relativistic velocities in special relativity, and Möbius's functional equation. Möbius's exponential equation is a functional equation defined by

$$f(a \oplus_M b) = f(a)f(b),$$

where $\oplus_M$ is Möbius addition given by $a \oplus_M b = \dfrac{a+b}{1+\bar{a}b}$ for all complex numbers a and b of modulus less than one, and the product $f(a)f(b)$ is taken in the field of complex numbers. We indicate that, in some sense, Möbius's exponential equation is an extension of Cauchy's exponential equation. We also exhibit a one-to-one correspondence between the irreducible linear representations of an abelian group on a complex vector space and the solutions of Cauchy's exponential equation and extend this to the case of Möbius's exponential equation. We then give the complete family of Borel measurable solutions to Cauchy's exponential equation with domain as the group of one-dimensional relativistic velocities under the restriction of Möbius addition.

Cauchy's Functional Equations

Recall that Cauchy's exponential equation is a functional equation of the form

$$f(x+y) = f(x)f(y). \tag{CE}$$

T. Suksumran (✉)
Center of Excellence in Mathematics and Applied Mathematics, Department of Mathematics, Faculty of Science, Chiang Mai University, Chiang Mai, Thailand
e-mail: teerapong.suksumran@cmu.ac.th

N. J. Daras, Th. M. Rassias (eds.), *Modern Discrete Mathematics and Analysis*, Springer Optimization and Its Applications 131,
https://doi.org/10.1007/978-3-319-74325-7_20

In the present article, we will primarily be interested in a solution f of (CE) from an *abelian* group to the complex numbers, $\mathbb{C}$. In this section, we indicate how solutions of Cauchy's exponential equation are connected to irreducible linear representations of an abelian group on a complex vector space.

Recall that a linear representation of a group G on a *complex* vector space V is simply a group homomorphism from G into the general linear group $\mathrm{GL}(V)$, which consists of the invertible linear transformations on V. In other words, a map $\pi: G \to \mathrm{GL}(V)$ is a linear representation of G on V if

$$\pi(gh) = \pi(g) \circ \pi(h) \tag{1}$$

for all $g, h \in G$, where $\circ$ denotes composition of functions. Let π be a linear representation of G on V. A subspace W of V is said to be *invariant* if $\pi(g)(W) \subseteq W$ for all $g \in G$. A linear representation (V, π) is said to be *irreducible* if the only invariant subspaces of V are $\{0\}$ and V itself. Roughly speaking, irreducible linear representations serve as building blocks for other linear representations.

Let G be an *abelian* group. As a consequence of Schur's lemma in representation theory, if (V, π) is an irreducible linear representation of G over $\mathbb{C}$, then for each $g \in G$, there is a complex scalar λ for which $\pi(g) = \lambda I_V$, that is, $\pi(g)v = \lambda v$ for all $v \in V$. It follows that every irreducible linear representation of G over $\mathbb{C}$ is one dimensional. Hence, one can represent G on $\mathbb{C}$, which is made into a one-dimensional vector space over itself. This results in a one-to-one correspondence between the irreducible linear representations of G on $\mathbb{C}$ and the solutions of Cauchy's exponential equation. More precisely, suppose that π is an irreducible linear representation of G on $\mathbb{C}$. As noted earlier, for each $g \in G$, there exists a unique complex scalar λ_g such that

$$\pi(g)z = \lambda_g z \qquad \text{for all } z \in \mathbb{C}.$$

Since $\pi(g + h) = \pi(g) \circ \pi(h)$ for all $g, h \in G$, it follows that $\lambda_{g+h} = \lambda_g \lambda_h$ for all $g, h \in G$. Define a function f by

$$f(g) = \lambda_g, \qquad g \in G.$$

Then f is a function from G into the set of nonzero complex numbers, denoted by $\mathbb{C}^\times$, that satisfies

$$f(g + h) = f(g)f(h)$$

for all $g, h \in G$. This shows f is a solution of Cauchy's exponential equation (CE). Conversely, suppose that $f: G \to \mathbb{C}^\times$ is a solution of (CE). For each $g \in G$, define

$$\pi(g)z = f(g)z, \qquad z \in \mathbb{C}.$$

It is easy to see that $\pi(g)$ is an invertible linear transformation on $\mathbb{C}$ for all $g \in G$ (this is because $f(g) \neq 0$). Define π by $g \mapsto \pi(g)$, $g \in G$. Then $\pi(g+h) = \pi(g) \circ \pi(h)$ for all $g, h \in G$ because $f(g+h) = f(g)f(h)$ for all $g, h \in G$. Hence, π defines a linear representation of G on $\mathbb{C}$. Since $\dim \mathbb{C}_{\mathbb{C}} = 1$, π must be irreducible. Therefore, we obtain the following one-to-one correspondence:

$$\{\text{irreducible representations of } G \text{ on } \mathbb{C}\} \quad \longleftrightarrow \quad \{f\colon G \to \mathbb{C}^{\times} \text{ satisfying (CE)}\},$$

and summarize this in the following theorem.

Theorem 1 *Let G be an abelian group. Then there is a one-to-one correspondence between the family of irreducible linear representations of G on $\mathbb{C}$ and the family of solutions $f\colon G \to \mathbb{C}^{\times}$ to Cauchy's exponential equation* (CE).

In the next section, we are going to apply Theorem 1 to a particular example of an abelian group, namely, the group of one-dimensional relativistic velocities.

One-Dimensional Special Relativity

In one-dimensional special relativity, where the speed of light c in vacuum is normalized to be 1, *relativistic* velocities u and v (which means $-1 < u, v < 1$) do not add in the same way as for ordinary velocities. Rather, they add according to the rule

$$u \oplus v = \frac{u+v}{1+uv}, \tag{2}$$

where the symbol + on the right-hand side of Eq. (2) is ordinary addition of real numbers; see, for instance, [5, p. 36]. From a mathematical point of view, the set of one-dimensional relativistic velocities

$$\{v \in \mathbb{R}\colon |v| < 1\} = (-1, 1)$$

forms an abelian group under addition given by (2). According to Theorem 1, it is possible to determine all the irreducible linear representations of the one-dimensional relativistic velocities by determining the solutions of Cauchy's exponential equation, which in turn by determining the solutions of Cauchy's additive equation:

$$f(x+y) = f(x) + f(y). \tag{CA}$$

First, let us mention the general solutions of Cauchy's exponential equation (CE). By Theorem 1 on p. 54 of [1], the nontrivial general solutions $L\colon \mathbb{R} \to \mathbb{C}$ of (CE) are given by

$$L(x) = \exp\left(A(x) + iB(x)\right), \qquad x \in \mathbb{R}, \tag{3}$$

where $A: \mathbb{R} \to \mathbb{R}$ is an arbitrary solution of Cauchy's additive equation (CA) and $B: \mathbb{R} \to \mathbb{R}$ is an arbitrary solution of

$$B(x+y) \equiv B(x) + B(y) \pmod{2\pi}, \qquad x, y \in \mathbb{R}. \tag{4}$$

In the case when Lebesgue measurability is assumed, the solutions of (CE) reduce to

$$L(x) = e^{\lambda x}, \qquad x \in \mathbb{R}, \tag{5}$$

for some complex constant λ (Aczél and Dhombres, [1, Theorem 4 on p. 56]). For simplicity, we will limit the solutions of (CE) with domain as the group of one-dimensional relativistic velocities to the *Borel* measurable ones (it will become clear why Borel measurability is assumed instead of Lebesgue measurability). This will lead us to the family of irreducible linear representations of the one-dimensional relativistic velocities that corresponds to the family of Borel measurable solutions to the functional equation:

$$f(u \oplus v) = f(u)f(v), \qquad u, v \in (-1, 1). \tag{6}$$

Here, f is a function from $(-1, 1)$ to $\mathbb{C}$.

In order to accomplish our goal, we will impose an appropriate σ-algebra on the one-dimensional relativistic velocities. Let $\mathcal{B}$ and $\mathcal{B}_{\mathbb{C}}$ be the Borel σ-algebras on $\mathbb{R}$ and on $\mathbb{C}$, respectively. Set $G = (-1, 1)$. Then G is made into a measurable space by introducing the σ-algebra induced by $\mathcal{B}$. More precisely, G is equipped with the σ-algebra:

$$\mathcal{M} = \{G \cap E : E \in \mathcal{B}\}. \tag{7}$$

Unless explicitly mentioned, G will be equipped with the σ-algebra given by (7). When $f: (G, \mathcal{M}) \to (\mathbb{C}, \mathcal{B}_{\mathbb{C}})$ is a measurable function, it is customary to abuse the terminology and say that $f: G \to \mathbb{C}$ is "Borel" measurable. Denote by tanh the *hyperbolic tangent* function on $\mathbb{R}$. Recall that tanh is a bijective continuous function from $\mathbb{R}$ to G. The following theorem is a standard result in measure theory.

Theorem 2 *If $f: G \to \mathbb{C}$ is a Borel measurable function, then $L = f \circ \tanh: \mathbb{R} \to \mathbb{C}$ is a Borel measurable function; in particular, L is Lebesgue measurable.*

We are now in a position to determine the nontrivial solutions of (6) that are Borel measurable. Suppose that $f: G \to \mathbb{C}$ is Borel measurable and is a solution of (6). Define $L = f \circ \tanh$. By Theorem 2, L is Lebesgue measurable. Furthermore, L is a solution of (CE). In fact, suppose that $x, y \in \mathbb{R}$. As $\tanh x$ and $\tanh y$ belong to G, one obtains by inspection that

$$
\begin{aligned}
L(x+y) &= f(\tanh(x+y)) \\
&= f\left(\frac{\tanh x + \tanh y}{1 + (\tanh x)(\tanh y)}\right) \\
&= f(\tanh x \oplus \tanh y) \\
&= f(\tanh x) f(\tanh y) \\
&= L(x)L(y).
\end{aligned}
$$

As mentioned at the beginning of this section, there exists a complex constant λ for which

$$
L(x) = e^{\lambda x} \qquad \text{for all } x \in \mathbb{R}.
$$

From this it follows that

$$
f(v) = e^{\lambda \tanh^{-1} v} \qquad \text{for all } v \in G. \tag{8}
$$

To see this, let $v \in G$. Then $v = \tanh x$ for some $x \in \mathbb{R}$ since tanh is surjective and so

$$
f(v) = f(\tanh x) = L(x) = e^{\lambda x} = e^{\lambda \tanh^{-1} v}
$$

for tanh is one-to-one. Conversely, suppose that f is of the form (8). Let $u, v \in G$. Then $u = \tanh x$ and $v = \tanh y$ for some $x, y \in \mathbb{R}$. Direct computation shows that

$$
\begin{aligned}
f(u \oplus v) &= e^{\lambda \tanh^{-1}(u \oplus v)} \\
&= \exp\left(\lambda \tanh^{-1}\left(\frac{u+v}{1+uv}\right)\right) \\
&= \exp\left(\lambda \tanh^{-1}\left(\frac{\tanh x + \tanh y}{1 + (\tanh x)(\tanh y)}\right)\right) \\
&= e^{\lambda \tanh^{-1}(\tanh(x+y))} \\
&= e^{\lambda(x+y)} \\
&= e^{\lambda x} e^{\lambda y} \\
&= e^{\lambda \tanh^{-1} u} e^{\lambda \tanh^{-1} v} \\
&= f(u) f(v).
\end{aligned}
$$

Thus, f is indeed a solution of (6). In summary, we have proved the following theorem:

Theorem 3 *The family of Borel measurable solutions to Cauchy's exponential equation* (6) *with domain G is*

$$\{v \mapsto e^{\lambda \tanh^{-1} v} : \lambda \in \mathbb{C}\},$$

where λ is a parameter for the family.

Möbius's Functional Equations

There is a natural extension of one-dimensional relativistic velocity addition described by (2) to the open unit disk of the complex plane. Let $\mathbb{D}$ be the complex open unit disk, that is,

$$\mathbb{D} = \{z \in \mathbb{C} \colon |z| < 1\}.$$

Möbius addition [4] is defined on $\mathbb{D}$ by

$$a \oplus_M b = \frac{a+b}{1+\bar{a}b}, \qquad a, b \in \mathbb{D}, \tag{9}$$

where the bar denotes complex conjugation. Unfortunately, the system $(\mathbb{D}, \oplus_M)$ does not form a group because $\oplus_M$ is not associative. Nevertheless, $(\mathbb{D}, \oplus_M)$ is rich in structure and possesses many algebraic properties like groups, called by Ungar a *gyrocommutative gyrogroup* [7] and by others a *Bruck loop* [2] or *K-loop* [3]. The system $(\mathbb{D}, \oplus_M)$ is known as the *Möbius* gyrogroup and has been studied extensively. For instance, the authors of [6] develop the elementary theory of linear representations of a gyrogroup, following the classical study of linear representations of a group. They also determine the irreducible linear representations of $(\mathbb{D}, \oplus_M)$ on a complex vector space, using a gyrogroup version of Schur's lemma.

Roughly speaking, a gyrogroup consists of a nonempty set G and a binary operation $\oplus$ on G that comes with a family of automorphisms of $(G, \oplus)$, called *gyroautomorphisms* $\mathrm{gyr}[a, b]$, $a, b \in G$, in which an identity exists and each element has an inverse. In general, gyrogroups fail to satisfy the associative law. Rather, they satisfy a weak form of associativity:

$$a \oplus (b \oplus c) = (a \oplus b) \oplus \mathrm{gyr}[a, b]c \qquad \text{(left gyroassociative law)}$$

$$(a \oplus b) \oplus c = a \oplus (b \oplus \mathrm{gyr}[b, a]c), \qquad \text{(right gyroassociative law)}$$

called the *gyroassociative* law. In addition, any gyrogroup satisfies the *loop* property:

$$\text{gyr}[a \oplus b, b] = \text{gyr}[a, b] \qquad \text{(left loop property)}$$

$$\text{gyr}[a, b \oplus a] = \text{gyr}[a, b]. \qquad \text{(right loop property)}$$

The loop property turns out to be extremely useful, forcing rigid structure on a gyrogroup. It also gives a strong connection between gyrogroups and loops [3, Theorem 6.4]. As an example, the lack of associativity in $(\mathbb{D}, \oplus_M)$ is repaired by the gyroautomorphisms as follows:

$$a \oplus_M (b \oplus_M c) = (a \oplus_M b) \oplus_M \text{gyr}[a, b]c$$

$$(a \oplus_M b) \oplus_M c = a \oplus_M (b \oplus_M \text{gyr}[b, a]c),$$

where $\text{gyr}[a, b]$ is an automorphism of $(\mathbb{D}, \oplus_M)$ given by

$$\text{gyr}[a, b]z = \frac{1 + a\bar{b}}{1 + \bar{a}b}z$$

for all $a, b, z \in \mathbb{D}$. Since $\dfrac{1 + a\bar{b}}{1 + \bar{a}b}$ is a complex number of modulus one, $\text{gyr}[a, b]$ represents a rotation of $\mathbb{D}$. This justifies the use of the prefix "gyro."

A natural generalization of the functional equation (6) is, of course, the functional equation of the form

$$f(a \oplus_M b) = f(a)f(b). \qquad \text{(ME)}$$

Here, f is a function from $\mathbb{D}$ into $\mathbb{C}$. A functional equation of this type is called *Möbius's exponential equation* and was studied in some detail in [6]. Following the same steps as in section "Cauchy's Functional Equations", one can show that if $f: \mathbb{D} \to \mathbb{C}$ satisfies (ME), then the composite $g = f \circ \tanh: \mathbb{R} \to \mathbb{C}$ satisfies Cauchy's exponential equation. This combined with the assumption of f being Borel measurable forces either $f = 0$ or there is a complex constant λ such that

$$f(r) = e^{\lambda \tanh^{-1} r} \qquad (10)$$

for all $r \in \mathbb{R}$ with $-1 < r < 1$ [6, Theorem 1.2]. Equation (10) gives us part of the Borel measurable solutions to Möbius's exponential equation on $(-1, 1)$. However, it is extremely difficult to extend the solutions on $(-1, 1)$ to the whole disk $\mathbb{D}$. Moreover, Möbius's exponential equation does not have nontrivial *holomorphic* solutions, as proved in [6]. This is because the identity $\tanh^{-1}(a \oplus_M b) = \tanh^{-1} a + \tanh^{-1} b$ is missing when the hyperbolic tangent function is extended to the complex plane.

It is still true that the nonzero solutions of Möbius's exponential equation (ME) and the irreducible linear representations of $(\mathbb{D}, \oplus_M)$ on a complex vector space are in bijective correspondence. In fact, as a consequence of a gyrogroup version of Schur's lemma [6, Theorem 3.13], we can represent $(\mathbb{D}, \oplus_M)$ on $\mathbb{C}$, which is viewed as a complex vector space over itself, and obtain the following one-to-one correspondence:

$$\{\text{nonzero irreducible representations of } \mathbb{D} \text{ on } \mathbb{C}\} \quad \longleftrightarrow \quad \operatorname{Hom}_M(\mathbb{D}, \mathbb{C}^\times),$$

where $\operatorname{Hom}_M(\mathbb{D}, \mathbb{C}^\times) = \{f : \mathbb{D} \to \mathbb{C}^\times \text{ satisfying (ME)}\}$. Hence, knowing the solutions of (ME) amounts to knowing the building blocks for linear representations of $(\mathbb{D}, \oplus_M)$ on a complex vector space.

There is another type of functional equation involving Möbius addition, namely,

$$f(a \oplus_M b) = f(a) + f(b), \tag{MA}$$

called *Möbius's additive equation.* As usual, if $f : \mathbb{D} \to \mathbb{C}$ is a solution of (MA), then $F(z) = e^{f(z)}$, $z \in \mathbb{D}$, is a solution of (ME). We end this section with the remark that concrete solutions of (ME) and (MA) are unknown.

Acknowledgements The author would like to thank Keng Wiboonton for his useful comments. This research was supported by Chiang Mai University.

References

1. Aczél, J., Dhombres, J.: Functional Equations in Several Variables. Encyclopedia of Mathematics and its Applications, vol. 31. Cambridge University Press, Cambridge (1989)
2. Baumeister, B., Stein, A.: The finite Bruck loops. J. Algebra **330**, 206–220 (2011)
3. Kiechle, H.: Theory of K-Loops. Lecture Notes in Mathematics, vol. 1778. Springer, Berlin (2002)
4. Kinyon, M.K., Ungar, A.A.: The gyro-structure of the complex unit disk. Math. Mag. **73**(4), 273–284 (2000)
5. Mould, R.: Basic Relativity. Springer, New York (1994)
6. Suksumran, T., Wiboonton, K.: Möbius's functional equation and Schur's lemma with applications to the complex unit disk. Aequat. Math. **91**(3), 491–503 (2017)
7. Ungar, A.A.: Analytic Hyperbolic Geometry and Albert Einstein's Special Theory of Relativity. World Scientific, Hackensack (2008)

Plane-Geometric Investigation of a Proof of the Pohlke's Fundamental Theorem of Axonometry

Thomas L. Toulias

Abstract Consider a bundle of three given coplanar line segments (radii) where only two of them are permitted to coincide. Each pair of these radii can be considered as a pair of two conjugate semidiameters of an ellipse. Thus, three concentric ellipses E_i, $i = 1, 2, 3$, are then formed. In a proof by G.A. Peschka of Karl Pohlke's fundamental theorem of axonometry, a parallel projection of a sphere onto a plane, say, $\mathbb{E}$, is adopted to show that a new concentric (to E_i) ellipse E exists, "circumscribing" all E_i, i.e., E is simultaneously tangent to all $E_i \subset \mathbb{E}$, $i = 1, 2, 3$. Motivated by the above statement, this paper investigates the problem of determining the form and properties of the circumscribing ellipse E of E_i, $i = 1, 2, 3$, exclusively from the analytic plane geometry's point of view (unlike the sphere's parallel projection that requires the adoption of a three-dimensional space). All the results are demonstrated by the actual corresponding figures as well as with the calculations given in various examples.

Introduction

This work is motivated by a work of Müller and Kruppa [4], where they suggested a proof (out of many in the literature) of Karl Pohlke's theorem, also known as the fundamental theorem of axonometry. Recalling Pohlke's theorem, we consider a bundle of three arbitrary chosen coplanar line segments, say, OP, OQ, and OR, where only one of them can be of zero length (nondegenerate segments). Under the assumption that the points $O, P, Q, R \in \mathbb{E}$ are not collinear (belonging to a plane $\mathbb{E}$), their corresponding line segments, as above, can always be considered as parallel projections of three other line segments of the three-dimensional space, say, O^*P^*, O^*Q^*, and O^*R^*, respectively, having equal length and being orthogonal with each other, i.e., $|O^*P^*| = |O^*Q^*| = |O^*R^*|$ with $\measuredangle(O^*P^*, O^*Q^*) =$

T. L. Toulias (✉)
Technological Educational Institute of Athens, Egaleo, Athens, Greece

N. J. Daras, Th. M. Rassias (eds.), *Modern Discrete Mathematics and Analysis*,
Springer Optimization and Its Applications 131,
https://doi.org/10.1007/978-3-319-74325-7_21

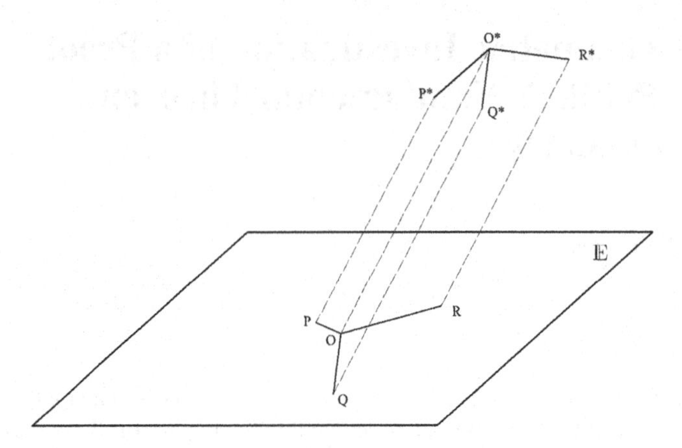

Fig. 1 Parallel projection for Pohlke's theorem

$\measuredangle(O^*Q^*, O^*R^*) = \measuredangle(O^*R^*, O^*P^*) = \pi/2$; see Fig. 1. The orthogonal projection can also be considered as a special case of the above parallel projection. For further reading about Pohlke's theorem, see also [1] and [6].

The specific method in [4] of the proof of Pohlke's theorem (which consists of finding the orthogonal segments O^*P^*, O^*Q^*, and O^*R^*) is based on the adoption of a parallel projection, say, $\mathscr{P}$, applied on an appropriate sphere S onto plane $\mathbb{E}$ (on which OP, OQ, and OR lie). The following proposition of three concentric ellipses, proven in [4, pg. 244] through the use of $\mathscr{P}$, was an intermediate step toward the proof of Pohlke's theorem.

Proposition 1 *Consider four noncollinear points O, P, Q, R on a plane, forming a bundle of three line segments OP, OQ, and OR, where only two of them are permitted to coincide. If the pairs (OP, OR), (OQ, OR), and (OP, OQ) are considered to be pairs of conjugate semidiameters defining the ellipses, say, E_1, E_2, and E_3, respectively, then a new ellipse E (concentric to E_i) exists and is tangent to all E_i, $i = 1, 2, 3$.*

The use of sphere, for the proving of the above proposition, appears for the first time in a work by J.W. v. Deschwanden and subsequently by G.A. Peschka in his elementary proof of Pohlke's fundamental theorem of axonometry; see [5]. Under this parallel projection $\mathscr{P}$, a cylindrical surface is created, tangent to sphere S and around its maximum circle, say, $k \subset S$; see Fig. 2. Hence, k is the contour of sphere S through $\mathscr{P}$, and it is parallel-projected (via $\mathscr{P}$) onto an ellipse E of plane $\mathbb{E}$, while the orthogonal line segments O^*P^*, O^*Q^*, and O^*R^* are parallel-projected onto radii OP, OQ and OR of the ellipses E_i, $i = 1, 2, 3$. Therefore, E is circumscribing all E_i, $i = 1, 2, 3$. According to the proof of Proposition 1, these ellipses are defined by three pairs of conjugate semidiameters, or conjugate radii, (OP, OR), (OQ, OR), and (OP, OQ), respectively, which are parallel projections, through $\mathscr{P}$, of the corresponding three maximum circles on sphere S. These maximum circles belong to planes (perpendicular to each other)

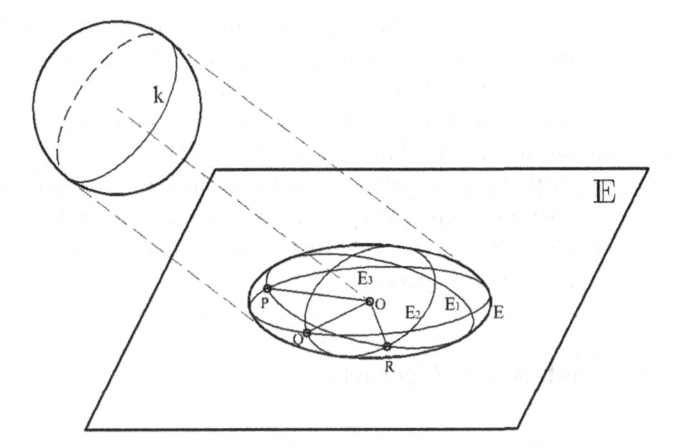

Fig. 2 Pohlke's theorem through a parallel-projected sphere onto plane $\mathbb{E}$

Fig. 3 Representation of the four ellipses problem

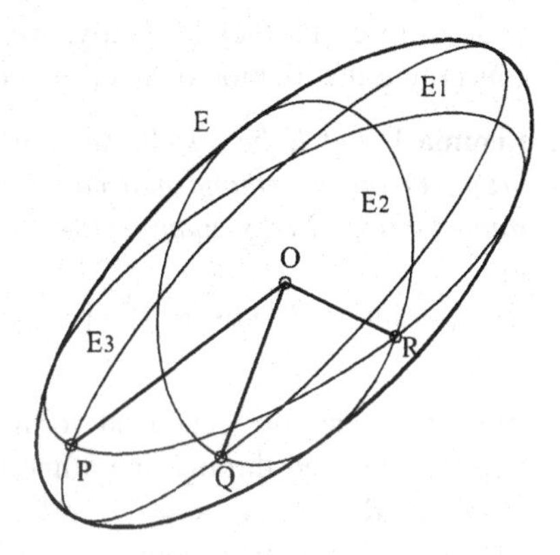

spanned, respectively, by $\{O^*P^*, O^*R^*\}$, $\{O^*Q^*, O^*R^*\}$, and $\{O^*P^*, O^*Q^*\}$. Figure 2 demonstrates the above projections (used by G.A. Peschka) for the proof of Proposition 1.

One can notice that the problem in Proposition 1 is, by its nature, a plane-geometric (two-dimensional) problem. The discrete mathematical problem of finding the E_i's "circumscribing" ellipse E, exclusively in terms of analytic plane geometry, shall be called hereafter as the *four ellipses problem*, while E shall be called as a *common tangential ellipse* (c.t.e.) of the ellipses E_i, $i = 1, 2, 3$. A visualization of the four ellipses problem is given in Fig. 3.

A construction method of the enveloping ellipse E was proposed in [3], employing synthetic projective geometry of the plane. Note that the authors in [2] studied also the discrete mathematics problem of the containment orders of concentric circumscribing (tangent) ellipses.

The present paper, in particular, provides an investigation of the four ellipses problem, i.e., the two-dimensional problem of finding a c.t.e. of three given concentric ellipses, where each one of these three ellipses is defined by two conjugate radii which belong to a given bundle of three (coplanar) nondegenerated line segments (where only two of them can coincide). Useful formulas are also provided and various examples demonstrate the results.

Affine transformations play a key role in this investigation. A study of certain affine transformations is carried out in section "Ellipses Through Affine Geometry," which helps simplify the corresponding formulations. In section "The Four Ellipses Problem," these transformations first apply for the special case where one of the ellipses E_i is assumed to be a circle, and then the investigation is generalized for the case where all E_i are ellipses in general.

Ellipses Through Affine Geometry

In this section, a certain affine transformation is studied transforming circles into ellipses, which is needed for the development of our study in section "The Four Ellipses Problem." Firstly, we consider the following preliminary Lemma concerning the rotation of an ellipse around its center.

Lemma 1 *Let E be an ellipse with principal radii $0 < b < a$, centered at the origin O of an orthonormal coordinate system. The analytical expression of the rotated ellipse E_θ by an angle $\theta \in [-\pi, \pi]$ around its center O is then given by*

$$E_\theta : \left(a^2 \sin^2\theta + b^2 \cos^2\theta\right)x^2 - \left(a^2 - b^2\right)(\sin 2\theta)xy + \left(a^2 \cos^2\theta + b^2 \sin^2\theta\right)y^2 = a^2 b^2. \tag{1}$$

Proof Consider the canonical form of $E : (x/a)^2 + (y/b)^2 = 1$. Then, the result is obtained through the affine transformation (rotation) R_θ defined by $\mathbf{x}' = R_\theta(\mathbf{x}) := \mathbf{R}_\theta \mathbf{x}$, where $\mathbf{x} := (x, y)^{\mathrm{T}} \in \mathbb{R}^{2\times 1}$ and $\mathbf{x}' := (x', y')^{\mathrm{T}} \in \mathbb{R}^{2\times 1}$ denote the coordinates' vectors of points X and X', while

$$\mathbf{R}_\theta := \begin{pmatrix} \cos\theta & -\sin\theta \\ \sin\theta & \cos\theta \end{pmatrix} \in \mathbb{R}_\perp^{2\times 2}, \tag{2}$$

is the usual rotation (orthonormal) matrix which defines the rotation $X' := R_\theta(X)$ of the Euclidean plane $\mathbb{R}^2$ by an angle θ around its origin O. Thus, applying R_θ to the matrix form of ellipse E, i.e., $E : \mathbf{a}\mathbf{x}^{\mathrm{T}} = 1$, $\mathbf{a} := (a, b) \in \mathbb{R}^2$, we obtain the rotated ellipse $E_\theta := R_\theta(E)$, expressed by (1).

A given circle can be transformed into an ellipse through an axis-invariant affine transformation which shall be called *axis shear*. Specifically, we shall call *x-shear* the axis shear that preserves the horizontal axis $x'Ox$ of a given coordinate system (c.s.), while *y-shear* shall be called the axis shear which preserves the corresponding c.s.'s vertical axis $y'Oy$. Hence, the y-shear transforms the orthonormal vector base $\mathscr{B} = \{\mathbf{e}_1, \mathbf{e}_2\}$ into $\mathscr{B}' = \{\mathbf{u}, \mathbf{e}_2\}$, while the x-shear transforms the vector base $\mathscr{B}$ into $B' = \{\mathbf{e}_1, \mathbf{u}\}$, $\mathbf{u} \in \mathbb{R}^{2\times 1} \setminus \{\mathbf{0}\}$, where $\mathbf{e}_1$ and $\mathbf{e}_2$ denote the usual orthonormal vectors $\mathbf{e}_1 := (1, 0)^{\mathrm{T}}$ and $\mathbf{e}_2 := (0, 1)^{\mathrm{T}}$ of $\mathbb{R}^2$. Note that the c.s. spanned by the vector base $\mathscr{B}$ is an orthonormal c.s. (o.s.c.). The vector $\mathbf{u} \neq \mathbf{0}$ shall be called as the *shearing vector*, while the $\mathbf{u}$'s angle (with respect to the adopted o.c.s.) $\omega := \measuredangle(\mathbf{e}_1, \mathbf{u}) \in (-\pi, \pi)$ shall often be called as the *shearing angle*. In particular, the x-shear, say, S_x, is an affine transformation defined through its matrix representation $\mathbf{x}' = S_x(\mathbf{x}) := \mathbf{N}_x\mathbf{x}$, where the transformation matrix $\mathbf{N}_x$ is given by

$$\mathbf{N}_x = (\mathbf{e}_1, \mathbf{u}) := \begin{pmatrix} 1 & u\cos\omega \\ 0 & u\sin\omega \end{pmatrix}, \tag{3}$$

with $\omega \in (-\pi, \pi)$ being the shearing angle of the x-shear S_x. Respectively, the y-shear, say, S_y, is defined by $\mathbf{x}' = S_y(\mathbf{x}) := \mathbf{N}_y\mathbf{x}$, where the transformation matrix $\mathbf{N}_y$ is given by

$$\mathbf{N}_y = (\mathbf{u}, \mathbf{e}_2) := \begin{pmatrix} u\cos\omega & 0 \\ u\sin\omega & 1 \end{pmatrix}, \tag{4}$$

with $\omega \in (-\pi, \pi)$ being the corresponding S_y's shearing angle, which is the same as in (3) (as S_y is referring to the same sharing vector $\mathbf{u}$ as S_x). It is clear that the x-shears preserve the horizontal axis, while the y-shears preserve the vertical axis, i.e., $S_k(A_k) = A_k$, $k \in \{x, y\}$, where $A_x = \{r\mathbf{e}_1\}_{r\in\mathbb{R}}$ and $A_y = \{r\mathbf{e}_2\}_{r\in\mathbb{R}}$. The x- or y-shears are therefore completely defined through a given shearing vector $\mathbf{u} \in \mathbb{R}^{2\times 1}$, i.e., by its length $u \in \mathbb{R}_+$, and its angle $\omega \in (-\pi, \pi)$ (with respect to some c.s.).

For the ellipse derived from a y-shear of a circle, we consider the following. Note that with the term *directive angle* of an ellipse, we shall refer hereafter to the angle formed by the ellipse's major axis with respect to the horizontal $x'Ox$ axis of an adopted c.s. In general, the directive angle of an ellipse with respect to a given line shall refer to the angle between the ellipse's major axis and this given line.

Lemma 2 *Let C be a circle of radius $\rho > 0$ centered at the origin O (of an o.c.s.). The y-sheared circle $S_y(C)$ corresponds to an ellipse, centered also at the origin O, with analytical expression*

$$S_y(C):\ (1+u^2\sin^2\omega)x^2 - u^2(\sin 2\omega)xy + u^2\left(\cos^2\omega\right)y^2 = \rho^2u^2\cos^2\omega, \tag{5}$$

where $u > 0$ and $\omega \in [-\pi, \pi]$ are the corresponding S_y's shearing vector length and angle. or equivalently, when $\omega \in [-\pi, \pi] \setminus \{\pm\pi/2\}$,

$$S_y(C):\ \left(1 + u^2\tan^2\omega\right)x^2 - 2u^2(\tan\omega)xy + u^2y^2 = \rho^2u^2. \tag{6}$$

The principal radii $0 < b < a$ of $S_y(C)$ are given by

$$a, b = \frac{\sqrt{2}\rho u|\cos\omega|}{\sqrt{u^2 + 1 \pm \sqrt{u^4 + 1 - 2u^2\cos 2\omega}}}, \tag{7}$$

where the minus sign corresponds to the major radius a, while the plus sign to the minor radius b. Note that the surface area of $S_y(C)$ *adopts the compact form* $A = \pi u\rho^2|\cos\omega|$*. Moreover, it holds that* $|\omega| < |\theta|$*, where* θ *is the directive angle of the ellipse* $S_y(C)$*, which is given through the* $S_y(C)$*'s major axis slope* $\tan\theta$ *by*

$$\tan\theta = \frac{u^2\sin 2\omega}{u^2\cos 2\omega - 1 + \sqrt{u^4 + 1 - 2u^2\cos 2\omega}}. \tag{8}$$

See Appendix for the proof.

Working similarly for the x-shear of the circle C, the following holds.

Lemma 3 *Let* C *be a circle of radius* $\rho > 0$ *centered at the origin* O *(of an o.c.s.). The* x*-sheared circle* $S_x(C)$ *corresponds to an ellipse centered also at the origin* O*, with analytical expression*

$$S_x(C):\ u^2\left(\sin^2\omega\right)x^2 - u^2(\sin 2\omega)xy + \left(1 + u^2\cos^2\omega\right)y^2 = \rho^2u^2\sin^2\omega, \tag{9}$$

where $u > 0$ *and* $\omega \in [-\pi,\ \pi]$ *are the corresponding shearing vector's length and angle, or equivalently, when* $\omega \in [-\pi,\ \pi] \setminus \{\pm\pi/2\}$*,*

$$S_x(C):\ u^2\left(\tan^2\omega\right)x^2 - 2u^2(\tan\omega)xy + \left(u^2 + 1\right)y^2 = \rho^2u^2\tan^2\omega. \tag{10}$$

The principal radii $0 < b < a$ *of* $S_x(C)$ *are given by*

$$a, b = \frac{\sqrt{2}\rho u|\sin\omega|}{\sqrt{u^2 + 1 \pm \sqrt{u^4 + 1 + 2u^2\cos 2\omega}}}, \tag{11}$$

where the minus sign corresponds to the major radius a, while the plus sign to the minor radius b. Note that the surface area of $S_x(C)$ *adopts the compact form* $A = \pi u\rho^2|\sin\omega|$*. Moreover, it holds that* $|\omega| > |\theta|$ *where* θ *is the directive angle of the ellipse* $S_x(C)$*, which is given, through the* $S_x(C)$*'s major axis slope* $\tan\theta$*, by*

$$\tan\theta = \frac{u^2\sin 2\omega}{1 + u^2\cos 2\omega + \sqrt{u^4 + 1 + 2u^2\cos 2\omega}}. \tag{12}$$

The proof of the above lemma is similar to the proof of Lemma 2 in the Appendix.

The following example clarifies the x- and y-shear transformations of a circle, which provides two ellipses having a common radius. It is also the basis for some other examples provided hereafter.

Example 1 Let C be a circle of radius $\rho := 10$ centered at the origin O of an o.c.s. spanned by the orthonormal vector base $\mathscr{B} = \{\mathbf{e}_1, \mathbf{e}_2\}$, and its points $P, Q \in C$, with $P(\rho, 0)$, $Q(0, \rho)$, which define C's orthogonal radii OP and OQ. These

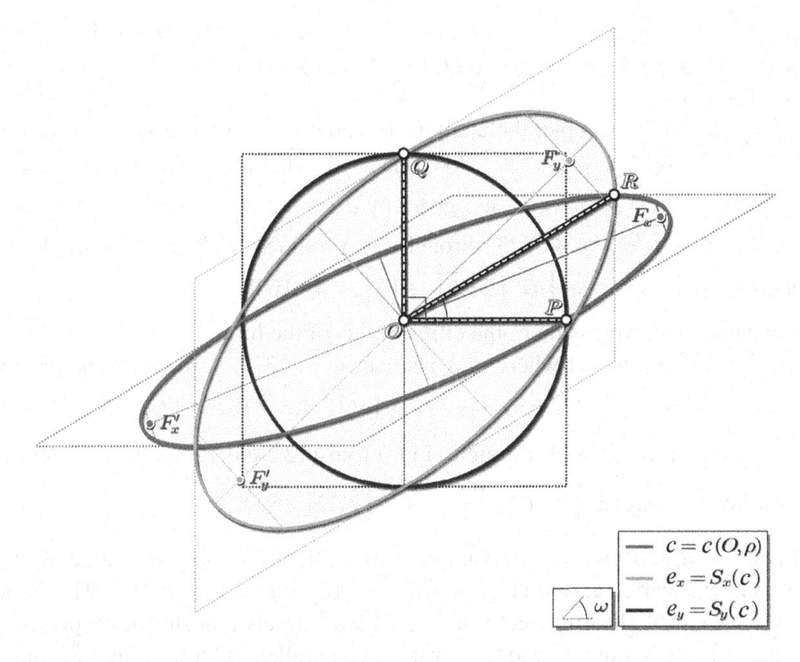

Fig. 4 Circle C and its x- and y-shears E_x and E_y

radii can be considered as C's trivially conjugate radii, with $OP \perp OQ$ and $|OP| = |OQ| = \rho = 10$. Let also a third point R with $|OR| = r := 15$ and angle $\measuredangle(\mathbf{e}_1, OR) = \omega := \pi/6$ ($= 30°$).

The y-shear transformation S_y which transforms point $P(\rho, 0) \in C$ into R, while preserving point $Q \in C$, is the one having the shearing vector $\mathbf{u} = \rho^{-1}OR$ (of length $u = |\mathbf{u}| = r/\rho = 3/2$ and angle $\omega = \pi/6$). Indeed, it can be shown through (4) that $S_y(Q) = Q$ and $S_y(P) = R$, with u and ω values as above. Therefore, the y-sheared circle corresponds to a concentric ellipse, say, $E_y := S_y(C)$, for which OR and OQ are its two conjugate radii. The conjugality of radii OP and OQ is derived from the fact that the y-shear (as well as x-shears) is, in principle, an affine transformation and, as such, it preserves the parallelism on the plane; see Fig. 4 where the circumscribing square frame of the circle C is transformed through S_y into a parallelogram circumscribing $E_y = S_y(C)$.

Working similarly, the x-shear transformation S_x which transforms $Q(\rho, 0) \in C$ into R, while preserving point $P \in C$, is the one having the same shearing vector $\mathbf{u}$ as S_y. From (3), it holds that $S_x(Q) = R$ and $S_x(P) = R$. Thus, the x-sheared circle corresponds to a concentric ellipse, say, $E_x := S_x(C)$, for which OR and OP are its two conjugate radii; see also Fig. 4.

Therefore, the given bundle of the three line segments OP, OQ, and OR corresponds to a bundle of three common (conjugate per pair) radii which define the ellipses E_x, Ey and, trivially, the circle C. In particular, we may write: $E_x = E_x(OP, OR)$, $E_y = E_y(OQ, OR)$, and $C = C(OP, OQ)$, meaning that E_x is defined by (its conjugate radii) OP and OQ, E_y is defined by QP and OR, while C is trivially defined by OP and OQ. Figure 4 provides again a clarification of the above discussion.

For the specific example, the analytical expression of ellipse E_y is of the form $E_y\colon\ 25x^2 - 18\sqrt{3}xy + 27y^2 = 2700$, due to (5) where it was set $u := 3/2$, while its principal radii $0 < b_y < a_y$ are given by $a_y = \frac{5}{2}\sqrt{26 + 2\sqrt{61}} \approx 16.1285$ and $b_y = \frac{5}{2}\sqrt{26 - 2\sqrt{61}} \approx 8.0543$, through (7). From (8), E_y's directive angle θ_y is then given by $\theta_y = \arctan\left\{\frac{\sqrt{3}}{27}\left(2\sqrt{61} - 1\right)\right\} \approx 43.1648°$.

The analytical expression of the ellipse E_x is of the form $E_x\colon\ 9x^2 - 18\sqrt{3}xy + 43y^2 = 900$, due to (9) where was also set $u := 3/2$, while its principal radii $0 < b_x < a_x$ are given by $a_x = \frac{5}{2}\left(\sqrt{19} + \sqrt{7}\right) \approx 17.5116$ and $b_x = \frac{5}{2}\left(\sqrt{19} - \sqrt{7}\right) \approx 4.2829$, through (11). From (12), the E_x's directive angle θ_x is given by $\theta_x = \arctan\left\{\frac{\sqrt{3}}{27}\left(2\sqrt{133} - 17\right)\right\} \approx 21.2599°$.

Figure 4 visualizes exactly Example 1 by depicting the circle C together with its x- and y-shears, i.e., the ellipses $E_x = S_x(C)$ and $E_y = S_y(C)$. The S_x and S_y axis shears (of shearing vector $\mathbf{u} = \rho^{-1}OR$) are also illustrated by presenting the square frame around C and its transformed parallelograms, through S_x and S_y, around the ellipses E_x and E_y, respectively. The corresponding foci F_x and F_y are also depicted.

The Four Ellipses Problem

In this section we deliver the main results concerning the so-called Four Ellipses problem, which is the problem of finding a common tangential ellipse around three given *mutually conjugate* ellipses which are described as follows:

Definition 1 Three concentric and coplanar ellipses shall be called mutually conjugate (with each other) when each of them is defined by a pair of two conjugate radii taken from a bundle of three given (nondegenerated) line segments, where only two of these segments are permitted to coincide. These line segments as above shall be called as the three *mutually conjugate radii* corresponding to the three mutually conjugate ellipses.

Recall Example 1 where E_x, E_y, and C (all centered at O) are indeed three mutually conjugate ellipses defined by their three given mutually conjugate radii OP, OQ, and OR, such that $E_x := E_x(OP, OR)$, $E_y := E_y(OQ, OR)$, and $C := C(OP, OQ)$. Furthermore, we shall extend the Four Ellipses problem, in the sense that we shall derive (not one but) all the common tangential ellipses (of three given mutually conjugate ellipses) that can exist. This section is divided into two subsections regarding the following cases:

- *The circular case*, in which the existence of a common tangential ellipse of a given circle and two ellipses, mutually conjugate with each other, is investigated, and
- *The noncircular case*, based on the circular one, where the existence of a common tangential ellipse around three given noncircular mutually conjugate ellipses is investigated.

Before these subsections we state and prove the following Lemma which is needed for our study. This Lemma investigates the form of a *tangential ellipse* of a given ellipse E, i.e., a concentric (to E) ellipse, say, $\overline{E}$, which is tangent to E. The ellipses $\overline{E}$ and E are then intersect with each other at two (in total) distinct diametrical (contact) points, on which their corresponding two tangent lines coincide.

Lemma 4 *Consider a tangential ellipse $\overline{E}$ of a given ellipse E and $\theta \in [-\pi, \pi]$ be a given angle between the major axes of E and $\overline{E}$. When $\overline{E}$'s minor radius $\bar{b}$ is given, then its corresponding major radius $\bar{a}$ satisfies the relation*

$$A\bar{a}^4 + B\bar{a}^2 + C = 0, \tag{13}$$

where

$$A := 4\bar{b}^2\lambda^{-2}\left(\bar{b}^2 - a^2\right)\cos^2\theta + 4a^2\bar{b}^2\left(\lambda^{-2} - 1\right)\cos^4\theta, \tag{14a}$$

$$B := a^2\bar{b}^4\left(1 - \lambda^{-2}\right)\cos^2 2\theta + 2\bar{b}^2\left(a^4 - \lambda^{-2}\bar{b}^4\right)\cos 2\theta \\ + 2\bar{b}^2\Big[a^4 + \lambda^{-2}\bar{b}^4 - a^2\bar{b}^2\left(1 + \lambda^{-2}\right)\Big], \tag{14b}$$

$$C := a^2\bar{b}^6\left(1 - \lambda^{-2}\right)\sin^2 2\theta + 4a^2\bar{b}^4\left(a^2 - \bar{b}^2\right)\sin^2\theta, \tag{14c}$$

with $0 < b < a$ being the E's given principal radii and λ being its corresponding aspect ratio, i.e., $\lambda := b/a$. The analytical expression of the requested $\overline{E}$, on an o.c.s. spanned by the principal axes of the given ellipse E, is then formulated by

$$\overline{E}: \left(\bar{a}^2\sin^2\theta + \bar{b}^2\cos^2\theta\right)x^2 - \left(\bar{a}^2 - \bar{b}^2\right)(\sin 2\theta)xy \\ + \left(\bar{a}^2\cos^2\theta + \bar{b}^2\sin^2\theta\right)y^2 = \bar{a}^2. \tag{15}$$

See Appendix for the proof.

The Circular Case

We consider the following lemma which investigates the existence of a c.t.e. of a circle and its two x- and y-shear transformations.

Lemma 5 *Let C be a circle of radius $\rho > 0$ centered at the origin O of an o.c.s. The circle C as well as the ellipses E_x and E_y, produced by x- and y-shears of C with the same given shearing vector $\mathbf{u}$ of length $u > 0$ and angle $\omega \in [-\pi, \pi]$, can always adopt a c.t.e. E in the direction of the vector $\mathbf{u}$ (i.e., E's directive angle is ω). The principal radii $0 < b < a$ of the requested c.t.e. E are then given by $a = \rho\sqrt{u^2+1}$ and $b = \rho$, with E's foci semi-distance (or linear eccentricity) is being $f = u\rho$, as E is analytically expressed (in the adopted o.c.s.) by*

$$E : \Big[\Big(u^2{+}1\Big)\sin^2\omega{+}\cos^2\omega\Big]x^2{-}u^2(\sin 2\omega)xy + \Big[\Big(u^2+1\Big)\cos^2\omega+\sin^2\omega\Big]y^2 = \rho^2\Big(u^2+1\Big). \tag{16}$$

See Appendix for the proof.

The following Theorem proves the existence of a c.t.e. of a circle and two ellipses mutually conjugate with each other.

Theorem 1 *Consider a circle C of radius $\rho > 0$ centered at point O and two ellipses E_1 and E_2 such that E_1, E_2, and C correspond to three given mutually conjugate ellipses. Hence, these ellipses are defined by a bundle of three given mutually conjugate radii, say, OP, OQ, and OR, such that $C = C(OP, OQ)$, $E_1 = E_1(OP, OR)$, and $E_2 = E_2(OQ, OR)$, with $\rho := |OP| = |OQ|$ and $OP \perp OQ$, as C is a circle. These radii are fully determined by the given length $r := |OR|$ and angle $\omega := \measuredangle(OP, OR) \in [-\pi, \pi]$. Then, a common tangential ellipse E, of E_1, E_2, and C, always exists in the direction of the non-orthogonal radius OR (i.e., the major radius of E is spanned by OR), while point R is one of E's foci. The principal radii $0 < b < a$ of E are then given by $a = \sqrt{\rho^2+r^2}$ and $b = \rho$, while its eccentricity and foci semi-distance are $\varepsilon = r/\sqrt{\rho^2+r^2}$ and $f = r$, respectively. This c.t.e. E is analytically expressed by*

$$E : \Big[\Big(\rho^2+r^2\Big)\sin^2\omega+\rho^2\cos^2\omega\Big]x^2 + \Big[\Big(\rho^2+r^2\Big)\cos^2\omega+\rho^2\sin^2\omega\Big]y^2 - r^2(\sin 2\omega)xy = \rho^2\Big(\rho^2+r^2\Big), \tag{17}$$

in an o.c.s. spanned by the orthogonal radii OP and OQ.

The two diametrical (common) contact points $T_1(x_1, y_1)$, $x_1 > 0$, and $T_1'(-x_1, -y_1)$ between E_1 and its tangential ellipse E are then given by

$$x_1 = \sqrt{\rho^2 + r^2\cos^2\omega} \quad \textit{and} \quad y_1 = \frac{r^2 \sin 2\omega}{2\sqrt{\rho^2+r^2\cos^2\omega}}, \tag{18}$$

while their corresponding two common tangent lines t_1 and t_1' (at points T_1 and T_1', respectively) are being parallel to OQ (which spans the o.c.s.'s vertical axis $y'Oy$), i.e., $t_1\colon\ x = x_1$ and $t_1'\colon\ x = -x_1$.

For the two diametrical contact points $T_2(x_2, y_2)$, $y_2 > 0$, and $T_1'(-x_2, -y_2)$ between E_2 and its tangential ellipse E, we have

$$x_2 = \frac{r^2 \sin 2\omega}{2\sqrt{\rho^2 + r^2 \sin^2 \omega}} \quad \text{and} \quad y_2 = \sqrt{\rho^2 + r^2 \sin^2 \omega}, \tag{19}$$

while their corresponding two common tangent lines t_2 and t_2' (at points T_2 and T_2', respectively) are being parallel to OP (which spans the o.c.s.'s horizontal axis $x'Ox$), i.e., $t_2\colon\ y = y_2$ and $t_2'\colon\ y = -y_2$.

Finally, for the last two contact points $T_3(x_3, y_3)$ and $T_3'(-x_3, -y_3)$ between circle C and its tangential ellipse E, it holds that

$$x_3 = -\rho \sin \omega \quad \text{and} \quad y_3 = \rho \cos \omega, \tag{20}$$

while their corresponding two tangent lines t_3 and t_3' (at points T_3 and T_3', respectively) are being parallel to the non-orthogonal radii OR, as they are given by

$$t_3 :\ y \cos \omega - x \sin \omega = \rho \quad \text{and} \quad t_3' :\ x \sin \omega - y \cos \omega = \rho. \tag{21}$$

See Appendix for the proof.

Example 2 Consider the bundle of three line segments OP, OQ, and OR as in Example 1, i.e., $\rho = |OP| = |OQ| := 10$ and $r = |OR| := 15$, with $OP \perp OQ$ and $\omega = \measuredangle(OP, OR) := \pi/6\ (= 30°)$. These three given line segments correspond to three mutually conjugate radii which define three mutually conjugate ellipses such that $E_1 = E_1(OP, OR)$, $E_2 = E_2(OQ, OR)$, and $E_3 = E_3(OP, OQ)$; see also Fig. 5. Notably, E_3 is a circle of radius $\rho = 10$ as defined by the pair (OP, OQ) of its orthogonal (and equal) conjugate radii. Consider also point O as the origin of an o.c.s. with its horizontal and vertical axes spanned, respectively, by orthogonal radii OP and OQ.

Ellipses E_1 and E_2 can be expressed as the x- and y-shears of circle E_3, with shearing vector $\mathbf{u} := \rho^{-1}OR$ of length $u := r/\rho = 1.5$. Therefore, E_1 and E_2 are then given in Example 1, where E_x and E_y correspond now to E_1 and E_2, respectively, while their principal radii $0 < b_i < a_i$, $i = 1, 2$, and their directive angles θ_i, $i = 1, 2$, are given, respectively, by the values of a_x, b_x, a_y, b_y, and θ_x, θ_y in Example 1. Applying now Theorem 1, a c.t.e. E of the given three mutually conjugate ellipses E_i, $i = 1, 2, 3$, always exists, and its major axis is spanned by (non-orthogonal common radius) OR, with point R being one of its foci. The corresponding E's principal radii are then given by $a = \sqrt{\rho^2 + r^2} = 5\sqrt{13} \approx 18.0278$ and $b = \rho = 10$, while eccentricity, foci distance, and surface area are given, respectively, by $\varepsilon = \frac{3}{13}\sqrt{13} \approx 0.832$, $f = 30$, and $A = 50\pi\sqrt{13} \approx 566.359$. As E's focal point, say, F, is identical to R, then $F = R(r\cos\omega, r\sin\omega) = R\big(\frac{15}{2}\sqrt{3}, \frac{15}{2}\big)$, and $F'\big(-\frac{15}{2}\sqrt{3}, -\frac{15}{2}\big)$ then correspond to the two diametrical foci of E. The analytical expression of c.t.e. E is given, through (17), by $E\colon\ 25x^2 - 18\sqrt{3}xy + 43y^2 = 5200$.

The two contact points $T_1(x_1, y_1)$, $y_1 > 0$, and $T_1'(-x_1, -y_1)$ between ellipse E_1 and its tangential E are calculated through (18), i.e., $x_1 = \frac{5}{2}\sqrt{43} \approx 16.394$ and $y_1 = \frac{45}{86}\sqrt{129} \approx 5.943$, with their corresponding two tangent lines t_1 and t_1' (at their contact points T_1 and T_1') to be parallel to OQ (which spans o.c.s.'s vertical axis $y'Oy$), i.e., $t_1\colon\ x = x_1 = \frac{5}{2}\sqrt{43}$ and $t_1'\colon\ x = -x_1 = -\frac{5}{2}\sqrt{43}$.

The other two contact points $T_2(x_2, y_2)$, $y_2 > 0$, and $T_2'(-x_2, -y_2)$ between ellipse E_2 and its tangential E are calculated through (19), i.e., $x_2 = \frac{9}{2}\sqrt{13} \approx 7.7942$ and $y_2 = 25/2 = 12.5$, with their corresponding two tangent lines t_2 and t_2' (at their contact points T_2 and T_2') to be parallel to OP (which spans o.c.s.'s horizontal axis $x'Ox$), i.e., $t_2\colon\ y = y_2 = 25/2$ and $t_2'\colon\ y = -y_2 = -25/2$.

Finally, the last two contact points $T_3(x_3, y_3)$, $y_3 > 0$, and $T_3'(-x_3, -y_3)$ between circle E_3 and its tangential E are calculated through (20), i.e., $x_3 = -5$ and $y_3 = 5\sqrt{3} \approx 8.6603$, with their corresponding two tangent lines t_3 and t_3' (at their contact points T_3 and T_3') obtained through (20), i.e., $t_3\colon\ \sqrt{3}y - x = 20$ and $t_3'\colon\ x - \sqrt{3}y = 20$, which are parallel to the non-orthogonal radii OR, as their slopes are both $\tan(\pi/6) = \sqrt{3}/3$.

Figure 5 visualizes exactly Example 2 by presenting the three mutually conjugate ellipses $E_1 = E_1(OP, OR)$, $E_2 = E_2(OQ, OR)$, and $E_3 = E_3(OP, OQ)$, with their defining three given mutually conjugate radii as well as their foci points. The c.t.e. E of E_i, $i = 1, 2, 3$, is also presented, together with their six common tangent lines at their six corresponding contact points. The intersection angles of the ellipses E_i, $i = 1, 2, 3$, at their common points P, Q, and R, are also denoted, while $\vartheta := \pi/2 - \omega$ and $\varphi := \measuredangle(OP, OQ) = \pi/2$.

The Noncircular Case

We shall now extend the results of the circular case, as discussed in section "The Circular Case," to the general case of three mutually conjugate noncircular ellipses.

From the circular case, investigated in Theorem 1, we concluded that there always exists a c.t.e. (at the direction of the non-orthogonal radius) around three given mutually conjugate ellipses, when one of them is a circle. Having the above in mind, we consider the following definition.

Definition 2 Let E_i, $i = 1, 2, 3$, be three mutually conjugate ellipses. We can always consider a coordinate system in which one of the ellipses is expressed as a circle, and hence two of their three mutually conjugate radii, which define E_i, $i = 1, 2, 3$, are of equal length and orthogonal with each other. A c.t.e. E of E_i,

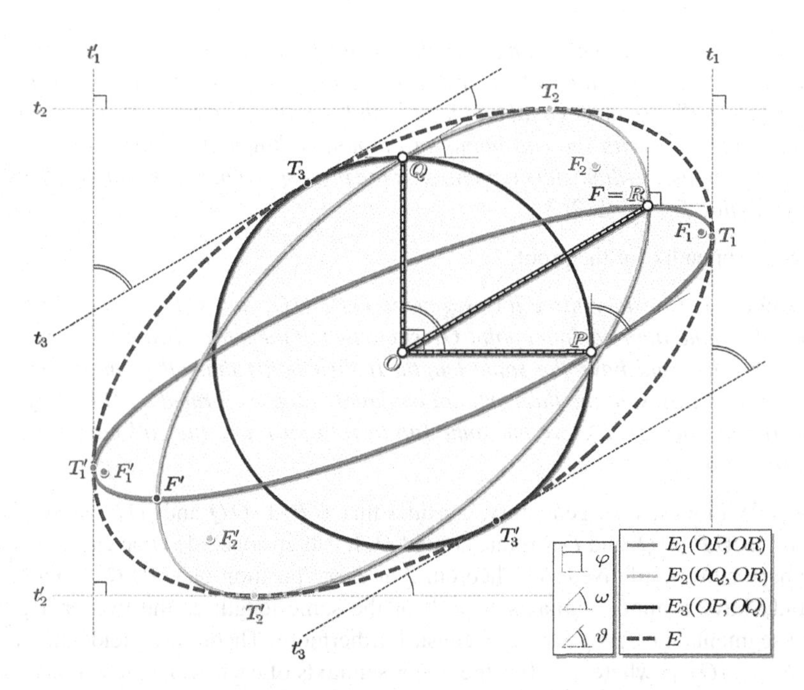

Fig. 5 Graphs of the three mutually conjugate ellipses E_i, $i = 1, 2, 3$, of Example 2, and their c.t.e. E together with their common tangent lines

$i = 1, 2, 3$ shall then be called primary when its major semiaxis coincides with the non-orthogonal radius from the corresponding three mutually conjugate radii (defining $E_i, i = 1, 2, 3$).

The following is the main theorem and investigates the existence of a common tangential ellipse around three mutually conjugate ellipses, and therefore extends Peschka's Proposition 1. It also provides a plane-geometric proof of it.

Theorem 2 *Consider three mutually conjugate ellipses E_i, $i = 1, 2, 3$, defined by a bundle of three given line segments which correspond to E_i's three mutually conjugate radii. These given radii, say, OP, OQ, and OR, are determined by their lengths $p, q, r > 0$, respectively, and the two angles $\varphi := \measuredangle(OP, OQ) \in (0, \pi)$ and $\omega := \measuredangle(OP, OR) \in (0, \pi)$. Let also $\vartheta := \angle(OR, OQ) \in (0, \pi) = \varphi - \omega$. Then, there always exists a primary c.t.e. E of E_i, $i = 1, 2, 3$, with their corresponding common tangent lines, between each E_i, $i = 1, 2, 3$, and their c.t.e. E, being parallel to each of the three given conjugate radii. The analytical expression of the primary common tangential ellipses E can be calculated, in the vector base where E_3 is expressed as a circle $C(O, p)$, by setting $r := r'$ and $\omega := \omega'$ both into* (17)*, where r' and ω' are given by*

$$r' = \tfrac{r}{q}(\csc\varphi)\sqrt{p^2 \sin^2\omega + q^2 \sin^2\vartheta} \;\; \text{and} \;\; \omega' = \arctan\left(\frac{p \sin\omega}{q \sin\vartheta}\right), \tag{22}$$

while ω' is the E's directive angle (in this vector base). In case one of the mutually conjugate ellipses is reduced to a circle (and therefore two out of three E_i's mutually conjugate radii are orthogonal and have equal length), then the foci of the primary c.t.e. E coincide with the end points of the non-orthogonal diameter of the three mutually conjugate diameters (spanned by the three given mutually conjugate radii) which define E_i, $i = 1, 2, 3$.

See Appendix for the proof.

Corollary 1 *Consider three line segments OP, OQ, and OR that can be freely rotated around their common point O, where two of them are always perpendicular to each other and have the same length. It then holds that all primary common tangential ellipses, of the three mutual conjugate ellipses formed by OP, OQ, and OR (as in Theorem 2), are the same (up to rotation), i.e., they all have the same aspect ratio.*

Proof Without loss of generality, consider that $OP \perp OQ$ and $|OP| = |OQ| =$ const. as OP, OQ, and OR rotate around O. It is then concluded that each common primary c.t.e. E (derived via Theorem 2 at every position of OP, OQ, and OR) should have a minor semiaxis $b > 0$ of the same length as the two orthogonal line segments, i.e., $\beta = |OP| =$ const. Furthermore, Theorem 2 yields that $a^2 = |OP|^2 + |OR|^2$ where $a > 0$ is the major semiaxis of each E, i.e., $a =$ const., and hence corollary has been proved.

Discussion

Consider the problem of finding a concentric common tangential ellipse around three given coplanar and concentric ellipses, say, E_i, $i = 1, 2, 3$, each one defined by a pair of two conjugate semidiameters which are taken from a bundle of three given line segments. This plane-geometric problem was first stated and addressed by G.A. Peschka in [5], in his proof of Pohlke's fundamental theorem of axonometry. However, his proof is based on a parallel projection of an appropriate sphere S onto E_i's common plane, say, $\mathbb{E}$. According to [4, 5], the common tangential ellipse, say, E, of all E_i, $i = 1, 2, 3$, is then the parallel projection of the sphere's contour onto $\mathbb{E}$. This parallel projection method has been used in literature for the proof of Pohlke's fundamental theorem of axonometry.

With this present paper, the above Peschka's problem is investigated and addressed in terms of analytic plane geometry, and some properties were derived. The exact direction of the requested E was calculated as well as its characteristics. The examples and figures provided in this paper are demonstrating these results.

Acknowledgements The author would like to thank Prof. G.E. Lefkaditis for posing the problem and providing the historical background and Prof. C.P. Kitsos for the useful discussions during the preparation of this article.

Appendix

Proof (of Lemma 2) For the y-shear S_y, defined by $\mathbf{x}' = S_y(\mathbf{x}) := \mathbf{N}_y\mathbf{x}$, with transformation matrix $\mathbf{N}_y$ as in (4), we have that $\mathbf{x} = S_y^{-1}(\mathbf{x}') = \mathbf{N}_y^{-1}\mathbf{x}'$, i.e.,

$$\begin{pmatrix} x \\ y \end{pmatrix} =: \mathbf{x} = \mathbf{N}_y^{-1}\mathbf{x}' = \begin{pmatrix} (u\cos\omega)^{-1} & 0 \\ -\tan\omega & 1 \end{pmatrix} \begin{pmatrix} x' \\ y' \end{pmatrix} = \begin{pmatrix} \frac{x'}{u\cos\omega} \\ y' - x'\tan\omega \end{pmatrix}, \tag{23}$$

and thus the matrix representation form $C: \mathbf{x}^{\mathrm{T}}\mathbf{x} = \rho^2$ of the circle $C: x^2 + y^2 = \rho^2$ implies that $S_y(C): \mathbf{x}'^{\mathrm{T}}(\mathbf{N}_y^{-1})^{\mathrm{T}}\mathbf{N}_y^{-1}\mathbf{x}' = \rho^2$ or, equivalently, $S_y(C): \mathbf{x}^{\mathrm{T}}(\mathbf{N}_y^{-1})^{\mathrm{T}}\mathbf{N}_y^{-1}\mathbf{x} = \rho^2$ (as $S_y(C)$ is referred again to the adopted o.c.s.). Therefore, $S_y(C): x^2 + u^2(y\cos\omega - x\sin\omega)^2 = \rho^2u^2\cos^2\omega$, and hence (5) and (6) are obtained. The ellipse $S_y(C)$ is centered also at the origin O. This is true because x- and y-shears preserve the center O as it belongs to both the preserved coordinates' axes $x'Ox$ and $y'Oy$.

In order to obtain the principal radii $0 < b < a$, the ellipse $S_y(C)$ has to be rotated by its directive angle, say, θ, around its center O until its canonical form is obtained (on the adopted o.c.s.), i.e., until $S_y(C)$ is transformed into $E_y: \widetilde{\mathbf{x}}^{\mathrm{T}}\widetilde{\mathbf{D}}\widetilde{\mathbf{x}} = 1$, $\widetilde{\mathbf{x}} \in \mathbb{R}^{2\times1}$, where $\widetilde{\mathbf{D}} \in \mathbb{R}^{2\times2}_{\text{diag}}$ is a real diagonal 2×2 matrix. For this purpose, we consider the rotation transform

$$\widetilde{\mathbf{x}} = R_\theta(\mathbf{x}) := \mathbf{R}_\theta\mathbf{x}, \quad \mathbf{x} \in \mathbb{R}^{2\times1}, \tag{24}$$

with rotation matrix $\mathbf{R}_\theta \in \mathbb{R}^{2\times2}_{\perp}$ as in (2) that provides the ellipse $S_y(C)$, as in (5), from its requested canonical form $E_y: \widetilde{\mathbf{x}}^{\mathrm{T}}\widetilde{\mathbf{D}}\widetilde{\mathbf{x}} = 1$, i.e., $R_\theta(E_y) = S_y(C)$ or $E_y = R_\theta^{-1}\left(S_y(C)\right)$. Recall the matrix representation of the ellipse $S_y(C)$ (mentioned earlier), i.e.,

$$S_y(C): \mathbf{x}^{\mathrm{T}}\mathbf{N}\mathbf{x} = \rho^2, \quad \mathbf{N} := (\mathbf{N}_y^{-1})^{\mathrm{T}}\mathbf{N}_y^{-1} \in \mathbb{R}^{2\times2}_{\text{sym}}. \tag{25}$$

Then, (24) implies that $\mathbf{x} = \mathbf{R}_\theta^{-1}\mathbf{x} = \mathbf{R}_\theta^{\mathrm{T}}\widetilde{\mathbf{x}} = \mathbf{R}_{-\theta}\widetilde{\mathbf{x}}$, and by substitution to (25), it holds that

$$E_y = R_\theta^{-1}\left(S_y(C)\right) = \left(R_\theta^{-1} \circ S_y\right)(C): \widetilde{\mathbf{x}}^{\mathrm{T}}\mathbf{R}_{-\theta}\mathbf{N}\mathbf{R}_\theta\widetilde{\mathbf{x}} = \rho^2. \tag{26}$$

Recall that the inverse of an orthogonal matrix equals to the transpose of the matrix, i.e., $\mathbf{M}^{-1} = \mathbf{M}^{\mathrm{T}}$ for $\mathbf{M} \in \mathbb{R}^{n\times n}_{\perp}$, $n \in \mathbb{N}$. By orthogonal decomposition of the symmetric matrix $\mathbf{N} \in \mathbb{R}^{2\times2}_{\text{sym}}$, we obtain that $\mathbf{N} = \mathbf{V}\mathbf{D}\mathbf{V}^{\mathrm{T}}$, where $\mathbf{D} := \operatorname{diag}(d_1, d_2) \in \mathbb{R}^{2\times2}_{\text{diag}}$ is the spectral matrix of $\mathbf{N}$ (i.e., the diagonal 2×2 matrix of the eigenvalues d_1 and d_2 of $\mathbf{N}$) and $\mathbf{V} \in \mathbb{R}^{2\times2}_{\perp}$ is the orthonormal 2×2 matrix of the eigenvectors corresponding to eigenvalues d_1 and d_2. Thus, $\mathbf{D} = \mathbf{V}^{\mathrm{T}}\mathbf{N}\mathbf{V}$, and by setting $\mathbf{R}_\theta := \mathbf{V}$ (i.e., we adopt as orthonormal rotation matrix $\mathbf{R}_\theta$ the (orthonormal) eigenvalues matrix $\mathbf{V}$), the relation (26) implies the canonical form of $S_y(C)$, i.e., $E_y : \tilde{\mathbf{x}}^{\mathrm{T}} \left(\rho^{-2}\mathbf{D}\right) \tilde{\mathbf{x}} = 1$. Hence, the major and minor radius of the ellipse $E_y : (d_1\tilde{x}/\rho)^2 + (d_2\tilde{y}/\rho)^2 = 1$ (which are the same for the rotated ellipse $S_y(C)$) are then given by $a = \rho/\sqrt{\max\{d_1, d_2\}}$ and $b = \rho/\sqrt{\min\{d_1, d_2\}}$, respectively.

The eigenvalues d_1 and d_2 of the symmetric matrix $\mathbf{N}$ can be calculated through the roots of the $\mathbf{N}$'s characteristic polynomial $P_{\mathbf{N}}(d) := |\mathbf{N} - d\mathbb{I}_2|, d \in \mathbb{R}, i = 1, 2,$ where $\mathbb{I}_2$ being the unitary 2×2 matrix. After some algebra, we derive that

$$d_i = \frac{\sec^2 \omega}{2u^2} \left[\left(u^2 + 1\right) + (-1)^{i-1}\sqrt{u^4 + 1 - 2u^2 \cos 2\omega} \right], \quad i = 1, 2, \tag{27}$$

with $0 < (u^2 - 1)^2 = u^4 - 2u^2 + 1 < u^4 + 1 - 2u^2 \cos 2\omega$ for every $u > 0$, and therefore, the requested major and minor radius of the ellipse $S_y(C)$ correspond to the eigenvalues d_2 and d_1, respectively (as $d_1 > d_2$), and hence given by (7).

Moreover, the non-unitary orthogonal eigenvectors $\mathbf{v}_i = (v_{i;1}, v_{i;2})^{\mathrm{T}} \in \mathbb{R}^{2\times 1}$ that correspond to its eigenvalues d_i, $i = 1, 2$, i.e., $\mathbf{V} = (\mathbf{v}_1, \mathbf{v}_2) \in \mathbb{R}_\perp^{2\times 2}$, are then calculated through (27) and (after some algebra) are found to be

$$\mathbf{v}_i = \left(-\frac{u^2 + 1 - 2u^2 \cos^2 \omega + (-1)^{i-1}\sqrt{u^4 + 1 - 2u^2 \cos 2\omega}}{u^2 \sin 2\omega}, 1 \right)^{\mathrm{T}}, \quad i = 1, 2.$$

Hence, as the major radius a corresponds to the eigenvalue d_2 (shown earlier), the directive angle θ of the ellipse $S_y(C)$ is then of the form $\theta = \angle(\mathbf{e}_1, \mathbf{v}_2) = \arctan(v_{2;2}/v_{2;1})$, and it is given by (8).

Notice also the fact that the assumption $\tan\theta < \tan\omega$ yields, through (8), that $u^2 + 1 > \sqrt{u^4 + 1 - 2u^2 \cos 2\omega}$, for $\omega > 0$, which cannot hold (as it would then implies that $\cos^2 \omega < 0$). Therefore, it holds that $\tan\theta > \tan\omega$ when $\omega > 0$ while, similarly, $\tan\theta < \tan\omega$ when $\omega < 0$. Hence $|\theta| > |\omega|$.

Proof (of Lemma 4) We consider the ellipse E centered at the origin O of an o.c.s. which is spanned by its principal axes, i.e., the E's major and minor axis form, respectively, the o.c.s.'s horizontal and vertical axes. Hence, the ellipse E (in this o.c.s.) adopts its canonical form

$$E : (x/a)^2 + (y/b)^2 = 1. \tag{28}$$

The tangential ellipse $\overline{E}$ of E, which (its major axis) forms a given angle θ with (the major axis of) E, adopts two diametrical points P and P' in common with E. Therefore, $\overline{E}$ can be expressed in the form of the ellipse (centered at the origin O) as in (1), and thus (15) holds, with $0 < \bar{b} < \bar{a}$ being the $\overline{E}$'s principal radii. Notice that θ is indeed the directive angle of $\overline{E}$ (with respect to the adopted o.c.s. as above) as it coincides with the given angle θ between the major axes of E and $\overline{E}$.

Assuming now that the $\overline{E}$'s minor radius $\bar{b}$ is given, we shall provide in the following the $\overline{E}$'s major axis $\bar{a}$, and thus the tangential ellipse $\overline{E}$ of a given ellipse E (in the direction of the given angle θ) can then be calculated easily through (15). Let $\lambda := b/a$ and $\bar{\lambda} := \bar{b}/\bar{a}$ be the aspect ratios of the ellipses E and $\overline{E}$, respectively. We consider the contact point $P(x_0, y_0) \in E \cap \overline{E}$, which lies on the o.c.s.'s upper semi-plane, i.e., $y_0 \geq 0$. Solving (28) and (15), with respect to y, we obtain

$$y_0 = y_E(x_0) := \lambda\sqrt{a^2 - x_0^2}, \quad x_0 \in [-a, a], \text{ and} \tag{29a}$$

$$y_0 = y_{\overline{E}}(x_0) := \frac{\frac{1}{2}\left(\bar{\lambda}^{-2} - 1\right) x_0 \sin 2\theta + \bar{\lambda}^{-1}\sqrt{\bar{\lambda}^{-2}\bar{b}^2\cos^2\theta + \bar{b}^2\sin^2\theta - x_0^2}}{\bar{\lambda}^{-2}\cos^2\theta + \sin^2\theta}, \tag{29b}$$

where, for the latter function $y_{\overline{E}}(x_0)$, it is assumed that $x_0 \in [-t, t]$ with the value $t := \bar{b}^2\left(\bar{\lambda}^{-2}\cos^2\theta + \sin^2\theta\right)$. Equating the right-hand side of (29a) and (29b), it holds that

$$\lambda K\sqrt{a^2 - x_0^2} = Lx_0 + \bar{\lambda}^{-1}\bar{b}^2\sqrt{K - x_0^2}, \tag{30}$$

where

$$K := \bar{\lambda}^{-2}\bar{b}^2\cos^2\theta + \bar{b}^2\sin^2\theta = \bar{b}^2\left[1 + \left(\bar{\lambda}^{-2} - 1\right)\cos^2\theta\right] > 0 \text{ and} \tag{31a}$$

$$L := \tfrac{1}{2}\bar{b}^2\left(\bar{\lambda}^{-2} - 1\right)\sin 2\theta = \left(K - \bar{b}^2\right)\tan\theta. \tag{31b}$$

In order the curves $y = y_E(x)$ and $y = y_{\overline{E}}(x)$, as defined in (29a) and (29b), to be tangent to each other at their common point $P(x_0, y_0)$ (and hence to have a common tangent line on P), the derivatives of $y_E = y_E(x)$ and $y_{\overline{E}} = y_{\overline{E}}(x)$ must coincide at $x = x_0$ with x_0 satisfying (30). The derivative of (30), with respect to $x = x_0$, yields

$$\lambda K\frac{x_0}{\sqrt{a^2 - x_0^2}} = -L + \bar{b}^2\frac{x_0}{\bar{\lambda}\sqrt{K - x_0^2}}. \tag{32}$$

Solving (30) and (32) in terms of $\sqrt{K - x_0^2}$, we get

$$\frac{\lambda\bar{\lambda}K}{\bar{b}^2}\sqrt{a^2 - x_0^2} - \frac{L\bar{\lambda}x_0}{\bar{b}^2} = \sqrt{K - x_0^2} = \frac{\bar{\lambda}^{-1}\bar{b}^2 x_0\sqrt{a^2 - x_0^2}}{\lambda K x_0 + L\sqrt{a^2 - x_0^2}}. \tag{33}$$

By multiplication of the left and right side of (33), we obtain the squared middle expression of (33) of the form

$$K - x_0^2 = \frac{\left(\lambda K\sqrt{a^2 - x_0^2} - Lx_0\right) x_0\sqrt{a^2 - x_0^2}}{\lambda K x_0 + L\sqrt{a^2 - x_0^2}} \quad \text{or}$$

$$\sqrt{a^2 - x_0^2} = \lambda \frac{a^2 - K}{L} x_0. \tag{34}$$

Moreover, relation (32) can be written, through (32), as

$$\lambda\bar{\lambda}a^2\sqrt{K - x_0^2} = \bar{b}^2\sqrt{a^2 - x_0^2}, \tag{35}$$

while substituting (30) again to the right-hand side of (35), we obtain

$$\sqrt{K - x_0^2} = \frac{\bar{b}^2}{L\bar{\lambda}a^2}\left(a^2 - K\right) x_0. \tag{36}$$

Applying (34) and (36) into (30), we have $[Kb^2 - (\bar{\lambda}^{-1}\bar{b}^2)^2]\left(a^2 - K\right) = L^2a^2$, and using (31b),

$$K^2b^2\left(1 + \lambda^{-2}\tan^2\theta\right) - K\left[\left(\bar{\lambda}^{-1}\bar{b}^2\right)^2 + a^2b^2 + 2a^2\bar{b}^2\tan^2\theta\right]$$
$$+ a^2\bar{b}^4\left(\bar{\lambda}^{-2} + \tan^2\theta\right) = 0. \tag{37}$$

Finally, by substitution of (31a) into the above (37), we obtain the biquadratic polynomial relation (13) with respect to $\overline{E}$'s requested major radius $\bar{a}$.

Proof (of Lemma 5) Consider an orthonormal vector base $\mathscr{B}$ of an o.c.s. of origin O, and let S_x and S_y be the x- and y-shear transformations with the same shearing vector (of length $u > 0$ and angle ω). Then, according to Lemmas 3 and 2, the x- and y-sheared circles $E_x := S_x(C)$ and $E_y := S_y(C)$ are ellipses centered also at O with their corresponding principal radii $0 < b_x < a_x$ and $0 < b_y < a_y$ given by (11) and (7). The directive angles $\theta_x, \theta_y \in [-\pi, \pi]$ of the tangential ellipses E_x and E_y (with respect to the o.c.s.'s coordinate axes) are then given by (12) and (8), respectively. Let also $\overline{E}_x$ and $\overline{E}_y$ be the (concentric) tangential ellipses of E_x and E_y, respectively, adopted in the same direction as the common shearing vector (i.e., their directive angles are both ω), and having principal radii $0 < \bar{b}_x < \bar{a}_x$ and $0 < \bar{b}_y < \bar{a}_y$, respectively.

We shall now investigate the form of tangential ellipses $\overline{E}_x$ and $\overline{E}_y$ of E_x and E_y in the direction of ω, such that both $\overline{E}_x$ and $\overline{E}_y$ are both tangent to the circle C. Thus, their minor radii $\bar{b}_x = \bar{b}_y = \rho$, while their directive angles $\bar{\theta}_x = \bar{\theta}_y = \omega$. In order to apply Lemma 4, we consider as angle θ (which is the angle between the major axes of E_x and $\overline{E}_x$) the difference $\theta := \bar{\theta}_x - \theta_x = \omega - \theta_x$, while for the angle between (the major axes of) E_y and $\overline{E}_y$, we must consider, respectively, $\theta := \bar{\theta}_y - \theta_y = \omega - \theta_y$.

Figure 4 clarifies also the above discussion. Let $k := \bar{b}_x/b_x$, while $\lambda_x := b_x/a_x$ and $\bar{\lambda}_x := \bar{b}_x/\bar{a}_x$ denote the aspect ratios of E_x and $\overline{E}_x$, respectively. Relation (13), through (31a), yields

$$\bar{\lambda}_x^{-2}k^4 - \Big[\left(\bar{\lambda}_x^{-2} + \lambda_x^{-2}\right)\cos^2\theta + \left(1 + \bar{\lambda}_x^{-2}\lambda_x^{-2}\right)\sin^2\theta\Big]k^2 + \lambda_x^{-2} = 0.$$

Substituting $k = \bar{b}_x/b_x = \rho/b_x$ (as $\bar{b}_x = \rho$ were assumed), the above polynomial (38) can be solved with respect to $\bar{\lambda}_x^{-2}$ as

$$\bar{\lambda}_x^{-2} = b_x^2\rho^{-2}\frac{b_x^2\lambda_x^{-2} - \rho^2 - \rho^2\left(\lambda_x^{-2} - 1\right)\cos^2\theta}{b_x^2\lambda_x^{-2} - \rho^2 - b_x^2\left(\lambda_x^{-2} - 1\right)\cos^2\theta}. \tag{38}$$

From (11) we have

$$b_x^2\lambda_x^{-2} = \frac{8u^4\rho^2\sin^4\omega}{\delta_x^2\left(u^2 + 1 + \sqrt{u^4 + 1 + 2u^2\cos 2\omega}\right)} = 2\delta_x^{-1}u^2\rho^2\sin^2\omega, \tag{39}$$

where

$$\delta_x := u^2 + 1 - \sqrt{u^4 + 1 + 2u^2\cos 2\omega}, \tag{40}$$

and thus, by substitution of (39) into (38) and then applying b_x as in (11), we obtain

$$\bar{\lambda}_x^2 = \frac{\rho^2\delta_x\Big[u^2 - 1 + \left(u^2 + 1 - \delta_x\right)\left(1 - 2\cos^2\theta\right)\Big]}{2\rho^2u^2\delta_x\sin^2\omega - \rho^2\delta_x^2\sin^2\theta - 4\rho^2u^2\sin^2\omega\cos^2\theta.} \tag{41}$$

Moreover, substituting θ_x from (12) to the relation

$$\cos^2\theta = \frac{1}{1 + \tan^2(\omega - \theta_x)} = \frac{(1 + \tan\omega\tan\theta_x)^2}{(1 + \tan\omega\tan\theta_x)^2 + (\tan\omega - \tan\theta_x)^2},$$

as $\theta := \omega - \theta_x$, we obtain that through (40),

$$\cos^2\theta = \frac{\left(2u^2 + 2 - \delta_x\right)^2}{\left(2u^2 + 2 - \delta_x\right)^2 + \left(\tan^2\omega\right)(2 - \delta_x)^2}. \tag{42}$$

By substitution of the above $\cos^2\theta$ into (41), we obtain (after some algebra) that

$$\bar{\lambda}_x^2 = \frac{\delta_x\left(u^2 - 1 - \delta_x\right)\Big[2\left(u^2 + 1\right) - \delta_x\Big]^2\left(1 - \sin^2\omega\right) + \delta_x\left(u^2 - 1 + \delta_x\right)g}{\Big[\delta_x\left(\delta_x - 2\right)\left(2u^2\sin^2\omega - \delta_x\right) + 2u^2\Big[2\left(u^2 + 1\right) - \delta_x\Big]^2\left(1 - \sin^2\omega\right)\Big]g}, \tag{43}$$

where $g := (\delta_x - 2)^2 \sin^2 \omega$. From (40) we have

$$\sin^2 \omega = 1 - \cos^2 \omega = 1 - \tfrac{1}{2}(1 + \cos 2\omega) = 1 + \tfrac{1}{4u^2}(\delta_x - 2)\left(2u^2 - \delta_x\right), \tag{44}$$

as it is easy to see, through (40), that $0 < \delta_x < 2$ and $\delta_x < 2u^2$. Substituting (44) into (43), we finally derive (after a series of simplifications) that $\bar{\lambda}_x^2 = u^2 + 1$, and hence the major radius of $\overline{E}_x$ is given by $\bar{a}_x = \rho\sqrt{u^2+1}$, as $\bar{b}_x = \rho$ was assumed.

Working similarly for the case of the tangential ellipse $\overline{E}_y$, we obtain its aspect ratio

$$\bar{\lambda}_y^2 = \frac{\rho^2\delta_y\left[u^2 - 1 + \left(u^2 + 1 - \delta_y\right)\left(1 - 2\cos^2\theta\right)\right]}{2\rho^2u^2\delta_y\cos^2\omega - \rho^2\delta_y^2\sin^2\theta - 4\rho^2u^2\cos^2\omega\cos^2\theta}, \quad \text{where} \tag{45}$$

$$\delta_y := u^2 + 1 - \sqrt{u^4 + 1 - 2u^2\cos 2\omega}. \tag{46}$$

Then, after some algebra, it holds also that $\bar{\lambda}_y^2 = u^2 + 1$, and hence the major radius of $\overline{E}_y$ is given by $\bar{a}_y = \rho\sqrt{u^2+1}$, as $\bar{b}_y = \rho$ was also assumed.

Therefore, the principal radii of $\overline{E}_x$ and $\overline{E}_y$ coincide, as $\bar{a}_x = \bar{a}_y = \rho\sqrt{u^2+1}$ and $\bar{b}_x = \bar{b}_y = \rho$, and hence the ellipses $\overline{E}_x$ and $\overline{E}_y$ are of the same shape. Moreover, as their directive angles $\bar{\theta}_x$ and $\bar{\theta}_y$ are both assumed to be ω, it is clear that $\overline{E}_x = \overline{E}_y$. We can then denote with $E := \overline{E}_x = \overline{E}_y$ the c.t.e. of E_x, E_y and C. The directive angle θ of E is ω, and its principal radii $0 < b < a$ are of the form $a = \rho\sqrt{u^2+1}$ and $b = \rho$. The foci semi-distance of E is formulated by $f := \sqrt{a^2 - b^2} = u\rho$. The analytical expression of the c.t.e. E is finally given by setting $\theta := \omega$ into (1), and thus (16) is finally derived.

Proof (of Theorem 1) Consider an o.c.s. with origin O where its horizontal and vertical axes are spanned by the given orthogonal vectors OP and OQ. Let $\mathscr{B} = \{\mathbf{e}_1, \mathbf{e}_2\}$ be the corresponding orthonormal vector base. Hence $P_{\mathscr{B}}(\rho, 0)$ and $Q_{\mathscr{B}}(0, \rho)$. Let S_x be the x-shear transformation with shearing vector $\mathbf{u} := \rho^{-1}OR$, which transforms point $Q \in C$ into R. Indeed, from (3) and setting $u := |\mathbf{u}| = r/\rho$, we obtain $S_x(Q) = R$. Notice also that $S_x(P) = P$ (as the horizontal axis $x'Ox$ is invariant under S_x), and hence $P \in C \cap E_1$. Therefore, $S_x(C) = E_1$ as the ellipse E_1 is (by assumption) defined by its two conjugate radii OP and OR. Let also S_y be the y-shear with the same shearing vector $\mathbf{u}$ which (similar to S_x) transforms point $P \in C$ also into R. Indeed, from (4) and setting again $u := |\mathbf{u}| = r/\rho$, we obtain $S_y(P) = R$. Also $S_y(Q) = Q$ (as the vertical axis $y'Oy$ is invariant under S_y), and hence $Q \in C \cap E_2$. Thus $S_y(C) = E_2$ as the ellipse E_2 (by assumption) is defined by its two conjugate radii OQ and OR. From the above discussion, point R is an intersecting point of the ellipses E_1 and E_2, i.e., $R \in E_1 \cap E_2$.

The given pairs of line segments (OP, OR) and (OQ, OR) indeed correspond to pairs of conjugate radii for E_1 and E_2, respectively, as these segments are affine transformations (recall S_x and S_y) of the C's orthogonal radii OP and OQ. As affinity preserves parallelism, the tangent lines of E_1 and E_2 at their points P and Q, respectively, are parallel to OR, while the tangent lines of E_1 and E_2 at their point R are parallel to OP and OQ, respectively (because the tangent lines of the circle C at its points P and Q are, trivially, parallel to its orthogonal radii OQ and OP). Figure 4 clarifies also the above discussion (on which the referred ellipses E_x and E_y correspond to E_1 and E_2, respectively) as the square frame around C is transformed, through S_x and S_y, into the parallelograms around E_1 and E_2.

Lemma 5 can now be applied, where we have to replace E_x and E_y with E_1 and E_2, respectively, and set $u := r/\rho$. Therefore, a c.t.e. E of E_1, E_2, and C always exists with directive angle ω (in the adopted o.c.s.), and hence point R lies onto E's major semiaxis. Moreover, through Lemma 5, the foci separation of the c.t.e. E is then $f = 2u\rho$, i.e., $f = 2r = 2|OR|$ (as $u := r/\rho$). Thus, the intersection point $R \in E_1 \cap E_2$ is indeed a focal point of the common tangential E. Moreover, the E's principal radii are then given by $a = \rho\sqrt{u^2+1} = \sqrt{\rho^2+r^2}$ and $b = \rho$, while its eccentricity $\varepsilon := a^{-1}\sqrt{a^2-b^2} = u/\sqrt{u^2+1} = r/\sqrt{\rho^2+r^2}$. The c.t.e. E can then be analytically expressed (in the adopted o.c.s.) as (16), where we set $u := r/\rho$, and therefore (17) holds.

For the calculations of the two contact points $T_1(x_1, y_1)$ and $T_1'(-x_1, -y_1)$ between E_1 and its tangential ellipse E, as well as for their corresponding tangent lines t_1 and t_1' (at T_1 and T_1'), we consider the following approach. The analytical expression of E_1 is given by (9), as $E_1 = S_x(C)$, where we set $u := r/\rho$. Let now $x = x_0$ be a vertical tangent line of E_1 at some point $T_0(x_0, y_0) \in E_1$, $y_0 > 0$. Then, setting $x := x_0$ into (9), a trinomial with respect to y is derived, say, $\eta_1 = \eta_1(y;\, x_0) = 0$. Because we expect $\eta_1(y) = 0$ to have one real (double) root $y = y_0$ (due to the fact that line $x = x_0$ was assumed to be a tangent line of E_1 at T_0), its discriminant must be zero, i.e., $4u^2\sin^2\omega\,(u^2\rho^2\cos^2\omega + \rho^2 - x_0^2) = 0$, or equivalently

$$x_0 = \pm\rho\sqrt{1+u^2\cos^2\omega} = \pm\sqrt{\rho^2+r^2\cos^2\omega},$$

while setting the x_0 value, as above, to the trinomial $\eta(y;\, x_0) = 0$, its double real roots would then be given by

$$y_0 = \pm\frac{u^2\rho\sin 2\omega}{2\sqrt{1+u^2\cos^2\omega}} = \tfrac{1}{2}x_0^{-1}r^2\sin 2\omega.$$

Adopting the positive value for x_0, from the above relation, we obtain

$$x_0 = \sqrt{\rho^2+r^2\cos^2\omega} \;\text{ and }\; y_0 = \frac{r^2\sin 2\omega}{2\sqrt{\rho^2+r^2\cos^2\omega}}. \tag{47}$$

Similarly, if $x = x_0'$ is now assumed to be also a vertical tangent line of E at some point $T_0'(x_0', y_0') \in E$, then by setting $x := x_0'$ into (17), a trinomial with respect also to y is derived, say, $\eta_1' = \eta_1'(y;\, x_0') = 0$. As we expect $\eta_1'(y) = 0$ to have also one real (double) root $y = y_0'$ (due to the fact that line $x = x_0'$ was assumed to be a tangent line of E at T_0'), its discriminant must again be zero. The new calculations yield that x_0' and y_0' expressions are exactly the same as x_0 and y_0 in (47), i.e., $x_0' = x_0$ and $y_0' = y_0$. Therefore, the vertical tangent lines of E_1 and E coincide at point $T_0 = T_0'$, i.e., the contact point $T_1(x_1, y_1)$ of E_1 and E is indeed given by (47), where x_0 and y_0 notations were replaced by x_1 and y_1, respectively, and hence (18) holds. The corresponding tangent line t_1 at point T_1 is thus the vertical tangent line $x = x_1$ $(= x_0 = x_0')$ which is proved to be a common tangent line between E_1 and E. For clarification, see also Fig. 5.

For the two contact points $T_2(x_2, y_2)$, $y_2 > 0$, and $T_2'(-x_2, -y_2)$ between E_2 and E, as well as for their corresponding tangent lines t_2 and t_2' (at T_2 and T_2'), we consider the analytical expression of E_2 given in (5), as $E_2 = S_y(C)$, where we also have to set $u := r/\rho$. Let $y = y_0$ be a horizontal tangent line of E_2 at some point $S_0(\chi_0, \psi_0) \in E_2$, $\psi_0 > 0$. Then, by setting $y := \psi_0$ into (5), a trinomial with respect to x is derived, say, $\eta_2 = \eta_2(x;\ \psi_0) = 0$. Because we expect $\eta_2(x) = 0$ to have one real (double) root $x = \chi_0$ (due to the fact that line $y = \psi_0$ was assumed to be a tangent line of E_2 at S_0), its discriminant must be zero, i.e., $4u^2 \cos^2\omega\, (u^2\rho^2 \sin^2\omega + \rho^2 - y_0^2) = 0$, or equivalently

$$\psi_0 = \pm\rho\sqrt{1 + u^2 \sin^2\omega} = \pm\sqrt{\rho^2 + r^2 \sin^2\omega}, \tag{48}$$

while adopting the positive value for ψ_0 as above, and setting it to the trinomial $\eta_2(x;\ \psi_0) = 0$, its double real root is then given by

$$\chi_0 = \frac{u^2\rho \sin 2\omega}{2\sqrt{1 + u^2 \sin^2\omega}} = \frac{r^2 \sin 2\omega}{2\sqrt{\rho^2 + r^2 \sin^2\omega}}. \tag{49}$$

Similarly, if $y = \psi_0'$ is assumed to be a horizontal tangent line of E at some point $S_0'(\chi_0', \psi_0') \in E$, $\psi_0' > 0$, then, by setting $y := \psi_0'$ into (17), a trinomial with respect also to x is derived, say, $\eta_2' = \eta_2'(x;\ \psi_0') = 0$. As we expect the trinomial $\eta_2'(x) = 0$ to have one real (double) root $x = \chi_0'$ (due to the fact that line $y = \psi_0'$ was assumed to be a tangent line of E at S_0'), its discriminant must again be zero. The calculations yield that the values of χ_0' and ψ_0' coincide to the χ_0 and ψ_0 values as in (49), i.e., $\chi_0' = \chi_0$ and $\psi_0' = \psi_0$. Therefore, the horizontal tangent lines of E_2 and E coincide at point $S_0 = S_0'$, i.e., the contact point $T_2(x_2, y_2)$, $y_2 > 0$, between E_2 and E, is indeed $S_0(\chi_0, \psi_0)$, and hence (19) holds. The corresponding tangent line t_2 at point T_2 then coincides with the horizontal tangent line $y = y_2$ $(= \psi_0 = \psi_0')$ which is proved to be a common tangent line of E_2 and E. For clarification, see also Fig. 5.

For the last two contact points $T_3(x_3, y_3)$ and $T_3'(-x_3, -y_3)$ between circle C and its tangential ellipse E, it holds that $x_3^2 + y_3^2 = \rho^2$. Moreover, the slope of diameter T_3T_3' is given by $\tan(\omega + \pi/2) = y_3/x_3$. This is due to the fact that ellipse E is a tangential ellipse to C. Hence, the minor axis of E (orthogonal to the E's major axis which forms an angle ω with o.c.s.'s horizontal axis) is assumed to be equal to the radius of C, i.e., $\rho = b$; see also Fig. 5. The above two relations imply that

$$x_3 = -y_3 \tan\omega \text{ and } y_3^2 = \frac{\rho^2}{1+\tan^2\omega} = \rho^2\cos^2\omega, \tag{50}$$

and therefore (20) is derived. The corresponding two tangent lines t_3 and t_3' of C at their contact points T_3 and T_3', respectively, are given in their usual forms t_3 : $x_3x+y_3y = \rho^2$ and t_3': $x_3x+y_3y = -\rho^2$ (on the adopted o.c.s.), as they are tangent lines of the circle C, and hence by substitution of (20) into them, relations (21) hold. Therefore, the tangent lines t_3 and t_3' are parallel to the non-orthogonal radii OR, as their slopes are both $\tan\omega$.

Proof (of Theorem 2) Consider an o.c.s. where its horizontal and vertical axes are spanned by radii OP and OQ', respectively, where $OQ' \perp OP$ and $|OQ'| = |OP|$. Hence, the corresponding orthonormal vector base is given by $\mathscr{B} := \{\mathbf{e}_1 := OP/|OP|, \mathbf{e}_2 := OQ'/|OQ'|\}$, i.e., $P_{\mathscr{B}}(p, 0)$ and $Q'_{\mathscr{B}}(0, p)$, or simply $P(p, 0)$ and $Q'(0, p)$. For point Q it then holds that $Q(q\cos\varphi, q\sin\varphi)$. Let also $E_1 := E_1(OP, OR)$, $E_2 := E_2(OQ, OR)$, and $E_3 := E_3(OP, OQ)$ be three mutually conjugate ellipses defined by their corresponding three mutually conjugate radii OP, OQ, and OR. As pointed out in Definition 2, a new c.s. can be adopted so that one of the ellipses E_i can be expressed as a circle. Without loss of generality, we may choose the c.s. in which ellipse E_3 is expressed as circle $C(O, p)$. It then holds that $OP \perp OQ$ and $|OP| = |OQ|$, i.e., $\varphi = \pi/2$ and $p = q$. Hence, Theorem 1 can be applied in order to derive the primary c.t.e. E of E_i, $i = 1, 2, 3$. Moreover, due to the fact that $0 < \vartheta = \varphi - \omega$, or $\omega < \varphi$, the given radius OR lies "between" OP and OQ, as $\omega \in (0, \varphi) \subset (0, \pi)$. The meaning of "between" indicates that the angles between the non-orthogonal (mutually conjugate) radius OR and the other two orthogonal (mutually conjugate) radii OP and OQ are not exceeding $\pi/2$; see also Fig. 5 where OR lies between OP and OQ.

For the construction of this new c.s., in which E_3 corresponds to circle $C(O, p)$, we consider the following: Let C be a circle of radius p centered at O and its points $P(p, 0)$ and $Q'(0, p)$. Let $S_x : \mathbf{x}' = \mathbf{N_u x}$ be a x-shear transformation with shearing vector $\mathbf{u} := p^{-1}OQ$ in the initially adopted o.c.s. (of vector base $\mathscr{B}$). Then, S_x transforms point the $Q'(0, p)$ into $Q(q\cos\varphi, q\sin\varphi)$. Indeed, substituting the shearing vectors' length $u := |\mathbf{u}| = q/p$ and its angle $\omega := \varphi$ into (3), the x-shear transformation matrix is then given by

$$\mathbf{N_u} = \begin{pmatrix} 1 & u\cos\omega \\ 0 & u\sin\omega \end{pmatrix} = \begin{pmatrix} 1 & \frac{q}{p}\cos\varphi \\ 0 & \frac{q}{p}\sin\varphi \end{pmatrix}, \tag{51}$$

and hence it can be easily verified that $S_x(Q') = Q$. Moreover, it holds that $S_x(P) = P$, as the horizontal axis $x'Ox$ (in $\mathscr{B}$) is an S_x-invariant. Therefore, the x-sheared circle C, i.e., $S_x(C)$, is essentially the ellipse E_3, as the orthogonal pair of radii (OP, OQ') is transformed into the pair of conjugate radii (OP, OQ) which defines ellipse E_3, as $E_3 = E_3(OP, OQ)$ is assumed, i.e., $S_x(C) = E_3$ with $P \in E_3 \cap E_1$ and $Q \in E_3 \cap E_2$.

Consider now a new vector base, say, $\mathscr{B}'$, in which ellipse E_3 is expressed as circle $C(O, p)$ or, equivalently, $OP \perp_{\mathscr{B}'} OQ$ and $|OP|_{\mathscr{B}'} = |OQ|_{\mathscr{B}'}$. Essentially, the affine transformation S_x (as defined above) corresponds to a vector base change, from $\mathscr{B}$ to $\mathscr{B}'$, of the form $\mathbf{x} = \mathbf{N_u}\mathbf{x}'$, in which $P_{\mathscr{B}'}(p, 0)$ and $Q_{\mathscr{B}'}(0, p)$. The length $r' := |OR|_{\mathscr{B}'}$ and angle $\omega' := \measuredangle_{\mathscr{B}'}(OP, OR)$ can now be calculated in order to use them instead of values $r := |OP|$ and $\omega := \measuredangle(OP, OR)$, for the application of Theorem 1. The components' vector $\mathbf{r}$ of OR is also given, in the initial vector base $\mathscr{B}$, by $\mathbf{r} = (r_1, r_2)^{\mathrm{T}} := (r\cos\omega, r\sin\omega)^{\mathrm{T}} \in \mathbb{R}^{2\times 1}$. When OR is expressed in $\mathscr{B}'$, its new components' vector, say, $\mathbf{r}'$, is then given by $\mathbf{r}' = (r_1', r_2')^{\mathrm{T}} := (r'\cos\omega', r'\sin\omega')^{\mathrm{T}} \in \mathbb{R}^{2\times 1}$. Applying the vector base change $\mathbf{x} = \mathbf{N_u}\mathbf{x}'$ on the vector $\mathbf{r}$, we obtain that $\mathbf{r} = \mathbf{N_u}\mathbf{r}'$, or

$$\left(r_1', r_2'\right)^{\mathrm{T}} = \mathbf{r}' = \mathbf{N_u}^{-1}\mathbf{r} = r(\csc\varphi)\left(\sin(\varphi - \omega), \tfrac{p}{q}\sin\omega\right)^{\mathrm{T}}, \tag{52}$$

and thus $r' = |OR|_{\mathscr{B}'} = |\mathbf{r}'|_{\mathscr{B}'} = \sqrt{r_1'^2 + r_2'^2}$, while OR's corresponding angle ω' is then given by $\omega' = \measuredangle_{\mathscr{B}'}(OP, OR) = \arctan(r_2'/r_1') \in (-\pi, \pi)$. Therefore, from (52), we derive, after some algebra, the expressions as in (22).

Recall the mutually conjugate ellipses $E_1 = E_1(OP, OR)$, $E_2 = E_2(OQ, OR)$, and $E_3 = E_3(OP, OQ)$, with $P_{\mathscr{B}'}(p, 0)$, $Q_{\mathscr{B}'}(0, p)$ and $R_{\mathscr{B}'}\left(r_1', r_2'\right)$. Therefore, in vector base $\mathscr{B}'$, where the ellipse E_3 corresponds to circle $C(O, \rho)$, Theorem 1 can be applied, and hence the primary c.t.e. E of E_i, $i = 1, 2, 3$ (as in Definition 2) always exists. Recalling Theorem 1, the non-orthogonal radii OR coincides with E's major semiaxis, while the foci of E are essentially the end points of (non-orthogonal) diameter spanned by radius OR. In vector base $\mathscr{B}'$, Theorem 1 shows also that each pair of (diametrical) common tangent lines (t_i, t_i') between ellipse E_i and its primary c.t.e. E, for $i = 1, 2, 3$, are being parallel to each of the three given mutually conjugate radii OP, OQ, and OR. In particular, t_1 and t_1' are parallel to OQ, t_2 and t_2' are parallel to OP, while t_3 and t_3' are parallel to OR. Therefore, as S_x preserves parallelism, we conclude that also in the initially adopted vector base $\mathscr{B}$, the common tangent lines between each E_i and E are parallel to each (of the three) mutually conjugate radii that define E_i, $i = 1, 2, 3$.

Finally, we point out that the analytical expression of the primary common tangential ellipses E can be calculated (in the vector base $\mathscr{B}'$) by setting $r := r'$ and $\omega := \omega'$ both into (17), where r' and ω' are as in (22).

References

1. Emch, A.: Proof of Pohlke's theorem and its generalizations by affinity. Am. J. Math. **40**(2), 366–374 (1918)
2. Fishburn, P.C., Trotter, W.T.: Containment orders for similar ellipses with a common center. Discret. Math. **256**, 129–136 (2002)
3. Lefkaditis, G.E., Toulias, T.L., Markatis, S.: The four ellipses problem. Int. J. Geom. **5**(2), 77–92 (2016)
4. Müller, E., Kruppa, E.: Lehrbuch der Darstellenden Geometrie. Springer, Wien (1961)
5. Peschka, G.A.: Elementarer beweis des Pohlke'schen fundamentalsatzes der Axonometrie. Stzgsb. Math. Nat., Akad. Wien **LXXVIII, II Abth**, 1043–1054 (1878)
6. Sklenáriková, Z., Pémová, M.: The Pohlke-Schwarz theorem and its relevancy in the didactics of mathematics. Quaderni di Ricerca in Didattica **17** (2007). http://math.unipa.it/~grim/quad17_sklenarikova-pemova_07.pdf

Diagonal Fixed Points of Geometric Contractions

Mihai Turinici

Abstract A geometric Meir-Keeler extension is given for the diagonal fixed point result in Ćirić and Presić (Acta Math Comenianae 76:143–147, 2007).

Keywords Metric space · Diagonal fixed point · Ćirić-Presić contraction · Geometric Meir-Keeler relation · Matkowski admissible function

Introduction

Let X be a nonempty set. Call the subset Y of X, *almost singleton* (in short *asingleton*), provided $[y_1, y_2 \in Y$ implies $y_1 = y_2]$; and *singleton* if, in addition, Y is nonempty; note that in this case $Y = \{y\}$, for some $y \in X$.

Take a *metric* $d : X \times X \to R_+ := [0, \infty[$ over X; the couple (X, d) will be referred to as a *metric space*. Further, let $S \in \mathcal{F}(X)$ be a self-map of X. [Here, for each couple A, B of nonempty sets, $\mathcal{F}(A, B)$ stands for the class of all functions from A to B; when $A = B$, we write $\mathcal{F}(A)$ in place of $\mathcal{F}(A, A)$]. Denote $\mathrm{Fix}(S) = \{x \in X; x = Sx\}$; each point of this set is referred to as *fixed* under S. The determination of such points is to be performed in the context below, comparable with the one in Rus [38, Ch 2, Sect 2.2]:

(pic-0) We say that S is *fix-asingleton*, provided $\mathrm{Fix}(S)$ is an asingleton; and *fix-singleton*, provided $\mathrm{Fix}(S)$ is a singleton

(pic-1) We say that S is a *Picard operator* (modulo d) if, for each $x \in X$, the iterative sequence $(S^n x; n \geq 0)$ is d-convergent

(pic-2) We say that S is a *strong Picard operator* (modulo d) if, for each $x \in X$, $(S^n x; n \geq 0)$ is d-convergent with $\lim_n(S^n x) \in \mathrm{Fix}(S)$.

M. Turinici (✉)
A. Myller Mathematical Seminar, A. I. Cuza University, Iaşi, Romania
e-mail: mturi@uaic.ro

N. J. Daras, Th. M. Rassias (eds.), *Modern Discrete Mathematics and Analysis*,
Springer Optimization and Its Applications 131,
https://doi.org/10.1007/978-3-319-74325-7_22

The basic result in this area is the 1922 one due to Banach [3]. Call $S \in \mathcal{F}(X)$ *Banach* $(d; \alpha)$*-contractive* (where $\alpha \geq 0$), if

(B-contr) $d(Sx, Sy) \leq \alpha d(x, y)$, for all $x, y \in X$.

Theorem 1 *Assume that* S *is Banach* $(d; \alpha)$*-contractive, for some* $\alpha \in [0, 1[$. *In addition, let* X *be* d*-complete. Then,*

(11-a) S *is fix-singleton; i.e.,* $\text{Fix}(S) = \{z\}$, *for some* $z \in X$
(11-b) S *is a strong Picard operator:*
$S^n x \xrightarrow{d} z$ *as* $n \to \infty$, *for each* $x \in X$.

This result (referred to as Banach's fixed point theorem) found some basic applications to the operator equations theory. Consequently, a multitude of extensions for it were proposed. From the perspective of this exposition, the following ones are of interest.

(I) Metrical type extensions: the initial Banach contractive property is taken in a *(geometric) implicit* way, as

(ge-im) $(d(Sx, Sy), d(x, y), d(x, Sx), d(y, Sy), d(x, Sy), d(Sx, y)) \in \mathcal{M}$, for all $x, y \in X$,

where $\mathcal{M} \subseteq R_+^6$ is a (nonempty) subset. In particular, for the *functional implicit* version of this

(fct-im) $(t_1, t_2, t_3, t_4, t_5, t_6) \in \mathcal{M}$ iff $H(t_1, t_2, t_3, t_4, t_5, t_6) \leq 0$

(where $H : R_+^6 \to R$ is a function), certain technical aspects have been considered by Leader [25] and Turinici [47]. And, for the *explicit* case of this, expressed as

(fct-ex) $(t_1, t_2, t_3, t_4, t_5, t_6) \in \mathcal{M}$ iff $t_1 \leq K(t_2, t_3, t_4, t_5, t_6)$

(where $K : R_+^5 \to R_+$ is a function), some consistent lists of such contractions may be found in the survey papers by Rhoades [35], Collaco and E Silva [10], Kincses and Totik [23], as well as the references therein.

(II) Diagonal type extensions: the initial self-map S is to be viewed as the *diagonal part* of a mapping $T : X^k \to X$ (for some $k \geq 1$); i.e.,

$Sx = T(x^k), x \in X$.

(Here, for each element $u \in X$, we introduced the convention

u^k=the vector $(z_0, \ldots, z_{k-1}) \in X^k$, where $(z_j = u, 0 \leq j \leq k-1)$;

note that in this case, $u^1 = u$.) Then, the fixed points of S may be approximated by k-step iterative processes attached to T. The first result of this type seems to have been obtained in 1965 by Presić [33]; a refinement of it is to be found in 2007 by Cirić and Presić [8].

Having these precise, it is our aim in the following to extend—from a geometric functional perspective—this last class of results. The obtained facts may be compared with some related statements in Rus [37]; further aspects will be delineated in a separate paper.

Dependent Choice Principles

Throughout this exposition, the axiomatic system in use is Zermelo-Fraenkel's (ZF), as described by Cohen [9, Ch 2]. The notations and basic facts to be considered are standard; some important ones are discussed below.

(**A**) Let X be a nonempty set. By a *relation* over X, we mean any (nonempty) part $\mathcal{R} \subseteq X \times X$; then, $(X, \mathcal{R})$ will be referred to as a *relational structure*. Note that $\mathcal{R}$ may be regarded as a mapping between X and $\exp[X]$ (=the class of all subsets in X). In fact, let us simplify the string $(x, y) \in \mathcal{R}$ as $x\mathcal{R}y$; and put, for $x \in X$,

$$X(x, \mathcal{R}) = \{y \in X; x\mathcal{R}y\} \text{ (the } \textit{section} \text{ of } \mathcal{R} \text{ through } x);$$

then, the desired mapping representation is $(\mathcal{R}(x) = X(x, \mathcal{R}); x \in X)$. A basic example of such object is

$$\mathcal{I} = \{(x, x); x \in X\} \text{ [the } \textit{identical relation} \text{ over } X].$$

Given the relations $\mathcal{R}$, $\mathcal{S}$ over X, define their *product* $\mathcal{R} \circ \mathcal{S}$ as

$$(x, z) \in \mathcal{R} \circ \mathcal{S}, \text{ if there exists } y \in X \text{ with } (x, y) \in \mathcal{R}, (y, z) \in \mathcal{S}.$$

Also, for each relation $\mathcal{R}$ in X, denote

$$\mathcal{R}^{-1} = \{(x, y) \in X \times X; (y, x) \in \mathcal{R}\} \text{ (the } \textit{inverse} \text{ of } \mathcal{R}).$$

Finally, given the relations $\mathcal{R}$ and $\mathcal{S}$ on X, let us say that $\mathcal{R}$ is *coarser* than $\mathcal{S}$ (or, equivalently, $\mathcal{S}$ is *finer* than $\mathcal{R}$), provided

$$\mathcal{R} \subseteq \mathcal{S}; \text{ i.e., } x\mathcal{R}y \text{ implies } x\mathcal{S}y.$$

Given a relation $\mathcal{R}$ on X, the following properties are to be discussed here:

(P1) $\mathcal{R}$ is *reflexive*: $\mathcal{I} \subseteq \mathcal{R}$
(P2) $\mathcal{R}$ is *irreflexive*: $\mathcal{I} \cap \mathcal{R} = \emptyset$
(P3) $\mathcal{R}$ is *transitive*: $\mathcal{R} \circ \mathcal{R} \subseteq \mathcal{R}$
(P4) $\mathcal{R}$ is *symmetric*: $\mathcal{R}^{-1} = \mathcal{R}$
(P5) $\mathcal{R}$ is *antisymmetric*: $\mathcal{R}^{-1} \cap \mathcal{R} \subseteq \mathcal{I}$.

This yields the classes of relations to be used; the following ones are important for our developments:

(C0) $\mathcal{R}$ is *amorphous* (i.e., it has no specific properties)
(C1) $\mathcal{R}$ is a *quasi-order* (reflexive and transitive)

(C2) $\mathcal{R}$ is a *strict order* (irreflexive and transitive)
(C3) $\mathcal{R}$ is an *equivalence* (reflexive, transitive, symmetric)
(C4) $\mathcal{R}$ is a *(partial) order* (reflexive, transitive, antisymmetric)
(C5) $\mathcal{R}$ is the *trivial* relation (i.e., $\mathcal{R} = X \times X$).

(B) A basic example of relational structure is to be constructed as below. Let

$$N = \{0, 1, 2, \ldots\}, \text{ where } (0 = \emptyset, 1 = \{0\}, 2 = \{0, 1\}, \ldots)$$

denote the set of *natural* numbers. Technically speaking, the basic (algebraic and order) structures over N may be obtained by means of the *(immediate) successor* function $\text{suc} : N \to N$ and the following Peano properties (deductible in our axiomatic system (ZF)):

(pea-1) $0 \in N$ and $0 \notin \text{suc}(N)$
(pea-2) suc(.) is injective ($\text{suc}(n) = \text{suc}(m)$ implies $n = m$)
(pea-3) if $M \subseteq N$ fulfills $[0 \in M]$ and $[\text{suc}(M) \subseteq M]$, then $M = N$.

(Note that, in the absence of our axiomatic setting, these properties become the well-known Peano axioms, as described in Halmos [15, Ch 12]; we do not give details). In fact, starting from these properties, one may construct, in a recurrent way, an *addition* $(a, b) \mapsto a + b$ over N, according to

$$(\forall m \in N)\text{: } m + 0 = m;\ m + \text{suc}(n) = \text{suc}(m + n).$$

This, in turn, makes possible the introduction of a relation $(\leq)$ over N, as

$$(m, n \in N)\text{: } m \leq n \text{ iff } m + p = n, \text{ for some } p \in N.$$

Concerning the properties of this structure, the most important one writes

$(N, \leq)$ is well ordered:
any (nonempty) subset of N has a first element;

hence (in particular), $(N, \leq)$ is (partially) ordered. Denote, for $r \in N$,

$$N(r, \leq) = \{n \in N; r \leq n\} = \{r, r + 1, \ldots, \}, r \geq 0,$$
$$N(r, >) = \{n \in N; r > n\} = \{0, \ldots, r - 1\}, r \geq 1;$$

the latter one is referred to as the *initial interval* (in N) induced by r. Any set P with $N \sim P$ (in the sense: there exists a bijection from N to P) will be referred to as *effectively denumerable*. In addition, given some natural number $n \geq 1$, any set Q with $N(n, >) \sim Q$ will be said to be *n-finite*; when n is generic here, we say that Q is *finite*. Finally, the (nonempty) set Y is called (at most) *denumerable* iff it is either effectively denumerable or finite.

By a *sequence* in X, we mean any mapping $x : N \to X$, where $N = \{0, 1, \ldots\}$ is the set of *natural* numbers. For simplicity reasons, it will be useful to denote it as $(x(n); n \geq 0)$, or $(x_n; n \geq 0)$; moreover, when no confusion can arise, we further simplify this notation as $(x(n))$ or (x_n), respectively. Also, any sequence $(y_n := x_{i(n)}; n \geq 0)$ with

$(i(n); n \geq 0)$ is *strictly ascending* (hence, $i(n) \to \infty$ as $n \to \infty$)

will be referred to as a *subsequence* of $(x_n; n \geq 0)$.

(C) Remember that an outstanding part of (ZF) is the *Axiom of Choice* (AC); which, in a convenient manner, may be written as

(AC) For each couple (J, X) of nonempty sets and each function $F : J \to \exp(X)$, there exists a (selective) function $f : J \to X$, with $f(\nu) \in F(\nu)$, for each $\nu \in J$.

(Here, $\exp(X)$ stands for the class of all nonempty subsets in X.) Sometimes, when the ambient set X is endowed with denumerable type structures, the case of $J = N$ will suffice for all choice reasoning to be used there; and the existence of such a selective function may be determined by using a weaker form of (AC), called *Dependent Choice* principle (DC). Let us say that the relation $\mathcal{R}$ over X is *proper*, when

$$(X(x, \mathcal{R}) =)\mathcal{R}(x) \text{ is nonempty, for each } x \in X.$$

Then, $\mathcal{R}$ is to be viewed as a mapping between X and $\exp(X)$; and the couple $(X, \mathcal{R})$ will be referred to as a *proper relational structure*. Further, given $a \in X$, we term the sequence $(x_n; n \geq 0)$ in X, $(a; \mathcal{R})$*-iterative* provided

$$x_0 = a \text{ and } x_n \mathcal{R} x_{n+1} \text{ (i.e., } x_{n+1} \in \mathcal{R}(x_n)\text{), for all } n.$$

Proposition 1 *Let the relational structure $(X, \mathcal{R})$ be proper. Then, for each $a \in X$, there is at least an $(a; \mathcal{R})$-iterative sequence in X.*

This principle—proposed, independently, by Bernays [5] and Tarski [43]—is deductible from (AC), but not conversely (cf. Wolk [53]). Moreover, by the developments in Moskhovakis [29, Ch 8] and Schechter [39, Ch 6], the *reduced system* (ZF-AC+DC) is comprehensive enough so as to cover the "usual" mathematics; see also Moore [28, Appendix 2].

Let $(\mathcal{R}_n; n \geq 0)$ be a sequence of relations on X. Given $a \in X$, let us say that the sequence $(x_n; n \geq 0)$ in X is $(a; (\mathcal{R}_n; n \geq 0))$*-iterative*, provided

$$x_0 = a, \text{ and } x_n \mathcal{R}_n x_{n+1} \text{ (i.e., } x_{n+1} \in \mathcal{R}_n(x_n)\text{), for all } n.$$

The following *Diagonal Dependent Choice* principle (DDC) is available.

Proposition 2 *Let $(\mathcal{R}_n; n \geq 0)$ be a sequence of proper relations on X. Then, for each $a \in X$, there exists at least one $(a; (\mathcal{R}_n; n \geq 0))$-iterative sequence in X.*

Clearly, (DDC) includes (DC) to which it reduces when $(\mathcal{R}_n; n \geq 0)$ is constant. The reciprocal of this is also true. In fact, letting the premises of (DDC) hold, put $P = N \times X$; and let $\mathcal{S}$ be the relation over P introduced as

$$\mathcal{S}(i, x) = \{i + 1\} \times \mathcal{R}_i(x), \ (i, x) \in P.$$

It will suffice applying (DC) to $(P, \mathcal{S})$ and $b := (0, a) \in P$ to get the conclusion in our statement; we do not give details.

Summing up, (DDC) is provable in (ZF-AC+DC). This is valid as well for its variant, referred to as the *Selected Dependent Choice* principle (SDC).

Proposition 3 *Let the map $F : N \to \exp(X)$ and the relation $\mathcal{R}$ over X fulfill*

($\forall n \in N$): $\mathcal{R}(x) \cap F(n+1) \neq \emptyset$, for all $x \in F(n)$.

Then, for each $a \in F(0)$, there exists a sequence $(x(n); n \geq 0)$ in X, with

$$x(0) = a, \ x(n) \in F(n), \ x(n+1) \in \mathcal{R}(x(n)), \ \forall n.$$

As before, (SDC) $\Longrightarrow$ (DC) ($\Longleftrightarrow$ (DDC)); just take $(F(n) = X; n \geq 0)$. But, the reciprocal is also true, in the sense (DDC) $\Longrightarrow$ (SDC). This follows from

Proof (Proposition 3) Let the premises of (SDC) be true. Define a sequence of relations $(\mathcal{R}_n; n \geq 0)$ over X as: for each $n \geq 0$,

$\mathcal{R}_n(x) = \mathcal{R}(x) \cap F(n+1)$, if $x \in F(n)$,
$\mathcal{R}_n(x) = \{x\}$, otherwise ($x \in X \setminus F(n)$).

Clearly, $\mathcal{R}_n$ is proper, for all $n \geq 0$. So, by (DDC), it follows that for the starting $a \in F(0)$, there exists an $(a, (\mathcal{R}_n; n \geq 0))$-iterative sequence $(x(n); n \geq 0)$ in X. Combining with the very definition above, one derives the desired fact.

In particular, when $\mathcal{R} = X \times X$, the regularity condition imposed in (SDC) holds. The corresponding variant of the underlying statement is just (AC(N)) (=the *denumerable axiom of choice*). Precisely, we have

Proposition 4 *Let $F : N \to \exp(X)$ be a function. Then, for each $a \in F(0)$, there exists a function $f : N \to X$ with $f(0) = a$ and $f(n) \in F(n)$, $\forall n \in N$.*

As a consequence of the above facts, (DC) $\Longrightarrow$ (AC(N)) in (ZF-AC). A direct verification of this is obtainable by taking $A = N \times X$ and introducing the relation $\mathcal{R}$ over it, according to

$\mathcal{R}(n, x) = \{n+1\} \times F(n+1), \ n \in N, x \in X;$

we do not give details. The reciprocal of the written inclusion is not true; see, for instance, Moskhovakis [29, Ch 8, Sect 8.25].

Conv-Cauchy Structures

Let X be a nonempty set; and $\mathcal{S}(X)$ stand for the class of all sequences (x_n) in X. By a (sequential) *convergence structure* on X, we mean any part $\mathcal{C}$ of $\mathcal{S}(X) \times X$, with the properties (cf. Kasahara [19]):

(conv-1) ($\mathcal{C}$ is hereditary)
$((x_n); x) \in \mathcal{C} \Longrightarrow ((y_n); x) \in \mathcal{C}$, for each subsequence (y_n) of (x_n)

(conv-2) ($\mathcal{C}$ is reflexive)
$(\forall u \in X)$: the constant sequence $(x_n = u; n \geq 0)$ fulfills $((x_n); u) \in \mathcal{C}$.

For each sequence (x_n) in $\mathcal{S}(X)$ and each $x \in X$, we simply write $((x_n); x) \in \mathcal{C}$ as $x_n \xrightarrow{\mathcal{C}} x$; this reads

(x_n), $\mathcal{C}$-converges to x (also referred to as x is some *limit* of (x_n)).

The set of all such x is denoted as $\lim_n(x_n)$; when it is nonempty, we say that (x_n) is $\mathcal{C}$-*convergent*. The following condition is to be optionally considered here:

(conv-3) $\mathcal{C}$ is separated:
$\lim_n(x_n)$ is an asingleton, for each sequence (x_n);

when it holds, $x_n \xrightarrow{\mathcal{C}} z$ will be also written as $\lim_n(x_n) = z$.

Further, let us say that the subset $Cauchy(X) \subseteq \mathcal{S}(X)$ is a (sequential) *Cauchy structure* on X, provided

(Cauchy-1) $(Cauchy)(X)$ is hereditary)
$(x_n) \in Cauchy(X)$ implies $(y_n) \in Cauchy(X)$, for each subsequence $(y_n; n \geq 0)$ of $(x_n; n \geq 0)$
(Cauchy-2) ($Cauchy(X)$ is reflexive)
$(\forall u \in X)$: the constant sequence $(x_n = u; n \geq 0)$ fulfills $(x_n) \in Cauchy(X)$.

As before, an optional condition to be considered here is

(Cauchy-3) $Cauchy(X)$ is $\mathcal{C}$-compatible:
any $\mathcal{C}$-convergent sequence is in $Cauchy(X)$.

A standard way of introducing such structures is the *(pseudo) metrical* one. By a *pseudometric* over X, we shall mean any map $d : X \times X \to R_+$. Fix such an object; with, in addition,

(r-s) d *reflexive sufficient*: $x = y \Longleftrightarrow d(x, y) = 0$;

in this case, (X, d) is called a *rs-pseudometric space*. Given the sequence (x_n) in X and the point $x \in X$, we say that (x_n) d-*converges* to x (written as $x_n \xrightarrow{d} x$) provided $d(x_n, x) \to 0$ as $n \to \infty$, i.e.,

$\forall \varepsilon > 0, \exists i = i(\varepsilon)$: $i \leq n \Longrightarrow d(x_n, x) < \varepsilon$.

By this very definition, we have the hereditary and reflexive properties:

(d-conv-1) ($(\xrightarrow{d})$ is hereditary)
$x_n \xrightarrow{d} x$ implies $y_n \xrightarrow{d} x$, for each subsequence (y_n) of (x_n)
(d-conv-2) ($(\xrightarrow{d})$ is reflexive)
$(\forall u \in X)$: the constant sequence $(x_n = u; n \geq 0)$ fulfills $x_n \xrightarrow{d} u$.

As a consequence, $(\xrightarrow{d})$ is a sequential convergence on X. The set of all such limit points of (x_n) will be denoted as $\lim_n(x_n)$; if it is nonempty, then (x_n) is called *d-convergent*. Finally, note that $(\xrightarrow{d})$ is not separated, in general; i.e., there may be sequences (x_n) in X, for which $\lim_n(x_n)$ is not an asingleton. However, this property holds, provided (in addition)

(tri) d is triangular: $d(x, y) \leq d(x, z) + d(z, y), \forall x, y, z \in X$
(sym) d is *symmetric*: $d(x, y) = d(y, x)$, for all $x, y \in X$;

i.e., when d becomes a *metric* on X. Further, call the sequence (x_n) (in X) *d-Cauchy* when $d(x_m, x_n) \to 0$ as $m, n \to \infty$, $m < n$, i.e.,

$$\forall \varepsilon > 0, \exists j = j(\varepsilon): \ j \leq m < n \Longrightarrow d(x_m, x_n) < \varepsilon;$$

the class of all these will be denoted as $Cauchy(X, d)$. As before, we have the hereditary and reflexive properties:

(d-Cauchy-1) ($Cauchy(X, d)$ is hereditary)
(x_n) is d-Cauchy implies (y_n) is d-Cauchy,
for each subsequence (y_n) of (x_n)
(d-Cauchy-2) ($Cauchy(X, d)$ is reflexive)
$(\forall u \in X)$: the constant sequence $(x_n = u; n \geq 0)$ is d-Cauchy;

hence, $Cauchy)(X, d)$ is a Cauchy structure on X. Note that, by the imposed (upon d) conditions, $Cauchy)(X, d)$ is not $(\xrightarrow{d})$-compatible, in general; i.e., there may be d-convergent sequences in X that are not endowed with the d-Cauchy property. But, when d is triangular symmetric, this happens as it can be directly seen.

We close this section with a few remarks involving convergent real sequences. For each sequence (r_n) in R, and each element $r \in R$, denote

$$r_n \to r+ \text{ (resp., } r_n \to r-\text{), when } r_n \to r \text{ and } [r_n > r \text{ (resp., } r_n < r\text{)}, \forall n].$$

Proposition 5 *Let the sequence $(r_n; n \geq 0)$ in R and the number $\varepsilon \in R$ be such that $r_n \to \varepsilon+$. Then, there exists a subsequence $(r_n^* := r_{i(n)}; n \geq 0)$ of $(r_n; n \geq 0)$ with*

$(r_n^*; n \geq 0)$ strictly descending and $r_n^* \to \varepsilon+$.

Proof Put $i(0) = 0$. As $\varepsilon < r_{i(0)}$ and $r_n \to \varepsilon+$, we have that

$A(i(0)) := \{n > i(0); r_n < r_{i(0)})\}$ is not empty;
hence, $i(1) := \min(A(i(0)))$ is an element of it, and $r_{i(1)} < r_{i(0)}$.

Likewise, as $\varepsilon < r_{i(1)}$ and $r_n \to \varepsilon+$, we have that

$A(i(1)) := \{n > i(1); r_n < r_{i(1)})\}$ is not empty;
hence, $i(2) := \min(A(i(1)))$ is an element of it, and $r_{i(2)} < r_{i(1)}$.

This procedure may continue indefinitely and yields (without any choice technique) a strictly ascending rank sequence $(i(n); n \geq 0)$ (hence, $i(n) \to \infty$ as $n \to \infty$) for which the attached subsequence $(r_n^* := r_{i(n)}; n \geq 0)$ of $(r_n; n \geq 0)$ fulfills

$r_{n+1}^* < r_n^*$, for all n; hence, (r_n^*) is (strictly) descending.

On the other hand, by this very subsequence property,

$(r_n^* > \varepsilon, \forall n)$ and $\lim_n r_n^* = \lim_n r_n = \varepsilon$.

Putting these together, we get the desired fact.

A bidimensional counterpart of these facts may be given along the lines below. Let $\pi(t, s)$ (where $t, s \in R$) be a logical property involving couples of real numbers.

Given the couple of real sequences $(t_n; n \geq 0)$ and $(s_n; n \geq 0)$, call the subsequences $(t_n^*; n \geq 0)$ of (t_n) and $(s_n^*; n \geq 0)$ of (s_n), *compatible* when

$(t_n^* = t_{i(n)}; n \geq 0)$ and $(s_n^* = s_{i(n)}, n \geq 0)$, for the same strictly ascending rank sequence $(i(n); n \geq 0)$.

Proposition 6 *Let the couple of real sequences $(t_n; n \geq 0)$, $(s_n; n \geq 0)$ and the couple of real numbers (a, b) be such that*

$t_n \to a+$, $s_n \to b+$ as $n \to \infty$ and ($\pi(t_n, s_n)$ is true, $\forall n$).

There exists then a couple of subsequences $(t_n^*; n \geq 0)$ of $(t_n; n \geq 0)$ and $(s_n^*; n \geq 0)$ of $(s_n; n \geq 0)$, respectively, with

$(t_n^*; n \geq 0)$ and $(s_n^*; n \geq 0)$ are strictly descending and compatible
($t_n^* \to a+$, $s_n^* \to b+$, as $n \to \infty$) and ($\pi(t_n^*, s_n^*)$ holds, for all n).

Proof By the preceding statement, the sequence $(t_n; n \geq 0)$ admits a subsequence $(T_n := t_{i(n)}; n \geq 0)$, with

$(T_n; n \geq 0)$ is strictly descending and ($T_n \to a+$, as $n \to \infty$).

Denote $(S_n := s_{i(n)}; n \geq 0)$; clearly,

$(S_n; n \geq 0)$ is a subsequence of $(s_n; n \geq 0)$ with $S_n \to b+$ as $n \to \infty$.

Moreover, by this very construction,

$\pi(T_n, S_n)$ holds, for all n.

Again by that statement, there exists a subsequence $(s_n^* := S_{j(n)} = s_{i(j(n))}; n \geq 0)$ of $(S_n; n \geq 0)$ (hence of $(s_n; n \geq 0)$ as well), with

$(s_n^*; n \geq 0)$ is strictly descending and ($s_n^* \to b+$, as $n \to \infty$).

Denote further $(t_n^* := T_{j(n)} = t_{i(j(n))}; n \geq 0)$; this is a subsequence of $(T_n; n \geq 0)$ (hence of $(t_n; n \geq 0)$ as well), with

$(t_n^*; n \geq 0)$ strictly descending and ($t_n^* \to a+$, as $n \to \infty$).

Finally, by this very construction (and a previous relation),

$\pi(t_n^*, s_n^*)$ holds, for all n.

Summing up, the couple of subsequences $(t_n^*; n \geq 0)$ and $(s_n^*; n \geq 0)$ has all needed properties; and the conclusion follows.

Admissible Functions

Let $R_+^0 :=]0, \infty[$ stand for the class of strictly positive real numbers. Denote by $\mathcal{F}(re)(R_+^0, R)$ the family of all functions $\varphi \in \mathcal{F}(R_+^0, R)$ with

φ is *regressive*: $\varphi(t) < t$, for all $t \in R_+^0$.

For each $\varphi \in \mathcal{F}(re)(R_+^0, R)$, let us introduce the sequential properties:

(n-d-a) φ is *non-diagonally admissible*:
there are no strictly descending sequences $(t_n; n \geq 0)$ in R_+^0
and no elements $\varepsilon \in R_+^0$ with $t_n \to \varepsilon+$, $\varphi(t_n) \to \varepsilon+$
(M-a) φ is *Matkowski admissible*:
for each $(t_n; n \geq 0)$ in R_+^0 with $(t_{n+1} \leq \varphi(t_n), \forall n)$, we have $\lim_n t_n = 0$
(str-M-a) φ is *strongly Matkowski admissible*:
for each $(t_n; n \geq 0)$ in R_+^0 with $(t_{n+1} \leq \varphi(t_n), \forall n)$, we have $\sum_n t_n < \infty$.

(The conventions (M-a) and (str-M-a) are taken from Matkowski [26] and Turinici [50], respectively.) Clearly, for each $\varphi \in \mathcal{F}(re)(R_+^0, R)$,

strongly Matkowski admissible implies Matkowski admissible;

but the converse is not in general true.

To get concrete circumstances under which such properties hold, we need some conventions. Given $\varphi \in \mathcal{F}(re)(R_+^0, R)$, let us introduce the global properties:

(R-a) φ is *Rhoades admissible*:
for each $\varepsilon > 0$, there exists $\delta > 0$, such that
$t, s > 0, t \leq \varphi(s)$, and $\varepsilon < s < \varepsilon + \delta$ imply $t \leq \varepsilon$
(MK-a) φ is *Meir-Keeler admissible*:
for each $\varepsilon > 0$, there exists $\delta > 0$, such that
$\varepsilon < s < \varepsilon + \delta \Longrightarrow \varphi(s) \leq \varepsilon$.

Proposition 7 *For each $\varphi \in \mathcal{F}(re)(R_+^0, R)$, we have in (ZF-AC+DC)*

(41-1) (MK-a) $\Longrightarrow$ (R-a) $\Longrightarrow$ (M-a)
(41-2) (M-a) $\Longrightarrow$ (n-d-a) $\Longrightarrow$ (MK-a).

Hence, for each $\varphi \in \mathcal{F}(re)(R_+^0, R)$, the properties (MK-a), (R-a), (M-a), (n-d-a) are equivalent to each other.

Proof

(i) Suppose that φ is Meir-Keeler admissible; we claim that φ is Rhoades admissible, with the same couple (ε, δ). In fact, let $\varepsilon > 0$ be given and $\delta > 0$ be assured by the Meir-Keeler admissible property. Take the numbers $t, s > 0$ with $t \leq \varphi(s)$ and $\varepsilon < s < \varepsilon + \delta$. As φ is Meir-Keeler admissible, this yields $\varphi(s) \leq \varepsilon$, wherefrom (as $t \leq \varphi(s)$) we must have $t \leq \varepsilon$; hence the claim.

(ii) Suppose that φ is Rhoades admissible; we have to establish that φ is Matkowski admissible. Let $(s_n; n \geq 0)$ be a sequence in R_+^0 with the property $(s_{n+1} \leq \varphi(s_n); n \geq 0)$. Clearly, (s_n) is strictly descending in R_+^0; hence, $\sigma := \lim_n s_n$ exists in R_+. Suppose by contradiction that $\sigma > 0$; and let $\rho > 0$ be given by the Rhoades admissible property of φ. By the above convergence relations, there exists some rank $n(\rho)$, such that

$n \geq n(\rho)$ implies $\sigma < s_n < \sigma + \rho$.

But then, under the notation $(t_n := \varphi(s_n); n \geq 0)$, we get (for the same ranks)

$$\sigma < s_{n+1} \leq t_n < s_n < \sigma + \rho;$$

in contradiction with the Rhoades admissible property. Hence, necessarily, $\sigma = 0$; and conclusion follows.

(iii) Suppose that φ is Matkowski admissible; we assert that φ is non-diagonally admissible. If φ is not endowed with such a property, there must be a strictly descending sequence $(t_n; n \geq 0)$ in R_+^0 and an $\varepsilon > 0$, such that

$t_n \to \varepsilon+$ and $\varphi(t_n) \to \varepsilon+$, as $n \to \infty$.

Put $i(0) = 0$. As $\varepsilon < \varphi(t_{i(0)})$ and $t_n \to \varepsilon+$, we have that

$A(i(0)) := \{n > i(0); t_n < \varphi(t_{i(0)})\}$ is not empty;
hence, $i(1) := \min(A(i(0)))$ is an element of it, and $t_{i(1)} < \varphi(t_{i(0)})$.

Likewise, as $\varepsilon < \varphi(t_{i(1)})$ and $t_n \to \varepsilon+$, we have that

$A(i(1)) := \{n > i(1); t_n < \varphi(t_{i(1)})\}$ is not empty;
hence, $i(2) := \min(A(i(1)))$ is an element of it, and $t_{i(2)} < \varphi(t_{i(1)})$.

This procedure may continue indefinitely and yields (without any choice technique) a strictly ascending rank sequence $(i(n); n \geq 0)$ (hence, $i(n) \to \infty$ as $n \to \infty$), for which the attached subsequence $(s_n := t_{i(n)}; n \geq 0)$ of (t_n) fulfills

$s_{n+1} < \varphi(s_n)(< s_n)$, for all n.

On the other hand, by this very subsequence property,

$(s_n > \varepsilon, \forall n)$ and $\lim_n s_n = \lim_n t_n = \varepsilon$.

The obtained relations are in contradiction with the Matkowski property of φ; hence, the working condition cannot be true; and we are done.

(iv) Suppose that φ is non-diagonally admissible; we show that, necessarily, φ is Meir-Keeler admissible. If φ is not endowed with such a property, we must have (for some $\varepsilon > 0$)

$$H(\delta) := \{t \in R^0_+; \varepsilon < t < \varepsilon + \delta, \varphi(t) > \varepsilon\} \text{ is not empty, for each } \delta > 0.$$

Taking a strictly descending sequence $(\delta_n; n \geq 0)$ in R^0_+ with $\delta_n \to 0$, we get by the denumerable axiom of choice (AC(N)) [deductible, as precise, in (ZF-AC+DC)], a sequence $(t_n; n \geq 0)$ in R^0_+, so as

$(\forall n)$: t_n is an element of $H(\delta_n)$;

or equivalently (by the very definition above and φ=regressive)

$(\forall n)$: $\varepsilon < \varphi(t_n) < t_n < \varepsilon + \delta_n$;
hence, in particular: $\varphi(t_n) \to \varepsilon+$ and $t_n \to \varepsilon+$.

By a previous result, there exists a subsequence $(r_n := t_{i(n)}; n \geq 0)$ of $(t_n; n \geq 0)$, such that

(r_n) is strictly descending and $r_n \to \varepsilon+$; hence, necessarily, $\varphi(r_n) \to \varepsilon+$.

But, this last relation is in contradiction with the non-diagonal admissible property of our function. Hence, the assertion follows; and we are done.

In the following, some sequential counterparts of the Rhoades admissible and Meir-Keeler admissible properties are provided.

For any $\varphi \in \mathcal{F}(re)(R^0_+, R)$ and any $s \in R^0_+$, put

$\Lambda^+\varphi(s) = \inf_{\varepsilon>0} \Phi(s+)(\varepsilon)$, where $\Phi(s+)(\varepsilon) = \sup \varphi(]s, s+\varepsilon[)$, $\varepsilon > 0$
(the *right superior limit* of φ at s).

From the regressiveness of φ,

$$-\infty \leq \Lambda^+\varphi(s) \leq s, \quad \forall s \in R^0_+;$$

but the case of these extremal values being attained cannot be avoided.

The following properties of this concept will be useful.

Proposition 8 *Let $\varphi \in \mathcal{F}(re)(R^0_+, R)$ and $s \in R^0_+$ be arbitrary fixed. Then,*

(42-1) $\limsup_n(\varphi(t_n)) \leq \Lambda^+\varphi(s)$,
for each sequence (t_n) in R^0_+ with $t_n \to s+$
(42-2) there exists a strictly descending sequence (r_n) in R^0_+ with $r_n \to s+$ and $\varphi(r_n) \to \Lambda^+\varphi(s)$.

Proof Denote, for simplicity,

$\alpha = \Lambda^+\varphi(s)$; hence, $\alpha = \inf_{\varepsilon>0} \Phi(s+)(\varepsilon)$, and $-\infty \leq \alpha \leq s$.

(i) Given $\varepsilon > 0$, there exists a rank $p(\varepsilon) \geq 0$ such that $s < t_n < s + \varepsilon$, for all $n \geq p(\varepsilon)$; hence

$$\limsup_n(\varphi(t_n)) \leq \sup\{\varphi(t_n); n \geq p(\varepsilon)\} \leq \Phi(s+)(\varepsilon).$$

Passing to infimum over $\varepsilon > 0$ yields (see above)

$$\limsup_n(\varphi(t_n)) \leq \inf_{\varepsilon>0} \Phi(s+)(\varepsilon) = \alpha;$$

and the claim follows.

(ii) Define $(\beta_n := \Phi(s+)(2^{-n}); n \geq 0)$; clearly,

$-\infty < \beta_n \leq s + 2^{-n}$
(because $\varphi(]s, s+2^{-n}[) \subseteq]-\infty, s+2^{-n}[)$, for all n;
(β_n) is descending, $(\beta_n \geq \alpha, \ \forall n)$, $\inf_n \beta_n = \alpha$; hence $\lim_n \beta_n = \alpha$.

By these properties, there may be constructed a sequence $(\gamma_n; n \geq 0)$ in R, with

$$\gamma_n < \beta_n, \ \forall n; \ \lim_n \gamma_n = \lim_n \beta_n = \alpha.$$

(For example, we may take $(\gamma_n = \beta_n - 3^{-n}; n \geq 0)$; but this is not the only possible choice). Let $n \geq 0$ be arbitrary fixed. By the supremum definition, there exists $t_n \in]s, s+2^{-n}[$ such that $\varphi(t_n) > \gamma_n$; moreover (again by definition), $\varphi(t_n) \leq \beta_n$; so, putting these together,

$(s < t_n < s + 2^{-n}, \gamma_n < \varphi(t_n) \leq \beta_n, \forall n)$;
whence $t_n \to s+$ and $\varphi(t_n) \to \alpha$.

By a previous result, there exists a subsequence $(r_n := t_{i(n)}; n \geq 0)$ of $(t_n; n \geq 0)$, in such a way that

(r_n) is strictly descending and $r_n \to \varepsilon+$ as well as $\varphi(r_n) \to \alpha$.

In other words the obtained sequence $(r_n; n \geq 0)$ has all needed properties; wherefrom the conclusion follows.

Call $\varphi \in \mathcal{F}(re)(R^0_+, R)$, *Boyd-Wong admissible* [6], if

(BW-a) $\Lambda^+\varphi(s) < s$, for all $s > 0$.

In particular, $\varphi \in \mathcal{F}(re)(R^0_+, R)$ is Boyd-Wong admissible provided

φ is *upper semicontinuous at the right* on R^0_+:
$\Lambda^+\varphi(s) \leq \varphi(s)$, for each $s \in R^0_+$.

This, e.g., is fulfilled when φ is *continuous at the right* on R^0_+; for, in such a case,

$\Lambda^+\varphi(s) = \varphi(s)$, for each $s \in R^0_+$.

A sequential counterpart of this convention may be described along the lines below. Call $\varphi \in \mathcal{F}(re)(R^0_+, R)$, *sequential Boyd-Wong admissible* provided

(s-BW-a) for each strictly descending sequence $(t_n; n \geq 0)$ in R^0_+
and each $\varepsilon > 0$ with $t_n \to \varepsilon+$, we have $\limsup_n \varphi(t_n) < \varepsilon$.

Finally, call $\varphi \in \mathcal{F}(re)(R^0_+, R)$ *sequentially Rhoades admissible*, provided

(s-R-a) for each strictly descending sequence $(t_n; n \geq 0)$ in R^0_+ and each $\varepsilon > 0$ with $t_n \to \varepsilon+$, we have $\liminf_n \varphi(t_n) < \varepsilon$.

Proposition 9 *The following inclusions are valid in (ZF-AC+DC), for a generic function* $\varphi \in \mathcal{F}(re)(R^0_+, R)$

(43-1) (BW-a) $\Longrightarrow$ *(s-BW-a)* $\Longrightarrow$ *(BW-a); hence, (BW-a)* $\Longleftrightarrow$ *(s-BW-a)*
(43-2) (s-BW-a) $\Longrightarrow$ *(s-R-a)* $\Longrightarrow$ *(s-BW-a); so, (s-BW-a)* $\Longleftrightarrow$ *(s-R-a)*
(43-3) (s-R-a) $\Longrightarrow$ *(n-d-a), (s-R-a)* $\Longrightarrow$ *(R-a), (s-R-a)* $\Longrightarrow$ *(MK-a)*
(43-4) (s-BW-a) $\Longrightarrow$ *(n-d-a), (s-BW-a)* $\Longrightarrow$ *(R-a), (s-BW-a)* $\Longrightarrow$ *(MK-a).*

Proof

(i) The first half is immediate, by the preceding statement. This is also true for the second half of the same; however, for completeness reasons, we shall supply an argument. Suppose that (BW-a) is not true. As φ=regressive, we have

$\Lambda^+\varphi(\varepsilon) = \varepsilon$, for some $\varepsilon > 0$.

By the quoted auxiliary fact, there exists a strictly descending sequence (r_n) in R^0_+, with the properties

$r_n \to \varepsilon+$ and $(\limsup_n \varphi(r_n) =) \lim_n \varphi(r_n) = \Lambda^+\varphi(\varepsilon) = \varepsilon$.

But then, φ does not satisfy (s-BW-a), a contradiction; hence the assertion.

(ii) The first half of this chain is immediate, by definition. For the second part of the same, one may proceed as below. Suppose by contradiction that $\varphi \in \mathcal{F}(re)(R^0_+, R)$ is not sequentially Boyd-Wong admissible; hence,

$\limsup_n \varphi(t_n) = \varepsilon$, for some strictly descending sequence $(t_n; n \geq 0)$ in R^0_+ and some $\varepsilon > 0$ with $t_n \to \varepsilon+$.

Combining with

$\varepsilon = \limsup_n \varphi(t_n) \leq \Lambda^+\varphi(\varepsilon) \leq \varepsilon$ (see above),

one derives $\Lambda^+\varphi(\varepsilon) = \varepsilon$. This, by a previous auxiliary fact, yields

$(\liminf_n \varphi(r_n) =) \lim_n \varphi(r_n) = \varepsilon$,
for some strictly descending sequence $(r_n; n \geq 0)$ in R^0_+ with $r \to \varepsilon+$.

But then, the sequential Rhoades admissible property of φ will be contradicted. Hence, our working assumption is not true; and the assertion follows.

(iii) Let us establish the first relation in this series. Suppose by contradiction that (n-d-a) is not holding. By definition, there must be a strictly descending sequence $(t_n; n \geq 0)$ in R^0_+ and a number $\varepsilon > 0$, such that

$t_n \to \varepsilon+$, $\varphi(t_n) \to \varepsilon+$, as $n \to \infty$.

This, along with

$\liminf_n \varphi(t_n) = \limsup_n \varphi(t_n) = \varepsilon$, contradicts the property (s-R-a). Hence, our working assumption is not true; and the conclusion is clear. The remaining relations follow at once in view of

(n-d-a) $\Longleftrightarrow$ (R-a) $\Longleftrightarrow$ (MK-a) (see above);

however, for completeness we provide a direct proof of the second one. Suppose that $\varphi \in \mathcal{F}(re)(R^0_+, R)$ is sequentially Rhoades admissible; we claim that φ is Rhoades admissible. If this does not hold, then (for some $\varepsilon > 0$)

$H(\delta) := \{(t,s) \in R^0_+ \times R^0_+; t \leq \varphi(s), \varepsilon < s < \varepsilon + \delta, t > \varepsilon\}$
is not empty, for each $\delta > 0$.

Taking a strictly descending sequence $(\delta_n; n \geq 0)$ in R^0_+ with $\delta_n \to 0$, we get by the denumerable axiom of choice (AC(N)) [deductible, as precise, in (ZF-AC+DC)], a sequence $((t_n, s_n); n \geq 0)$ in R^0_+, so as

$(\forall n)$: (t_n, s_n) is an element of $H(\delta_n)$;

or, equivalently (by the very definition above)

$(\forall n)$: $\varepsilon < t_n \leq \varphi(s_n) < s_n < \varepsilon + \delta_n$.

As a direct consequence of this, we have

$t_n \to \varepsilon+$, $s_n \to \varepsilon+$, as $n \to \infty$.

On the other hand,

$(\varepsilon < \varphi(s_n) < \varepsilon + \delta_n, \forall n)$ implies $\varepsilon = \lim_n \varphi(s_n) = \liminf_n \varphi(s_n)$.

By a previous result, there exists a subsequence $(s^*_n := s_{i(n)}; n \geq 0)$ of $(s_n; n \geq 0)$, such that

(s^*_n) is strictly descending and $s^*_n \to \varepsilon+$;
hence, necessarily, $(\liminf_n \varphi(s^*_n) =) \lim_n \varphi(s^*_n) = \varepsilon$;

in contradiction with the sequential Rhoades admissible property of φ. The remaining inclusions are obtainable in a similar way.

In the following, a special version of these facts is formulated, in terms of increasing functions from $\mathcal{F}(R^0_+, R)$.

Let $\mathcal{F}(re, in)(R^0_+, R)$ stand for the class of all $\varphi \in \mathcal{F}(re)(R^0_+, R)$, with

φ is increasing on R^0_+ ($0 < t_1 \leq t_2$ implies $\varphi(t_1) \leq \varphi(t_2)$).

The following characterization of our previous Matkowski properties imposed to φ is now available. Given $t > 0$, let

$$\varphi^0(t) = t, \varphi^1(t) = \varphi(t), \ldots, \varphi^{n+1}(t) = \varphi(\varphi^n(t)) \ (n \geq 0)$$

stand for the iterates sequence of φ at this point. Note that such a construction may be noneffective; for, e.g.,

$\varphi^2(t) = \varphi(\varphi(t))$ is undefined whenever $\varphi(t) \leq 0$.

Proposition 10 *For each $\varphi \in \mathcal{F}(re, in)(R^0_+, R)$, we have*

(44-1) *φ is Matkowski admissible, iff*
$(\forall t > 0)$: $\lim_n \varphi^n(t) = 0$, whenever $(\varphi^n(t); n \geq 0)$ exists
(44-2) *φ is strongly Matkowski admissible, iff*
$(\forall t > 0)$: $\sum_n \varphi^n(t) < \infty$, whenever $(\varphi^n(t); n \geq 0)$ exists.

The proof is immediate, by the increasing property of φ and

$$(\forall t > 0)\text{: } (\varphi^n(t); n \geq 0) \text{ exists iff } \{\varphi^n(t); n \geq 0\} \subseteq R^0_+;$$

so, further details are not necessary.

As before, we need sufficient conditions (involving the class $\mathcal{F}(re, in)(R^0_+, R)$) under which this property holds. Two situations occur.

The former one concerns the Matkowski admissible property. For each $\varphi \in \mathcal{F}(re, in)(R^0_+, R)$, denote

$$\varphi(s+0) := \lim_{t \to s+} \varphi(t),\ s \in R^0_+ \text{ (the *right limit* of } \varphi \text{ at } s);$$

clearly, the following evaluation holds

$$\varphi(s) \leq \varphi(s+0) \leq s, \text{ for all } s > 0.$$

Proposition 11 *Suppose that the function $\varphi \in \mathcal{F}(re, in)(R^0_+, R)$ fulfills*

φ is strongly regressive: $\varphi(s+0) < s$, for each $s > 0$.

Then, φ is Matkowski admissible.

Proof Given $s_0 > 0$, assume that

the iterative sequence $(s_n = \varphi^n(s_0); n \geq 0)$ exists (in R^0_+).

By the regressive property of φ, (s_n) is strictly descending; hence, $s := \lim_n s_n$ exists, with (in addition) $s_n > s$, for all n. Suppose by contradiction that $s > 0$. Combining with the immediate property

$$\varphi(s+0) = \lim_n \varphi(s_n) = \lim_n s_{n+1}$$

yields $\varphi(s+0) = s$, a contradiction. Hence, $s > 0$; and we are done.

Remark 1 Concerning the reverse inclusion, let us consider the particular function $\varphi \in \mathcal{F}(re, in)(R^0_+, R)$, according to (for some $r > 0$):

$$(\varphi(t) = t/2, \text{ if } t \leq r), (\varphi(t) = r, \text{ if } t > r).$$

Clearly, φ is Matkowski admissible, as it can be directly seen. On the other hand,

$\varphi(r+0) = r$; whence φ is not strongly regressive;

and this proves our claim. For an extended example of this type, see Turinici [48] and the references therein.

Now, it is natural to establish the connection between the introduced class and the Meir-Keeler one. An appropriate answer to this is contained in

Proposition 12 *Under these conventions we have, generically,*

(for each $\varphi \in \mathcal{F}(re, in)(R_+^0, R)$*):*
φ *is Matkowski admissible iff* φ *is Meir-Keeler admissible.*

Proof The verification of this fact is entirely deductible from the one developed in a related statement; however, for completeness reasons, we shall give a (different) reasoning for it.

(i) (cf. Jachymski [17]). Assume that $\varphi \in \mathcal{F}(re, in)(R_+^0, R)$ is Matkowski admissible; we have to establish that it is Meir-Keeler admissible. If the underlying property fails, then (for some $\gamma > 0$):

$$\forall \beta > 0, \exists t \in]\gamma, \gamma + \beta[, \text{ such that } \varphi(t) > \gamma.$$

As φ is increasing, this yields (by the arbitrariness of β)

$$(\varphi(t) > \gamma, \forall t > \gamma); \text{ whence, by induction, } (\varphi^n(t) > \gamma, \forall n, \forall t > \gamma).$$

Taking some $t > \gamma$ and passing to limit as $n \to \infty$; one gets $0 \geq \gamma$; a contradiction.

(ii) Assume that $\varphi \in \mathcal{F}(re, in)(R_+^0, R)$ is Meir-Keeler admissible; we have to establish that φ is Matkowski admissible. Given $s_0 > 0$, suppose that

the iterative sequence $(s_n = \varphi^n(s_0); n \geq 0)$ exists (in R_+^0).

By the regressive property of φ, (s_n) is strictly descending; hence, $s := \lim_n s_n$ exists, with (in addition) $s_n > s$, for all n. Suppose by contradiction that $s > 0$; and let $r > 0$ be the number assured by the Meir-Keeler admissible property of φ. By definition, there exists a rank $n(r) \geq 0$, such that

$$n \geq n(r) \text{ implies } s < s_n < s + r.$$

This, by the underlying property, gives (for the same ranks)

$$s < s_{n+1} = \varphi(s_n) \leq s; \text{ a contradiction.}$$

Hence, $s = 0$; wherefrom φ is Matkowski admissible.

Passing to the strong Matkowski admissible property, the following simple criterion will be in effect. Given $\varphi \in \mathcal{F}(re)(R_+^0, R)$, let us associate it the function $g \in \mathcal{F}(R_+^0, R)$, according to

$$g(t) = t/(t - \varphi(t)), t > 0.$$

Clearly, by the progressive condition imposed upon φ, we have

g is strictly positive: $g(t) > 0$ when $t > 0$ (hence, $g \in \mathcal{F}(R_+^0)$).

Assume in the following that

$g(.)$ is decreasing on R_+^0;
referred to as φ is *relatively decreasing*.

In this case, the associated primitive

$$G(t) := \int_0^t g(\xi)d\xi = \lim_{s\to 0+} \int_s^t g(\xi)d\xi, t > 0$$

is well defined as a function from R_+^0 to $R_+^0 \cup \{\infty\}$; moreover,

$$G(.,.) \text{ is strictly increasing on } R_+^0 \ (0 < t_1 \le t_2 \Longrightarrow G(t_1) < G(t_2)).$$

Proposition 13 *Let the function $\varphi \in \mathcal{F}(re)(R_+^0, R)$ be relatively decreasing and the associated (to $g(.)$) primitive $G(.)$ be such that*

$$G(t) < \infty, \textit{ for each } t \in R_+^0.$$

Then, φ is strongly Matkowski admissible (see above).

For a direct proof, we refer to Altman [1]. However, for completeness reasons, we shall supply an argument; which differs, in part, from the original one.

Proof By the very definitions above,

$$G(.) \text{ is continuous on } R_+^0 \text{ and } G(0+) := \lim_{t\to 0+} G(t) = 0.$$

Let the sequence $(t_n; n \ge 0)$ in R_+^0 be such that

$$t_{n+1} \le \varphi(t_n), \text{ for all } n \ge 0;$$

clearly, $(t_n; n \ge 0)$ is strictly descending. Further, let $i \ge 0$ be arbitrary fixed. By the above choice,

$$(0 <)t_i - \varphi(t_i) \le t_i - t_{i+1}; \text{ whence, } 1 \le (t_i - t_{i+1})/(t_i - \varphi(t_i)).$$

Combining with the decreasing property of $g(.)$ (and definition of $G(.)$), we get

$$t_i \le (t_i - t_{i+1})g(t_i) \le G(t_i) - G(t_{i+1}).$$

This gives (under the notation $t_\infty = \lim_n t_n$) an evaluation like

$$\sum_n t_n \le G(t_0) - G(t_\infty+) < \infty;$$

i.e., the series $\sum_n t_n$ converges. The proof is complete.

Technically speaking, this result is not the most general one in the area; but, it may be useful in practice. Further aspects may be found in Timofte [46].

Meir-Keeler Relations

Let $\Omega \subseteq R_+^0 \times R_+^0$ be a relation over R_+^0; as a rule, we write $(t, s) \in \Omega$ as $t\Omega s$. The following basic (global) properties upon this object are considered

(u-diag) Ω is *upper diagonal*: $t\Omega s$ implies $t < s$
(1-decr) Ω is *first variable decreasing*:
$t_1, t_2, s \in R_+^0, t_1 \ge t_2$, and $t_1\Omega s$ imply $t_2\Omega s$

(2-incr) Ω is *second variable increasing*:
$t, s_1, s_2 \in R^0_+$, $s_1 \le s_2$, and $t\Omega s_1$ imply $t\Omega s_2$.

The class of all upper diagonal relations will be denoted as udiag(R^0_+). Note that all subsequent constructions are being handled in this setting. In particular, the following sequential properties for upper diagonal relations Ω are considered:

(matk-1) Ω in *Matkowski admissible*:
$(t_n; n \ge 0)$ in R^0_+ and $(t_{n+1}\Omega t_n, \forall n)$ imply $\lim_n t_n = 0$
(matk-2) Ω in *strongly Matkowski admissible*:
$(t_n; n \ge 0)$ in R^0_+ and $(t_{n+1}\Omega t_n, \forall n)$ imply $\sum_n t_n < \infty$.

The following local condition involving our relations is—in particular—useful for discussing the former of these properties

(g-mk) (Ω has the *geometric Meir-Keeler property*)
$\forall \varepsilon > 0, \exists \delta > 0$: $t\Omega s, \varepsilon < s < \varepsilon + \delta \Longrightarrow t \le \varepsilon$.

(This condition is related to the developments in Meir and Keeler [27]; we do not give details). Precisely, the following auxiliary fact is available.

Proposition 14 *For each upper diagonal relation $\Omega \subseteq R^0_+ \times R^0_+$, we have*

Ω has the geometric Meir-Keeler property implies
Ω is Matkowski admissible.

Proof Suppose that Ω has the geometric Meir-Keeler property; we have to establish that Ω is Matkowski admissible. Let $(t_n; n \ge 0)$ be a sequence in R^0_+, with

$t_{n+1}\Omega t_n$, for all n.

By the upper diagonal property, we get

$t_{n+1} < t_n$, for all n

i.e., (t_n) is strictly descending. As a consequence, $\tau = \lim_n t_n$ exists in R_+; with, in addition, $(t_n > \tau$, for all $n)$. Let $\sigma > 0$ be the number assured by the geometric Meir-Keeler property. By definition, there exists an index $n(\sigma)$, with

$(t_{n+1}\Omega t_n$ and$)$ $\tau < t_n < \tau + \sigma$, for all $n \ge n(\sigma)$.

This, by the quoted property, gives (for the same ranks)

$\tau < t_{n+1} \le \tau$, a contradiction.

Hence, necessarily, $\tau = 0$; and the conclusion follows.

In the following, a sufficient (sequential) condition is given for the global property attached to our (relational) concepts above. Given the upper diagonal relation Ω over R^0_+, let us introduce the (asymptotic type) convention

(a-mk) Ω has the *asymptotic Meir-Keeler property*:
there are no strictly descending sequences (t_n) and (s_n) in R^0_+

and no elements ε in R_+^0, with
$((t_n, s_n) \in \Omega, \forall n)$ and $(t_n \to \varepsilon+, s_n \to \varepsilon+)$.

Here, for each sequence (r_n) in R and each element $r \in R$, we denoted

$$r_n \to r+ \text{ (resp., } r_n \to r-\text{), when } r_n \to r \text{ and } [r_n > r \text{ (resp., } r_n < r), \forall n].$$

Proposition 15 *The following generic relationship is valid (for an arbitrary upper diagonal relation $\Omega \in \mathrm{udiag}(R_+^0)$) in the reduced system (ZF-AC+DC):*

asymptotic Meir-Keeler property implies geometric Meir-Keeler property.

Proof Let $\Omega \in \mathrm{udiag}(R_+^0)$ be asymptotic Meir-Keeler; but—contrary to the conclusion—assume that Ω does not have the geometric Meir-Keeler property; i.e. (for some $\varepsilon > 0$)

$$H(\delta) := \{(t, s) \in \Omega; \varepsilon < s < \varepsilon + \delta, t > \varepsilon\} \neq \emptyset, \text{ for each } \delta > 0.$$

Taking a zero converging sequence $(\delta_n; n \geq 0)$ in R_+^0, we get by the denumerable axiom of choice (AC(N)) [deductible, as precise, in (ZF-AC+DC)], a sequence $((t_n, s_n); n \geq 0)$ in Ω, so as

$$(\forall n)\text{: } (t_n, s_n) \text{ is an element of } H(\delta_n);$$

or, equivalently (by definition and upper diagonal property)

$$((t_n, s_n) \in \Omega \text{ and}) \ \varepsilon < t_n < s_n < \varepsilon + \delta_n, \text{ for all } n.$$

Note that, as a direct consequence,

$$((t_n, s_n) \in \Omega, \text{ for all } n), \text{ and } t_n \to \varepsilon+, s_n \to \varepsilon+, \text{ as } n \to \infty.$$

By a previous result, there exists a compatible couple of subsequences $(t_n^* := t_{i(n)}; n \geq 0)$ of $(t_n; n \geq 0)$ and $(s_n^* := s_{i(n)}; n \geq 0)$ of $(s_n; n \geq 0)$, with

$(t_n^* \Omega s_n^*$, for all $n)$, $(t_n^*; n \geq 0)$, $(s_n^*; n \geq 0)$ are strictly descending, and
$(t_n^* \to \varepsilon+, s_n^* \to \varepsilon+$, as $n \to \infty)$.

This, however, is in contradiction with respect to our hypothesis; wherefrom, the assertion follows.

In the following, some basic examples of (upper diagonal) Matkowski admissible and strongly Matkowski admissible relations are given. The general scheme of constructing these may be described as follows.

Let $R(\pm\infty) := R \cup \{-\infty, \infty\}$ stand for the set of all *extended real numbers*. For each relation Ω over R_+^0, let us associate a function $\Lambda : R_+^0 \times R_+^0 \to R(\pm\infty)$, as

$$\Lambda(t, s) = 0, \text{ if } (t, s) \in \Omega; \ \Lambda(t, s) = -\infty, \text{ if } (t, s) \notin \Omega.$$

It will be referred to as the *function* generated by Ω; clearly,

$$(t, s) \in \Omega \text{ iff } \Lambda(t, s) \geq 0.$$

Conversely, given a function $\Lambda : R^0_+ \times R^0_+ \to R(\pm\infty)$, we may associate it a relation Ω over R^0_+ as

$\Omega = \{(t, s) \in R^0_+ \times R^0_+; \Lambda(t, s) \geq 0\}$; in short: $\Omega = [\Lambda \geq 0]$;
referred to as the *the positive section* of Λ.

Note that the correspondence between the function Λ and its associated relation $\Omega = [\Lambda \geq 0]$ (given as before) is not injective because, for the function $\Delta := \alpha\Lambda$ (where $\alpha > 0$), its associated relation $[\Delta \geq 0]$ is identical with the relation $[\Lambda \geq 0]$ attached to Λ.

Now, under the geometric model above, the following basic (global) properties upon a relational function $\Lambda : R^0_+ \times R^0_+ \to R(\pm\infty)$ are considered

(u-diag) Λ is *upper diagonal*: $\Lambda(t, s) \geq 0$ implies $t < s$
(1-decr) Λ is *first variable decreasing*:
$t_1, t_2, s \in R^0_+$, $t_1 \geq t_2$ and $\Lambda(t_1, s) \geq 0$ imply $\Lambda(t_2, s) \geq 0$
(2-incr) Λ is *second variable increasing*:
$t, s_1, s_2 \in R^0_+$, $s_1 \leq s_2$, and $\Lambda(t, s_1) \geq 0$ imply $\Lambda(t, s_2) \geq 0$.

As before, all subsequent constructions are being considered within the upper diagonal setting. In particular, the following sequential properties for upper diagonal functions Λ are considered:

(matk-1) Λ in *Matkowski admissible*:
$(t_n; n \geq 0)$ in R^0_+ and $(\Lambda(t_{n+1}, t_n) \geq 0, \forall n)$ imply $\lim_n t_n = 0$
(matk-2) Λ is *strongly Matkowski admissible*:
$(t_n; n \geq 0)$ in R^0_+ and $(\Lambda(t_{n+1}, t_n) \geq 0, \forall n)$ imply $\sum_n t_n < \infty$.

The following local condition involving our functions is—in particular—useful for discussing the former of these properties

(g-mk) Λ has the *geometric Meir-Keeler property*:
$\forall \varepsilon > 0, \exists \delta > 0$: $\Lambda(t, s) \geq 0, \varepsilon < s < \varepsilon + \delta \Longrightarrow t \leq \varepsilon$.

The relationship between this geometric condition and the Matkowski one attached to upper diagonal functions is nothing else than a simple translation of the previous one involving upper diagonal relations; we do not give details.

Summing up, any concept (like the ones above) about (upper diagonal) relations over R^0_+ may be written as a concept about (upper diagonal) functions in the class $\mathcal{F}(R^0_+ \times R^0_+, R(\pm\infty))$. For the rest of our exposition, it will be convenient working with relations over R^0_+ and not with functions in $\mathcal{F}(R^0_+ \times R^0_+, R(\pm\infty))$; this, however, is but a methodology question.

Having these precise, we may now pass to the description of some particular objects in this area.

Part Case (I) The former of these corresponds to the (functional) choice

$\Lambda(t, s) = \chi(s) - t$, $t, s \in R^0_+$, where $\chi \in \mathcal{F}(R^0_+, R)$.

Precisely, let χ be a function in $\mathcal{F}(re)(R_+^0, R)$. Define the (associated) *Matkowski* relation $\Omega := \Omega[\chi]$ over R_+^0, as

$$(t, s) \in \Omega \text{ iff } t \leq \chi(s);$$

note that, as χ=regressive, Ω is upper diagonal. The introduced terminology is suggested by the developments in Matkowski [26].

Proposition 16 *Under the precise conventions, the following conclusions (involving our upper diagonal relation $\Omega := \Omega[\chi]$) are valid:*

(53-1) Ω is first variable decreasing
(53-2) Ω is second variable increasing when χ is increasing
(53-3) Ω has the geometric Meir-Keeler property whenever χ is Rhoades admissible (or, equivalently, Matkowski admissible)
(53-4) Ω is Matkowski admissible when χ is Matkowski admissible (or, equivalently, Meir-Keeler admissible)
(53-5) Ω is strongly Matkowski admissible whenever χ is strongly Matkowski admissible (see above).

Proof Evident, by definition.

Part Case (II) The second particular case of our investigations corresponds to the (functional) choice

$$\Lambda(t, s) = \psi(s) - \psi(t) - \varphi(s), \text{ where } \psi, \varphi \in \mathcal{F}(R_+^0, R).$$

So, let (ψ, φ) be a pair of functions in $\mathcal{F}(R_+^0, R)$, fulfilling

(R-couple) (ψ, φ) is a *Rhoades couple*:
ψ is increasing and φ is *strictly positive* ($\varphi(R_+^0) \subseteq R_+^0$).

Define the (associated) *Rhoades* relation $\Omega = \Omega[\psi, \varphi]$ in $R_+^0 \times R_+^0$, as

$$(t, s) \in \Omega \text{ iff } \psi(t) \leq \psi(s) - \varphi(s).$$

Note that, by the Rhoades couple condition, Ω is upper diagonal. In fact, let $t, s \in R_+^0$ be such that

$$(t, s) \in \Omega, \text{ i.e., } \psi(t) \leq \psi(s) - \varphi(s).$$

By the strict positivity of φ, one gets $\psi(t) < \psi(s)$; and this, along with the increasing property of ψ, shows that $t < s$; whence the assertion. The introduced convention is related to the developments in Rhoades [36]. (Some related aspects may be found in Wardowski [52]; see also Dutta and Choudhury [13]).

Now, further properties of this associated relation are available under some extra sequential conditions upon (ψ, φ). The former of these writes

(a-pos) φ is *asymptotic positive*: for each strictly descending sequence $(t_n; n \geq 0)$ in R_+^0 and each $\varepsilon > 0$ with $t_n \to \varepsilon+$, we must have $\limsup_n(\varphi(t_n)) > 0$.

In this context, it will be useful to introduce the combined notion

(a-R-couple) (ψ, φ) is *asymptotic Rhoades couple*:
(ψ, φ) is a Rhoades couple with φ=asymptotic positive.

For the next one, we need some conventions relative to the Rhoades couple (ψ, φ). Namely, let us say that (ψ, φ) is *finitely asymptotic regular*, provided

(f-a-reg-1) $\psi(0+0)(=\lim_{t\to 0+}\psi(t)) > -\infty$
(f-a-reg-2) φ is *asymptotic expansive*:
for each strictly descending sequence $(t_n; n \geq 0)$ in R_+^0
with $\sum_n \varphi(t_n) < \infty$, we have $\sum_n t_n < \infty$.

Also, let us say that (ψ, φ) is *infinitely asymptotic regular*, provided

(i-a-reg-1) $\psi(0+0) = -\infty$ and φ is increasing
(i-a-reg-2) $\exists \mu > 1$, such that (ψ, φ) is *asymptotic μ-dominating*, i.e.,
for each strictly descending sequence $(t_n; n \geq 0)$ in R_+^0 with $\lim_n t_n = 0$ and $\sum_n[\varphi(t_n)/(1+\psi(t_0)-\psi(t_{n+1}))]^\mu < \infty$, we have $\sum_n t_n < \infty$.

Proposition 17 *Let (ψ, φ) be a Rhoades couple and $\Omega := \Omega[\psi, \varphi]$ stand for its associated relation. Then*

(54-1) Ω is first variable decreasing
(54-2) Ω is second variable increasing if $\psi - \varphi$ is increasing
(54-3) Ω is asymptotic Meir-Keeler (hence, geometric Meir-Keeler; hence, Matkowski admissible), if φ is asymptotic positive
(54-4) Ω is strongly Matkowski admissible if (ψ, φ) is finitely asymptotic regular, or infinitely asymptotic regular.

Proof (i), (ii) Evident.

(iii) Suppose that Ω does note have the asymptotic Meir-Keeler property: there exist strictly descending sequences (t_n) and (s_n) in R_+^0 and elements ε in R_+^0, with

$(t_n \Omega s_n, \forall n)$ and $(t_n \to \varepsilon+, s_n \to \varepsilon+)$.

By the former of these, we get

$$(0 <)\varphi(s_n) \leq \psi(s_n) - \psi(t_n),\ \forall n.$$

Passing to the superior limit as $n \to \infty$, and noting that

$$\lim_n \psi(s_n) = \lim_n \psi(t_n) = \psi(\varepsilon + 0),$$

one gets $\limsup_n \varphi(t_n) = 0$; in contradiction with the asymptotic positive property of φ. So, necessarily, Ω has the asymptotic Meir-Keeler property.

(iv) Let $(t_n; n \geq 0)$ be a sequence in R_+^0, with

$t_{n+1}\Omega t_n$ (i.e., $\psi(t_{n+1}) \leq \psi(t_n) - \varphi(t_n)$), for all n.

There are two alternatives to be discussed.

Alter 1 Suppose that (ψ, φ) is finitely asymptotic regular. By the posed hypothesis about (t_n), one gets

$$\varphi(t_n) \leq \psi(t_n) - \psi(t_{n+1}), \text{ for all } n.$$

On the other hand,

$$\psi(0+0) > -\infty; \text{ whence, } \sum_n [\psi(t_n) - \psi(t_{n+1})] < \infty.$$

Combining these, we have

$$\sum_n \varphi(t_n) < \infty; \text{ whence, } \sum_n t_n < \infty,$$

if we remember that φ is asymptotic expansive.

Alter 2 Suppose that (ψ, φ) is infinitely asymptotic regular. By the same working hypothesis about (t_n), one gets

$$\varphi(t_n) \leq \psi(t_n) - \psi(t_{n+1}), \text{ for all } n.$$

Adding the first $(n + 1)$ relations, we derive (via φ=increasing)

$$(n+1)\varphi(t_n) \leq \psi(t_0) - \psi(t_{n+1}) \leq 1 + \psi(t_0) - \psi(t_{n+1}), \text{ for all } n \geq 0.$$

By the upper diagonal property of Ω, $(t_n; n \geq 0)$ is strictly descending; hence, $\tau := \lim_n t_n$ exists. We claim that

$$\tau = 0; \text{ i.e., } t_n \to 0 \text{ as } n \to \infty.$$

In fact, assume by contradiction that

$$\tau > 0; \text{ whence } \varphi(\tau) > 0 \text{ (as } \varphi \text{ is strictly positive).}$$

From the above inequality, one gets (via φ=increasing)

$$(n+1)\varphi(\tau) \leq \psi(t_0) - \psi(\tau), \text{ for all } n.$$

Passing to limit as $n \to \infty$, we derive

$$\infty \leq \psi(t_0) - \psi(\tau) < \infty; \text{ a contradiction;}$$

hence the claim.

Now, by an appropriate multiplication of the (additively generated) inequality above, one gets

$$[\varphi(t_n)/(1 + \psi(t_0) - \psi(t_{n+1}))]^\mu \leq 1/(n+1)^\mu, \text{ for all } n \geq 0.$$

As $\sum_n 1/(n+1)^\mu < \infty$, it follows that

$$\sum_n [\varphi(t_n)/(1 + \psi(t_0) - \psi(t_{n+1}))]^\mu < \infty; \text{ whence, } \sum_n t_n < \infty,$$

if we remember that (ψ, φ) is asymptotic μ-dominating; hence the claim.

Remark 2 Concerning the finitely asymptotic regular property of (ψ, φ), the following simple facts are useful for applications. Let (ψ, φ) be a couple of functions over $\mathcal{F}(R^0_+, R)$, with

(ψ, φ) a Rhoades couple with $\psi(0+0) > -\infty$.

The following cases are to be discussed.

Case 1 Take the function $\varphi \in \mathcal{F}(R_+^0, R)$ according to

$\varphi(t) = \alpha$, $t > 0$; where $\alpha > 0$ is a constant.

Clearly, φ is strictly positive and asymptotic positive. Hence, by the above statement, the associated Rhoades relation $\Omega := \Omega[\psi, \varphi]$ is upper diagonal and fulfills

Ω is asymptotic Meir-Keeler; hence, geometric Meir-Keeler;
hence, Matkowski admissible.

But, this last property is holding in a trivial way; precisely,

there are no sequences $(t_n; n \geq 0)$ in R_+^0, with $(t_{n+1}\Omega t_n$, for all $n)$.

For, if $(t_n; n \geq 0)$ is such a sequence, then

$\alpha = \varphi(t_n) \leq \psi(t_n) - \psi(t_{n+1})$, for all n,

wherefrom, passing to limit as $n \to \infty$, one gets $\alpha = 0$; a contradiction. Moreover, the obtained nonexistence relation also tells us that Ω is strongly Matkowski admissible in a trivial way. An indirect argument for verifying this is based on the immediate observation that

there are no sequences $(t_n; n \geq 0)$ in R_+^0 with $\sum_n \varphi(t_n) < \infty$.

Note that, as a direct consequence, constructions like in Wardowski [52] are not working here.

Case 2 Summing up, the case of φ=constant is noneffective for us. To get genuine examples, let us start from the class of functions $\varphi : R_+^0 \to R_+^0$ given as

φ is *strongly asymptotic expansive*:
$(\varphi(t) \geq \alpha t$, if $\varphi(t) \leq \beta)$, where $\alpha, \beta > 0$ are constants.

Note that any such function is asymptotic expansive. In fact, let the strictly descending sequence $(t_n; n \geq 0)$ in R_+^0 be such that $\sum_n \varphi(t_n) < \infty$. In particular, $\lim_n \varphi(t_n) = 0$; so, there must be some rank $n(\beta)$ such that

$\varphi(t_n) \leq \beta$, for all $n \geq n(\beta)$.

This, along with the strong asymptotic expansive property, gives

$\varphi(t_n) \geq \alpha t_n$, for all $n \geq n(\beta)$, wherefrom $\sum_n t_n < \infty$;

and the claim follows.

By the statement above, one derives that

$\Omega := \Omega[\psi, \varphi]$ is strongly Matkowski admissible
(hence, Matkowski admissible).

However, for practical reasons, we want that Ω be second variable increasing as well. To reach this property, let $\varphi : R_+^0 \to R_+^0$ be such that

φ is *admissible* (in the sense: φ is increasing and strongly asymptotic expansive).

Note that, in such a case,

φ is asymptotic positive as well.

In fact, for each strictly descending sequence $(t_n; n \geq 0)$ in R_+^0 and each $\varepsilon > 0$ with $t_n \to \varepsilon+$, we have

$$\limsup_n(\varphi(t_n)) = \lim_n \varphi(t_n) = \varphi(\varepsilon + 0) \geq \varphi(\varepsilon) > 0;$$

and the claim follows. The function $\psi : R_+^0 \to R_+^0$ is introduced as

$$\psi(t) = (\chi(t) + 1)\varphi(t), t \in R_+^0$$

where $\chi : R_+^0 \to R_+^0$ is increasing, fulfills

ψ is increasing, $\psi(0+0) \geq 0 > -\infty$ and $\psi - \varphi = \chi\varphi$ is increasing.

Summing up, the couple (ψ, φ) fulfills all general/specific conditions imposed above; so, its associated Rhoades relation Ω is endowed with the following properties:

Ω is first variable decreasing, second variable increasing,
asymptotic Meir-Keeler, and strongly Matkowski admissible.

In particular, the function introduced as (for some $\nu \in]0, 1]$)

$$\varphi = \Gamma_\nu; \text{ where } \Gamma_\nu(t) = t^\nu, t > 0$$

is admissible. Evidently, Γ_ν is asymptotic positive, (strictly) increasing, and strongly asymptotic expansive (just take $\alpha = \beta = 1$ in the corresponding definition), hence the conclusion. So, with respect to the function

$$\psi_\nu(t) = (\chi(t) + 1)\Gamma_\nu(t), t \in R_+^0 \text{ (where } \chi : R_+^0 \to R_+^0 \text{ is increasing)}$$

the associated relation $\Omega := \Omega[\psi_\nu, \Gamma_\nu]$ has the above listed properties.

Remark 3 Passing to the discussion of infinitely asymptotic regular properties, let (ψ, φ) be a couple of functions over $\mathcal{F}(R_+^0, R)$, with

(ψ, φ) is a Rhoades couple, $\psi(0+0) = -\infty$, and $\varphi =$ increasing.

In particular, this last property holds when $\varphi \in \mathcal{F}(R_+^0, R)$ is defined as

$\varphi(t) = \alpha, t > 0$, where $\alpha > 0$ is a constant;

note that, in such a case, φ is strictly positive and asymptotic positive. Moreover, the domination property becomes

there exists $\mu > 1$, such that ψ is *asymptotic μ-dominating*:
for each strictly descending sequence $(t_n; n \geq 0)$ in R_+^0 with $\lim_n t_n = 0$ and $\sum_n [1 + \psi(t_0) - \psi(t_{n+1})]^{-\mu} < \infty$, we have $\sum_n t_n < \infty$.

A basic particular case when this property holds is

there exists $\lambda \in]0, 1[$, such that
ψ is *λ-polynomial*: $\lim_{t\to 0+} \psi(t)t^\lambda = 0$.

In fact, put

$\mu := 1/\lambda$ (hence, $\mu > 1$);

and let the sequence $(t_n; n \geq 0)$ in R^0_+ be as in the premise above. From the posed assumption, we have (by direct calculations)

$\lim_n [1 + \psi(t_0) - \psi(t_{n+1})]t^\lambda_{n+1}=0$;
and then (under the precise convention)
$\lim_n [1 + \psi(t_0) - \psi(t_{n+1})]^\mu t_{n+1}=0$.

The sequence in the left-hand side of this last relation is therefore bounded; so, there must be some $C > 0$, with

$[1 + \psi(t_0) - \psi(t_{n+1})]^\mu t_{n+1} \leq C$, for all n,

or, equivalently,

$t_{n+1}/C \leq [1 + \psi(t_0) - \psi(t_{n+1})]^{-\mu}$, for all n.

But, from this, the desired assertion is clear, hence the conclusion.

Remark 4 The construction of Matkowski relation $\Omega := \Omega[\chi]$ involving a certain function $\chi \in \mathcal{F}(re)(R^0_+, R)$ is nothing else than a particular case of this one, corresponding to the choice

$\psi(t) = t$, $\varphi(t) = t - \chi(t)$, $t \in R^0_+$.

In fact, it is not hard to see that (under these notations)

(i) φ is strictly positive (as χ belongs to $\mathcal{F}(re)(R^0_+, R)$)
(ii) φ is asymptotic positive iff χ is sequentially Rhoades admissible.

Since the verification is immediate, we do not give details.

Part Case (III) The third particular case of our investigations may be viewed as a bidimensional version of the preceding one, corresponding to the choice

$\Lambda(t, s) = \psi(s) - \psi(t) - \Phi(t, s)$, $t, s > 0$,
where $\psi \in \mathcal{F}(R^0_+, R)$, $\Phi \in \mathcal{F}(R^0_+ \times R^0_+, R)$.

Precisely, let $(\psi : R^0_+ \to R, \Phi : R^0_+ \times R^0_+ \to R)$ be a pair of functions, with

(p-R-couple) ψ increasing and Φ *strictly positive*
($\Phi(t, s) > 0, \forall t, s > 0$) (referred to as: (ψ, Φ) is a *plane Rhoades couple*).

Define the (associated) *plane Rhoades* relation $\Omega = \Omega[\psi, \Phi]$ in $R^0_+ \times R^0_+$, as

$(t, s) \in \Omega$ iff $\psi(t) \leq \psi(s) - \Phi(t, s)$.

Note that, by the plane Rhoades couple condition, Ω is upper diagonal. In fact, let $t, s \in R^0_+$ be such that

$$(t, s) \in \Omega; \text{ i.e., } \psi(t) \le \psi(s) - \Phi(t, s).$$

By the strict positivity of Φ, one gets $\psi(t) < \psi(s)$; and this, along with the increasing property of ψ, shows that $t < s$; whence the assertion. The introduced convention is related to the developments in Khojasteh et al. [22] (see below).

Now, further properties of this associated relation are available under some extra sequential conditions upon (ψ, Φ). The first group of these writes

(d-a-pos) Φ is *diagonal asymptotic positive*:
for each strictly descending sequence $(t_n; n \ge 0)$ in R^0_+
and each $\varepsilon > 0$ with $t_n \to \varepsilon+$, we must have $\limsup_n \Phi(t_{n+1}, t_n) > 0$
(a-pos) Φ is *asymptotic positive*:
for each couple of strictly descending sequences (t_n) and (s_n) in R^0_+ and each $\varepsilon > 0$ with $t_n \to \varepsilon+$, $s_n \to \varepsilon+$, we have $\limsup_n \Phi(t_n, s_n) > 0$.

Clearly,

Φ is asymptotic positive implies Φ is diagonal asymptotic positive.

The reciprocal inclusion is not in general true. To verify this, fix $\alpha > 0$; and let the function $g = g_\alpha : R^0_+ \to R^0_+$ be defined as

$$g(t) = 1, \text{ if } 0 < t \le \alpha; \; g(t) = t - \alpha, \text{ if } t > \alpha.$$

Then, let $\Phi = \Phi_g : R^0_+ \times R^0_+ \to R^0_+$ be introduced according to

$$\Phi(t, s) = \max\{t, s\}, \text{ if } t \ne s, t, s > 0; \; \Phi(r, r) = g(r), r > 0.$$

Clearly, Φ is *diagonal asymptotic positive*; given the strictly descending sequence $(t_n; n \ge 0)$ in R^0_+ and the number $\varepsilon > 0$ with $t_n \to \varepsilon+$, we have (by definition)

$\Phi(t_{n+1}, t_n) = \max\{t_{n+1}, t_n\} = t_n > \varepsilon$, for all n;
whence $\limsup_n \Phi(t_{n+1}, t_n) \ge \varepsilon > 0$.

On the other hand, Φ is not asymptotic positive. In fact, let the couple of strictly descending sequences (t_n) and (s_n) in R^0_+ be such that

$(t_n = s_n = r_n$, for all $n \ge 0)$, where
$(r_n; n \ge 0)$ is a strictly descending sequence in R^0_+, with $r_n \to \alpha+$.

Then (again by definition)

$\Phi(t_n, s_n) = g(r_n) = r_n - \alpha \to 0$, as $n \to \infty$,
whence $\limsup_n \Phi(t_n, s_n) = \lim_n \Phi(t_n, s_n) = 0$;

and this proves our assertion.

In this context, it will be useful to introduce the combined notion

(d-a-p-R-couple) (ψ, Φ) is *diagonal asymptotic plane Rhoades couple*:
(ψ, Φ) is a plane Rhoades couple with Φ=diagonal asymptotic positive

(a-p-R-couple) (ψ, Φ) is *asymptotic plane Rhoades couple*:
(ψ, Φ) is a plane Rhoades couple with Φ=asymptotic positive.

As before, we have

(ψ, Φ) is asymptotic plane Rhoades couple implies
(ψ, Φ) is diagonal asymptotic plane Rhoades couple.

But (cf. our previous discussion), the reciprocal of this is not in general true.

For the second group of conditions, we need some conventions relative to the plane Rhoades couple (ψ, Φ). Namely, call (ψ, Φ), *finitely asymptotic regular* if

(f-a-reg-1) $\psi(0+0)(= \lim_{t\to 0+} \psi(t)) > -\infty$
(f-a-reg-2) Φ is *diagonal asymptotic expansive*:
for each strictly descending sequence $(t_n; n \geq 0)$ in R_+^0
with $\sum_n \Phi(t_{n+1}, t_n) < \infty$, we have $\sum_n t_n < \infty$.

Also, let us say that (ψ, Φ) is *infinitely asymptotic regular*, provided

(i-a-reg-1) $\psi(0+0) = -\infty$ and Φ is increasing in each variable
(i-a-reg-2) $\exists \mu > 1$, such that (ψ, Φ) is *asymptotic* μ*-dominating*; i.e.,
for each strictly descending sequence $(t_n; n \geq 0)$ in R_+^0 with $\lim_n t_n = 0$ and $\sum_n [\Phi(t_{n+1}, t_n)/(1 + \psi(t_0) - \psi(t_{n+1}))]^\mu < \infty$, we have $\sum_n t_n < \infty$.

Proposition 18 *Let (ψ, Φ) be a plane Rhoades couple; and $\Omega := \Omega[\psi, \Phi]$ stand for its associated relation. Then*

(55-1) Ω is first variable decreasing if
$t \mapsto \psi(t) + \Phi(t, s)$ is increasing, for each $s \in R_+^0$
(55-2) Ω is second variable increasing if
$s \mapsto \psi(s) - \Phi(t, s)$ is increasing, for each $t \in R_+^0$
(55-3) Ω is asymptotic Meir-Keeler (hence, geometric Meir-Keeler; hence, Matkowski admissible), if Φ is asymptotic positive
(55-4) Ω is strongly Matkowski admissible if (ψ, Φ) is finitely asymptotic regular, or infinitely asymptotic regular.

Proof (i), (ii) Evident.

(iii) Suppose that Ω does note have the asymptotic Meir-Keeler property: there exist strictly descending sequences (t_n) and (s_n) in R_+^0 and elements ε in R_+^0, with

$(t_n \Omega s_n, \forall n)$ and $(t_n \to \varepsilon+, s_n \to \varepsilon+)$.

By the former of these, we get

$$(0 <)\Phi(t_n, s_n) \leq \psi(s_n) - \psi(t_n), \ \forall n.$$

Passing to the superior limit as $n \to \infty$, and noting that

$$\lim_n \psi(s_n) = \lim_n \psi(t_n) = \psi(\varepsilon + 0),$$

one gets $\limsup_n \Phi(t_n, s_n) = 0$, in contradiction with the asymptotic positive property of Φ. So, necessarily, Ω has the asymptotic Meir-Keeler property.

(iv): Let $(t_n; n \geq 0)$ be a sequence in R^0_+, with

$$t_{n+1}\Omega t_n \text{ (i.e., } \psi(t_{n+1}) \leq \psi(t_n) - \Phi(t_{n+1}, t_n)\text{), for all } n.$$

There are two cases to be discussed.

Case 1 Suppose that (ψ, Φ) is finitely asymptotic regular. By the posed hypothesis about (t_n), one gets

$$\Phi(t_{n+1}, t_n) \leq \psi(t_n) - \psi(t_{n+1}), \text{ for all } n.$$

On the other hand,

$$\psi(0+0) > -\infty; \text{ whence, } \sum_n [\psi(t_n) - \psi(t_{n+1})] < \infty.$$

Combining these, we have

$$\sum_n \Phi(t_{n+1}, t_n) < \infty; \text{ whence, } \sum_n t_n < \infty,$$

if we remember that Φ is diagonal asymptotic expansive.

Case 2 Suppose that (ψ, Φ) is infinitely asymptotic regular. By the same working hypothesis about (t_n), one gets

$$\Phi(t_{n+1}, t_n) \leq \psi(t_n) - \psi(t_{n+1}), \text{ for all } n.$$

Adding the first $(n + 1)$ relations, we derive (via Φ=increasing in each variable)

$$(n+1)\Phi(t_{n+1}, t_n) \leq \psi(t_0) - \psi(t_{n+1}) \leq 1 + \psi(t_0) - \psi(t_{n+1}), \text{ for all } n \geq 0.$$

By the upper diagonal property of Ω, $(t_n; n \geq 0)$ is strictly descending; whence, $\tau := \lim_n t_n$ exists. We claim that

$$\tau = 0; \text{ i.e., } t_n \to 0 \text{ as } n \to \infty.$$

In fact, assume by contradiction that

$$\tau > 0; \text{ wherefrom } \Phi(\tau, \tau) > 0 \text{ (as } \Phi \text{ is strictly positive).}$$

By the above inequality, one gets

$$(n+1)\Phi(\tau, \tau) \leq \psi(t_0) - \psi(\tau), \text{ for all } n \geq 0.$$

Passing to limit as $n \to \infty$, we derive

$$\infty \leq \psi(t_0) - \psi(\tau) < \infty, \text{ a contradiction;}$$

hence the claim.

Now, by an appropriate multiplication of the (additively generated) inequality above, one gets

$$[\Phi(t_{n+1}, t_n)/(1 + \psi(t_0) - \psi(t_{n+1}))]^\mu \leq (1/(n+1))^\mu, \text{ for all } n \geq 0.$$

As $\sum_n [1/(n+1)]^\mu < \infty$, it follows that

$$\sum_n [\Phi(t_{n+1}, t_n)/(1 + \psi(t_0) - \psi(t_{n+1}))]^\mu < \infty; \text{ whence, } \sum_n t_n < \infty,$$

if we remember that (ψ, Φ) is asymptotic μ-dominating, hence the claim.

Remark 5 Concerning the finitely asymptotic regular property of (ψ, Φ), the following simple facts are useful for applications. Let $\psi : R_+^0 \to R$, $\Phi : R_+^0 \times R_+^0 \to R$ be a couple of functions with

(ψ, Φ) is a plane Rhoades couple with $\psi(0+0) > -\infty$.

The following alternatives are to be discussed.

Alter 1 Take the function $\Phi : R_+^0 \times R_+^0 \to R$ according to

$\Phi(t, s) = \alpha, t, s > 0$, where $\alpha > 0$ is a constant.

Clearly, Φ is strictly positive and asymptotic positive. Hence, by the above statement, the plane Rhoades associated relation $\Omega := \Omega[\psi, \Phi]$ fulfills

Ω is asymptotic Meir-Keeler; hence, geometric Meir-Keeler;
hence, Matkowski admissible.

But, this last property is holding in a trivial way; precisely,

there are no sequences $(t_n; n \geq 0)$ in R_+^0, with $(t_{n+1} \Omega t_n$, for all $n)$.

For, if $(t_n; n \geq 0)$ is such a sequence, then

$\alpha = \Phi(t_{n+1}, t_n) \leq \psi(t_n) - \psi(t_{n+1})$, for all n;

wherefrom, passing to limit as $n \to \infty$, one gets $\alpha = 0$; a contradiction. Moreover, the obtained nonexistence relation also tells us that Ω is strongly Matkowski admissible in a trivial way. An indirect argument for verifying this is based on the immediate observation that

there are no sequences $(t_n; n \geq 0)$ in R_+^0 with $\sum_n \Phi(t_{n+1}, t_n) < \infty$.

Note that, as a direct consequence, constructions like in Wardowski [52] are not working here.

Alter 2 Summing up, the case of Φ=constant is noneffective for us. To get genuine examples, let us start from the class of functions $\Phi : R_+^0 \times R_+^0 \to R$, with

Φ is strongly asymptotic expansive:
$(\Phi(t, s) \geq \alpha(t + s)$, whenever $\Phi(t, s) \leq \beta)$, where $\alpha, \beta > 0$ are constants.

Note that any such function is diagonal asymptotic expansive. In fact, let the strictly descending sequence $(t_n; n \geq 0)$ in R_+^0 be such that $\sum_n \Phi(t_{n+1}, t_n) < \infty$. In particular, $\lim_n \Phi(t_{n+1}, t_n) = 0$; so, there must be some rank $n(\beta)$ such that

$\Phi(t_{n+1}, t_n) \leq \beta$, for all $n \geq n(\beta)$.

This, along with the strong asymptotic expansive property, gives

$\Phi(t_{n+1}, t_n) \geq \alpha(t_{n+1} + t_n) \geq 2\alpha t_{n+1}$, for all $n \geq n(\beta)$,

wherefrom $\sum_n t_n < \infty$; and the claim follows.

By the statement above, one derives that (with ψ taken as precise)

$\Omega := \Omega[\psi, \Phi]$ is strongly Matkowski admissible
(hence, Matkowski admissible).

However, for practical reasons, we want that Ω be second variable increasing as well. To reach this property, a separated version of the above condition is needed:

Φ is *separated strongly asymptotic expansive*: $\Phi(t, s) = \varphi_1(t) + \varphi_2(s)$,
where $\varphi_1, \varphi_2 \in \mathcal{F}(R^0_+)$ are strongly asymptotic expansive (see above).

[Note that any such function is strongly asymptotic expansive, as it can be directly seen]. In addition, suppose that

φ_2 is increasing (hence, asymptotic positive; see above).

Clearly, the introduced function Φ has the additional property:

Φ is asymptotic positive; hence diagonal asymptotic positive.

In fact, for each couple of strictly descending sequences (t_n) and (s_n) in R^0_+ and each $\varepsilon > 0$ with $t_n \to \varepsilon+$, $s_n \to \varepsilon+$, we have

$\limsup_n \Phi(t_n, s_n) \geq \limsup_n \varphi_2(s_n) > 0$;

and the claim follows. But then, the function $\psi : R^0_+ \to R^0_+$ is introduced as

$\psi(s) = (\chi(s) + 1)\varphi_2(s)$, $s \in R^0_+$

where $\chi : R^0_+ \to R^0_+$ is increasing, fulfills

ψ is increasing, $\psi(0 + 0) \geq 0 > -\infty$ and $\psi - \varphi_2 = \chi\varphi_2$ is increasing;
hence, $s \mapsto \psi(s) - \Phi(t, s)$ is increasing, for each $t > 0$.

Summing up, the couple (ψ, Φ) fulfills all general/specific conditions imposed above; so, its associated plane Rhoades relation $\Omega := \Omega[\psi, \Phi]$ is endowed with the following regularity properties:

Ω is second variable increasing, asymptotic Meir-Keeler,
and strongly Matkowski admissible.

In particular, the function $\Phi_{\lambda,\mu} : R^0_+ \times R^0_+ \to R$ introduced as (for some couple $\lambda, \mu \in]0, 1]$)

$\Phi_{\lambda,\mu}(t, s) = \Gamma_\lambda(t) + \Gamma_\mu(s)$, $t, s > 0$ (where $\Gamma_\nu(r) = r^\nu$, $r > 0$, $\nu > 0$)

is separated strongly asymptotic expansive; moreover, it is evident that

$\Phi_{\lambda,\mu}$ is asymptotic positive and (strictly) increasing in each variable.

So, with respect to the function $\psi_\mu : R_+^0 \to R_+^0$ introduced as

$\psi_\mu(s) = (\chi(s) + 1)\Gamma_\mu(s)$, $s > 0$ (where $\chi : R_+^0 \to R_+^0$ is increasing),

the associated plane Rhoades relation $\Omega := \Omega[\psi_\mu, \Phi_{\lambda,\mu}]$ has the properties

Ω is second variable increasing, asymptotic Meir-Keeler,
and strongly Matkowski admissible.

Remark 6 Passing to the infinitely asymptotic regular property, the following simple fact is useful. Let $\psi : R_+^0 \to R$, $\Phi : R_+^0 \times R_+^0 \to R$ be a couple of functions, where

(ψ, Φ) is a plane Rhoades couple, $\psi(0+0) = -\infty$, and
Φ is increasing in each variable.

In particular, this last property holds when $\Phi : R_+^0 \times R_+^0 \to R$ is defined as

$\Phi(t, s) = \alpha$, $t, s > 0$, where $\alpha > 0$ is a constant.

Note that, in such a case, Φ is strictly positive and continuous (hence, asymptotic positive). Moreover, the domination property becomes

there exists $\mu > 1$, such that ψ is *asymptotic μ-dominating*:
for each strictly descending sequence $(t_n; n \geq 0)$ in R_+^0 with $\lim_n t_n = 0$ and $\sum_n [1 + \psi(t_0) - \psi(t_{n+1})]^{-\mu} < \infty$, we have $\sum_n t_n < \infty$.

A basic particular case when this property holds is

there exists $\lambda \in]0, 1[$, such that
ψ is *λ-polynomial*: $\lim_{t\to 0+} \psi(t)t^\lambda = 0$.

In fact, put

$\mu := 1/\lambda$ (hence, $\mu > 1$);

and let the sequence $(t_n; n \geq 0)$ in R_+^0 be as in the premise above. From the posed assumption, we have (combining with $\lim_n t_n = 0$)

$\lim_n [1 + \psi(t_0) - \psi(t_{n+1})]t_{n+1}^\lambda = 0$; and then
$\lim_n [1 + \psi(t_0) - \psi(t_{n+1})]^\mu t_{n+1} = 0$.

The sequence in the left-hand side of this relation is therefore bounded; so, there must be some $C > 0$, with

$[1 + \psi(t_0) - \psi(t_{n+1})]^\mu t_{n+1} \leq C$, for all n

or, equivalently,

$t_{n+1}/C \leq [1 + \psi(t_0) - \psi(t_{n+1})]^{-\mu}$, for all n.

But, from this, the desired assertion is clear, hence the conclusion.

Remark 7 The construction in the preceding step (involving a certain couple $(\psi, \varphi \in \mathcal{F}(R^0_+, R))$ is nothing else than a particular case of this one, (relative to the couple $(\psi \in \mathcal{F}(R^0_+, R), \Phi \in \mathcal{F}(R^0_+ \times R^0_+, R))$, corresponding to the choice

$(\psi$=as before) and $\Phi(t,s) = \varphi(s), t, s > 0$.

In fact, it is not hard to see that (under these notations)

(i) Φ is diagonal asymptotic positive if φ is asymptotic positive
(ii) Φ is asymptotic positive if φ is asymptotic positive
(iii) Φ is diagonal asymptotic expansive if φ is asymptotic expansive
(iv) $(\forall \mu > 1)$: (ψ, Φ) is asymptotic μ-dominating when (ψ, φ) is asymptotic μ-dominating;

since the verification is immediate, we do not give details.

Part-Case (IV) Given the functions $\psi : R^0_+ \to R$ and $\xi : R^0_+ \times R^0_+ \to R$, we say that (ψ, ξ) is a *simulation couple*, if

(diag) ψ is increasing and ξ is *ψ-diagonal*:
$\xi(t,s) < \psi(s) - \psi(t)$, for each $t, s > 0$.

In this case, the relation $\Omega[\xi]$ introduced as

$t\Omega[\xi]s$ iff $\xi(t,s) \geq 0$

will be referred to as *associated* to (ψ, ξ). This concept may be viewed as an extension of the related one defined (over pseudometric structures) in the survey paper by Turinici [51]; we do not give details.

As before, some useful properties of this associated relation are available under extra sequential conditions upon (ψ, ξ). The first group of these writes

(d-a-neg) ξ is *diagonal asymptotic negative*:
for each strictly descending sequence (t_n) in R^0_+ and each $\varepsilon > 0$ with $t_n \to \varepsilon+$, we have $\liminf_n \xi(t_{n+1}, t_n) < 0$
(a-pos) ξ is *asymptotic negative*:
for each couple of strictly descending sequences (t_n) and (s_n) in R^0_+ and each $\varepsilon > 0$ with $t_n \to \varepsilon+$, $s_n \to \varepsilon+$, we have $\liminf_n \xi(t_n, s_n) < 0$.

In this context, it will be useful to introduce the combined notion

(d-a-sim-couple) (ψ, ξ) is a *diagonal asymptotic simulation couple*:
(ψ, ξ) is a simulation couple with ξ=diagonal asymptotic negative
(a-sim-couple) (ψ, ξ) is an *asymptotic simulation couple*:
(ψ, ξ) is a simulation couple with ξ=asymptotic negative.

Now, to get the relationships between these, the following *duality* principle will be useful. Given the couple (ψ, ξ) where $\psi : R^0_+ \to R$ and $\xi : R^0_+ \times R^0_+ \to R$, let us construct another couple (ψ, Φ), where ψ is the precise one and $\Phi : R^0_+ \times R^0_+ \to R$ is taken as

$\Phi(t,s) = \psi(s) - \psi(t) - \xi(t,s), t, s > 0;$

we say that (ψ, Φ) is the *dual* to (ψ, ξ) couple. As usual, given such a couple, we denote by $\Omega[\psi, \Phi]$ the associated relation

$t\Omega[\psi, \Phi]s$ iff $\psi(t) \leq \psi(s) - \Phi(t, s)$.

Conversely, given the couple (ψ, Φ) where $\psi : R^0_+ \to R$ and $\Phi : R^0_+ \times R^0_+ \to R$, let us construct another couple (ψ, ξ), where ψ is as before and $\xi : R^0_+ \times R^0_+ \to R$ is introduced as

$\xi(t, s) = \psi(s) - \psi(t) - \Phi(t, s), t, s > 0;$

we say that (ψ, ξ) is the *dual* to (ψ, Φ) couple.

Proposition 19 *Let these conventions be in use. Then,*

(56-1) *If (ψ, ξ) is a (diagonal) asymptotic simulation couple (i.e., a simulation couple where ξ is (diagonal) asymptotic negative), then the dual couple (ψ, Φ) is a (diagonal) asymptotic Rhoades couple (i.e., a Rhoades couple with Φ=(diagonal) asymptotic positive); moreover, the associated relations $\Omega[\xi]$ and $\Omega[\psi, \Phi]$ are identical, that is,*

$(\forall t, s > 0)$: $t\Omega[\xi]s$ iff $t\Omega[\psi, \Phi]s$.

(56-2) *If (ψ, Φ) is a (diagonal) asymptotic Rhoades couple (i.e., a Rhoades couple where Φ is (diagonal) asymptotic positive), then the dual couple (ψ, ξ) is a (diagonal) asymptotic simulation couple (i.e., a simulation couple with ξ=(diagonal) asymptotic negative); moreover, the associated relations $\Omega[\psi, \Phi]$ and $\Omega[\xi]$ are identical, that is,*

$(\forall t, s > 0)$: $t\Omega[\psi, \Phi]s$ iff $t\Omega[\xi]s$.

Proof There are two stages to be completed.

(i) By the very definition of simulation couple

$\Phi(t, s) = \psi(s) - \psi(t) - \xi(t, s) > 0$, for all $t, s > 0$;
whence, Φ is strictly positive;

and this, along with ψ=increasing, tells us that (ψ, Φ) is a Rhoades couple. Moreover, by these conventions, one has for each $t, s > 0$

$t\Omega[\xi]s$ iff $\xi(t, s) \geq 0$
$t\Omega[\psi, \Phi]s$ iff $\psi(t) \leq \psi(s) - \Phi(t, s)$.

This, along with

$\xi(t, s) \geq 0$ iff $\psi(s) - \psi(t) - \Phi(t, s) \geq 0$,

gives us the identity between these relations.

Finally, we have two alternatives to discuss.

Alter i-1 Given the strictly descending sequence $(t_n; n \geq 0)$ in R^0_+ and the number $\varepsilon > 0$ with $t_n \to \varepsilon+$, we must have (as ξ=diagonal asymptotic negative)

$$\limsup_n \Phi(t_{n+1}, t_n) = \limsup_n[\psi(t_n) - \psi(t_{n+1}) - \xi(t_{n+1}, t_n)] =$$
$$\psi(\varepsilon + 0) - \psi(\varepsilon + 0) - \liminf_n \xi(t_{n+1}, t_n) > 0;$$

whence, Φ is diagonal asymptotic positive; so, (ψ, Φ) is a diagonal asymptotic Rhoades couple.

Alter i-2 For each couple of strictly descending sequences (t_n) and (s_n) in R^0_+ and each $\varepsilon > 0$ with $t_n \to \varepsilon+$, $s_n \to \varepsilon+$, we have (as ξ=asymptotic negative)

$$\limsup_n \Phi(t_n, s_n) = \limsup_n[\psi(s_n) - \psi(t_n) - \xi(t_n, s_n)] =$$
$$\psi(\varepsilon + 0) - \psi(\varepsilon + 0) - \liminf_n \xi(t_n, s_n) > 0;$$

whence, Φ is asymptotic positive; so, (ψ, Φ) is an asymptotic Rhoades couple.

(ii) By the very definition of a Rhoades couple

$$\xi(t, s) = \psi(s) - \psi(t) - \Phi(t, s) < \psi(s) - \psi(t), \text{ for all } t, s > 0;$$

and this, along with ψ=increasing, tells us (ψ, ξ) is a simulation couple. Moreover, by these conventions, one has for each $t, s > 0$

$$t\Omega[\psi, \Phi]s \text{ iff } \psi(t) \le \psi(s) - \Phi(t, s),$$
$$t\Omega[\xi]s \text{ iff } \xi(t, s) \ge 0.$$

This, along with

$$\psi(s) - \psi(t) - \Phi(t, s) \ge 0 \text{ iff } \xi(t, s) \ge 0,$$

gives us the identity between these relations.

Finally, as before, we have two alternatives to discuss

Alter ii-1 Given the strictly descending sequence $(t_n; n \ge 0)$ in R^0_+ and the number $\varepsilon > 0$ with $t_n \to \varepsilon+$, we must have (as Φ=diagonal asymptotic positive)

$$\liminf_n \xi(t_{n+1}, t_n) = \liminf_n[\psi(t_n) - \psi(t_{n+1}) - \Phi(t_{n+1}, t_n)] =$$
$$\psi(\varepsilon + 0) - \psi(\varepsilon + 0) - \limsup_n \Phi(t_{n+1}, t_n) < 0;$$

whence, ξ is diagonal asymptotic positive; so, (ψ, ξ) is a diagonal asymptotic simulation couple.

Alter ii-2 For each couple of strictly descending sequences (t_n) and (s_n) in R^0_+ and each $\varepsilon > 0$ with $t_n \to \varepsilon+$, $s_n \to \varepsilon+$, we have (as Φ=asymptotic positive)

$$\liminf_n \xi(t_n, s_n) = \liminf_n[\psi(s_n) - \psi(t_n) - \Phi(t_n, s_n)] =$$
$$\psi(\varepsilon + 0) - \psi(\varepsilon + 0) - \limsup_n \Phi(t_n, s_n) < 0;$$

whence, ξ is asymptotic positive so (ψ, ξ) is an asymptotic simulation couple.

Remark 8 As a consequence of this, we may give an appropriate answer to the comparison problem between these two classes of simulation couples. Precisely, given the simulation couple (ψ, ξ), where $\psi : R^0_+ \to R$, $\xi : R^0_+ \times R^0_+ \to R$, we have

(ψ, ξ) is an asymptotic simulation couple implies
(ψ, ξ) is a diagonal asymptotic simulation couple;

The converse inclusion is not in general true. To verify this, let $\psi : R^0_+ \to R$ be an increasing function. Fix some $\alpha > 0$, and take the function $g = g_\alpha : R^0_+ \to R^0_+$, as

$g(t) = 1$, if $0 < t \leq \alpha$; $g(t) = t - \alpha$, if $t > \alpha$.

Then, let $\Phi = \Phi_g : R^0_+ \times R^0_+ \to R^0_+$ be introduced according to

$\Phi(t, s) = \max\{t, s\}$, if $t \neq s$, $t, s > 0$; $\Phi(r, r) = g(r)$, $r > 0$.

As precise in a previous place,

Φ is diagonal asymptotic positive, but not asymptotic positive.

Now, let (ψ, ξ) be the dual to (ψ, Φ) couple; that is,

(ψ=as before) and $\xi(t, s) = \psi(s) - \psi(t) - \Phi(t, s)$, $t, s > 0$.

By the previous developments, (ψ, ξ) is a simulation couple, with, in addition,

ξ diagonal asymptotic negative, but not asymptotic negative.

This, according to our conventions, means

(ψ, ξ) is a diagonal asymptotic simulation couple;
but not an asymptotic simulation couple;

and concludes our assertion.

Remark 9 Let $I(.)$ stand for the identity function of $\mathcal{F}(R^0_+)$:

$I(t) = t$, $t > 0$ (hence, $I(.)$ is increasing).

Further, let $\xi : R^0_+ \times R^0_+ \to R$ be a function. According to a general convention, we say that (I, ξ) is an *asymptotic simulation couple* when

(asc-1) ξ is *I-diagonal*: $\xi(t, s) < s - t$, for each $t, s > 0$;
(whence, (I, ξ) is a *simulation couple*)
(asc-2) ξ is *asymptotic negative*:
for each couple of strictly descending sequences (t_n) and (s_n) in R^0_+ and each $\varepsilon > 0$ with $t_n \to \varepsilon+$, $s_n \to \varepsilon+$, we have $\liminf_n \xi(t_n, s_n) < 0$.

A variant of this construction is as follows. Let us say that (I, ξ) is a *strong asymptotic simulation couple* when

(sasc-1) (I, ξ) is a simulation couple (see above)
(sasc-2) ξ is *strong asymptotic negative*:

for each couple of sequences (t_n) and (s_n) in R^0_+ and each $\varepsilon > 0$ with $(t_n \to \varepsilon, s_n \to \varepsilon)$, we have $\limsup_n \xi(t_n, s_n) < 0$.

This concept (also written as $\xi : R^0_+ \times R^0_+ \to R$ is a *simulation function*) has been introduced by Khojasteh et al. [22]. According to these authors, it may be viewed as a unifying tool for many metrical fixed point theorems; so, a discussion of its basic lines may be of interest. Clearly,

(I, ξ)=strong asymptotic simulation couple implies
(I, ξ)=asymptotic simulation couple.

Concerning the converse inclusion, the natural problem to be posed is as follows:

for each asymptotic simulation couple (ψ, ξ), we have—or not—that (I, ξ) is a strong asymptotic simulation couple.

The answer to this is negative, as shown in the example below.

Let $\psi : R^0_+ \to R$ be an increasing discontinuous function; hence,

$$\alpha := \sup\{\psi(t+0) - \psi(t-0); t \in R^0_+\} > 0.$$

Let $\beta \in]0, \alpha[$ be fixed; note that, by the very definition above,

there exists $\theta \in R^0_+$ such that $\psi(\theta+0) - \psi(\theta-0) > \beta$.

Given the couple (ψ, β), define a function $\xi : R^0_+ \times R^0_+ \to R$ as

$$\xi(t, s) = \psi(s) - \psi(t) - \beta, \ t, s > 0.$$

(i) We claim that (ψ, ξ) is an asymptotic simulation function. In fact, as $\beta > 0$,

(ψ, ξ) is a simulation couple (as ξ is ψ-diagonal).

On the other hand, ξ is asymptotic negative. Let the couple of strictly descending sequences (t_n) and (s_n) in R^0_+ and the number $\varepsilon > 0$ be such that $t_n \to \varepsilon+$, $s_n \to \varepsilon+$. Then, by definition

$$\lim_n \xi(t_n, s_n) = \psi(\varepsilon+0) - \psi(\varepsilon+0) - \beta = -\beta < 0;$$

and the assertion follows.

(ii) Further, we claim that ξ is not strong asymptotic negative. Indeed, let the couple of sequences (t_n) and (s_n) in R^0_+ and the number $\theta > 0$ (described as before) be such that $t_n \to \theta-$, $s_n \to \theta+$. Then, by definition

$$\lim_n \xi(t_n, s_n) = \psi(\theta+0) - \psi(\theta-0) - \beta > 0;$$

hence the assertion.

Part Case (V) Given the functions $\psi : R^0_+ \to R$ and $\eta : R^0_+ \times R^0_+ \to R$, we say that (ψ, η) is an *asymptotic manageable couple*, if

(amc-1) (ψ, η) is a *simulation couple*; i.e. (see above), ψ is increasing and η is ψ-*diagonal* ($\eta(t, s) < \psi(s) - \psi(t)$, for each $t, s > 0$)
(amc-2) η is *asymptotic subunitary*:

for each couple of strictly descending sequences (t_n) and (s_n) in R^0_+ and each $\varepsilon > 0$ with $t_n \to \varepsilon+$, $s_n \to \varepsilon+$, we have $\liminf_n [t_n + \eta(t_n, s_n)]/s_n < 1$.

The relationships between this concept and the preceding one (of asymptotic simulation couple) are described in the statement below.

Proposition 20 *Let the functions $\psi : R^0_+ \to R$ and $\eta : R^0_+ \times R^0_+ \to R$ be such that (ψ, η) is an asymptotic manageable couple. Then,*

(57-1) *We necessarily have that η is asymptotic negative; hence, (ψ, η) is an asymptotic simulation couple*

(57-2) *The associated relation $\Omega := \Omega[\eta]$ on R^0_+, introduced as*

$$t\Omega s \text{ iff } \eta(t, s) \geq 0,$$

is upper diagonal and asymptotic Meir-Keeler (hence, geometric Meir-Keeler; hence, Matkowski admissible).

Proof There are two steps to be passed.

(i) Let the couple of strictly descending sequences (t_n) and (s_n) in R^0_+ and the number $\varepsilon > 0$ be such that $t_n \to \varepsilon+$, $s_n \to \varepsilon+$. By the imposed condition, we have

$$\liminf_n [t_n + \eta(t_n, s_n)]/s_n < 1.$$

On the other hand, for each $n \geq 0$ we have

$$\eta(t_n, s_n)/s_n = (t_n + \eta(t_n, s_n))/s_n + (-t_n/s_n).$$

Passing to lim inf as $n \to \infty$, we have (by direct calculations)

$$\liminf_n \eta(t_n, s_n)/s_n < 1 - 1 = 0;$$

and this yields (via definition of inferior limit)

$\liminf_n \eta(t_n, s_n) < 0$; i.e., η is asymptotic negative.

Combining with (ψ, η) being a simulation function, one derives that (ψ, η) is an asymptotic simulation couple, as claimed.

(ii) The assertion is clear, in view of a preceding statement involving asymptotic simulation couples; however, for completeness reasons, we will provide an argument for this.

Step ii-1 Let $t, s > 0$ be such that

$t\Omega s$, i.e., $\eta(t, s) \geq 0$.

By the simulation couple property,

$\psi(t) - \psi(s) < -\eta(t, s) \leq 0$; whence, $\psi(t) < \psi(s)$.

Combining with ψ=increasing yields $t < s$; so, Ω is upper diagonal.

Step ii-2 Suppose by contradiction that there exists a couple of strictly descending sequences (t_n) and (s_n) in R_+^0 and a number $\varepsilon > 0$, with

$$t_n \to \varepsilon+, s_n \to \varepsilon+, \text{ and } t_n \Omega s_n \text{ (i.e., } \eta(t_n, s_n) \geq 0\text{), for each } n.$$

Then,

$$\liminf_n [t_n + \eta(t_n, s_n)]/s_n \geq \liminf_n t_n/s_n = 1,$$

in contradiction with η being asymptotic subunitary. Hence, Ω has the asymptotic Meir-Keeler property; and, from this, we are done.

Remark 10 Let $I(.)$ stand for the identity function of $\mathcal{F}(R_+^0)$:

$I(t) = t, t > 0$ (hence, $I(.)$ is increasing).

Further, let $\eta : R_+^0 \times R_+^0 \to R$ be a function. We say that (I, η) is a *strong asymptotic manageable couple* when

(samc-1) η is *I-diagonal*: $\eta(t, s) < s - t$, for each $t, s > 0$
(whence: (I, η) is a *simulation couple*)
(samc-2) η is *strong asymptotic subunitary*:
for each bounded sequence (t_n) in R_+^0 and each decreasing sequence (s_n) in R_+^0, we have $\limsup_n [t_n + \eta(t_n, s_n)]/s_n < 1$.

This concept (also written as $\eta : R_+^0 \times R_+^0 \to R$ is a *manageable function*) is due to Du and Khojasteh [12]. We claim that it is a particular case of the preceding one:

(I, η) is a strong asymptotic manageable couple that implies
(I, η) is an asymptotic manageable couple.

In fact, let the couple of strictly descending sequences (t_n) and (s_n) in R_+^0 and the number $\varepsilon > 0$ be such that $t_n \to \varepsilon+$, $s_n \to \varepsilon+$. Clearly, $(t_n; n \geq 0)$ is a bounded sequence; and then, by the imposed hypothesis,

$$\liminf_n [t_n + \eta(t_n, s_n)]/s_n \leq \limsup_n [t_n + \eta(t_n, s_n)]/s_n < 1;$$

so, η is an asymptotic subunitary. This, added to (I, η) being a simulation couple, proves our claim.

Part-Case (VI) Let $\psi \in \mathcal{F}(R_+^0, R)$ and $\Delta \in \mathcal{F}(R)$ be a couple of functions. The following regularity condition involving these objects will be considered here:

(BV-c) (ψ, Δ) is a *Bari-Vetro couple*:
ψ is increasing and Δ is regressive ($\Delta(r) < r$, for all $r \in R$).

In this case, by definition,

$$\varphi(t) := \psi(t) - \Delta(\psi(t)) > 0, \text{ for all } t > 0;$$

so, (ψ, φ) is a Rhoades couple of functions in $\mathcal{F}(R_+^0, R)$. Let $\Omega := \Omega[\psi, \Delta]$ be the (associated) *Bari-Vetro relation* over R_+^0, introduced as

$$t\Omega s \text{ iff } \psi(t) \leq \Delta(\psi(s)).$$

(This convention is related to the developments in Di Bari and Vetro [11]; we do not give details). As Δ=regressive, Ω is an upper diagonal relation over R^0_+. It is natural then to ask under which extra assumptions about our data we have that Ω is an asymptotic Meir-Keeler relation. The simplest one may be written as

(a-reg) Δ is *asymptotic regressive*:
for each descending sequence (r_n) in R and each $\alpha \in R$ with $r_n \to \alpha$, we have that $\liminf_n \Delta(r_n) < \alpha$.

Note that, by the non-strict character of the descending property above,

Δ is asymptotic regressive implies that Δ is regressive.

Proposition 21 *Let the functions ($\psi \in \mathcal{F}(R^0_+, R)$, $\Delta \in \mathcal{F}(R)$) be such that*

(ψ, Δ) is an asymptotic Bari-Vetro couple; i.e.,
ψ is increasing and Δ is asymptotic regressive.

Then,

(58-1) *The couple (ψ, φ) (where φ is the one above) is asymptotic Rhoades*
(58-2) *The associated relation Ω is upper diagonal and asymptotic Meir-Keeler (hence, geometric Meir-Keeler; hence, Matkowski admissible).*

Proof There are two steps to be passed.

(i) By a previous observation, (ψ, φ) is a Rhoades couple; so, it remains only to establish that φ is asymptotic positive. Let the strictly descending sequence $(t_n; n \geq 0)$ in R^0_+ and the number $\varepsilon > 0$ be such that $t_n \to \varepsilon+$; we must derive that $\limsup_n(\varphi(t_n)) > 0$. Denote, for simplicity,

$$r_n = \psi(t_n), n \geq 0; \alpha = \psi(\varepsilon + 0).$$

By the imposed conditions (and ψ=increasing)

(r_n) is descending and $r_n \to \alpha$ as $n \to \infty$.

In this case,

$$\limsup_n \varphi(t_n) = \limsup_n [r_n - \Delta(r_n)] = \alpha - \liminf_n \Delta(r_n) > 0;$$

hence the claim.

(ii) The assertion follows at once from (ψ, φ) being an asymptotic Rhoades couple and a previous remark involving these objects. However, for simplicity reasons, we provide an argument for this.

Step ii-1 Let $t, s > 0$ be such that

$t\Omega s$, i.e., $\psi(t) \leq \Delta(\psi(s))$.

As Δ is regressive,

$\psi(t) < \psi(s)$; whence, $t < s$ (in view of ψ=increasing);

so, Ω is upper diagonal.

Step ii-2 Suppose by contradiction that there exists a couple of strictly descending sequences (t_n) and (s_n) in R^0_+, and a number $\varepsilon > 0$, with

$t_n \to \varepsilon+$, $s_n \to \varepsilon+$, and $t_n \Omega s_n$ [i.e., $\psi(t_n) \leq \Delta(\psi(s_n))$], for each n.

From the increasing property of ψ, we get (under $\alpha := \psi(\varepsilon + 0)$)

$(u_n := \psi(t_n))$ and $(v_n := \psi(s_n))$ are descending sequences in R, with $u_n \to \alpha$, $v_n \to \alpha$, as $n \to \infty$;

so, passing to lim inf as $n \to \infty$ in the relation above [i.e., $u_n \leq \Delta(v_n)$, $\forall n$], one derives (via Δ=asymptotic regressive)

$\alpha = \liminf_n u_n \leq \liminf_n \Delta(v_n) < \alpha$; a contradiction.

Hence, our working assumption is not acceptable; and the conclusion follows.

In particular, when ψ and Δ are continuous, our statement reduces to the one in Jachymski [18]; see also Suzuki [42]. Nevertheless, it is to be stressed that the proposed geometric techniques cannot help us—in general—to handle contractive conditions like in Khan et al. [21]; so, we may ask of under which conditions is this removable. Further aspects will be delineated elsewhere.

Main Result

Let X be a nonempty set. Given the couple of vectors

$a = (a_1, \ldots, a_p) \in X^p$, $b = (b_1, \ldots, b_q) \in X^q$ (where $p, q \geq 1$),

define the *concatenation* vector $(a, b) \in X^{p+q}$ as

$(a, b) = (c_1, \ldots, c_{p+q})$, where $(c_i = a_i, 1 \leq i \leq p)$, $(c_{p+j} = b_j, 1 \leq j \leq q)$.

For simplicity reasons, we also use the equivalent writings for this object

$(a, b) = (a_1, \ldots, a_p, b) = (a, b_1, \ldots, b_q)$.

This operation may be extended to a finite number, $r(\geq 1)$, of such vectors. In particular (when $p = q = 1$, $a = b = u$), we introduce the convention: for each $u \in X$,

u^r=the vector $(z_0, \ldots, z_{r-1}) \in X^r$, where $(z_j = u, 0 \leq j \leq r - 1)$;

note that, in this case, $u^1 = u$.

Let (X, d) be a metric space and $k \geq 1$ be a natural number. Further, let $T : X^k \to X$ be a map and $S \in \mathcal{F}(X)$ be defined as

$$Sx = T(x^k),\ x \in X \text{ (the } \textit{diagonal} \text{ map attached to } T).$$

Then, let us put

$$\text{Fixd}(T) = \text{Fix}(S)(= \{z \in X; z = Sz\});$$

each element of this set will be called a *diagonal fixed point* of T.

The existence of such points is to be discussed in the context below. Given $U_0 := (u_0, \ldots, u_{k-1}) \in X^k$, the sequence $(u_n; n \geq 0)$ in X given as

(dia-ip) u_n=the above one, $0 \leq n \leq k-1$; $u_n = T(u_{n-k}, \ldots, u_{n-1}), n \geq k$

will be called the *(diagonal) iterative process* generated by U_0; and denoted [for simplicity reasons] as $(u_n = T^n U_0; n \geq 0)$.

(pp-0) Let us say that T is *fixd-asingleton*, if $\text{Fixd}(T) = \text{Fix}(S)$ is an asingleton, and *fixd-singleton*, provided $\text{Fixd}(T) = \text{Fix}(S)$ is a singleton

(pp-1) Call $U_0 = (u_0, \ldots, u_{k-1}) \in X^k$ a *Presić point* (modulo (d, T)), provided the associated iterative sequence $(u_n := T^n U_0; n \geq 0)$ (see above) is d-convergent. If each $U_0 = (u_0, \ldots, u_{k-1}) \in X^k$ is a Presić point (modulo (d, T)), then T will be referred to as a *Presić operator* (modulo d)

(pp-2) Call $U_0 = (u_0, \ldots, u_{k-1}) \in X^k$ a *strong Presić point* (modulo (d, T)), provided the associated iterative sequence $(u_n := T^n U_0; n \geq 0)$ is d-convergent and $z := \lim_n(u_n)$ is an element of $\text{Fixd}(T) = \text{Fix}(S)$. If each $U_0 = (u_0, \ldots, u_{k-1}) \in X^k$ is a strong Presić point (modulo (d, T)), then T will be referred to as a *strong Presić operator* (modulo d).

As a basic completion of these, we have to introduce the specific (diagonal) metric type contractions. Denote, for each i, j with $i < j$,

$$A(z_i, \ldots, z_j) = \max\{d(z_i, z_{i+1}), \ldots, d(z_{j-1}, z_j)\},\ (z_i, \ldots, z_j) \in X^{j-i+1}.$$

Further, let $\Omega \subseteq R_+^0 \times R_+^0$ be a relation over R_+^0; supposed to satisfy

Ω is upper diagonal: $t\Omega s$ implies $t < s$.

We say that the mapping T is *(d, Ω)-contractive*, provided

(d-Om-con) $d(T(y_0, \ldots, y_{k-1}), T(y_1, \ldots, y_k))\Omega A(y_0, \ldots, y_{k-1}, y_k)$,
for all $(y_0, \ldots, y_{k-1}, y_k) \in X^{k+1}$ with
$d(T(y_0, \ldots, y_{k-1}), T(y_1, \ldots, y_k)) > 0$, $A(y_0, \ldots, y_{k-1}, y_k) > 0$.

Some concrete examples of such contractions will be given a bit further. For the moment, we shall be interested to derive a nonexpansive property of such objects.

Proposition 22 *Suppose that T is (d, Ω)-contractive, where [as precise] Ω is upper diagonal. Then, necessarily,*

(61-1) T is d-nonexpansive:
$d(T(y_0,\ldots,y_{k-1}),T(y_1,\ldots,y_k)) \le A(y_0,\ldots,y_{k-1},y_k)$,
for all $(y_0,\ldots,y_{k-1},y_k) \in X^{k+1}$.

Proof Let $(y_0,\ldots,y_{k-1},y_k) \in X^{k+1}$ be arbitrary fixed. Two alternatives occur.

(I) Suppose that $A(y_0,\ldots,y_{k-1},y_k) > 0$. If $d(T(y_0,\ldots,y_{k-1}),T(y_1,\ldots,y_k)) = 0$, then the d-nonexpansive property is clear. Suppose further that

$$d(T(y_0,\ldots,y_{k-1}),T(y_1,\ldots,y_k)) > 0.$$

By the (d,Ω)-contractive property, one gets (via Ω=upper diagonal)

$$d(T(y_0,\ldots,y_{k-1}),T(y_1,\ldots,y_k)) < A(y_0,\ldots,y_{k-1},y_k);$$

and the d-nonexpansive property is again clear.

(II) Suppose that $A(y_0,\ldots,y_{k-1},y_k) = 0$. By definition, this gives

$$y_0 = \ldots = y_{k-1} = y_k = a, \text{ for some } a \in X.$$

But then evidently,

$$T(y_0,\ldots,y_{k-1}) = T(a^k), T(y_1,\ldots,y_k) = T(a^k);$$
$$\text{whence } d(T(y_0,\ldots,y_{k-1}),T(y_1,\ldots,y_k)) = 0;$$

which tells us that the d-nonexpansive property holds. The proof is complete.

Finally, let us say that S is *d-strictly nonexpansive*, provided

(d-s-nex) $d(Sx,Sy) < d(x,y)$, whenever $x \neq y$.

Our main result in this exposition (referred to as Cirić-Presić relational fixed point statement (CP-rela)) is to be stated as follows.

Theorem 2 *Suppose that T is (d,Ω)-contractive, for some upper diagonal relation Ω over R^0_+, with the additional properties:*

(61-i) Ω is second variable increasing
(61-ii) Ω is strongly Matkowski admissible.

In addition, let X be d-complete. Then, the following conclusions hold:

(61-a) *If (in addition) S is d-strictly nonexpansive, then*

$$T \text{ is fixd-asingleton; i.e., } \mathrm{Fixd}(T) = \mathrm{Fix}(S) \text{ is an asingleton}$$

(61-b) *T is strong Presić (modulo d); i.e., for each $U_0 := (u_0,\ldots u_{k-1}) \in X^k$, the iterative sequence $(u_n := T^n U_0; n \ge 0)$ given according to*

$$u_n\text{=the above ones, } 0 \le n \le k-1;\ u_n = T(u_{n-k},\ldots,u_{n-1}),\ n \ge k$$

fulfills $u_n \xrightarrow{d} z$, for some diagonal fixed point $z \in \mathrm{Fixd}(T) = \mathrm{Fix}(S)$.

Proof The first half of this statement is immediate, as results from

Step 0 Let $z_1, z_2 \in X$ be a couple of points in $\text{Fixd}(T) = \text{Fix}(S)$ (i.e., $z_1 = Sz_1$, $z_2 = Sz_2$); and—contrary to the written conclusion—suppose that $z_1 \neq z_2$. By the d-strict nonexpansive property of S,

$d(z_1, z_2) = d(Sz_1, Sz_2) < d(z_1, z_2)$; a contradiction.

Hence, necessarily, $z_1 = z_2$; and the claim follows.

It remains now to establish that T is strong Presić (modulo d). There are several steps to be passed.

Step 1 Put, for simplicity,

$\rho_n = d(u_n, u_{n+1}), n \geq 0.$

Note that, by a previous convention, we have for each i, j with $i < j$,

$A(u_i, \ldots, u_j) = \max\{\rho_i, \ldots, \rho_{j-1}\}.$

For technical reasons, it would be useful to denote, for each $n \geq k$,

$B_n = A(u_{n-k}, \ldots, u_{n-1}, u_n)$, i.e., $B_n = \max\{\rho_{n-k}, \ldots, \rho_{n-1}\}$.

Proposition 23 *Suppose that*

$B_n = 0$, *for some* $n \geq k$.

Then

(62-1) *The subsequence* $(y_i := u_{n-k+i}; i \geq 0)$ *is constant, i.e.,*

$$u_{n-k} = u_{n-k+1} = \ldots = a, \text{ for some } a \in X.$$

(62-2) *In addition, we have*

$$a = T(a^k) = Sa;$$

hence, $a \in X$ *is a diagonal fixed point of* T.

Proof (Proposition 23) Let $n \geq k$ be such that $B_n = 0$. Then, by definition,

$$u_{n-k} = \ldots = u_{n-1} = u_n = a, \text{ for some } a \in X.$$

Combining with the iterative procedure, we also get

$$a = u_n = T(u_{n-k}, \ldots, u_{n-1}) = T(a^k);$$

so, $a \in X$ is a diagonal fixed point of T. Finally, we have

$$u_{n+1} = T(u_{n-k+1}, \ldots, u_{n-1}, u_n) = T(a^k) = a,$$
$$u_{n+2} = T(u_{n-k+2}, \ldots, u_n, u_{n+1}) = T(a^k) = a, \ldots;$$

and, from this, $(u_{n-k+i} = a; i \geq 0)$.

As a consequence of these remarks, it follows that, whenever

$B_n = 0$, for some $n \geq k$,

we are done; so, without loss, one may assume that

(str-pos) $B_n > 0$, for all $n \geq k$.

The following auxiliary fact concentrates the "deep" part of our argument.

Proposition 24 *Under these conditions, we have*

(63-1) $\rho_n \Omega B_n$ *(hence,* $\rho_n < B_n$*), for each* $n \geq k$ *with* $\rho_n > 0$*;*
whence $\rho_n < B_n$*, for each* $n \geq k$
(63-2) $B_{n+1} \leq B_n$*,* $\forall n \geq k$*; so,* $(B_{k+i}; i \geq 0)$ *is descending*
(63-3) $B_{n+k} \Omega B_n$*,* $\forall n \geq k$ *whence* $B_{(i+1)k} \Omega B_{ik}$*, for each* $i \geq 1$.

Proof

(i) By definition, we have for each $n \geq k$

$$A(u_{n-k}, \ldots, u_{n-1}, u_n) = \max\{\rho_{n-k}, \ldots, \rho_{n-1}\} = B_n.$$

On the other hand, from our iterative construction,

$$\rho_n = d(u_n, u_{n+1}) = d(T(u_{n-k}, \ldots, u_{n-1}), T(u_{n-k+1}, \ldots, u_{n-1}, u_n));$$

and this, by the contractive property, gives (from the working hypothesis about our sequence $(B_{k+i}; i \geq 0)$)

$\rho_n \Omega B_n$, for each $n \geq k$ with $\rho_n > 0$.

Finally, for each $n \geq k$ with $\rho_n = 0$, we derive

$\rho_n < B_n$ (in view of $0 < B_n$);

so, our assertion follows.

(ii) From the representation above, one has, for each $n \geq k$,

$B_{n+1} = \max\{\rho_{n-k+1}, \ldots, \rho_{n-1}, \rho_n\} \leq \max\{B_n, \rho_n\}$;

and this, along with $\rho_n < B_n$, yields the written relation.

(iii) Let $n \geq k$ be arbitrary fixed. By definition,

$B_{n+k} = \max\{\rho_n, \ldots, \rho_{n+k-1}\}$;

so, $B_{n+k} = \rho_{n+j}$, for some $j \in \{0, \ldots, k-1\}$,
with, in addition, $\rho_{n+j} > 0$ (because $B_{n+k} > 0$).

From the preceding step, we have

$\rho_{n+j} \Omega B_{n+j}$; that is, $B_{n+k} \Omega B_{n+j}$.

As $B_{n+j} \leq B_n$ (see above), one gets

$B_{n+k} \Omega B_n$ (as Ω is second variable increasing);

hence, the desired assertion follows.

Having these established, we may now pass to the final part of our argument.

Step 2 From the evaluations above, we have

$B_{(i+1)k} \Omega B_{ik}$, for all $i \geq 1$.

As Ω is strongly Matkowski admissible, this yields

$\sum_{i \geq 0} B_{(i+1)k} (= B_k + B_{2k} + \ldots) < \infty$.

On the other hand, by definition, one gets (for all ranks $i \geq 0$)

$\rho_{ik}, \ldots, \rho_{(i+1)k-1} \leq B_{(i+1)k}$; whence $\rho_{ik} + \ldots + \rho_{(i+1)k-1} \leq k B_{(i+1)k}$;

so (by simply adding these inequalities)

$\sum_{n \geq 0} \rho_n \leq k \sum_{i \geq 0} B_{(i+1)k} < \infty$,

which tells us that the iterative sequence $(u_n; n \geq 0)$ is d-Cauchy. As X is d-complete, we necessarily have

$u_n \xrightarrow{d} z$ as $n \to \infty$, for some $z \in X$.

Step 3 We now claim that the obtained limit $z \in X$ is a diagonal fixed point for T; i.e., $z = T(z^k)(= Sz)$. To verify this, it will suffice proving that

$u_n \xrightarrow{d} Sz(= T(z^k))$, as $n \to \infty$.

In fact, by the triangle inequality and d-nonexpansive property (deductible—as precise—from our contractive condition), we have for each $n \geq k$

$$\begin{aligned}
&d(u_n, Sz) = d(T(u_{n-k}, \ldots, u_{n-1}), T(z^k)) \leq \\
&d(T(u_{n-k}, \ldots, u_{n-2}, u_{n-1}), T(u_{n-k+1}, \ldots, u_{n-1}, z)) + \\
&d(T(u_{n-k+1}, \ldots, u_{n-2}, u_{n-1}, z), T(u_{n-k+2}, \ldots, u_{n-1}, z^2)) + \ldots + \\
&d(T(u_{n-1}, z^{k-1}), T(z^k)) \leq \\
&\max\{\rho_{n-k}, \ldots, \rho_{n-2}, d(u_{n-1}, z)\} + \\
&\max\{\rho_{n-k+1}, \ldots, \rho_{n-2}, d(u_{n-1}, z)\} + \ldots + d(u_{n-1}, z);
\end{aligned}$$

and this yields

$$d(u_n, Sz) \leq k \max\{\rho_{n-k}, \ldots, \rho_{n-2}, d(u_{n-1}, z)\}, \ \forall n \geq k.$$

Passing to limit as $n \to \infty$, we get

$$\lim_n d(u_n, Sz) = 0; \text{ so (as } d\text{=separated), } z = Sz(= T(z^k)).$$

The proof is thereby complete.

Note that further extensions of this result are possible, in the framework of quasi-metric spaces, taken as in Hitzler [16]; see also Turinici [49]. Some abstract counterparts of these facts may be found in Kurepa [24].

Particular Versions

Let again (X, d) be a metric space and $k \geq 1$ be a natural number. Further, let $T : X^k \to X$ be a mapping; and $S : X \to X$ be its associated diagonal map. Remember that our main problem is that of establishing (via diagonal iterative processes) some existential results about the set

Fixd$(T) =$ Fix(S) (the *diagonal fixed points* of T).

The basic directions under which a determination of such points is to be made were already sketched in a previous place; and, as an appropriate answer to the posed problem, the main result of this exposition—referred to as the Ćirić-Presić relational fixed point statement (CP-rela) – has been stated and proved. It is our aim in the following to give some useful versions of the underlying statement (CP-rela), by starting from the classes of (strongly) Matkowski admissible (upper diagonal) relations on R^0_+ we just discussed. Remember that, for each i, j with $i < j$, we denoted (for simplicity)

$A(z_i, \ldots, z_j) = \max\{d(z_i, z_{i+1}), \ldots, d(z_{j-1}, z_j)\}, (z_i, \ldots, z_j) \in X^{j-i+1};$

moreover, we agreed to say that the self-map S is *d-strictly nonexpansive*, provided

$d(Sx, Sy) < d(x, y)$, whenever $x \neq y$.

Part-Case (I) Let $\mathcal{F}(re)(R^0_+, R)$ stand for the class of all $\varphi \in \mathcal{F}(R^0_+, R)$ with

φ *regressive* (in the sense: $\varphi(t) < t$, for all $t > 0$).

For each $\varphi \in \mathcal{F}(re)(R^0_+, R)$, we introduced the properties:

(m-ad) φ is *Matkowski admissible*:
for each $(t_n; n \geq 0)$ in R^0_+ with $(t_{n+1} \leq \varphi(t_n); n \geq 0)$, we have $\lim_n t_n = 0$
(s-m-ad) φ is *strongly Matkowski admissible*:
for each $(t_n; n \geq 0)$ in R^0_+ with $(t_{n+1} \leq \varphi(t_n); n \geq 0)$, we have $\sum_n t_n < \infty$.

Clearly, for each $\varphi \in \mathcal{F}(re)(R^0_+, R)$,

strongly Matkowski admissible implies Matkowski admissible;

but the converse is not in general true. A basic technical aspect of these conventions is related to the increasing property of such functions. Precisely, let $\mathcal{F}(re, in)(R_+^0, R)$ stand for the class of all $\varphi \in \mathcal{F}(re)(R_+^0, R)$, with

φ increasing on R_+^0 ($0 < t_1 \leq t_2$ implies $\varphi(t_1) \leq \varphi(t_2)$).

The following characterization of our Matkowski properties imposed upon φ is now available. Denote, for each $t \geq 0$,

$$\varphi^0(t) = t, \varphi^1(t) = \varphi(t), \ldots, \varphi^{n+1}(t) = \varphi(\varphi^n(t)), n \geq 0;$$

it is just the iterations sequence of φ. Then, for each $\varphi \in \mathcal{F}(re, in)(R_+^0, R)$, we have

(m-ad-in) φ is Matkowski admissible, iff
$(\forall t > 0)$: $\lim_n \varphi^n(t) = 0$, whenever $(\varphi^n(t); n \geq 0)$ exists
(s-m-ad-in) φ is strongly Matkowski admissible, iff
$(\forall t > 0)$: $\sum_n \varphi^n(t) < \infty$, whenever $(\varphi^n(t); n \geq 0)$ exists.

Now, given $\varphi \in \mathcal{F}(R_+^0, R)$, let us say that T is (d, φ)-*contractive*, provided

$d(T(y_0, \ldots, y_{k-1}), T(y_1, \ldots, y_k)) \leq \varphi(A(y_0, \ldots, y_{k-1}, y_k))$,
for all $(y_0, \ldots, y_{k-1}, y_k) \in X^{k+1}$ with
$d(T(y_0, \ldots, y_{k-1}), T(y_1, \ldots, y_k)) > 0$, $A(y_0, \ldots, y_{k-1}, y_k) > 0$.

As a first application of our main result, the following (practical) fixed point principle (referred to as Cirić-Presić-Matkowski fixed point statement (CP-Matkowski)) is available.

Theorem 3 *Suppose that T is (d, φ)-contractive, for some $\varphi \in \mathcal{F}(R_+^0, R)$, with*

φ regressive, increasing, and strongly Matkowski admissible.

In addition, let X be d-complete. Then, the following conclusions hold:

(71-a) *If (in addition) S is d-strictly nonexpansive, then*

T is fixd-asingleton, i.e., $\mathrm{Fixd}(T) = \mathrm{Fix}(S)$ *is an asingleton*

(71-b) *T is strong Presić (modulo d); i.e., for each $U_0 := (u_0, \ldots u_{k-1}) \in X^k$, the iterative sequence $(u_n := T^n U_0; n \geq 0)$ given according to*

u_n=the above ones, $0 \leq n \leq k-1$; $u_n = T(u_{n-k}, \ldots, u_{n-1})$, $n \geq k$

fulfills $u_n \overset{d}{\longrightarrow} z$, for some diagonal fixed point $z \in$ $\mathrm{Fixd}(T) = \mathrm{Fix}(S)$.

Proof Let $\Omega := \Omega[\varphi]$ stand for the associated Matkowski relation over R_+^0

$(t, s \in R_+^0)$: $t\Omega s$ iff $t \leq \varphi(s)$.

We have to establish that the main result is applicable with respect to this relation.

Step (1) By the imposed contractive condition, one derives that T is (d, Ω)-contractive, in the sense we already precise.

Step (2) By the increasing property of φ, it is clear that

$t, s_1, s_2 \in R_+^0$, $t\Omega s_1$, and $s_1 \le s_2$ imply $t\Omega s_2$

or, in other words, Ω is second variable increasing.

Step (3) Finally, by the strongly Matkowski admissible property of φ, it results that Ω is strongly Matkowski admissible:

$(t_n; n \ge 0)$ in R_+^0 and $(t_{n+1}\Omega t_n, \forall n)$ imply $\sum_n t_n < \infty$.

Summing up, the Ćirić-Presić relational (CP-rela) fixed point statement is indeed applicable to these data; so, we are done.

In particular, when φ is linear, i.e.,

$\varphi(t) = \beta t, t \in R_+$ (for some $\beta \in [0, 1[$),

the obtained result is just the one in Ćirić and Presić [8]. On the other hand, when φ is continuous, the corresponding version of our Ćirić-Presić -Matkowski fixed point statement (CP-Matkowski) includes a related 1981 result in Rus [37]; we shall discuss this fact elsewhere. Further technical aspects may be found in Khan et al. [20]; see also Shahzad and Shukla [40].

Part-Case (II) Let (ψ, φ) be a pair of functions in $\mathcal{F}(R_+^0, R)$. The following regularity conditions are to be used here:

(R-couple) (ψ, φ) is a *Rhoades couple*:
ψ is increasing, and φ is *strictly positive* ($\varphi(t) > 0, \forall t > 0$)
(a-pos) φ is *asymptotic positive*:
for each sequence $(t_n; n \ge 0)$ in R_+^0 and each $\varepsilon > 0$ with $t_n \to \varepsilon+$,
we must have $\limsup_n(\varphi(t_n)) > 0$.

In addition, we need a lot of conventions relative to the Rhoades couple (ψ, φ). Precisely, define the properties:

(f-a-r) (ψ, φ) is *finitely asymptotic regular*: $\psi(0+0) > -\infty$ and φ is *asymptotic expansive* (for each strictly descending sequence $(t_n; n \ge 0)$ in R_+^0 with $\sum_n \varphi(t_n) < \infty$, we have $\sum_n t_n < \infty$)
(i-a-r) (ψ, φ) is *infinitely asymptotic regular*: $\psi(0 + 0) = -\infty$, φ is increasing, and there exists $\mu > 1$ such that (ψ, φ) is *asymptotic μ-dominating* (for each strictly descending sequence $(t_n; n \ge 0)$ in R_+^0 with $t_n \to 0$ as $n \to \infty$ and $\sum_n[\varphi(t_n)/(1 + \psi(t_0) - \psi(t_{n+1}))]^\mu < \infty$, we have $\sum_n t_n < \infty$).

Given the couple of functions (ψ, φ) over $\mathcal{F}(R_+^0, R)$, let us say that T is *$(d; \psi, \varphi)$-contractive*, provided

$\psi(d(T(y_0, \ldots, y_{k-1}), T(y_1, \ldots, y_k))) \le$
$\psi(A(y_0, \ldots, y_{k-1}, y_k)) - \varphi(A(y_0, \ldots, y_{k-1}, y_k))$,
for all $(y_0, \ldots, y_{k-1}, y_k) \in X^{k+1}$ with
$d(T(y_0, \ldots, y_{k-1}), T(y_1, \ldots, y_k)) > 0, A(y_0, \ldots, y_{k-1}, y_k) > 0$.

As a second application of our main result, the following (practical) fixed point principle (referred to as Cirić-Presić-Rhoades fixed point statement (CP-Rhoades)) is available.

Theorem 4 *Suppose that T is $(d;\psi,\varphi)$-contractive, for some Rhoades couple (ψ,φ) over $\mathcal{F}(R^0_+,R)$, with $(\psi-\varphi)$ = increasing, fulfilling one of the conditions:*

(72-i) *(ψ,φ) is finitely asymptotic regular*
(72-ii) *(ψ,φ) is infinitely asymptotic regular.*

Then, the following conclusions hold:

(72-a) *If (in addition) S is d-strictly nonexpansive, then*

$$T \text{ is fixd-asingleton; i.e., } \mathrm{Fixd}(T)=\mathrm{Fix}(S) \text{ is an asingleton}$$

(72-b) *T is strong Presić (modulo d); i.e., for each $U_0:=(u_0,\dots u_{k-1})\in X^k$, the iterative sequence $(u_n:=T^nU_0;\ n\geq 0)$ given according to*

$$u_n\text{=the above ones, } 0\leq n\leq k-1;\ u_n=T(u_{n-k},\dots,u_{n-1}),\ n\geq k$$

fulfills $u_n\xrightarrow{d} z$, for some diagonal fixed point $z\in\mathrm{Fixd}(T)=\mathrm{Fix}(S)$.

Proof Let $\Omega:=\Omega[\psi,\varphi]$ stand for the Rhoades relation over R^0_+ attached to the Rhoades couple (ψ,φ), i.e.,

$$(t,s\in R^0_+)\colon t\Omega s \text{ iff } \psi(t)\leq\psi(s)-\varphi(s).$$

We have to establish that the main result is applicable with respect to this relation.

Step (1) By the imposed contractive condition, one derives that T is (d,Ω)-contractive, in the sense we already precise.

Step (2) From the increasing property of $\psi-\varphi$, it is clear that

$$t,s_1,s_2\in R^0_+,\ t\Omega s_1, \text{ and } s_1\leq s_2 \text{ imply } t\Omega s_2;$$

or, in other words, Ω is second variable increasing.

Step (3) Finally, by the posed regularity assumptions about (ψ,φ), it results (see above) that Ω is strongly Matkowski admissible:

$$(t_n;n\geq 0) \text{ in } R^0_+ \text{ and } (t_{n+1}\Omega t_n,\forall n) \text{ imply } \textstyle\sum_n t_n<\infty.$$

Summing up, the Cirić-Presić relational fixed point statement (CP-rela) is indeed applicable to these data; so, we are done.

Concerning the asymptotic regularity properties of (ψ,φ), the following particular cases are of interest. Let (ψ,φ) be a Rhoades couple of functions over $\mathcal{F}(R^0_+,R)$.

(I) Suppose, in addition, that

$$(\psi(0+0)>-\infty) \text{ and } (\varphi(t)\geq\alpha t, \text{ whenever } \varphi(t)\leq\beta),$$

where $\alpha, \beta > 0$ are constants. Then, (ψ, φ) is finitely asymptotic regular; and the conclusion in the statement above is retainable. For example, the function $\varphi \in \mathcal{F}(R^0_+)$, taken according to

$$\varphi = \Gamma_\nu, \text{ for some } \nu \in]0, 1] \text{ (where } \Gamma_\nu(t) = t^\nu, t > 0)$$

fulfills this last property, with $\alpha = \beta = 1$; moreover, it is evident that

φ is (strictly positive and) increasing (hence, asymptotic positive).

On the other hand, the function $\psi \in \mathcal{F}(R^0_+)$ introduced as

$$\psi(t) = (\chi(t) + 1)\varphi(t), t \in R^0_+,$$

where $\chi \in \mathcal{F}(R^0_+)$ is increasing and φ is taken as before, fulfills

ψ is strictly positive and increasing and $\psi - \varphi = \chi\varphi$ is increasing.

This yields a corresponding version of our statement with practical significance.

(II) Suppose, in addition, that

$\psi(0 + 0) = -\infty$, φ is constant ($\varphi(t) = \gamma$, $t > 0$, where $\gamma > 0$ is a constant), and there exists $\mu > 1$ such that ψ is asymptotic μ-dominating (for each strictly descending sequence $(t_n; n \geq 0)$ in R^0_+ with $t_n \to 0$ as $n \to \infty$ and $\sum_n [1 + \psi(t_0) - \psi(t_{n+1})]^{-\mu} < \infty$, we have $\sum_n t_n < \infty$).

Then, (ψ, φ) is infinitely asymptotic regular; and the conclusion in the statement above is retainable.

A basic particular case when this last property holds is

there exists $\lambda \in]0, 1[$, such that
ψ is *λ-polynomial*: $\lim_{t \to 0+} \psi(t)t^\lambda = 0$.

The obtained statement may be then viewed as a diagonal type counterpart of some developments in Wardowski [52]. Further aspects may be found in Abbas et al. [2].

Cirić-Presić Approach

Let X be a nonempty set. Given the couple of vectors

$$a = (a_1, \ldots, a_p) \in X^p, b = (b_1, \ldots, b_q) \in X^q \text{ (where } p, q \geq 1),$$

define the *concatenation* vector $(a, b) \in X^{p+q}$ as

$$(a, b) = (c_1, \ldots, c_{p+q}), \text{ where } (c_i = a_i, 1 \leq i \leq p), (c_{p+j} = b_j, 1 \leq j \leq q).$$

For simplicity reasons, we also use the equivalent writings for this object

$$(a, b) = (a_1, \dots, a_p, b) = (a, b_1, \dots, b_q).$$

This operation may be extended to a finite number (r say) of such vectors. In particular (when $p = q = 1, a = b = u$), we introduce the convention: for each $u \in X$

$$u^r\text{=the vector } (z_0, \dots, z_{r-1}) \in X^p, \text{ where } (z_j = u, 0 \le j \le r-1),$$

note that, in this case, $u^1 = u$.

Let (X, d) be a metric space and $k \geq 1$ be a natural number. Further, let $T : X^k \to X$ be a map and $S \in \mathcal{F}(X)$ be defined as

$$Sx = T(x^k), x \in X \text{ (the } \textit{diagonal} \text{ map attached to } T).$$

Then, let us put

$$\text{Fixd}(T) = \text{Fix}(S)(= \{z \in X; z = Sz\});$$

each element of this set will be called a *diagonal fixed point* of T. In the following, the existence of such points is to be discussed in the linear context of Ćirić-Presić-Matkowski fixed point statement (CP-Matkowski), characterized as

$$\varphi(t) = \beta t, t \in R_+, \text{ for some } \beta \in [0, 1[.$$

The basic result of this type is the 2007 one in Ćirić and Presić [8].

Given $\beta \geq 0$, let us say that T is *Ćirić-Presić $(d; \beta)$-contractive*, provided

$$\text{(CP-contr) } d(T(x_0, \dots, x_{k-1}), T(x_1, \dots, x_k)) \le \beta \max\{d(x_0, x_1), \dots, d(x_{k-1}, x_k)\}, \text{ for each } (x_0, \dots, x_k) \in X^{k+1}.$$

The following result (referred to as Ćirić-Presić standard fixed point statement (CP-st)) is now available.

Theorem 5 *Assume that T is Ćirić-Presić $(d; \beta)$-contractive, for a certain $\beta \in [0, 1[$. In addition, let X be d-complete. Then, the following conclusions hold:*

(81-a) *T is fixd-asingleton, whenever*

> *(d-s-nex) S is d-strictly nonexpansive:*
> $d(Sx, Sy) < d(x, y), \forall x, y \in X, x \neq y$

(81-b) *T is strong Presić (modulo d); i.e., for each $U_0 := (u_0, \dots u_{k-1}) \in X^k$, the iterative sequence $(u_n = T^n U_0; n \geq 0)$ given as*

$$u_n\text{=the above ones, } 0 \le n \le k-1; \; u_n = T(u_{n-k}, \dots, u_{n-1}), n \ge k$$

fulfills $u_n \xrightarrow{d} z$ *as* $n \to \infty$, *for some* $z \in \text{Fixd}(T) = \text{Fix}(S)$.

As already precise, the Ćirić-Presić standard fixed point statement (CP-st) is nothing else than a particular version of Ćirić-Presić-Matkowski fixed point statement (CP-Matkowski), when the underlying function $\varphi \in \mathcal{F}(R_+^0, R)$ is linear (see above). However, for a number of technical reasons, we provide a proof of this result; which differs, in part, from the original one.

Proof It will suffice considering the case $\beta > 0$; hence, $\beta \in]0, 1[$. There are several parts to be passed.

Part 1 Let $z_1, z_2 \in X$ be such that

$$z_1 = Sz_1, z_2 = Sz_2, z_1 \neq z_2.$$

Then, by the d-strict nonexpansive property of S,

$$d(z_1, z_2) = d(Sz_1, Sz_2) < d(z_1, z_2); \text{ a contradiction.}$$

Hence, $z_1 = z_2$, which tells us that $\text{Fix}(S) = \text{Fixd}(T)$ is an asingleton.

Part 2 Denote for simplicity

$$(\rho_n := d(u_n, u_{n+1}); n \geq 0), \theta = \beta^{1/k} \text{ (hence, } \theta^k = \beta);$$
$$B = \max\{\rho_i/\theta^i; 0 \leq i \leq k-1\} \text{ (hence, } \rho_i \leq B\theta^i, \text{ for } 0 \leq i \leq k-1).$$

We prove, by induction, that

$$(\forall i \geq 0)\colon \rho_i \leq B\theta^i \text{ [where } \theta \in]0, 1[\text{ is the above number].}$$

The case of $i \in \{0, \ldots, k-1\}$ is clear, by definition. Suppose that the assertion is true for all $i \in \{0, \ldots, j\}$, where $j \geq k-1$; we want to establish that it holds as well for $i = j+1$. From the contractive property (and imposed convention),

$$\rho_{j+1} \leq \beta \max\{\rho_{j+1-k}, \ldots, \rho_j\} \leq$$
$$B\beta \max\{\theta^{j+1-k}, \ldots, \theta^j\} = B\beta\theta^{j+1-k} = B\theta^{j+1};$$

and the claim follows. As a consequence of this (and $0 < \theta < 1$),

$$\sum_n \rho_n \text{ converges; whence, } (u_n) \text{ is } d\text{-Cauchy.}$$

Combining with X being d-complete, one derives

$$x_n \stackrel{d}{\longrightarrow} z \text{ as } n \to \infty, \text{ for some } z \in X.$$

Part 3 We now claim that the obtained limit z is a diagonal fixed point for T, in the sense $z = T(z^k)(= Sz)$. To do this, it will suffice proving that

$$u_n \stackrel{d}{\longrightarrow} Sz(= T(z^k)), \text{ as } n \to \infty.$$

In fact, the triangle inequality and contractive condition yield, for each $n \geq k$,

$$d(u_n, Sz) = d(T(u_{n-k}, \ldots, u_{n-1}), T(z^k)) \leq$$
$$d(T(u_{n-k}, \ldots, u_{n-2}, u_{n-1}), T(u_{n-k+1}, \ldots, u_{n-1}, z))+$$
$$d(T(u_{n-k+1}, \ldots, u_{n-2}, u_{n-1}, z), T(u_{n-k+2}, \ldots, u_{n-1}, z^2)) + \ldots +$$
$$d(T(u_{n-1}, z^{k-1}), T(z^k)) \leq$$
$$\beta \max\{\rho_{n-k}, \ldots, \rho_{n-2}, d(u_{n-1}, z)\}+$$
$$\beta \max\{\rho_{n-k+1}, \ldots, \rho_{n-2}, d(u_{n-1}, z)\} + \ldots + \beta d(u_{n-1}, z).$$

Passing to limit as $n \to \infty$ gives

$\lim_n d(u_n, Sz) = 0$; whence (as d=separated), $z = Sz$;

and this yields the desired fact. The proof is thereby complete.

A basic particular case of these developments may be constructed as below. Given $\Gamma := (\gamma_0, \ldots, \gamma_{k-1}) \in R_+^k$, let us say that T is *Presić* $(d; \Gamma)$*-contractive*, if

(P-contr) $d(T(x_0, \ldots x_{k-1}), T(x_1, \ldots, x_k)) \leq$
$\gamma_0 d(x_0, x_1) + \ldots + \gamma_{k-1} d(x_{k-1}, x_k)$, for each $(x_0, \ldots, x_k) \in X^{k+1}$.

The regularity condition imposed upon the vector Γ appearing here is

(norm-sub) Γ is *norm subunitary*: $\beta := \gamma_0 + \ldots + \gamma_{k-1} < 1$.

The following 1965 fixed point result obtained by Presić [33] is now available:

Theorem 6 *Suppose that T is Presić $(d; \Gamma)$-contractive, where $\Gamma = (\gamma_0, \ldots, \gamma_{k-1}) \in R_+^k$ is norm subunitary. In addition, let X be d-complete. Then,*

(82-a) *T is fixd-singleton:* $\mathrm{Fixd}(T) = \mathrm{Fix}(S) = \{z\}$, *for some $z \in X$*

(82-b) *T is a strong Presić operator (modulo d): for each (starting point) $U_0 := (u_0, \ldots u_{k-1}) \in X^k$, the iterative sequence $(u_n = T^n U_0; n \geq 0)$ given as*

u_n=the above ones, $0 \leq n \leq k-1$; $u_n = T(u_{n-k}, \ldots, u_{n-1})$, $n \geq k$

fulfills $u_n \xrightarrow{d} z$ as $n \to \infty$.

Proof Clearly, from the Presić $(d; \Gamma)$-contractive property, it results the Cirić-Presić one, where β is introduced as above. Hence, the Cirić-Presić standard fixed point statement (CP-st) is indeed applicable here; and, from this, we are done.

Note, finally, that further extensions—to the coincidence point setting—of these facts are possible, under the lines in George and Khan [14], Păcurar [31, 32], and Pathak et al. [30]; we shall discuss them elsewhere. For direct applications of these developments to convergence questions involving real sequences, we refer to Chen [7]; see also Berinde and Păcurar [4].

Further Aspects

In the following, we show that the 2007 Ćirić-Presić standard fixed point statement (CP-st) we just presented (cf. [8]) is "almost" deductible from an early 1976 result in Tasković [44].

(A) Given $k \geq 1$, let $f : R_+^k \to R_+$ be a function and $\Gamma := (\gamma_0, \ldots, \gamma_{k-1})$ be an element in R_+^k. We say that (f, Γ) is *admissible*, provided

(adm-1) f is *increasing*:
$u_i \leq v_i, i \in \{1, \ldots, k\}$ imply $f(u_1, \ldots, u_k) \leq f(v_1, \ldots, v_k)$
(adm-2) f is *semi-homogeneous*:
$f(\lambda x_1, \ldots, \lambda x_k) \leq \lambda f(x_1, \ldots, x_k), \forall (x_1, \ldots, x_k) \in R_+^k, \forall \lambda \geq 0$
(adm-3) the *associated* map $g : R_+ \to R_+$ introduced as
$(g(t) = f(\gamma_0 t, \ldots, \gamma_{k-1} t^k); t \in R_+)$ is continuous at $t = 1$.

Further, let us introduce the conditions

(sub) (f, Γ) is *subunitary*: $\beta := f(\gamma_0, \ldots, \gamma_{k-1}) < 1$
(tele-sub) (f, Γ) is *telescopic subunitary*:
$\alpha := f(\gamma_0, 0, \ldots, 0) + \ldots + f(0, \ldots, 0, \gamma_{k-1}) < 1$.

Having these precise, let us say that $T : X^k \to X$ is *Tasković $(d; f, \Gamma)$-contractive* (where $f \in \mathcal{F}(R_+^k, R_+)$, $\Gamma := (\gamma_0, \ldots, \gamma_{k-1}) \in R_+^k$), provided

(T-contr) $d(T(x_0, \ldots, x_{k-1}), T(x_1, \ldots, x_k)) \leq$
$f(\gamma_0 d(x_0, x_1), \ldots, \gamma_{k-1} d(x_{k-1}, x_k))$, for each $(x_0, \ldots, x_k) \in X^{k+1}$.

The following 1976 fixed point statement in Tasković [44] (referred to as *Tasković partial result* (T-part)) is our starting point. (See also Tasković [45], for a lot of related facts.)

Theorem 7 *Assume that T is Tasković $(d; f, \Gamma)$-contractive, where f is a mapping in $\mathcal{F}(R_+^k, R_+)$ and $\Gamma = (\gamma_0, \ldots, \gamma_{k-1})$ is a vector in R_+^k. In addition, let X be d-complete. Then, the following conclusions hold:*

(91-a) *If the couple (f, Γ) is telescopic subunitary (see above), then T is fixd-asingleton, in the sense*

$$\text{Fixd}(T) = \text{Fix}(S) \textit{ is an asingleton.}$$

(91-b) *If the couple (f, Γ) is admissible and subunitary, then T is strong Presić (modulo d); i.e., for each $U_0 := (u_0, \ldots u_{k-1}) \in X^k$, the iterative sequence $(u_n = T^n U_0; n \geq 0)$ given according to*

$$u_n = \textit{the above ones, } 0 \leq n \leq k-1;\ u_n = T(u_{n-k}, \ldots, u_{n-1}),\ n \geq k$$

fulfills $u_n \xrightarrow{d} z$, for some diagonal fixed point $z \in \text{Fixd}(T) = \text{Fix}(S)$.

Before passing to the verification of this result, some technical remarks are in order. Given $f \in \mathcal{F}(R_+^k, R_+)$, call it *homogeneous* provided

$$f(\lambda x_1, \ldots, \lambda x_k) = \lambda f(x_1, \ldots, x_k), \forall (x_1, \ldots, x_k) \in R_+^k, \forall \lambda \geq 0.$$

Clearly, any such function is semi-homogeneous. But, the reciprocal inclusion is also true, as it results from

Proposition 25 *For each function $f \in \mathcal{F}(R_+^k, R_+)$, we have*

(91-1) semi-homogeneous $\Longrightarrow$ homogeneous
(91-2) semi-homogeneous $\Longleftrightarrow$ homogeneous.

Proof Suppose that $f : R_+^k \to R_+$ is semi-homogeneous, i.e.,

$$f(\lambda x_1, \ldots, \lambda x_k) \leq \lambda f(x_1, \ldots, x_k), \forall (x_1, \ldots, x_k) \in R_+^k, \forall \lambda \geq 0.$$

Putting $\lambda = 0$ in this relation gives

$$0 \leq f(0, \ldots, 0) \leq 0; \text{ whence, } f(0, \ldots, 0) = 0,$$

i.e., the homogeneous property is fulfilled in case of $\lambda = 0$. It remains then to verify the underlying property in case of $\lambda > 0$. Denote for simplicity $\mu := 1/\lambda$; and let $(x_1, \ldots, x_k) \in R_+^k$ be arbitrary fixed. By the semi-homogeneous property, we have

$$f(x_1, \ldots, x_k) = f(\mu\lambda x_1, \ldots, \mu\lambda x_k) \leq \mu f(\lambda x_1, \ldots, \lambda x_k)$$
or, equivalently (by our notation): $\lambda f(x_1, \ldots, x_k) \leq f(\lambda x_1, \ldots, \lambda x_k)$.

This, along with the semi-homogeneous inequality, gives

$$f(\lambda x_1, \ldots, \lambda x_k) = \lambda f(x_1, \ldots, x_k), \forall (x_1, \ldots, x_k) \in R_+^k, \forall \lambda > 0$$

and proves our claim.

As a consequence of this, the semi-homogeneous property imposed to f is, ultimately, a homogeneous property of the same. On the other hand, the third admissible condition (involving the associated function $g(.)$) may be removed, as we shall see. Summing up, the following simplified variant of (T-part) (referred to as Tasković refined statement (T-refi)) is to be considered. Given $k \geq 1$, let $f : R_+^k \to R_+$ be a function and $\Gamma := (\gamma_0, \ldots, \gamma_{k-1})$ be an element in R_+^k. We say that (f, Γ) is *quasi admissible*, provided

(qadm-1) f is increasing:
$u_i \leq v_i, i \in \{0, \ldots, k-1\}$ imply $f(u_0, \ldots, u_{k-1}) \leq f(v_0, \ldots, v_{k-1})$
(qadm-2) f is homogeneous:
$f(\lambda x_0, \ldots, \lambda x_{k-1}) = \lambda f(x_0, \ldots, x_{k-1}), \forall (x_0, \ldots, x_{k-1}) \in R_+^k, \forall \lambda \geq 0.$

Further, let us introduce the conditions

(sub) (f, Γ) is *subunitary*: $\beta := f(\gamma_0, \ldots, \gamma_{k-1}) < 1$.
(tele-sub) (f, Γ) is *telescopic subunitary*:
$\alpha := f(\gamma_0, 0, \ldots, 0) + \ldots + f(0, \ldots, 0, \gamma_{k-1}) < 1$.

Theorem 8 *Assume that T is Tasković $(d; f, \Gamma)$-contractive, where f is a mapping in $\mathcal{F}(R_+^k, R_+)$ and $\Gamma = (\gamma_0, \ldots, \gamma_{k-1})$ is a vector in R_+^k. In addition, let X be d-complete. Then, the following conclusions hold:*

(92-a) *If the couple (f, Γ) is telescopic subunitary (see above), then T is fixd-asingleton, in the sense*

$$\text{Fixd}(T) = \text{Fix}(S) \text{ is an asingleton}$$

(92-b) *If the couple (f, Γ) is quasi admissible and subunitary, then T is strong Presić (modulo d); i.e., for each $U_0 := (u_0, \ldots u_{k-1}) \in X^k$, the iterative sequence $(u_n = T^n U_0; n \geq 0)$ given according to*

$$u_n\text{=the above ones, } 0 \leq n \leq k-1;\ u_n = T(u_{n-k}, \ldots, u_{n-1}),\ n \geq k$$

fulfills $u_n \xrightarrow{d} z$, *for some diagonal fixed point* $z \in \text{Fixd}(T) = \text{Fix}(S)$

Concerning the relationships between this last result and the standard Ciric-Presić one, the following answer is available.

Proposition 26 *Under the above conventions,*

(92-1) *(T-refi) is deductible from (CP-st)*
(92-2) *(CP-st) (the second half) follows from (T-refi) (the second half)*
(92-3) *(CP-st) (the first half) is not in general deductible from (T-refi)*

(the first half).

Proof

(i) Let the mapping $T : X^k \to X$, the function $f : R_+^k \to R_+$, and the vector $\Gamma = (\gamma_0, \ldots, \gamma_{k-1})$ in R_+^k be such that T is Tasković $(d; f, \Gamma)$-contractive.

(i-1) Assume that the telescopic subunitary condition in the first half of (T-refi) holds, i.e.,

$$\alpha := f(\gamma_0, 0, \ldots, 0) + \ldots + f(0, \ldots, 0, \gamma_{k-1}) < 1.$$

We claim that

(the diagonal map) S is $(d; \alpha)$-contractive;
hence, all the more, d-strictly nonexpansive.

In fact, let $x, y \in X$ be arbitrary fixed with $x \neq y$. From the Tasković contractive condition (and f=homogeneous), one derives

$$\begin{aligned}
&d(Sx, Sy) = d(T(x^k), T(y^k)) \leq \\
&d(T(x^k), T(x^{k-1}, y)) + \ldots + d(T(x, y^{k-1}), T(y^k)) \leq \\
&f(\gamma_0 d(x, y), 0, \ldots, 0) + \ldots + f(0, \ldots, 0, \gamma_{k-1} d(x, y)) = \\
&\alpha d(x, y) < d(x, y);
\end{aligned}$$

hence, (CP-st) (the first half) implies (T-refi) (the first half).

(i-2) Suppose that (f, Γ) is quasi admissible and subunitary (see above). From the imposed properties, we have (under our notations)

$$f(\gamma_0 t_0, \ldots, \gamma_{k-1} t_{k-1}\} \leq \beta \max\{t_0, \ldots, t_{k-1}\},$$
$$\text{for each } (t_0, \ldots, t_{k-1}) \in R_+^k.$$

Consequently, the mapping T in (T-refi) is Cirić-Presić $(d; \beta)$-contractive; and this tells us that (CP-st) (the second half) implies (T-refi) (the second half). Putting these together gives our desired fact.

(ii) Evident, by simply taking the couple (f, Γ) as

$$f(t_0, \ldots, t_{k-1}) = \max\{t_0, \ldots, t_{k-1}\}, (t_0, \ldots, t_{k-1}) \in R_+^k;$$
$$\Gamma = (\gamma_0, \ldots, \gamma_{k-1}) \in R_+^k \colon \gamma_i = \beta, 0 \leq i \leq k-1.$$

(iii) The deduction of (CP-st) (the first half) from (T-refi) (the first half) means, ultimately, a deduction of telescopic subunitary condition (tele-sub) in (T-refi) from the strict nonexpansive condition (d-s-nex) in (CP-st). But, as precise, the telescopic subunitary condition (tele-sub) in (T-refi) yields a Banach contractive property for the diagonal operator S. Hence, a deduction of telescopic subunitary condition (tele-sub) from the strict nonexpansive condition (d-s-nex) amounts to a deduction of Banach contractive property for S from the strict nonexpansive property of the same, which—in general—is not possible.

Summing up, (T-refi) may be viewed as a particular case of (CP-st); but the reciprocal is not in general true. However, we stress that, by the above developments, (T-refi) covers the most important part of (CP-st). Further aspects may be found in Rao et al. [34]; see also Shukla et al. [41].

References

1. Altman, M.: An integral test for series and generalized contractions. Am. Math. Mon. **82**, 827–829 (1975)
2. Abbas, M., Ilić, D., Nazir, T.: Iterative approximation of fixed points of generalized weak Presić type k-step iterative methods for a class of operators. Filomat **29**, 713–724 (2015)
3. Banach, S.: Sur les opérations dans les ensembles abstraits et leur application aux équations intégrales. Fundam. Math. **3**, 133–181 (1922)
4. Berinde, V., Păcurar, M.: An iterative method for approximating fixed points of Presić nonexpansive mappings. Rev. Anal. Numér. Théor. Approx. **38**, 144–153 (2009)
5. Bernays, P.: A system of axiomatic set theory: Part III. Infinity and enumerability analysis. J. Symb. Log. **7**, 65–89 (1942)
6. Boyd, D.W., Wong, J.S.W.: On nonlinear contractions. Proc. Am. Math. Soc. **20**, 458–464 (1969)
7. Chen, Y.-Z.: A Presić type contractive condition and its applications. Nonlinear Anal. **71**, 2012–2017 (2009)
8. Cirić, L.B., Presić, S.B.: On Presić type generalization of the Banach contraction mapping principle. Acta Math. Comenianae **76**, 143–147 (2007)
9. Cohen, P.J.: Set Theory and the Continuum Hypothesis. Benjamin, New York (1966)
10. Collaco, P., E Silva, J.C.: A complete comparison of 25 contractive definitions. Nonlinear Anal. **30**, 441–476 (1997)

11. Di Bari, C., Vetro, C.: Common fixed point theorems for weakly compatible maps satisfying a general contractive condition. Int. J. Math. Math. Sci. **2008**, Article ID 891375, 8 (2008)
12. Du, W.-S., Khojasteh, F.: New results and generalizations for approximate fixed point property and their applications. Abstr. Appl. Anal. **2014**, Article ID 581267, 9 (2014)
13. Dutta, P.N., Choudhury, B.S.: A generalization of contraction principle in metric spaces. Fixed Point Theory Appl. **2008**, Article ID 406368, 8 (2008)
14. George, R., Khan, M.S.: On Presić type extension of Banach contraction principle. Int. J. Math. Anal. **5**, 1019–1024 (2011)
15. Halmos, P.R.: Naive Set Theory. Van Nostrand Reinhold Co., New York (1960)
16. Hitzler, P.: Generalized metrics and topology in logic programming semantics. PhD Thesis, Natl. Univ. Ireland, Univ. College Cork (2001)
17. Jachymski, J.: Common fixed point theorems for some families of mappings. Indian J. Pure Appl. Math. **25**, 925–937 (1994)
18. Jachymski, J.: Equivalent conditions for generalized contractions on (ordered) metric spaces. Nonlinear Anal. **74**, 768–774 (2011)
19. Kasahara, S.: On some generalizations of the Banach contraction theorem. Publ. Res. Inst. Math. Sci. Kyoto Univ. **12**, 427–437 (1976)
20. Khan, M.S., Berzig, M., Samet, B.: Some convergence results for iterative sequences of Presić type and applications. Adv. Differ. Equ. **2012**, 38 (2012)
21. Khan, M.S., Swaleh, M., Sessa, S.: Fixed point theorems by altering distances between the points. Bull. Aust. Math. Soc. **30**, 1–9 (1984)
22. Khojasteh, F., Shukla, S., Radenović, S.: Formulization of many contractions via the simulation functions, 13 Aug 2013. arXiv:1109-3021-v2
23. Kincses, J., Totik, V.: Theorems and counterexamples on contractive mappings. Math. Balkanica **4**, 69–99 (1999)
24. Kurepa, D.: Fixpoint approach in mathematics. In: Rassias, Th.M. (ed.) Constantin Carathéodory: An International Tribute, vol. 1–2, pp. 713–767, World Scientific, Singapore (1991)
25. Leader, S.: Fixed points for general contractions in metric spaces. Math. Jpn. **24**, 17–24 (1979)
26. Matkowski, J.: Integrable Solutions of Functional Equations. Dissertationes Mathematicae, vol. 127, Polish Scientific Publishers, Warsaw (1975)
27. Meir, A., Keeler, E.: A theorem on contraction mappings. J. Math. Anal. Appl. **28**, 326–329 (1969)
28. Moore, G.H.: Zermelo's Axiom of Choice: Its Origin, Development and Influence. Springer, New York (1982)
29. Moskhovakis, Y.: Notes on Set Theory. Springer, New York (2006)
30. Pathak, H.K., George, R., Nabway, H.A., El-Paoumi, M.S., Reshma, K.P.: Some generalized fixed point results in a b-metric space and application to matrix equations. Fixed Point Theory Appl. **2015**, 101 (2015)
31. Păcurar, M.: Approximating common fixed points of Presić-Kannan type operators by a multi-step iterative method. An. Şt. Univ. "Ovidius" Constanţa (Mat.) **17**, 153–168 (2009)
32. Păcurar, M.: Fixed points of almost Presić operators by a k-step iterative method. An. Şt. Univ. "Al. I. Cuza" Iaşi (Mat.) **57**, 199–210 (2011)
33. Presić, S.B.: Sur une classe d'inéquations aux différences finies et sur la convergence de certaines suites. Publ. Inst. Math. (Beograd) (N.S.) **5**(19), 75–78 (1965)
34. Rao, K.P.R., Ali, M.M., Fisher, B.: Some Presic type generalizations of the Banach contraction principle. Math. Moravica, 15, 41–47 (2011)
35. Rhoades, B.E.: A comparison of various definitions of contractive mappings. Trans. Am. Math. Soc. **226**, 257–290 (1977)
36. Rhoades, B.E.: Some theorems on weakly contractive maps. Nonlinear Anal. **47**, 2683–2693 (2001)
37. Rus, I.A.: An iterative method for the solution of the equation $x = f(x, \ldots, x)$. Mathematica (Rev. Anal. Numer. Theory Approx.) **10**, 95–100 (1981)

38. Rus, I.A.: Generalized Contractions and Applications. Cluj University Press, Cluj-Napoca (2001)
39. Schechter, E.: Handbook of Analysis and Its Foundation. Academic Press, New York (1997)
40. Shahzad, N., Shukla, S.: Set-valued G-Presić operators on metric spaces endowed with a graph and fixed point theorems. Fixed Point Theory Appl. **2015**, 24 (2015)
41. Shukla, S., Radenović, S., Pantelić, S.: Some fixed point theorems for Presić-Hardy-Rogers type contractions in metric spaces. J. Math. **2013**, Article ID 295093, 8 (2013)
42. Suzuki, T.: Discussion of several contractions by Jachymski's approach. Fixed Point Theory Appl. **2016**, 91 (2016)
43. Tarski, A.: Axiomatic and algebraic aspects of two theorems on sums of cardinals. Fundam. Math. **35**, 79–104 (1948)
44. Tasković, M.R.: Some results in the fixed point theory. Publ. Inst. Math. (N.S.) **20**(34), 231–242 (1976)
45. Tasković, M.R.: On a question of priority regarding a fixed point theorem in a Cartesian product of metric spaces. Math. Moravica **15**, 69–71 (2011)
46. Timofte, V.: New tests for positive iteration series. Real Anal. Exch. **30**, 799–812 (2004/2005)
47. Turinici, M.: Fixed points of implicit contraction mappings. An. Şt. Univ. "Al. I. Cuza" Iaşi (S I-a, Mat.) **22**, 177–180 (1976)
48. Turinici, M.: Nonlinear contractions and applications to Volterra functional equations. An. Şt. Univ. "Al. I. Cuza" Iaşi (S I-a, Mat.) **23**, 43–50 (1977)
49. Turinici, M.: Pseudometric versions of the Caristi-Kirk fixed point theorem. Fixed Point Theory. **5**, 147–161 (2004)
50. Turinici, M.: Wardowski implicit contractions in metric spaces, 15 Sep 2013. arxiv:1211-3164-v2
51. Turinici, M.: Contraction maps in pseudometric structures. In: Rassias, T.M., Pardalos, P.M. (eds.) Essays in Mathematics and Its Applications, pp. 513–562. Springer, Cham (2016)
52. Wardowski, D.: Fixed points of a new type of contractive mappings in complete metric spaces. Fixed Point Theory Appl. **2012**, 94 (2012)
53. Wolk, E.S.: On the principle of dependent choices and some forms of Zorn's lemma. Can. Math. Bull. **26**, 365–367 (1983)

A More Accurate Hardy–Hilbert-Type Inequality with Internal Variables

Bicheng Yang

Abstract By the use of the way of weight coefficients, the technique of real analysis, and Hermite-Hadamard's inequality, a more accurate Hardy–Hilbert-type inequality with internal variables and a best possible constant factor is given. The equivalent forms, the reverses, the operator expressions with the norm, and some particular cases are also considered.

Keywords Hardy–Hilbert-type inequality · Weight coefficient · Equivalent form · Reverse · Operator

Introduction

If $p > 1, \frac{1}{p} + \frac{1}{q} = 1$, $a_m, b_n \geq 0, a = \{a_m\}_{m=1}^{\infty} \in l^p, b = \{b_n\}_{n=1}^{\infty} \in l^q$, $||a||_p = (\sum_{m=1}^{\infty} a_m^p)^{\frac{1}{p}} > 0$, and $||b||_q > 0$, then we have the following Hardy–Hilbert's inequality with the best possible constant factor $\frac{\pi}{\sin(\pi/p)}$ (cf. [1], Th. 315):

$$\sum_{m=1}^{\infty} \sum_{n=1}^{\infty} \frac{a_m b_n}{m+n} < \frac{\pi}{\sin(\pi/p)} ||a||_p ||b||_q. \tag{1}$$

The more accurate inequality of (1) was given as follows (cf. [1], Th. 323):

$$\sum_{m=1}^{\infty} \sum_{n=1}^{\infty} \frac{a_m b_n}{m+n-\alpha} < \frac{\pi}{\sin(\pi/p)} ||a||_p ||b||_q \tag{2}$$

B. Yang (✉)
Department of Mathematics, Guangdong University of Education, Guangzhou, Guangdong 510303, People's Republic of China
e-mail: bcyang@gdei.edu.cn; bcyang818@163.com

N. J. Daras, Th. M. Rassias (eds.), *Modern Discrete Mathematics and Analysis*, Springer Optimization and Its Applications 131,
https://doi.org/10.1007/978-3-319-74325-7_23

$(0 \le \alpha \le 1)$, which is an extension of (1). Also we have the following Hardy-Littlewood-Polya's inequality with the best possible constant factor $[\frac{\pi}{\sin(\pi/p)}]^2$ (cf. [1], Th. 342):

$$\sum_{m=1}^{\infty}\sum_{n=1}^{\infty}\frac{\ln\left(\frac{m}{n}\right)}{m-n}a_m b_n < \left[\frac{\pi}{\sin(\pi/p)}\right]^2 ||a||_p||b||_q. \tag{3}$$

Inequalities (1)–(3) are important in analysis and its applications (cf. [1–3]).

Suppose that $\mu_i, \upsilon_j > 0\,(i, j \in \mathbf{N} = \{1, 2, \dots\})$,

$$U_m := \sum_{i=1}^{m}\mu_i,\ V_n := \sum_{j=1}^{n}\upsilon_j\ (m, n \in \mathbf{N}), \tag{4}$$

we have the following Hardy–Hilbert-type inequality with the internal variables (cf. [1], Th. 321, replacing $\mu_m^{1/q}a_m$ and $\upsilon_n^{1/p}b_n$ by a_m and b_n) :

$$\sum_{m=1}^{\infty}\sum_{n=1}^{\infty}\frac{a_m b_n}{U_m + V_n} < \frac{\pi}{\sin(\frac{\pi}{p})}\left(\sum_{m=1}^{\infty}\frac{a_m^p}{\mu_m^{p-1}}\right)^{\frac{1}{p}}\left(\sum_{n=1}^{\infty}\frac{b_n^q}{\upsilon_n^{q-1}}\right)^{\frac{1}{q}}. \tag{5}$$

For $\mu_i = \upsilon_j = 1\ (i, j \in \mathbf{N})$, inequality (5) reduces to (1).

In 2015, Yang [4] gave an extension of (5) as follows: If $0 < \lambda_1, \lambda_2 \le 1, \lambda_1 + \lambda_2 = \lambda$, $\{\mu_m\}_{m=1}^{\infty}$ and $\{\upsilon_n\}_{n=1}^{\infty}$ are decreasing with $U_\infty = V_\infty = \infty$, then we have the following inequality with the same internal variables and the best possible constant factor $B(\lambda_1, \lambda_2)$:

$$\sum_{m=1}^{\infty}\sum_{n=1}^{\infty}\frac{a_m b_n}{(U_m + V_n)^\lambda} < B(\lambda_1, \lambda_2)\left[\sum_{m=1}^{\infty}\frac{U_m^{p(1-\lambda_1)-1}a_m^p}{\mu_m^{p-1}}\right]^{\frac{1}{p}}\left[\sum_{n=1}^{\infty}\frac{V_n^{q(1-\lambda_2-1)}b_n^q}{\upsilon_n^{q-1}}\right]^{\frac{1}{q}}, \tag{6}$$

where, $B(u, v)$ is the beta function indicated by (cf. [5])

$$B(u, v) := \int_0^{\infty}\frac{t^{u-1}}{(1+t)^{u+v}}dt\ (u, v > 0). \tag{7}$$

Recently, a strengthened Mulholland-type inequality with the internal variables as (4) was given by [6], and some other Hilbert-type inequalities were provided by [7–19].

In this paper, by the use of the way of weight coefficients, the technique of real analysis and Hermite-Hadamard's inequality, a Hardy–Hilbert-type inequality with the same internal variables and a best possible constant factor $[\frac{\pi}{\sin(\pi/p)}]^2$ is given as follows: If $\{\mu_m\}_{m=1}^{\infty}$ and $\{\upsilon_n\}_{n=1}^{\infty}$ are decreasing with $U_\infty = V_\infty = \infty$, $a_m, b_n \geq 0$, $\sum_{m=1}^{\infty} \frac{a_m^p}{\mu_{m+1}^{p-1}} \in \mathbf{R}_+$, $\sum_{n=1}^{\infty} \frac{b_n^q}{\upsilon_{n+1}^{q-1}} \in \mathbf{R}_+$, then an extension of (3) is given as follows:

$$\sum_{m=1}^{\infty}\sum_{n=1}^{\infty} \frac{\ln\left(\frac{U_m}{V_n}\right)}{U_m - V_n} a_m b_n < \left[\frac{\pi}{\sin(\pi/p)}\right]^2 \left(\sum_{m=1}^{\infty} \frac{a_m^p}{\mu_{m+1}^{p-1}}\right)^{\frac{1}{p}} \left(\sum_{n=1}^{\infty} \frac{b_n^q}{\upsilon_{n+1}^{q-1}}\right)^{\frac{1}{q}}, \tag{8}$$

where, the constant factor $[\frac{\pi}{\sin(\pi/p)}]^2$ is the best possible. Moreover, the more accurate inequality of (8) with a few parameters and a best possible constant factor is obtained. The equivalent forms, the reverses, the operator expressions with the norm, and some particular cases are also considered.

Some Lemmas

In the following, we make appointment that $p \neq 0, 1, \frac{1}{p} + \frac{1}{q} = 1, \lambda_1, \lambda_2 > 0$, $\lambda_1 + \lambda_2 = \lambda \leq 1$, $\mu_i, \upsilon_j > 0$ $(i, j \in \mathbf{N})$, $\mu(t) := \mu_m, t \in (m-1, m](m \in \mathbf{N})$; $\upsilon(t) := \upsilon_n, t \in (n-1, n](n \in \mathbf{N})$,

$$U(x) := \int_0^x \mu(t)dt, V(y) := \int_0^y \upsilon(t)dt \ (x, y \geq 0). \tag{9}$$

U_m and V_n are defined by (4), $0 \leq \alpha \leq \frac{\mu_1}{2}$, $0 \leq \beta \leq \frac{\upsilon_1}{2}$, $a_m, b_n \geq 0$, $||a||_{p,\Phi_\lambda} := (\sum_{m=1}^{\infty} \Phi_\lambda(m) a_m^p)^{\frac{1}{p}}$ and $||b||_{q,\Psi_\lambda} := (\sum_{n=1}^{\infty} \Psi_\lambda(n) b_n^q)^{\frac{1}{q}}$, where,

$$\Phi_\lambda(m) := \frac{(U_m - \alpha)^{p(1-\lambda_1)-1}}{\mu_{m+1}^{p-1}}, \ \Psi_\lambda(n) := \frac{(V_n - \beta)^{q(1-\lambda_2)-1}}{\upsilon_{n+1}^{q-1}} \ (m, n \in \mathbf{N}). \tag{10}$$

Lemma 1 *If $a \in \mathbf{R}$, $f(x)$ in continuous in $[a-\frac{1}{2}, a+\frac{1}{2}]$, $f'(x)$ is strictly increasing in $(a-\frac{1}{2}, a)$ and $(a, a+\frac{1}{2})$, respectively, and*

$$\lim_{x \to a-} f'(x) = f'(a-0) \leq f'(a+0) = \lim_{x \to a+} f'(x),$$

then we have the following Hermite-Hadamard's inequality (cf. [20]):

$$f(a) < \int_{a-\frac{1}{2}}^{a+\frac{1}{2}} f(x)dx. \tag{11}$$

Proof Since $f'(a-0)(\leq f'(a+0))$ is finite, we set function $g(x)$ as follows:

$$g(x) := f'(a-0)(x-a) + f(a), x \in \left[a - \frac{1}{2}, a + \frac{1}{2}\right].$$

In view of $f'(x)$ being strictly increasing in $(a - \frac{1}{2}, a)$, then for $x \in (a - \frac{1}{2}, a)$, $(f(x) - g(x))' = f'(x) - f'(a-0) < 0$. Since $f(a) - g(a) = 0$, it follows that $f(x) - g(x) > 0, x \in (a - \frac{1}{2}, a)$. In the same way, we can obtain $f(x) - g(x) > 0, x \in (a, a + \frac{1}{2})$. Hence, we find

$$\int_{a-\frac{1}{2}}^{a+\frac{1}{2}} f(x)dx > \int_{a-\frac{1}{2}}^{a+\frac{1}{2}} g(x)dx = f(a),$$

namely, (11) follows.

The lemma is proved.

Example 1 If $\{\mu_m\}_{m=1}^{\infty}$ and $\{\upsilon_n\}_{n=1}^{\infty}$ are decreasing, $U(x)$ and $V(y)$ are defined by (9), then it follows that $U(m) = U_m$, $V(n) = V_n$, $U(\infty) = U_\infty$, $V(\infty) = V_\infty$ and

$$U'(x) = \mu(x) = \mu_m, x \in (m-1, m),$$
$$V'(y) = \upsilon(y) = \upsilon_n, y \in (n-1, n) \ (m, n \in \mathbf{N}).$$

For fixed $m, n \in \mathbf{N}$, we also set function $f(x)$ as follows:

$$f(x) := \frac{(V(x) - \beta)^{\lambda_2 - 1} \ln\left(\frac{U_m - \alpha}{V(x) - \beta}\right)}{(U_m - \alpha)^\lambda - (V(x) - \beta)^\lambda}, x \in \left[n - \frac{1}{2}, n + \frac{1}{2}\right].$$

Then $f(x)$ in continuous in $[n - \frac{1}{2}, n + \frac{1}{2}]$. We set

$$h_1(u) := \frac{\ln\left(\frac{U_m - \alpha}{u}\right)}{(U_m - \alpha)^\lambda - u^\lambda}, h_2(u) := \frac{1}{u^{1-\lambda_2}} \ (u \in \mathbf{R}_+).$$

Then for $0 < \lambda_2 < \lambda \le 1$, it follows that (cf. [3]) $(-1)^j h_i^{(j)}(u) > 0 \, (i, j = 1, 2)$, and $f(x) = h_1(V(x) - \beta) h_2(V(x) - \beta) > 0$.

For $x \in (n - \frac{1}{2}, n)(n \in \mathbf{N})$, we find that $V'(x) = \upsilon_n$ and

$$f'(x) = [h_1'(V(x) - \beta) h_2(V(x) - \beta) + h_1(V(x) - \beta) h_2'(V(x) - \beta)]\upsilon_n < 0,$$

$$f''(x) = [h_1''(V(x) - \beta) h_2(V(x) - \beta) + h_1'(V(x) - \beta) h_2'(V(x) - \beta)$$
$$+ h_1'(V(x) - \beta) h_2'(V(x) - \beta) + h_1(V(x) - \beta) h_2''(V(x) - \beta)]\upsilon_n^2 > 0.$$

It follows that $f'(x)$ is strictly increasing in $(n - \frac{1}{2}, n)$ and

$$\lim_{x\to n-} f'(x) = f'(n-0) = [h_1'(V_n-\beta)h_2(V_n-\beta) + h_1(V_n-\beta)h_2'(V_n-\beta)]\upsilon_n.$$

In the same way, for $x \in (n, n+\frac{1}{2})$, we find that

$$f'(x) = [h_1'(V(x)-\beta)h_2(V(x)-\beta) + h_1(V(x)-\beta)h_2'(V(x)-\beta)]\upsilon_{n+1} < 0,$$

$$f''(x) = [h_1''(V(x)-\beta)h_2(V(x)-\beta) + h_1'(V(x)-\beta)h_2'(V(x)-\beta) \\ +h_1'(V(x)-\beta)h_2'(V(x)-\beta) + h_1(V(x)-\beta)h_2''(V(x)-\beta)]\upsilon_{n+1}^2 > 0,$$

and $f'(x)$ is strict increasing in $(n, n+\frac{1}{2})$.

In view of $\upsilon_{n+1} \leq \upsilon_n$, it follows that

$$\lim_{x\to n+} f'(x) = [h_1'(V_n-\beta)h_2(V_n-\beta) + h_1(V_n-\beta)h_2'(V_n-\beta)]\upsilon_{n+1} \\ = f'(n+0) \geq f'(n-0).$$

Then by (10), for $m, n \in \mathbf{N}$, we have

$$f(n) < \int_{n-\frac{1}{2}}^{n+\frac{1}{2}} f(x)dx = \int_{n-\frac{1}{2}}^{n+\frac{1}{2}} \frac{(V(x)-\beta)^{\lambda_2-1}\ln\left(\frac{U_m-\alpha}{V(x)-\beta}\right)}{(U_m-\alpha)^\lambda - (V(x)-\beta)^\lambda}dx. \tag{12}$$

Definition 1 For $m, n \in \mathbf{N}$, we define the following weight coefficients:

$$\omega(\lambda_2, m) := \sum_{n=1}^{\infty} \frac{\upsilon_{n+1}\ln\left(\frac{U_m-\alpha}{V_n-\beta}\right)}{(U_m-\alpha)^\lambda - (V_n-\beta)^\lambda}\frac{(U_m-\alpha)^{\lambda_1}}{(V_n-\beta)^{1-\lambda_2}}, \tag{13}$$

$$\varpi(\lambda_1, n) := \sum_{m=1}^{\infty} \frac{\mu_{m+1}\ln\left(\frac{U_m-\alpha}{V_n-\beta}\right)}{(U_m-\alpha)^\lambda - (V_n-\beta)^\lambda}\frac{(V_n-\beta)^{\lambda_2}}{(U_m-\alpha)^{1-\lambda_1}}. \tag{14}$$

Lemma 2 *If $\{\mu_m\}_{m=1}^\infty$ and $\{\upsilon_n\}_{n=1}^\infty$ are decreasing with $U_\infty = V_\infty = \infty$, setting*

$$k(\lambda_1) := \frac{1}{\lambda^2}\int_0^\infty \frac{\ln t}{t-1}t^{\frac{\lambda_2}{\lambda}-1}dt,$$

then it follows that (cf. [1])

$$k(\lambda_1) = \frac{1}{\lambda^2}\int_0^\infty \frac{\ln t}{t-1}t^{\frac{\lambda_1}{\lambda}-1}dt = \left[\frac{\pi}{\lambda\sin(\frac{\pi\lambda_1}{\lambda})}\right]^2 \in \boldsymbol{R}_+,$$

and for $m, n \in \mathbf{N}$, we have the following inequalities:

$$\omega(\lambda_2, m) < k(\lambda_1), \tag{15}$$

$$\varpi(\lambda_1, n) < k(\lambda_1). \tag{16}$$

Proof For $x \in (n - \frac{1}{2}, n + \frac{1}{2})\backslash\{n\}$, since $\upsilon_{n+1} \le V'(x)$, by (12), we find

$$\begin{aligned}
\omega(\lambda_2, m) &< \sum_{n=1}^{\infty} \int_{n-\frac{1}{2}}^{n+\frac{1}{2}} \frac{\upsilon_{n+1} \ln\left(\frac{U_m-\alpha}{V(x)-\beta}\right)}{(U_m - \alpha)^{\lambda} - (V(x) - \beta)^{\lambda}} \frac{(U_m - \alpha)^{\lambda_1} dx}{(V(x) - \beta)^{1-\lambda_2}} \\
&\le \sum_{n=1}^{\infty} \int_{n-\frac{1}{2}}^{n+\frac{1}{2}} \frac{(U_m - \alpha)^{\lambda_1} \ln\left(\frac{U_m-\alpha}{V(x)-\beta}\right)}{(U_m - \alpha)^{\lambda} - (V(x) - \beta)^{\lambda}} \frac{V'(x) dx}{(V(x) - \beta)^{1-\lambda_2}} \\
&= \int_{\frac{1}{2}}^{\infty} \frac{(U_m - \alpha)^{\lambda_1} \ln\left(\frac{U_m-\alpha}{V(x)-\beta}\right)}{(U_m - \alpha)^{\lambda} - (V(x) - \beta)^{\lambda}} \frac{V'(x)}{(V(x) - \beta)^{1-\lambda_2}} dx.
\end{aligned}$$

Setting $t = \frac{(V(x)-\beta)^{\lambda}}{(U_m-\alpha)^{\lambda}}$, since $V(\frac{1}{2}) - \beta = \frac{\upsilon_1}{2} - \beta \ge 0$, we find $V'(x)dx = \frac{1}{\lambda}(U_m - \alpha)t^{\frac{1}{\lambda}-1}dt$, and

$$\omega(\lambda_2, m) < \frac{1}{\lambda^2} \int_0^{\infty} \frac{\ln t}{t - 1} t^{\frac{\lambda_2}{\lambda}-1} dt = k(\lambda_1). \tag{17}$$

Hence, we obtain (15). In the same way, we obtain (16).

The lemma is proved.

Note. For example, $\mu_n, \upsilon_n = \frac{1}{n^{\sigma}}$ $(0 \le \sigma \le 1)$ are satisfied the assumptions of Lemma 2.

Lemma 3 *With regards the assumptions of Lemma 2, (i) for $m, n \in \mathbf{N}$, we have*

$$k(\lambda_1)(1 - \theta(\lambda_2, m)) < \omega(\lambda_2, m), \tag{18}$$

$$k(\lambda_1)(1 - \vartheta(\lambda_1, n)) < \varpi(\lambda_1, n), \tag{19}$$

where,

$$\begin{aligned}
\theta(\lambda_2, m) :&= \frac{1}{k(\lambda_1)} \frac{(\upsilon_1 - \beta)^{\lambda_2} \ln\left(\frac{U_m-\alpha}{\upsilon_1\theta(m)-\beta}\right)}{\lambda_2\left[1 - \left(\frac{\upsilon_1\theta(m)-\beta}{U_m-\alpha}\right)^{\lambda}\right]} \frac{1}{(U_m - \alpha)^{\lambda_2}} \\
&= O\left(\frac{1}{(U_m - \alpha)^{\lambda_2/2}}\right) \in (0, 1) \left(\theta(m) \in \left(\frac{\beta}{\upsilon_1}, 1\right)\right),
\end{aligned} \tag{20}$$

$$\vartheta(\lambda_1, n) := \frac{1}{k(\lambda_1)} \frac{(\mu_1 - \alpha)^{\lambda_1} \ln\left(\frac{V_n - \beta}{\mu_1 \vartheta(n) - \alpha}\right)}{\lambda_1\left[1 - \left(\frac{\mu_1 \vartheta(n) - \alpha}{V_n - \beta}\right)^{\lambda}\right]} \frac{1}{(V_n - \beta)^{\lambda_1}}$$

$$= O\left(\frac{1}{(V_n - \beta)^{\lambda_1/2}}\right) \in (0, 1) \ \left(\vartheta(n) \in \left(\frac{\alpha}{\mu_1}, 1\right)\right); \quad (21)$$

(ii) for any $c > 0$, *we have*

$$\sum_{m=1}^{\infty} \frac{\mu_{m+1}}{(U_m - \alpha)^{1+c}} = \frac{1}{c}\left[\frac{1}{(\mu_1 - \alpha)^c} + cO_1(1)\right], \quad (22)$$

$$\sum_{n=1}^{\infty} \frac{\upsilon_{n+1}}{(V_n - \beta)^{1+c}} = \frac{1}{c}\left[\frac{1}{(\upsilon_1 - \beta)^c} + cO_2(1)\right]. \quad (23)$$

Proof In view of $0 \leq \beta \leq \frac{\upsilon_1}{2} < \upsilon_1$, it follows that $0 \leq \frac{\beta}{\upsilon_1} < 1$. Since by Example 1, $f(x)$ is strictly decreasing in $[n, n+1]$, then for $m \in \mathbf{N}$, we find

$$\omega(\lambda_2, m) > \sum_{n=1}^{\infty} \int_n^{n+1} \frac{\upsilon_{n+1} \ln\left(\frac{U_m - \alpha}{V(x) - \beta}\right)}{(U_m - \alpha)^{\lambda} - (V(x) - \beta)^{\lambda}} \frac{(U_m - \alpha)^{\lambda_1} dx}{(V(x) - \beta)^{1-\lambda_2}}$$

$$= \int_1^{\infty} \frac{(U_m - \alpha)^{\lambda_1} \ln\left(\frac{U_m - \alpha}{V(x) - \beta}\right)}{(U_m - \alpha)^{\lambda} - (V(x) - \beta)^{\lambda}} \frac{V'(x) dx}{(V(x) - \beta)^{1-\lambda_2}}$$

$$= \int_{\frac{\beta}{\upsilon_1}}^{\infty} \frac{(U_m - \alpha)^{\lambda_1} \ln\left(\frac{U_m - \alpha}{V(x) - \beta}\right)}{(U_m - \alpha)^{\lambda} - (V(x) - \beta)^{\lambda}} \frac{V'(x) dx}{(V(x) - \beta)^{1-\lambda_2}}$$

$$- \int_{\frac{\beta}{\upsilon_1}}^{1} \frac{(U_m - \alpha)^{\lambda_1} \ln\left(\frac{U_m - \alpha}{V(x) - \beta}\right)}{(U_m - \alpha)^{\lambda} - (V(x) - \beta)^{\lambda}} \frac{V'(x) dx}{(V(x) - \beta)^{1-\lambda_2}}.$$

Setting $t = \frac{(V(x) - \beta)^{\lambda}}{(U_m - \alpha)^{\lambda}}$, we have $V(\frac{\beta}{\upsilon_1}) - \beta = \upsilon_1 \frac{\beta}{\upsilon_1} - \beta = 0$, and

$$\omega(\lambda_2, m) > \frac{1}{\lambda^2} \int_0^{\infty} \frac{\ln t}{t - 1} t^{\frac{\lambda_2}{\lambda} - 1} dt$$

$$- \int_{\frac{\beta}{\upsilon_1}}^{1} \frac{(U_m - \alpha)^{\lambda_1} \ln\left(\frac{U_m - \alpha}{V(x) - \beta}\right)}{(U_m - \alpha)^{\lambda} - (V(x) - \beta)^{\lambda}} \frac{V'(x) dx}{(V(x) - \beta)^{1-\lambda_2}}$$

$$= k(\lambda_1)(1 - \theta(\lambda_2, m)) > 0.$$

where,

$$\theta(\lambda_2, m) = \frac{1}{k(\lambda_1)} \int_{\frac{\beta}{\upsilon_1}}^{1} \frac{(U_m - \alpha)^{\lambda_1} \ln\left(\frac{U_m - \alpha}{V(x) - \beta}\right)}{(U_m - \alpha)^{\lambda} - (V(x) - \beta)^{\lambda}} \frac{V'(x)dx}{(V(x) - \beta)^{1-\lambda_2}} \in (0, 1).$$

In view of the integral mid-value theorem and

$$\int_{\frac{\beta}{\upsilon_1}}^{1} \frac{d(V(x) - \beta)}{(V(x) - \beta)^{1-\lambda_2}} dx = \frac{1}{\lambda_2}(V(x) - \beta)^{\lambda_2}\Big|_{\frac{\beta}{\upsilon_1}}^{1} = \frac{(\upsilon_1 - \beta)^{\lambda_2}}{\lambda_2},$$

there exists a $\theta(m) \in \left(\frac{\beta}{\upsilon_1}, 1\right)$, such that

$$\begin{aligned}
\theta(\lambda_2, m) &= \frac{1}{k(\lambda_1)} \frac{(U_m - \alpha)^{\lambda_1} \ln\left(\frac{U_m - \alpha}{V(\theta(m)) - \beta}\right)}{(U_m - \alpha)^{\lambda} - (V(\theta(m)) - \beta)^{\lambda}} \int_{\frac{\beta}{\upsilon_1}}^{1} \frac{V'(x)}{(V(x) - \beta)^{1-\lambda_2}} dx \\
&= \frac{1}{k(\lambda_1)} \frac{(\upsilon_1 - \beta)^{\lambda_2} (U_m - \alpha)^{\lambda_1} \ln\left(\frac{U_m - \alpha}{\upsilon_1 \theta(m) - \beta}\right)}{\lambda_2[(U_m - \alpha)^{\lambda} - (\upsilon_1 \theta(m) - \beta)^{\lambda}]} \\
&= \frac{1}{k(\lambda_1)} \frac{(\upsilon_1 - \beta)^{\lambda_2} \ln\left(\frac{U_m - \alpha}{\upsilon_1 \theta(m) - \beta}\right)}{\lambda_2\left[1 - \left(\frac{\upsilon_1 \theta(m) - \beta}{U_m - \alpha}\right)^{\lambda}\right]} \frac{1}{(U_m - \alpha)^{\lambda_2}}.
\end{aligned}$$

Since we obtain

$$\frac{1}{(U_1 - \alpha)^{\lambda_2/2}} \cdot \frac{\ln\left(\frac{U_1 - \alpha}{\upsilon_1 \theta(1) - \beta}\right)}{1 - \left(\frac{\upsilon_1 \theta(1) - \beta}{U_1 - \alpha}\right)^{\lambda}} \in \mathbf{R}_+,$$

$$\frac{1}{(U_m - \alpha)^{\lambda_2/2}} \cdot \frac{\ln\left(\frac{U_m - \alpha}{\upsilon_1 \theta(m) - \beta}\right)}{1 - \left(\frac{\upsilon_1 \theta(m) - \beta}{U_m - \alpha}\right)^{\lambda}} \to 0 \ (m \to \infty),$$

there exists a constant $L > 0$, such that

$$0 < \theta(\lambda_2, m) \leq \frac{(\upsilon_1 - \beta)^{\lambda_2} L}{\lambda_2 k(\lambda_1)} \frac{1}{(U_m - \alpha)^{\lambda_2/2}},$$

namely, $\theta(\lambda_2, m) = O\left(\frac{1}{(U_m - \alpha)^{\lambda_2/2}}\right)$, then we have (18) and (20). In the same way, we obtain (19) and (21).

For any $c > 0$, we find

$$\sum_{m=1}^{\infty} \frac{\mu_{m+1}}{(U_m - \alpha)^{1+c}} \leq \sum_{m=1}^{\infty} \frac{\mu_m}{(U_m - \alpha)^{1+c}}$$

$$= \frac{\mu_1}{(U_1-\alpha)^{1+c}} + \sum_{m=2}^{\infty} \frac{\mu_m}{(U_m-\alpha)^{1+c}}$$

$$= \frac{\mu_1}{(\mu_1-\alpha)^{1+c}} + \sum_{m=2}^{\infty} \int_{m-1}^{m} \frac{U'(x)}{(U_m-\alpha)^{1+c}} dx$$

$$< \frac{\mu_1}{(\mu_1-\alpha)^{1+c}} + \sum_{m=2}^{\infty} \int_{m-1}^{m} \frac{U'(x)}{(U(x)-\alpha)^{1+c}} dx$$

$$= \frac{\mu_1}{(\mu_1-\alpha)^{1+c}} + \int_{1}^{\infty} \frac{U'(x)}{(U(x)-\alpha)^{1+c}} dx$$

$$= \frac{\mu_1}{(\mu_1-\alpha)^{1+c}} + \frac{1}{c(\mu_1-\alpha)^{c}} = \frac{1}{c}\left[\frac{1}{(\mu_1-\alpha)^{c}} + \frac{c\mu_1}{(\mu_1-\alpha)^{1+c}}\right],$$

$$\sum_{m=1}^{\infty} \frac{\mu_{m+1}}{(U_m-\alpha)^{1+c}} \geq \sum_{m=1}^{\infty} \int_{m}^{m+1} \frac{U'(x)}{(U(x)-\alpha)^{1+c}} dx$$

$$= \int_{1}^{\infty} \frac{U'(x)dx}{(U(x)-\alpha)^{1+c}} = \frac{1}{c(\mu_1-\alpha)^{c}}.$$

Hence we obtain (22). In the same way, we obtain (23).

The lemma is proved.

Main Results and Operator Expressions

We set

$$\widetilde{\Phi}_{\lambda}(m) := \frac{\omega(\lambda_2, m)}{\mu_{m+1}^{p-1}} (U_m-\alpha)^{p(1-\lambda_1)-1},$$

$$\widetilde{\Psi}_{\lambda}(n) := \frac{\varpi(\lambda_1, n)}{\upsilon_{n+1}^{q-1}} (V_n-\beta)^{q(1-\lambda_2)-1} \quad (m, n \in \mathbf{N}). \tag{24}$$

Theorem 1

(i) For $p > 1$, we have the following equivalent inequalities:

$$I := \sum_{n=1}^{\infty} \sum_{m=1}^{\infty} \frac{\ln\left(\frac{U_m-\alpha}{V_n-\beta}\right)}{(U_m-\alpha)^{\lambda} - (V_n-\beta)^{\lambda}} a_m b_n \leq ||a||_{p,\widetilde{\Phi}_{\lambda}} ||b||_{q,\widetilde{\Psi}_{\lambda}}, \tag{25}$$

$$J := \left\{\sum_{n=1}^{\infty} \frac{\upsilon_{n+1}(V_n-\beta)^{p\lambda_2-1}}{(\varpi(\lambda_1,n))^{p-1}} \left[\sum_{m=1}^{\infty} \frac{\ln\left(\frac{U_m-\alpha}{V_n-\beta}\right)a_m}{(U_m-\alpha)^{\lambda}-(V_n-\beta)^{\lambda}}\right]^p\right\}^{\frac{1}{p}}$$
$$\leq ||a||_{p,\tilde{\Phi}_\lambda}; \tag{26}$$

(ii) for $0 < p < 1$ (or $p < 0$), we have the equivalent reverses of (25) and (26)

Proof

(i) By Hölder's inequality with weight (cf. [20]) and (14), we have

$$\left[\sum_{m=1}^{\infty} \frac{\ln\left(\frac{U_m-\alpha}{V_n-\beta}\right)}{(U_m-\alpha)^{\lambda}-(V_n-\beta)^{\lambda}} a_m\right]^p$$
$$= \left[\sum_{m=1}^{\infty} \frac{\ln\left(\frac{U_m-\alpha}{V_n-\beta}\right)}{(U_m-\alpha)^{\lambda}-(V_n-\beta)^{\lambda}}\right.$$
$$\left.\times \left(\frac{\upsilon_{n+1}^{1/p}(U_m-\alpha)^{(1-\lambda_1)/q}}{\mu_{m+1}^{1/q}(V_n-\beta)^{(1-\lambda_2)/p}} a_m\right)\left(\frac{\mu_{m+1}^{1/q}(V_n-\beta)^{(1-\lambda_2)/p}}{\upsilon_{n+1}^{1/p}(U_m-\alpha)^{(1-\lambda_1)/q}}\right)\right]^p$$
$$\leq \sum_{m=1}^{\infty} \frac{\upsilon_{n+1}\ln\left(\frac{U_m-\alpha}{V_n-\beta}\right)}{(U_m-\alpha)^{\lambda}-(V_n-\beta)^{\lambda}} \frac{(U_m-\alpha)^{(1-\lambda_1)(p-1)}}{\mu_{m+1}^{p/q}(V_n-\beta)^{1-\lambda_2}} a_m^p$$
$$\times \left[\sum_{m=1}^{\infty} \frac{\mu_{n+1}\ln\left(\frac{U_m-\alpha}{V_n-\beta}\right)}{(U_m-\alpha)^{\lambda}-(V_n-\beta)^{\lambda}} \frac{(V_n-\beta)^{(1-\lambda_2)(q-1)}}{\upsilon_{n+1}^{q-1}(U_m-\alpha)^{1-\lambda_1}}\right]^{p-1}$$
$$= \frac{(\varpi(\lambda_1,n))^{p-1}}{\upsilon_{n+1}(V_n-\beta)^{p\lambda_2-1}}$$
$$\times \sum_{m=1}^{\infty} \frac{\upsilon_{n+1}\ln\left(\frac{U_m-\alpha}{V_n-\beta}\right)}{(U_m-\alpha)^{\lambda}-(V_n-\beta)^{\lambda}} \frac{(U_m-\alpha)^{(1-\lambda_1)(p-1)}}{\mu_{m+1}^{p-1}(V_n-\beta)^{1-\lambda_2}} a_m^p. \tag{27}$$

In view of (13), we find

$$J \leq \left[\sum_{n=1}^{\infty}\sum_{m=1}^{\infty} \frac{\upsilon_{n+1}\ln\left(\frac{U_m-\alpha}{V_n-\beta}\right)}{(U_m-\alpha)^{\lambda}-(V_n-\beta)^{\lambda}} \frac{(U_m-\alpha)^{(1-\lambda_1)(p-1)}}{\mu_{m+1}^{p-1}(V_n-\beta)^{1-\lambda_2}} a_m^p\right]^{\frac{1}{p}}$$
$$= \left[\sum_{m=1}^{\infty}\sum_{n=1}^{\infty} \frac{\upsilon_{n+1}\ln\left(\frac{U_m-\alpha}{V_n-\beta}\right)}{(U_m-\alpha)^{\lambda}-(V_n-\beta)^{\lambda}} \frac{(U_m-\alpha)^{(1-\lambda_1)(p-1)}}{\mu_{m+1}^{p-1}(V_n-\beta)^{1-\lambda_2}} a_m^p\right]^{\frac{1}{p}}$$
$$= \left[\sum_{m=1}^{\infty} \frac{\omega(\lambda_2,m)}{\mu_{m+1}^{p-1}}(U_m-\alpha)^{p(1-\lambda_1)-1} a_m^p\right]^{\frac{1}{p}}, \tag{28}$$

and then (26) follows.

By Hölder's inequality (cf. [20]), we have

$$I = \sum_{n=2}^{\infty}\left[\frac{\upsilon_{n+1}^{1/p}(V_n-\beta)^{\lambda_2-\frac{1}{p}}}{(\varpi(\lambda_1,n))^{\frac{1}{q}}}\sum_{m=1}^{\infty}\frac{a_m}{\ln^{\lambda}[(U_m-\alpha)(V_n-\beta)]}\right]$$
$$\times\left[(\varpi(\lambda_1,n))^{\frac{1}{q}}\frac{(V_n-\beta)^{\frac{1}{p}-\lambda_2}}{\upsilon_{n+1}^{1/p}}b_n\right] \le J||b||_{q,\widetilde{\Psi}_\lambda}. \tag{29}$$

Then by (26), we have (25).

On the other hand, assuming that (25) is valid, we set

$$b_n := \frac{\upsilon_{n+1}(V_n-\beta)^{p\lambda_2-1}}{(\varpi(\lambda_1,n))^{p-1}}\left[\sum_{m=1}^{\infty}\frac{\ln\left(\frac{U_m-\alpha}{V_n-\beta}\right)}{(U_m-\alpha)^{\lambda}-(V_n-\beta)^{\lambda}}a_m\right]^{p-1}, n\in\mathbf{N}. \tag{30}$$

Then we find $J^p = ||b||_{q,\widetilde{\Psi}_\lambda}^q$. If $J=0$, then (26) is trivially valid; if $J=\infty$, then by (28), (26) takes the form of equality. Suppose that $0<J<\infty$. By (25), it follows that

$$||b||_{q,\widetilde{\Psi}_\lambda}^q = J^p = I \le ||a||_{p,\widetilde{\Phi}_\lambda}||b||_{q,\widetilde{\Psi}_\lambda}, \tag{31}$$

$$||b||_{q,\widetilde{\Psi}_\lambda}^{q-1} = J \le ||a||_{p,\widetilde{\Phi}_\lambda}, \tag{32}$$

and then (26) follows, which is equivalent to (25).

(ii) For $0<p<1$ (or $p<0$), by the reverse Hölder's inequality with weight (cf. [20]) and (14), we obtain the reverse of (27) (or (27)), then we have the reverse of (28), and then the reverse of (26) follows. By Hölder's inequality (cf. [20]), we have the reverse of (29) and then by the reverse of (26), the reverse of (25) follows.

On the other hand, assuming that the reverse of (25) is valid, we set b_n as (30). Then we find $J^p = ||b||_{q,\widetilde{\Psi}_\lambda}^q$. If $J=\infty$, then the reverse of (26) is trivially valid; if $J=0$, then by the reverse of (28), (26) takes the form of equality ($=0$). Suppose that $0<J<\infty$. By the reverse of (25), it follows that the reverses of (31) and (32) are valid, and then the reverse of (26) follows, which is equivalent to the reverse of (25).

The theorem is proved.

Theorem 2 *If $p>1$, $\{\mu_m\}_{m=1}^{\infty}$ and $\{\upsilon_n\}_{n=1}^{\infty}$ are decreasing with $U_\infty = V_\infty = \infty$, $||a||_{p,\Phi_\lambda}\in\mathbf{R}_+$ and $||b||_{q,\Psi_\lambda}\in\mathbf{R}_+$, then we have the following equivalent inequalities:*

$$\sum_{n=1}^{\infty}\sum_{m=1}^{\infty}\frac{\ln\left(\frac{U_m-\alpha}{V_n-\beta}\right)}{(U_m-\alpha)^{\lambda}-(V_n-\beta)^{\lambda}}a_m b_n < k(\lambda_1)||a||_{p,\Phi_\lambda}||b||_{q,\Psi_\lambda}, \tag{33}$$

$$J_1 := \left\{\sum_{n=1}^{\infty}\frac{\upsilon_{n+1}}{(V_n-\beta)^{1-p\lambda_2}}\left[\sum_{m=1}^{\infty}\frac{\ln\left(\frac{U_m-\alpha}{V_n-\beta}\right)a_m}{(U_m-\alpha)^{\lambda}-(V_n-\beta)^{\lambda}}\right]^p\right\}^{\frac{1}{p}}$$
$$< k(\lambda_1)||a||_{p,\Phi_\lambda}, \tag{34}$$

where, the constant factor $k(\lambda_1)=\left[\frac{\pi}{\lambda\sin(\frac{\pi\lambda_1}{\lambda})}\right]^2$ *is the best possible.*

Proof Using (15) and (16) in (25) and (26), we obtain equivalent inequalities (33) and (34).

For $\varepsilon\in(0,p\min\{\lambda_1,1-\lambda_2\})$, we set $\widetilde{\lambda}_1=\lambda_1-\frac{\varepsilon}{p}$ $(\in(0,1))$, $\widetilde{\lambda}_2=\lambda_2+\frac{\varepsilon}{p}$ $(\in(0,1))$, and

$$\widetilde{a}_m := \mu_{m+1}(U_m-\alpha)^{\widetilde{\lambda}_1-1},\ \widetilde{b}_n := \upsilon_{n+1}(V_n-\beta)^{\widetilde{\lambda}_2-\varepsilon-1}. \tag{35}$$

Then by (22), (23), and (19), we have

$$\begin{aligned}||\widetilde{a}||_{p,\Phi_\lambda}||\widetilde{b}||_{q,\Psi_\lambda} &= \left[\sum_{m=1}^{\infty}\frac{\mu_{m+1}}{(U_m-\alpha)^{1+\varepsilon}}\right]^{\frac{1}{p}}\left[\sum_{n=1}^{\infty}\frac{\upsilon_{n+1}}{(V_n-\beta)^{1+\varepsilon}}\right]^{\frac{1}{q}}\\ &= \frac{1}{\varepsilon}\left[\frac{1}{(\mu_1-\alpha)^{\varepsilon}}+\varepsilon O_1(1)\right]^{\frac{1}{p}}\left[\frac{1}{(\upsilon_1-\beta)^{\varepsilon}}+\varepsilon O_2(1)\right]^{\frac{1}{q}},\end{aligned}$$

$$\begin{aligned}\widetilde{I} &:= \sum_{n=1}^{\infty}\sum_{m=1}^{\infty}\frac{\ln\left(\frac{U_m-\alpha}{V_n-\beta}\right)}{(U_m-\alpha)^{\lambda}-(V_n-\beta)^{\lambda}}\widetilde{a}_m\widetilde{b}_n\\ &= \sum_{n=1}^{\infty}\left[\sum_{m=1}^{\infty}\frac{\mu_{m+1}\ln\left(\frac{U_m-\alpha}{V_n-\beta}\right)}{(U_m-\alpha)^{\lambda}-(V_n-\beta)^{\lambda}}\frac{(V_n-\beta)^{\widetilde{\lambda}_2}}{(U_m-\alpha)^{1-\widetilde{\lambda}_1}}\right]\frac{\upsilon_{n+1}}{(V_n-\beta)^{\varepsilon+1}}\\ &= \sum_{n=1}^{\infty}\varpi(\widetilde{\lambda}_1,n)\frac{\upsilon_{n+1}}{(V_n-\beta)^{\varepsilon+1}}\\ &\geq k(\widetilde{\lambda}_1)\left[\sum_{n=1}^{\infty}\frac{\upsilon_{n+1}}{(V_n-\beta)^{\varepsilon+1}}-\sum_{n=1}^{\infty}O\left(\frac{\upsilon_{n+1}}{(V_n-\beta)^{\frac{1}{2}(\lambda_1-\frac{\varepsilon}{p})+\varepsilon+1}}\right)\right]\\ &= \frac{1}{\varepsilon}k(\widetilde{\lambda}_1)\left[\frac{1}{(\upsilon_1-\beta)^{\varepsilon}}+\varepsilon(O_2(1)-O(1))\right].\end{aligned}$$

If there exists a positive constant $K \le k(\lambda_1)$, such that (33) is valid when replacing $k(\lambda_1)$ by K, then in particular, we have $\varepsilon\widetilde{I} < \varepsilon K||\widetilde{a}||_{p,\Phi_\lambda}||\widetilde{b}||_{q,\Psi_\lambda}$, namely,

$$k\left(\lambda_1 - \frac{\varepsilon}{p}\right)\left[\frac{1}{(\upsilon_1-\beta)^\varepsilon} + \varepsilon(O_2(1) - O(1))\right]$$
$$< K\left[\frac{1}{(\mu_1-\alpha)^\varepsilon} + \varepsilon O_1(1)\right]^{\frac{1}{p}}\left[\frac{1}{(\upsilon_1-\beta)^\varepsilon} + \varepsilon O_2(1)\right]^{\frac{1}{q}}.$$

It follows that $k(\lambda_1) \le K$ $(\varepsilon \to 0^+)$. Hence, $K = k(\lambda_1)$ is the best possible constant factor of (33).

Similarly to (29), we still can find that

$$I \le J_1||b||_{q,\Psi_\lambda}. \tag{36}$$

Hence, we can prove that the constant factor $k(\lambda_1)$ in (34) is the best possible. Otherwise, we would reach a contradiction by (36) that the constant factor in (33) is not the best possible.

The theorem is proved.

Remark 1

(i) For $\alpha = \beta = 0$ in (33) and (34), setting

$$\varphi_\lambda(m) := \frac{U_m^{p(1-\lambda_1)-1}}{\mu_{m+1}^{p-1}},\ \psi_\lambda(n) := \frac{V_n^{q(1-\lambda_2)-1}}{\upsilon_{n+1}^{q-1}} (m, n \in \mathbf{N}),$$

we have the following equivalent Hardy–Hilbert-type inequalities (cf. [21]):

$$\sum_{n=1}^{\infty}\sum_{m=1}^{\infty}\frac{\ln\left(\frac{U_m}{V_n}\right)a_m b_n}{U_m^\lambda - V_n^\lambda} < k(\lambda_1)||a||_{p,\varphi_\lambda}||b||_{q,\psi_\lambda}, \tag{37}$$

$$\left\{\sum_{n=1}^{\infty}\upsilon_{n+1}V_n^{p\lambda_2-1}\left[\sum_{m=1}^{\infty}\frac{\ln\left(\frac{U_m}{V_n}\right)a_m}{U_m^\lambda - V_n^\lambda}\right]^p\right\}^{\frac{1}{p}} < k(\lambda_1)||a||_{p,\varphi_\lambda}, \tag{38}$$

which are extensions of (8) and the following inequality (for $\lambda = 1, \lambda_1 = \frac{1}{q}, \lambda_2 = \frac{1}{p}$ in (38)):

$$\left\{\sum_{n=1}^{\infty}\upsilon_{n+1}\left[\sum_{m=1}^{\infty}\frac{\ln\left(\frac{U_m}{V_n}\right)a_m}{U_m - V_n}\right]^p\right\}^{\frac{1}{p}} < \left[\frac{\pi}{\sin(\frac{\pi}{p})}\right]^2\left(\sum_{m=1}^{\infty}\frac{a_m^p}{\mu_{m+1}^{p-1}}\right)^{\frac{1}{p}}. \tag{39}$$

(ii) For $\mu_i = \upsilon_j = 1(i, j \in \mathbf{N})$, $0 \le \beta = \alpha \le \frac{1}{2}$, $\lambda = 1$, $\lambda_1 = \frac{1}{q}$, $\lambda_2 = \frac{1}{p}$, (33) reduces to the following more accurate Hardy-Littlewood-Polya's inequality:

$$\sum_{m=1}^{\infty}\sum_{n=1}^{\infty} \frac{\ln(\frac{m-\alpha}{n-\alpha}) a_m b_n}{m-n} < \left[\frac{\pi}{\sin(\frac{\pi}{p})}\right]^2 ||a||_p ||b||_q. \tag{40}$$

For $p > 1$, $\Psi_\lambda^{1-p}(n) = \upsilon_{n+1}(V_n - \beta)^{p\lambda_2 - 1}$, we define the following normed spaces:

$$l_{p,\Phi_\lambda} := \{a = \{a_m\}_{m=1}^{\infty}; ||a||_{p,\Phi_\lambda} < \infty\},$$
$$l_{q,\Psi_\lambda} := \{b = \{b_n\}_{n=1}^{\infty}; ||b||_{q,\Psi_\lambda} < \infty\},$$
$$l_{p,\Psi_\lambda^{1-p}} := \{c = \{c_n\}_{n=1}^{\infty}; ||c||_{p,\Psi_\lambda^{1-p}} < \infty\}.$$

Assuming that $a = \{a_m\}_{m=1}^{\infty} \in l_{p,\Phi_\lambda}$, setting

$$c = \{c_n\}_{n=1}^{\infty}, c_n := \sum_{m=1}^{\infty} \frac{\ln\left(\frac{U_m - \alpha}{V_n - \beta}\right)}{(U_m - \alpha)^\lambda - (V_n - \beta)^\lambda} a_m, n \in \mathbf{N},$$

we can rewrite (34) as follows:

$$||c||_{p,\Psi_\lambda^{1-p}} < k(\lambda_1)||a||_{p,\Phi_\lambda} < \infty,$$

namely, $c \in l_{p,\Psi_\lambda^{1-p}}$.

Definition 2 Define a Hardy–Hilbert-type operator $T : l_{p,\Phi_\lambda} \to l_{p,\Psi_\lambda^{1-p}}$ as follows: For any $a = \{a_m\}_{m=1}^{\infty} \in l_{p,\Phi_\lambda}$, there exists a unique representation $Ta = c \in l_{p,\Psi_\lambda^{1-p}}$. Define the formal inner product of Ta and $b = \{b_n\}_{n=1}^{\infty} \in l_{q,\Psi_\lambda}$ as follows:

$$(Ta, b) := \sum_{n=1}^{\infty}\left[\sum_{m=1}^{\infty} \frac{\ln\left(\frac{U_m - \alpha}{V_n - \beta}\right)}{(U_m - \alpha)^\lambda - (V_n - \beta)^\lambda} a_m\right] b_n. \tag{41}$$

Then we can rewrite (33) and (34) as follows:

$$(Ta, b) < k(\lambda_1)||a||_{p,\Phi_\lambda}||b||_{q,\Psi_\lambda}, \tag{42}$$
$$||Ta||_{p,\Psi_\lambda^{1-p}} < k(\lambda_1)||a||_{p,\Phi_\lambda}. \tag{43}$$

Define the norm of operator T as follows:

$$||T|| := \sup_{a(\neq\theta)\in l_{p,\Phi_\lambda}} \frac{||Ta||_{p,\Psi_\lambda^{1-p}}}{||a||_{p,\Phi_\lambda}}.$$

Then by (43), we find $||T|| \leq k(\lambda_1)$. Since the constant factor in (43) is the best possible, we have

$$||T|| = k(\lambda_1) = \left[\frac{\pi}{\lambda \sin(\frac{\pi\lambda_1}{\lambda})}\right]^2. \tag{44}$$

Some More Accurate Reverses

In the following, we also set

$$\widetilde{\Omega}_\lambda(m) := \frac{1-\theta(\lambda_2, m)}{\mu_{m+1}^{p-1}}(U_m - \alpha)^{p(1-\lambda_1)-1},$$

$$\widetilde{\Theta}_\lambda(n) := \frac{1-\vartheta(\lambda_1, n)}{\upsilon_{n+1}^{q-1}}(V_n - \beta)^{q(1-\lambda_2)-1} \ (m, n \in \mathbf{N}). \tag{45}$$

For $0 < p < 1$ or $p < 0$, we still use the formal symbols $||a||_{p,\Phi_\lambda}$, $||b||_{q,\Psi_\lambda}$, $||a||_{p,\widetilde{\Omega}_\lambda}$ and $||b||_{q,\widetilde{\Theta}_\lambda}$, et al.

Theorem 3 *If* $0 < p < 1$, $\{\mu_m\}_{m=1}^\infty$ *and* $\{\upsilon_n\}_{n=1}^\infty$ *are decreasing with* $U_\infty = V_\infty = \infty$, $||a||_{p,\Phi_\lambda} \in \mathbf{R}_+$ *and* $||b||_{q,\Psi_\lambda} \in \mathbf{R}_+$, *then we have the following equivalent inequalities with the best possible constant factor* $k(\lambda_1)$:

$$\sum_{n=1}^\infty \sum_{m=1}^\infty \frac{\ln\left(\frac{U_m-\alpha}{V_n-\beta}\right)}{(U_m-\alpha)^\lambda - (V_n-\beta)^\lambda} a_m b_n > k(\lambda_1)||a||_{p,\widetilde{\Omega}_\lambda}||b||_{q,\Psi_\lambda}, \tag{46}$$

$$\left\{\sum_{n=1}^\infty \frac{\upsilon_{n+1}}{(V_n-\beta)^{1-p\lambda_2}}\left[\sum_{m=1}^\infty \frac{\ln\left(\frac{U_m-\alpha}{V_n-\beta}\right)a_m}{(U_m-\alpha)^\lambda - (V_n-\beta)^\lambda}\right]^p\right\}^{\frac{1}{p}} > k(\lambda_1)||a||_{p,\widetilde{\Omega}_\lambda} \tag{47}$$

Proof Using (18) and (16) in the reverses of (25) and (26), since

$$(\omega(\lambda_2, m))^{\frac{1}{p}} > (k(\lambda_1))^{\frac{1}{p}}(1-\theta(\lambda_2, m))^{\frac{1}{p}} \ (0 < p < 1),$$

$$(\varpi(\lambda_1, n))^{\frac{1}{q}} > (k(\lambda_1))^{\frac{1}{q}} \ (q < 0),$$

and

$$\frac{1}{(k(\lambda_1))^{p-1}} > \frac{1}{(\varpi(\lambda_1, n))^{p-1}} \ (0 < p < 1),$$

we obtain equivalent inequalities (46) and (47).

For $\varepsilon \in (0, p\min\{\lambda_1, 1 - \lambda_2\})$, we set $\widetilde{\lambda}_1, \widetilde{\lambda}_2, \widetilde{a}_m$, and $\widetilde{b}_n$ as (35). Then by (22), (23), and (16), we find

$$\begin{aligned}
&||a||_{p,\widetilde{\Omega}_\lambda}||b||_{q,\Psi_\lambda}\\
&= \left[\sum_{m=1}^{\infty} \frac{(1-\theta(\lambda_2, m))\mu_{m+1}}{(U_m - \alpha)^{1+\varepsilon}}\right]^{\frac{1}{p}} \left[\sum_{n=1}^{\infty} \frac{\upsilon_{n+1}}{(V_n - \beta)^{1+\varepsilon}}\right]^{\frac{1}{q}}\\
&= \left[\sum_{m=1}^{\infty} \frac{\mu_{m+1}}{(U_m - \alpha)^{1+\varepsilon}} - \sum_{m=1}^{\infty} O\left(\frac{\mu_{m+1}}{(U_m - \alpha)^{1+\frac{\lambda_2}{2}+\varepsilon}}\right)\right]^{\frac{1}{p}} \left[\sum_{n=1}^{\infty} \frac{\upsilon_{n+1}}{(V_n - \beta)^{1+\varepsilon}}\right]^{\frac{1}{q}}\\
&= \frac{1}{\varepsilon}\left[\frac{1}{(\mu_1 - \alpha)^{\varepsilon}} + \varepsilon(O_1(1) - O(1))\right]^{\frac{1}{p}} \left[\frac{1}{(\upsilon_1 - \beta)^{\varepsilon}} + \varepsilon O_2(1)\right]^{\frac{1}{q}},
\end{aligned}$$

$$\begin{aligned}
\widetilde{I} &:= \sum_{n=1}^{\infty}\sum_{m=1}^{\infty} \frac{\ln\left(\frac{U_m-\alpha}{V_n-\beta}\right)}{(U_m - \alpha)^{\lambda} - (V_n - \beta)^{\lambda}} \widetilde{a}_m \widetilde{b}_n\\
&= \sum_{n=1}^{\infty}\left[\sum_{m=1}^{\infty} \frac{\mu_{m+1}\ln\left(\frac{U_m-\alpha}{V_n-\beta}\right)}{(U_m - \alpha)^{\lambda} - (V_n - \beta)^{\lambda}} \frac{(V_n-\beta)^{\widetilde{\lambda}_2}}{(U_m - \alpha)^{1-\widetilde{\lambda}_1}}\right] \frac{\upsilon_{n+1}}{(V_n - \beta)^{\varepsilon+1}}\\
&= \sum_{n=1}^{\infty} \varpi(\widetilde{\lambda}_1, n)\frac{\upsilon_{n+1}}{(V_n - \beta)^{\varepsilon+1}} \le k(\widetilde{\lambda}_1)\sum_{n=1}^{\infty} \frac{\upsilon_{n+1}}{(V_n - \beta)^{\varepsilon+1}}\\
&= \frac{1}{\varepsilon}k(\widetilde{\lambda}_1)\left[\frac{1}{(\upsilon_1 - \beta)^{\varepsilon}} + \varepsilon O_2(1)\right].
\end{aligned}$$

If there exists a positive constant $K \ge k(\lambda_1)$, such that (46) is valid when replacing $k(\lambda_1)$ by K, then in particular, we have $\varepsilon\widetilde{I} > \varepsilon K||\widetilde{a}||_{p,\widetilde{\Omega}_\lambda}||\widetilde{b}||_{q,\Psi_\lambda}$, namely,

$$\begin{aligned}
&k\left(\lambda_1 - \frac{\varepsilon}{p}\right)\left[\frac{1}{(\upsilon_1 - \beta)^{\varepsilon}} + \varepsilon O_2(1)\right]\\
&> K\left[\frac{1}{(\mu_1 - \alpha)^{\varepsilon}} + \varepsilon(O_1(1) - O(1))\right]^{\frac{1}{p}} \left[\frac{1}{(\upsilon_1 - \beta)^{\varepsilon}} + \varepsilon O_2(1)\right]^{\frac{1}{q}}.
\end{aligned}$$

It follows that $k(\lambda_1) \ge K$ $(\varepsilon \to 0^+)$. Hence, $K = k(\lambda_1)$ is the best possible constant factor of (46).

The constant factor $k(\lambda_1)$ in (47) is still the best possible. Otherwise, we would reach a contradiction by the reverse of (36) that the constant factor in (46) is not the best possible.

The theorem is proved.

Remark 2 For $\alpha = \beta = 0$ in (46) and (47), setting

$$\begin{aligned}\widetilde{\theta}(\lambda_2, m) &= \frac{1}{k(\lambda_1)} \frac{\upsilon_1^{\lambda_2} \ln\left(\frac{U_m}{\upsilon_1 \theta(m)}\right)}{\lambda_2\left[1 - \left(\frac{\upsilon_1 \theta(m)}{U_m}\right)^{\lambda}\right]} \frac{1}{U_m^{\lambda_2}} \\ &= O\left(\frac{1}{U_m^{\lambda_2/2}}\right) \in (0,1)\ (\theta(m) \in (0,1)),\end{aligned}$$

$$\widetilde{\varphi}_\lambda(m) := \frac{1 - \widetilde{\theta}(\lambda_2, m)}{\mu_{m+1}^{p-1}} U_m^{p(1-\lambda_1)-1},$$

we have the following equivalent inequalities:

$$\sum_{n=1}^{\infty} \sum_{m=1}^{\infty} \frac{\ln\left(\frac{U_m}{V_n}\right) a_m b_n}{U_m^{\lambda} - V_n^{\lambda}} > k(\lambda_1) ||a||_{p,\widetilde{\varphi}_\lambda} ||b||_{q,\psi_\lambda}, \tag{48}$$

$$\left\{\sum_{n=1}^{\infty} \upsilon_{n+1} V_n^{p\lambda_2-1} \left[\sum_{m=1}^{\infty} \frac{\ln\left(\frac{U_m}{V_n}\right) a_m}{U_m^{\lambda} - V_n^{\lambda}}\right]^p\right\}^{\frac{1}{p}} > k(\lambda_1) ||a||_{p,\widetilde{\varphi}_\lambda}, \tag{49}$$

where, the constant factor $k(\lambda_1)$ is still the best possible.

Theorem 4 *If $p < 0$, $\{\mu_m\}_{m=1}^{\infty}$ and $\{\upsilon_n\}_{n=1}^{\infty}$ are decreasing with $U_\infty = V_\infty = \infty$, $||a||_{p,\Phi_\lambda} \in \mathbf{R}_+$ and $||b||_{q,\Psi_\lambda} \in \mathbf{R}_+$, then we have the following equivalent inequalities with the best possible constant factor $k(\lambda_1)$:*

$$\sum_{n=1}^{\infty} \sum_{m=1}^{\infty} \frac{\ln\left(\frac{U_m-\alpha}{V_n-\beta}\right)}{(U_m - \alpha)^{\lambda} - (V_n - \beta)^{\lambda}} a_m b_n > k(\lambda_1) ||a||_{p,\Phi_\lambda} ||b||_{q,\widetilde{\Theta}_\lambda}, \tag{50}$$

$$\begin{aligned}J_2 &:= \left\{\sum_{n=1}^{\infty} \frac{\upsilon_{n+1}(V_n - \beta)^{p\lambda_2-1}}{(1 - \vartheta(\lambda_1, n))^{p-1}} \left[\sum_{m=1}^{\infty} \frac{\ln\left(\frac{U_m-\alpha}{V_n-\beta}\right) a_m}{(U_m - \alpha)^{\lambda} - (V_n - \beta)^{\lambda}}\right]^p\right\}^{\frac{1}{p}} \\ &> k(\lambda_1) ||a||_{p,\Phi_\lambda}.\end{aligned} \tag{51}$$

Proof Using (15) and (19) in the reverses of (25) and (26), since

$$(\omega(\lambda_2, m))^{\frac{1}{p}} > (k(\lambda_1))^{\frac{1}{p}} \ (p < 0),$$
$$(\varpi(\lambda_1, n))^{\frac{1}{q}} > (k(\lambda_1))^{\frac{1}{q}}(1 - \vartheta(\lambda_1, n))^{\frac{1}{q}} \ (0 < q < 1),$$

and

$$\left[\frac{1}{(k(\lambda_1))^{p-1}(1 - \vartheta(\lambda_1, n))^{p-1}}\right]^{\frac{1}{p}} > \left[\frac{1}{(\varpi(\lambda_1, n))^{p-1}}\right]^{\frac{1}{p}} \ (p < 0),$$

we obtain equivalent inequalities (50) and (51).

For $\varepsilon \in (0, q\min\{\lambda_2, 1 - \lambda_1\})$, we set $\widetilde{\lambda}_1 = \lambda_1 + \frac{\varepsilon}{q}$ $(\in (0, 1))$, $\widetilde{\lambda}_2 = \lambda_2 - \frac{\varepsilon}{q}$ $(\in (0, 1))$, and

$$\widetilde{a}_m := \mu_{m+1}(U_m - \alpha)^{\widetilde{\lambda}_1-\varepsilon-1}, \widetilde{b}_n = \upsilon_{n+1}(V_n - \beta)^{\widetilde{\lambda}_2-1}.$$

Then by (22), (23), and (15), we have

$$\begin{aligned}
&||\widetilde{a}||_{p,\Phi_\lambda}||\widetilde{b}||_{q,\widetilde{\Theta}_\lambda} \\
&= \left[\sum_{m=1}^{\infty}\frac{\mu_{m+1}}{(U_m - \alpha)^{\varepsilon+1}}\right]^{\frac{1}{p}}\left[\sum_{n=1}^{\infty}\frac{(1 - \vartheta(\lambda_1, n))\upsilon_{n+1}}{(V_n - \beta)^{\varepsilon+1}}\right]^{\frac{1}{q}} \\
&= \left[\sum_{m=1}^{\infty}\frac{\mu_{m+1}}{(U_m-\alpha)^{\varepsilon+1}}\right]^{\frac{1}{p}}\left[\sum_{n=1}^{\infty}\frac{\upsilon_{n+1}}{(V_n-\beta)^{\varepsilon+1}} - \sum_{n=1}^{\infty}O\left(\frac{\upsilon_{n+1}}{(V_n-\beta)^{1+\frac{1}{2}\left(\lambda_1+\frac{\varepsilon}{q}\right)+\varepsilon}}\right)\right]^{\frac{1}{q}} \\
&= \frac{1}{\varepsilon}\left[\frac{1}{(\mu_1 - \alpha)^{\varepsilon}} + \varepsilon O_1(1)\right]^{\frac{1}{p}}\left[\frac{1}{(\upsilon_1 - \beta)^{\varepsilon}} + \varepsilon(O_2(1) - O(1))\right]^{\frac{1}{q}},
\end{aligned}$$

$$\begin{aligned}
\widetilde{I} &= \sum_{m=1}^{\infty}\sum_{n=1}^{\infty}\frac{\ln\left(\frac{U_m-\alpha}{V_n-\beta}\right)}{(U_m - \alpha)^{\lambda} - (V_n - \beta)^{\lambda}}\widetilde{a}_m\widetilde{b}_n \\
&= \sum_{m=1}^{\infty}\left[\sum_{n=1}^{\infty}\frac{(U_m - \alpha)^{\widetilde{\lambda}_1}\ln\left(\frac{U_m-\alpha}{V_n-\beta}\right)}{(U_m - \alpha)^{\lambda} - (V_n - \beta)^{\lambda}}\frac{\upsilon_{n+1}}{(V_n - \beta)^{1-\widetilde{\lambda}_2}}\right]\frac{\mu_{m+1}}{(U_m - \alpha)^{\varepsilon+1}} \\
&= \sum_{m=1}^{\infty}\omega(\widetilde{\lambda}_2, m)\frac{\mu_{m+1}}{(U_m - \alpha)^{\varepsilon+1}} \le k(\widetilde{\lambda}_1)\sum_{n=1}^{\infty}\frac{\mu_{m+1}}{(U_m - \alpha)^{\varepsilon+1}} \\
&= \frac{1}{\varepsilon}k(\widetilde{\lambda}_1)\left[\frac{1}{(\mu_1 - \alpha)^{\varepsilon}} + \varepsilon O_1(1)\right].
\end{aligned}$$

If there exists a positive constant $K \geq k(\lambda_1)$, such that (50) is valid when replacing $k(\lambda_1)$ by K, then in particular, we have $\varepsilon\widetilde{I} > \varepsilon K||\widetilde{a}||_{p,\Phi_\lambda}||\widetilde{b}||_{q,\widetilde{\Theta}_\lambda}$, namely,

$$k\left(\lambda_1+\frac{\varepsilon}{q}\right)\left[\frac{1}{(\mu_1-\alpha)^\varepsilon}+\varepsilon O_1(1)\right]$$
$$> K\left[\frac{1}{(\mu_1-\alpha)^\varepsilon}+\varepsilon O_1(1)\right]^{\frac{1}{p}}\left[\frac{1}{(\upsilon_1-\alpha)^\varepsilon}+\varepsilon(O_2(1)-O(1))\right]^{\frac{1}{q}}.$$

It follows that $k(\lambda_1) \geq K$ ($\varepsilon \to 0^+$). Hence, $K = k(\lambda_1)$ is the best possible constant factor of (50).

Similarly to the reverse of (29), we still can find that

$$I \geq J_2||b||_{q,\widetilde{\Theta}_\lambda}. \tag{52}$$

Hence the constant factor $k(\lambda_1)$ in (51) is still the best possible. Otherwise, we would reach a contradiction by (52) that the constant factor in (50) is not the best possible.

The theorem is proved.

Remark 3 For $\alpha = \beta = 0$ in (50) and (51), setting

$$\widetilde{\vartheta}(\lambda_1, n) := \frac{1}{k(\lambda_1)}\frac{\mu_1^{\lambda_1}\ln\left(\frac{V_n}{\mu_1\vartheta(n)}\right)}{\lambda_1\left[1-\left(\frac{\mu_1\vartheta(n)}{V_n}\right)^\lambda\right]}\frac{1}{V_n^{\lambda_1}}$$
$$= O\left(\frac{1}{V_n^{\lambda_1/2}}\right) \in (0,1)\ (\vartheta(n) \in (0,1)),$$

$$\widetilde{\psi}_\lambda(n) := \frac{1-\widetilde{\vartheta}(\lambda_1, n)}{\upsilon_{n+1}^{q-1}}V_n^{q(1-\lambda_2)-1},$$

we have the following equivalent inequalities:

$$\sum_{n=1}^\infty\sum_{m=1}^\infty \frac{\ln\left(\frac{U_m}{V_n}\right)}{U_m^\lambda - V_n^\lambda}a_m b_n > k(\lambda_1)||a||_{p,\varphi_\lambda}||b||_{q,\widetilde{\psi}_\lambda}, \tag{53}$$

$$\left\{\sum_{n=1}^\infty \frac{\upsilon_{n+1}V_n^{p\lambda_2-1}}{(1-\widetilde{\vartheta}(\lambda_1,n))^{p-1}}\left[\sum_{m=1}^\infty \frac{\ln\left(\frac{U_m}{V_n}\right)}{U_m^\lambda - V_n^\lambda}a_m\right]^p\right\}^{\frac{1}{p}} > k(\lambda_1)||a||_{p,\varphi_\lambda}, \tag{54}$$

where, the constant factor $k(\lambda_1)$ is still the best possible.

Acknowledgements This work is supported by the National Natural Science Foundation (Nos. 61370186, 61640222) and Appropriative Researching Fund for Professors and Doctors, Guangdong University of Education (No. 2015ARF25). We are grateful for their help.

References

1. Hardy, G.H., Littlewood, J.E., Pólya, G.: Inequalities. Cambridge University Press, Cambridge (1934)
2. Mitrinović, D.S., Pečarić, J.E., Fink, A.M.: Inequalities Involving Functions and Their Integrals and Derivatives. Kluwer Academic Publishers, Boston (1991)
3. Yang, B.C.: Discrete Hilbert-Type Inequalities. Bentham Science Publishers Ltd., Sharjah (2011)
4. Yang, B.C.: An extension of a Hardy-Hilbert-type inequality. J. Guangdong Univ. Edu. **35**(3), 1–7 (2015)
5. Wang, D., Guo, D.: Introduction to Spectral Functions. Science Press, Beijing (1979)
6. Wang, A., Huang, Q., Yang, B.: A strengthened Mulholland-type inequality with parameters. J. Inequal. Appl. **2015**, 329 (2015)
7. Perić, I., Vuković, P.: Multiple Hilbert's type inequalities with a homogeneous kernel. Banach J. Math. Anal. **5** (2), 33–43(2011)
8. Krnić, M., Pečarić, J.: General Hilbert's and Hardy's inequalities. Math. Inequal. Appl. **8**(1), 29–51 (2005)
9. Rassias, M.Th., Yang, B.C.: On half-discrete Hilbert's inequality. Appl. Math. Comput. **220**, 75–93 (2013)
10. Rassias, M.Th., Yang, B.C.: A multidimensional half-discrete Hilbert-type inequality and the Riemann zeta function. Appl. Math. Comput. **225**, 263–277 (2013)
11. Rassias, M.Th., Yang, B.C.: On a multidimensional half-discrete Hilbert-type inequality related to the hyperbolic cotangent function. Appl. Math. Comput. **242**, 800–813 (2014)
12. Rassias, M.Th., Yang, B.C.: On a multidimensional Hilbert–type integral inequality associated to the gamma function. Appl. Math. Comput. **249**, 408–418 (2014)
13. Adiyasuren, V., Batbold, T., Krnić, M.: Multiple Hilbert-type inequalities involving some differential operators. Banach J. Math. Anal. **10**(2), 320–337 (2016)
14. Rassias, M.Th., Yang, B.C.: A Hilbert-type integral inequality in the whole plane related to the hyper geometric function and the beta function. J. Math. Anal. Appl. **428**(2), 1286–1308 (2015)
15. Rassias, M.Th., Yang, B.C.: On a Hardy–Hilbert-type inequality with a general homogeneous kernel. Int. J. Nonlinear Anal. Appl. **7**(1), 249–269 (2016)
16. Rassias, M.Th., Yang, B.C.: A more accurate half-discrete Hardy–Hilbert-type inequality with the best possible constant factor related to the extended Riemann–Zeta function. Int. J. Nonlinear Anal. Appl. **7**(2), 1–27 (2016)
17. He, B.: A multiple Hilbert-type discrete inequality with a new kernel and best possible constant factor. J. Math. Anal. Appl. **431**, 990–902 (2015)
18. Gu, Z.H., Yang, B.C.: A Hilbert-type integral inequality in the whole plane with a non-homogeneous kernel and a few parameters. J. Inequal. Appl. **2015**, 314 (2015)
19. Agarwal, R.P., O'Regan, D., Saker, S.H.: Some Hardy-type inequalities with weighted functions via Opial type inequalities. Adv. Dyn. Syst. Appl. **10**, 1–9 (2015)
20. Kuang, J.C.: Applied Inequalities. Shangdong Science Technic Press, Jinan (2004)
21. Huang, Q.: A new extension of Hardy-Hilbert-type inequality. J. Inequal. Appl. **2015**, 397 (2015)

An Optimized Unconditionally Stable Approach for the Solution of Discretized Maxwell's Equations

Theodoros T. Zygiridis

Abstract Explicit finite-difference time-domain (FDTD) methods are well-established grid-based computational approaches for the solution of the time-dependent Maxwell's equations that govern electromagnetic problems. As they describe time-evolving phenomena, iterative procedures that calculate the necessary field values at successive time instants constitute fundamental procedures of the corresponding algorithms. In order for these time-marching processes to remain stable and provide reliable results, the size of the selected time step (i.e., the distance between successive time instants) should not exceed a known stability limit. Fulfillment of this condition may render the implementation of explicit FDTD approaches inefficient in certain types of simulations, where a large number of iterations are necessary to be executed. Appropriate unconditionally stable FDTD techniques have been developed throughout the years for modeling such classes of problems, which enable the selection of larger time steps without sacrificing stability. These methods are implicit, but they can perform efficiently as well, since they require fewer iterations for a given time period. In this work, we present an optimized version of a four-stage split-step unconditionally stable algorithm that is characterized by minimized errors at selected frequency bands, depending on the problem under investigation. Unlike conventional FDTD schemes, the spatial operators do not comply with the standard, Taylor series-based definitions but are designed in a special fashion, so that they minimize an error formula that represents the inherent space-time errors. In this way, the performance of the unconditionally stable algorithm is upgraded by simply modifying the values of some constant coefficients, thus producing a more efficient alternative without augmentation of the involved computational cost. The properties of the proposed algorithm are assessed theoretically as well as via numerical simulations, and the improved performance ensured by the new spatial operators is verified.

T. T. Zygiridis (✉)
Department of Informatics and Telecommunications Engineering, University of Western Macedonia, Kozani, Greece
e-mail: tzygiridis@uowm.gr

N. J. Daras, Th. M. Rassias (eds.), *Modern Discrete Mathematics and Analysis*,
Springer Optimization and Its Applications 131,
https://doi.org/10.1007/978-3-319-74325-7_24

Introduction

Computational electromagnetics is the scientific discipline that pertains to the development and implementation of computational techniques for the solution of electromagnetic (EM) problems, governed by Maxwell's equations. In the case of high-frequency time-dependent problems, methodologies based on finite-difference approximations are among the most popular choices, as they display various attractive features such as relatively simple implementation, ability to model a variety of different excitations and materials, low computational complexity, etc. Specifically, the finite-difference time-domain (FDTD) algorithm [1, 2] is a widely accepted methodology with numerous applications in diverse problems. In essence, the original version of the FDTD method discretizes Maxwell's equations in their differential form and employs second-order finite differences on a dual (one for electric and one for magnetic field components), structured, orthogonal grid in space, in a fashion that is also compatible with the integral form of Maxwell's equations. Discretization of the time derivatives complies with a similar pattern as well (the so-called leapfrog scheme), resulting in an explicit updating procedure of the involved field components (i.e., no matrix inversions are required). The update procedure is conditionally stable, in the sense that the magnitude of the time step (i.e., the distance between successive time instants) is bounded by the Courant-Friedrichs-Lewy (CFL) stability condition [3], which in the case of 3D problems has the form

$$\Delta t \leq \Delta t_{\text{Yee}} = \frac{1}{c_0\sqrt{\frac{1}{\Delta x^2} + \frac{1}{\Delta y^2} + \frac{1}{\Delta z^2}}} \tag{1}$$

where c_0 is the speed of light in free space and Δx, Δy, Δz are the spatial steps of the grid along the x, y, z axes, respectively.

The direct connection between the spatial and temporal steps may become an undesirable factor that diminishes the efficiency of the FDTD method, in cases where dense spatial sampling of the computational space is required. For example, simulations of configurations involving fine geometric features call for higher than usual mesh densities, in order to guarantee proper modeling of the corresponding details. Evidently, the selection of small spatial steps produces small time steps as well. Hence, more iterations are required for simulating a given time period. To deal with this deficiency, FDTD variants free from the aforementioned time-step constraint have been proposed, enabling the selection of the Δt size according to accuracy criteria only. Among these unconditionally stable techniques, some representative approaches are the alternating direction implicit (ADI) FDTD scheme [4, 5] and the locally one-dimensional (LOD) FDTD algorithm [6], which are members of the more general family of split-step approaches. A common feature of such methodologies is their implicitness, i.e., systems of equations need to be solved in each iteration. Consequently, the involved computational cost increases, compared to explicit approaches. Nevertheless, this is usually compensated by the reduced number of iterations needed, thanks to the larger Δt selected.

Due to the discretization process in time, the selection of large time steps has a nontrivial impact on the produced temporal errors, which may constitute a detrimental factor for the reliability of the simulations. In the standard FDTD method, spatial and temporal errors tend to balance each other, when the time step is set equal to the corresponding stability limit [2]. Given that the normal implementation of unconditionally stable approaches entails the selection of larger time steps (otherwise, no gain over explicit schemes can be anticipated), it becomes clear that the abovementioned balance between space-time flaws tends to be disrupted. In addition, common solutions for accuracy improvement, such as the implementation of higher-order spatial approximations, become inefficient [7]. Consequently, any methodology that enables the reduction of the pertinent discretization errors in unconditionally stable FDTD schemes is deemed important, as it may extend their efficiency and applicability range to more challenging problems.

In this work, we present an unconditionally stable FDTD-based approach for three-dimensional (3D) problems that is also optimized, in terms of spectral accuracy. Starting from a standard second-order split-step scheme [8], we introduce parametric spatial operators with extended stencils. Then, we determine the involved coefficients in the aforementioned finite-difference approximations by minimizing a proper error estimator that is based on the scheme's numerical dispersion relation. Similar approaches have been extensively considered in explicit finite-difference methodologies [9–14] but are rarely implemented in implicit schemes [15–17]. The resulting algorithm preserves the structure of the original methodology; hence, no augmentation of the involved computational burden is observed. The properties of the modified computational approach are assessed theoretically and numerically and compared against similar techniques with conventional operators. The results verify that the proposed optimized design improves the method's performance significantly even at large time steps, without modifying the computational complexity of the method with fourth-order operators. It is also confirmed that "clever" algorithmic designs have the potential to outperform standard solutions based on conventional finite-difference formulae, even in the case of unconditionally stable methodologies.

Development of the Proposed Methodology

The Basic Algorithm

We are interested in studying wave propagation phenomena in an isotropic lossless medium with electric permittivity ε and magnetic permeability μ. The electric field ($\mathbf{E}$) and magnetic field ($\mathbf{H}$) intensities are governed by Maxwell's equations

$$\nabla \times \mathbf{E} = -\mu \frac{\partial \mathbf{H}}{\partial t} \tag{2}$$

$$\nabla \times \mathbf{H} = \varepsilon \frac{\partial \mathbf{E}}{\partial t} \tag{3}$$

when sources are absent. An equivalent representation of Maxwell's system is

$$\frac{\partial}{\partial t}[\mathbf{f}] = [\mathbf{S}][\mathbf{f}] \tag{4}$$

where $[\mathbf{f}] = [E_x\ E_y\ E_z\ H_x\ H_y\ H_z]^{\mathrm{T}}$ is the vector comprising all field components (T denotes the transpose) and

$$[\mathbf{S}] = \begin{bmatrix} 0 & 0 & 0 & 0 & -\frac{1}{\varepsilon}\frac{\partial}{\partial z} & \frac{1}{\varepsilon}\frac{\partial}{\partial y} \\ 0 & 0 & 0 & \frac{1}{\varepsilon}\frac{\partial}{\partial z} & 0 & -\frac{1}{\varepsilon}\frac{\partial}{\partial x} \\ 0 & 0 & 0 & -\frac{1}{\varepsilon}\frac{\partial}{\partial y} & \frac{1}{\varepsilon}\frac{\partial}{\partial x} & 0 \\ 0 & \frac{1}{\mu}\frac{\partial}{\partial z} & -\frac{1}{\mu}\frac{\partial}{\partial y} & 0 & 0 & 0 \\ -\frac{1}{\mu}\frac{\partial}{\partial z} & 0 & \frac{1}{\mu}\frac{\partial}{\partial x} & 0 & 0 & 0 \\ \frac{1}{\mu}\frac{\partial}{\partial y} & -\frac{1}{\mu}\frac{\partial}{\partial x} & 0 & 0 & 0 & 0 \end{bmatrix} \tag{5}$$

As known, the solution of (4) subject to the initial condition $[\mathbf{f}]_{t=0} = [\mathbf{f}_0]$ has the form $[\mathbf{f}] = e^{[\mathbf{S}]t}[\mathbf{f}_0]$. Considering a discretization of the time axis using the time nodes $t = n\Delta t$, $n = 0, 1, 2, \ldots$, the aforementioned solution can be written in the form of an update equation between successive instants, as

$$[\mathbf{f}]^{n+1} = e^{[\mathbf{S}]\Delta t}[\mathbf{f}]^n \tag{6}$$

where $[\mathbf{f}]^k$ denotes the solution at the k-th time instant.

For our purposes, the—spatial—matrix $[\mathbf{S}]$ can be decomposed according to $[\mathbf{S}] = [\mathbf{A}] + [\mathbf{B}]$, where

$$[\mathbf{A}] = \begin{bmatrix} 0 & 0 & 0 & 0 & 0 & \frac{1}{\varepsilon}\frac{\partial}{\partial y} \\ 0 & 0 & 0 & \frac{1}{\varepsilon}\frac{\partial}{\partial z} & 0 & 0 \\ 0 & 0 & 0 & 0 & \frac{1}{\varepsilon}\frac{\partial}{\partial x} & 0 \\ 0 & \frac{1}{\mu}\frac{\partial}{\partial z} & 0 & 0 & 0 & 0 \\ 0 & 0 & \frac{1}{\mu}\frac{\partial}{\partial x} & 0 & 0 & 0 \\ \frac{1}{\mu}\frac{\partial}{\partial y} & 0 & 0 & 0 & 0 & 0 \end{bmatrix}, \quad [\mathbf{B}] = -\begin{bmatrix} 0 & 0 & 0 & 0 & \frac{1}{\varepsilon}\frac{\partial}{\partial z} & 0 \\ 0 & 0 & 0 & 0 & 0 & \frac{1}{\varepsilon}\frac{\partial}{\partial x} \\ 0 & 0 & 0 & \frac{1}{\varepsilon}\frac{\partial}{\partial y} & 0 & 0 \\ 0 & 0 & \frac{1}{\mu}\frac{\partial}{\partial y} & 0 & 0 & 0 \\ \frac{1}{\mu}\frac{\partial}{\partial z} & 0 & 0 & 0 & 0 & 0 \\ 0 & \frac{1}{\mu}\frac{\partial}{\partial x} & 0 & 0 & 0 & 0 \end{bmatrix} \tag{7}$$

The unconditionally stable approximate scheme that we use in this work is described by the following four-stage split-step approach [8]:

$$\left([\mathbf{I}] - \frac{\Delta t}{4}[\mathbf{A}]\right)[\mathbf{f}]^{n+1/4} = \left([\mathbf{I}] + \frac{\Delta t}{4}[\mathbf{A}]\right)[\mathbf{f}]^n \tag{8}$$

$$\left([\mathbf{I}] - \frac{\Delta t}{4}[\mathbf{B}]\right)[\mathbf{f}]^{n+2/4} = \left([\mathbf{I}] + \frac{\Delta t}{4}[\mathbf{B}]\right)[\mathbf{f}]^{n+1/4} \tag{9}$$

$$\left([\mathbf{I}] - \frac{\Delta t}{4}[\mathbf{B}]\right)[\mathbf{f}]^{n+3/4} = \left([\mathbf{I}] + \frac{\Delta t}{4}[\mathbf{B}]\right)[\mathbf{f}]^{n+2/4} \tag{10}$$

$$\left([\mathbf{I}] - \frac{\Delta t}{4}[\mathbf{A}]\right)[\mathbf{f}]^{n+1} = \left([\mathbf{I}] + \frac{\Delta t}{4}[\mathbf{A}]\right)[\mathbf{f}]^{n+3/4} \tag{11}$$

where $[\mathbf{I}]$ is the 6×6 identity matrix. Note that other popular unconditionally stable approaches also have split-step structures. For instance, we mention the LOD-FDTD method [6], which is described by

$$\left([\mathbf{I}] - \frac{\Delta t}{2}[\mathbf{A}]\right)[\mathbf{f}]^{n+\frac{1}{2}} = \left([\mathbf{I}] + \frac{\Delta t}{2}[\mathbf{A}]\right)[\mathbf{f}]^{n} \tag{12}$$

$$\left([\mathbf{I}] - \frac{\Delta t}{2}[\mathbf{B}]\right)[\mathbf{f}]^{n+1} = \left([\mathbf{I}] + \frac{\Delta t}{2}[\mathbf{B}]\right)[\mathbf{f}]^{n+\frac{1}{2}} \tag{13}$$

and the ADI-FDTD scheme [4, 5], which is based on the following equations:

$$\left([\mathbf{I}] - \frac{\Delta t}{2}[\mathbf{A}]\right)[\mathbf{f}]^{n+\frac{1}{2}} = \left([\mathbf{I}] + \frac{\Delta t}{2}[\mathbf{B}]\right)[\mathbf{f}]^{n} \tag{14}$$

$$\left([\mathbf{I}] - \frac{\Delta t}{2}[\mathbf{B}]\right)[\mathbf{f}]^{n+1} = \left([\mathbf{I}] + \frac{\Delta t}{2}[\mathbf{A}]\right)[\mathbf{f}]^{n+\frac{1}{2}} \tag{15}$$

Although they appear similar, the ADI-FDTD approach features second-order temporal accuracy, whereas the LOD-FDTD method has only first-order accuracy in time. Note, also, that the scheme used herein applies the LOD update at time instants $n + \frac{1}{4}$ and $n + \frac{1}{2}$ and the LOD update with inverse matrix order at $n + \frac{3}{4}$ and $n + 1$.

For reference, the update equations concerning the first sub-step of the considered methodology are

$$E_x|^{n+1/4} = E_x|^{n} + \frac{\Delta t}{4\varepsilon} D_y H_z|^{n} + \frac{\Delta t}{4\varepsilon} D_y H_z|^{n+1/4} \tag{16}$$

$$E_y|^{n+1/4} = E_y|^{n} + \frac{\Delta t}{4\varepsilon} D_z H_x|^{n} + \frac{\Delta t}{4\varepsilon} D_z H_x|^{n+1/4} \tag{17}$$

$$E_z|^{n+1/4} = E_z|^{n} + \frac{\Delta t}{4\varepsilon} D_x H_y|^{n} + \frac{\Delta t}{4\varepsilon} D_x H_y|^{n+1/4} \tag{18}$$

$$H_x|^{n+1/4} = H_x|^{n} + \frac{\Delta t}{4\mu} D_z E_y|^{n} + \frac{\Delta t}{4\mu} D_z E_y|^{n+1/4} \tag{19}$$

$$H_y|^{n+1/4} = H_y|^{n} + \frac{\Delta t}{4\mu} D_x E_z|^{n} + \frac{\Delta t}{4\mu} D_x E_z|^{n+1/4} \tag{20}$$

$$H_z|^{n+1/4} = H_z|^{n} + \frac{\Delta t}{4\mu} D_y E_x|^{n} + \frac{\Delta t}{4\mu} D_y E_x|^{n+1/4} \tag{21}$$

where the implicit nature of the calculations becomes evident. In the aforementioned equations, D_x, D_y, D_z denote finite-difference approximations for the $\frac{\partial}{\partial x}, \frac{\partial}{\partial y}, \frac{\partial}{\partial z}$ operators, respectively. By substituting (21) in (16), (19) in (17), and (20) in (18), we obtain

$$E_x|^{n+1/4} - \frac{\Delta t^2}{16\mu\varepsilon} D_y D_y\, E_x|^{n+1/4} = E_x|^n + \frac{\Delta t}{2\varepsilon} D_y\, H_z|^n + \frac{\Delta t^2}{16\mu\varepsilon} D_y D_y\, E_x|^n \tag{22}$$

$$E_y\Big|^{n+1/4} - \frac{\Delta t^2}{16\mu\varepsilon} D_z D_z\, E_y\Big|^{n+1/4} = E_y\Big|^n + \frac{\Delta t}{2\varepsilon} D_z\, H_x|^n + \frac{\Delta t^2}{16\mu\varepsilon} D_z D_z\, E_y\Big|^n \tag{23}$$

$$E_z|^{n+1/4} - \frac{\Delta t^2}{16\mu\varepsilon} D_x D_x\, E_z|^{n+1/4} = E_z|^n + \frac{\Delta t}{2\varepsilon} D_x\, H_y\Big|^n + \frac{\Delta t^2}{16\mu\varepsilon} D_x D_x\, E_z|^n \tag{24}$$

which formulate three distinct systems to be solved. In case of second-order finite-difference approximations, the $D_x D_x = D_{xx}$ operator corresponds to the second partial derivative with respect to the x variable and has the form

$$D_{xx}\, f|^n_{i,j,k} = \frac{1}{\Delta x^2}\left(f|^n_{i+1,j,k} - 2\, f|^n_{i,j,k} + f|^n_{i-2,j,k} \right) \tag{25}$$

and similar expressions are used for the D_{yy}, D_{zz} operators as well.[1] This means that the emerging systems have tridiagonal form and can be solved efficiently.

Once $E_x|^{n+\frac{1}{4}}$, $E_y\Big|^{n+\frac{1}{4}}$, $E_z|^{n+\frac{1}{4}}$ have been obtained from (22) to (24), the corresponding magnetic field components can be calculated in an explicit fashion from (19) to (21). Similar procedures are performed for the remaining three sub-steps $n + \frac{1}{2}, n + \frac{3}{4}, n + 1$. In the following subsection of this work, we explore the possibility to improve the accuracy of the described algorithm, simply by modifying the D_x, D_y, D_z operators.

Performance Improvement

Our aim is to explore the potential of the aforementioned discretization scheme to accurately solve electromagnetic problems using time steps that are larger than the common stability limit. At the same time, we would like to avoid increasing the temporal accuracy of the method, as this necessitates the increase of the number of split steps [8], which has an unavoidable impact on the involved computational

[1]In this work we adopt the notation $f|^n_{i,j,k} = f(i\Delta x, j\Delta y, k\Delta z, n\Delta t)$.

burden. Instead, we choose to modify the applied spatial approximations in such a manner, so that the combined space-time error is reduced. We adopt a dispersion relation-preserving strategy to achieve our goal, which dictates the utilization of error formulae that do not coincide with the standard approach based on maximizing the order of cutoff terms. As mentioned in the Introduction section, such approaches have been proven in the past to be very useful in demanding electromagnetic simulations, where wave propagation over long distances or extended time periods needed to be explored [9–14].

To add extra design degrees of freedom that will facilitate performance optimization, we introduce the following 1D approximations for spatial derivatives:

$$\left.\frac{\partial f}{\partial u}\right|_i \simeq D_u f|_i = \frac{1}{\Delta u}\left[C_1^u\left(f|_{i+\frac{1}{2}} - f|_{i-\frac{1}{2}}\right) + C_2^u\left(f|_{i+\frac{3}{2}} - f|_{i-\frac{3}{2}}\right)\right] \tag{26}$$

where C_1^u, C_2^u are constant coefficients. As known, the above symmetric formula is also used for conventional (Taylor series-based) expressions: in case of second-order approximations, we have $C_1^u = 1, C_2^u = 0$, while it is $C_1^u = \frac{9}{8}, C_2^u = -\frac{1}{24}$ in case of fourth-order approximations. However, we avoid implementing a high-order approximation, as this would correct only spatial flaws, leaving the dominant temporal errors unaffected. Instead of adopting the aforementioned values, which minimize the corresponding cutoff terms according to Taylor expansions, we treat C_1^u, C_2^u as unknowns and determine the values that minimize other (alternative) error estimators.

At this point, the numerical scheme's discrete dispersion relation plays the key role of the required accuracy indicator, which is derived by requiring the existence of plane-wave solutions:

$$f(\mathbf{r}, t) = A_0 e^{\mathrm{j}(\omega t - \tilde{\mathbf{k}}\cdot\mathbf{r})} \tag{27}$$

where ω is the angular frequency, $\mathrm{j} = \sqrt{-1}$, $\tilde{\mathbf{k}} = \tilde{k}_x\hat{\mathbf{x}} + \tilde{k}_y\hat{\mathbf{y}} + \tilde{k}_z\hat{\mathbf{z}}$ is the numerical (discrete) wave vector, and $\mathbf{r}$ is the position vector. The requirement that the considered four-stage split-step method admits nontrivial solutions leads to the condition

$$\det\left\{e^{\mathrm{j}\omega\Delta t}[\mathbf{I}] - \left([\mathbf{I}] - \frac{\Delta t}{4}[\mathscr{A}]\right)^{-1}\left([\mathbf{I}] + \frac{\Delta t}{4}[\mathscr{A}]\right)\left([\mathbf{I}] - \frac{\Delta t}{4}[\mathscr{B}]\right)^{-1}\right.$$
$$\left([\mathbf{I}] + \frac{\Delta t}{4}[\mathscr{B}]\right)\left([\mathbf{I}] - \frac{\Delta t}{4}[\mathscr{B}]\right)^{-1}\left([\mathbf{I}] + \frac{\Delta t}{4}[\mathscr{B}]\right)\left([\mathbf{I}] - \frac{\Delta t}{4}[\mathscr{A}]\right)^{-1}$$
$$\left.\left([\mathbf{I}] + \frac{\Delta t}{4}[\mathscr{A}]\right)\right\} = 0 \tag{28}$$

where the $[\mathscr{A}], [\mathscr{B}]$ are obtained from the $[\mathbf{A}], [\mathbf{B}]$ matrices, after replacing the D_x, D_y, D_z operators with

$$\mathscr{X} = -\frac{2\mathrm{j}}{\Delta x}\left[C_1^x \sin\left(\frac{1}{2}\tilde{k}_x \Delta x\right) + C_1^x \sin\left(\frac{3}{2}\tilde{k}_x \Delta x\right)\right] \tag{29}$$

$$\mathscr{Y} = -\frac{2\mathrm{j}}{\Delta y}\left[C_1^y \sin\left(\frac{1}{2}\tilde{k}_y \Delta y\right) + C_1^y \sin\left(\frac{3}{2}\tilde{k}_y \Delta y\right)\right] \tag{30}$$

$$\mathscr{Z} = -\frac{2\mathrm{j}}{\Delta z}\left[C_1^z \sin\left(\frac{1}{2}\tilde{k}_z \Delta z\right) + C_1^z \sin\left(\frac{3}{2}\tilde{k}_z \Delta z\right)\right] \tag{31}$$

respectively. Specifically, the dispersion relation can be written as

$$\left(\frac{\sin(\omega \Delta t)}{c_0 \Delta t}\right)^2 - \frac{\Xi}{\mathrm{Z}} = 0 \tag{32}$$

where $c_0 = 1/\sqrt{\mu\varepsilon}$ is the free-space speed of light and

$$\Xi = -4194304\delta_1 - 16384(c_0\Delta t)^4(\delta_4 - 12\delta_3) - 64(c_0\Delta t)^8\delta_2\delta_3 \tag{33}$$

$$\begin{aligned} \mathrm{Z} = {} & 16777216 - 2097152(c_0\Delta t)^2\delta_1 \\ & + 65536(c_0\Delta t)^4\left(\delta_1^2 + 2\delta_2\right) - 8192(c_0\Delta t)^6(\delta_4 + 4\delta_3) \\ & + 256(c_0\Delta t)^8(\delta_5 + 4\delta_1\delta_3) - 32(c_0\Delta t)^{10}\delta_2\delta_3 + (c_0\Delta t)^{12}\delta_3^2 \end{aligned} \tag{34}$$

$$\delta_1 = \mathscr{X}^2 + \mathscr{Y}^2 + \mathscr{Z}^2 \tag{35}$$

$$\delta_2 = \mathscr{X}^2\mathscr{Y}^2 + \mathscr{Y}^2\mathscr{Z}^2 + \mathscr{X}^2\mathscr{Z}^2 \tag{36}$$

$$\delta_3 = \mathscr{X}^2\mathscr{Y}^2\mathscr{Z}^2 \tag{37}$$

$$\delta_4 = \mathscr{X}^2\mathscr{Y}^4 + \mathscr{X}^2\mathscr{Z}^4 + \mathscr{Y}^2\mathscr{X}^4 + \mathscr{Y}^2\mathscr{Z}^4 + \mathscr{Z}^2\mathscr{X}^4 + \mathscr{Z}^2\mathscr{Y}^4 \tag{38}$$

$$\delta_5 = \mathscr{X}^4\mathscr{Y}^4 + \mathscr{Y}^4\mathscr{Z}^4 + \mathscr{X}^4\mathscr{Z}^4 \tag{39}$$

Given that $\tilde{k}_x = \tilde{k}\sin\theta\cos\phi$, $\tilde{k}_y = \tilde{k}\sin\theta\sin\phi$, $\tilde{k}_z = \tilde{k}\cos\theta$, where θ, ϕ are the azimuth and polar angles in spherical coordinates, the numerical dispersion relation (32) can be compactly written as $\mathscr{E}(\omega, \tilde{k}, \theta, \phi) = 0$. Following other dispersion relation-preserving schemes in the pertinent literature, the discretization error can be represented by the formula $\mathscr{E}(\omega, k, \theta, \phi)$, where $k = \omega/c_0$ is the exact value of the wave number (i.e., the magnitude of the wave vector). Apparently, it is $\mathscr{E}(\omega, k, \theta, \phi) \neq 0$, and the magnitude of $\mathscr{E}$ is directly related to the accuracy of the underlying discretization scheme (ideally, a computationally exact methodology would yield $\mathscr{E}(\omega, k, \theta, \phi) = 0$ for all frequencies and angles). Note than in case of standard spatial approximation formulae, the proposed error expression displays the expected behavior. For instance, if we consider second-order spatial accuracy ($C_1^x = C_1^y = C_1^z = 1$, $C_2^x = C_2^y = C_2^z = 0$), application of Taylor expansions to (32) results in

$$\mathcal{E} = \left(\frac{\omega}{c_0}\right)^2 - \frac{1}{12}\left(\frac{\omega}{c_0}\right)^4 (c_0\Delta t)^2 + \ldots - \left[(k_x)^2 + \left(k_y\right)^2 + (k_z)^2\right]$$
$$+\frac{1}{24}\left\{2\left[(k_x)^4\Delta x^2 + \left(k_y\right)^4\Delta y^2 + (k_z)^4\Delta z^2\right] + 3\left[(k_x)^4 + \left(k_y\right)^4 + (k_z)^4\right](c_0\Delta t)^2\right.$$
$$\left. +6\left[\left(k_x k_y\right)^2 + \left(k_y k_z\right)^2 + (k_z k_x)^2\right](c_0\Delta t)^2\right\} + \ldots \quad (40)$$

which verifies that the exact dispersion relation $k_x^2 + k_y^2 + k_z^2 = \left(\frac{\omega}{c_0}\right)^2$ is approximated to second order, with respect to spatial and temporal increments. If, on the other hand, spatial operators with fourth-order accuracy are implemented ($C_1^x = C_1^y = C_1^z = \frac{9}{8}, C_2^x = C_2^y = C_2^z = -\frac{1}{24}$), then the error descriptor can be written as

$$\mathcal{E} = \left(\frac{\omega}{c_0}\right)^2 - \frac{1}{12}\left(\frac{\omega}{c_0}\right)^4 (c_0\Delta t)^2 + \ldots - \left[(k_x)^2 + \left(k_y\right)^2 + (k_z)^2\right]$$
$$-\frac{1}{8}\left[(k_x)^4 + \left(k_y\right)^4 + (k_z)^4 + 2\left(k_x k_y\right)^2 + 2\left(k_y k_z\right)^2 + 2(k_z k_x)^2\right](c_0\Delta t)^2$$
$$+\frac{1}{1280}\left\{-12\left[(k_x)^6\Delta x^4 + \left(k_y\right)^6\Delta y^4 + (k_z)^6\Delta z^4\right]\right.$$
$$+15\left[(k_x)^6 + (k_z)^6 + (k_z)^6\right](c_0\Delta t)^4$$
$$+40\left[(k_x)^4\left(k_y\right)^2 + (k_x)^2\left(k_y\right)^4 + (k_x)^4(k_z)^2\right.$$
$$\left.\left. +\left(k_y\right)^4(k_z)^2 + (k_x)^2(k_z)^4 + \left(k_y\right)^2(k_z)^4\right](c_0\Delta t)^4\right\} + \ldots \quad (41)$$

As seen, the second-order temporal and fourth-order spatial accuracy of the specific discretization scheme are verified by the suggested $\mathcal{E}$ estimator.

In order to determine now the unknown coefficients involved in the expressions of the spatial operators, we pursue the minimization of the average—with respect to the propagation directions—discretization error at a specific (selected) frequency. In essence, such an error indicator can be represented by

$$\Lambda_1 = \frac{1}{4\pi}\int_0^{2\pi}\int_0^{\pi} |\mathcal{E}(\omega_0, k_0, \theta, \phi)| \sin\theta \, d\theta \, d\phi \quad (42)$$

where ω_0 denotes the selected design frequency point, and $k_0 = \omega_0/c_0$. Evidently, the rather complicated form of $\mathcal{E}$ renders the calculation of the aforementioned double integral a difficult task. Instead, we search for those operator coefficients that minimize the quantity

$$L_1 = \sum_{i=1}^{N} |\mathcal{E}(\omega_0, k_0, \theta_i, \phi_i)| \sin\theta_i \quad (43)$$

where (θ_i, ϕ_i), $i = 1, \ldots, N$ are N pairs of propagation angles that cover the rectangle $[0, \pi] \times [0, 2\pi]$. The computation of the aforementioned minimization problem can be performed via standard optimization techniques, e.g., the Nelder-Mead method [18].

To complete the description of the proposed algorithm, it is reminded that half of the update equations necessitate implicit calculation, via systems that incorporate finite-difference expressions of second partial derivative operators. Once the optimized formulae for the first partial derivatives have been obtained, the remaining approximations can be easily determined if the formula (26) is expressed as a linear combination of shift operators:

$$D_u f|_i = \frac{1}{\Delta u} \left[C_1^u \left(\mathrm{S}_{\frac{1}{2}} - \mathrm{S}_{-\frac{1}{2}} \right) + C_2^u \left(\mathrm{S}_{\frac{3}{2}} - \mathrm{S}_{-\frac{3}{2}} \right) \right] f|_i \tag{44}$$

where the shift operators S are defined according to their action on f:

$$\mathrm{S}_{i_0} f|_i = f|_{i+i_0} \tag{45}$$

Considering that the multiplication of two shift operators can be defined as $\mathrm{S}_{i_0} \cdot \mathrm{S}_{j_0} = \mathrm{S}_{i_0+j_0}$, the following expression for the D_{xx} operator is obtained:

$$\begin{aligned} \Delta x^2 D_{xx} f|_i = & -2 \left[(C_1^x)^2 + (C_2^x)^2 \right] f|_i + \left[(C_1^x)^2 - 2C_1^x C_2^x \right] (f|_{i+1} + f|_{i-1}) \\ & + 2C_1^x C_2^x (f|_{i+2} + f|_{i-2}) + (C_2^x)^2 (f|_{i+3} + f|_{i-3}) \end{aligned} \tag{46}$$

and similar formulae are obtained for D_{yy}, D_{zz} as well.

Of course, it should be kept in mind that the advantage of main time-domain algorithms (such as the one of this work) over frequency-domain methods is their ability to provide results for a wide frequency range. Therefore, it is not certain—at least at this point—that the aforementioned selection of a single design frequency preserves this characteristic feature to a sufficient degree. Consequently, a detailed study of the method's characteristics appears to be necessary, and such an investigation is performed in the following section.

Theoretical Assessment

In order to estimate the level of improvement that the proposed design strategy may accomplish, we examine the values of the numerical phase velocity $\tilde{c}$ in free space, calculated as $\tilde{c} = \omega/\tilde{k}$, that are predicted by computational schemes implementing standard or optimized finite-difference approximations. The phase velocity can be used as a reliable indicator of the numerical accuracy, as its exact value in homogeneous media is constant (equal to $\frac{1}{\sqrt{\mu\varepsilon}} \simeq 2.998 \times 10^8$ m/s in free space). It is also reminded that the numerical wave number is computed from the corresponding dispersion relation (32).

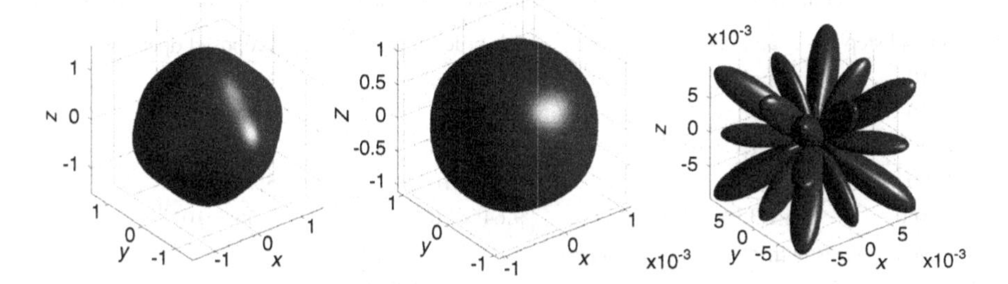

Fig. 1 Error % in numerical phase velocity for all propagation directions, in case of second-order (left figure), fourth-order (middle figure), and optimized spatial operators (right figure), when $Q = 4$ and $\Delta x = \Delta y = \Delta z = \lambda/20$

First, we consider the case of cubic mesh cells ($\Delta x = \Delta y = \Delta z = \lambda/20$, where λ is the free-space wavelength), when a time step that is four times larger than the common stability criterion is selected ($Q = \Delta t/\Delta t_{\text{Yee}} = 4$). The % error in $\tilde{c}$ for all possible propagation angles is given in Fig. 1, when the unconditional updating is combined with second-order, fourth-order, or optimized spatial approximations. It can be easily observed that there are no significant differences between the schemes with the conventional operators, despite their different formal orders of accuracy. Apart from the error of the second-order case being slightly more anisotropic than the error of the fourth-order approximations, a quite uniform distribution is evident, which is primarily due to the—isotropic—temporal error, emanating from the implementation of a large time step. On the other hand, the implementation of the proposed spatial approximations yields significantly lower error values, as the different scaling of the plot axes implies. Additionally, the proposed design procedure results in the total vanishing of the error at various propagations angles, which gives rise to the more complex spatial distribution depicted in the corresponding figure.

To better assess the accuracy improvement that the new spatial operators accomplish, we calculated the average of the absolute error in phase velocity, defined as

$$\text{overall error} = \frac{1}{4\pi}\int_0^{2\pi}\int_0^{\pi}\left|\frac{c_0-\tilde{c}}{c_0}\right|\sin\theta\, d\theta\, d\phi \tag{47}$$

The aforementioned formula takes into account the contributions over all possible directions of propagation. The computed values are presented in Table 1 for different grid densities and time steps defined by $Q = 3$. As expected, the fourth-order operators do not improve the accuracy of the method significantly, compared to the second-order approximations, mainly due to the domination of the larger temporal errors. In addition, both standard schemes display a second-order convergence, while the proposed approach is characterized by a fourth-order convergence at the

Table 1 Overall error for different mesh resolutions, considering various cases of spatial operators

Spatial step	2nd-order operators	4th-order operators	Proposed operators
$\lambda/10$	3.62×10^{-2}	2.62×10^{-2}	3.03×10^{-3}
$\lambda/20$	8.73×10^{-3}	6.26×10^{-3}	1.55×10^{-5}
$\lambda/40$	2.16×10^{-3}	1.55×10^{-3}	9.39×10^{-7}
$\lambda/80$	5.40×10^{-4}	3.86×10^{-4}	5.71×10^{-8}
$\lambda/160$	1.35×10^{-4}	9.64×10^{-5}	3.56×10^{-9}

Time-step sizes with $Q = 3$ are considered

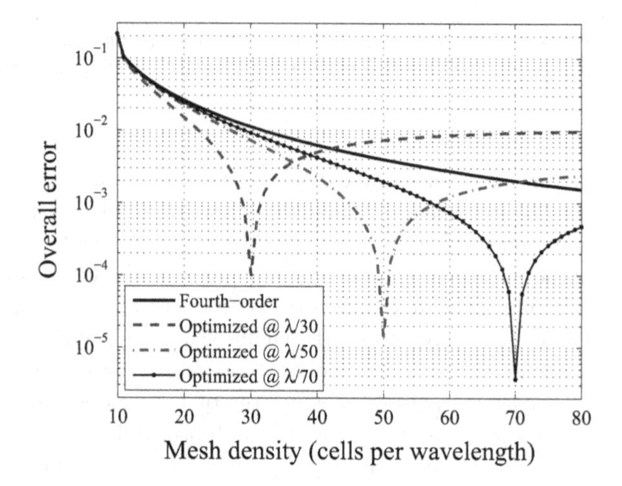

Fig. 2 Overall error in phase velocity versus discretization spatial density for standard fourth-order and proposed spatial approximations. The time-step size is determined by $Q = 6$

design frequency. This behavior verifies the efficient treatment of the combined space-time errors by the proposed discretization approach, despite the low-order accuracy in time.

It should be kept in mind that the above-described excellent behavior of the modified operators is accomplished at the selected design frequency point. Therefore, similar benefits should be expected in case of time-harmonic problems. In order, however, to predict the performance of the proposed approach in problems with wideband excitations, the corresponding error should be examined throughout an extended frequency range. Figure 2 plots the overall error over a frequency band, where spatial density extends from 10 to 80 cells per wavelength. Unlike the conventional scheme with fourth-order operators, the proposed one exhibits a sharp minimum at the selected optimization point. Furthermore, the accomplished improvement is not strictly narrowband, but it extends over several other frequencies. This feature renders the optimized scheme suitable for wideband simulations as well, provided that the optimization point is selected carefully. The deterioration of the performance at denser discretizations is, of course, a drawback, but it should not be too alarming, as these points correspond to lower frequencies, where usually the phase flaws are less likely to produce significant harmful results.

Numerical Validation

Apart from the theoretical assessment of the previous section, we also proceed to the validation of the performance of the proposed algorithm via numerical simulations. Specifically, we consider a $4 \times 4 \times 4\,\text{cm}^3$ cavity with perfectly conducting walls filled with air, which is known to support distinct modes at specific frequencies. This is a problem that has a known analytic solution; hence, the error produced by any numerical scheme can be easily computed. Selected electromagnetic modes can be excited by imposing the exact field solution throughout the computational domain at the first time instant of the simulation. As the walls are considered to be perfectly conducting, they can be modeled by setting the tangential electric field and the normal magnetic field components to zero at all time steps. In addition, in order to facilitate the implementation of the extended spatial stencils near the boundaries, the computational domain is augmented by three cells in all directions, and proper symmetry/antisymmetry conditions are applied for the computation of the field values at these buffer cells. Evidently, the formulation of the equation system for the implicit updates requires special attention at the points near the boundaries. Taking into account the symmetry of the electric field due to the presence of the perfectly conducting walls, the system's matrix takes the form

$$\begin{bmatrix} 1 & 0 & 0 & 0 & 0 & 0 & 0 & \dots \\ \zeta_1 & \zeta_0 - \zeta_2 & \zeta_1 - \zeta_3 & \zeta_2 & \zeta_3 & 0 & 0 & \dots \\ \zeta_2 & \zeta_1 - \zeta_3 & \zeta_0 & \zeta_1 & \zeta_2 & \zeta_3 & 0 & \dots \\ \zeta_3 & \zeta_2 & \zeta_1 & \zeta_0 & \zeta_1 & \zeta_2 & \zeta_3 & \dots \\ \vdots & \vdots & \vdots & \vdots & \vdots & \vdots & \vdots & \ddots \end{bmatrix} \tag{48}$$

where $\zeta_0, \zeta_1, \zeta_2, \zeta_3$ are the coefficients obtained from the left-hand sides of e.g., (22)–(24). For instance, in the case of (22), which is used for the update of $E_x|^{n+\frac{1}{4}}$, we have

$$\zeta_0 = 1 + \frac{(c_0 \Delta t)^2}{8\Delta y^2}\left[\left(C_1^y\right)^2 + \left(C_2^y\right)^2\right] \tag{49}$$

$$\zeta_1 = -\frac{(c_0 \Delta t)^2}{16\Delta y^2}\left[\left(C_1^y\right)^2 - 2C_1^y C_2^y\right] \tag{50}$$

$$\zeta_2 = -\frac{(c_0 \Delta t)^2}{8\Delta y^2} C_1^y C_2^y \tag{51}$$

$$\zeta_3 = -\frac{(c_0 \Delta t)^2}{16\Delta y^2}\left(C_2^y\right)^2 \tag{52}$$

Similar expressions are extracted for all the other cases as well. Note that the first line of the matrix in (48) is used to set to zero the component that is tangential to the boundary.

Starting from a single-frequency simulation, a selected mode is excited within the cavity, which is characterized by the indices $(\ell, m, n) = (2, 1, 1)$. The latter determine the corresponding resonant frequency, according to

$$f_{\ell,m,n} = \frac{c_0}{2d}\sqrt{\ell^2 + m^2 + n^2} \tag{53}$$

where d is the edge length of the cubic cavity. In this case, it is $f_{2,1,1} = 9.179$ GHz. We perform two different simulations. In the first one, which is executed for 2000 time steps, an $80 \times 80 \times 80$ mesh is used with the time-step size set by $Q = 2$, and fourth-order spatial operators are applied. Regarding the second simulation, the new approximations are implemented in a $40 \times 40 \times 40$ grid with $Q = 5$. Due to the larger spatial increments and time step in the second case, only 400 iterations need to be carried out now for the same time period. The performance of the two schemes is shown in Fig. 3, where the L_2 error with respect to E_z, defined as

$$L_2(n\Delta t) = \sqrt{\frac{\sum_{i=1}^{i_{\max}}\sum_{j=1}^{j_{\max}}\sum_{k=1}^{k_{\max}}\left(E_z|^n_{i,j,k+\frac{1}{2}} - E_z^{\text{exact}}|^n_{i,j,k+\frac{1}{2}}\right)^2}{i_{\max}j_{\max}k_{\max}}} \tag{54}$$

is plotted for the two cases. It becomes evident that the optimized approach produces significantly more accurate results, as the maximum L_2 error is reduced by 17.46 times. What is more important is the fact that this accuracy upgrade is also accompanied by a nontrivial modification of the necessary computational cost, as the simulation's duration is reduced from 1933 s to only 29 s, i.e., almost 67×

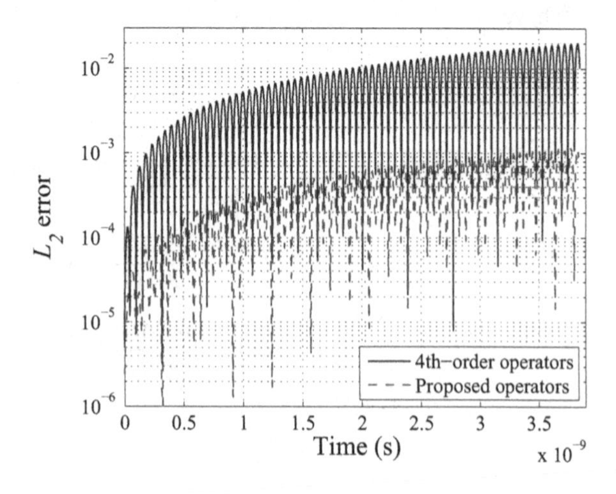

Fig. 3 L_2 error versus time for the cubic cavity problem supporting a single mode

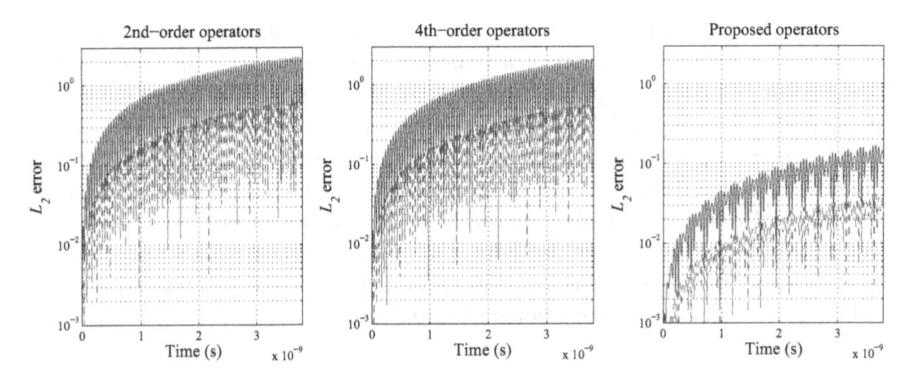

Fig. 4 L_2 error versus time for the problem with two cavity modes at different frequencies. The blue continuous lines correspond to $40 \times 40 \times 40$ grids, and the red dashed lines correspond to $80 \times 80 \times 80$ grids

acceleration of the involved computations. Consequently, through this numerical demonstration, we validate the potential of the proposed numerical scheme to provide credible, unconditionally stable computational models for high-frequency electromagnetic problems.

We also examine the case where the aforementioned cavity supports, at the same time, two distinct modes at different frequencies, namely, 9.179 ((2, 1, 1) indices) and 11.242 ((2, 2, 1) indices) GHz. This is a wideband problem; hence, a careful selection of the optimization frequency for the proposed method is required; otherwise, there is no guarantee that precision improvement will be accomplished. Considering the impact of dispersion errors at high frequencies, the design point is selected at 11 GHz. The time evolution of the L_2 error for different methods and two grid densities (mesh size of either $40 \times 40 \times 40$ or $80 \times 80 \times 80$ cells) is plotted in Fig. 4. The time step is determined by $Q = 4$, and 500 iterations are performed in the case of the coarse grid. In agreement with previous conclusions, increasing the spatial order of the spatial operators does not result in significant accuracy improvement. Specifically, the maximum L_2 error is reduced from 2.3 to 2.102 using the first mesh and from 0.678 to 0.591 in the case of the second grid. On the other hand, the new methodology provides considerable enhancement, as the maximum L_2 values observed at the two simulations are 0.167 and 4.218×10^{-2}, respectively. In other words, the proposed modification of the approximations' coefficients lowers the specific error norm by 12.6–14 times, thus validating the suitability of the new approach for broadband simulations as well.

Conclusions

As the study of electromagnetic problems involving wave propagation in discretized domains can be challenging in diverse aspects, the implementation of unconditionally stable time-domain methodologies becomes more and more popular, because they permit calculations that are free from time-step restrictions. In this work, we presented an improved version of a four-stage split-step FDTD approach, which features single-frequency optimization of the spatial operators and, at the same time, is capable of handling wideband simulations efficiently. The theoretical assessment of the scheme's performance verified its superiority over the algorithm's versions employing standard finite-difference expressions, and numerical results displayed the practical gains over existing approaches. The findings of this works verify the potential of alternative design procedures to produce reliable grid-based, unconditionally stable methods for electromagnetic problems that outperform conventional solutions without extra computational requirements.

References

1. Yee, K.: Numerical solution of initial boundary value problems involving Maxwell's equations in isotropic media. IEEE Trans. Antennas Propag. **14**(3), 302–307 (1966)
2. Taflove, A., Hagness, S.C.: Computational Electrodynamics: The Finite-Difference Time-Domain Method. Artech House, Norwood (2005)
3. Courant, R., Friedrichs, K., Lewy, H.: On the partial difference equations of mathematical physics. IBM J. Res. Dev. **11**(2), 215–234 (1967)
4. Namiki, T.: A new FDTD algorithm based on alternating-direction implicit method. IEEE Trans. Microwave Theory Tech. **47**(10), 2003–2007 (1999)
5. Zheng, F., Chen, Z., Zhang, J.: A finite-difference time-domain method without the courant stability conditions. IEEE Microwave Guided Wave Lett. **9**(11), 441–443 (1999)
6. Shibayama, J., Muraki, M., Yamauchi, J., Nakano, H.: Efficient implicit FDTD algorithm based on locally one-dimensional scheme. Electron. Lett. **41**(19), 1046–1047 (2005)
7. Staker, S.W., Holloway, C.L., Bhobe, A.U., Piket-May, M.: Alternating-direction implicit (ADI) formulation of the finite-difference time-domain (FDTD) method: algorithm and material dispersion implementation. IEEE Trans. Electromagn. Compat. **45**(2), 156–166 (2003)
8. Lee, J., Fornberg, B.: A split step approach for the 3-d Maxwell's equations. J. Comput. Appl. Math. **158**(2), 485–505 (2003)
9. Tam, C.K.W., Webb, J.C.: Dispersion-relation-preserving finite difference schemes for computational acoustics. J. Comput. Phys. **107**(2), 262–281 (1993)
10. Wang, S., Teixeira, F.L.: Dispersion-relation-preserving FDTD algorithms for large-scale three-dimensional problems. IEEE Trans. Antennas Propag. **51**(8), 1818–1828 (2003)
11. Sun, G., Trueman, C.W.: Suppression of numerical anisotropy and dispersion with optimized finite-difference time-domain methods. IEEE Trans. Antennas Propag. **53**(12), 4121–4128 (2005)
12. Zygiridis, T.T., Tsiboukis, T.D.: Optimized three-dimensional FDTD discretizations of Maxwells equations on cartesian grids. J. Comput. Phys. **226**(2), 2372–2388 (2007)
13. Finkelstein, B., Kastner, R.: Finite difference time domain dispersion reduction schemes. J. Comput. Phys. **221**(1), 422–438 (2007)
14. Sheu, T.W.H., Chung, Y.W., Li, J.H., Wang, Y.C.: Development of an explicit non-staggered scheme for solving three-dimensional Maxwells equations. Comput. Phys. Commun. **207**, 258–273 (2016)
15. Liang, F., Wang, G., Lin, H., Wang, B.Z.: Numerical dispersion improved three-dimensional locally one-dimensional finite-difference time-domain method. IET Microwaves Antennas Propag. **5**(10), 1256–1263 (2011)
16. Zhou, L., Yang, F., Ouyang, J., Yang, P., Long, R.: ADI scheme of a nonstandard FDTD method and its numerical properties. IEEE Trans. Antennas Propag. **64**(10), 4365–4373 (2016)

17. Zygiridis, T.T.: Improved unconditionally stable FDTD method for 3-d wave-propagation problems. IEEE Trans. Microwave Theory Tech. **65**(6), 1921–1928 (2017)
18. Nelder, J.A., Mead, R.: A simplex method for function minimization. Comput. J. **7**, 308–313 (1965)

Author Correction to: Moment Generating Functions and Moments of Linear Positive Operators

Vijay Gupta, Neha Malik, and Themistocles M. Rassias

Author Correction to: Chapter 8 in: N. J. Daras, Th. M. Rassias (eds.), *Modern Discrete Mathematics and Analysis*, Springer Optimization and Its Applications 131, https://doi.org/10.1007/978-3-319-74325-7_8

The original version of the chapter was inadvertently published with some errors. The chapter has now been corrected and approved by the authors.

The updated online version of this chapter can be found at
https://doi.org/10.1007/978-3-319-74325-7_8

N. J. Daras, Th. M. Rassias (eds.), *Modern Discrete Mathematics and Analysis*,
Springer Optimization and Its Applications 131,
https://doi.org/10.1007/978-3-319-74325-7_25

Printed in the United States
By Bookmasters

Printed in the United States
By Bookmasters